SOIL ENGINEERING

Series in Civil Engineering

Series Editor—Russell C. Brinker
New Mexico State University

Bouchard and Moffitt—*Surveying*, 6th edition
Brinker and Wolf—*Elementary Surveying*, 6th edition
Clark, Viessman and Hammer—*Water Supply and Pollution Control*, 3rd edition
Jumikis—*Foundation Engineering*
McCormac—*Structural Analysis*, 3rd edition
McCormac—*Structural Steel Design*, 2nd edition
Meyer—*Route Surveying and Design*, 4th edition
Moffitt—*Photogrammetry*, 2nd edition
Salmon and Johnson—*Steel Structures: Design and Behavior*
Spangler and Handy—*Soil Engineering*, 3rd edition
Viessman, Harbaugh, and Knapp—*Introduction to Hydrology*, 2nd edition
Wang—*Computer Methods in Advanced Structural Analysis*
Wang and Salmon—*Reinforced Concrete Design*, 2nd edition

SOIL ENGINEERING

THIRD EDITION

Merlin G. Spangler
Richard L. Handy
Iowa State University

INTEXT
HARPER & ROW, PUBLISHERS
New York Hagerstown San Francisco London

SOIL ENGINEERING, Third Edition

 For information address Harper & Row, Publishers, Inc., 10 East 53rd Street, New York, N.Y. 10022.

Library of Congress Cataloging in Publication Data

Spangler, Merlin Grant, 1894–
Soil engineering.
(Intext series in civil engineering)
Includes bibliographical references.
1. Soil mechanics. I. Handy, Richard Lincoln, 1929– joint author. II. Title.
TA710.S7 1973 624'.151 73-8
ISBN 0-7002-2402-5

Contents

Preface to the Third Edition

While this edition differs from the two previous editions in several respects, including the addition of a second author, the objective remains the same: to present the material simply, and to direct it toward the undergraduate level of instruction and the beginner in this field.

The scope of the book has been expanded into geotechnical areas to offer an introductory appreciation of landforms, weathering, and clay mineralogy in relation to soil occurrences, properties, and uses. This approach derives from a concept developed some years ago by the authors and their associates at that time, Dr. T. Y. Chu, Professor W. W. Williams, and the late Dr. D. T. Davidson: that every soil has a past, a present, and a future, each stage depending on those preceding. The concept may be further elaborated by dividing the history of a soil into a geological past during which the rocks and sediments were formed, and a pedological past whereby these were weathered into soils. History determines the present composition of the soil—its structure, particle sizes, mineralogy, etc., which in turn determine its present engineering properties such as strength, compressibility, and plasticity. These latter directly influence the usefulness of the soil in engineering.

In our own experience we have found that information ordered within this past-present-future framework becomes more understandable and is more easily remembered. The facts about the origin and mineralogy of a soil frequently suggest potential problems and indicate which engineering tests should be pertinent. Insufficient recognition of what are important facts can easily lead to stupid blunders, with costs measured in thousands of dollars and sometimes tens or hundreds of lives. In some countries the authors of such blunders have been found criminally liable, a rather sobering thought when one considers the extreme variability of soil as an engineering material.

The authors have tried to emphasize that while soil mechanics is an invaluable academic foundation it is not the same as soil engineering, but rather bears about the same relationship to it as English does to journalism. Soil engineering includes many uses of soils which as yet have no theoretical basis in mechanics and are therefore derived from empirical information. Furthermore, it seems irrelevant to the authors to present and argue minute details of theoretical soil mechanics when by far the most important problem facing the soil engineer is his evaluation of the soil. In practically all failures, soil mechanics, the science, was not at fault although the mechanics may have been misapplied; almost inevitably there was some important factor or sampling error which was not recognized. Our emphasis is therefore on statistical site study and *in situ* testing, the latter to reduce the disturbance in "undisturbed" sampling.

The authors wish to thank and acknowledge valuable and otherwise unrewarded assistance in their areas of specialty from Mr. John W. (Jack) Guinnee, Engineer of Soils, Geology and Foundations of the Highway Research Board; Dr. Don Kirkham of the Department of Agronomy, Iowa State University; Professor James M. Hoover, Dr. Turgut Demirel, Dr. Jack Mickle, Dr. Robert Lohnes, and other members of the Soils and Materials staff of the Department of Civil Engineering, Iowa State University, as well as encouragement and clerical assistance from the department.

Preface to the Second Edition

All branches of engineering are constantly undergoing change, both in subject matter and in emphasis upon various components, and certainly soil engineering is not an exception in this regard. Since the first edition of this text became available in 1951, a number of changes in this field have occurred, such as the development of the Unified System of soil classification. Also, certain techniques in the determination of allowable bearing capacity of foundation soils have grown in popularity and usefulness, and a textbook on soil engineering must incorporate a presentation of these modern techniques if it is to be considered up-to-date. These and other less extensive, though no less important, changes have dictated the necessity for preparing this second edition.

The basic philosophy of the first edition has been retained; that is, the objective has been to present the material in simple language aimed primarily toward the undergraduate level of instruction and the beginner in this field. In spite of the contemporary trend toward inclusion of more of the "sciences" and less of the "arts" of engineering practice in undergraduate curricula, the author has chosen to retain the practical aspects of the text and to expand them where possible. Too much emphasis cannot be placed upon the value of the sciences as background material, but the necessity for developing "know how" and "common sense" beyond the limits of basic science is inescapable. It is hoped that this text will provide the student with a foundation in soil science which will enable him to understand what he observes in practice and to develop that priceless commodity sometimes referred to as "horse sense."

The author is grateful to a number of readers of the first edition—students, teachers, and practicing engineers—who have offered criticism and constructive suggestions from time to time. Many of their ideas have been incorporated in the second edition. Also, appreciation is extended to Stephen T. Mikochik, Associate Professor of Civil Engineering at Brooklyn Polytechnic Institute, who read the new manuscript critically and offered a number of highly valued suggestions.

Preface to the First Edition

Soil Engineering has been written for the beginner in this field, which in the past quarter century has assumed a position of commanding importance in the civil, architectural, and agricultural branches of engineering. It is intended for use by undergraduate students at the junior and senior levels and by practicing engineers who may not have had formal training in soils.

An attempt has been made to present the material in the simplest possible language. The aim has been to lead the reader through appropriate phases of the basic sciences of geology, pedology, soil physics, and physical chemistry and to illustrate the applications of these sciences in soil mechanics and soil engineering. It has been the intent of the author to emphasize the fact that soil is an engineering material and needs to be studied and handled in the engineering manner wherever it is encountered in connection with the design, construction, and maintenance of highways, airports, foundations, underground structures, earth dams, and similar structures. At the same time the cultural aspects of the soil have not been neglected, and it is the author's hope that the reader of the book not only will be assisted in his engineering practice, but will be better equipped to enjoy a walk or ride in the country and to appreciate the significance of the various landforms and features of the landscape which he observes.

The author wishes to express his great sense of gratitude to Professor Lowell O. Stewart, whose friendly interest and encouragement created a very stimulating atmosphere during the period of growth and preparation of the manuscript. Appreciation is also extended to Mr. Harold Allen and Mr. E. S. Barber of the U.S. Bureau of Public Roads, who read the manuscript and gave the author the benefit of their criticism. Thanks are also due to Dr. Donald T. Davidson and Professor Wilfred T. Hosmer of the Iowa State College, who used preliminary drafts of the manuscript in the classroom and made valuable suggestions for improvement of the material and the manner of its presentation.

SOIL ENGINEERING

1

Introduction

1.1. BRANCHES OF MECHANICS. The term *mechanics* is defined as "that part of physical science which treats of the action of forces on bodies." There are many branches or subdivisions of mechanics, each of which is pertinent to a particular kind of body or classification of matter. Mechanics pertaining to astronomical bodies is called *celestial mechanics*. The mechanics of gaseous bodies is called *pneumatics*; of water, *hydraulics*; of solid bodies in motion, *dynamics*; of solid bodies at rest, *statics*; of heat-energy transfer, *thermodynamics*, and so on. Similarly, the branch of mechanics which deals with the action of forces on soil masses may be called *soil mechanics*.

1.2. TYPES OF PROBLEMS IN SOIL MECHANICS. The soil which occurs at or near the surface of the earth is one of the most widely encountered materials in the civil, architectural, and agricultural branches of engineering; and soil ranks high in the scale of importance and frequency of contact among the numerous materials encountered and used in engineering. The practice of engineering which involves the application of the principles of soil mechanics may be called *soil engineering*.

Soil is used as a construction material in many engineering structures, such as highway pavements, earth embankments, dams, and levees. All structures (except those founded on solid rock), regardless of the material of which they are constructed, rest ultimately upon the soil. Consequently, the design of the foundations and the settlement behavior of the finished structure depend on the character of the underlying soil and on its action under the stresses imposed by the foundation. In many of the types of structures just mentioned, the absorption, retention, and flow of water through the soil—in the capillary and film phases, as moisture vapor, or as gravity water—cause important stresses in the soil and changes in its stress-resisting properties which may affect the structure.

In the case of sewers, culverts, tunnels, and other types of underground structures, and in partially embedded structures, such as retaining walls, revetments, and bulkheads, soil is important not only as the material upon which the structures are founded but also as the major source of the loads to which they are subjected in service and which they must be designed to carry. Also, the enveloping and overlying soil is the medium through which loads applied at the surface—from truck and airplane wheels, adjacent building foundations, railroad tracks, and the like—are transmitted to these underground and partially embedded structures. The action of water and wind upon soil is an important aspect in the design of stream-control, soil-erosion, and beach-control works.

1.3. COMPARISON OF SOIL WITH OTHER STRUCTURAL MATERIALS. Soil engineering is, in the broadest sense, a phase or subdivision of structural engineering, since it deals either with the soil as the foundation material upon which structures rest or with the soil when used as a structural material. It differs from conventional structural engineering, however, in several important respects. For example, in the creation of a steel structure the structural engineer needs to devote only a very minor part of his total effort to a determination of the kind and quality of the material with which he builds. Steel is a manufactured material whose physical and chemical properties can be very accurately controlled during the manufacturing process. It is only necessary, therefore, for a designer to specify the kind of steel which he wishes to use; and, by providing adequate inspection facilities, he is certain to obtain the desired material for his structure. On the contrary, soils are natural materials which occur in infinite variety over the earth and whose engineering properties may vary widely from place to place within the relatively small confines of a single engineering project. Also, because of the great bulk of soil involved in most projects, it is not feasible to transport the material for more than relatively short distances. Nor is it possible to manufacture or process the soil, except to a very limited extent. Generally speaking, soil must be used in the locality and in the condition in which it is found. A large part of soil-engineering practice, therefore, must be devoted to the location of the various soils encountered on a project, the determination of their engineering properties, the correlation of those properties with the engineering requirements of the job, and the selection of the best of the available soils for use in the various elements of the project.

Furthermore, steel is a material whose properties remain unchanged during the life of a structure, whereas the properties of soils are continuously changing as the amount of moisture fluctuates and other environmental influences vary. One of the most perplexing problems which faces a soil engineer is the fact that soil, being an assemblage of more or less movable particles plus variable amounts of water and air, may change dramatically under load. Two common examples are: (a) loading commonly increases soil density and strength by pushing the particles closer together, so long as the pore fluid can escape from between the particles; (b) on the other hand, if the pore fluid cannot escape, the soil may be drastically weakened by loading because the load is transferred to the pore water, which has zero resistance to shearing. The first behavior is utilized by artificially compacting soil to improve its engineering characteristics, whereas the latter behavior is frequently a factor in landslides and bearing capacity failures. The soil engineer therefore attempts to anticipate soil behavior under the anticipated plan of loading—whether the load builds up gradually and remains, as in the case of building construction, or is applied and released almost instantaneously, as in the case of highways and airfields.

Symptomatic of the difficulties inherent in soil engineering are the generous "factors of safety" frequently used, whereby the measured soil strength is divided by a suitable factor, usually 2 to 5, to insure being on the safe side for design. Thus "factor of safety" is a misnomer which tends to build overconfidence; "factor of ignorance" may be more accurate, but not too acceptable to the profession. Where occasional failures are not disastrous and repairs are cheaper than gross overdesign, as in highway pavement construction, a bare minimum "factor of safety" may be used.

1.4. ANCIENT APPLICATIONS OF SOIL ENGINEERING. Man's first contact with soil as an engineering material is lost in antiquity, along with the origin of mankind itself; but it is certain that he encountered problems in soil engineering very early. We know that excellent paved highways existed in Egypt several thousand years before the Christian era and were used by the pyramid builders for transportation of the construction materials for those huge structures. Remnants of various types of underground conduits which served ancient people as drains, tunnels, and aqueducts, and many other kinds of structures which involved the application of principles of mechanics to the soil, have been unearthed at the sites of early civilizations. It is certain from these evidences that engineers in ancient times encountered and solved in some manner many problems in soil engineering, even though we have no record of the methods by which they obtained their results.

1.5. EARLY LITERATURE ON SOIL ENGINEERING. Some of the first contributions to modern engineering literature deal with subject matter which is now embraced in the field of soil engineering. As early as 1687 a French military engineer named Vauban set forth certain empirical rules and formulas for the design and construction of revetments to withstand the overturning and translating forces exerted by the soil behind them. These rules of Vauban were recommended by Wheeler in his "Manual of Civil Engineering for U.S. Military Cadets" as late as 1877, nearly 200 years after their publication.

Bullet (1691), of the French Royal Academy of Architecture, recorded the earliest theory of lateral earth pressures based on the principles of mechanics and was the first to introduce the idea of a "sliding wedge." He assumed that all the soil above a 45° plane through the heel of a retaining wall tends to slide downward and must be resisted by the wall. Coulomb (*9*) in 1773 applied the principle of maxima and minima to the sliding-wedge theory introduced by Bullet. He determined the slope of the sliding wedge which would produce the maximum pressure on a retaining wall, and showed that this slope is dependent on the characteristics of the soil. He did not pretend to determine the distribution of pressures on a wall in a rational manner but merely assumed them to be quasi-hydrostatic in character with the resultant acting at the upper limit of the lower one-third of the height of the wall.

Coulomb is also credited with having made important contributions to our understanding of the nature of shearing resistance of soils. He introduced the concept that shearing resistance is composed of two components, namely, cohesion and friction. His empirical formula embodying these components is universally used in soil-engineering practice today.

1.6. NINETEENTH-CENTURY DEVELOPMENTS. Rankine (*14*), in his treatise "On the Stability of Loose Earth" in 1856, applied the theory of elasticity to the retaining-wall problem by assuming the soil backfill to be homogeneous, elastic, isotropic, and of indefinite extent. His analysis indicated a quasi-hydrostatic distribution of pressure and checked Coulomb's assumption in that respect. By the application of the theory of conjugate stresses he arrived at the conclusion that the resultant pressure on a wall is always parallel to the surface of the backfill. The contributions of both Rankine and Coulomb to the subject of soil pressures are regarded as classic and have served engineers well throughout the years, although it is generally recognized that they represent only special limiting cases which are seldom encountered in actual engineering practice.

In 1856 two other concepts which play an important role in soil engineering were introduced. These are Darcy's law relative to the flow of gravitational water through porous media such as soils (*10*) and the Stokes law of the velocity of fall of solid particles precipitated in liquids (*15*).

Another important contribution in the field of soil pressures is that of Sir Benjamin Baker (*2*) in 1881. Baker gave the results of many years of extensive experience in the construction of retaining walls, revetments, and other types of underground structures. He pointed out a number of weaknesses in the Rankine and Coulomb theories and was especially critical of the theory that, when a bank of soil fails, the surface of sliding is a plane passing through the heel of the wall or the base of an unconfined bank. His observations showed that surface cracks which develop in a soil mass when failure is imminent are oriented vertically and continue in this direction to a comparatively great depth before bending toward the base of the mass. Baker's observations served to focus attention upon the value of experimental evidence and service performance of soil structures.

Another nineteenth-century contribution to science, which was destined to become extremely useful in modern soil engineering, was the solution by Boussinesq (*3*) in 1885 of the problem of stress distribution in a semi-infinite elastic medium when acted upon by a point load applied at the surface of the mass. Although soil is far from being an elastic material, as that term is ordinarily employed in mechanics, and soil masses are seldom if ever isotropic and homogeneous, experience has indicated that the classical Boussinesq solution is very valuable as a guide in estimating stresses in the undersoil due to foundation loads applied near the soil surface. It is also useful when determining pressures transmitted to underground structures by loads applied at the soil surface such as those from truck and airplane wheels. A modified form of the Boussinesq solution may be employed for computing the magnitude and distribution of the lateral pressure on a retaining wall caused by loads applied at the surface of the soil backfill.

1.7. STRAHAN'S STUDIES. In the field of highway engineering, Dr. C. M. Strahan (*16*), while a county engineer in Georgia, in 1906 began a systematic study of the distribution of particle sizes in gravel road-surface mixtures to correlate the quality and performance of such roads with their mechanical analysis. He selected numerous stretches of gravel roads which had performed satisfactorily and analyzed samples from each one. His early work provided the basis for the far-reaching studies in soil stabilization which today play such an important part in the design and construction of highways and airport runways. It is worth recording that, although tremendous advances in soil stabilization have been made in the past 50 years, the empirical method of approach used by Strahan is still the most powerful tool of the highway engineer at the present time.

1.8. U.S. DEPARTMENT OF AGRICULTURE STUDIES. Early in the twentieth century several soil scientists connected with the U.S. Department of Agriculture were active in the study of the mechanics of soil moisture. Among these was Briggs (*4*), who suggested a useful classification of soil moisture and the centrifuge-moisture-equivalent technique for studying certain water-holding relationships in soils. Concurrently Buckingham (*5*) proposed the concept of capillary potential and conductivity, which has led to a better understanding of the forces responsible for the retention and movement of capillary water in soils.

1.9. MARSTON'S EXPERIMENTS. In 1908, Anson Marston (*12*) of the Iowa State University began his studies of actual loads on sewer pipes and drains in ditches and of methods of testing and laying such conduits to insure that they will adequately withstand the loads to which they are subjected in service. Later his work was extended to include those conduits, such as railway and highway culverts, which are constructed at the surface of the ground and are then covered with soil embankments. He also studied the transmission of surface loads, such as those of truck wheels and other traffic units, to underground structures and, at the suggestion of Griffith, correlated these measured loads with those calculated by the Boussinesq solution mentioned in Section 1.6. The subject of Marston's studies had been considered by many engineers to be too complex for orderly classification and solution, but he succeeded in developing a theoretical approach to the problem which was amply supported by extensive and carefully obtained experimental evidence. It is widely known as "Marston's Theory of Loads on Underground Conduits."

1.10. ATTERBERG'S TESTS. The Swedish scientist, Atterberg (*1*), suggested two simple tests in 1911 for determining the moisture content at the upper and lower limits of the moisture range within which a soil exhibits the properties of a plastic solid. These tests for *liquid limit* and *plastic limit* are widely used to identify soils and give an indication of certain properties, such as plasticity, cohesiveness, and bonding characteristics. The numerical difference between the liquid limit and the plastic limit is called the *plasticity index* of the soil, and knowledge of this property has become a necessity in judging the character and quality of an engineering soil. Prior to about 1925 the abbreviation "P.I." in engineering jargon referred only to the point of intersection of two tangents on a survey line; but today it is widely used to refer to the plasticity index of a soil.

1.11. TERZAGHI'S CONTRIBUTIONS. With the publication in 1925 of "Erdbaumechanik" by Dr. Karl Terzaghi (*18*), soil engineering became recognized as a distinct branch of engineering science. Terzaghi's contributions in this field, both technically and in focusing the attention of engineers on the necessity of studying the soil, have been outstanding, and he is fittingly called the "father of soil mechanics." Probably his greatest technical contribution has been the theory of consolidation of soils in which the effects of load, degree of permeability, and time on settlements of foundations resting on soil are definitely set forth. This theory has materially advanced the science of foundations, and has provided a rational basis for the interpretation of actual settlement observations.

1.12. PROCTOR'S WORK. Another and more recent contribution of outstanding importance in the field of soil engineering is that of Proctor (*13*), who in 1933 published the principles of soil compaction and showed the relationship between compacting energy, moisture content, and the density of the soil. By application of these principles in the construction of soil structures, such as embankments, levees, earth dams, and subgrades for pavements for highways and airport runways, and by the development of the sheep's-foot tamping roller and other efficient types of compacting equipment, it has become possible to attain a greater degree of compaction, and consequently greater strength and less settlement of finished structures than formerly.

1.13. PORTLAND CEMENT ASSOCIATION STUDIES. Stabilization of soils with portland cement and with hydrated lime was tried in the 1920s but did not become a successful technique until the 1930s, following adoption of Proctor's concepts of compaction. In 1935 the U.S. Portland Cement Association undertook extensive tests to establish design criteria for soil-cement. This research, conducted by Miles Catton (*8*) and his associates, literally "paved the way" for use of soil stabilization in highways and airfields. This use has increased exponentially since the 1940s and continues to grow as construction methods improve and aggregate supplies become depleted. Simultaneously the theoretical basis for soil stabilization was pioneered by Dr. Hans F. Winterkorn while a professor at the University of Missouri and later at Princeton University.

1.14. RECENT CONTRIBUTIONS. As research continues and expands, milestones will continue to be laid, through a combination of intensive effort, luck, and unusual insight and perception. Starting in the 1940s Professor Donald Burmister of Columbia University (*6*) successfully applied a Boussinesq type solution to layered soil systems commonly encountered in highway and foundations work. Dr. Arthur Casagrande (*7*) of Harvard University, renowned teacher of soil mechanics, devised a new soil classification system which has come into wide use. Dr. T. W. Lambe of M.I.T. and L. Bjerrum and his co-workers at the Norwegian Geotechnical Institute early recognized and demonstrated the importance of clay minerals and clay structure on engineering properties. In granular soils, the energy expenditures due to volume change during shearing were discussed in the textbook by Dr. Donald W. Taylor (*17*) in 1948 and recalculated and elaborated by Bishop, Rowe, and others. In the 1950s Skempton demonstrated valuable new relationships from old and well-known test methods. In the 1960s the behavior of soils under cyclic loading, as occurs in earthquakes, was intensively studied by Dr. H. B. Seed and his co-workers at the University of California. The volume of research and the number of highly competent researchers in the U.S., England, Norway, Australia, Russia, etc., are rapidly expanding, indicative of a new, vital, and growing field of engineering.

1.15. SOIL ENGINEERING AS A PROFESSION. The brief history noted above has necessarily overlooked many important contributions and contributors to the history of soil engineering and is by no means intended to be complete. The important fact is that soil engineering is well established in engineering practice today. Practically every engineering organization maintains a soil laboratory and a staff of soil engineers whose function is to render advice concerning the engineering performance of the soils encountered on various projects. Final locations of many airports, highways, earth dams, and bridges often are established largely on the basis of study of the soils involved at various alternate sites. Land-use maps for development of urban areas depend heavily on soil engineering evaluations.

The acceptance of soil engineering has not been altogether spontaneous; rather it came partly as a result of sad experiences and loss of life where it had been ignored. In almost any community, buildings may be found showing cracks, sagging, tilting, or otherwise displaying signs of distress. Soil engineers are frequently called on to suggest remedial measures for such conditions; however their primary job is to prevent such problems in future construction. As architectural, structural, and legal requirements become more complicated and construction progresses into less suitable areas, soil engineering becomes even more essential.

While one or two courses in soil engineering now are in practically every undergraduate civil engineering curriculum, the term "soil engineer" usually designates a person having a Masters degree or the equivalent in this specialty. In recent years many practicing professional engineers engaged in soil work have taken advanced degrees or refresher courses in soil engineering to improve their effectiveness in practice.

PROBLEMS

1.1. Define: (a) mechanics; (b) soil mechanics; (c) soil engineering.

1.2. Name several types of engineering structures or projects which involve the practice of soil engineering.

1.3. What contributions to modern soil engineering were made by Coulomb in the eighteenth century?

1.4. Name six scientists and engineers, other than Coulomb, whose work is directly useful in the practice of soil engineering and state the nature of the contribution made by each.

1.5. Discuss two important features involved in the practice of soil engineering which differ from similar features in conventional structural engineering.

REFERENCES

1. Atterberg, A. "Über die physikalische Bodenuntersuchung, und über die Plastizität der Tone." *Internationale Mitteilungen für Bodenkunde* **1** (1911).
2. Baker, Benjamin. "The Actual Lateral Pressure of Earthwork." *Proc. Inst. of Civ. Engrs.* (London) **65,** Part 3 (1880–81).
3. Boussinesq, J. *Application des Potentiels á l'Étude de l'Équilibre et du Mouvement des Solids Élastiques.* Gauthier-Villars, Paris, 1885.
4. Briggs, Lyman J., and John W. McLane. "The Moisture Equivalent of Soils." Bul. 45. *U.S. Bureau of Soils*, Washington, D.C., 1907.
5. Buckingham, E. "Studies on the Movement of Soil Moisture." Bul. 38, *U.S. Bureau of Soils*, Washington, D.C., 1907.
6. Burmister, Donald M. "The Theory of Stresses in Layered Systems and Applications to Design of Airport Runways." *Proc. Highway Research Board* **23,** 126 (1943).
7. Casagrande, Arthur. "Classification and Identification of Soils." *Trans. American Society of Civil Engineers* **113,** 901 (1948).
8. Catton, Miles D. "Research on the Physical Relations of Soil and Soil-cement Mixtures." *Proc. Highway Research Board* **20,** 821 (1940).
9. Coulomb, C. A. "Essai sur une Application des Règles de Maximis et Minimis à Quelques Problèmes de Statique Relatives à l'Architecture." *Royale Acadèmie des Sciences* (Paris) **7** (1776).
10. Darcy, H. *Les Fontaines Publiques de la Ville de Dijon.* Dijon, Paris, 1856.
11. Feld, Jacob. "History of the Development of Lateral Earth Pressure Theories." *Proc. Brooklyn Engineer's Club,* 61–104 (January, 1928).
12. Marston, Anson. "The Theory of External Loads on Closed Conduits in the

Light of the Latest Experiments." Bul. 96, Iowa Engineering Experiment Station, Ames, Iowa, 1930.

13. Proctor, R. R. "Fundamental Principles of Soil Compaction." *Engineering News-Record* (Aug. 31 and Sept. 7, 21, and 28, 1933).
14. Rankine, W. J. Macquorn. "On the Stability of Loose Earth." *Phil Transactions, Royal Society*, London (1857).
15. Stokes, C. G. "On the Effect of the Internal Friction of Fluids on the Motion of Pendulums." *Transactions Cambridge Philosophical Society* **9,** Part 2 (1856).
16. Strahan, C. M. "Research Work on Semi-Gravel, Topsoil and Sand-Clay, and Other Road Materials in Georgia." *Bulletin, Univ. of Georgia* **22,** No. 5-a (June 1932).
17. Taylor, Donald W. *Fundamentals of Soil Mechanics.* John Wiley & Sons, New York, 1948.
18. Terzaghi, Karl. "Erdbaumechanik auf bodenphysikalischer Grundlage." *Deuticke* (1925).

2

The Soil Profile

2.1. NATURE OF SOIL PROFILE. Rock or soil material, derived or laid down by one or more geological process, is subjected to physical and chemical changes brought about by the climate and other factors prevalent in the locale of the soil and in the geological era subsequent to its deposition. Vegetation gains a foothold on the newly deposited mineral debris; rainfall begins the processes of topographical development and leaching and eluviation of the surface of the soil material; and profound changes gradually take place in the character of the soil with the passage of geological time. This development brings about the soil profile.

The soil profile, or weathering profile as it is sometimes called, is a natural succession of zones or strata below the ground surface and represents the alterations in the original soil material which have been brought about by weathering processes. It may extend to various depths, and each stratum may have various thicknesses. There are, generally speaking, three distinct strata or horizons in a natural soil profile. However, this number may be increased to five, or possibly more, in soils which are very old and mature or in which the weathering processes have been unusually intense.

The upper layer of the profile, which is often designated as the A horizon or *epipedon* ("over-soil"), usually is rich in humus and organic plant residues. The subsoil immediately below the topsoil layer is called the B horizon, and it can usually be distinguished from the topsoil by a contrast in color. The relatively unweathered parent material lying below the A and B horizons and from which they were derived is designated as the C horizon. The horizon zonation of soil profiles is shown in Fig. 2-1.

The layered nature of soil profiles is of prime importance in soil engineering because properties of each layer usually differ from those of the others. For example, the uppermost layer, or topsoil, is usually soft and is seldom used for construction. Similarly, deeper layers will have a varying suitability, depending on their properties and environment. Equally important is to be able to relate test data to individual soil layers, to allow meaningful evaluation of a site, rather than scrambling "apples and oranges." If data from several layers of widely varying strength are inadvertently averaged the result may be extremely misleading and dangerous, for example, by obscuring the reasons for choosing between shallow foundations and piling.

2.1.1. *The A Horizon.* In regions of humid or semi-humid climate, the A horizon is usually eluviated and leached; that is, the ultrafine colloidal material and the soluble mineral salts have been washed out of this

Fig. 2-1. (a) Gray A_2 topsoil horizon over a clayey B horizon over C horizon silt, or loess. Range pole for scale. (b) Subangular blocky B horizon. Shiny coatings on the individual peds are clay skins.

horizon by percolating water. It is frequently darker in color than the underlying strata, because of the accumulation of organic matter, or it may be gray or white owing to intensive acid leaching. The thickness of this top layer typically ranges from a few inches to about 2 ft.

Because of its high humus content, the A horizon often exhibits undesirable engineering characteristics, such as high compressibility and elasticity, high resistance to compaction, and variable plasticity. In engineering operations this horizon has its greatest value as a topping-out material where seeding, sodding, and planting are desirable for erosion control or aesthetic purposes. When used in this manner, the soil is stripped and stockpiled in the early construction phases and is later spread over the areas to be planted, such as highway slopes and ditches and the areas between runways of airports.

Organic debris such as leaves, vegetation, etc., on top of the A horizon is termed an O horizon. It is not used in construction and should be removed prior to placement of fill soil or it will constitute a weak layer under the fill.

2.1.2. The B Horizon. In many soils the colloidal material which has migrated from the A horizon by eluviation is deposited in the subsoil layer, or B horizon, sometimes referred to as the zone of accumulation. Clay particles deposited in the B horizon are oriented with their flat surfaces parallel to the surface of deposition, whether that surface is a capillary channel, a vertical or horizontal tension crack, or a single grain of another material. Such "clay skins" are readily viewed in a small, thinly ground cross section or "thin section" under a polarizing microscope, or can be seen by careful inspection with a 10× hand lens. Sometimes they are sufficiently well developed to produce a shiny appearance on soil blocks or peds [Fig. 2-1(b)], and they frequently are darker in color than the nonoriented interior of peds.

B horizons containing expansive clays often show a blocky or columnar structure caused by periodic drying and shrinkage of the clay. Oriented clay skins developing in the open cracks then preserve the structure by offering poor surfaces for bonding, a factor significant in engineering. Columnar structure is most prominent in sodium montmorillonite clays. A faint columnar structure may be seen in the lower B horizon in Fig. 2-1(a).

Under very acid conditions all humus is removed from the A horizon, and the B horizon contains concentrations of humus plus iron and amorphous aluminum hydroxides. The latter do not deposit as clay skins or produce a blocky structure, but the iron may form a hard, cemented layer or *ortstein*. The kind of A and B horizons formed depends on climate and several other variables, which will be discussed later.

Still another type of B horizon is found only in severely weathered tropical soils. It has a high clay content that is not the result of clay deposition but rather of removal of the nonclay minerals, which are more weatherable. The relative ease of weathering of different minerals is discussed near the end of this chapter.

Since the B horizon often contains more clay and thus is more surface-chemically active and unstable than the soil either above it or below it, the B horizon is important in highway and airfield design and construction, and in other work in which foundations are located near the ground surface.

2.1.3. The C Horizon. The A and B horizons together usually constitute the soil scientists' "solum," or true soil, and the underlying C horizon has been referred to as the "parent material." The C horizon may be weathered rock, or it may be any of the variety of unconsolidated sediments discussed in Chapter 3. If of weathered rock it exhibits the structure of the rock rather than the blocky or columnar structure of a solum. Nevertheless the C horizon by definition has been weathered to a soft material. If of sediment, it is already soft, and often oxidized and leached of soluble carbonates in the upper part. In engineering, the C is the most important of the three horizons; it furnishes the bulk of the material of which large soil structures, such as earth dams, levees, and embankments, are constructed, and is frequently used for foundations of buildings, highways, etc.

2.1.4. Subhorizons and Pans. The A and B horizons may be subdivided into subhorizons which have distinctive physical and chemical characteristics and which are designated by subscripts or numbers on the principal horizon letter, such as A_1, A_2, B_1, and B_2. The contact between two horizons or subhorizons is not a sharply defined line or plane; rather, the change from one to another occurs through a zone of some thickness. Also, in some soils, the upper horizons may extend into lower horizons in the form of tongues or sharply defined intrusions. Selected horizon nomenclature is given in Table 2-1.

The term *hardpan* is frequently used to describe any buried, hard, impervious layer occurring in the A, B, or C horizons. Shallow hardpans or *duripans* (L. *durus* means hard) are common in arid and semiarid areas, where they represent soil materials cemented by calcium carbonate which has been deposited through gradual evaporation of ground water. Such horizons are designated A_{ca}, B_{ca}, or C_{ca}, the subscript indicating calcium carbonate, or chalk. In the southwestern U.S. these layers are referred to as *caliche* and are crushed and used to pave secondary roads.

TABLE 2-1. Soil Horizon Designations[a]

Horizon	Subhorizon	Characteristics	Other Common Horizons: Designation	Other Common Horizons: Characteristics
O	O	Organic horizon (leaves, plant remains, etc.)		
A	A_1	Eluviated surface horizon with humus accumulation (common black or brown loamy topsoil)	A_p	Plowed A
	A_2	Eluviated surface horizon with humus removed (usually a gray or white, ashy-appearing layer)	A_{ca}, B_{ca}, C_{ca}	Calcium carbonate accumulations in A, B, C (caliche)
	A_3	Transitional to the B } = AB if not readily differentiated		
B	B_1	Transitional to the A } = AB if not readily differentiated	B_x, C_x	Fragipan in B or C
	B_2	Layer most clearly showing B horizon characteristics (the heart of a clayey or blocky subsoil)		
	B_3	Transitional to the C	A_g, B_g, C_g	Gleyed A, B, C
C	C	Mineral layer showing appreciable chemical alteration but unlike the B	B_{ir}	Iron accumulation in B
	IIC	Roman numeral signifies change in lithology from C	B_m	Cemented B (laterite, ortstein)
	III A_b, III B_b, III C	Buried A and B horizons (paleosol); Roman numeral signifies change in lithology from II	B_n	Humus accumulation in B (acid soil)
			B_t	Clay accumulation in B
	R	Bedrock	C_{cs}	Gypsum accumulation in C

[a]Adapted from the Seventh Approximation (U.S.D.A., 1960).

Fig. 2-2. Perched water table and septic tank failure due to an underlying fragipan, a hard, impermeable subsoil. North Carolina.

In more humid climates where the direction of water movement is more often downward, the dissolved calcium carbonates are deposited deeper within the sediment or bedrock, and are referred to as *concretions.* Concretions are usually ellipsoidal or nodular and irregular and rounded in shape, may be hollow, and sometimes occur in layers. They may be very hard, resembling limestone.

Another type of pan which occurs within the soil profile is termed a *fragipan,* meaning brittle pan. Fragipans usually start at 1.5 to 2 ft depth, fade away at a depth of 4 or 5 ft, and occur mainly in silty-clay soils leached of carbonates. The main factor contributing hardness is a high bulk density, but the cause of the high density is not understood. Fragipans are very slowly permeable and may contribute to a locally high, or *perched* water table (Fig. 2-2).

2.1.5. Deeper Zones of Weathering. The distinct changes in a soil profile with depth are a result of more biological activity and more intense weathering at shallower depths, in addition to vertical movements of solutions and minerals. That is, minerals occurring in the A horizon show that it is more intensely weathered than those in the B, and the B is more weathered than the C.

Similarly deeper in the profile, the upper part of the C horizon is more weathered than farther down; it may be leached of carbonates, reducing the density and strength compared to deeper, unleached C horizon material. The depth of leaching is important to recognize in soil reconnaissance and may be detected with hydrochloric acid diluted about 1:10 with water—when drops of acid are applied to the soil with an eyedropper, soil containing carbonates effervesces or "fizzes."

Below the limits of the soil profile, sediments such as glacial till and alluvium may be gray, greenish or bluish, which was their color when deposited. Weathering of such materials initiates upon exposure to air or oxygen-laden water, and the soil color changes to tan or brown. The occurrence of gray or blue colored soils found by deep drilling therefore does not necessarily indicate a different geological material, but only a penetration below the depth of weathering. In fact, samples of such materials may be observed to change color in a matter of hours or even minutes after exposure to air. Nevertheless such colors indicative of a lack of weathering may be highly significant in engineering, since this relates either to continuous saturation below the water table, or youth, density, and impermeability of the deposit.

2.2. FIELD IDENTIFICATION OF SOIL BY COLOR. The color of a soil stratum is an important means for identification of soils which are shown on the plans of a project. The color by which a soil is designated should be that which it has when moist and in its natural state. The color designation of a stratum should give the range of color included, such as black to dark brown or dark brown to reddish brown. In using such compound terms as grayish brown or pinkish gray, the adjective is recognized as the modifying term. Thus, a grayish-brown soil is a brown soil with a grayish cast sufficiently noticeable to require recognition; a pinkish-gray soil is a gray soil with a pinkish cast.

A soil which has spots, streaks, or splotches of one or more colors against a background of another predominant color is called a *mottled* soil. For example, a soil may be described as "olive gray, mottled with brown." In a mottled soil the colors are not mixed and blended, but each color is more or less distinct in relation to the general background color. When two or more distinct colors occur mixed throughout the mass in approximately equal amounts, the soil is said to be *marbled*. In a marbled soil there is not a general or predominant background color, as in the case of a mottled soil. The terms *spotted*, *speckled*, *streaked*, or *variegated* may be used in a soil designation when the generally accepted meaning of the term clearly describes the color distributions that occur in the soil.

2.3. COLOR STANDARDS. The U.S. Department of Agriculture uses the Munsell color chart in pedological soil-survey operations. In this system colors are designated by the following three variables: hue, value, and chroma. Hue is the dominant spectral or rainbow color. Value refers to the relative lightness or darkness of color. Chroma, sometimes called saturation, is the relative purity or strength of the hue; it increases with decreasing grayness. The relation among hue, value, and chroma is illustrated in Fig. 2-3.

In the Munsell notation of color, each of the three components—hue, value, and chroma—is measured on a scale from 0 to 10. For example, in the notation for value the number 0 stands for absolute black, and 10 stands for absolute white. A color to which the number 5 is assigned is visually midway between these extremes. An example of the Munsell notation is "5YR 4/4." This means a yellowish-red soil of hue 5 for which the value is 4 and the chroma is 4.

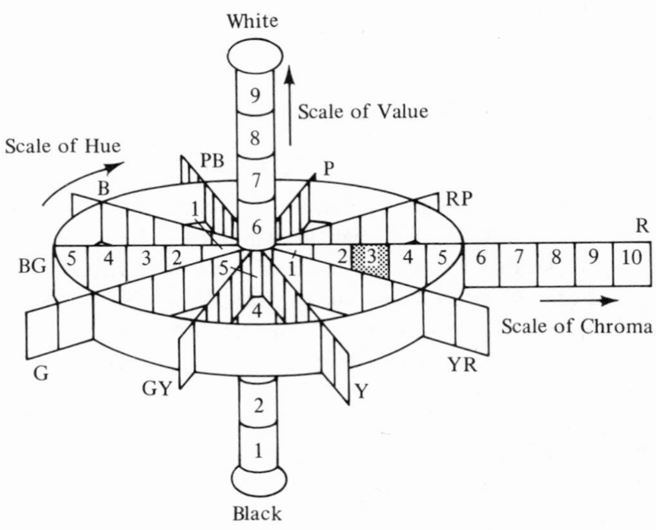

Fig. 2-3. The Munsell system of color. *Hue,* or spectral color, is indicated by number and letter: R—red; YR—yellow-red; G—green; etc. *Value,* or relative darkness, is shown on the vertical scale, and *chroma,* or intensity of the color, on the radiating horizontal scale. The shaded block is 10R 5/3, 10 indicating clockwise scale of the hue. Colors along the central axis are designated N (for neutral).

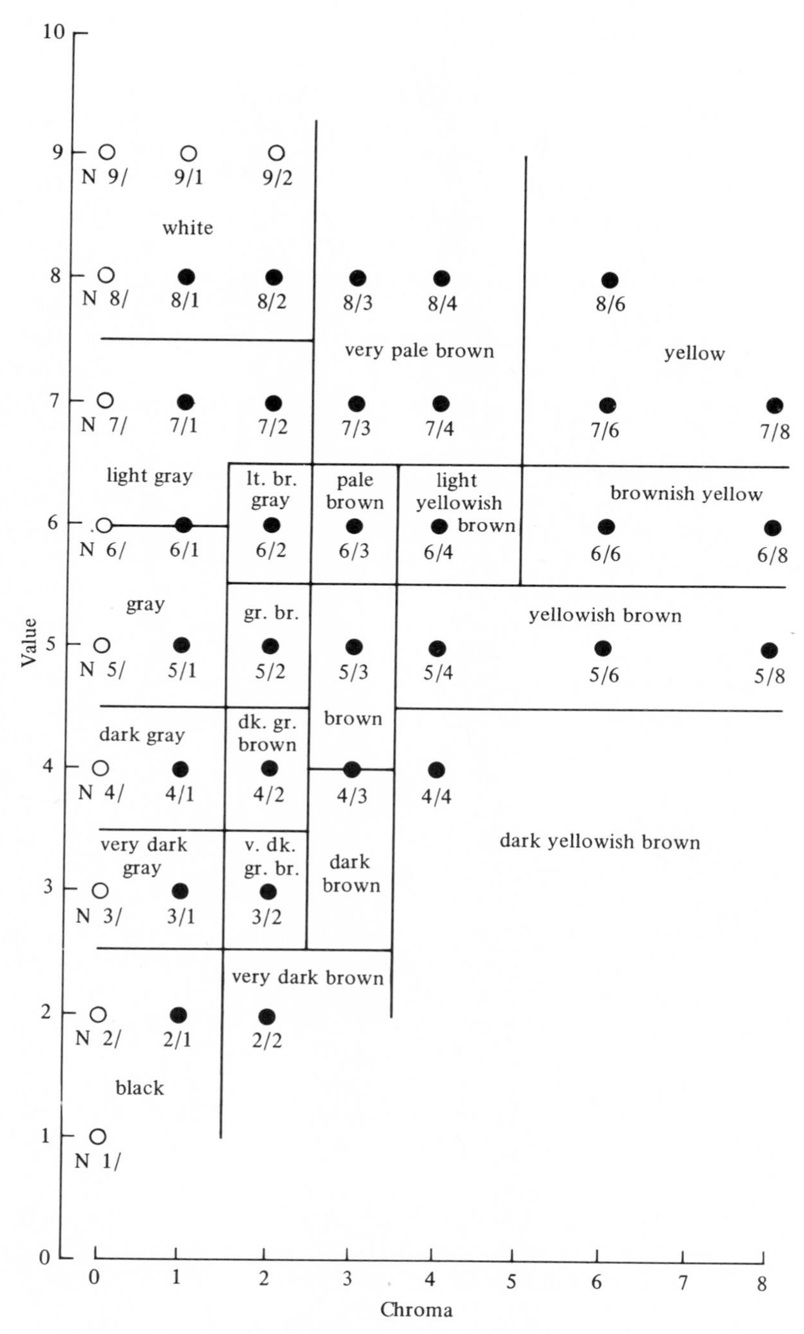

Fig. 2-4. Soil color names for several combinations of value and chroma and hue 10YR. Neutral colors are at the left.

Soil colors are most conveniently and accurately determined by comparison with a color chart. The chart used by the U.S.D.A. consists of 175 colored cards or chips systematically arranged by hue, value, and chroma and marked by their Munsell notations. As a rule, engineering usage does not require that the soil color be determined with the degree of refinement which is possible by means of the U.S.D.A. chart. Ordinarily, an engineering organization supplies each of its soil-survey parties with a smaller number of color cards, say 12 to 24. The actual number depends on the range of soil colors encountered in the state or region in which the work is being done. These cards, however, should match with selected colors on the U.S.D.A. chart, in order that a satisfactory degree of uniformity in color terminology may be achieved. The relationship of color name to Munsell notation for a single hue is shown in Fig. 2-4.

2.4. IMPLICATIONS OF SOIL COLOR; GLEYING. Soil color is mainly a function of grain surface coatings which constitute only a small percentage of the soil. The implications of some colors are: Intense rusty red-browns signify iron oxide (hematite) coatings; yellow browns are caused by minor amounts of iron hydroxides; black usually means humus or less commonly, trace amounts of manganese oxide; a white, crusty appearance, calcium carbonate. Soils without grain coatings are usually light gray or white, as in A_2 horizons.

Dark gray or green or blue hues indicate reducing conditions or gleying, which occurs below a permanent water table. These colors relate to ferrous iron content, and rapidly change upon exposure to air. Since gleying indicates lack of oxygen, organic materials such as timber pile may be preserved for centuries in gleyed soils with little change, explaining why bodies of Roman soldiers have been found more or less intact in British peat bogs (but not explaining how they got there). Mottled gray and brown colors often indicate proximity to a water table or zone of fluctuating water table. Gleying also occurs in soil adjacent to gas pipeline leaks and may be used to locate or confirm old leaks.

If the chroma is 0 there is no color, and the hue is designated by N (neutral): N1/0 designates a very low reflectance neutral shade (black) and N10/0 represents absolute white.

2.5. TRUNCATED AND SUPERIMPOSED PROFILES. Because of unconformities which may have developed in the soil profile at a particular place, all parts of a normal profile do not always exist. For example, in an extensively eroded area the A horizon may be entirely missing, and borings in such an area will reveal only the B and C horizons. Within the flood plain adjacent to a stream, sufficient time may not have elapsed since

Fig. 2-5. Layer sequence from the top: A and B horizons (dark) developed in C horizon (medium), leached, loess; II C horizon (light, unleached loess); III B horizon (thin, dark), a paleosol developed in III C horizon (medium), glacial till; road level.

deposition of the material for a soil profile to have developed, and borings may reveal only a heterogeneous mass of water-sorted, stratified alluvium with a weak A horizon. This is referred to as an AC profile, indicative of immature soil development. Alternately, several soil profiles may be superimposed one upon another. For instance, a deep cut in a glacial area may reveal two or more complete profiles because profiles developed during interglacial periods may have been buried by a later glacial deposit, upon which a new profile developed (Fig. 2-5). In a volcanic region, a soil profile may have developed from the parent lava which in the geological past may have buried an ancient soil profile; hence, two distinct profiles exist, one above the other.

2.6. MORPHOLOGY OF SOIL. The profile of a soil results from the combined influence of all the past events and environment in the history of the soil. The morphology or form of a soil is expressed by a complete description of the texture, structure, color, and other characteristics of the various horizons, and by their thicknesses and depths in the soil profile. These parts or properties of a soil are called its internal characteristics. In addition, each soil has certain external characteristics, such as climate, native vegetation, landscape (as it influences internal and external drainage), and age or maturity.

Starting with freshly weathered rock or recently deposited parent material, the soil gradually acquires a characteristic morphology, the details of which depend on the kind of material, the native vegetation, the climate, and the relief. Thus, a soil may change gradually from a mass of mineral debris to an individual body composed of distinctive horizons almost entirely unlike the original mineral material. These processes of change are called processes of soil genesis.

2.7. AGRONOMIC SOIL MAPS. Governmental agencies in the U.S. and in many other countries in the world are actively engaged in mapping soils as an aid to land evaluation and in order to improve soil uses. While the primary goal is agriculture, soil surveys in the U.S. now include engineering data and are an essential part of the library of practicing soil engineers. In the U.S., surveys and maps are prepared on a county basis and are available from state universities or county or state offices of the U.S.D.A. Soil Conservation Service.

The basic agronomic soil mapping unit is called a *soil series* whose members have the same weathering profile and genetic history. That is, the soils have the same number of horizons of similar depth; the horizons were developed from the same kind of parent material under the same climate and native vegetation; they have essentially the same slope and landscape; and they are of approximately equal age. Each soil series is given a name, usually the proper name of a lake, city, stream, or other geographical entity near the place where the series was first found and scientifically described. Examples are the Webster series in Iowa, and the Hilo series in Hawaii. After a soil series has been discovered, appropriately described, and named, that name is applied to all soils fitting the description, wherever they may be encountered.

Soil series names are obviously of little use unless one is familiar with local soil names and their respective profiles. Series having similar but not identical characteristics are classified together into families; similar families into subgroups, etc., through groups, suborders, and orders. One of the most useful soil classification categories above series is the *great soil group*, which traditionally embraces some key climatic, ground water, parent material, or other soil-forming factor believed to be most influential in development of the soils in a particular group. Unfortunately, the common great soil group names such as Brunizem and Chernozem are not very descriptive. These are Russian in origin and mean brown earth and black earth, respectively, in reference to the A horizon. However, many other soils are also brown or black.

2.8. THE SEVENTH APPROXIMATION. Recently the U.S. Department of Agriculture has developed a classification system tied more to soil properties than to inferences regarding soil genesis, and utilizing a new nomenclature consisting of coined words that describe important characteristics. It is called the Seventh Approximation, since it was reviewed before publication seven times by soil scientists around the world. The most important category is the *order,* the 10 orders being as follows, arranged more or less according to increasing degree of weathering:

2.8.1. *Histosol.* Literally, "tissue soil." Bog soils; peat and organic muck. Severe engineering problems.

2.8.2. *Entisol.* "Recent soil." Little or no genetic horizonation. Usually an AC profile, but may have an A_1, A_2, or A_p, and sometimes a very weak B horizon. Suborders are defined for wet soils (Aquents), sandy soils (Psammants), arid soils (Ustents), and humid soils (Udents). Under Aquents, the great group Cryaquent (cold-wet-recent) includes tundra soils, indicative of permafrost.

2.8.3. *Inceptisol.* "Beginning soil." Horizons mainly reflect changes in color and pH, with little textural change (no clay illuviation, etc.). Suborders for wet soils (Aquepts), predominantly volcanic ash or allophane soils (Andepts), dark-colored A horizon soils (Umbrepts), and light-colored A horizon soils (Ochrepts). Most important in engineering are Aquepts because they are wet, and Andepts because they may be sensitive to shock or vibrations. Cryaquepts (cold-wet-beginning soil) are on permafrost.

2.8.4. *Aridisol.* "Arid soil." Light-colored A horizon and a variety of B horizons with salt, gypsum, or calcium carbonate duripan. Suborders for nonclayey (Orthid) and clayey (Argid) soils; the latter are expansive clays (Na or Ca montmorillonite), discussed in Chapter 4.

2.8.5. *Mollisol.* "Soft soil," referring to the dark-colored, crumbly A horizon. May have A_2, illuvial B, etc. Developed under grass, these are prime agricultural soils. Suborders indicate development on limestones (Rendolls), white A_2 horizon (Albolls), wetness plus gleying or mottling (Aqualls), cool or high-altitude soils (Altolls), humid soils (Udalls), and hot, dry soils (Ustolls and Xerols). For engineering purposes, the A horizon is stripped away; B horizon is montmorillonitic and blocky and may be very clayey. Formerly called Brunizem, Chernozem, Chestnut, etc., soils.

2.8.6. *Alfisol.* "Aluminum-iron soil." Light colored A horizon over a clayey B. Developed under boreal or deciduous forest, usually in calcareous deposits of the Pleistocene (i.e., those containing calcium carbonate). Suborders for wet conditions (Aqualfs), high, cool climate (Altalfs), humid climate (Udolfs), and hot, dry climate (Ustalfs and Xeralfs). B horizon well developed, blocky, montmorillonitic. Formerly called Gray-Brown Podzolic, Gray Wooded, and similar soils.

2.8.7. *Vertisol.* "Inverted soil." Black A_1 horizon, sometimes extending to 3 to 5 ft depth. Vertical mixing due to seasonal shrinkage and swelling; usually an AC profile. Contains over 30% montmorillonitic clay. Vertical shrinkage cracks fill by sloughing in from above; re-expansion often causes shear failures within the soil mass, indicated by *slickensides* on inclined shear planes. Cracking sometimes creates a hummocky microrelief called *gilgai*. Parent material usually alluvial clay, weathered limestone, or basalt. Climate includes a dry season and may be temperate to tropical, mixing apparently preventing normal profile development and weathering of the montmorillonite to more stable clay minerals. Severe engineering problems due to shrinkage and expansion; tilting and tearing of buildings, pulling of pile, etc. Formerly called Grummosol, Regur, or Black Cotton Soils.

2.8.8. *Spodosol.* "Wood ash (Gr., *spodos*) soil"; contains illuvial accumulation of free oxides or humus or both. Formerly called Podzols, Brown Podzolics, and Ground Water Podzols. These have an ash-colored A_2 horizon (hence the name). Soils of the humid, usually coniferous forest areas, they do not develop on clayey parent material (>30% clay). Suborders for wet conditions (Aquods), humus accumulation (Humods), iron plus humus accumulation (Orthods), and iron accumulation (Ferrods).

2.8.9. *Ultisol.* "Ultimate soil." Variable A horizon over a red or yellow clayey B, often containing iron-cemented nodules or a layer termed *plinthite* (laterite). This is a highly weathered acid soil developed under forest, grass, or marsh in pre-Pleistocene deposits. B horizons very well developed, contain kaolinite with some montmorillonite.

2.8.10. *Oxisol.* "Oxide soils." Red B horizon of low density, weakly cemented silt and clay which are mostly oxide minerals plus kaolinite (described in Chapter 4). The more readily weathered minerals have been removed. Occurrence restricted to tropical and subtropical climates and older (pre-Pleistocene) land surfaces. Often have soft or hard iron accumulation layer or plinthite. Suborders for wet conditions (Aquox), extreme weathering (Acrox), humid climate (Udox), hot, dry climate (Ustox), and arid climate (Idox).

2.8.11. Great Groups and Subgroups. Great soil group names in the Seventh Approximation are suborder names plus a prefix designating a prime soil property.

For example, "Argalboll" designates a clay layer (arg)–white layer (alb)–soft layer (oll) soil, or a Mollisol with an A_1, A_2 and clayey B horizon. (This was formerly called a Planosol, meaning flat soil.) Similarly, subgroups names are added in front to indicate whether a soil fits the central concept for that group, termed "Orthic," or is an intergrade into another great group, suborder, or order. This is essential since soils form a continuum and are not isolated species.

In summary, the soil names from the Seventh Approximation should be deciphered from the back. For example, in Durustalf, *alf* is order Alfisol (aluminum-iron soil); suborder prefix *ust* means hot, dry climate; great group prefix *dur* signifies duripan.

2.9. LAND FILLS. A very good reason for one to become familiar with natural soil profiles and weathering is to better recognize an unnatural sequence, such as is imposed by man-made land fills. As urban areas develop, more and more filled land is utilized for building sites. Also, reconstruction on or near an old site almost invariably involves earth fill which was spoil from the earlier excavations. Since random fill is practically never compacted to a satisfactory density or strength, it may pose a serious construction problem that reaches literally devastating dimensions if it goes unrecognized. Furthermore the last step in most land-fill operations is to cover the random bricks, concrete, trees, brush, car bodies, tires, garbage, etc., with a layer of soil to seal it off and improve the appearance. A most frequent clue to filled land therefore is the occurrence of C horizon material or materials of recent origin (bricks, glass, etc.), where there should be a weathered profile, or the occurrence of A over C with no intermediate B. If available, old contour maps of the unfilled area can be a valuable aid in estimating the extent and thickness of filling. Even old random soil fills, as from basement excavations or filling of ravines, are seldom satisfactory to support loads since they are essentially in equilibrium support of their own weight, and any additional weight will induce considerable settlement.

2.10. SOIL ASSOCIATION AREAS. While, as shown in Fig. 2-6, particular soil orders tend to dominate in particular areas of the world, other orders also commonly occur in the same local area. Furthermore each order may be represented by more than one series. A *soil association area* is a geographical area which, because of geological and climatic uniformity, is characterized by several soil series. The soil association area thus crosses classification boundaries. One such area is shown in Fig. 2-7, and includes soil orders ranging from Alfisol to Mollisol and Entisol.

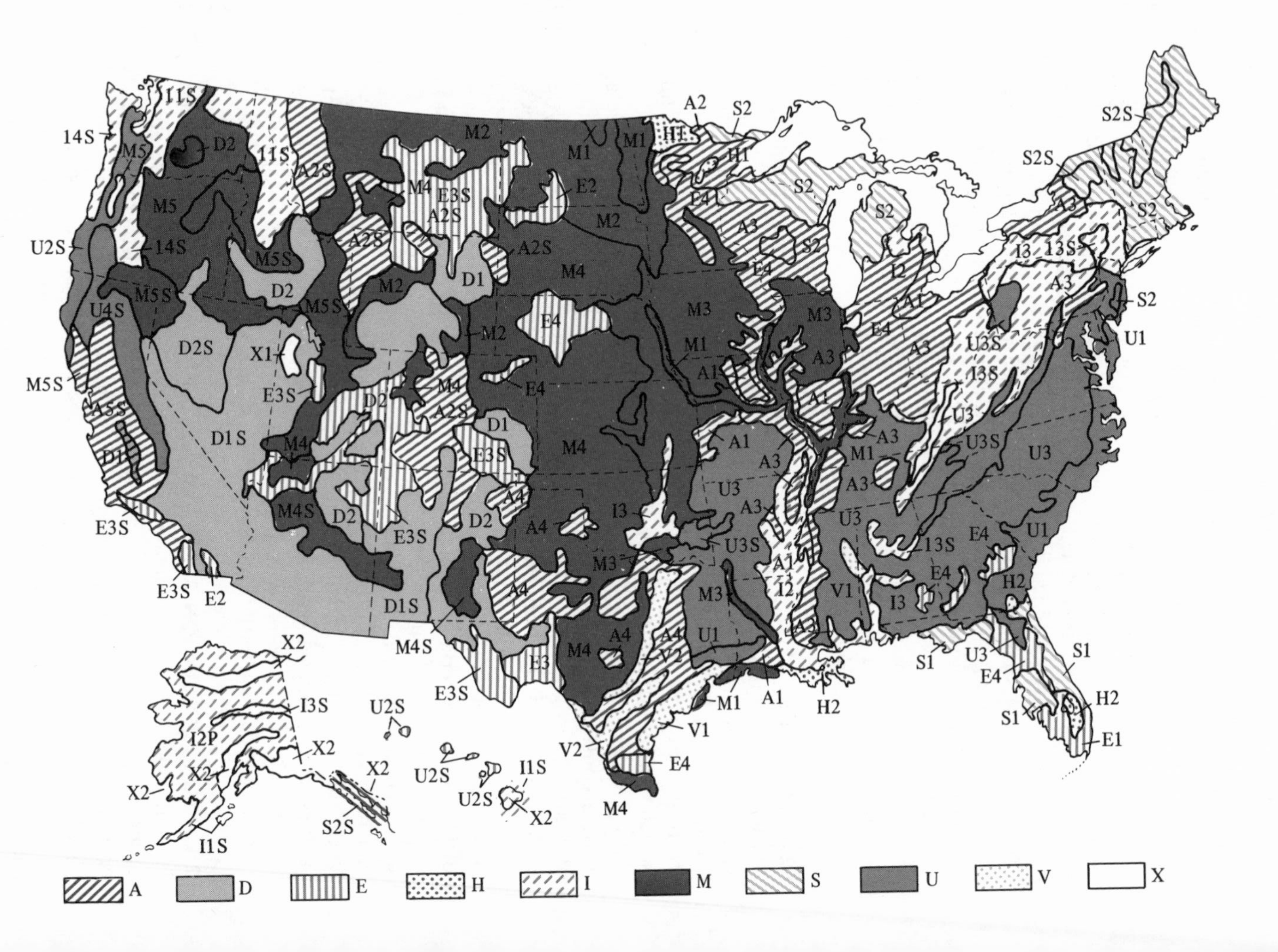
A
D
E
H
I
M
S
U
V
X

(A) Alfisol (Al-Fe soil, clay accumulation)	*(D) Aridisol (Dry more than 6 months/yr)*	*(E) Entisol (No pedogenic horizons)*	*(H) Histosol (Organic soil)*	*(I) Inceptisol (No accumulation horizons)*
A1 Aqualfs (seasonally water-saturated) A2 Boralfs (cool or cold) A3 Udalfs (temperate, moist) A4 Ustalfs (dry seasons) A5S[a] Xeralfs (long, dry summer)	D1 Argids (horizon of clay accumulation) D2 Orthids (no clay accumulation horizon)	El Aquents (seasonally water-saturated) E2 Orthents, deep (loamy or clayey) E3 Orthents, shallow (loamy or clayey) E4 Psamments (sandy)	H1 Fibrists (Fibrous: peat) H2 Saprists (Decomposed: muck)	I1S Andepts (with amorphous clay or volcanic glass) I2 Aquepts (seasonally water-saturated) I3 Ochrepts (light-colored surface horizons) I4S Umbrepts (dark-colored surface horizons)
(M) Mollisol (Dark surface horizon, high base supply)	*(S) Spodosal (Amorphous accumulations)*	*(U) Ultisol (Clay accumulation, low base supply)*	*(V) Vertisol (Expansive clay, seasonal cracking)*	*(X) Areas with little soil*
M1 Aquolls (seasonally water-saturated) M2 Boralls (cool or cold) M3 Udolls (temperate, moist) M4 Ustolls (dry seasons) M5 Xerolls (long, dry summer)	S1 Aquods (seasonally water-saturated) S2 Orthods (surface accumulation of Fe, A1, and organic matter)	U1 Aquults (seasonally water-saturated) U2S Humults (high organic) U3 Udults (temperate, moist) U4S Xerults (long, dry summer)	V1 Uderts (cracks open less than 3 months/yr) V2 Usterts (cracks open more than 3 months/yr, open twice each year)	X1 Salt flats X2 Rock, ice fields

[a]S after number, sloping to steep. P after number, permafrost.

Fig. 2-6. Dominant soil orders and suborders (modified from U.S.D.A.–SCS).

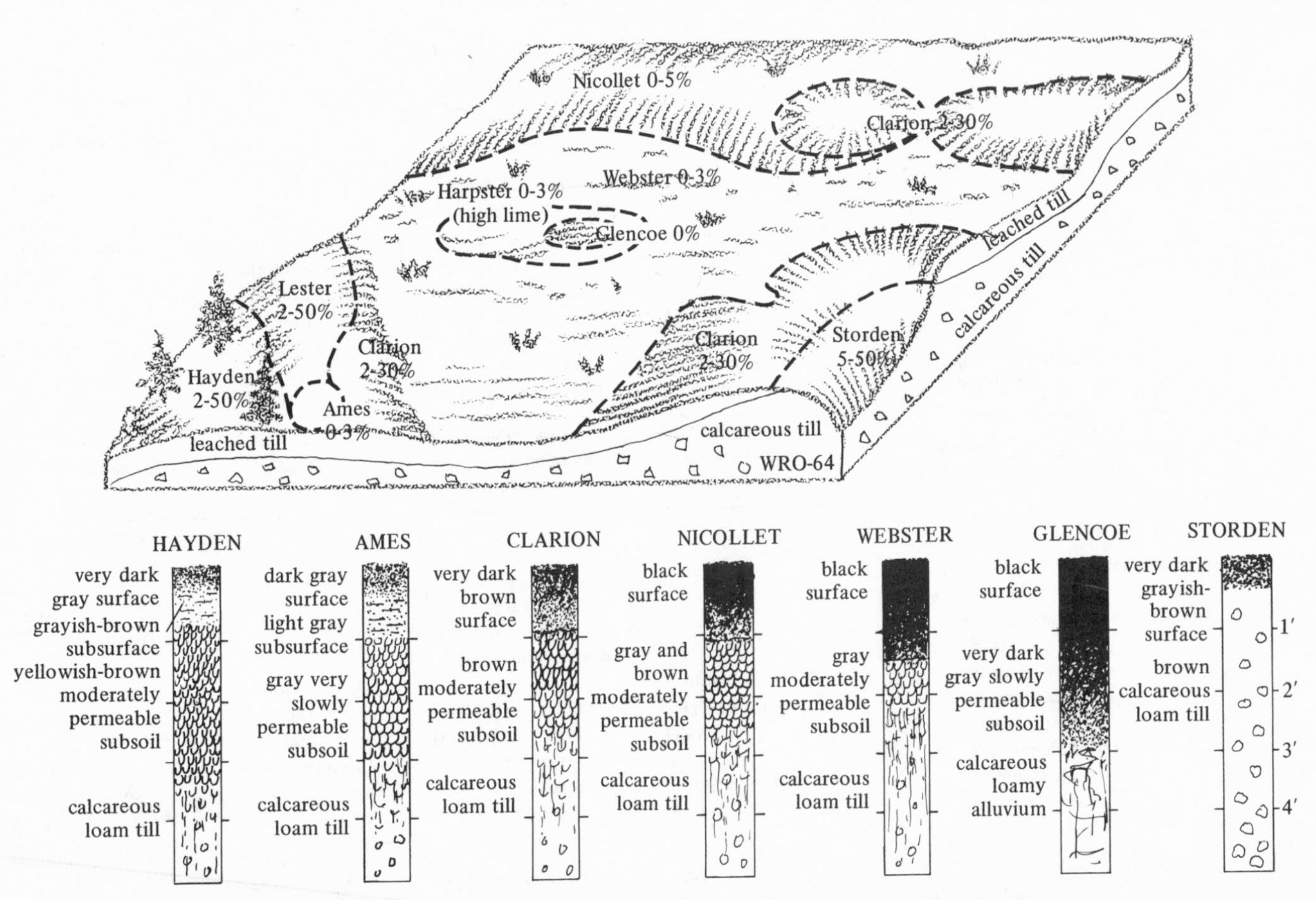

Fig. 2-7. Relationship of slope, vegetation, and parent material to soils of the Clarion-Nicollet-Webster soil association area.

PROBLEMS

2.1. (a) Since weathering is assumed to be most intense in the A horizon, changing primary minerals into clay minerals, where does the clay go? (b) Give evidences for the above.

2.2. Strength tests show that a dry, clayey B_{2t} behaves like a gravel. (a) Explain. (b) How will clay skins affect this behavior?

2.3. A soil is described as having A over C_g horizons. (a) What can you infer regarding water conditions? (b) What are the order and suborder?

2.4. The C_{ca} horizon occurring in the southwestern U.S. is sometimes crushed and used to gravel roads. (a) What is the cementing component in this material? (b) What is another name for this horizon? (c) What is the order?

2.5. The iron-cemented or laterite layer of tropical soils is frequently used as a road metal. (a) What is the horizon designation? (b) What is the soil order?

2.6. What are the probable orders for soil series profiles shown at the bottom in Fig. 2.7? (Note the A_2 horizons in the Hayden and Ames series.)

2.7. (a) What clay mineral dominates in Vertisols (or Black Cotton Soils)? (b) Discuss engineering problems and suggest remedies or preventative measures associated with the use of (1) pile, (2) shallow foundations, in such soils.

2.8. Identify hue, value, and chroma, and give color name: (a) 10YR 5/4; (b) 10YR 2/2; (c) N 9/0.

2.9. Drilling indicates the following sequences of horizons from the top down. Give one or more possible explanations for each sequence: (a) A_1, C; (b) A, B, C, B, A; (c) A, B, C, A, B, C; (d) B, C.

2.10. Compare the suitability for foundation bearing: B_{2t}, B_x, B_m.

2.11. (a) Examine soil horizons exposed in a road cut, and identify A_1, A_2, B, and C horizons if present. (b) Do the same by auger boring.

REFERENCES

1. *The Munsell Book of Color.* Munsell Color Co., Inc., 10 East Franklin Street, Baltimore, Maryland 21202.
2. Soil Survey Staff, Soil Conservation Service, U.S.D.A. *Soil Classification—A Comprehensive System—Seventh Approximation.* U.S. Government Printing Office, Washington, D.C., 1960.
3. Soil Survey Staff, U.S.D.A. *Soil Survey Manual.* U.S.D.A. Handbook No. 18. U.S. Government Printing Office, Washington, D.C., 1951.
4. Sowers, George F. "Foundation Problems in Sanitary Land Fills." *ASCE J. San. Eng. Div.* **94** (5A1), 103–116 (1968).

3

Airphoto Interpretation of Soils

3.1. SOIL PARENT MATERIALS. As indicated in the preceding chapter, the surficial A and B horizons are important in engineering, but the underlying C horizons are equally important and frequently more important. Agricultural soil maps are useful as a guide to deeper soils, but are not intended to adequately describe or predict properties of the deeper C horizons. Engineers therefore also rely on other information sources such as published state and federal geological survey reports, airphotos, and ultimately soil borings.

3.2. GEOLOGICAL CYCLE. In a broad sense, soil may be thought of as an incidental material in the vast geological cycle which has been going on continuously and relentlessly throughout the hundreds of millions of years of geological time. This cycle may be considered as consisting of a number of phases, illustrated in Fig. 3-1.

The first step in the cycle is represented by igneous rocks—that is, rocks that have solidified from molten magma. Igneous rocks include the oldest rocks found on earth and represent the original or primordial sources for soils. For example, grains of quartz in ordinary river sand, even though occurring far from an igneous source both in distance and in time, almost invariably contain microscopic bubbles indicative of their igneous origin. Although most igneous activity occurred in past geologic

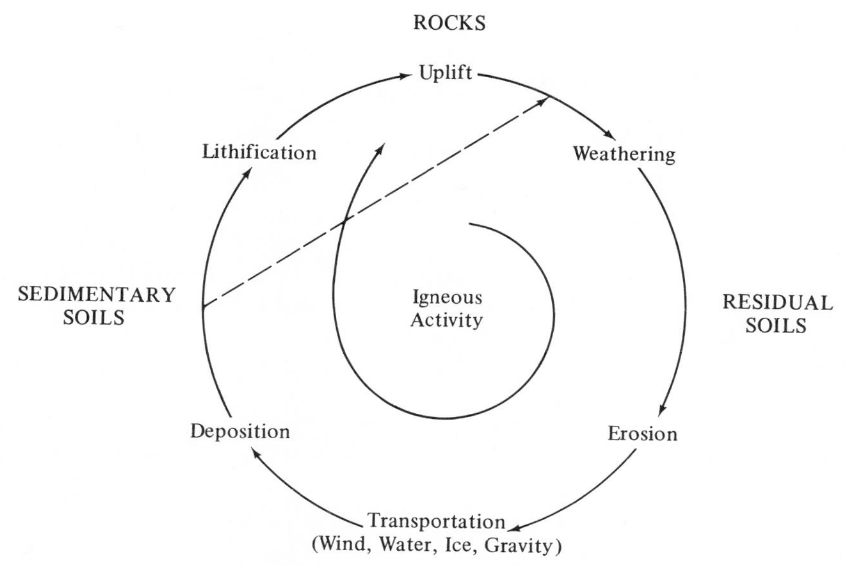

Fig. 3-1. The geological cycle.

eras, volcanoes and associated earthquakes are evidence that such activity still continues.

Igneous activity which involves uplift and exposure to the atmosphere initiates another step in the cycle, slow chemical degradation or weathering. The gradual breakdown of hard rock into soil results in *residual soils* indicated in Fig. 3-1. Residual soils may be very thick in areas of intense weathering such as the tropics, or they may be thin or absent in areas of rapid erosion such as steep slopes of mountains. Residual soils are usually clayey, and their properties are related to climate (discussed in Chapter 2).

In addition to exposing rocks to weathering, geologic uplift also initiates the forces of erosion. This is caused mostly by running water, but also by wind or moving ice (glaciers).

Erosion and transportation are followed by deposition in a different locale. The character of the resulting deposit closely reflects the modes of transportation and deposition, and of course the source material. For example, aeolian dunes consist of sand-size material. The composition of dune sand depends on the source material: As an illustration, the White Sands of New Mexico are dune sands uniquely composed of the mineral gypsum.

Genetically the unconsolidated deposits from wind, water, or ice are sediments. Sedimentary soils assume a highly exaggerated engineering importance because nearly all major cities are located at least in part on floodplains, deltas, and coastal plains, and other extensive industrialized areas are located in regions of continental glaciation. Soil engineers therefore should be familiar with the various sedimentation processes. Deposits by water include alluvial floodplains, coastal plains, and beaches; deposits by wind include sand dunes and loess; and deposits by melting ice include glacial till and outwash. Each of these materials has important behavioral idiosyncracies dependent on geological origin, and a geological name such as loess conveys much useful engineering information. In fact, sometimes the geological information is more useful to engineers than engineering tests, since tests are conducted on soil samples that are never truly "undisturbed."

After deposition, most sediments on the continents immediately reenter the weathering-erosion cycle, indicated by the dashed line in Fig. 3-1. However, sediments deposited in the sea or deeply buried under other sediments gradually undergo a process of *lithification* to form solid rock. In past eras of geological time many loose sediments were lithified into rock. For example, sands became sandstones; clays and silts became shales; and calcareous lime "oozes" became limestones. Many of these rock layers subsequently were uplifted and re-entered the weathering cycle.

The relative importance of past erosion cycles forming sedimentary rocks and sediments is indicated by the abundance of such materials on the continents: Sedimentary rocks constitute a thin "skin" which covers about three-fourths of the area of the continents. Much of the exposed igneous remainder is mountainous and of relatively lesser importance in soil engineering.

3.3. AIRPHOTO INTERPRETATION. Airphoto interpretation is a valuable tool for reconnaissance, for example, in preliminary highway, airfield, or dam location studies, urban planning, etc., where the use of airphotos allows one to literally "get the big picture" with a minimum of cost and effort. Airphotos often reveal soil boundaries with such clarity that they now are used as base maps for U.S.D.A. County Soil Surveys and topographic maps of the U.S. Geological Survey. Airphotos are an invaluable aid for materials prospecting and military uses, provided one knows how to interpret them.

Airphoto interpretation is a skill requiring specialized training and experience, but also is a valuable discipline for those who do not plan to become professional photo interpreters. This is because a systematic study of geologic landforms in relation to soils and rocks is invaluable for

Fig. 3-2. Infrared photograph of Malaspina Glacier, Alaska. (Authors' photos unless otherwise noted.)

understanding their occurrences and in many instances their properties. Furthermore, recognition of soil patterns which are obvious from the air helps one to recognize patterns not-so-obvious on the ground. For example, one who has studied limestone sink topography on airphotos will have little difficulty in recognizing such features and their engineering implications on the ground, in an area where perhaps one person in ten would notice anything unusual. The study of geologic landforms is termed *geomorphology;* airphoto interpretation in this sense is an example of applied geomorphology.

Recent trends in photo interpretation involve use of special films such as infrared (Fig. 3-2), color, etc., and remote sensing techniques such as side-looking airborne radar (SLAR) and infrared imagery, the latter to identify "hot spots" of volcanic or ground water activity or thermal pollution. Photos made from earth satellites offer advantages of wide simultaneous coverage, showing gross terrain and weather patterns.

I. SEDIMENTS

3.4. THE PLEISTOCENE AND RECENT. Whereas most rocks are many millions to several billions of years old, the Pleistocene Epoch of continental glaciation includes only the last million years and ended only about 10,000 years ago. In fact there is no widespread agreement that it has really ended. Even though Pleistocene deposits are geologically young, much weathering and erosion have occurred, and much of the earth's landscape reached its present form during the Pleistocene. Following the Pleistocene is the Recent, which began after the last glacial retreat. Deposits of the Recent include extensive river floodplains, beaches, sand dunes, and gravity deposits such as landslides. A geological classification of transported or residual soils which were formed for the most part during the Pleistocene and Recent is given in Table 3-1.

TABLE 3-1. Geological Classification of Soil

Position Group	*Transporting Agent*	*Geological Class*
Residual soils	None	Residual
Transported soils or sediments	Water	Alluvial
		Deltaic
		Marine
		Lacustrine
	Ice	Glacial drift
	Wind	Loess
		Dune sand
	Gravity	Colluvial

3.5. GLACIATION. The maximum limits of glaciation in the U.S. (Fig. 3-3) are very roughly bounded by the Missouri River on the west and the Ohio River to the east, overrunning the Missouri and in some places not reaching the Ohio. In addition, all of New England, New York, and the northern parts of Pennsylvania and New Jersey were glaciated. In Europe, most of the British Isles, Scandinavia, and northern parts of Germany, Poland, and Russia were glaciated. An illustration of spreading by glacial ice is shown in Fig. 3-2. In other areas mountain glaciers also advanced during the Pleistocene, leaving their deposits. Glacial soils are therefore of major importance in soil engineering.

Glaciation was a grand leveler of topography. The basal ice incorporated rocks and soils and became a glacial "sandpaper," grinding off hills and filling in valleys of pre-existing landscape. Bedrock in glaciated areas often shows long fluted gouges from ice movement, and boulders in glacial deposits frequently show fluted and striated faces, evidences of the tremendous grinding action occurring within the creeping masses of soil-laden ice (Fig. 3-4). Glacial erosion stripped many northern areas of soil, leaving irregular elongated lakes due to plucking and gouging out of the bedrock. The Great Lakes are in part due to glacial

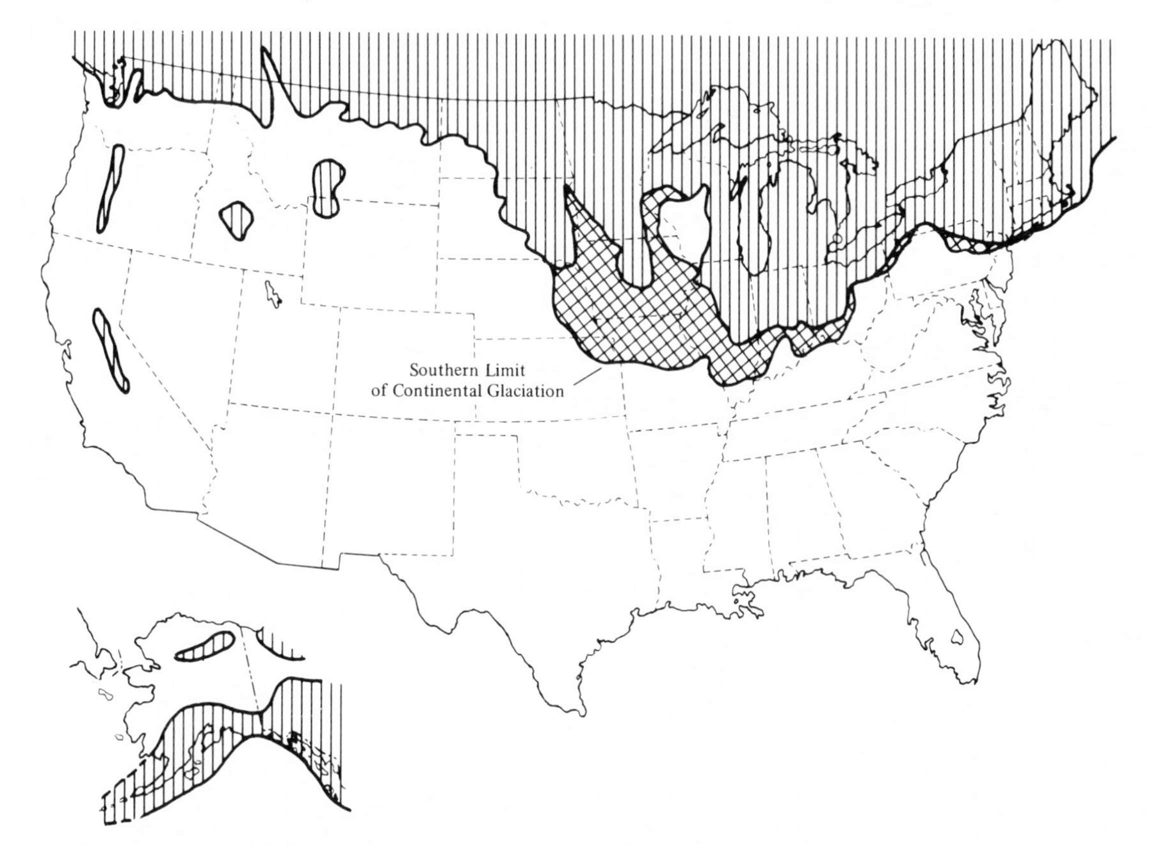

Fig. 3-3. Major glacial deposits of the U.S. Modified from "Glacial Map of the U.S. East of the Rocky Mountains," Geological Society of America, 1959, and "Retreat of the Wisconsin and Recent Ice in North America," Geological Survey Canada Map 1257A, 1969. Key: vertical shading, Wisconsin Age moraine; crossed shading, Kansan and Illinoian Age moraine (weathered, mostly loess covered); unshaded, nonglaciated and Nebraskan Age remnants.

Fig. 3-4. Striated glacial boulder in the Midwestern U.S.

scour action and suggest the lobate nature of the continental glaciers. Glacial activity in the U.S. was more a matter of deposition than of erosion, and the glaciated areas for the most part are mantled with the pulverized rock debris from melting glaciers.

3.5.1. Till, Drift, and Moraine. Glacial deposits are collectively known as glacial drift. The most abundant kind of drift is glacial till, a nonstratified mixture of all particle sizes (Fig. 3-5). Composition of the drift reflects the source area, which for the most part is less than a few hundred miles away. As in the case of other materials discussed in this chapter, glacial drift and till are identified from airphotos on a basis of landform. The most common glacial landforms, termed moraines, are classified as follows:

Ground moraine has a gently to moderately rolling topography sometimes termed swell-and-swale. Drainage is poor and for the most part directed into the swales, often resulting in a generous scattering of lakes or peat bogs. The swells sometimes show a subparallel "minor

Fig. 3-5. Active deposition of glacial till, an unstratified mixture of clay, silt, sand, and boulder. Matanuska Glacier, Alaska.

moraine" pattern first described by Gwynne (*6*), outlining former positions of the glacial front. In older ground moraine the swales are mostly filled with dark topsoil eroded from the adjacent swells, giving a mottled appearance to air photographs. Filled swales are avoided for engineering purposes because of the occurrence of peat and soft alluvial clay.

End moraines (terminal and recessional) are more hilly and usually higher than ground moraine. They are curvilinear (Fig. 3-2) and continue for miles, and probably indicate a temporarily stationary ice front. That is, these moraines piled up when the rate of melting equaled the rate of ice advance, partly because of impedance to ice movement by the moraine itself. Presumably a uniform rate of retreat of the ice front gives a more uniform ground moraine. The often-used term "ice retreat" is a misnomer, since the ice did not back up; instead the ice front retreated when the rate of ice melting exceeded the rate of advance. Sometimes ice masses stagnated, the rate of movement locally becoming zero.

End moraines as well as ground moraines often contain kettles, which are small, deep lakes marking the positions of stagnant ice blocks buried in drift. The blocks subsequently melted, leaving lakes. Kettles are most commonly associated with stratified drift, but also occur in till.

Glacial till in ground moraines is often very compact and hard, presumably from consolidation under the weight of the ice. In contrast, till in the end moraines is much looser and less desirable from an engineering standpoint. The two types of till have been referred to as "lodgement till" and "ablation till," respectively, but no definitive criteria have been set up for their identification. "Superglacial till," an accumulation of stones on the top of the glacial ice, is commonly observed on active glaciers and is presumed to contribute to ablation till. However, stony surfaces or "stone lines" also develop after till is deposited, owing to surficial erosion with selective removal of the fines.

3.5.2. Stratified Drift. Continuous melting of ice during the several glacial advances and retreats modified most major drainage systems, not so much because of the additional runoff water as because of the tremendous additional loads of sediments (Fig. 3-6). We know from modern glaciers that glacial meltwater streams are extremely muddy and loaded to their carrying capacity with all particle sizes up to and including small boulders. As a result the streams do not meander lazily down a floodplain; they race downhill and occupy most of the floodplain with a wild series of interconnecting, rapidly shifting, channels. This is a *braided stream pattern,* shown in Fig. 3-6. Deposits from braided streams are

Fig. 3-6. The Matanuska River, Alaska, a braided stream carrying glacial outwash.

primarily sand and gravel. Stratified drift thus has considerable commercial importance in engineering. Airphotos are a valuable aid in prospecting for such deposits.

Most stratified drift is *outwash,* deposited by glacial meltwater streams beyond the existing glacier margin. The excessive sediment load plus glacial lowering of sea level repeatedly modified equilibrium stream gradients, so outwash in stream valleys now occurs either in terraces above present floodplains or as gravelly substratam beneath the modern floodplain alluvium. Both are frequently exploited as sources for sand and gravel. The terraces, unless covered by younger deposits, usually exhibit a braided stream pattern on airphotos. Outwash terraces occur along the Ohio, Mississippi, and Missouri River floodplains and smaller tributary rivers in these systems.

Another common occurrence of outwash is as *outwash plains,* which occur close to the glacial margins. They are usually fan-shaped with the apex or apices at the glacier. Long Island and Cape Cod contain such a series of fans.

Although outwash extends far beyond glacial margins, it also may be deposited within a glaciated area after retreat of the glaciers. Other types of stratified drift appear to have been deposited under or on top of the ice. *Eskers* are long, sinuous sand-gravel ridges which probably represent beds of subglacial streams near the ice margin. *Kames* are sand-gravel mounds which apparently originated either as crevasse-fillings in the ice or outwash features against the ice, subsequently collapsing when the ice melted. *Kame terraces* occur along glaciated valleys, but have an irregular topography and included kettles as a result of similar collapse. Kames and kettles frequently occur together, constituting "kame-and-kettle" topography. Eskers and kames are identified on airphotos by their light tone, indicative of good drainage.

Ice marginal lakes were common as the ice melted, particularly where natural drainage to the north was prevented by the presence of the ice. Some of these former lakes were very extensive; for example, in Canada, Lake Agassiz, which extends into eastern North Dakota and western Minnesota, was larger than all the present Great Lakes combined. Because of the availability of outwash, such lakes rapidly silted up. Now drained, they are flat, nearly featureless plains. The lake beds usually show regular layering which has been presumed to be annual; these are referred to as *varved clays.* The lakes also show other characteristic lacustrine deposits, namely beaches and deltas.

3.5.3. Loess. An important by-product of glaciation was widespread dust storms arising from winds over the outwash. Modern examples exist

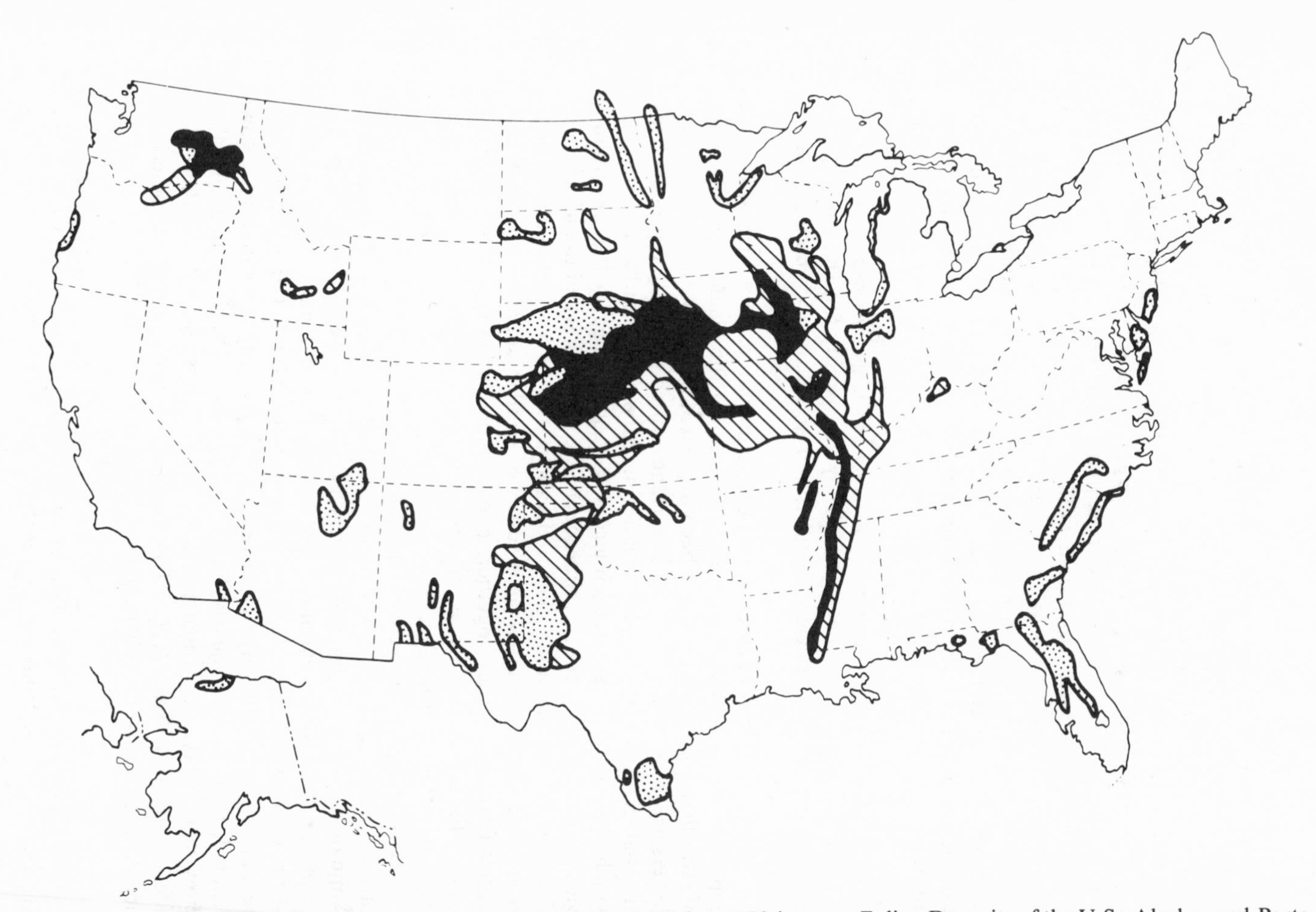

Fig. 3-7. Major eolian sand and silt deposits of the U.S. Modified from "Pleistocene Eolian Deposits of the U.S., Alaska, and Parts of Canada." Geological Society of America, 1952. Key: dots, dune sand (mostly stable); shading, loess (more than 16 ft thick); lines, loess, 4–16 ft thick.

along glacial outwash streams in Alaska. Some of the most monumental dust storms of all time apparently occurred during the Wisconsin Glacial Stage, and covered the central U.S., Europe, China, and Russia with thick deposits of silt called loess.

Loess is thickest close to former outwash areas such as the Platte, Missouri, Mississippi, and other river floodplains, and it thins exponentially with distance from these presumed sources. The maximum thickness in the U.S. is about 150 ft, and the loess cover extends from Nebraska and Kansas eastward to southern Ohio, and southward along the Mississippi floodplain into Tennessee and Mississippi (Fig. 3-7). Other small areas also occur, the most notable being the Palouse loess in southeast Washington.

Where the loess is adjacent to a source, the combined effects of easy erosion plus the low base level nearby cause deep gullying, giving almost a badlands topography (Fig. 3-8). However within a few miles the loess usually thins sufficiently that it merely mantles the older topography. Dissection of the thick loess and gullying of the thick or thin loess constitute the major hints on airphotos that loess is present, since the gullies are usually steep-sided to a considerable depth. The clay content and degree of weathering increase with distance from a source, causing a gradual transition from a friable silt to a less erodible sticky clay soil.

Fig. 3-8. Thick loess with deep gullying and catsteps. Valleys are occupied by loess-derived silty alluvium.

A very important engineering characteristic of loess is its low density and resulting high permeability. In areas of moderate rainfall much of the loess has never been saturated with water. If ever it does become saturated, as from canals, downspouts, etc., its strength often decreases sufficiently that its structure collapses and it consolidates under its own weight. Saturated loess is very weak and is a frequent cause of foundation problems.

Loess is perhaps most famous for its mystical ability to stand in high, steep bluffs; however, conventional soil mechanics appears to satisfactorily explain away much of the mystique, as discussed in later chapters.

3.6. DUNES. One of the more dramatic depositional landforms is sand dunes. While active dunes are easily recognized, few people appreciate the extent of older dunes that are now stable and covered with vegetation (see Fig. 3-9).

Sand dunes become stable and vegetated if something happens to shut off the supply of sand. Thus stable "cliff-head" dunes commonly

Fig. 3-9. Stable sand dunes, Nebraska. Photograph width, 1.25 mi. (U.S.D.A. Air Photograph.)

border former outwash-carrying alluvial floodplains. "Shore dunes" are common adjacent to sandy beaches. They may be either active or stable, depending on whether the beach is modern or an older, abandoned beach or "strand line." Prominent examples of the latter occur around the Great Lakes and in the Carolinas.

A third major occurrence of dunes is in deserts, where they are mainly derived from extensive alluvial fan deposits discussed below. Unfortunately to many people "desert" connotes only sand dunes, which actually comprise at most about one-fourth to one-third of the desert areas.

Unlike silt, which is carried in suspension by the wind, sand grains settle rapidly enough that they bounce along on the ground, sand-blasting exposed rocks and clipping off old fence posts close to the ground. Because of this transport by saltation, sand dunes gradually build out from a sand source and submerge the older landscape.

Dunes may exhibit many shapes indicative of directions of the prevailing winds. *Transverse dunes,* shown in Fig. 3-9, resemble gigantic ripples, being wavy and irregularly interconnecting, and transverse to the main direction of sand-laden wind. Farther downwind or at other places where the supply of sand is not sufficient, the transverse dunes are often partitioned by the wind into *barchans,* the classic cresent-shaped dunes with the tails extending downwind. Barchans usually occur *en echelon,* in staggered rows, a tail from one dune pointing to the head of the next dune.

According to Bagnold (*3*), occurrence of two dominant wind directions causes one tail of a barchan to grow longer, giving a *seif.* In seifs the long tails sometimes link dunes into a continuous chain constituting a *longitudinal dune* running more or less parallel with the major wind.

Because of dependence of dunes upon saltation, dune tracts always attach to the source area either directly or by means of *desert pavement,* a protective mosaic of stones too large to be removed by wind. Such pavements are practically devoid of vegetation because of the sand moving across. On otherwise featureless desert pavements and sand plains, thin, very straight sand strips or streaks resulting from a single storm may sometimes be seen running parallel to the wind direction.

One characteristic of all large dunes regardless of shape is a slip face on the leeward side. This is the primary area of sand deposition, the sand grains, having bounced up the windward side and over the dune crest, whence they sprinkled onto the slip face. Sand on the slip face is therefore very loose and continually readjusts to the angle of repose (30–35°) discussed in Chapter 19. Since dunes build or migrate downwind, their bulk is comprised of old slip faces and the interior sand is rather soft.

Sand bombardment erodes the windward side of unattached dunes, keeping the dune size down and causing them to migrate. Bombardment by sand grains also densifies the windward surface into a crust and causes the crust to creep uphill and act as a protective skin on the dune. Nevertheless since the windward side is eroding and the leeward side is depositing, the entire dune migrates downwind with little or no change in size or shape.

Dune migration sometimes poses a very difficult engineering problem, and construction should be avoided on the leeward and apparently "safe" side of active dunes. Larger dunes migrate more slowly because more sand is required to build a large slip face a given distance. While data are scarce, measurements of several barchans in Egypt gave an average rate of advance of about $180 \div H$ m/yr, where H is the dune height in meters, up to a maximum rate of 20 m/yr. These figures may be expected to vary widely depending on the availability of sand and frequency and velocity of winds. Where precise, such data could be used for past or future dating. For example, if the most extensive dune tract in the U.S., the Sand Hills area of NW Nebraska, is taken as 125 mi (200,000 m) across from NW to SE, and the dune migration was at the maximum rate of 10 m/yr, a minimum of 200,000/20 = 10,000 years were required to cover this area. For comparison the Wisconsin loess was deposited contemporaneously over a period of about 15,000 years. This may be only an interesting coincidence, since the origin of the Sand Hills is complex and debatable.

Shore and cliff-head dunes are analogous to snow drifts and do not migrate except by gradual transgression of the slip face. They therefore may be quite large. Excellent examples may be seen along the south shore of Lake Michigan from airplane flights into Chicago. In arid areas where there is no vegetative anchor, barchans tend to form and break away to the leeward and are true "children" of such dunes.

Although active dunes for the most part are in sparsely populated areas they pose difficult engineering problems. Stabilization with vegetation is effective, particularly if the vegetative growth can keep pace with slow submergence by sand. Sprinkling with pebbles or covering with asphalt in effect make a desert pavement, preventing erosion of the windward side of migrating dunes and retarding their migration. Where encroachment is imminent and protection paramount, a timber or steel tunnel may be built over a road or railroad, and the dune allowed to pass over it. Pressures on such a structure would be calculated by conduit theory, discussed in Chapter 25.

3.7. ALLUVIUM. Although somewhat limited in areal extent, deposits from streams and rivers are among the most important to the soil engineer, because many highways, most major cities, and practically all major bridges are located near rivers. Most rivers occupy *floodplains*—low, fairly level plains composed of alluvial sediments and sometimes subject to flooding.

Rivers with floodplains are said to be *mature*, meaning that downcutting is being prevented by a *base level*, and the river excess energies have been diverted to lateral cutting which makes the floodplain. Hard rock outcrops may form local base levels, but the ultimate and most important base level is sea level. That is, a stream cannot incise vertically much below sea level or it would have to run uphill. Streams that have not cut to base level, such as mountain streams, gullies, and small tributaries that make up most of a drainage pattern, are called *youthful;* they occupy V-shaped valleys and have no floodplain except perhaps for a narrow smear of poorly sorted localized alluvium.

Mature rivers are sometimes *rejuvenated*, or become youthful owing to a sudden lowering of base level. The river then entrenches, eventually cutting a new floodplain and leaving the old floodplain remnants as a terrace.

The opposite effect also occurs; a base level may rise and cause the river to aggrade or build up its floodplain to occupy a much wider area than would otherwise be possible. This is common because the last melting of the continental glaciers raised sea level an estimated 300–400 ft. Measurements of this amount differ because of contemporaneous movements of the earth's crust. However, the rise in sea level some 10,000 years ago created many *estuaries* (drowned river valleys), *fjords* (drowned glacial valleys), and anomalously broad alluvial floodplains which rapidly narrow upstream, such as the Mississippi floodplain south of Cairo, Illinois.

Mature streams are of two general types: braided and meandering. Some rivers such as the Missouri are borderline, sometimes meandering, sometimes braided, depending on local conditions. After channel straightening the modern channel may be braided whereas channels occupied 50 or 100 years ago were meandering, indicative of changes wrought by man.

3.7.1. Braided Stream Deposits. Braided streams have a high gradient, are rapidly flowing when they are flowing, and are characterized by a number of dividing, recombining, shifting channels enclosing almond-shaped sand bars (Fig. 3-6). Braided streams carry a large sediment load for their amount of water, such that small velocity changes cause deposition and plugging of a channel and shifting to a new channel. Because of their high gradient they can be very erosive, actively broadening the floodplain in spite of the heavy sediment load. The high gradient also dictates the deposits, which are usually sand and gravel since the water velocity is seldom slow enough to deposit silt and clay. Braided streams were and are a major source for dune sand and loess, the loess being derived from thin silt deposits left on bars during waning stages of the river.

Braided streams occur in two natural environments: arid and semiarid areas, and as glacial outwash. Both environments contribute excessive sediment for the amount of water. In arid and semiarid areas the streams are intermittent and may flow only every few years. In such areas bridges may not be used because of the tremendous energy and channel instability of the "flash floods" which wash bridges and roads out.

A special case of braiding is the *alluvial fan*, caused by a loaded youthful stream issuing onto a plain (Fig. 3-10). The sudden decrease in gradient causes deposition to occur. Since the stream is free to shift radially, a broad fan-shaped deposit is built up, the handle of the fan being the contributing stream valley. As the fan builds, some sorting ac-

Fig. 3-10. Alluvial fans encroaching onto a river floodplain. Rio Grande Valley, New Mexico. (U.S.A.F. Photograph.)

tion occurs, coarser particles being deposited first at the apex, and fines being deposited farther out or removed if the stream flows into a larger river. In closed desert basins the fines are often carried to one or more central lakes or *playas,* where they form an intermittently dry clay deposit.

Where adjacent fans coalesce, the resulting sloping plain is called a *bajada.* In many desert areas fans are so extensive they are submerging adjacent mountains. The most formidable examples in the U.S. are in the Basin and Range Province of Nevada, Utah, and Southern California, and include Death Valley in California. The desert lakes were much larger during the Pleistocene because of glacial meltwaters and greater rainfall. The largest and best known in the U.S. is Lake Bonneville, Great Salt Lake being the modern remnant.

Deposits from braided streams (including glacial outwash) are mainly sand and gravel, depending in part on what is available. Because of the erratically shifting manner of deposition the deposits are usually cross-bedded, with curved bedding planes truncating those underneath. Individual beds or deposits tend to be finer-grained at the top since deposition occurs in waning river stages with gradually decreasing water velocities. The resulting gradual transition from coarse to fine is termed *graded bedding.* In late flood stages a thin layer of silt is sometimes deposited over bars and constitutes the major source for loess.

Braided streams and their floodplain deposits are easily recognized on airphotos from the many branching channels enclosing almond-shaped bars. Older deposits occurring in terraces are less conspicuous because of masking by vegetation and younger flood deposits, loess, etc., and processes of weathering and erosion. Braided streams and alluvial fans may be seen in miniature after a rain in roadside ditches adjacent to bare, eroding soil slopes.

Since rivers draining glaciated areas were braided during the Pleistocene, remnants of these former wide, braided river beds are valuable deposits of sand and gravel for engineering construction. Because the stream gradient was steeper than at present, such deposits usually occur as terraces higher than and bordering the modern floodplain close to the glaciated areas, but downstream the terraces gradually dip below and are submerged by the modern floodplain. Near the sea the deposits are several hundred feet below the floodplain because of lowering of sea level during the Pleistocene. Even where buried, such deposits are important for support of foundations and piling, since they are typically much harder than the more recent alluvium.

3.7.2. Meander Belt Deposits. For some reason nature rejects the notion of a straight river, in spite of heroic and bullish efforts of man to

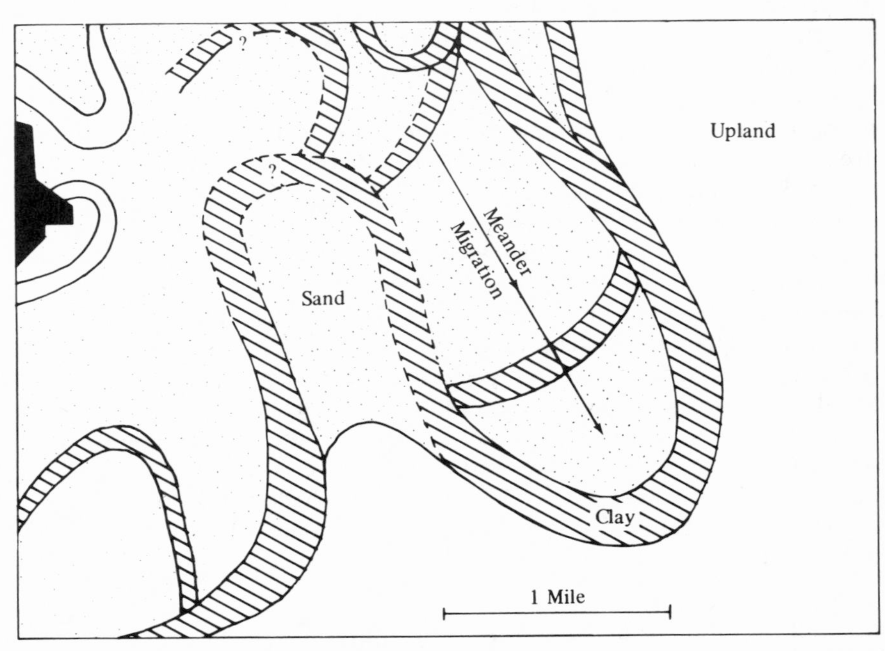

Fig. 3-11. Airphoto interpretation of a heavily forested floodplain in Alabama. (U.S.D.A. Photograph.)

straighten them. Straight channels soon develop a series of riffles, or gravel bars, spaced downstream at an interval of five to seven river widths. Many rivers meander, individual meander arcs being separated by shallow reaches which occur at approximately the same interval as the gravel bars.

While the causes for meandering are not understood, the effects are easily seen. As the river flows around a meander, linear momentum of the flowing water carries it to the outside, increasing the velocity, channel depth, and bank erosion around the outside of the meander. In uniform material the erosion of outer downstream banks coupled with sand deposition along inner upstream banks causes the entire channel meander pattern to gradually shift downstream. The deposits within meander loops are therefore sandy *point bars* (Fig. 3-11). Since bank erosion and downstream shifts mainly occur during late flood stages, point bars contain arcuate ridges, indicating former river margins.

Whenever a meander loop migration is slowed or halted, the next loop will catch up, reducing the point bar between the two loops to a narrow neck. Eventually this leads to a *neck cutoff* and abandonment of a meander loop. Since the loop represents a considerable detour for the river flow, the major flow is diverted through the cutoff and the ends of the abandoned loop rapidly fill with silty sand, isolating it as an *oxbow lake*. Gradual filling of the oxbow by sedimentation from trapped flood waters builds up a soft clay deposit known as a *clay plug* (Fig. 3-11). Clay plugs, being hard to erode, in turn are the most common resistance to meander migration, and hence are the cause of neck cutoffs, renewing the cycle.

Active meandering and channel changing often occur within only a portion of wide floodplains because of confinement on either side by old clay plugs. In this case the active area is called the *meander belt*. Deposits in the meander belt are mainly point bar sands extending to the maximum depth of scour, and clay plugs whose thickness reflects river channel geometry when the cutoff occurred during high water with attendant scour. Up to several feet of flood-deposited clay will mantle older point bars, obscuring the arcuate ridge pattern.

3.7.3. Overbank Deposits. The term floodplain implies a plain which floods, a major factor in most zoning ordinances. During high water the sediment load in a river usually increases, since floods arise from an unusually high surface runoff that also accelerates soil erosion. When the river is in flood, i.e., out of its banks, flow of the flood water is fast in the channel because of a better hydraulic geometry, whereas water beyond the banks flows slower, both because it is shallower and because of roughness afforded by vegetation, etc. Reduction of the water velocity causes deposition of sediment, the coarser material (fine sand and silt) close to the river channel, and the finer-grained remnant farther out. The silt deposits form low, broad ridges along the river bank, termed *natural levees;* the clay deposits gradually mantle the floodplain beyond the meander belt or channel belt, converting it to *backswamp.* Natural levees are preferred sites for buildings and highways since they flood less frequently than backswamp and consist of less clayey materials. Some rivers such as the loess-laden Yellow River in China develop natural levees that contain the river during normal stages, holding the river level higher than the adjacent floodplain, advantageous for irrigation, but precarious for flooding.

When floodwaters breach a natural or artificial levee the sudden outrush carries considerable sand that is deposited in a thin veneer having a braided pattern. Such *sand splays* may be misinterpreted as bona fide braided stream deposits suitable for foundations or gravel sources, when actually they are too thin to be valuable. Their main value lies in indicating where rapid overflows have occurred and may be expected in the future.

3.7.4. Deltas. A river flowing into a large body of water such as a lake or the sea quickly loses velocity and deposits its sediment load. Finer materials settle slowly from suspension and are widely distributed, constituting relatively thin *bottomset beds*, or bottom deposits. The main portion of the delta ideally consists of *foreset beds*, materials presumably deposited at the angle of repose. On top of these are a relatively thin layer of *topset beds.* The foreset beds in fresh-water deltas are mostly sand and silt, but in marine deltas they commonly include clays which have been flocculated by the salt water and settle rapidly. Topset beds resemble the river deposits, being extensions of the channel sands and backswamp clays.

Deltas are commonly arcuate in shape. Bird's-foot deltas such as that of the Mississippi River are less common, and resemble natural levee extensions into the sea. Estuarine deltas are quite common because of the postglacial rise in sea level, and vary from a drowned river valley to a shallow valley with islands and tidal mud flats to a completely alluviated and filled valley.

3.7.5. Beaches. Wave erosion of deltas provides a major source for beach sands, which are distributed laterally by *longshore currents.* The latter result from onshore winds oriented at an angle with the coast line.

Waves involve a near-circular movement of water particles in a vertical plane, the diameter of the circle being the height of the waves. This movement induces a series of orbital motions in underlying water. When a wave progresses into shallows, the orbits are distorted by frictional drag, and the wave breaks. Breaking ordinarily occurs when the water depth is about $1\frac{1}{4}$ times the wave height. Since breakers dissipate energy, sand deposition occurs to the landward side of the break zone, building a submerged *offshore bar*. The bar may be built above the usual water level by storm waves, isolating the shoreline and forming a *barrier beach* and *lagoon*. According to one theory, erosion on the seaward side of the barrier beach and deposition to the landward side causes it to migrate inland over the clayey lagoon sediments, reaching the shore to form a beach. Barrier beaches are common along the eastern and southern coasts of the U.S. and are readily noted even on small-scale maps.

Whereas modern beaches are obvious from the air or from the ground, older beaches resulting from an intermittent lowering of sea or lake levels are less obvious. Often there is a series of old beaches or strand lines. While active, each beach is a source for sand dunes, and dunes also are common on barrier beaches.

Materials in beaches reflect the continual reworking by water; they are well sorted (all one size), well rounded, and variable in size from gravel to sand.

3.8. LANDSLIDES. One of the more important engineering uses of airphoto interpretation is to show scars of old landslides. The land involved in the slide is usually an unstable jumble of wet rock and clay, very poor as a foundation material. Perhaps equally as important, all landslides indicate a possibility of future landslides if similar conditions exist in the immediate vicinity.

Landslides are devastating to life and property, and are usually preceded by slow creep and clicking "rock noises." The latter are inaudible to man but have been electronically detected, amplified, and recorded, and apparently serve as an early warning to animals which manage to scurry to safety well in advance of the actual slide. Where a landslide enters water it makes waves which spread the destructive energy for miles. The worst slide disaster in history involved a reservoir above the Vaiont Dam, Italy, in 1964. The slide area of about 1 × 1.5 mi filled the reservoir, created a wind strong enough to break windows over a mile away, and then created a wave which overtopped the dam some 300 ft and washed through the valley, killing over 2000 people. Ironically, landslides seldom occur without warning; at Vaiont engineers measured a gradually increasing creep rate two weeks prior to the disaster, and animals moved off a week prior to the disaster. As in many difficult matters, the main prob-

Fig. 3-12. Landslide in quick clay, Nicolet, Quebec. (Photograph courtesy of C. B. Crawford, National Research Council, Canada.)

lem appears to be human awareness; at Vaiont a precautionary drainage of the reservoir was started only one day before the slide, and by that time the creep was so fast that the water level rose instead of declining. Evacuation was not carried out since the slide was much larger than expected.

Natural causes of landslides include earthquakes, as in the Turnagain Heights landslide in 1964 in Anchorage, Alaska, and the leaching of flocculating salts, as in the quick clays of the St. Lawrence and Ottawa basins in Canada and marine clays of Norway. A landslide is shown in Fig. 3-12, and the soil mechanics of landslides will be discussed in detail in later chapters. Because of the economic and human factors, landslides pose a formidable problem in soil engineering.

II. ROCKS

3.9. IMPORTANCE OF ROCK IDENTIFICATION. Rock areas which are bare of sedimentary soils may still be weathered to a considerable depth, forming residual soils. The rock parent materials for such soils often show characteristic patterns on airphotos, and not only the rocks and soils but the patterns themselves may have considerable engineering significance. For example, limestone in humid areas tends to weather to form caverns and sinkholes which are readily apparent on airphotos and have a tremendous practical importance in foundation engineering and in the location of dams and reservoirs.

3.10. ROCK OCCURRENCES AND CONTINENTAL DRIFT. Within the span of a very few years the idea of continental drift has worked a quiet revolution in geological thinking, similar to the revolution initiated by Darwin a century ago. Although continental drift is at best an hypothesis and cannot be regarded as fact, it is useful because it fits observations into a readily recognizable pattern, much as the Periodic Table helps in defining and understanding chemical reactions.

Continental drift is a simplified model of earth crustal behavior which may explain the occurrences of mountain ranges, volcanic belts, midocean ridges, ocean deeps or trenches, wide coastal plains, etc. One should notice that these are all linear features instead of spot occurrences, and thus might be explained by either crumpling together or cracking apart of the earth's crust. Continental drift was popularized by the German geologist Wegener in the early 1900s, but was not widely accepted until new evidence was found in the 1960s.

Modern continental drift theory visualizes the earth's crust as a number of plates, some comprising continents and others part of ocean basins. Continental plates are less dense and consequently float higher on the underlying rock than do oceanic plates. For the last 200 million years the plates have been moving relative to one another at a rate varying from 1 to 10 cm/yr. The initiating action is believed to be extremely slow convection currents rising in deep-seated rocks in the earth's mantle (left, Fig. 3-13). The currents rise under the oceans, creating midocean ridges and rifts due to the lifting and separating action. The rock currents then spread laterally both ways away from the ridge, forming and carrying the ocean floor in an action that has been likened to a conveyor belt. Volcanoes also form along the ridge and become extinct as they are carried laterally, gradually sinking below sea level to become submarine "guyots."

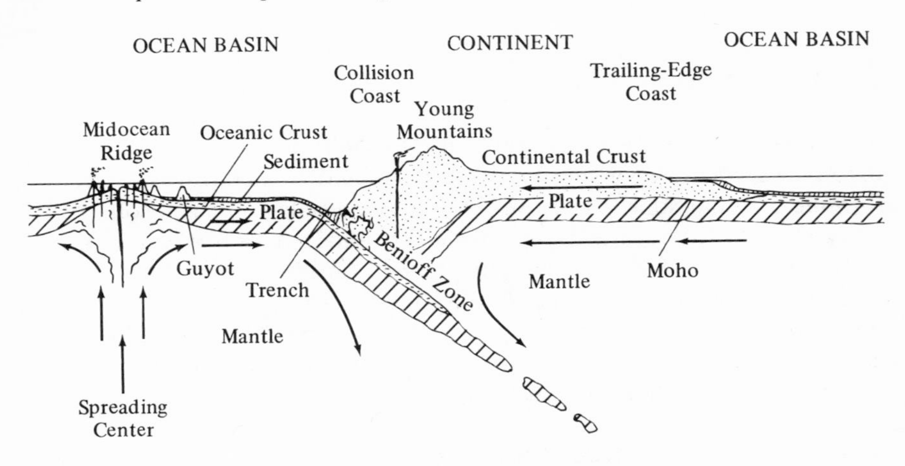

Fig. 3-13. Model for formation of mountain ranges by continental drift. After Inman and Nordstrom (*8*).

Early evidences for continental drift included "fit" of the continents now separated by the Atlantic Ocean, and similar rocks and fossils on both sides up to a certain geological time when the species diverge. Most convincing, however, is recent magnetic and radioactive dating of ocean bottom sediments, which shows that the youngest sediments occur at the midocean ridges, and sediments are progressively older farther away. The dating indicates that drifting began during the age of reptiles, about 200 million years ago, separating continental rocks as old as 3000 million years. For comparison, continental glaciers and early man came on the scene only about 1 million years ago.

The midocean ridges are extensively faulted and are focal points for earthquakes. The Atlantic Ridge runs south from Iceland, roughly halving the Atlantic Ocean. The Pacific-Indian Ocean Ridge extends southward from the San Andreas Fault in the Gulf of California, past Easter Island, westward between Australia and Antarctica, and northwest to the Red Sea. (If this theory is valid, the Gulf of California and the Red Sea may be the first stage of new ocean basins.)

The steady formation of new crustal plate material riding away from the ridges must be compensated by crustal shortening in other areas of the earth. Mountain ranges provide ample evidence for such shortening, the rock layers being folded and faulted over one another and crumpled up into heaps. In fact, the obvious crustal shortening led an earlier generation of geologists to conclude that the earth might be cooling and contracting.

The continental drift theory suggests that mountains form when plates collide or override one another. For example, when two continental plates collide the result is continental mountains, namely the Himalayas. Mountains also form where a continental and an oceanic plate collide wherein the latter, being denser, tends to plunge under the continental plate. The mountains then are at the edge of an ocean basin, as shown in Fig. 3-13. Volcanoes and earthquakes are frequent, as in the south coast of Europe and the west coasts of the Americas.

The remaining possibility, two oceanic plates colliding, results in a mutual plunge to form a deep ocean trench, backed by a line of volcanic islands or "island arc" such as the Aleutians, Japan, Indonesia, the Philippines, Solomons, etc. These also are active earthquake areas.

Continents in active collision with oceanic plates tend to tip away from the colliding edge, creating nonsymmetrical drainage and extensive plains and deltas along the trailing edge. Examples include the east side of the Americas. Continents not in active collision such as Africa have symmetrical drainage and less active stream erosion. Also relatively unaffected by continental drift are the coasts protected by offshore island areas, including the east coast of Asia and the Gulf of Mexico.

This theory accounts for the young and active mountain ranges and explains why they usually occur at continental margins. It does not so readily explain older mountain ranges such as the Appalachians, since the geological record of continental drift goes back only 200 million years, or only about 5% of all geological time. Presumably older ranges may have formed from previous cycles of separation and collision, and this may become a fruitful area for research.

Plate drift theory has some immensely practical implications in prediction and control of earthquakes and perhaps volcanism. For example, plates which are locally immobile may be building up stress which when released will cause a severe earthquake. The sliding friction phenomenon responsible for such action is called "stick-slip" (it is sometimes heard in the classroom as squeaking blackboard chalk). Current research indicates possible control of stick-slip by use of fluid pressure or temperature or both.

Another application of plate tectonic theory is for potential disposal of radioactive wastes. Bostrum and Sherif (*12*) have suggested that such wastes may be more or less permanently disposed by injecting into the trench (Fig. 3-13).

3.11. GRANITE. Granitic rocks, shown by dots in Fig. 3-13, are in effect a "slag" which has separated and floated up from subcrustal basaltic magmas in past geological time. As shown in Fig. 3-13, granites thus

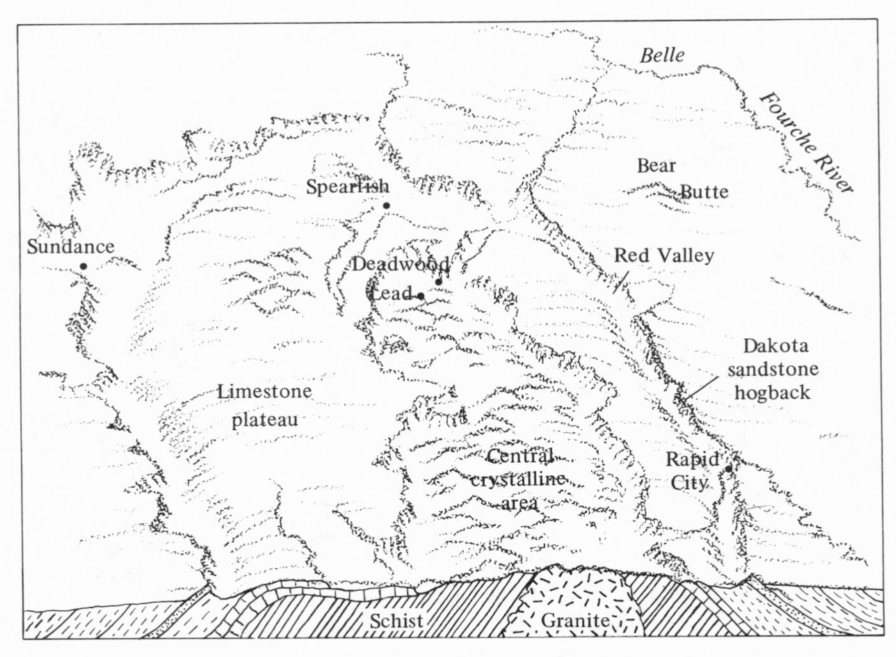

Fig. 3-14. Physiographic diagram of the northern half of the Black Hills, an eroded structural dome with a granite core flanked by metamorphic rocks and overlain by sedimentary rocks. The latter have been removed from the central area by erosion. (Reproduced from A. N. Strahler *Physical Geography*, John Wiley, New York, © 1960, 1969. By permission of the author and publishers.)

form the basement complex of continents and the cores of mountain ranges. They are therefore important as a primordial source for transported soils, which are derived from granite by weathering and erosion. For the most part the granite is covered by younger materials, and it outcrops only in areas of active uplift and erosion (Fig. 3-14). Outcrops of granite in North America are mainly associated with the Rocky Mountain System and an older, eroded system which comprises the Canadian Shield.

Granite on airphotos shows the wear-and-tear of being associated with mountains, particularly the tear. Granite areas are severely fractured in a criss-cross pattern that probably is related to directions of maximum shearing stresses (Fig. 3-15). The fractures may be apparent on a small scale on bare outcrops and make themselves felt on a large scale by influencing the drainage pattern. Instead of the normal fingerlike "dendritic" pattern, streams tend to follow fractures, giving obtuse junction angles and a "rectangular" drainage pattern. Granite outcrops weather along the intersecting fractures until the unweathered rock becomes rounded boulders, as shown in Fig. 3-16. Such large boulders are indicative of weathering, not rolling and rounding by a stream.

Granitic mountains in the United States include the Colorado Front Range, Sangre de Cristo Mts., Bighorns, Beartooth, Belt, Idaho Batholith, Sierra Nevada, the Coastal Range of SE Alaska and British Columbia, etc.

In most instances the granitic core of such ranges is flanked by sedimentary rocks which lap up onto the granite and erode into a mountainous series of linear ridges or hogbacks. The Black Hills are a small, idealized dome-type uplift of a granite core flanked by metamorphic rocks and sedimentary rock ridges (Fig. 3-14).

In the older and topographically less rugged Appalachian System, the Piedmont Plateau and the Blue Ridge Mts., granites and generous amounts of associated metamorphic rocks are included. Stone Mountain, Georgia, and the White Mts. in New England are also granitic, as is a small area designated the St. Francis Mts. of SE Missouri.

Fig. 3-15. Granitic area, northern Minnesota. Fracture patterns are readily visible. Elongated lakes are due to glacial scour. Photograph width, 1.25 mi. (U.S.D.A. Photograph.)

Fig. 3-16. Granite commonly forms rounded boulders, caused by weathering along fractures and exfoliation or spalling of the surfaces and corners.

The primary exposures of granitic rocks are therefore relatively inaccessible and of minor engineering interest, except as obstacles. However, prolonged erosion and uplift through long periods of geologic time expose more and more of the granitic core of mountain ranges, resulting in the second major occurrence of granite in large continental "shield areas." The major shield in North America is the Canadian Shield, which constitutes most of eastern Canada and extends southward into New York as the Adirondak Mts., and northern Minnesota and Wisconsin as the Superior Upland. As might be expected, such areas are geologically complex; furthermore the Canadian Shield has been glaciated. Figure 3-15 shows part of the Canadian Shield.

3.12. BASALT. Basalt is the black lava rock of volcanoes. Volcanoes are associated both with midocean ridges and with mountain ranges, as shown at the left and middle in Fig. 3-13. They also occur in areas paralleling ocean trenches, in effect forming a ring around the Pacific Ocean.

Lava flows emanating from a neck or fissure start to cool and form a surface crust while still moving, typically resulting in a very rough surface texture as the basalt crust continually breaks into blocks which change orientation as the underlying material flows. Bubble holes and voids are common in the lava, and occasionally the crust will form a bridge while the lava beneath flows out, leaving a tunnel. As a result the flows are very permeable and there may be little or no surface runoff. Basalts exposed in cliffs frequently show a columnar structure of vertical jointing presumed to be contraction cracks formed during cooling (Fig. 3-17).

The most prominent volcanic features in the continental U.S. are the Cascade Range, which includes Mts. Shasta, Rainier, Adams, Hood, etc. After eruption and cooling there may be subsidence around the crater to form a *caldera,* usually miles across and sometimes a location of continuing activity. Crater Lake, Oregon, is a well-known caldera; others are prominent in Hawaii. Extensive volcanic and plateau flows also occur in Utah, Arizona, Colorado, and New Mexico. Examples are San Francisco Mountain, Mt. Taylor, and the San Juan Mts. (including

Fig. 3-17. Lava flows sometimes show columar jointing due to cooling contraction cracking. North Shore, Lake Superior.

Fig. 3-18. Vertical airphoto of eroded successive lava flows, Eastern Washington. Width, 1.25 mi. (U.S.D.A. photograph.)

Uncompahgre). Older volcanoes in these and other areas are sometimes eroded until only the central neck is left standing (the cone is relatively easier to erode). Examples are prominent in Western movies.

The Aleutian Islands are a series of about 80 volcanoes in an island arc, some still active. These also include a number of calderas, including Katmai, which erupted violently in 1912, blowing out several cubic miles of volcanic ash.

The Hawaiian Islands all are of volcanic origin with many calderas and illustrate an interesting sequence of events. The volcanic cones or domes tend to sink very slowly, perhaps in part because of the additional weight imposed on the earth's crust, and farther west they are progressively older, lower, and smaller. Eventually only coral reefs remain, since reef-building tends to adjust to equal the rate of submergence. Thus the western islands in the Hawaiian chain except for Midway are a series of reefs inhabited only by birds. Where the reefs display a circular form reminiscent of the old volcanoes they are referred to as atolls.

Less dramatic but far more extensive deposits of basalt are those which exuded from fissures and form wide *basalt plains* such as the Columbia Plateau in Washington, Oregon, and Idaho (Fig. 3-18). Here a vast succession of flows, each averaging about 100 ft thick, extends horizontally for many tens of miles, covering an area of about 200,000 sq. mi. Another major example of a plateau basalt is the very extensive Dekkan Plateau in southwestern India.

3.13. SEDIMENTARY ROCKS. Sedimentary rocks are ultimately derived from igneous rocks such as granite and basalt and are for the most part deposited by water. They therefore are layered and form a relatively thin blanket which nevertheless covers about 75% of the area of the continents.

The three principle sedimentary rocks, shale, sandstone, and limestone, occur in layered sequences that may be continuous for hundreds of miles. If the layers are horizontal or nearly so, the pattern of streams is "dendritic" or branching in a series of acute angles (Fig. 3-19). The harder rocks tend to form a protective cap on flat-topped hills or *mesas,*

Fig. 3-19. Vertical photograph in Mesa County, Colorado. Dendritic drainage; sandstone and (smooth pattern) shale. Width, 1.8 mi. (U.S.D.A. photograph.)

Fig. 3-20. Limestone sinks, Indiana. Width, 1.25 mi. (U.S.D.A. photograph.)

or if there are no resistant rocks the hills tend to be rounded. Outcrop and vegetational patterns frequently give evidence of rock layering, although one must not be taken in by agricultural contour plowing or rice terraces.

One of the most distinctive airphoto patterns is that of limestone in humid areas (Fig. 3-20). In these areas limestone usually contains caverns which occur at or near the water table. Roofs of the caverns eventually collapse, giving a steep-sided pit at most a few hundred feet in diameter with internal drainage, appropriately termed a sink. Many examples are known of caverns collapsing to form sinks under existing structures, destroying the integrity of the structure. Sometimes sinks will contain lakes in the bottom, perhaps a result of plugging of underground drainage by the collapsed material. Where sinks are abundant or overlap, the land form is referred to as *karst topography*.

Sandstone and shale can often be differentiated because of the difference in density of the drainage pattern. Shale characteristically is much less permeable than sandstone, so there is a greater runoff and a higher *drainage density*, defined as the total stream length per unit area. Sometimes by measuring drainage density a fairly positive identification of a dominant rock type can be made even under heavy forest cover.

3.13.1. Tilted Sedimentary Rocks. Sedimentary rocks frequently are tilted sufficiently to modify the outcrop and the drainage pattern. A low dip angle, as in inland portions of a coastal plain, results in erosion to form a series of parallel outcrops, the more resistant rocks standing high as broad asymmetrical hills termed *cuestas*. A higher dip angle, as in the neighborhood of a mountain range, causes a steeper version called a *hogback* (Fig. 3-21). Between successive hogbacks or cuestas are the softer rocks, usually shale, and the streams. Some streams, called *consequent* streams, apparently were in position before the rocks tilted or folded and now cut across the beds; however most streams follow parallel to and between the ridges, and small tributary streams come in at right angles, flowing down the slopes of the ridges. This drainage is termed a *trellis pattern*.

Fig. 3-21. Tilted sedimentary rocks comprising a hogback. From top, shale, hard sandstone (hogback), shale (steep face), sandstone. San Rafael Swell, Utah. Width, 1.4 mi; relief 2000 ft. (U.S.G.S. photograph.)

3.14. FRACTURES AND FAULTS. Fractures were discussed in relation to granite and may be defined as breaks along which there have been no shear movements. Faults are fractures along which there has been movement and are much more serious from an engineering standpoint for two reasons: The shear displacement usually grinds up a considerable amount of rock which occupies the fault zone, and the fault may still be periodically active. Active faults, such as the San Andreas fault in California, are focal points for earthquakes. Structures located on top of an active fault comprise the original high-rent district.

Faults can be recognized by (a) displaced streams, (b) displaced outcropping beds, (c) the fault scarp or cliff itself, which is soon eroded into a series of triangular or trapezoidal facets (Fig. 3-22). A purely vertical movement can cause an apparent horizontal displacement of tilted beds, when the upthrown side erodes down to the same general elevation as the downthrown side.

A common igneous feature that may be misinterpreted as a fault is a dike, which is an igneous intrusion into a fracture. Since the dike usually

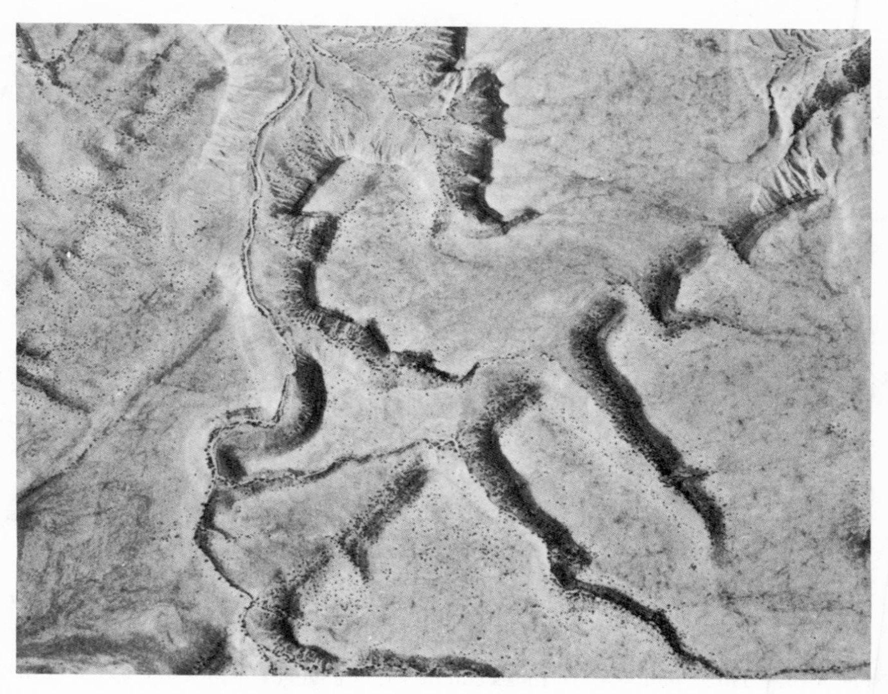

Fig. 3-22. Fault scarp dissected by streams, Utah. Width, 1.25 mi. (U.S.G.S. photograph, selected by R. G. Ray (*11*).

is harder than its surroundings it tends to occur as a linear ridge and may extend for miles. Dikes of almost pure quartz are not uncommon.

3.15. TALUS. Rock fragments falling in mountainous areas accumulate in a steep-sloping pile of loose rubble called a talus. Talus often poses a unique road construction problem since it lies at the angle of repose; therefore if anything is built on it or a cut is made in it, it tends to creep or slide. Roads constructed on a talus require frequent repair or replacement. Failures usually occur after heavy rain or snow melting, when the weight of the water adds to the weight of the loose rock, causing it to slide.

PROBLEMS

3.1. Starting with igneous activity, indicate all geological processes involved in the origin of each of the following: (a) residual soil on lava flows; (b) sand derived from weathered granite; (c) residual soil on sandstone; (d) dune sand; (e) limestone; (f) caverns in limestone.

3.2. Identify the major transportation agent for each of the following sedimentary soils: sand dunes; glacial till; alluvium; beach sand; volcanic ash; talus.

3.3. Indicate areas where one should find the following surficial features or deposits: (a) glacial till; (b) dune sand; (c) beach sand; (d) loess; (e) alluvium; (f) glacial erosion.

3.4. Radiocarbon dating shows Wisconsin Age loess in the U.S. was deposited during the period between 29,000 to 14,000 years B.P. (before present). Compare the average thickness accumulation per year of the thickest loess deposits with the accumulation of dust alongside a dusty road (as much as 0.1 in./yr).

3.5. Suggest two ways to stabilize sand dunes encroaching on a housing development, and indicate how they work.

3.6. The edge of a field of barchans 6 m high is located $1\frac{1}{2}$ mi upwind from an express highway. Estimate the time of arrival, and suggest and discuss ways to meet the problem.

3.7. (a) Name three major rivers in the U.S. which probably were braided during Pleistocene glaciation. (b) What leads you to this conclusion? (c) How is this pertinent to engineers?

3.8. Comment on the location of the railroad in Fig. 3-11.

3.9. Which should give the better prediction of maximum scour depth in a river: water depth in the river, or clay thickness in a nearby oxbow? Why? (Hint: What was the probable river stage when the oxbow was cut off?)

3.10. How does continental drift relate to the design of structures for earthquakes?

3.11. What is the major rock in the Red Valley area of Fig. 3-14? Name two towns located on this rock.

3.12. How might one explain the unusual hollowed-out appearance of the stream valleys in Fig. 3-18?

3.13. Where would you prospect for economical sources of (a) clean sand; (b) road aggregate, in Fig. 3-19?

3.14. Discuss problems in relation to (a) foundations; (b) dams in the area of Fig. 3-20.
3.15. (a) Explain the jagged appearance of drainage lines in the bottom area of Fig. 3-21. (b) Is this area probably a granite? Why (not)? (c) Identify the consequent streams.
3.16. Sketch outcrops of (a) hard sandstone; (b) medium sandstone; (c) shale, in Fig. 3-21.
3.17. Sketch outcrops of (a) sandstone cap rock, and (b) underlying shale, in Fig. 3-8. (c) Are any other faults visible in this photograph?

REFERENCES

1. American Geological Institute. *Glossary of Geology and Related Sciences.* 2nd ed. Natl. Acad. Sci., Natl. Res. Council, Washington, D.C., 1960.
2. American Society of Photogrammetry. *Manual of Photographic Interpretation.* Washington, D.C., 1960.
3. Bagnold, R. A. *The Physics at Blown Sand and Desert Dunes.* Mathven and Co., London, 1941. Reproduced by Dover Publ., New York, 1965.
4. Cadman, J. D., and R. E. Goodman. "Landslide Noise." *Science* **158,** 1182–1184 (Dec. 1967).
5. Flint, Richard Foster. *Glacial and Pleistocene Geology*, 3rd ed. John Wiley, New York, 1971.
6. Gwynne, C. S. "Swell and Swale Patterns of the Mankato Lobe of the Wisconsin Drift Plain in Iowa." *J. Geol.* **50,** 200–208 (1942).
7. Highway Research Board. *Remote Sensing and Its Application to Highway Engineering.* HRB Spec. Rept. 102. NRC-NAS-NAE Publ. 1640. Washington, D.C. 1969.
8. Inman, D. L., and C. E. Nordstrom. "On the Tectonic and Morphologic Classification of Coasts." *J. Geol.* **79,** 1–21 (1971).
9. Kiersch, George A. "Vaiont Reservoir Disaster." *ASCE Civil Engineering* **34** (3), 32–39 (1964).
10. Leopold, Luna B., M. Gordon Wolman, and John P. Miller. *Fluvial Processes in Geomorphology.* W. H. Freeman & Co., San Francisco, 1964.
11. Ray, R. G. *Aerial Photographs in Geologic Interpretation and Mapping.* U.S. Geol. Surv. Prof. Paper 373. U.S. Government Printing Office, Washington, D.C., 1960.
12. Sherif, M. A., Ed. *Proc. of the Int. Symp. on the Engr. Prop. of Sea-Floor Soils and Their Identification.* UNESCO-Univ. of Wash., Seattle, 1972.
13. Strahler, Arthur N. *The Earth Sciences.* Harper & Row, New York, 1963.

4

The Nature of Soil Constituents

4.1. DEFINITION OF SOIL. The word soil, in engineering terminology, refers to all of the fragmented mineral material at or near the surface of the earth, the moon, or other planetary body, plus the air, water, organic matter, and other substances which may be included therein. Soil is a nonhomogeneous, porous, earthen material whose engineering behavior is greatly influenced by changes in moisture content and density. It may range in character from the soft, spongy peat soils through the soft clays, loams, and silts to the coarse-grained sands and gravels, the feebly cemented conglomerates, the hard, stiff clays, and the semi-indurated shales. This definition of engineering soil does not include indurated rock in natural beds and extensive solidified masses. In a broad sense, however, the engineer may consider that bedrock is in a limiting state of solidification and represents an ideal condition for soil upon which to found the bridges, buildings, dams, and other load-producing structures which he builds.

The term *soil* has various meanings and connotations to different professional groups and practitioners. To the agronomist, as well as to the layman, the most appropriate definition is that given by the dictionary which defines soil as "finely divided rock material mixed with decayed vegetable and animal matter, constituting that portion of the earth's surface in which plants grow." Agronomic applications of soil science are directed toward the understanding and improvement of soils for growing the crops upon which mankind depends for sustenance. On the other hand, the design and construction of engineering structures often involve consideration of the soil at depths and in conditions which are far removed from those conducive to the growth of vegetation. The engineer is interested primarily in the determination and the improvement of the strength characteristics of soil. The properties of soil which are associated with high strength—i.e., high density, high internal friction and cohesion, low moisture content, etc.—are those which, in general, are associated with poor tilth and crop-growing conditions. For these reasons, many masses of material with which a soil engineer deals and in which he has a vital interest are so poor from the standpoint of plant growth that in all probability they would not even be called soil by an agronomist. Similarly, the term soil has different connotations in the fields of ceramics and geology.

As an extreme example of different interests, imagine a well-graded sand-gravel material which contains sufficient fines and inorganic colloidal matter having the proper plasticity characteristics to serve as a good binder. Such material, when compacted to a density approaching that of concrete, would serve as an excellent stabilized soil pavement, but it would have no value whatever as an agricultural soil and would hold but little professional interest for an agronomist.

It is to be emphasized that, although the basis of interest and the applied technologies may vary widely among the various professions which deal with the soil, the basic principles of soil science in physics, physical chemistry, petrology, pedology, etc., are the same for all fields of application. For this reason, engineers can learn and have learned much from these other professional groups who have, generally speaking, been studying the soil in the scientific manner for a longer time. Also, there is need for much closer liaison and interchange of thought among these groups. Such cooperation will certainly result in better understanding of the soil and improvement of practice in all these fields of human effort.

4.2. IGNEOUS ROCKS, PRIMORDIAL SOURCES FOR SOIL. As implied in the preceding chapter, the most common igneous rocks are granitic, occurring as the core of new or old mountain ranges, and basalt, occurring as lava flows. Basalt, having been extruded from the earth as lava, cools more rapidly and retains a much finer crystalline size. It frequently contains bubble holes due to exsolution of gases when pressure was released on the lava. An analogous phenomenon occurs when one opens a carbonated beverage. An extreme example of bubbly rock is *pumice* or rock froth, which is so vesicular it may float on water. Very rapid cooling, as

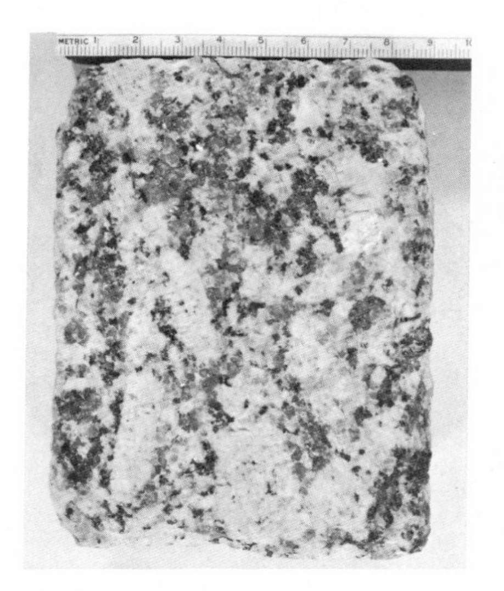

Fig. 4-1. The most abundant igneous rock types are granite (left), which is coarse-grained, and basalt (right), which is microcrystalline as a result of rapid cooling. In the granite, large white (or pink) crystals are feldspar, clear grains are quartz, and dark grains are iron-rich minerals. Specimens are from the Iowa State University Geological Museum.

when lava flows into water or is blown into the air, inhibits crystallization sufficiently to form black volcanic glass, or *obsidian*.

The two dominant types of igneous rocks, granite and basalt (shown in Fig. 4-1), also differ mineralogically. The major minerals in both types of igneous rocks are feldspars, but they differ in composition. Also quartz and mica are abundant in granite, whereas heavy, dark ferro-magnesian minerals abound in basalt.

4.3. CLASSES OF SEDIMENTARY ROCKS. The most abundant sedimentary rock is *shale*, which is a consolidated, thinly bedded mixture of silt and clay. Much less abundant are *sandstone*, which is cemented sand, and *limestone* ($CaCO_3$) (shown in Fig. 4-2).

Limestone may result from either chemical or biological deposition (shells), and usually contains some small amounts of silt and clay. Nodules or nodular beds of *chert* are frequently associated with limestone; chert is a hard, waxy appearing silica rock utilized by American Indians for arrowheads. (Colored varieties of chert may be referred to as flint, agate, or jasper). *Dolomite* occurs in geologically old limestones which

Fig. 4-2. Roadcut in limestone (right) with some thin beds of shale (dark streaks at left). Shale layers end abruptly at a fault. (Photograph by R. A. Lohnes.)

have undergone partial replacement of calcium by magnesium to make $Ca \cdot Mg(CO_3)_2$.

Sandstones close to a granitic source for sand have a granitic composition and appearance and are referred to as *arkose.* An example is rocks of the Garden of the Gods in Colorado. These rock beds are pink in color because of the high content of potash feldspar; they were derived from an adjacent granite source and later were folded up on end by a large thrust fault.

Sand is our most transient particle size in that transportation by wind or water involves "saltation," or bouncing along, which gradually degrades the feldspars and leaves quartz. Thus the eventual product of erosion-deposition cycles is a quartz-rich sand. Cemented, it becomes a *quartz sandstone*.

A third common variety of sandstone is called *graywacke;* this is a gray, fine-grained delta deposit usually associated with shale.

4.4. METAMORPHIC ROCKS. Metamorphic rocks are minor in occurrence, but are locally important, for example, in the Appalachian Mountain area. Metamorphism literally means change in form and occurs as a result of heat, pressure, and solution. Approximate metamorphic equivalents are shown in Table 4-1. Mica schist, which is the most common, is a shiny, micaceous rock that attracts children and illustrates the well-known epigram about all that glitters. Most commercial "marble" is not true marble but merely dense limestone that takes a polish.

TABLE 4-1. Common Metamorphic Rock Equivalents

Original Rock	*Low-Grade Metamorphism*	*High-Grade Metamorphism*
Granite	—	Gneiss
Basalt	—	Schist
Shale	Slate	Mica schist
Sandstone	—	Quartzite
Limestone	Marble	—

4.5. PYROCLASTIC ROCKS. Pyroclastic rocks are granular debris from exploding volcanoes. Such debris includes, in decreasing order of grain size: bombs, cinders, ash, and dust. Such materials result when hot or molten lava is exuberantly tossed in the air by an erupting volcano.

Bombs land close to the crater and usually have an ellipsoidal shape. *Cinders*, which are much more common, also land close, and these angular fragments comprise most or all of the volcanic cones. In contrast, *ash* may be carried tens of miles and constitutes the most lethal aspect of many eruptions; unlike lava flows, which move slowly, ash thrown aloft forms a hot dust cloud that rolls down the mountain much faster than people can escape. Deposits may be tens of feet thick. The ancient Roman town of Pompeii was buried by volcanic ash.

Volcanic ash is a pink to white powder composed of volcanic glass. Because of its lack of crystalline structure the ash quickly weathers to a nearly pure, sticky clay called *bentonite* or an amorphous clay called *allophane*. Bentonite is a nearly pure clay mineral called montmorillonite (discussed later) and is widely used as a base for drilling muds, and to some extent as a basement sealant.

Volcanic *dust* is carried on the wind for tens, hundreds, or even thousands of miles. When Krakatoa, an island in the Pacific, blew up in the early 1900s, dust girdled the earth and contributed an unusual brilliance to the sunsets for many months.

4.6. WEATHERING OF ROCKS. Weathering processes are often described as "physical," caused by thermal expansion, ice action, etc., or "chemical," involving chemical change. However, practically all rock weathering on earth is chemical or represents a combination of the physical and chemical. For example, strictly thermal-induced stresses are seldom sufficient to degrade a hard rock, and analysis of rock spalls invariably reveals some chemical change. Frost action is the most common example of physical weathering, and works by a combination of moisture movement and freezing expansion.

The major mineral constituent in igneous rocks is *feldspar*, which is actually a whole group of minerals. When feldspar weathers it hydrates and increases in volume such that when the outer layer of a boulder weathers slightly and expands, it tends to spall off or "exfoliate," as shown in Fig. 3-16.

Upon weathering, granite tends to separate into constitutent sand-sized grains of feldspar and quartz and is the principle source for sand. Examples of apparently competent "rotten" granite may be found in road cuts. Basalt, being finer grained, weathers directly to clay.

Major rock types and their weathered equivalents are indicated in Table 4-2. Shale probably represents the closest approach to pure physical weathering since separation of silt and clay grains sometimes occurs by wetting and drying. Sandstone weathers by loss of cementation, giving sand, and limestone weathers by dissolution, giving caverns and a residual soil which is clayey or cherty or both.

Caverns are extremely important in foundation engineering since the roof may collapse, giving a *sink*. Limestone areas also pose severe problems for the construction of dams, since the reservoirs frequently leak through caverns and fissures.

TABLE 4-2. Weathering Products

Rock	*Product*
Granite	Sand and clay
Basalt	Clay
Shale	Silt and clay
Sandstone	Sand
Limestone	Caverns and residual clay soil

4.7. NONCLAY MINERALS. Minerals may be defined as solid, homogeneous, naturally occurring materials. Frequently individual mineral crystals in a rock can be seen with the unaided eye, as on a broken or polished face on granite. Other mineral crystals are so small they can be seen only with an optical microscope, and still others require an electron microscope. In soils, the latter constitute a separate category known as clay minerals.

The coarser, nonclay minerals are characterized by physical properties such as color, hardness, and cleavage. Hardness is measured by scratching one mineral with another according to Mohs' relative hardness scale, each mineral in the scale being able to scratch those with lower hardness (Table 4-3). For rapid identification gypsum can be scratched with a fingernail, calcite with a penny, and feldspar with a hard knife or file.

TABLE 4-3. Mohs' Hardness Scale

1. Talc	6. Orthoclase feldspar (knife)
2. Gypsum (fingernail)	7. Quartz
3. Calcite (penny)	8. Topaz
4. Fluorite	9. Corundum
5. Apatite	10. Diamond

Cleavage is a unique property of minerals which causes them to break along one or more crystallographic planes. For example, mica has excellent cleavage in one direction, feldspar has two cleavage directions, and calcite three. Quartz has no cleavage and breaks in a concave rounded or conchoidal fracture. (Quartz may exhibit flat faces, but these occur because the crystals grew that way, not because they broke along planes.)

Engineering properties of minerals relate far more closely to crystalline structure than to chemical composition. This is reflected in mineral names, which often cover a variety of chemical compositions. Thus two soils with identical chemical compositions may differ widely in properties, depending on how the chemical elements are combined into minerals. Some examples of common minerals and their chemical compositions are given in Table 4-4.

TABLE 4-4. Common Nonclay Minerals

Mineral Name	*Composition*	*Rock Occurrences*
Quartz	SiO_2	Granite, sandstone, shale, etc.
Orthoclase feldspar	$KAlSi_3O_8$	Granite, sandstone, shale, etc.
Plagioclase feldspar	$NaAlSi_3O_8$–$CaAl_2Si_2O_8$	Granite, sandstone, shale, basalt
Muscovite (mica)	$KAl_2(Si_3Al)O_{10}(OH)_2$	Granite, sandstone, shale, schist
Calcite	$CaCO_3$	Limestone
Dolomite	$Ca \cdot Mg(CO_3)_2$	Dolomite
Gypsum	$CaSO_4 \cdot 2H_2O$	Gypsum, shale
Amphiboles and pyroxenes	Variable	The heavy dark minerals in granite, basalt, etc.

4.8. STRUCTURE OF THE CLAY MINERALS. The most important minerals in soil engineering are too fine-grained to be characterized by color, hardness, and cleavage, and in fact cannot be seen as particles without the aid of a powerful microscope. As the particle size decreases the particle surface area per gram increases exponentially. The properties of clays and clay soils therefore are in a large part due to their mineralogy and surface chemistry, particularly in regard to the interaction with water, as discussed in Chapter 9.

Clay minerals are structurally similar to and in many cases identical to micas—that is, they have a layer structure, each layer being composed of silica and alumina sublayers or sheets. Within the silica sheet each silicon ion is surrounded by four oxygen ions, comprising a tetrahedron; hence this is termed the *tetrahedral sheet*, sometimes abbreviated "A" because of the resemblance of a tetrahedron to the letter A. In the alumina sheet each aluminum ion is surrounded by six hydroxyl (OH) ions comprising an octahedron; hence this is termed the *octahedral sheet* or "B." The metallic ion in these sheets can vary widely in some minerals without markedly changing properties; for example, Mg^{+2} or Fe^{+2} may proxy for the aluminum, and aluminum may proxy for some of the silicon. These are referred to as *isomorphous substitutions*, since they do not change the mineral form; however, referring to a "B" sheet as an "alumina hydrate sheet" is incorrect, since the sheet may not contain any aluminum.

The method by which individual tetahedra or octahedra combine to form tetahedral or octahedral sheets is by sharing of corner ions, shown in Fig. 4-3. These sheets are the building blocks of clay minerals.

Usually one or more A sheets are joined to a B sheet by a further sharing of corner ions; that is, rings of apical O ions of the A sheet share with similar rings in the B, as shown in Fig. 4-3. The result is in Fig. 4-4, sharing being indicated by points of contact between the tetrahedra and octahedra, where the A joins the B. The layers extend indefinitely in the x and y directions, to crystal edges which therefore have unsatisfied ionic bond positions. A crystal therefore cannot be considered a "molecule" in a classic chemical sense.

In the Z direction, layer is stacked on layer but bonding between layers is relatively weak, so clay minerals are flakey minerals. This same layered structure occurs in mica, an easily recognized nonclay mineral which tends to cleave between layers to form flakes. The minimum thickness of such flakes theoretically is only 10 angstroms (Å), 1 Å equaling 10^{-8} cm.

Nearly all clay minerals have either the AB or ABA layer structure shown in Fig. 4-4. Because of the ratio of A to B, these clays are referred to and classified as 1:1 and 2:1 clay minerals, respectively. A classification of common clay minerals based on A to B ratios is shown in Table 4-5. It should be noted that the layer thickness, which equals the distance between corresponding ions in adjacent layers, closely relates to the structure: the thickness of the AB layer is 7.15 Å and of the ABA about 9.6 Å.

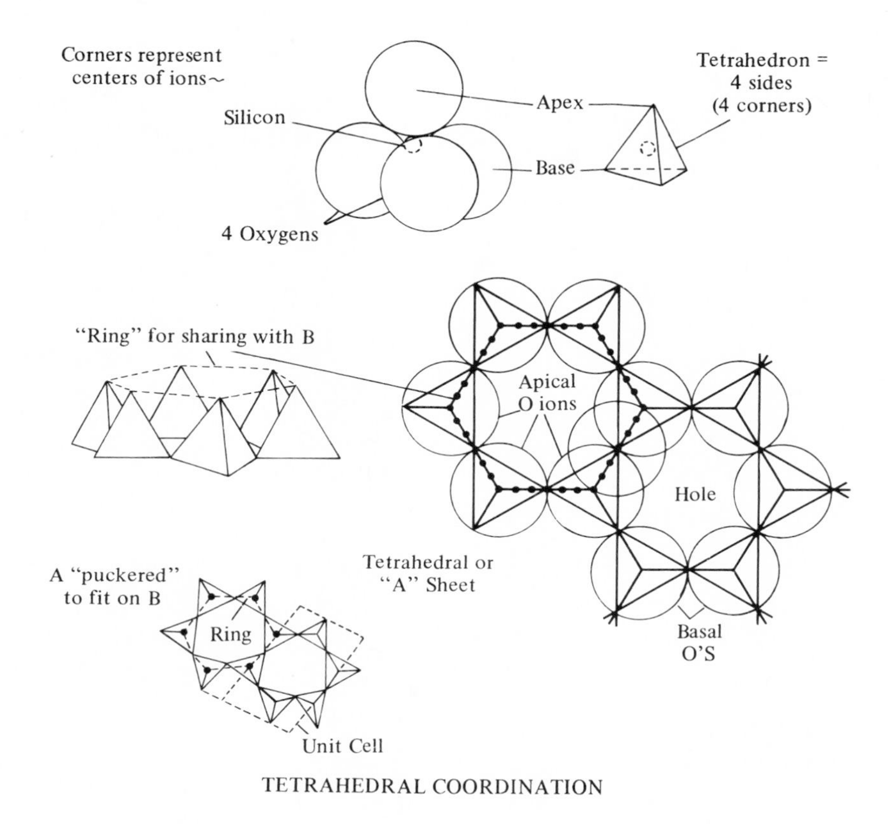

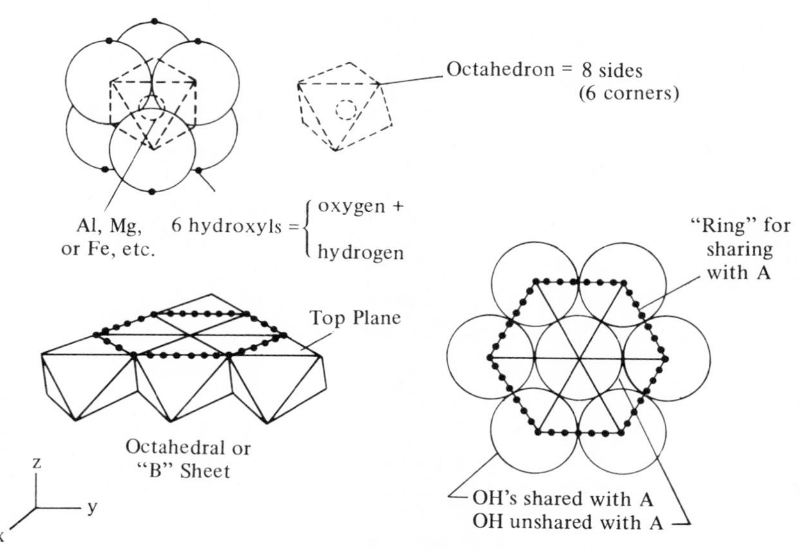

Fig. 4-3. Geometry of the A and B sheets in clays. [After Bradley and Grim in Brown (*3*).]

TABLE 4-5. Classification of Common Clay Minerals

Type	*Group*	*Approximate Negative Charge per Formula Unit*	*Common Minerals*	*Layer Thickness (Å)*	*Typical Occurrence*
1:1	Kaolinite	0	Kaolinite	7.15	Old, residual soils of tropics; SE U.S.
			Halloysite	10 (7.9 air-dry)	Same as kaolinite but under continuous moist conditions
2:1	Montmorillonite	0.5–1.3	Montmorillonite	Expansive: >10 (15 air-dry)	Bentonite; semiarid soils; western U.S.
	Vermiculite	1.2–1.8	Vermiculite	14	Common minor soil mineral
	Mica	1.2–2.0	Illite	10	Common soil mineral Also abundant in shale
		2.0	Muscovite	10	Common white shiny flakes in rocks and some soils
		2.0	Biotite	10	Black shiny flakes in rocks Weathers to vermiculite

2:1:1	Chlorite	0	Chlorite	14	Common in metamorphic rocks and shale
0:1	Alumina	0	Gibbsite	4.85	Common interlayer. Also in highly weathered bauxite (Al ore)
	Iron oxide minerals	0	Hematite	—	Red-brown } Concentrated in tropical laterites
			Lepidocrocite	—	Brown } Concentrated in tropical laterites
			Geothite	—	Yellow } Concentrated in tropical laterites
Variable 1:1 to 1:2	Amorphous clays	0	Allophane	Amorphous	Young, weathered volcanics under wet conditions; Hawaii
Mixed layer (examples)	Chlorite-vermiculite			14	Marine clay; deweathered montmorillonite
	Biotite-vermiculite			10–14	Weathered biotite
	Mica-montmorillonite (hydrous micas)			Expansive; >10	Weathered muscovite
	Illite-chlorite-montmorillonite			>10	Shale

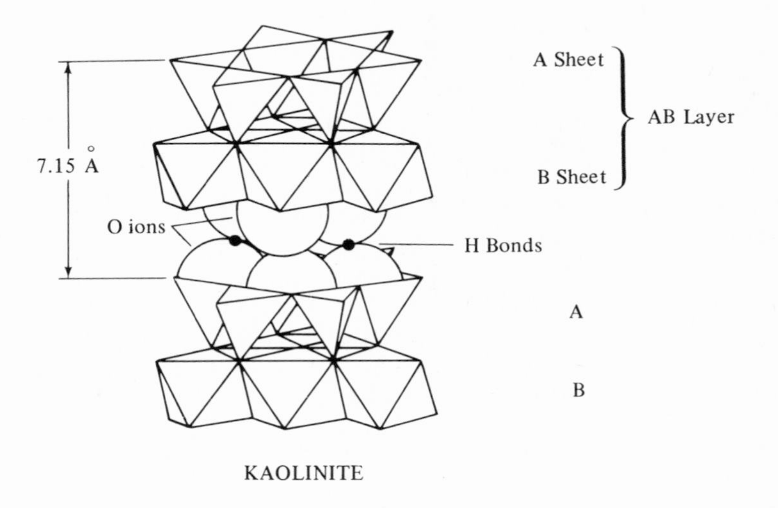

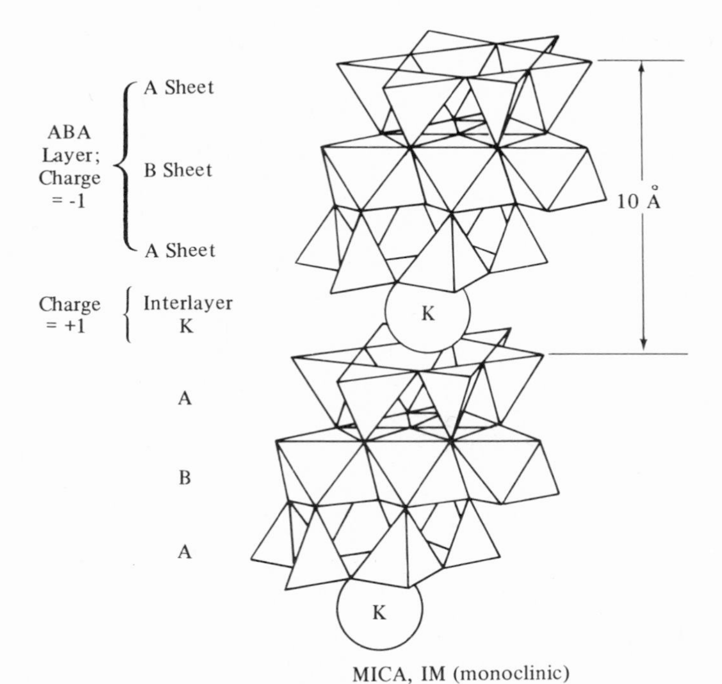

Fig. 4-4. (Top) An AB or 1:1 layer silicate, and (bottom), an ABA or 2:1 layer silicate. Note: Corners of polyhedra represent centers of ions.

To avoid confusion in the use of terms relating to clay sheets and layers, in 1968 the Clay Minerals Society proposed that "plane" refer to a planar arrangement of atoms, "sheet" to a single thickness of linked tetrahedra or octahedra, and "layer" to a combination of sheets (5). For example, a kaolinite AB *layer* is composed of A and B *sheets*, in turn made up of *planes* of oxygen, silicon, etc., ions. The reason atoms in minerals are referred to as *ions* is that they have (+) or (−) charges, which hold the crystal together.

4.9. COMMON CLAY MINERALS. Table 4-5 lists the common soil clay minerals now recognized. The earliest known mineral, kaolinite, when heated in its pure form makes porcelain. Kaolinite is common in soils, and is abundant in highly weathered soils of tropical and semitropical climates. Kaolinite crystals are fairly strong owing to hydrogen bonding between O's from the tetrahedral sheets and OH's from the next octahedral sheets, as shown in Fig. 4-4. That is, in the B sheets the hydrogen atoms have lost their electrons to adjacent oxygens, each OH pair constituting an OH^- ion. Hydrogen without an electron becomes a proton—a small, concentrated positive charge. The proton is mobile to the extent that it readily takes the outside of the exposed OH plane, close to exposed oxygens of the next AB layer. A proton so situated between two oxygen ions attracts both, and is referred to as a hydrogen bond.

Halloysite is a common soil mineral similar in occurrence and structur to kaolinite, but containing in addition to the AB structure one plane of interlayer water. This increases the total layer thickness to 10 Å. The water separation of layers and loss of hydrogen bonds allows the layers to curl and relieve stress caused by an imperfect fit of A to B, so halloysite exists as tiny tubes of clay, rolled up like a newspaper. Air drying causes halloysite to lose its interlayer water irreversibly, changing its properties. *Soil samples suspected of containing halloysite should not be allowed to dry prior to performance of engineering tests.*

The most common 2:1 layer clay mineral is montmorillonite, named after a town in France where it was first described. In this mineral the B sheet with its hydroxyls is sandwiched between two A sheets, giving no opportunity for hydrogen bonding between ABA layers. The result is that montmorillonite readily imbibes variable amounts of water between its ABA layers, expanding drastically and in some cases disastrously for structures supported on this clay. Montmorillonite thus is the major clay mineral in *expansive clay*, which readily becomes expensive clay if not recognized and apprehended by the sagacious soil engineer. Volcanic ash which has weathered to relatively pure montmorillonite is called *bentonite*.

The montmorillonite crystal structure usually contains more negative than positive ions, resulting in a net negative charge. The negatively charged crystals in turn attract and retain positive ions or cations which are held to mineral surfaces between and around the layers. Such ions are readily exchanged for other ions in solution and are referred to as *exchangeable cations*. The *cation exchange capacity* (c.e.c.) is the quantity of exchangeable ions per unit weight of soil, usually expressed in milliequivalents per 100 gm, and is an imperfect indication of montmorillonite content. Kaolinite has a small but measurable cation exchange capacity due to adsorption of both positive and negative ions at crystal edges.

Expandability and hence activity of montmorillonite is greatly influenced by the size and charge of the interlayer cation. Thus sodium or hydrogen montmorillonite is highly expandable; calcium or magnesium montmorillonite exhibits controlled expansion. Ammonium and potassium ions or polar organic molecules in the interlayer tend to stabilize the spacing.

Micas are very similar structurally to montmorillonite except that the negative charge is larger and is balanced by interlayer potassium ions (Fig. 4-4). Potassium ions are about the same size as oxygen, a nearly ideal size to fit in hexagonal surface holes and hold successive ABA layers together in a nonexpandable structure. Thus mica, like kaolonite, does not expand, but for a different reason.

Intermediate between mica and montmorillonite is illite, named after a well-known state in the U.S. Illite has less potassium than mica, is finer grained and less well crystallized, and is nonexpansive. Interstratified mixtures of montmorillonite and illite are common in soils. Illite also is common in marine shales as a result of "deweathering" of montmorillonite.

Another structurally similar mineral is vermiculite, which like montmorillonite intergrades to mica and occurs both as clay-size and as larger particles. Vermiculite has an ABA structure with a higher charge than montmorillonite. Magnesium ions and a discrete amount of water occur in the vermiculite interlayer. Vermiculite does not readily expand on wetting. Upon rapid heating, larger particles lose interlayer water explosively, producing an accordionlike expanded particle used for building insulation. In fact, the name vermiculite stems from the shape of expanded particles, *vermes* being Latin for worms.

Still another structurally similar mineral is chlorite, which occurs both as clay-size and as larger particles. Chlorite has an ABA-B-ABA structure with expansion prohibited by both hydrogen bonding and ionic attractions between ABA and B layers. The hydrogen bonds are similar to those in kaolinite, owing to the extra B layer which is unshared in the z direction. In addition the ABA has a negative charge of between 1.0

and 3.6, which is balanced by a positive charge in the isolated B layer. Chlorite therefore does not cleave so easily as other clay minerals and is a common nonclay as well as a clay mineral.

Such close structural similarities exist between the various clay minerals that it is not surprising that clay minerals in soils are frequently intimately mixed or interlayered, often as a result of partial weathering conversion of one mineral to another. For example, illite often represents a weathering stage of mica after removal of part of its interlayer potassium (K). As more K is removed, the mineral becomes interlayered illite and montmorillonite, the latter representing layers after a complete removal of their K. These and other interlayers are common and may be lateral interlayers, as when changes occur near crystal edges, or they may be stacking interlayers, where interlayer atoms are removed or replaced throughout a plane in a crystal.

Yellow, red, and brown colors of soils and rocks for the most part relate to iron oxides and hydroxides which are not shown in Table 4-5. These occur as clay-size particles and coatings on larger particles. Only a small percentage is sufficient to give color, although some deep reddish soils may contain as much as 20% readily removable or "free" iron oxides. The most common iron minerals in soils are hematite (Fe_2O_3) which gives a red color, goethite, which is yellow, and lepidocrocite, which is brown. The chemical composition of the last two is the same, $Fe_2O_3 \cdot H_2O$, but the crystal structures differ. The term "limonite" is frequently used for undifferentiated hydrated iron oxides.

Highly weathered soils of the tropics tend to be rich in iron oxides and hydroxides which frequently cement soil into a hard, brick-red layer called *laterite*. If sufficiently hard, laterite is sought as a road construction material.

4.10. IDENTIFICATION OF SOIL MINERALS. The importance of soil mineralogy for engineering behavior can scarcely be overemphasized. The mineralogical composition of a soil is indirectly indicated by routine engineering tests such as particle size gradation and plasticity. However, for an intelligent use of soil, i.e., to better anticipate problems and find solutions, major mineralogical components should be identified by analysis or by reference to available published reports.

Until recently, identification of soil minerals required a time-consuming examination by a skilled mineralogist using a polarizing microscope. The commercial availability of x-ray diffractometers has reduced soil mineral identification to a relatively rapid and easy procedure which is now routine in many laboratories. The American Society for Testing and Materials has assembled and published x-ray diffraction data which are an invaluable aid for identification of minerals (*1*).

x-ray diffraction offers a means for measuring the distance between repeating planes in a crystal structure, such as the basal spacings of clay minerals. Clay mineral basal spacings give very strong diffraction peaks, especially when the crystals are oriented flat and parallel to the sample surface. This orientation may be obtained by moistening the soil and troweling or smoothing the surface so it appears shiny.

Figure 4-5 shows the ideal orientation of mineral planes for diffraction in an x-ray diffractometer. A small portion of the x-ray light beam striking the sample is diffracted or "reflected" much as if the mineral crystals were mirrors, such that the two angles designated θ are equal (*2*). Such diffraction differs from a simple reflection in that it occurs only for specific combinations of crystal spacing d, x-ray wavelength λ, and diffraction angle θ. The relation between these is illustrated in Fig. 4-5(b). Although the following is not a strict proof of x-ray diffraction, it is termed the Bragg reflection analogy and the result is called the Bragg law.

In Fig. 4-5(b) the distances $AB + BC$ represent a difference in path length for two diffracted rays. Ions or atoms in a crystal become point sources for x-rays; therefore diffracted rays will arise from planes of ions, which coincide with crystallographic planes. As the angle θ is varied by scanning, the diffracted waves may either annul if they are out of phase, or reinforce if in phase. For reinforcement, the distance $AB + BC$ must equal a whole number of wavelengths n. Therefore

$$n\lambda = AB + BC \tag{4-1}$$

By mutual perpendiculars

$$\angle ADB = \theta = \angle BDC$$

and

$$\sin \theta = \frac{AB}{d} = \frac{BC}{d}$$

Solving for AB and BC and substituting in Eq. (4-1) gives

$$n\lambda = 2d \sin \theta \tag{4-2}$$

In x-ray crystallography n is usually included in the convention for designating crystal planes, and

$$\lambda = 2d_{\text{hkl}} \sin \theta \tag{4-3}$$

where d_{hkl} is the distance between corresponding crystal planes identified in the Miller index system. The basal spacing is refered to as d_{001}.

x-ray diffraction analysis utilizes a known λ and measures θ to obtain d. It very readily shows, for example, whether a clay contains kaolinite

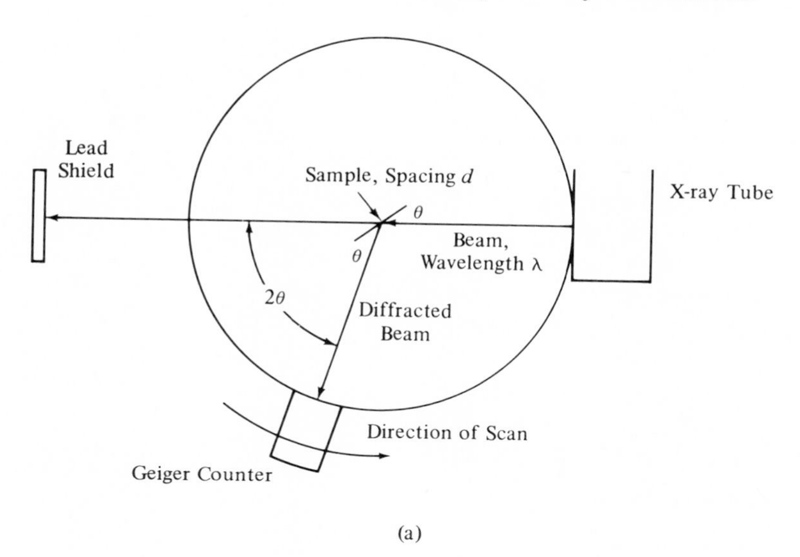

(a)

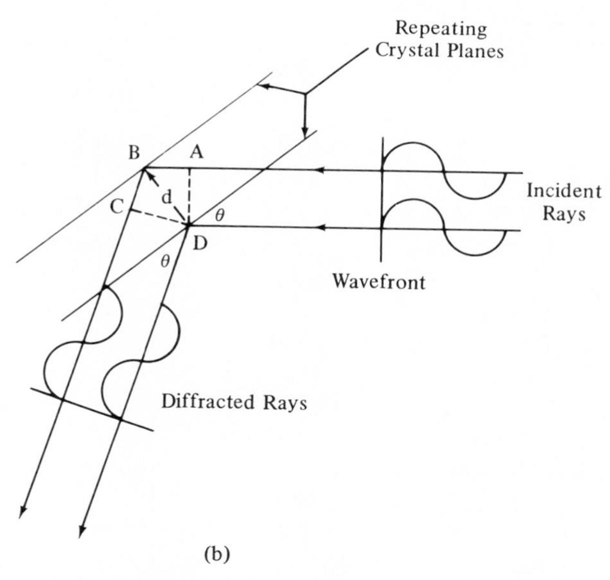

(b)

Fig. 4-5. (a) Schematic arrangement of an x-ray diffractometer, used for rapid identification of minerals in soils. (b) Geometry of diffraction illustrated by the Bragg reflection analogy. For diffraction to occur, emergent rays must be in phase.

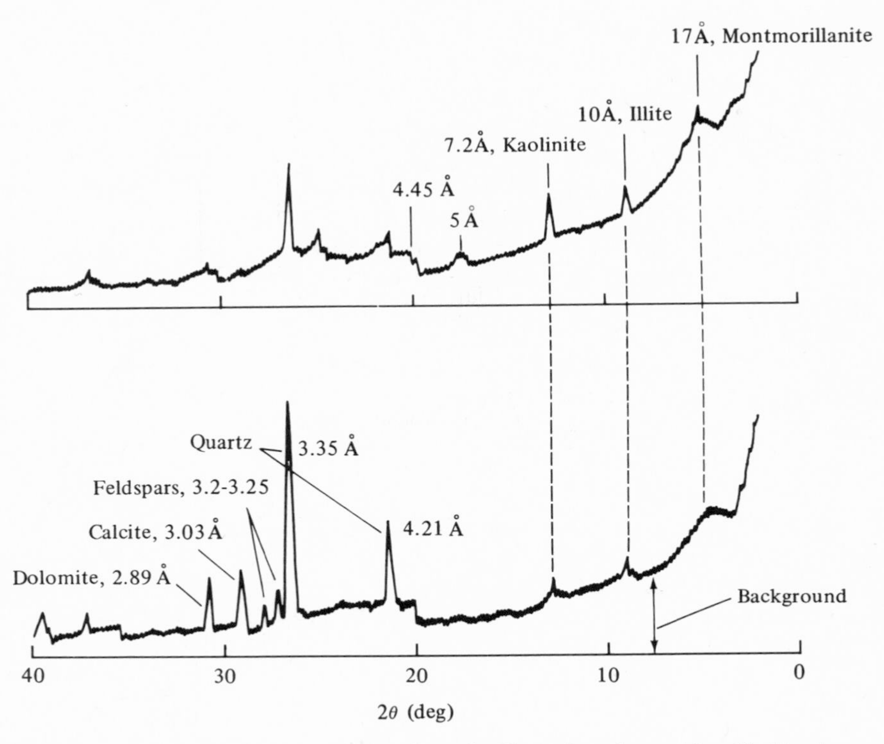

Fig. 4-6. x-ray diffraction analyses of a soil. Clay minerals are identified in the upper graph. Nonclay minerals in the lower one.

with a "reflection" at d_{001} = 7.15 Å, or illite with d_{001} = 10 Å, or montmorillonite (air dry) with d_{001} = 15.4 Å. x-ray diffraction also quickly shows any expansions on wetting, as when montmorillonite goes from 15 to 20 Å or more after simply wetting the soil with water.

Since any one mineral has a number of repeating planes running in many different directions through the crystal, each mineral gives a number of x-ray reflections. For example, in Fig. 4-6 the lower graph, an analysis of a whole soil sample, shows the two strongest peaks for quartz, 4.21 and 3.35 Å. In the upper graph the clay was concentrated to strengthen the clay mineral peaks, which nevertheless can be seen in the lower graph. In the upper graph the 5 Å peak represents a second-order (n = 2) reflection from 10 Å illite. The 4.45 Å peak measures a clay mineral spacing in the y direction and is not affected by basal spacing, and thus arises from all the clay minerals. This soil was mixed with ethylene glycol before x-raying as a convenience for identification of montmorillonite, since this chemical invades the interlayer, displacing the water and fixing the montmorillonite spacing at 17 Å.

4.11. OTHER SOLID CONSTITUENTS IN SOILS. Amorphous solid materials having little or no crystal structure occur in many soil clays and dominate in a few. These noncrystalline materials, termed allophane, are essentially hydrated mixtures of silica and alumina plus some iron, magnesia, etc., forming a gelatinous colloid or "glue." Commercial "water glass," a sodium silicate solution used for gluing flaps on cardboard boxes, is an example of an artificial amorphous hydrated silica material. Amorphous aluminosilicates dominate in certain weathered volcanic ash soils in Japan and other islands in the Pacific, including the wet volcanic summits in Hawaii. Because of the high degree of hydration—oven drying frequently reduces the weight by two-thirds—the allophane soils sometimes pose a perplexing engineering problem.

Allophane lacks crystal planes and hence does not give discrete x-ray peaks. However, it does give low, broad "humps" indicative of average interatomic spacings. Intensities are weak, and the sample must contain at least 25–50% allophane for its presence to be ascertained from x-ray diffraction. Allophane is soluble in alkaline solutions, and may be determined by weight loss or amount of silica and alumina removed after boiling in excess 0.5*N* NaOH and washing with water.

Organic matter is considered deleterious for soil engineering uses, since it contributes a spongy, unstable structure and is chemically reactive. Soil organic matter includes fresh plant and animal residues; black decomposed residues termed "humus"; and occasionally elemental carbon in the form of charcoal or coal. An extreme case of the first is peat, which is readily recognizable and is useless for support of structures. Far more common is humus or humic acids, which in amounts of 5% or less will aggregate and color a soil black and may contribute considerable acidity. Charcoal and coal are readily identified by inspection and although soft are not ordinarily deleterious unless they should happen to become ignited.

The percentage of organic matter may be approximated from weight loss after combustion at 400°C. Alternately the amount of evolved CO_2 may be measured to determine organic carbon, or the carbon may be oxidized with a potassium dichromate solution which is then titrated to determine the amount of remaining chromate ion. The weight of organic matter equals the weight of organic carbon times a factor of 1.9 to 2.5, or nominally 2.

A convenient method sometimes used to study clay and nonclay constituents of soils is *differential thermal analysis*, abbreviated DTA. In the

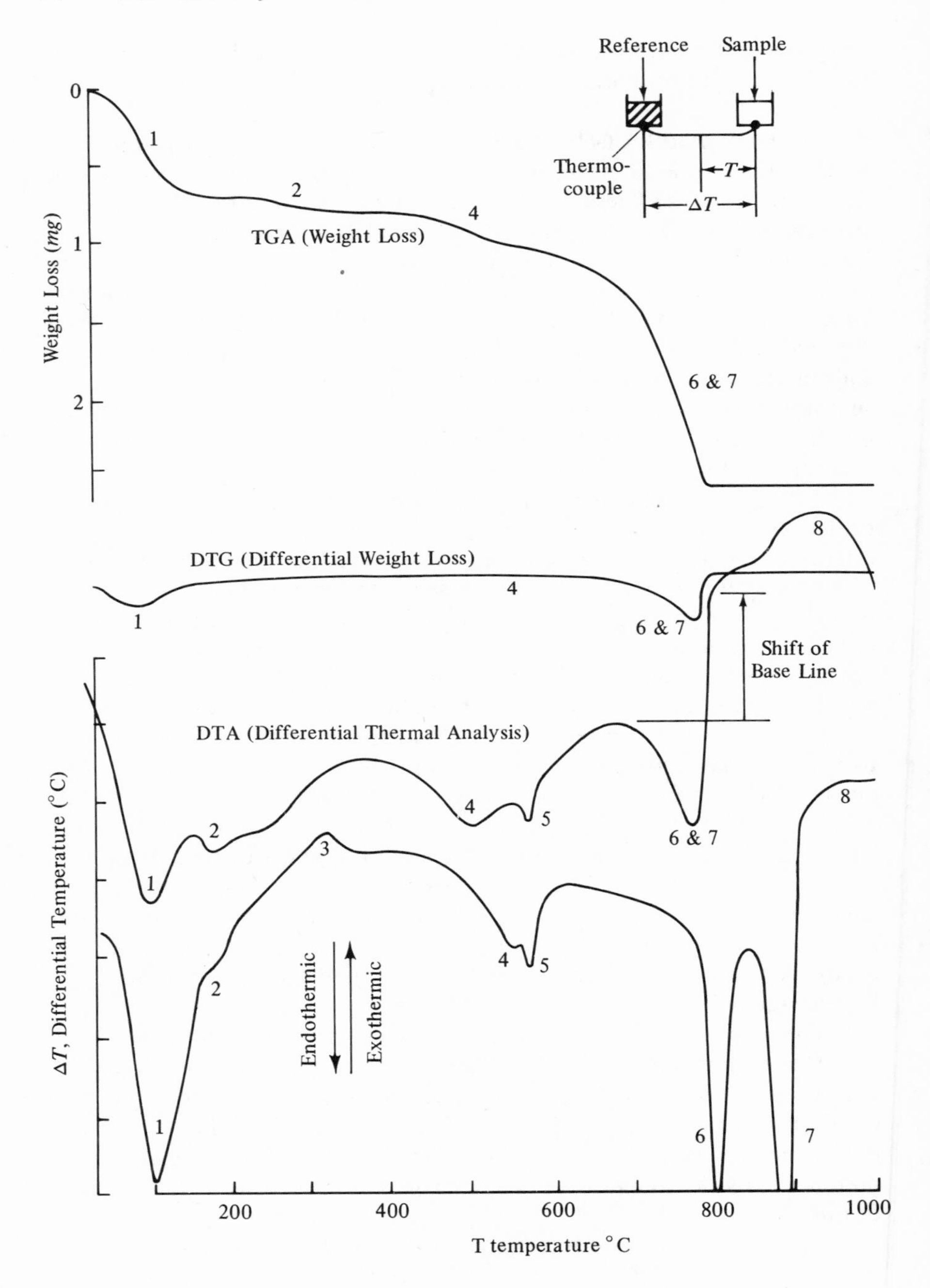

Reference
Sample
Thermo-couple
T
ΔT
0
1
2
Weight Loss (mg)
1
2
4
TGA (Weight Loss)
6 & 7
DTG (Differential Weight Loss)
1
4
6 & 7
8
Shift of Base Line
DTA (Differential Thermal Analysis)
1
2
4
5
6 & 7
8
3
4
5
2
1
6
7
Endothermic
Exothermic
ΔT, Differential Temperature (°C)
200
400
600
800
1000
T temperature °C

differential thermal method a small sample of soil and a reference sample of inert material such as Al_2O_3 are both heated at a uniform rate, and the minute temperature differences between the sample and reference are continuously recorded. When a reaction such as loss of water, burning of organic matter, etc., occurs in the soil, the heat either taken up or given off by the reaction causes a difference in temperature, recorded as an endothermic or exothermic peak, respectively, on the DTA curve (Fig. 4-7). As an illustration, consider free water present in the soil: The soil temperature cannot rise above 100°C until the free water is boiled away. Meanwhile the reference sample experiences no such thermal lag but continues heating, so there is a difference in temperature. The method is very powerful because the temperature differences are measured directly by a simple arrangement of the thermocouples, which is far more accurate than measuring the two temperatures separately and numerically subtracting.

Some instruments built to perform DTA also continuously record the weight of the sample, which helps to relate the DTA peaks to their causes. This method is called thermogravimetric analysis, or TGA. The slope or differential of a TGA curve is the weight-change analogy to the heat content-change DTA curve and is called a differential thermogravimetric or DTG curve.

4.12. MINERAL WEATHERABILITY SEQUENCE. The classification or occurrence of a soil can be a key factor in intelligent use of the soil without extensive mineralogical data. For example, an x-ray diffractometer analysis is not needed for one to expect that a red, tropical clay soil will be rich in kaolinite, or that an alluvial clay deposited downstream from a montmorillonitic soil area should be mostly montmorillonite. The relative weatherability of different minerals has been worked out from ob-

---

Fig. 4-7. (see p. 86) Thermogravimetric analysis, differential thermogravimetry, and differential thermal analysis curves for an Iowa loess soil. TGA serves to quantify the reactions involving weight losses, and DTG shows which DTA peaks have significant weight losses. Inset shows the DTA arrangement of thermocouples to give a precise measurement of the temperature differences ΔT between the sample and an inert reference material. Reactions are identified as follows: (1) hygroscopic water (note weight loss and DTG peak); (2) water held to Ca^{+2} exchangeable ions in the clay; (3) oxidation of trace organic matter; (4) OH water from clay minerals; (5) α to β quartz inversion; (6) CO_2 from magnesium carbonate in dolomite (note weight loss and DTG); (7) CO_2 from calcium carbonate; (8) recrystallization. The lower DTA curve is for the same soil tested in a deep sample holder which restricts the outflow of evolved gases, delaying and intensifying the peaks.

TABLE 4-6. Weatherability of Some Soil Minerals[a]

Nonclay Minerals			*Clay Minerals*	*Scale*
Halite (NaCl)				Very easily weatherable
Gypsum				
Calcite, dolomite				
Volcanic glass				Easily weatherable
Dark minerals, chlorite			Allophane	
Feldspars				
Muscovite				Moderately weatherable
	↓	↑	Illite	
	K loss	K gain	Mixed-layer, secondary chlorite, vermiculite	Moderately resistant
	↓	↑	Montmorillonite	
Quartz	SiO_2 loss ↓	SiO_2 gain	Kaolinite, halloysite	Resistant
			Gibbsite, hematite, limonite	Very resistant

[a]After Jackson *et al.* (*7*) and Jackson (*6*).

servations of different soil profiles and horizons, and is shown in Table 4-6. Unfortunately the data on actual soils are scattered and have not been rigorously summarized, so many of the inferences given below are just inferences.

Nevertheless from Table 4-6 one would not expect to encounter halite (common salt) in many soils, since it is easily and quickly removed by weathering. Progressively less readily weathered minerals are shown lower in the table. The high resistance of quartz in part accounts for its widespread abundance in soils.

The clay mineral weathering sequence involves first a removal of interlayer potassium (K) by leaching, changing the clay from a nonexpansive mica clay mineral to the very expansive and often troublesome montmorillonite. Upon further weathering, montmorillonite apparently loses a silica "A" sheet to become nonexpansive kaolinite.

Whereas illite and montmorillonite in moderately weathered soils incorporate considerable iron into their crystal lattices to give the clay a gray or brown color, kaolinite is better crystallized and rejects iron from its lattice. The iron then must occur as oxide coatings, giving the soil an intense red-brown color. A red color therefore indicates a kaolinitic soil if the soil exhibits some plasticity and the color is not inherited from the parent material.

Under very severe weathering conditions even kaolinite disappears, leaving the nonplastic oxide minerals to dominate and lend local color to the lateritic soils or Oxisols.

This weathering sequence applies to oxidized soils subjected to leaching. When soil materials are redeposited under water they tend to "de-weather," especially if ions are available to aid the transition. "De-weathering," indicated by upward arrows in Table 4-6, occurs for example in deltaic clays deposited in salt water, giving abundant illite and chlorite in older marine clays and shale. As another example, allophane, highly hydrated amorphous clay, forms quickly by weathering of volcanic ash in very wet conditions, after which weathering is more or less arrested by wet conditions and impermeability. Allophane eventually crystallizes into kaolinite, halloysite, or gibbsite, and does not readily fit into the weathering sequence for oxidized-and-leached conditions. Under drier conditions volcanic ash weathers instead to deposits of relatively pure montmorillonite, termed *bentonite.*

4.13. MINERALS IN RELATION TO SOIL ORDERS. As would be expected, soil minerals closely relate to soil classification, although precise relationships are yet to be resolved. Starting with the least weathered, Histosols and Entisols may be oxidized but otherwise are practically unweathered or (deweathered), so their minerals are inherited and reflect the source materials. Inceptisols are somewhat weathered, containing allophane or 2:1 clays including montmorillonite or both. The cool climate of Spodosols apparently arrests weathering at about the illite stage. Aridosols, Mollisols, and Alfisols contain 2:1 clays, usually with montmorillonite dominant, and may have minor kaolinite. Aridosols in addition contain secondary halite, gypsum, or calcite and are "deweathered" in this sense. Weathering of Vertisols is arrested at montmorillonite because of the continual vertical mixing due to shrinking and swelling. More intense weathering in the Ultisols causes kaolinite to dominate over 2:1 clays, but the latter may be present, and finally, Oxisols are mainly kaolinite and nonoxide minerals.

PROBLEMS

4.1. Classify as igneous, sedimentary, or metamorphic:
(a) slate; (b) chert; (c) arkose; (d) pumice; (e) schist.

4.2. Arrange in order of *increasing* scratch hardness: diamond, quartz, calcite, steel, feldspar.

4.3. Based on its crystal structure, would you expect mica to be hard, medium, or soft?

4.4. The atomic weight of Si is 28; of O, 16. What are the weight percentages of Si and O in quartz?

4.5. The ionic radius of Si in quartz is 0.4 Å; of O, 1.32 Å. What are the relative volume percentages of Si and O in quartz?

4.6. Identify the following clay minerals and give d_{001} spacings:

(a)	(b)	(c)
A	A	A
B	B	B
A	H_2O	A
A	A	Mg + H_2O
B	B	A
A	H_2O	B
		A
		Mg + H_2O

(d) Same as (b) without water.

4.7. An air-dry calcium montmorillonite has d_{001} = 15.4 Å, which expands to d_{001} = 19.6 Å upon wetting. Assuming that pores expand in proportion to the clay mineral expansion or that the pore size is negligible, calculate the following: (a) linear percent swelling along three axes, assuming perfect parallel orientation of clay plates; (b) volume percent swelling with orientation as in (a); (c) linear percent swelling with random orientation; (d) volume percent swelling with random orientation. (e) Discuss differences if any between (b) and (d).

4.8. Identify the following from their x-ray diffraction θ angles from a copper-target x-ray beam (λ = 1.54 Å):

2θ	θ	$\sin\theta$	d_{001}	Probable Mineral(s)
5°				
6.1°				
8.8°				
12.2°				

4.9. (a) What is the volume percentage of interlayer water in hydrated halloysite? (b) The specific gravity of kaolinite is 2.61. Calculate the specific gravity of hydrated halloysite, assuming the specific gravity of water is 1.0. (c) A dried (7.2 Å) halloysite is compacted in the laboratory to a density of 100 pcf. Using the specific gravity figures obtained in (b), what would be the density of 10 Å halloysite, assuming compaction to the same void ratio? (d) Which density in (c) above would be correct to specify for typical field compaction, or would either be correct? Why (not)?

4.10. Referring to Fig. 4-7, the initial sample weight was 58.4 mgm. (a) Calculate the hygroscopic moisture content. (Ans.: 1.2%.) (b) Calculate the CO_2 content. (c) Assuming the carbonate mineral is dolomite, $CaMg(CO_3)_2$, convert the answer in (b) to percentage of dolomite if the molecular weight of $2(CO_2)$ is 88; of $CaMg(CO_2)_2$, 152. (d) Do all DTA peaks involve a weight loss from the sample? Why (not)? (e) Explain the differences in the two DTA curves.

REFERENCES

1. American Society for Testing and Materials. *X-Ray Powder Data File.* Sets 1–5, Inorganic, 1960. Sets 6–10, Inorganic, 1967.
2. Black, C. A., Ed. *Methods of Soil Analysis.* American Society of Agronomists,

Inc., Madison, Wisconsin, 1965. (Note especially Chapter 45 on oxides, hydroxides, and amorphous aluminosilicates, and Chapters 89–90 on total and organic carbon.)

3. Brown, G., Ed. *The X-Ray Identification and Crystal Structures of Clay Minerals.* Mineralogical Society (Clay Minerals Group), London, 1961.
4. Carroll, Dorothy. "Clay Minerals: A Guide to their X-ray Identification." *Geol. Soc. Amer. Special Paper* 126. Geological Society of America, Boulder, Colorado, 1970.
5. Clay Minerals Society, Nomenclature Committee (1966–1967). *Clays and Clay Minerals* **16** (4), 322–324 (1968).
6. Jackson, M. L. "Frequency Distribution of Clay Minerals in Great Soil Groups as Related to the Factors of Soil Formation." In *"Clays and Clay Minerals," Proc. 6th Natl. Clay Conf.*, Pergamon Press, New York, 1959.
7. Jackson, M. L., et al. "Weathering Sequence of Clay-Size Minerals in Soils and Sediments." *J. Phys. Coll. Chem.* **52**, 1237–1260 (1948).
8. Krynine, D. P., and W. R. Judd. *Principles of Engineering Geology and Geotechnics.* McGraw-Hill, New York, 1957.
9. Leggett, Robert F. *Geology and Engineering*, 2nd ed. McGraw-Hill, New York, 1962.

5

Soil Reconnaissance and Geophysical Surveys

5.1. PURPOSE OF SOIL SURVEYS. Engineering soil surveys are preferably performed before final selection of a site, to obtain information on the kinds, location, and extent, both in plan and in profile, of soils which may be encountered or subjected to important stresses. In a few instances surveys are postponed until during or after construction, or whenever the trouble begins. Invariably the cost of remedial measures then far exceeds the cost of adequate original construction based on proper soil surveys and engineering analysis. For example, in some instances a very troublesome and expensive soil problem might be avoided by relocating a few tens of feet away. Similarly, the bargain building site may prove to be most expensive if it is located on soft soil, a landslide, or an area of subsidence. Most instances of sliding, differential or excessive settlement, or other severe soil-related problems stem from inadequate soil surveys, or in some instances, no on-site soil information at all. On the other hand, much construction made with little or no soil information does perform satisfactorily. Such construction of necessity must be overly conservative or run the risk of failure, and in general money can be saved and the worry factor greatly reduced by means of a good soil survey.

This chapter emphasizes shallow surveys to define the areal extent of soils. One problem in soil engineering is that no matter how exhaustive the exploratory drilling program, there can never be complete assurance that something bad does not exist between the bore holes. An appreciation of the site geology is therefore indispensible for interpolation, and in difficult areas geophysical methods of exploration form a valuable aid for interpolation between drill holes.

Some simple field tests are usually conducted during the soil survey to provide preliminary data for estimation of bearing capacity, and for later reidentification and correlation of soils from their strength properties. These tests are discussed in the next chapter.

5.2. PRELIMINARY STUDIES. Before any soil survey is undertaken, the surveyor should equip himself with all possible pertinent information concerning the surface geology and the soils in the vicinity of the project. Valuable information of this nature can often be obtained from geological reports of the region and from soil survey maps issued by the U.S. Department of Agriculture. Also, soil surveys made for other engineering projects in the vicinity may be helpful. In the case of highway- or railway-reconstruction projects, records of the maintenance department should be scanned for information concerning areas which have given trouble because of frost heave, seepage water, pumping of pavement joints, excessive or uneven settlement of the roadbed, bank slides, or any other difficulty associated with soil behavior. Knowing the locations of such areas in advance will enable the soil surveyor to more understandingly obtain information of the kind and extent needed by the designer to eliminate the difficulties previously experienced.

5.3. RECONNAISSANCE SURVEYS. Reconnaissance surveys are frequently made in connection with preliminary location and planning of extensive facilities, particularly when they are to be constructed in virgin or unknown territory. Obtaining general information relative to the soils in the various tentative locations under consideration is a vital part of the reconnaissance, and provision for obtaining this type of information should not be neglected. Whether the reconnaissance is made on foot, by automobile, or by aerial photography, extensive observations and notes regarding the soils are necessary. Such information can be obtained by observing the kind of vegetation, the extent and nature of rock outcrops, evidences of soil erosion and gullying, the character of the landscape, the presence or absence of boulders at the surface, and many other details of the terrain. Of course, observations should be made at road or railroad cuts and at stream-eroded banks. Aerial photography should not be overlooked as a valuable means of making reconnaissance soil surveys.

5.4. DEPTH OF SURVEY. The depth to which a soil should be investigated depends on the kind of engineering project or the size of the structure which is being contemplated. In the case of a highway or an airport, a good rule is to locate and map the various soil strata which are encountered within a depth of 6 ft below the finished grade line in cuts or within a zone 6 ft below the natural ground surface in fill areas. Exceptions to this rule should be made when the conditions are unusual, as where there is a deep bed of peat or where an unusually high fill is to be built. In the case of a very heavy structure which covers a relatively restricted area, such as a bridge pier, a high building, or a large earth dam, the soil investigation should extend to a depth equal to about $1\frac{1}{2}$ times the least horizontal dimension of the structure unless solid bedrock is encountered at a higher elevation. The value of $1\frac{1}{2}$ is approximately the maximum depth of shearing in a bearing capacity failure [Fig. 5-1(a)].

The above guidelines assume that the soil becomes stronger with greater depth, which is the usual case because of overburden pressures. Sometimes, however, weaker soils do occur under strong layers, in which

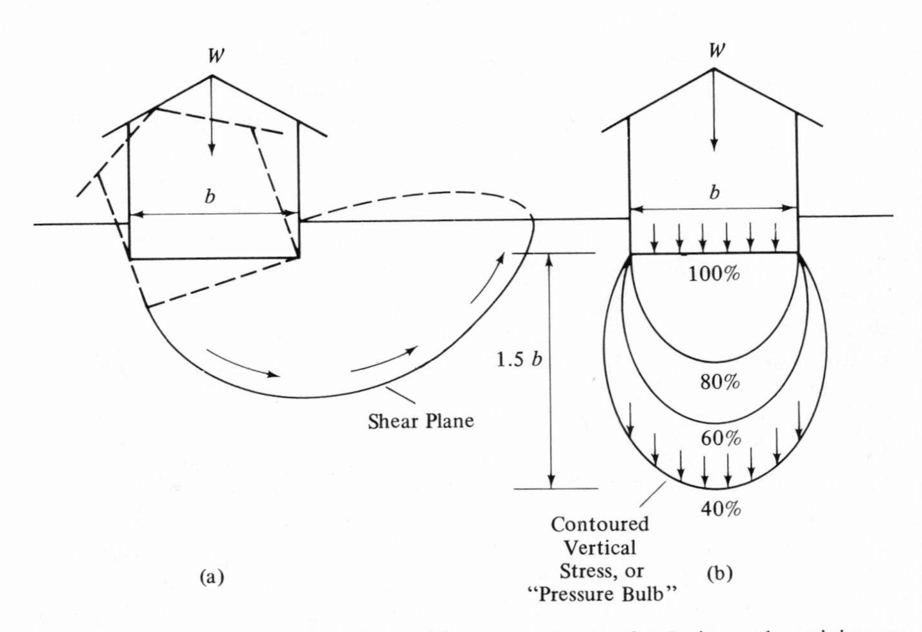

Fig. 5-1. Exploratory foundation drilling to a depth of 1.5 times the minimum width of the loaded area (which may be a footing) investigates soils which would be involved in a shearing bearing capacity failure (a) and soils within the zone of major vertical stress (b).

case deeper exploration is required to detect the weaker layers. For example, soil below a water table is weakened owing to buoyant forces. A fairly common occurrence is weaker soil under a genetic hardpan, discussed in the preceding chapter. In other instances weak clay layers are buried under many feet of gravel or sand, or shale layers occur under thin limestone rock. The occurrence of unmapped or unknown mines or caverns also must be suspected in likely areas. Often one or more of the exploration holes are extended to a depth of 100 ft or more to check for weak layers, even when a much shallower depth is used for most holes.

Deeper exploration also is needed for structures on deep, soft clays where settlement is an important limitation. As shown in Fig. 5-1(b), bearing stresses tend to dissipate with depth, but at the depth of $1.5b$, where b is the width of the bearing area, vertical stress is still about 40% of the applied surface stress. At a depth of $3b$ the stress is reduced to about 20% of the surface stress, so one or more exploration holes in soft clays may extend to a depth of $3b$ or even deeper. Terzaghi and Peck (*7*, p. 340) cite an example of a group of factory buildings built on a tidal flat; the maximum building width was 40 ft, and borings were made to 90 ft, which is $2.25b$. The buildings, founded on pile, subsequently settled more than 2 ft because of consolidation of a clay stratum at a depth of 115 to 145 ft, or about $3b$.

At the other extreme, very light structures supported by wall or column footings will usually require only relatively shallow drilling. In this case the drilling should extend sufficiently deep to reveal site geology, but the footing width may be substituted for b in the above guidelines. For example, a 2-ft footing width would indicate depth of drilling to 3 to 6 ft below the footing elevation, plus deeper holes to check for weak layers.

5.5. BORINGS FOR HIGHWAYS AND AIRFIELDS. After the reconnaissance survey of a proposed project has been completed and a temporary line and grade have been established, a detailed soil survey should be made. In the case of a highway, airport, railroad, or similar facility which extends over considerable length or area, the detailed soil survey is made by making borings at appropriate intervals along the line and at points laterally disposed from the control line. The points at which borings are made should be sufficiently close together to insure reasonably complete knowledge of the location and extent of each of the various soil horizons encountered on the project. For a single-line highway project, borings at each station on the center line will usually suffice, although some intermediate borings may be necessary to establish the limits of individual horizons which do not extend continuously between the bore holes at even stations. Also, some borings should be made at the right-of-way lines and perhaps also at the side-ditch lines at some stations in cut areas, in order to establish the lateral orientation of soil horizons and the water table. The frequency with which these cross sections are taken will depend on the apparent uniformity of the soil formations and is largely determined by the judgment of the soil surveyor.

Divided highways should be surveyed in a similar manner, but a line of holes should be put down along the center line of each lane. In the case of an airport, borings should be made along the center line and along each edge of each runway, especially in the cut areas, and also along the taxiways, apron, hardstands, and service drives. The areas between runways probably will not need to be surveyed, unless they are in cut areas which serve as borrow pits.

The borings for highway and airport soil surveys are usually relatively shallow and may conveniently be made by means of a 4-in. post-hole auger or mechanized helical flight auger. Additional sections may be added to the vertical shaft of the auger when the depth of the hole exceeds about 4 ft. A soil auger or probe $1-1\frac{1}{2}$ in. in diameter may be used for intermediate check holes when it is not deemed necessary to take soil samples at these points. Infrequently, it may be desirable to make a more detailed examination of a soil profile than is possible by means of a post-hole auger or a soil auger. If the need arises, an open pit with vertical sides may be dug. Such a pit will expose the profile and will permit careful examination of the morphology and structure of the soil. Obviously, open pits of this kind are relatively expensive and are seldom justified.

As the auger holes are dug, a log should be kept of the elevations and thicknesses of the various strata or horizons which are encountered. In addition, the color, structure, estimated textural classification, and any other identifying characteristics of the soil in each horizon should be noted. The designation of the textural class of a soil in the field is only a temporary identification procedure which will be modified or made permanent after the results of a laboratory mechanical analysis become available. However, the personnel of a soil-survey party may become quite proficient in estimating the textural classification by visual examination in the field.

The soil from a bore hole should be placed on a tarpaulin as it is excavated, the soil from each horizon should be kept in a separate pile, and the piles should be arranged in the order in which the horizons were encountered. These piles should be left for inspection by the chief of the soil-survey party, who will probably indicate from which horizons or piles laboratory samples should be taken. However, moisture samples should be taken from each horizon, at the time the soil is removed from the hole, by filling a 4 oz metal salve box. Place the lid on the box immediately, and seal the complete circumference of the joint by means of scotch tape or adhesive tape. Carefully identify the box and pack it for transportation to the laboratory. Doubled plastic bags also may be used to hold samples at their natural moisture content if the samples are not to be stored too long.

5.6. GROUND-WATER TABLE. If a ground-water table is encountered in a test hole, it is extremely important to determine and record its elevation. Information concerning the water table is vital in connection with the design of the project and may be very valuable to the contractor, particularly if material must be excavated near or below it. If reliable information relative to fluctuations of the water table can be secured from local residents or by examination of the local terrain, such information should be incorporated in the soil-survey notes.

The elevation of a water table is indicated by the elevation to which free water rises in a test hole. However, it should be realized that it takes time for the water to accumulate in the hole, the amount of time depending on the rate of flow of ground water from the adjacent soil. If the soil is pervious, the water in the test hole will rise to the water table within a very short time; but, if it is relatively impervious, considerable time may be required for the water to rise to the true water-table elevation. A good rule to follow in soil surveying is to allow the test hole to stand for about 24 hr after the hole has been dug before the elevation of the water is measured. Of course, if the soil in the region of the water table is known to be freely pervious, this time-lag period is not necessary.

5.7. DISTURBED SAMPLES. Disturbed samples are obtained where soil is to be excavated and used as a construction material, or for grain size analysis or classification tests. Disturbed samples are taken by hand scoop, by shovel, or from the auger borings, and are placed in cloth bags for transportation to the laboratory. No attempt is made to obtain the material in its natural state of structure or density. Such samples should be truly representative of the prototype material which they represent, however, and should be placed in clean bags to prevent contamination from samples previously contained therein. When samples are being taken from a stock pile or windrow, particular care must be exercised to make certain that the sample is representative, because of the fact that coarse material has a tendency to roll to the base of the pile, the finer material being left in the upper portion.

The size of a disturbed sample will depend on the kind and number of tests which are to be conducted. For a complete series of tests, including mechanical analysis, specific gravity, consistency limits, Proctor density, and California Bearing Ratio tests, a minimum of 40 lb of soil should be taken. If all these tests except the CBR tests are to be conducted, the sample may be cut down to about 10 lb. Some laboratories prefer to save a reserve supply of each sample for check purposes in case questions arise concerning the properties of the soil during construction of a project. If such a system is practiced, the size of samples should be increased accordingly. As the soils are sampled, great care should be exercised to see that the bags or containers are properly and completely labeled to show the location in plan and elevation from which the samples were taken. A convenient way to do this is to give each bag a number which should be recorded in the soil-survey notes along with a complete description of the sample contained in the bag.

5.8. UNDISTURBED SAMPLES FROM SHALLOW DEPTHS. Undisturbed soil samples are required when it is necessary to determine the natural or "cut" density of a soil stratum; the strength properties of the material, such as its shearing resistance and compressive strength; or its consolidation and permeability characteristics. It is much more difficult and expensive to obtain undisturbed samples than disturbed samples. The term "undisturbed," as it is used in this connection, is relative only; since it is physically impossible to take a sample of soil, transport it to a laboratory, and prepare it for testing without disturbing it to some degree. However, the objective of undisturbed sampling operations and subsequent handling of the samples should be to obtain test specimens which are as nearly as possible like the prototype soil in all respects. All apparatus should be designed, and all procedures should be carried out, with this objective constantly in mind. Several typical procedures will be described which will illustrate the principles to be observed, although the exact details of the operation may vary in different situations.

A hand-driven sampler may be used when the soil stratum to be sampled is reasonably cohesive in character and at a relatively shallow depth, as in the case of a highway or airport soil survey. The cores are about 3 or 4 in. in diameter and 6 or 8 in. long. Density of the soil may be determined by trimming the soil and measuring the weight of soil in the sampler and volume of the sampler, but this is not considered reliable because of compaction during driving. A removable liner may be used to retain the sample; in this case the liner should be sealed at both ends to prevent moisture loss.

Shallow-depth samples may also be taken from open pits dug to the level of the top of the stratum to be sampled. Mark out the top area of the sample, which may be either square or circular in shape, and carefully excavate around the sides to within $\frac{1}{2}$ in. or so of the area marked. Place a cutting edge of the size and shape of the sample on the top face and gently press downward, carefully trimming away the material outside the cutting edge. Continue this process until the full height of the sample has been exposed. A tight-fitting container such as an ice-cream carton may be lowered over the sample, and the sample cut off, capped, and sealed with tape or paraffin. Such samples should be stored in a laboratory humid room to inhibit evaporation of the natural moisture.

The density of an irregular-shaped soil mass may be determined in the laboratory by weighing, coating it with paraffin, and reweighing while submerged in water. Since the specific gravity of paraffin is close to that of water, 0.87 to 0.91, relatively little error results from considering it the volume equivalent of water. The loss of weight upon submergence represents the weight of water displaced by soil, and dividing by 62.4 gives the soil volume in cubic feet. Most commonly in engineering, however, densities are measured in the field by one of the methods in Chapter 8.

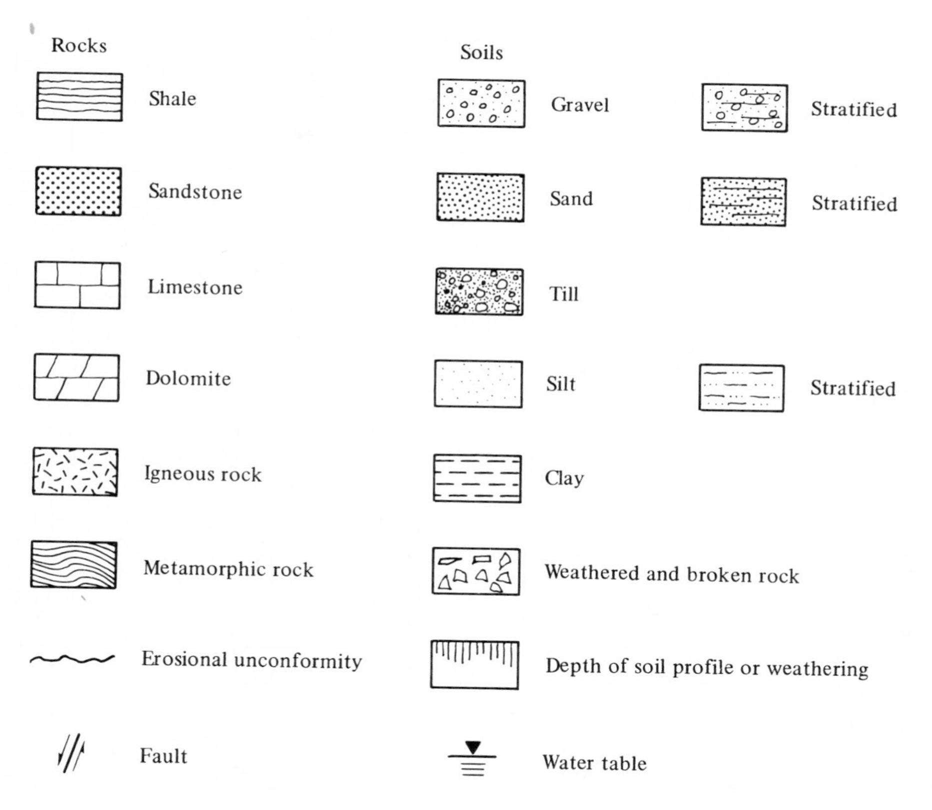

Fig. 5-2. Suggested symbols for cross sections.

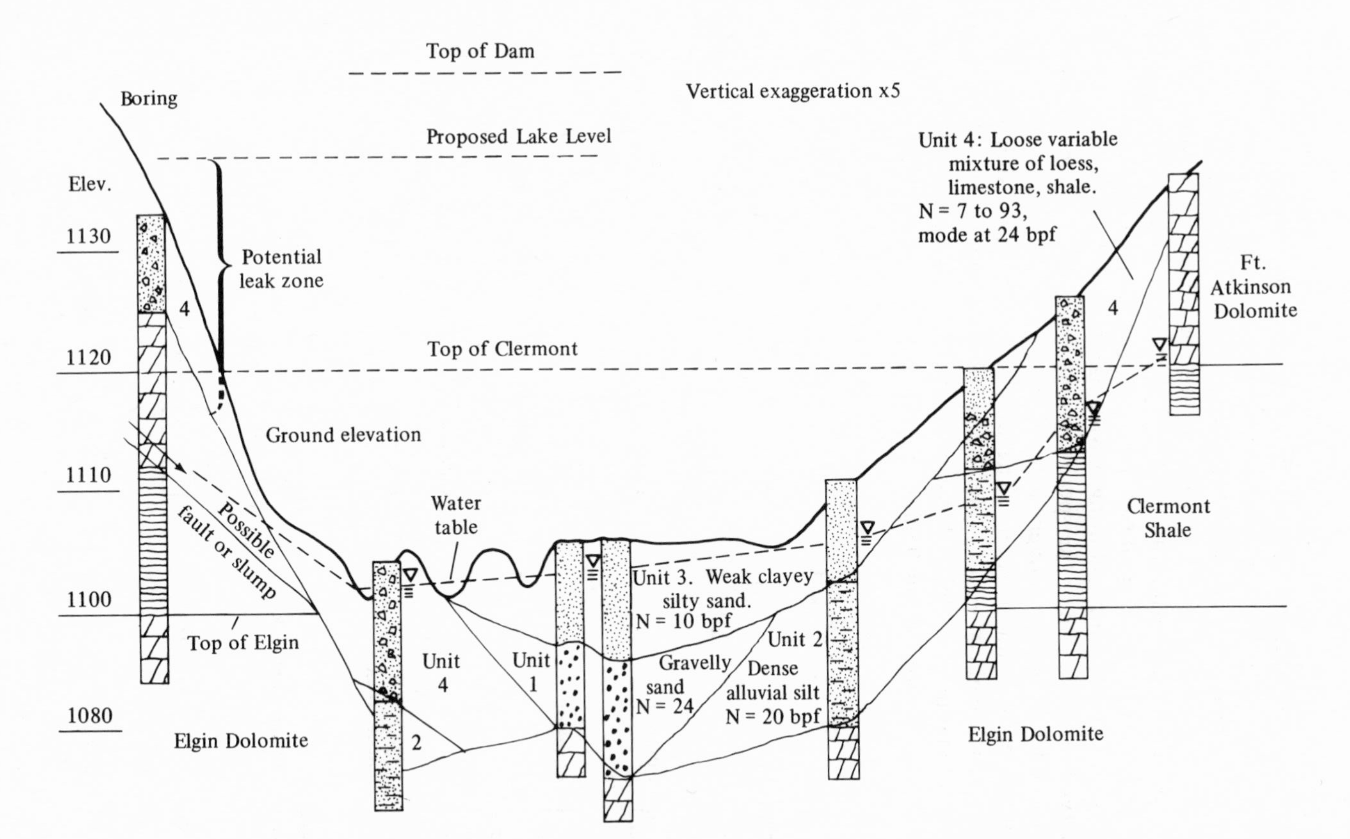

Fig. 5-3. Logs of exploratory drill holes along centerline of a proposed dam. A fault in the left abutment was inferred from the occurrence of dolomite at too low an elevation, and was later confirmed on excavating.

5.9. RECORD OF SOIL SURVEY. The information obtained in a soil survey may conveniently be shown graphically and should be made a part of the plans of a project. The various soil strata may be plotted on profile paper to a horizontal scale of 1 in. = 50 ft and a vertical scale of 1 in. = 5 ft. Each horizon of the profile is shown by a distinctive pattern (Fig. 5-2) and is labeled so as to indicate the color and engineering soil classification. The location and depth of each bore hole is shown by parallel vertical lines (Fig. 5-3), and the location and elevation of each undisturbed sample may be indicated on the tracing by a solid blacked-in symbol. The elevation of the water table may be shown by a dashed line. In the case of a highway or airport project, data relative to the natural moisture content, in-place density, standard (or modified) Proctor density, and optimum moisture content and any other pertinent information may be shown by figures and notes on the profile. Resistivity information also may be contoured on the profile. Sometimes the various strata may be drawn in to show a geological profile along the line of the structure.

5.10. HAND PENETROMETERS AND VANES. A preliminary evaluation of soil strength or trafficability may be made by jabbing soil with a boot heel, fist, or thumb, by pushing special rods called penetrometers, or by twisting an apparatus called a vane. The oldest test instrument is the thumb, which jabbed into the soil with vigor gives a measure of unconfined compressive strength or bearing capacity of clays. The rule-of-thumb is: buried to the thumbnail means 2 Tsf; to the first knuckle, 1 Tsf; and to the second, $\frac{1}{2}$ Tsf. This is considered very rough.

Somewhat more scientific is the spring-loaded pocket penetrometer, an inexpensive hand-operated device intended only for cohesive soils. Unconfined compressive strength in tons per square foot is read directly from the force required to push a 0.25 in. diameter plunger about 0.25 in. into the soil. Design of the penetrometer relates to Terzaghi bearing capacity theory, discussed in a later chapter. Under good conditions results agree to within ± 20 to $\pm 40\%$. However, since this is a near-surface test it is influenced by drying and sampling disturbances. The pocket penetrometer is sometimes used on drive samples obtained from the standard penetration test, but little or no reliance should be placed on such test results.

Other deeper penetrometers have been developed for degree of densification (Proctor penetrometer) or for a rapid indication of soil trafficability for military vehicles (U.S. Corps of Engineers penetrometer). The latter in particular has been extensively researched.

Recently hand-operated vane shear devices have been introduced.

These incorporate a round plate with thin metal projections that penetrate into the soil. The plate is pushed against the soil and torque applied until the soil shears. The torque is measured with a spring device and gives a measure of soil strength. The test is sensitive to the amount of pressure against the soil and is used only for soft clays.

5.11. GEOPHYSICAL RESISTIVITY SURVEYS. Geophysical methods are useful in reconnaissance surveys over large areas, and for interpolation between or extrapolation away from bore holes. The two common techniques used in soil engineering are the electrical resistivity and the seismic methods, measuring the resistance of soil to flow of an electrical current, or the velocity of sound waves, respectively.

Resistivity is defined as electrical resistance per unit length of a unit cross-sectional area. Resistivity of a soil primarily relates to its water content and the concentration of dissolved ions. The latter is particularly high in the case of clays. Dry soils or solid rocks therefore have a high resistivity; saturated granular soils have a moderate resistivity; and saturated clays have a low resistivity. Representative data are shown in Table 5-1.

TABLE 5-1. Representative Soil Resistivities[a]

$\rho(10^3$ *ohm-cm*$)$	*Soils*
0	
10	Clay, saturated silt
25	Sandy or silty clay
50	Clayey sand or saturated sand
150	Sand
500	Gravel
>500	Dry sand or bedrock

[a] After Barnes (*2*).

Resistivity surveys are used to locate or outline buried aquifers or gravel deposits, or find depths to a water table, bedrock, or a change in soil. Surveys also have been successfully conducted from boats on salt water in studies of bridge pier sites.

5.12. ELECTRODE ARRAYS. Resistivity surveys utilize four electrodes, commonly metal stakes spaced and driven into the ground at intervals along a straight line. A current is supplied through two electrodes by means of a battery or small generator, and a voltage drop or potential difference is measured between the other two. A non-60-cycle alternating or commutated current source is usually used to avoid both d.c. polarization and a.c. stray current effects.

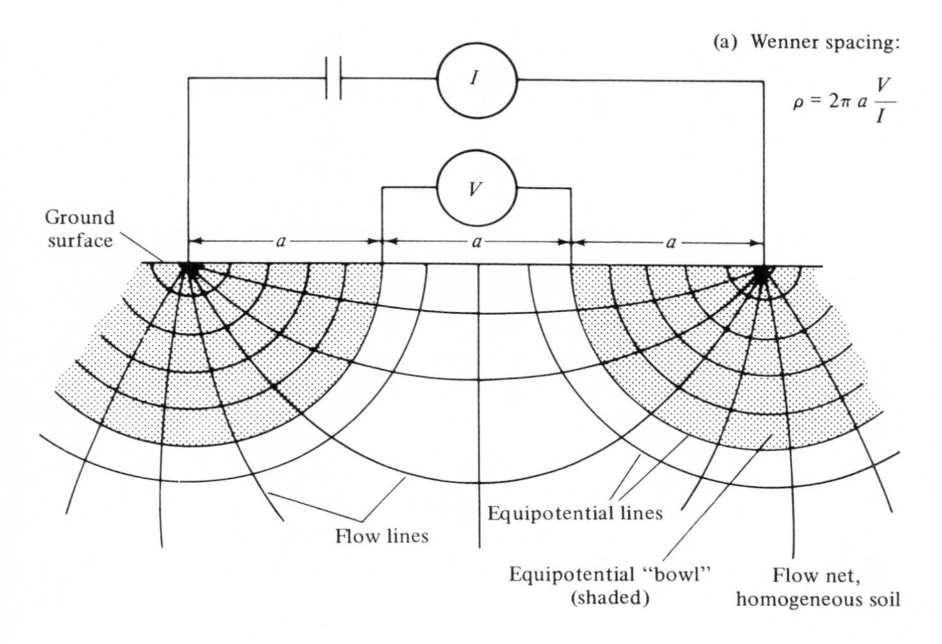

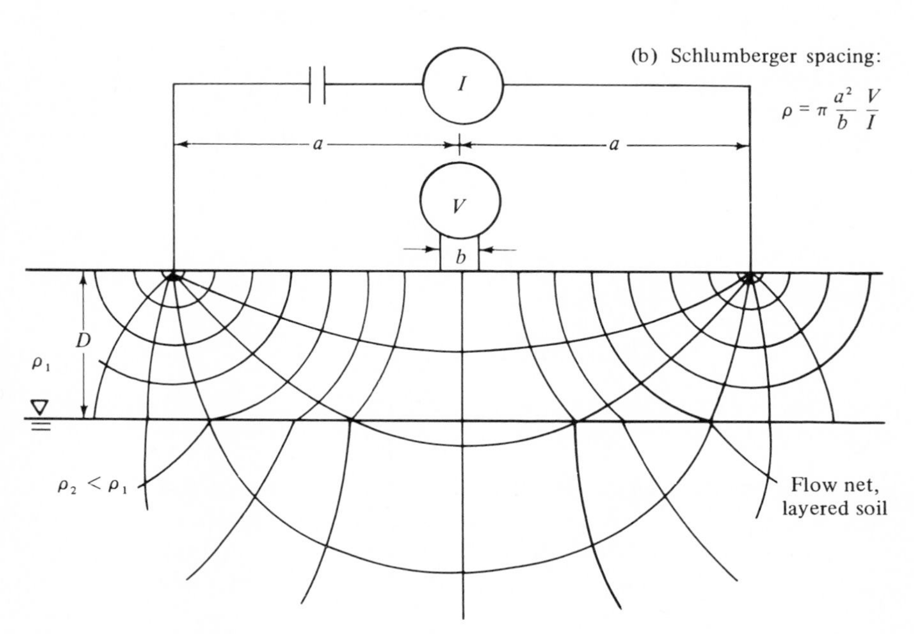

Fig. 5-4. Electrode spacings used in resistivity surveys.

The most common electrode arrangement for shallow surveys is called the Wenner spacing, shown in Fig. 5-4(a). The Schlumberger spacing, Fig. 5-4(b), differs somewhat in detail and is advantageous for deeper surveys. The flow of electrical current through soil is analogous to the flow of water through soil, voltage being the driving force, analogous to water head. The current paths and voltage contours therefore may be represented by a "flow net" such as shown in Fig. 5-4. The upper flow net in the figure is sketched for current flow in a homogeneous soil; in a layered soil the net will become distorted as shown in the lower sketch, because more flow lines will pass through the less resistive layer. As the electrode spacing a is increased, the ratio of measured voltage V to current I decreases in proportion if the soil is homogeneous, and not in proportion if the soil is layered, owing to changing distortion of the flow net.

Resististivity ρ is a property of a conductor, defined as electrical resistance per unit length of a unit cross-sectional area. For example, resistivity could be measured directly on a cubic block of soil 1 cm on a side, by applying a voltage to two opposite sides. From Ohm's law the resistance $R = V \div I$ where V is the voltage and I the current flow. Resistivity would therefore be $\rho = R \times$ (cross-sectional area/length) $= R \times 1\ \text{cm}^2/1\ \text{cm}$. If V is in volts and I is in amps, ρ will be in ohm-centimeters, abbreviated Ω cm.

The soil involved in resistivity measurements, as in Fig. 5-4, is far from being a cube, but it can be shown by use of calculus that with the Wenner arrangement the effective resistivity for any spacing a between two adjacent electrodes is

$$\rho = 2\pi a \frac{V}{I} \tag{5-1}$$

If a is in feet instead of meters,

$$\rho = 191 a \frac{V}{I} \tag{5-2}$$

where V is in volts, I is in amps, and ρ is in ohm-centimeters.

5.13. INTERPRETATION OF RESISTIVITY DATA. Equations have been derived to relate the total effective resistivity ρ at any spacing to the thickness and resistivity of individual layers, and books of graphs of log ρ versus log a are available in geophysical literature for interpreting data by visual comparison with ideal curves. For shallow depths in variable soils, two rapid approximation methods devised by engineers are in use in highway departments and consulting firms. Both utilize the Wenner spacing and involve a concept of equipotential "bowls" shown in Fig. 5-4(a).

Since by definition the voltage anywhere along an equipotential line is the same, the lines at spacing a from both electrodes define two nearly hemispherical surfaces which extend to a depth approximately equal to a. Therefore a voltage measurement V between the two "bowls" involves soil resistivity approximately to a depth a. By systematically increasing the spacing, the penetration depth can be gradually increased to give a variable depth survey. In highway practice a is often 3 ft plus intervals of 3 ft up to 15 ft, and then with added intervals of 5 or 6 ft up to 50 ft or more.

A method of interpretation proposed in 1945 by R. W. Moore (*6*) of the Bureau of Public Roads, is to plot $\sum \rho$ versus spacing a. If ρ is constant with depth the result is a straight line with a slope ρ_1/a (Fig. 5-5)

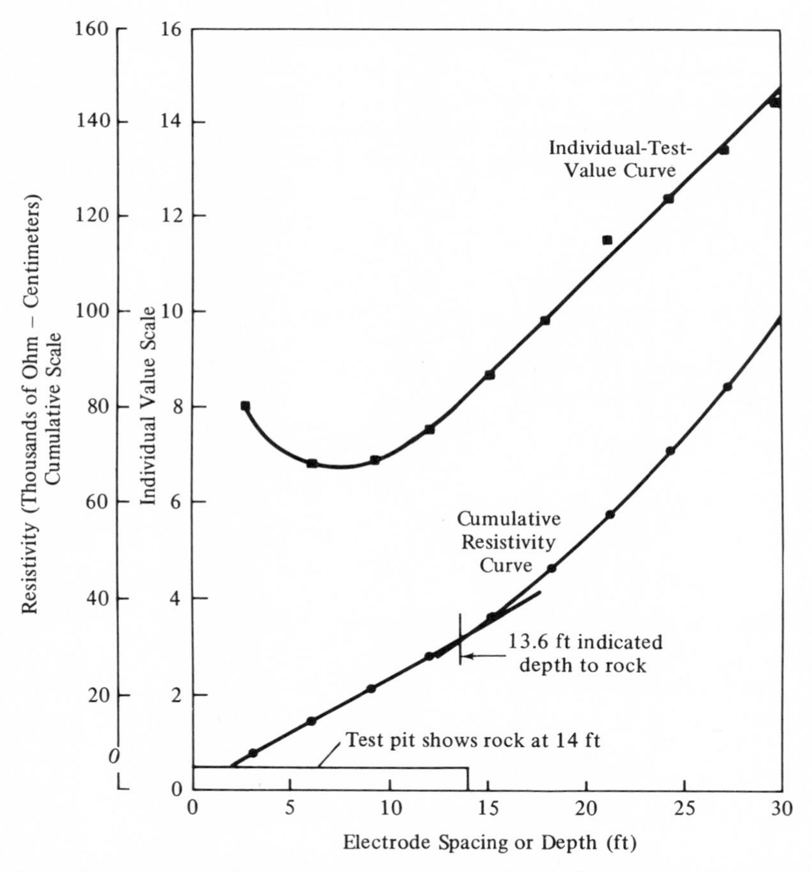

Fig. 5-5. Typical resistivity data and method of analysis using the cumulative resistivity curve. From Moore (*6*).

where ρ_1 is the resistivity of the upper layer. As the spacing is increased and a different material is encountered, a second line is formed with a slope ρ_2/a. The intersection of the two lines indicates depth of the boundary between the two layers, usually within 5 to 10%. The advantages over a simple plot of ρ versus depth can be seen by reference to the figure.

A method devised by H. E. Barnes (*2*) of the Michigan State Highway Department treats the soil as layers which act as resistors in parallel. In this case it can be shown that the conductance in mhos, or reciprocal of the resistance in ohms, of any layer may be found from

$$L_n = \sum L_n - \sum L_{n-1} \tag{5-3}$$

where L_n is defined as the conductance or reciprocal resistance of layer n, $\sum L_n$ is the conductance of all layers down to and including layer n and $\sum L_{n-1}$ is the conductance of all layers down to but not including layer n. For calculations one must therefore convert total resistance readings to total conductances, $\sum L$. Some meters are calibrated to read $\sum L$ directly in mhos, simplifying calculations. Successive $\sum L_n$ and $\sum L_{n-1}$ values are then subtracted to obtain individual L_n values. The layer conductance L_n is then converted to resistivity by use of a modified Wenner equation

$$\rho_n = \frac{191\, a_n}{L_n} \tag{5-4}$$

where a_n is the incremental increase in a in feet.

EXAMPLE 5-1. The following conductance readings were obtained at indicated electrode spacings:

A (ft)	$\sum L_n$ (mho)
3	0.0044
6	0.0143
9	0.0394

Calculate the conductance and resistivity of each layer.

SOLUTION. Subtracting successive conductances $\sum L$, one obtains the following:

Layer depth (ft)	L_n (mho)
0–3	0.0044
3–6	0.0099
6–9	0.0251

Substituting the values for L_n into Eq. (5-4) gives $\rho_n = 191(3)/L_n = 573/L_n$. Therefore,

0–3 ft	$\rho_1 = 130\ (10)^3\ \Omega$ cm
3–6	$\rho_2 = 58\ (10)^3$
6–9	$\rho_3 = 23\ (10)^3$

It is important to note that arbitrary layer boundaries determined by incremental changes in A will seldom coincide with precise depths to changes in strata. When a layer incorporates a boundary, the layer resistivity will be a weighted average for the two materials.

EXAMPLE 5-2. Let us assume the above example represents two geological layers. What are they and how thick are they?

SOLUTION. The values of ρ_1 and ρ_3 compared to data in Table 5-1 suggest the upper material is a sand and the lower is a sandy or silty clay. If d is the thickness of ρ_1 material in Layer 2, and $3 - d$ is the thickness of ρ_3 material, ρ_2 represents a weighted average, or

$$58(3) = 130d + 23(3 - d)$$

$$d = 1.0 \text{ ft}$$

$$\text{Thickness of the top stratum} = 3 + 1.0 = 4.0 \text{ ft}$$

Therefore the section is:

0
Sand, $\rho = 130(10)^3\ \Omega$ cm
4 ft
Sandy clay, $\rho = 23(10)^3$
9 ft
Unknown

Following a depth survey and interpretation by one of the methods indicated above, a common application of the resistivity method is to perform an area survey with a constant spacing a to outline the limits of a particular deposit by means of a resistivity contour map.

The constant a method also is useful to check lateral continuity of a buried firm layer that might be used for a building foundation and will, for example, outline the areal extent of caverns or pockets of weathering far more efficiently than is possible by drilling. One must be sure to recognize that resistivity varies with moisture content and indicates soils only indirectly, leading to occasional gross errors in interpretation. The method therefore should never be used without confirmatory drilling.

5.14. SEISMIC SURVEYS. In the seismic exploration method a sound pulse is introduced into soil, and the time of first arrival of the vibrations at several different horizontal distances is recorded. The sound is introduced by a small explosive charge or by artistically banging a steel

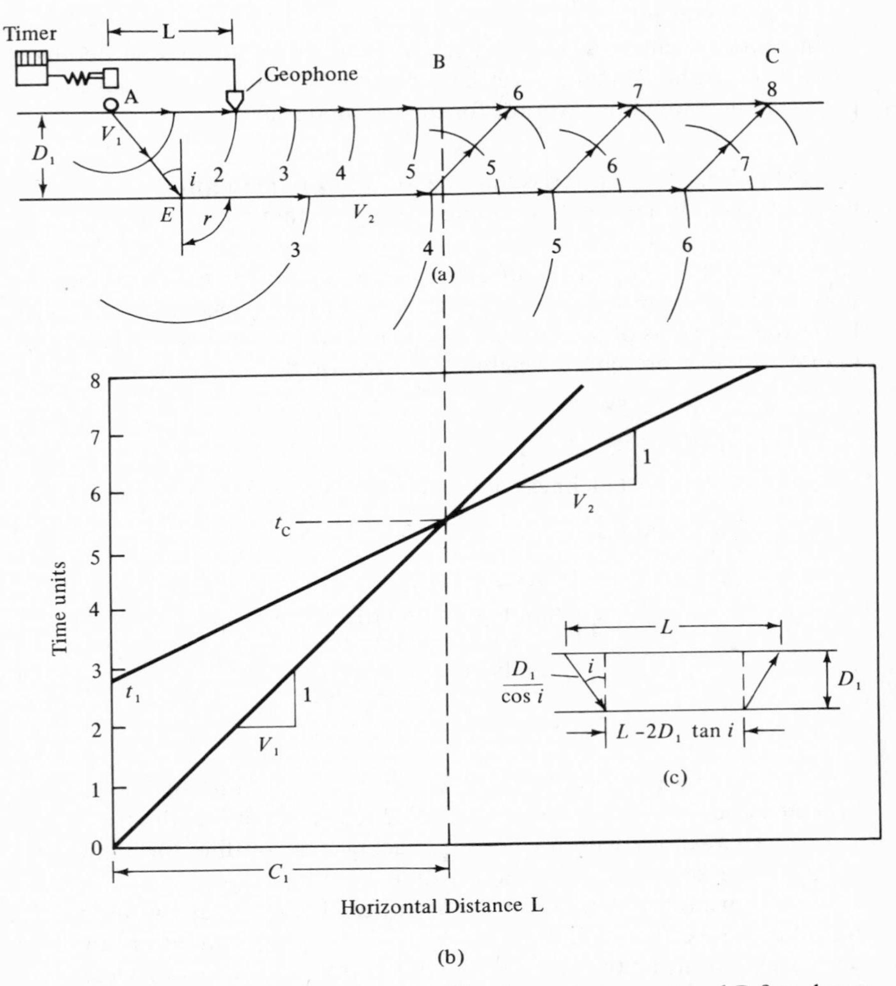

Fig. 5-6. Principle of seismic refraction. Geophones between A and B first detect sound coming through layer V_1, whereas between B and C the first arrivals occur by refraction through V_2. Note that velocities can be read directly from the slope of the graph.

plate or ball into the ground with a sledge hammer; the latter technique is limited to survey depths of about 50 ft or less. The sound is detected by a special vibration detector called a geophone. In some instruments the hammer blow starts an electric timer and the geophone stops it, and the time may be read directly from the timer. Most engineering seismographs utilize only one geophone, and either the geophone or the hammer is moved successively to different locations along a straight line. At each position the test is repeated to give the time-distance relationships, as shown in Fig. 5-6. When explosive charges must be used for deeper surveying, use of a series of geophones and a multichannel recorder may be more economical.

The pulse arrival time is plotted versus distance as in Fig. 5-6. The reciprocal slope of the line is velocity in feet per second. The line may not go through the origin because of a starting time error, but this should remain constant and thus will not affect slope.

As the survey distance is increased, usually by increments of 10 ft, the velocity may increase, indicated by a change in slope of the curve. Figure 5-6 shows two velocities labeled V_1 and V_2. The first, lowest, velocity could be soil, whereas the second velocity might represent bedrock. The buried higher velocity material shortens longer distance arrival times by a process of refraction, illustrated in the figure. Refraction also may occur through a buried pipe or boulder, or laterally through a building foundation, sometimes leading to erroneous interpretations. *Furthermore, refraction will not shorten arrival times if the surface layer has a higher velocity; therefore seismic exploration cannot be conducted through a pavement or frozen soil layer.* Seismic refraction therefore will not directly indicate a buried lower-velocity (and hence usually weaker) layer.

5.15. DEPTH TO A BURIED STRATUM. The approximate depth D_1 to a higher velocity layer can be calculated by assuming a rectangular travel path equal to $L + 2D_1$ where L is the distance between the source and the geophone. Dividing L and D_1 by the respective velocities gives the total travel time

$$t \approx \frac{L}{V_2} + \frac{2D_1}{V_1} \qquad (5\text{-}5)$$

The last term is a *delay time* due to twice traversing the lower velocity V_1 layer. The delay time, t_1, may be read directly on a t vs. L plot by extrapolating to $L = 0$.

$$t_1 \approx \frac{2D_1}{V_1}$$

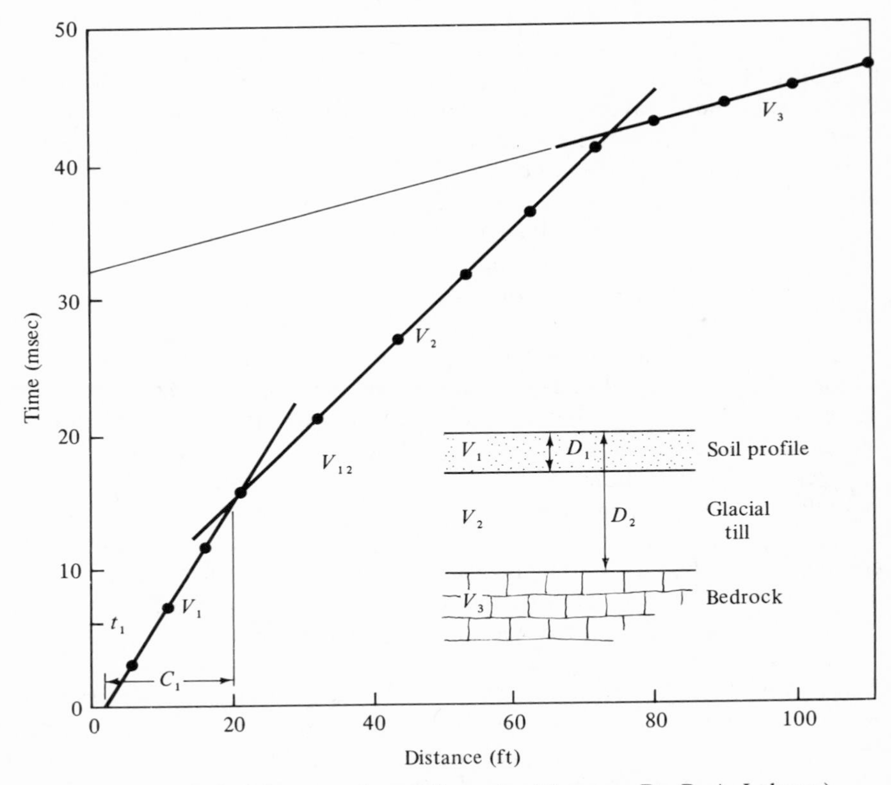

Fig. 5-7. Seismic data from northern Wisconsin. (Courtesy Dr. R. A. Lohnes.)

for which

$$D_1 \approx \frac{t_1 V_1}{2} \tag{5-6}$$

This equation is fairly accurate when V_1 and V_2 differ widely. However, when the difference is smaller the true travel path is trapezoidal, decreasing refraction travel times and t_1.

By utilizing the laws of refraction, Fig. 5-6(c), one can show that the precise expression for delay time is

$$t_1 = \frac{2D_1 \cos i}{V_1}$$

from which

$$D_1 = \frac{t_1 V_1}{2 \cos i} \tag{5-7}$$

where i is the angle of incidence. $\sin i = V_1/V_2$. A more convenient expression uses the intersection distance C_1 whereby the times for the direct and refracted waves are equal; that is,

$$\frac{C_1}{V_1} = \frac{C_1}{V_2} + \frac{2D_1 \cos i}{V_1}$$

Solving for D_1 and substituting for $\cos i$ gives

$$D_1 = \frac{C_1}{2}\sqrt{\frac{V_2 - V_1}{V_2 + V_1}} \tag{5-8}$$

This equation is particularly useful in soil engineering, where the main concern often is the depth to bedrock.

EXAMPLE 5-3. In Fig. 5-7, $C_1 = 18$ ft, $V_1 = 1200$ fps, and $V_2 = 2000$ fps. What is the thickness of the upper soil layer?

SOLUTION.

$$D_1 = \frac{18}{2}\sqrt{\frac{2000 - 1200}{2000 + 1200}} = \frac{18}{2}\sqrt{\frac{1}{4}} = 4.5 \text{ ft}$$

Figure 5-7 shows three velocities indicative of three different layers. The expressions for the thickness of deeper layers become rather complicated and are usually solved with the aid of nomographs. As a first approximation,

$$D_2 \approx \frac{C_2}{2}\sqrt{\frac{V_3 - V_2}{V_3 + V_2}} + 0.85\, D_1 \tag{5-9}$$

5.16. DIPPING STRATA. Equations (5-8) and (5-9) assume that strata are parallel to the ground surface. If such is not the case, the indicated velocities of buried refracting layers may be appreciably in error and may even become negative. As a check, a reversed seismic traverse should be performed over the same path, with the sound source or geophone distances stepped off from the opposite end of the traverse. As shown in Fig. 5-8, V_1 will be the same but the apparent V_2 velocities and intersection distances differ from one another. The apparent velocities and intersections may be used with Eq. (5-8) to calculate approximate depths to the change near respective ends of the traverse.

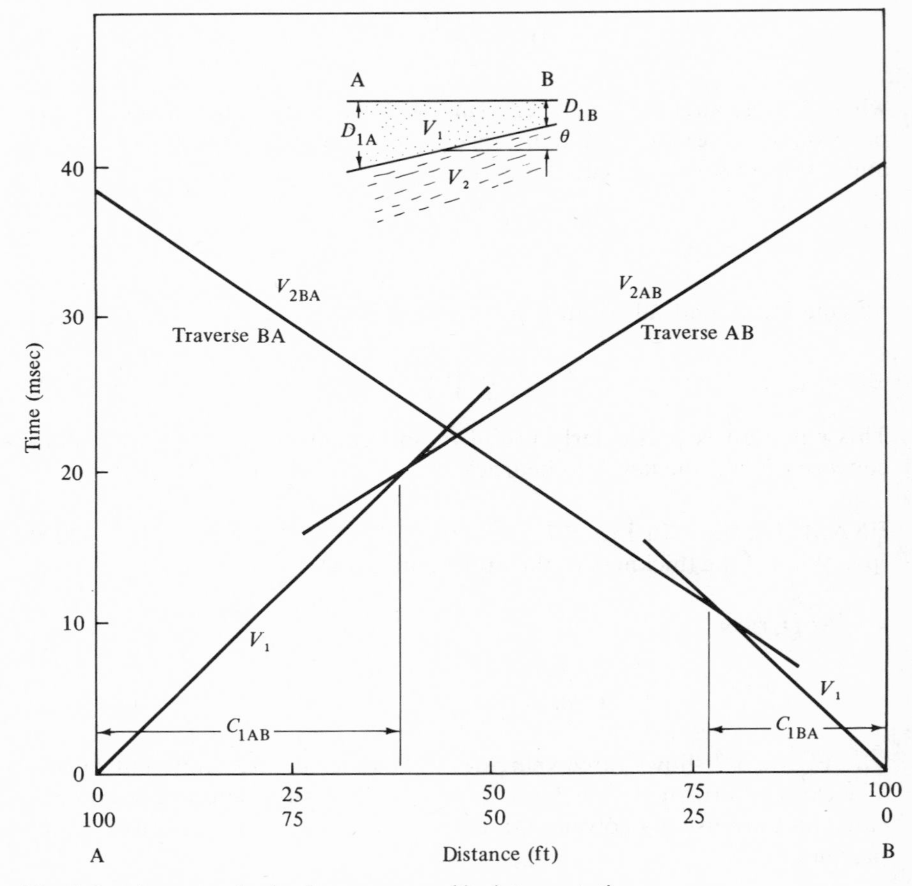

Fig. 5-8. A reversed seismic traverse and its interpretation.

EXAMPLE 5-4. The data of Fig. 5-8 give $V_1 = 2000$ fps, $V_{2AB} = 3100$ fps, $V_{2BA} = 2700$ fps, and indicated times and distances. Calculate the approximate D_1 at A and B.

SOLUTION.

$$D_{1A} \approx \frac{40}{2}\sqrt{\frac{3100 - 2000}{3100 + 2000}} = 9.3 \text{ ft}$$

$$D_{1B} \approx \frac{20}{2}\sqrt{\frac{2700 - 2000}{2700 + 2000}} = 3.0 \text{ ft}$$

5.17. SIGNIFICANCE OF SEISMIC VELOCITY. Seismic velocity is a valuable indicator of hardness. From the theory of elasticity the velocity of a compression wave in a solid is shown to be a function of mass, modulus of elasticity, and Poisson's ratio. The most important of these is the modulus of elasticity, which is very sensitive to continuity and grain-to-grain contact. As an approximation,

$$V_L \approx K\sqrt{E} \tag{5-10}$$

where E is in pounds per square foot, V_L is in feet per second, and K for soils is about 0.6 to 0.8.

Tests by the Caterpillar Tractor Company indicate that soils with seismic velocities below about 5000 fps can be ripped with a large crawler tractor, whereas velocities above about 6000 to 8000 fps indicate the need for blasting.

Typical seismic velocity data are shown in Table 5-2. The velocity of sound in air is about 1100 fps; in water, 5000 fps. Soil velocities below 1100 fps are difficult to measure with a seismic timer since the air wave may trip the geophone. This will give an erroneous value for V_1 and C_1 in the depth equations. Any soil velocity determined as about 1100 fps therefore should be suspect, and the test repeated while muffling the blow or reducing the electronic gain control.

TABLE 5-2. Seismic Velocity Data

Material	*Feet per second*
Soil	800–1800
Dense soil	1500–2000
Sand or gravel below water table	1500–4000
Cemented gravel	4000–7000
Clay, soft shale	3000–7000
Hard shale	6000–10,000
Sandstone	5000–10,000
Limestone, weathered	4000–8000
Limestone	8000–18,000
Basalt	8000–13,000
Granite	10,000–20,000
Frozen soil	4000–7000
Ice	10,000–12,000
Liquid water	5000
Air	1100

5.18. CONTINUOUS SEISMIC REFLECTION PROFILING. Seismic reflection or echo profiling is used extensively in oceanography to obtain not only bottom depths but also for discerning soil layers below the bottom, providing a valuable supplement to test borings for bridge piers, channel dredging, tunneling, etc. Perhaps the simplest example of bottom-depth recorders are "fish finders" which sell for about $100. These belong in a class called "pingers" because they use a piezoelectric transducer which emits a high-frequency pulse; after a time the echo then re-excites the transducer, causing an electrical pulse. The time between sending and arrival is indicative of the depth to the bottom, the accuracy depending on precision of the timing system. Depth surveys may be conducted from boats, or from the upper surface of ice if the water is frozen.

Pingers have the disadvantage of very little penetration into bottom sediments, and higher-energy acoustical "boomers" or electrical "sparkers" are preferred for engineering applications. The various transducers have a wide range of operating frequencies, as shown in Fig. 5-9. The higher the frequency the shorter the wavelength and higher the reso-

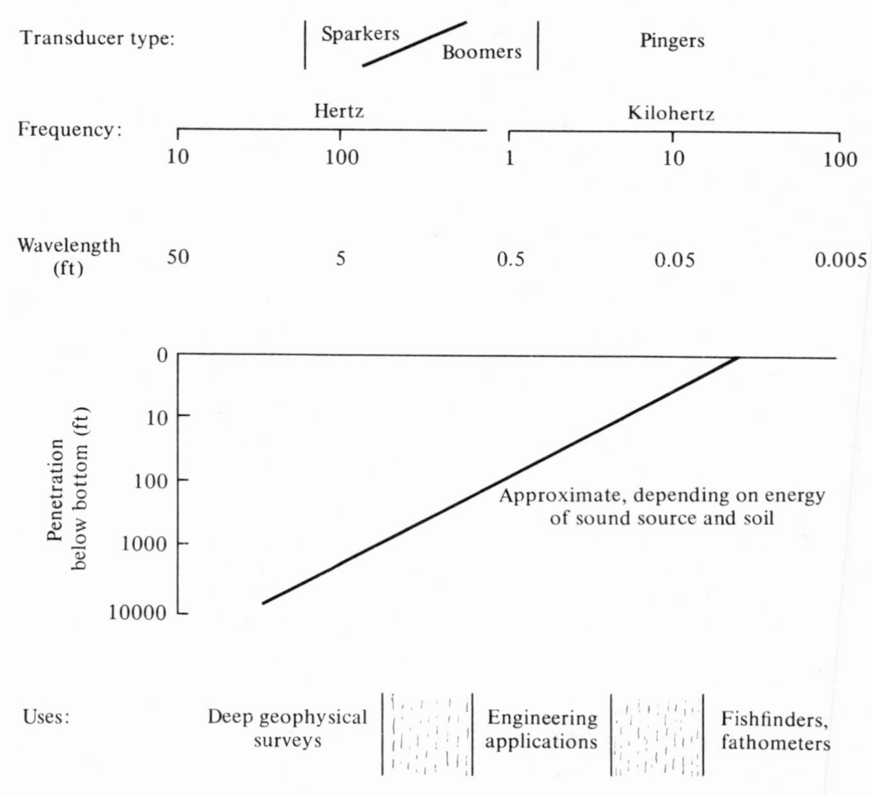

Fig. 5-9. Instruments for subbottom profiling.

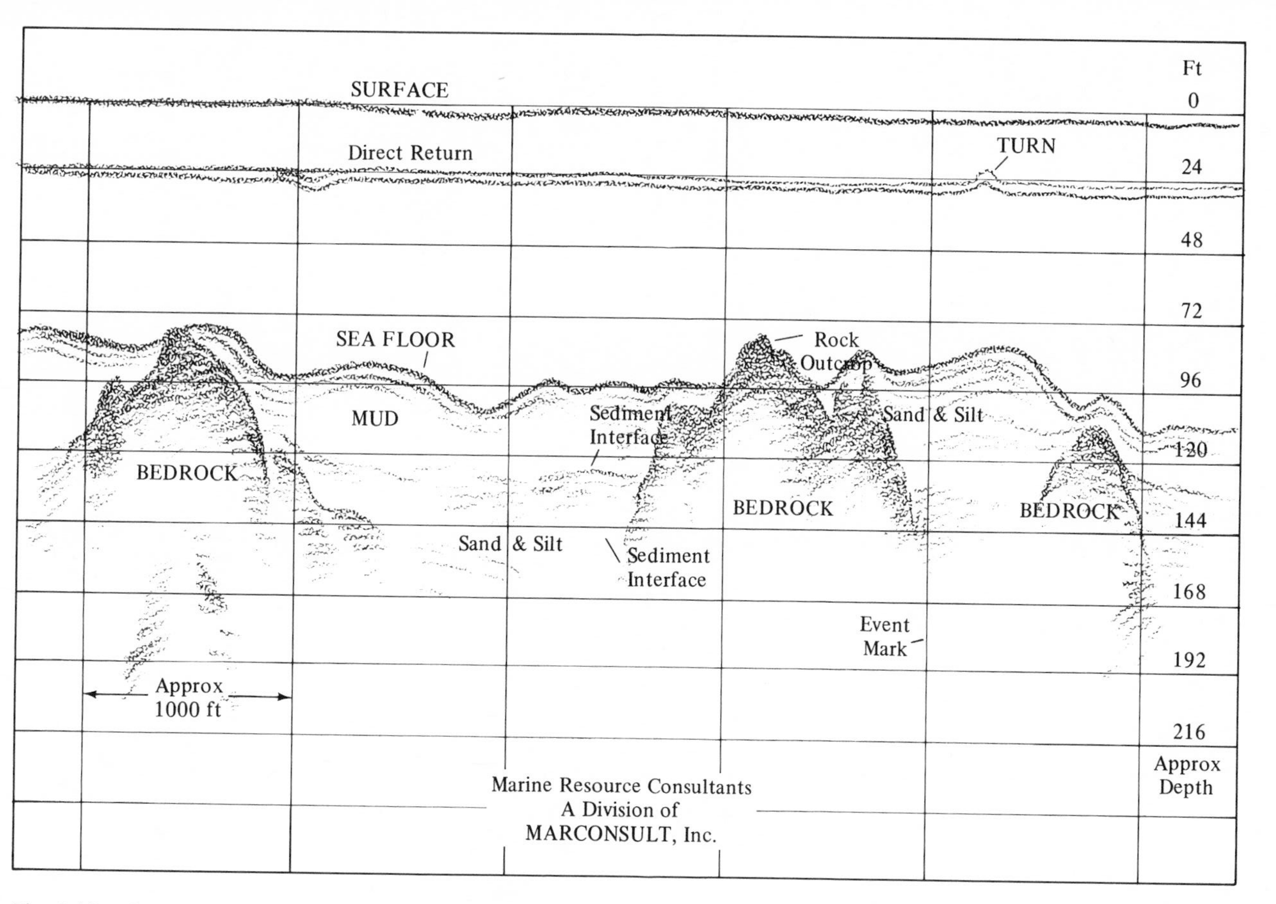

Fig. 5-10. Boomer subbottom profile. Reproduced from (*4*).

lution, but the lower the energy and lower the penetration. For engineering purposes sparkers or boomers operating in the frequency range around 1 kHz appear to effect the best compromise. The system is coupled to a high-speed recorder which gives a continuous subbottom profile as the boat moves. Since the velocity of sound in loose sediments is very close to that in water, approximate depths may be scaled directly on the profile. For example, in Fig. 5-10 the transducer and detector were towed behind a boat about 24 ft apart, giving a scale to convert time to distance.

The advantages of continuous profiling over usual spot drilling as performed on land are obvious from Fig. 5-10, where drilling at set intervals could easily miss the rock outcrops. It would be advantageous if similar techniques could be applied on land, but unfortunately the penetration is very poor into rock, dense sand, or unsaturated soils.

PROBLEMS

5.1. Suggest and discuss in class a site exploration program for a one-story warehouse building 100 ft × 800 ft, located on a level area; (a) where the site geology and soil maps suggest uniform soil conditions over the area; (b) where there is the possibility of limestone caverns or solution cavities filled with clay; (c) on the floodplain of a meandering river.

5.2. Interception of bedrock by the grade line of a highway or pipeline can easily increase excavation costs by a factor of 10. Suggest a minimal exploration program preparatory to bidding such a job.

5.3. Define goals of a site exploration program at a potential borrow area for 1,000,000 cu yd of earth fill.

5.4. Suggest an exploration program using a hand auger, at the site of a one-story 25 ft × 60 ft dwelling with 7 ft basement and a proposed footing width of 1.5 ft.

5.5. In Problem 5.4, what might be the significance of soil color?

5.6. Obtain a county soil survey report from a local county extension office, select two or three potential factory sites along a railway, sketch cross sections of anticipated soils, and suggest a drilling program to select the best site.

5.7. Why should clayey soils have so much lower a resistivity than saturated sand or gravel?

5.8. Reduce the following resistivity data, identifying geological layers and giving their thicknesses:

A (ft)	ΣL_n (mho × 10^3)	L_n	ρ_n
0	61		
3	133		
6	201		
9	215		

A (ft)	ΣL_n (mho × 10^3)	L_n	ρ_n
12	224		
15	234		
20	242		
25	246		
30	251		
35	256		
40	258		
45	259		
50	259		

Suggestion: Calculate ρ's and plot as a bar graph vs. depth. The approximate soil boundary position within a transitional layer can be estimated by connecting bars on either side with a straight line and noting at what depth the line cuts the intermediate bar.

5.9. A plot of seismic data gives velocities of 590 fps for a surface layer and 4800 fps for a subsurface layer, the two time-distance plotted lines intersecting at C = 16 ft. Calculate thickness of the surface layer.

5.10. In Fig. 5-7, t_1 = 6 msec. Compare D_1 from the example with D_1 calculated by means of Eq. (5-6).

5.11. In Fig. 5-7, C_2 = 71 ft and V_3 = 4000 fps. Calculate D_2.

5.12. Plot the following seismic data and calculate depths to the refracting layer:

Forward		*Reversed*	
L (ft)	*t (msec)*	*L (ft)*	*t (msec)*
0	0	100	60
10	9	90	56
20	18	80	51
30	26	70	49
40	36	60	44
50	44	50	40
60	48	40	35
70	50	30	27
80	53	20	18
90	56	10	10
100	59	0	0

5.13. Explain why on a forward and a reversed seismic traverse the end travel times should be equal.

5.14. Soil reports show an area having loess soil over weathered limestone. Calculate a minimum seismic traverse length to assure 30-ft penetration.

5.15. Calculate a minimum seismic traverse to prospect for depth to gravel under a soil cover, where economic considerations prohibit removal of overburden in excess of 15 ft.

5.16. In Problems 5.14 and 5.15 a higher V_2 has the effect of decreasing L for a desired penetration. Explain why.

5.17. In Fig. 5-7, V_1 = 1200 fps; V_2 = 2000 fps; V_3 = 4000 fps. (a) Are the geological interpretations appropriate for the indicated velocities? (b) Which materials can probably be ripped with ripper on a large crawler tractor? (c) Calculate approximate dynamic moduli of elasticity.

REFERENCES

1. Barkan, D. D. *Dynamics of Bases and Foundations* (translated from Russian by L. Drashevska). McGraw-Hill, New York, 1962.
2. Barnes, H. E. "Soil Investigation Employing a New Method of Layer-Value Determination for Earth-Resistivity Interpretation." *Highway Research Board Bull.* **65,** 26–36 (1952).
3. Bhattacharya, P. K., and H. P. Patra. *Direct Current Geoelectric Sounding.* Elsevier, New York, 1968. (Summary from a geophysicist's viewpoint.)
4. Herron, Robert F. "The Application of Acoustic Subbottom Profiling to Engineering Problems." *Proc. Intern. Symp. Engineering Properties of Sea-Floor Soils and Their Geophysical Identification,* pp. 350–359. Sponsored by UNESCO, NSF. University of Washington, Seattle, Washington, 1971.
5. Irving, F. R., N. Bigelow, Jr., W. R. Platts, D. F. Malott, and C. E. Lawson. "Geophysical Methods in Highway Engineering." *Highway Research Board Rec.* **81,** 2–48 (1965).
6. Moore, R. Woodward. "Geophysics Efficient in Exploring the Subsurface." *ASCE J. Soil Mechanics and Found. Division* **87** (SM3: Pt. 1), 69–100 (June 1961).
7. Terzaghi, Karl, and R. B. Peck. *Soil Mechanics in Engineering Practice.* 2nd ed. Wiley, New York, 1967.

6

Test Borings

6.1. BORINGS FOR HEAVY STRUCTURES. Borings for heavy buildings, bridge piers and abutments, and other heavy structures are conducted for the same basic purpose as soil surveys, i.e., to determine the character and location in plan and profile of the soils beneath the structure. The principal difference in the two situations is that borings should be carried to much greater depths in the case of heavy structures, in accordance with the criteria given in Section 5.4. Also, greater emphasis needs to be given to the determination of strength characteristics of the soils beneath a heavy foundation, and much more extensive sampling of the undisturbed undersoil is required.

As in soil surveys, the bore holes for a heavy structure should be sufficiently close together to give complete information concerning the thickness and extent of each of the various soil strata which are encountered. In the case of a bridge, one or two bore holes at each pier and each abutment will probably suffice. For a building, one hole at each corner and one at the center will provide a good start, and additional holes may be dug at such locations as appear to be necessary to complete the geological profile throughout the area of the site. In the case of a large earth dam, the borings will need to be deep and to cover a large area because of the large size and great weight of this type of structure. If rock is encountered in making the borings, drilling should continue well into the rock, and cores should be taken to determine its character and to make sure that it is not simply a boulder or thin ledge with little load-carrying capacity.

Since borings for heavy structures are usually carried much deeper than soil-survey borings, more elaborate drilling equipment usually will be required. Infrequently an ordinary post-hole auger with extensions may be employed. Also, open vertical shafts may sometimes be appropriately used to explore the undersoil. Usually, however, some type of well-drilling rig will be necessary, either with or without casing pipe, to maintain the integrity of the hole as the drilling progresses. Generally speaking, borings of this kind should be made by drilling contractors who have the necessary equipment and experience for handling the work. Payment for drilling operations is made on the basis of a unit price per foot of hole plus a stipulated sum for each soil sample taken, since it is often impossible to estimate in advance how much drilling or how many samples will be required.

The simplest and most rapid method for power drilling employs a helical screw "flight" auger (Fig. 6-1), which may be used continuously without periodic withdrawal from the hole. However, soil identifications and sampling should be performed by withdrawing the auger and removing soil from the tip, since soil samples otherwise become contaminated in a long circuitous journey up and out of the hole. It is better to periodically withdraw the flight auger and use a special sampler. In many instances the driller will be able to detect a change in soil type by a change in speed, sound, or "feel" of the drilling. Whenever this happens the depth should be noted and a sample obtained from the new layer.

Drill holes in sand below the ground-water table tend to cave in if the drilling tools are withdrawn, and may even cave in during the drilling operation owing to seepage of water into the drill hole, which causes a "quick" condition. This may impede penetration of the drill and result in excavation of an underground cavern. The unexpected collapse of such caverns may endanger nearby foundations. An experienced driller will be wary of this possibility and use a drilling mud consisting of a bentonite-water slurry. The mud is pumped down inside the drill stem to emerge at the drill and rise back to the surface between the drill stem and the walls of the hole. Since the drilling mud carries out the drilling cuttings, a continuous auger offers no advantage, and drilling is done with a "fishtail" tool or a rock bit attached to a drill pipe. Positive fluid pressure of the mud avoids ground water seepage into the drill hole, prevents quicksand, and causes a mud filter "cake" to build up on and reinforce the circumference of the hole.

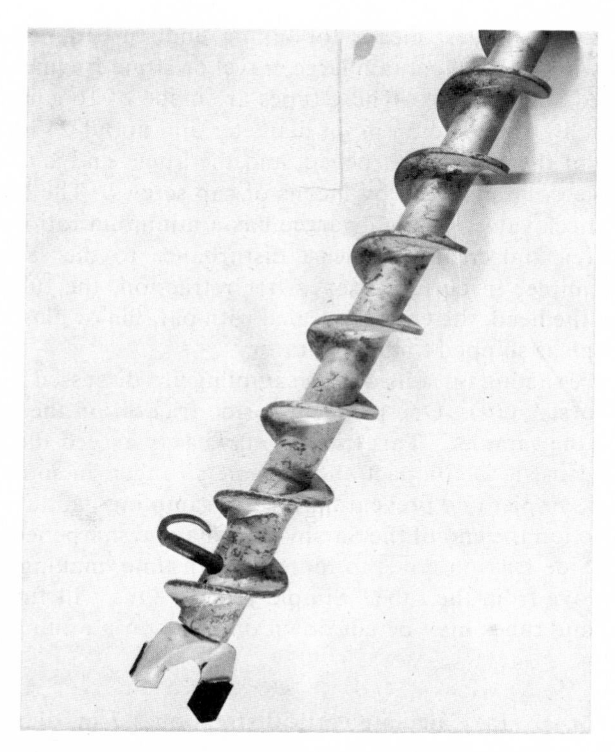

Fig. 6-1. Continuous flight auger with interchangeable carbide-faced bit attached by U-shaped drive pin (shown ready to drive).

6.2. DESIGN OF SAMPLERS. A number of different samplers have been designed by various engineers. No attempt will be made to describe them in detail. Essentially, each consists of a cylindrical barrel fitted at the bottom with a cutting edge which is slightly smaller in diameter than the inside of the barrel. At the top, the barrel is threaded into a transition section which connects with extra-strength drive pipe extending to the surface. Inside the transition section is a ball check valve which permits air or water to escape from the barrel as the sample enters from below. This check valve prevents the return of fluid from above and helps to hold the sample in place while the sampler is lifted from the hole. Some designs provide for a split barrel which facilitates removal of the sample after the cutting edge and the transition section are removed. In other designs, there is a removable, thin, metal or plastic cylindrical liner which encases the sample and facilitates its removal from the barrel and protects it during shipment to the laboratory.

Perhaps the simplest means for taking undisturbed samples of cohesive soils which do not contain large gravel or stone fragments is by use of thin-walled Shelby tubes. These tubes are made of 16-gauge steel, and they are usually from 2 to $5\frac{1}{2}$ in. in diameter and about 24 in. long. The cutting edge of the tube is sharpened, and the upper end is drilled for attachment to a coupling head by means of cap screws. The head is fitted with a ball check valve. A Shelby tube has a minimum ratio of wall area to sample area and creates the least disturbance to the sample of any drive-type sampler in current use. After retraction, the tube is disconnected from the head, the ends are sealed with paraffin or plastic end caps, and the sample is shipped to the laboratory.

The forces acting on soils during sampling are discussed in the classic work of Hvorslev (*10*). One problem is side friction on the sample as it slides inside the sampler. This friction may easily exceed the strength of the soil and cause it to compact and become stronger, in some cases even plugging the sampler and preventing further sampling. In an effort to reduce side friction the end of the Shelby tube may be sharpened and turned in slightly. Side friction tends to increase with time, making the soil difficult to remove from the tube. Simple jacking thus will further densify the sample, and tubes may be cut down one side on a milling machine to aid extrusion.

EXAMPLE 6-1. (a) Calculate vertical stress on a 3-in. diameter Shelby tube sample per inch of penetration, assuming the soil adheres to the tube with a skin friction of 0.5 psi. (b) What is the best way to extrude such a Shelby tube sample 20 in. long, by pushing from the bottom or from the top?

SOLUTION. The perimeter contact area per inch of penetration is 3π sq in./in. Multiplying by skin friction gives $3\pi \times 0.5 = 1.5\pi$ lb/in., and dividing by cross-sectional area gives $1.5\pi \div \pi(1.5)^2 = 0.66$ psi/in.

Pushing the tube 20 in. will cause a vertical stress ranging from 0 to a maximum of $20 \times 0.66 = 13.3$ psi on the bottom part of the sample. The sample therefore should be pushed from the bottom to avoid a similar compression on the top of the sample during extrusion.

Piston samplers offer an ingenious means to reduce side friction (Fig. 6-2). A short piston is fitted inside the sample tube and is attached to a rod which extends to the ground surface inside the drill rod. The piston rod is clamped to the drill rod so that the piston is flush with the cutting edge, while the sample tube is lowered to the bottom of the hole. As the tube is pushed into the soil, the piston is released. It is clamped again

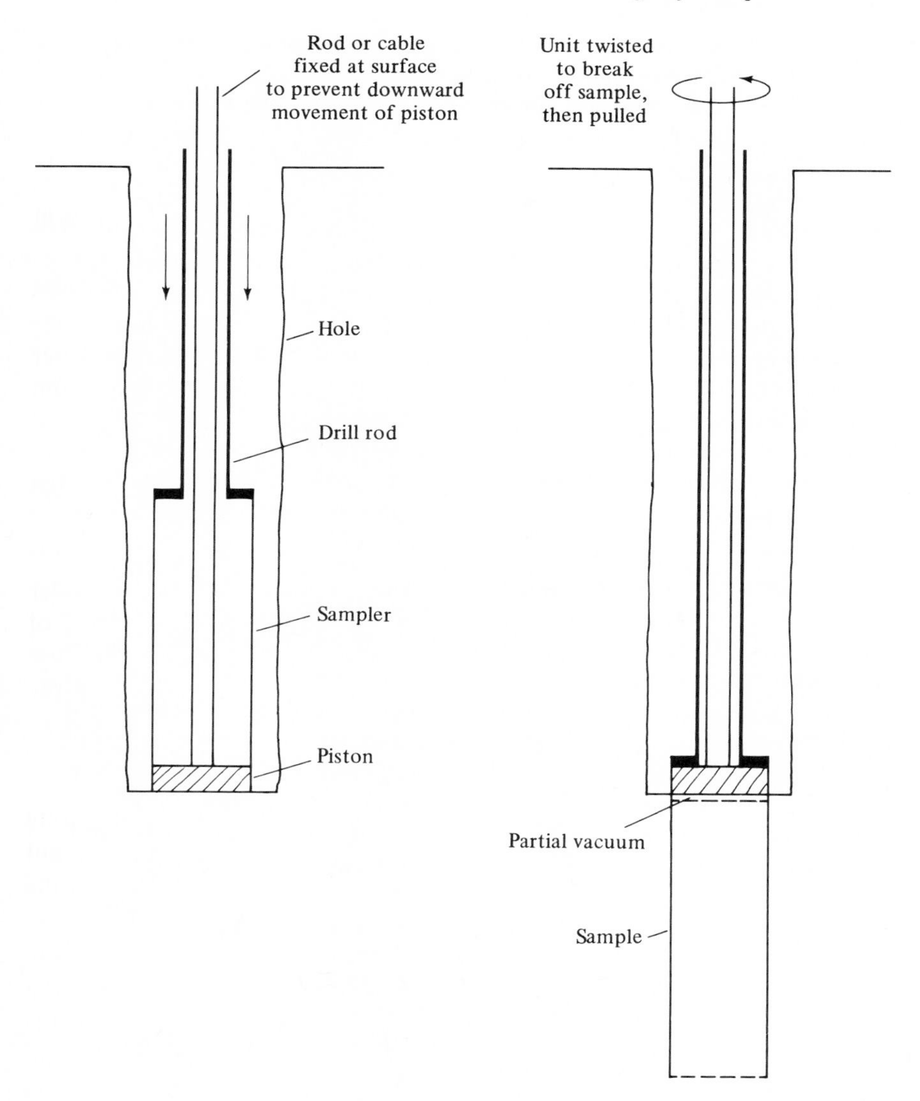

Fig. 6-2. Principle of the piston sampler.

during removal of the tube and sample. The piston prevents undesirable slurry and other foreign material from entering the sample tube and helps to retain the sample by vacuum. Piston samplers and techniques for their use have been refined to such a degree that, with the aid of drilling mud, it is often possible to recover samples of sand from below the water table.

In the Osterberg sampler the sampling tube is attached to a second piston and pushed by fluid pressure through the drill pipe.

The maximum vacuum, or pressure deficiency, in a piston sampler cannot exceed the total fluid pressure (water plus atmosphere) at the sampling depth. For example, at a depth 10 ft below the water table the water pressure is $(10 \times 62.4)/144 = 4.3$ psi, so the theoretical maximum vacuum would be $4.3 + 14.7 = 19$ psi. Piston samplers are therefore well adapted to ocean or lake bottom sampling, where there is a considerable hydrostatic pressure. The complications of such remote sampling are discussed by Richards and Parker (*14*).

Perhaps the most successful sampler for long undisturbed samples is the Swedish foil sampler [Kjellman *et al.* (*12*)], which is a piston sampler utilizing 16 rolled-up metal foil ribbons arranged around the periphery of the sampler. As the sample tube is pushed, the ribbons unroll and enclose the sample. Since the ribbons move with the sample, side friction on the soil becomes zero, and very soft soil samples as long as 60 ft have been obtained.

6.3. PENETRATION TESTS AND SAMPLING. Because of the elaborate precautions, equipment, and expense of undisturbed sampling, a different philosophy is to load or drive a sampler and evaluate strength from the resistance to penetration. In this case the "split-spoon" sampler (Fig. 6-3) is disassembled to remove the samples in the field. The samples are disturbed and properly used only for purposes of identification of strata, a representative portion being cut from each layer and placed in a sample bottle. The penetration resistance is the number of blows from a standardized hammer to cause a given amount of penetration. The hammer is usually lifted by use of a drill rig, but can be raised with a hand-operated winch and tripod.

The degree of disturbance of spoon samples is influenced by the re-

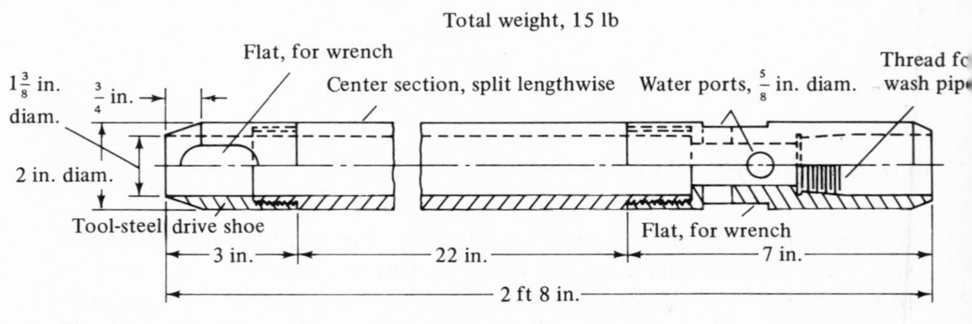

Fig. 6-3. Gow-type split-spoon sampler used in standard penetration test (Courtesy of Raymond Concrete Pile Co.)

lationship between the inside and outside cross-sectional areas of the spoon, in addition to the method of driving. The area ratio of a spoon may be expressed by the equation.

$$A_r = \frac{E_d^2 - I_d^2}{I_d^2} \times 100 \qquad (6\text{-}1)$$

in which

A_r = area ratio, in percent;
E_d = external diameter; and
I_d = internal diameter.

In the Standard Penetration Test (ASTM Designation D1586-67) the split spoon has an outside diameter of 2 in. and an inside diameter of $1\frac{3}{8}$ in., as illustrated in Fig. 6-3. This sampler is usually driven into the soil stratum by means of a 140 lb weight falling a distance of 30 in. and striking an anvil at the upper end of the drill rod or wash pipe to which the spoon is attached. When the sampler is so driven, the operation is known as the standard penetration test, and the number of blows of the hammer required to cause the sampler to penetrate 1 ft into the soil is called the standard penetration value or the N value. Since the sampler is 18 in. long, blows for the three 6-in. intervals are normally recorded in the field notes, and the engineer adds the last two for an N value. If the last count is abnormally high, indicating a different layer, the first two counts are used.

The Standard Penetration Test is best adapted to sands, which are very difficult to sample without disturbance. On the other hand Shelby tubes or the other methods for undisturbed sampling are best for clays. When the strata are not known, the two methods may be alternated at 5 ft depth intervals. One advantage of Standard Penetration split spoon sampling is that samples are examined immediately in the field and retain sufficient original structure to be very useful for identification, whereas Shelby tube samples can be examined only at the exposed ends.

The Dutch cone penetrometer, popular in Europe and gaining favor in the U.S., is a static penetration test involving pushing a 60° cone of 1.4 in. base diameter. Both penetration resistance of the cone and skin friction on a 6 in. casing tube are recorded. The ratio of skin friction to point bearing indicates the cohesive nature of the soil; for sand it is usually less than 2%, and for clays 2–10%. In sands the cone penetration value in tons per square foot can be converted to an equivalent N or N' value by dividing by a factor of from 2.5 to 6, averaging about 4 [Meyerhoff (*13*)].

Cone penetrometers may also be driven, and comprise perhaps the cheapest and simplest deep-hole test in that an elaborate or heavy hold-down apparatus is not needed for hydraulic pushing, and the penetrometer need not be withdrawn after every 18 in. of penetration. The latter can be very time-consuming, since the drill rod must be dissassembled and stored as it is pulled, then reassembled for the next test. Cone tests must be accompanied by drilling to identify the soil types, since penetration data without soil identifications are worthless. Conversion of a cone blow count to an equivalent N value requires extensive comparative tests in a given soil. As a rough guide, driving resistance is approximately proportional to the displacement area. Any penetrometers tend to give erratic values when the maximum soil particle size approaches the diameter of the penetrometer; therefore unusually high blow counts are usually omitted from calculated averages.

6.4. VANE SHEAR TEST. Because of inevitable disturbances to soil during sampling, increased emphasis is being placed on *in situ* tests in bore holes. One of the oldest and most widely used methods is the vane shear test, developed and investigated extensively in Sweden from the late 1940s. The essential element is a rod with radial fins or vanes. The rod is pushed into soil and twisted at 0.1°/sec while the torque to cause twisting is measured. The torque increases until it reaches a maximum value as the soil shears around a cylindrical surface (Fig. 6-4). Both a maximum and a constant residual torque are measured, to give a measure of sensitivity of the clay to remolding.

Vane shear torques are converted to soil shearing stresses by integrating across an assumed stress distribution shown in Fig. 6-4. The result for vanes of rectangular section and a length twice the diameter is

$$\tau = \frac{3T}{28\pi r^3} \tag{6-2}$$

in which T is the torque in inch-pounds, r the radius in inches, and τ the shearing stress in pounds per square inch. (Since the equation is dimensionally homogeneous, other units may be used.) Commercial vanes are available tapered at 45° on the ends to facilitate their use through drill casing. In this case the conversion factor differs and takes into account the diameter of the drill rod (cf. ASTM Designation D2573-67T). Vanes introduced by Aas (*1*) of Norway vary the length to diameter ratio in order to establish horizontal versus vertical shearing strength.

As will be discussed later, the shearing resistance of soils can be thought of as a sum of cohesion and internal friction. The vane shear test is adapted only to clayey soils, and is sometimes taken as a measure of cohesion, which is the shearing strength when there is no stress acting perpendicular or normal to the shear plane. As shown in Fig. 6-4, twisting a

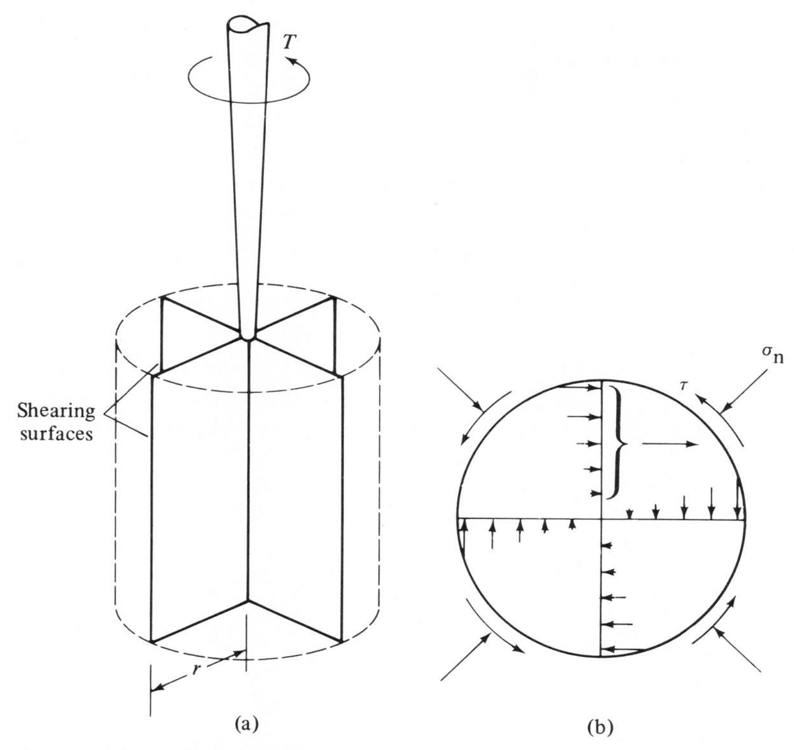

Fig. 6-4. Principle of the vane shear test. (a) Shearing involves one vertical cylindrical surface and two horizontal end surfaces. (b) Twisting of the vane induces normal stress σ_n on the shearing surface.

smooth-bladed vane produces major stress components normal to the shearing surface. In saturated clays much of this normal stress will be taken by pore water pressure, provided no water can drain away while the soil is stressed. Indirect evidence suggests that some water may drain away, allowing part of the σ_n to press soil grains together and increase friction. For example, Farrant (*6*) showed that if friction is zero, shearing strength should be independent of the number of blades, whereas he found that strength is higher with fewer blades. He then utilized the strength ratio from different bladed vanes to calculate a coefficient of friction. Unfortunately the method is rather insensitive compared to methods discussed in Chapter 19. One can appreciate the importance of friction in the vane test by trying it on sand, which drains well and allows a full effect of σ_n.

6.5. PRESSURE-EXPANSION TESTS. A relatively simple test involving complex theory was developed over the past decade by L. Menard of France. The test involves inflating a rubber membrane inside a bore hole and measuring the volume expansion as a function of applied pressure. The expanding membrane is contained axially in the hole by inflated guard cells (Fig. 6-5). During the early part of the test the soil is assumed to behave elastically. From theory of elasticity applied to thick cylinders,

$$\frac{\Delta V}{V_0} = 2p(1 + \nu)\frac{1}{E} \tag{6-3}$$

where $\Delta V/V_0$ is the change in volume per unit volume, p the applied pressure, ν Poisson's ration, and E the modulus of elasticity. Similar devices developed for rock mechanics are instrumented to measure the change in radius of the bore hole, $\Delta r/r_0$ being substituted for $\Delta V/V$ in the above equation.

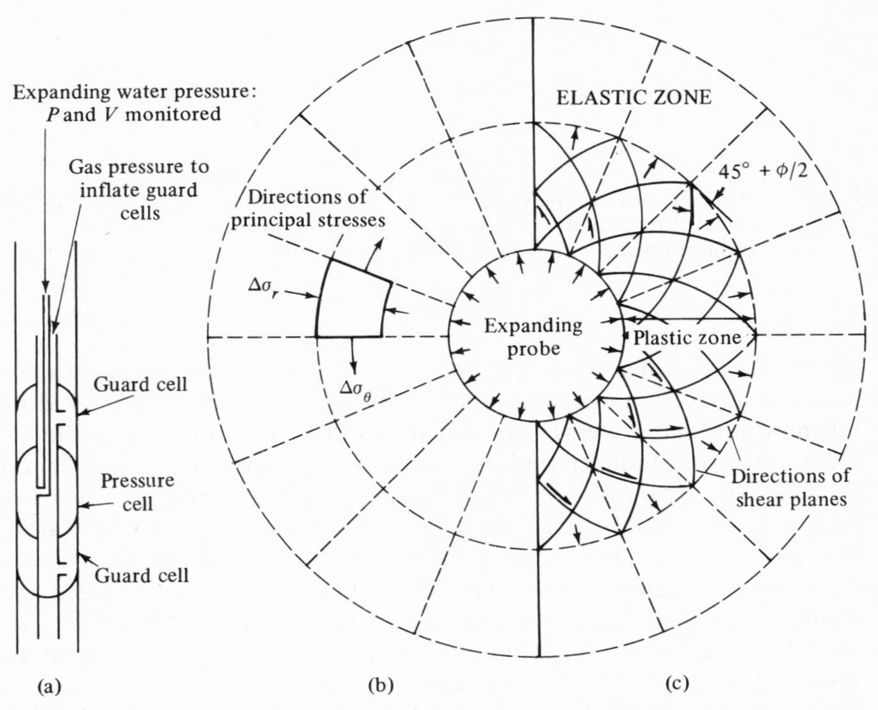

Fig. 6-5. (a) Menard pressuremeter. (b) Principal stresses induced around the probe. Note the tensile stress $\Delta\sigma_\phi$. (c) Higher pressure may initiate plastic shear failure with sliding along vertically oriented surfaces.

In soils, further expansion initiates plastic shear failure, as shown in Fig. 6-5, starting when the applied pressure equals the original horizontal pressure plus the soil cohesion. If the soil has no internal friction, a pressure theoretically will be reached when the hole will expand indefinitely. Calculation of cohesive shearing strength is based on several assumptions, and results differ from other test results by up to factor of 2 [Gibson and Anderson (*8*), Higgins (*9*)]. Pressure-expansion tests have been used to predict settlement from the evaluation of E, and to predict ultimate bearing capacity.

6.6. IN SITU STRENGTH VERSUS DEPTH. Standard penetration, vane shear, and pressure-expansion test strengths all tend to increase with increasing depth, more than do results of laboratory tests of samples from the same depths. This is a common argument for greater reliability of in situ tests, the inference being that higher deep strengths are a result of less sample disturbance. However, other factors are also important.

One factor is the increase in geostatic pressure with depth. For example, the vertical stress at 100 ft depth is 100 times the effective unit weight of the soil in pounds per cubic foot. The horizontal stress is a fraction of the vertical stress, ordinarily about 0.4 to 0.5. (This fraction is the coefficient of earth pressure at rest, K_0, discussed in a later chapter.)

Gibbs and Holtz (*7*) showed that standard penetration tests performed in sands are greatly influenced by overburden pressure. This is reasonable since sand strength is primarily from internal friction, which by definition is strength that is sensitive to confining pressure. As a result of their findings the U.S. Bureau of Reclamation recommends that N values be evaluated in terms of the graph in Fig. 6-6. The line for 40 psi overburden pressure approximately follows a widely used relationship suggested by Terzaghi and Peck (*16*). An equation to correct N values to the 40 psi reference pressure is

$$N' = N \frac{50}{p + 10} \tag{6-4}$$

where p is the effective overburden pressure in pounds per square inch, not to exceed 40 psi. If the wet unit weight γ is nominally 130 pcf,

$$N' = N \frac{50}{0.9h + 10} \tag{6-5}$$

where h is the depth in feet, or one-half the depth in feet if below the water table.

EXAMPLE 6-2. A blow count of 10 bpf was obtained at a depth of 30 ft in sand below the water table. Estimate the relative density and the equivalent N under 40 psi pressure.

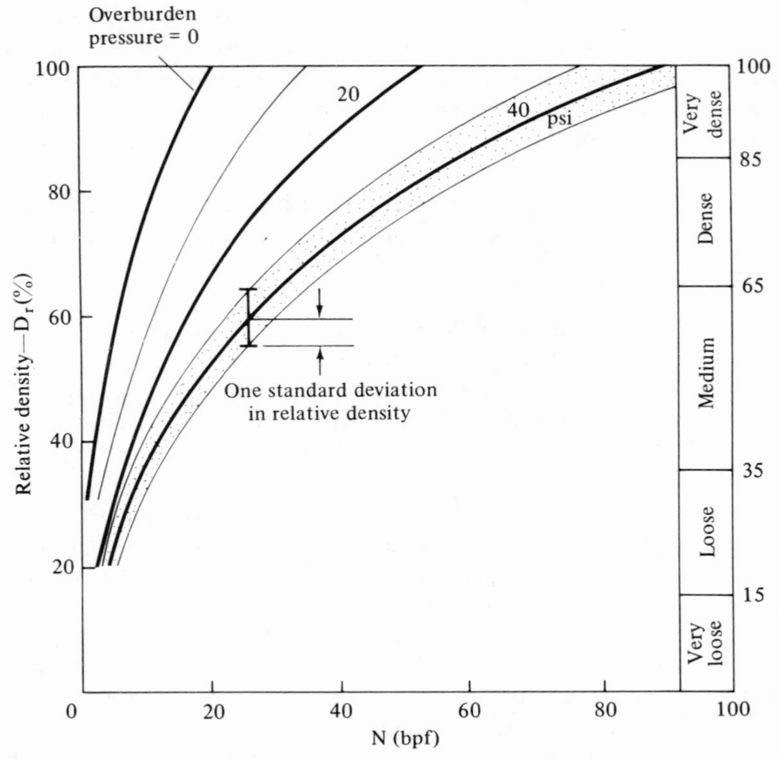

Fig. 6-6. Relative density of sand in relation to Standard Penetration Test and overburden pressure. After Gibbs and Holtz (*7*).

SOLUTION. From Eq. (6-5),

$$N' = 10 \frac{50}{0.9\,(15) + 10} = 21 \text{ bpf}$$

Terzaghi and Peck (*16*) recommend that if N values in very fine sand below the water table exceed 15, they be reduced as shown:

$$N' = 15 + \tfrac{1}{2}(N - 15) \tag{6-6}$$

This recognizes that dense, very fine sands tend to develop negative pore pressure or "suction" during rapid shearing, temporarily increasing their strength.

Vane shear tests, being applicable only to clays, would be expected to show less sensitivity to overburden pressure. Nevertheless in normally

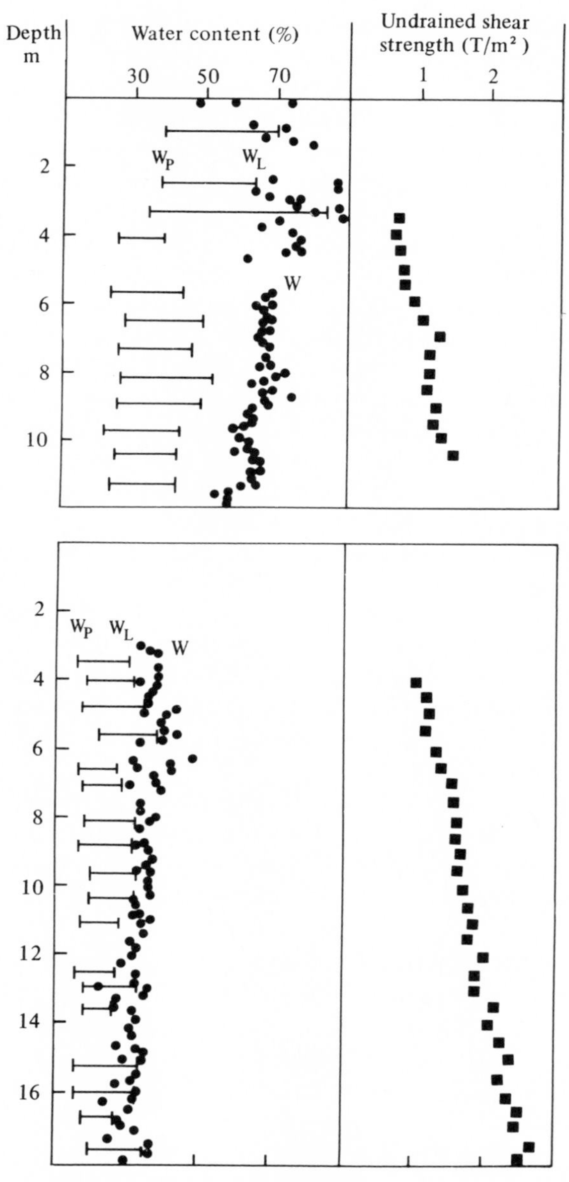

Fig. 6-7. Squares show the increase in vane shear strength with depth in two Norwegian clays. Natural water contents are shown on the left by circles. After Aas (*1*).

consolidated clays, that is those which have consolidated to equilibrium under their own weight, vane shear strengths increase linearly with depth, as shown in Fig. 6-7. The slope of this line has been designated the c/p' ratio, implying an increase in cohesion with depth. However, internal friction also may be important, particularly when we note that p' is a vertical pressure, but shearing takes place mainly on a vertical surface and is influenced by horizontal pressure.

Aas (*1*) and others have found that in normally consolidated clays the strength on a vertical plane is only $\frac{1}{2}$ to $\frac{2}{3}$ that on the horizontal plane, and attribute this to the difference in horizontal vs. vertical pressure. In clays subjected to an earlier loading where horizontal and vertical pressures are equal, vane shear strengths were found to be the same vertically and horizontally. This sensitivity of vane shear strength to pressure is a characteristic of internal friction, not cohesion alone, as suggested by the designation c/p' ratio. Karlson and Viberg (*11*) and Aas (*1*) found the c/p' ratio to vary between 0.1 and 1.0, and relate more to existing horizontal stress than to clay physical properties.

Strengths from pressuremeter tests in overconsolidated clays such as the London clay, where the horizontal pressure is high, increase faster with depth than do laboratory test data in such clays [Gibson and Anderson (*8*)]. In normally consolidated clays this is not true [Higgins (*9*)]. Since the pressuremeter causes shearing on vertical planes (Fig. 6-5), this also suggests an influence from horizontal geostatic pressure and internal friction, even though the latter is assumed to be zero in order to interpret tests in clays.

A field test devised by one of the authors to control horizontal pressure on the shear plane and separably determine cohesion and internal friction is described in Chapter 19. So far as is known, this test is insensitive to geostatic pressure except as it influences soil properties.

6.7. STATISTICAL TREATMENT OF DATA. Soil is variable, so obviously it is not enough to drill one hole or run one test and assume this information applies to an entire site or project. Instead many holes are drilled and many tests performed, and results expressed as averages or arithmetic means. An *average* or *mean* is the simplest statistic, readily calculated and understood.

In most instances the mean is not enough either, since the engineer's design must encompass not only the average, but also the range in properties. Thus both the mean and the range are important. However, the observed range will increase with an increasing number of tests: For example if the average thoracic measurement of freshmen girls is 32 in., one may require very extensive sampling to find a 40.

The *standard deviation* is defined to remove the effect of number of tests on the calculated range:

$$\sigma = \sqrt{\frac{\sum X^2}{n} - \bar{X}^2} \tag{6-7}$$

or

$$\sigma = \sqrt{\frac{\sum (X - \bar{X})^2}{n}} \tag{6-8}$$

where X represents an individual value, $\bar{X}$ is the mean, and n is the number of tests. The estimate of σ becomes more reliable with an increasing number of tests, but σ itself should show no systematic change.

One use of the standard deviation is to calculate how much of a population probably exceeds certain limits. This is done by assuming a normal distribution of data, as shown in Fig. 6-8. For example, 68% of the area under this curve, representing 68% of the population, is bounded by $\bar{X} \pm \sigma$, shown shaded in Fig. 6-8.

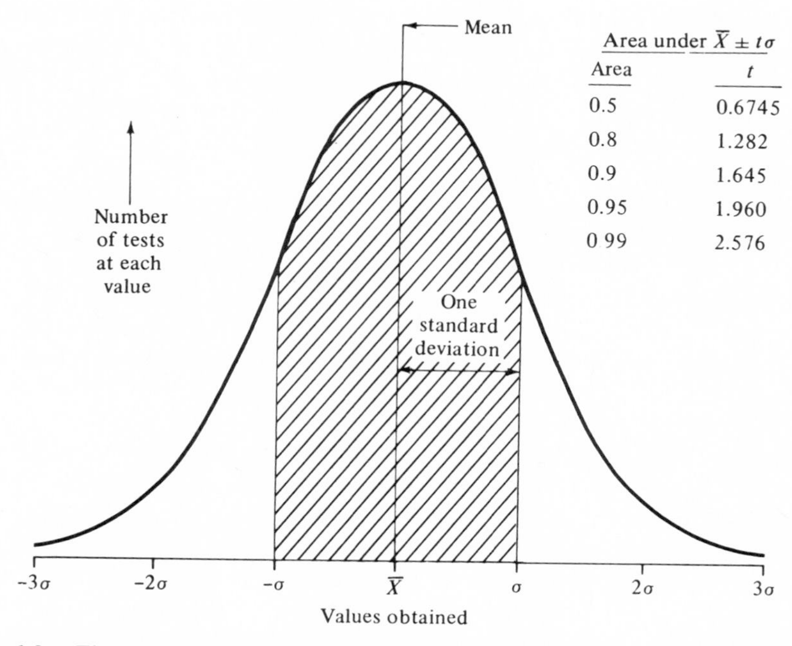

Area under $\bar{X} \pm t\sigma$	
Area	t
0.5	0.6745
0.8	1.282
0.9	1.645
0.95	1.960
0.99	2.576

Fig. 6-8. The normal distribution curve. The area under the curve limited by $\bar{X} \pm \sigma$ is shaded.

EXAMPLE 6-3. The following standard penetration test blow counts were obtained in 10 tests. Calculate the standard deviation and estimate the N value *exceeded* by 95% of the deposit. (Hint: Use 90% area boundaries since these will define 5% "tails" at both ends of the distribution.) Given: $N = 4, 12, 18, 9, 16, 6, 13, 10, 8, 14$.

SOLUTION. The mean, $\overline{N} = \sum N \div n$ (number of tests) $= 110 \div 10 = 11.0$. Values of $(N - \overline{N})$ are $-7, 1, 7, -2, 5, -5, 2, -1, -3, 3$. Therefore $\sum (N - \overline{N})^2 = 49 + 1 + 49 + 4 + 25 + 25 + 4 + 1 + 9 + 9 = 176$.

$$\sigma = \sqrt{\frac{176}{10}} = 4.2$$

or $\overline{N} = 11.0 \pm 4.2$ where the $\pm$ entry signifies one standard deviation. If we assume that $\overline{N}$ and σ are accurate estimates, from Fig. 6-8, 90% of the values will be between $11.0 \pm 1.645\,(4.2) = 4.1$ to 17.9. If the distribution is symmetrical, 5% will be less than 4.1 and 95% will exceed 4.1.

QUESTION. Why not use 95% boundaries in the above problem?

Another common use of standard deviation is to estimate validity of the calculated mean as representing the true population. This is referred to as *confidence limits on the mean* and is a function of σ and n. Note that it describes only the mean, and not the population or "spread." The larger the sample, the closer the confidence limits.

EXAMPLE 6-4. By use of Table 6-1 calculate the 95% confidence limits on $\overline{N} = 11.0$, $\sigma = 4.2$, and $n = 10$.

SOLUTION. The 95% confidence limits are $\overline{N} \pm 0.754\,(\sigma) = 11.0 \pm 0.75\,(4.2) = 11.0 \pm 3.2 = 7.8$ to 14.2. That is, from the above data we are 95% confident that the true average N for the entire population is between 7.8 and 14.2.

Before statistical measures are applied to data, the distribution should be plotted as a bar graph. Figure 6-9 shows a bar graph of 123 standard penetration tests conducted in glacial till at a hospital site. To plot such a graph an arbitrary interval is assigned on the horizontal axis, and each test value is drawn as a box in the appropriate interval. For example in Fig. 6-9 three tests have N values from 10 to 14, seven from 15 to 19, etc. This particular distribution is said to be *bimodal*, since it shows two *modes*, or population peaks which correlate very well with the field

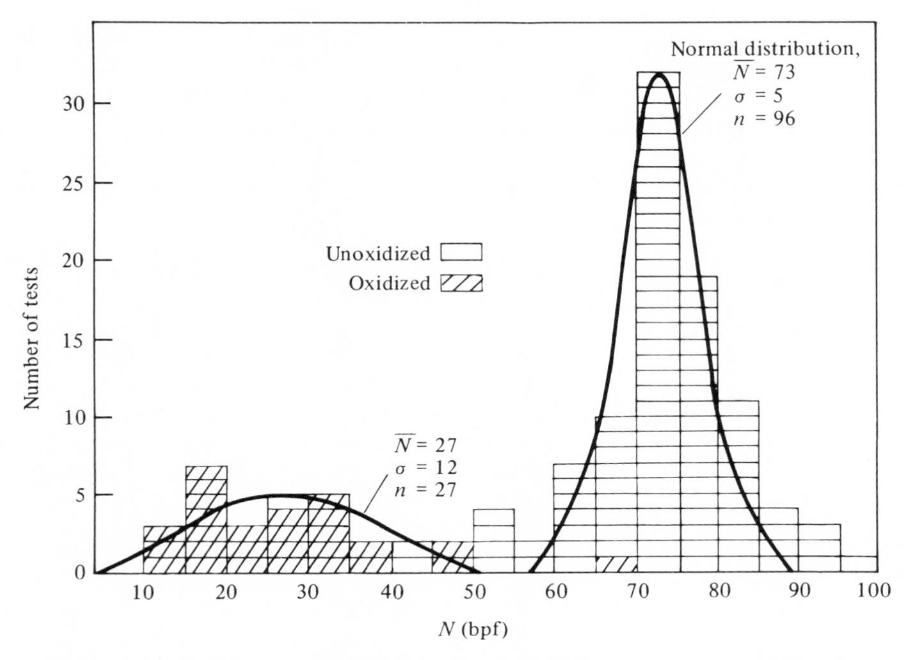

Fig. 6-9. Histogram of standard penetration blow counts in glacial till showing bimodal distribution related to weathering.

TABLE 6-1. 95% Confidence Limits on an Average $\overline{X} \pm a\sigma$

No. of Observations n	*Multiplier* a
4	1.84
5	1.39
6	1.15
7	1.00
8	0.89
9	0.82
10	0.75
12	0.66
14	0.60
16	0.55
18	0.51
20	0.48
25	0.42
>25	$1.96 \div \sqrt{n-3}$

description. The two populations therefore were treated separately to calculate means and standard deviations. The ideal normal distributions are shown by heavy lines. From this one may conclude that not only is the weathered till much weaker than the unoxidized till ($\bar{N}$ = 27 compared to 73 bpf), it also is more variable (σ = 12 compared to 5 bpf) and could have some very weak areas. The building was therefore supported on the unoxidized till.

PROBLEMS

6.1. Plot the total force from skin friction on a 2 in. diameter sample versus depth of penetration from 0 to 25 in. for the following: (a) Shelby tube sample. Assume friction equals 1 lb for each square inch of contact area. (b) Piston sampler. Assume the same friction, and a maximum vacuum of 14 psi on the upper end of the sample. (c) Swedish foil sampler.

6.2. Calculate area ratios for the following: (a) Standard Penetration Test sampler; (b) 2 in. o.d. Shelby tube, i.d. = $1\frac{7}{8}$ in.; (c) 3 in. o.d. Shelby tube, i.d. = $2\frac{7}{8}$ in.; (d) $4\frac{1}{2}$ in. o.d. Shelby tube, i.d. = $4\frac{3}{8}$ in.

6.3. Calculate a multiplying factor to convert blow counts from the Canadian 2.25 in. drive cone test to SPT N values, assuming the same driving energy and: (a) soil inside the split spoon offers no resistance to driving; (b) soil inside the split spoon resists driving such that the spoon behaves as a solid rod.

6.4. Develop an expression between rectangular shear vane height-to-radius ratio and the ratio of vertical to horizontal shearing surface.

6.5. A pressure-expansion test shows an increase in bore-hole diameter from 3.0 to 3.05 in. with an applied pressure of 80 psi. Calculate E, assuming the soil is incompressible. (ν = 0.5.)

6.6. Calculate the vertical and horizontal pressure at a depth of 40 ft in soil weighing 110 lb/ft^3. K_o = 0.4. Neglect the effect of a water table.

6.7. Explain why the c/p' ratio of vane shear strength to overburden pressure should be higher for overconsolidated than for normally consolidated clays.

6.8. Calculate the mean and standard deviation for the following data: N = 17, 16, 25, 20, 18, 37, 45, 26, 18, 22 bpf.

6.9. Plot a bar graph of the data in Problem 6.8. Would you be justified in deleting any values? If so, explain and recalculate $\bar{N}$ and σ.

6.10. Calculate 95% confidence intervals on the means from 6.8 and 6.9. Does this appear to justify the omission of data?

6.11. Assuming means in standard deviations in Fig. 6-9 are an accurate representation of the true conditions, 90% of each soil exceeds what N value?

6.12. Calculate 95% confidence intervals on the means of Fig. 6-9.

REFERENCES

1. Aas, G. "Vane Tests for Investigation of Anisotropy of Undrained Shear Strength of Clays." *Proc. Geotech. Conf. Oslo* **1,** 3–8 (1967).

2. *ASTM Book of Standards*, Part 11. American Society for Testing and Materials, Philadelphia, Pennsylvania, 1971.
3. ASTM Committee E-11. *ASTM Manual on Quality Control of Materials.* American Society for Testing and Materials, Philadelphia, Pennsylvania, 1951.
4. Cadling, Lyman, and Sten Odenstad. "The Vane Borer." *Proc. Roy. Swedish Geotech. Inst.* **2,** (1950).
5. Clark, K. R. "Mechanical Methods of Undisturbed Soil Sampling." *ASTM Spec. Tech. Publ.* **351,** 86–95 (1964).
6. Farrent, T. A. "The Interpretation of Vane Tests in Soils Having Friction." *Proc. 3rd Australia-New Zealand Conf. Soil Mechanics and Found. Eng.*, 81–86, (1960).
7. Gibbs, H. J., and W. G. Holtz. "Research on Determining the Density of Sands by Spoon Penetration Testing." *Proc. 4th Intern. Conf. on Soil Mechanics and Found. Eng.* **I,** 35 (1957).
8. Gibson, R. E., and W. F. Anderson. "In-situ Measurement of Soil Properties with the Pressuremeter." *Civ. Eng. and Publ. Works Rev.* **56** (568); 615–618 (1961).
9. Higgins, C. M. "Pressuremeter Correlation Study." *Highway Research Rec.* **284,** 51–62 (1969). Discussion by H. Y. Fang.
10. Hvorslev, M. J. "Subsurface Exploration and Sampling of Soils for Civil Engineering Purposes." WES, U.S. Army Corps of Engrs., Vicksburg, Mississippi, 1949. Reprinted in 1962 by Engineering Foundation, United Engineering Center, 347 E. 47th St., New York, N. Y. 10017.
11. Karlsson, R., and L. Viberg. "The Ratio c/p' in Relation to Liquid Limit and Plasticity Index, with Special Reference to Swedish Clays." *Proc. Geotech. Conf. Oslo* **1,** 43–47 (1967).
12. Kjellman, W., T. Kallstenius, and O. Wagner. "Soil Sampler with Metal Foils." *Proc. Royal Swedish Geotech. Inst. No.* **1,** 76 pp. 1950.
13. Meyerhoff, G. G. "Penetration Tests and Bearing Capacity of Cohesionless Soils." *ASCE J.* 82 SMF Paper 866. 1956.
14. Richards, Adrian F., and H. W. Parker. "Surface Coring for Shear Strength Measurements." *Proc. Conf. Civil Engr. in the Oceans*, pp. 445–488. ASCE 1968.
15. Schmertmann, John M. Static Cone Penetrometers for Soil Exploration. *ASCE Civil Engr.* 71–73 (June 1967).
16. Terzaghi, Karl, and R. B. Peck. *Soil Mechanics in Engineering Practice,* 2nd ed. John Wiley, New York, 1967.

7

Gradation, Texture, and Structure

7.1. ELEMENTS OF SOIL STRUCTURE. A soil may be visualized as an assemblage of mineral particles interspersed with open spaces called voids or pores. The void spaces may contain air, water, or water vapor in varying amounts and combinations. In some cases organic matter and bacteria may be present in the voids, particularly in surface soils which support vegetation or those in which plants have grown in the geological past. The solid mineral particles in soils vary widely in size, shape, mineralogical composition, and surface-chemical characteristics. This solid portion of the soil mass is often referred to as the soil skeleton, and the pattern of arrangement of the individual particles or of aggregations and combinations of particles within the skeleton is called the soil structure.

7.2. SIZE OF SOIL PARTICLES. The sizes of soil particles and the distribution of sizes throughout the soil mass are important factors which influence soil properties and performance. For convenience in expressing the size characteristics of the various soil fractions, a number of particle-size classifications have been proposed by different agencies interested in soil studies. These classifications are based, for the most part, upon purely arbitrary considerations, and they reflect the ideas and requirements of the various agencies which proposed them. Several widely recognized particle-size classifications are shown in Table 7-1.

TABLE 7-1. Size Classification of Soil Particles

Agency[a]	Particle Size Limits (mm)								
	Gravel	*Very Coarse Sand*	*Coarse Sand*	*Medium Sand*	*Fine Sand*	*Very Fine Sand*	*Silt-Size*	*Clay-Size*	*Colloidal-Size*
AASHO	76.2 (3 in.) to 2.0	—	2.0 to 0.42	—	0.42 to 0.075	—	0.075 to 0.002	<0.002	<0.001
ASTM	76.2 (3 in.) to 2.0	—	2.0 to 0.42	—	0.42 to 0.074	—	0.074 to 0.005	<0.005	<0.001
UNIFIED	Cobbles > 76.2 (3 in.) Coarse 76.2 to 19.05 Fine 19.05 to 4.76	—	4.76 to 2.0	2.0 to 0.42	0.42 to 0.074	—	Fines (silt or clay) <0.074		
USDA	>2.0	2.0 to 1.0	1.0 to 0.5	0.5 to 0.25	0.25 to 0.10	0.10 to 0.05	0.05 to 0.002	<0.002	—
MIT	>2.0	—	2.0 to 0.6	0.6 to 0.2	0.2 to 0.06	—	0.06 to 0.002	<0.002	—
ISSS	>2.0	—	2.0 to 0.2	—	0.2 to 0.02	—	0.02 to 0.002	<0.002	—

[a] AASHO—American Association of State Highway Officials—Designation: M146-70.
ASTM—American Society for Testing and Materials—Designation: D422-63.
UNIFIED—Corps of Engineers, U.S. Army, and United States Bureau of Reclamation.
USDA—United States Department of Agriculture.
MIT—Massachusetts Institute of Technology.
ISSS—International Society of Soil Science.

There are other grain-size classifications in existence in addition to those shown in Table 7-1, but they are used less widely. Because there are so many, it may be necessary at times to indicate the scale which one has in mind when using such terms as sand and silt. Probably the most important difference in the classifications is in the maximum diameter of particles included in the clay-size fraction. Since clay content is a very significant characteristic of soil, it is important that the meaning of the term clay be clearly understood. Therefore, it is advisable to use the term "2-micron clay" or "5-micron clay," as the case may be, when expressing clay content. Recent efforts at standardization initiated by the Soil Science Society of America indicate increasing acceptance of 0.002 mm for the upper limit of clay and 2.0 mm for the upper limit of sand, but little likelihood of a general agreement on the division between silt and sand, 0.074 mm being preferred in engineering [see (*2*)].

7.3. CALCULATION OF TWO-MICRON CLAY CONTENT. After clays are dispersed for a particle size analysis, 0.002 mm appears to be more valid than 0.005 mm to separate clay from nonclay minerals, since many soil engineering properties relate more closely to 0.002 mm clay content than to 0.005 mm clay content. Engineering organizations are in the process of changing to 0.002 mm as an upper boundary for clay, but 0.001 mm and 0.005 mm clay are more conveniently determined by hydrometer analysis, requiring about 1 hr and 1 day settling time, respectively. Measurement of 0.002 mm clay content requires something over 8 hr. The 0.002 mm clay content may be estimated by plotting and interpolating the particle-size accumulation curve. Alternately if both 0.001 and 0.005 mm clay contents are known, the 0.002 mm clay can be fairly estimated by assuming a straight-line semilogarithmic relationship, from which

$$P_{002} = 0.57\,P_{001} + 0.43\,P_{005} \approx 0.6\,P_{001} + 0.4\,P_{005} \qquad (7\text{-}1)$$

EXAMPLE 7-1. If $P_{005} = 36\%$, $P_{001} = 28\%$, then

$$P_{002} = 0.57\,(28) + 0.43\,(36) = 31.5\%$$

7.4. GRADING OF SOIL. The distribution of particle sizes throughout a mass of soil is known as the grading of the soil. For the sand and gravel fractions, the grading is determined by separating a representative sample of the soil into various size groups or separates by shaking it through a nest of sieves, each sieve having a different size of opening or mesh. The smaller size fractions are separated by measuring the settling velocity of the particles when the sample is thoroughly dispersed in a soil-water suspension. Since the larger particles settle out of suspension more rapidly than do the smaller particles, the time rate of settlement provides a measure of the relative size of the fine soil grains in accordance with the Stokes' law pertaining to the settling velocity of a sphere in a liquid.

7.5. MECHANICAL ANALYSIS OF SOIL. The process of separating a soil into particle-size groups, including both the sieve analysis of the coarser grains and the measurement of settling velocity of the fine grains, is called the mechanical analysis of the soil. The results of a mechanical analysis for determining the grading are often expressed by stating the percentage of the total weight of the dry soil particles which falls in each of the size classes, namely, gravel, sand, silt-size, clay-size, and colloidal-size. A typical analysis on this basis is illustrated in Table 7-2.

TABLE 7-2. Mechanical Analysis Based on Weight in Each Size Class

Size Fraction	*Percent by Weight*
Gravel	4
Sand	31
Coarse, 7%	
Fine, 24%	
Silt-size	44
Clay-size	21
Colloidal-size, 8%	

TABLE 7-3. Mechanical Analysis Based on Percent Finer than Size Indicated

Diameter of Particle or Sieve Number (mm)	*Percent Finer or Passing by Weight*
No. 4 (4.76)	100
No. 10 (2.0)	96
No. 20 (0.84)	92
No. 40 (0.42)	89
No. 60 (0.25)	82
No. 100 (0.147)	78
No. 200 (0.074)	65
0.025	52
0.010	31
0.005	21
0.002	13
0.001	8

Another common method of expressing the grading of a soil is to give the percentage of the total weight of the dry soil particles which is finer than each of a series of stated diameters from the smallest size up through the maximum size of particle contained in the soil. This latter method is illustrated in Table 7-3.

7.6. PARTICLE-SIZE DISTRIBUTION CURVE. The grading of a soil may be represented graphically by means of a particle-size distribution curve. Such a curve is drawn on a semilogarithmic scale and the percentages smaller than the various sizes (percent passing) as ordinates on the arithmetic scale, as shown in Fig. 7-1. The shape of such a curve shows at a glance the general grading characteristics of the soil.

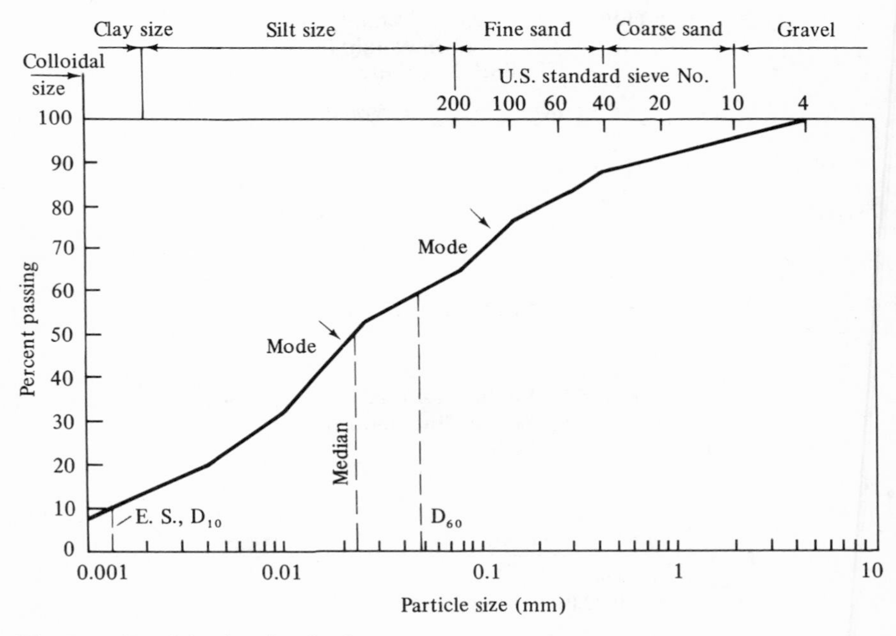

Fig. 7-1. Particle-size distribution curve.

7.7. MEDIAN, EFFECTIVE SIZE, AND UNIFORMITY COEFFICIENT. Although graphs such as Fig. 7-1 are excellent to describe a soil since they do not depend on arbitrary size grade definitions, they are not very convenient. Such graphs may be approximately described by statistical measures or textural descriptions.

The median grain size D_{50} may be defined as the 50% size, or the size which divides the distribution such that 50% of the soil by weight is finer and 50% is coarser than this size. The median is easily read from the 50% line on a particle-size distribution curve (Fig. 7-1).

While the median describes an average particle size, it says nothing about the range in particle sizes. Statistical measures such as the standard deviation have been used but are difficult to calculate and not particularly applicable in engineering. A measure proposed many years ago by Hazen to describe filter sand is the *effective size*, D_{10}, or the maximum diameter of the smallest 10%, by weight, of the soil particles. *Uniformity coefficient* is the quotient obtained by dividing the maximum diameter of the smallest 60% by weight, of the soil particles by the effective size. Thus, in Fig. 7-1 the abscissa of the point having an ordinate of 10% on the grading curve is the effective size, which in this case is 0.0012 mm. The maximum size of the smallest 60% of the soil particles, determined in the same manner, is 0.049 mm; and the uniformity coefficient is 0.049/0.0012 = 41.

It is customary in soil work to carry out the mechanical analysis only down to the 0.001-mm particle size. If the amount of material smaller than this size is greater than 10%, it is not possible to determine the effective size and the uniformity coefficient. The preferred practice in such cases is to report the effective size as less than 0.001 mm.

A low value of effective size indicates that the soil contains a relatively large amount of fine material, while a higher value indicates a relatively smaller percentage of fines. A low value of uniformity coefficient indicates a soil in which the grains are fairly uniform in size. A high value indicates that the size of grains is distributed over a wide range. For example, a wind-blown deposit of silt may have a uniformity coefficient in the neighborhood of 10 to 20, while a well-graded sand-gravel soil may have a uniformity coefficient in the range of 200 to 300 or more.

7.8. MODES, MIXTURES, AND GAP-GRADING. A measure descriptive of the most abundant particle size is the *mode*. A mode occurs wherever a particle size accumulation curve steepens, indicating a considerable change in percent passing within a given size range. A sample may have one or two or more modes; more than one indicating a mixture. The mixture may be due to geological or pedological processes, or to composite sampling of more than one material. For example, a B horizon sample may be bimodal owing to secondary enrichment by clay. Gravels are frequently bimodal—after the gravel is deposited and the water velocity is reduced, the voids are filled by sand. Engineers frequently specify mixtures of soil to obtain a better gradation for a particular use. For example, a sand-gravel mixture with sand filling the voids will be denser and stronger than pure gravel. Similarly clay may be added to improve cohesion, but too much clay will decrease strength.

Figure 7-1 shows two modes, indicated by arrows. The two materials can be visualized by drawing a horizontal dividing line where the curve flattens, at about 55 to 60% passing, and considering this as the 100% line for the lower graph and the 0% line for the upper.

Soils mixed for engineering uses such as in road base courses or concrete aggregate preferably have overlapping particle sizes so the modes do not give well-defined steps on the gradation curve. Where steps and modes are obvious, the mixture is said to be "gap-graded."

7.9. SHAPE OF MINERAL GRAINS IN SOIL. The physical properties of soils and their engineering performance may depend to a considerable degree on the shape and size of the mineral grains of which they are composed. With reference to shape, there are two major groups of soil particles: those which are bulky and those which are flat or flake-shaped. A third class, prismatic or needle-shaped, is much less common. Artificially crushed rock widely used as road and concrete aggregate is a distinctive mixture of highly angular shapes.

Geologists use the terms angular, subangular, rounded, etc., to describe roundness of grain corners rather than overall grain shape. Whereas corner roundness depends to a large extent on the degree of wear or abrasion by water or wind action, shape relates more to composition of the grains themselves. For example, micas and clay minerals are inevitably flaky because of their crystal structure, whereas quartz grains tend to be bulky. On the other hand very extensive rounding, as occurs in alluvial gravels because of their large grain mass and momentum on impact, also alters the shape.

A measure of grain shape is *sphericity*, a degree of closeness to a sphere. Sphericity is defined on the basis of surface areas, but may be approximated by

$$S = \text{intermediate grain diameter} \div \text{maximum grain diameter}$$

This formula is not satisfactory for plate-shaped particles, which have a maximum sphericity of 0.5 when the formula gives a value of 1.0. Representative sphericities are shown in the grain outline chart in Fig. 7-2. This chart may be used to estimate sphericity by assuming that grains rest on their shortest diameter. Usually about 100 random grains are selected, their sphericities estimated and averaged. Since sphericity and corner roundness vary depending on particle size, larger particles being more rounded, measurements for comparison purposes are often limited to the most abundant size grade.

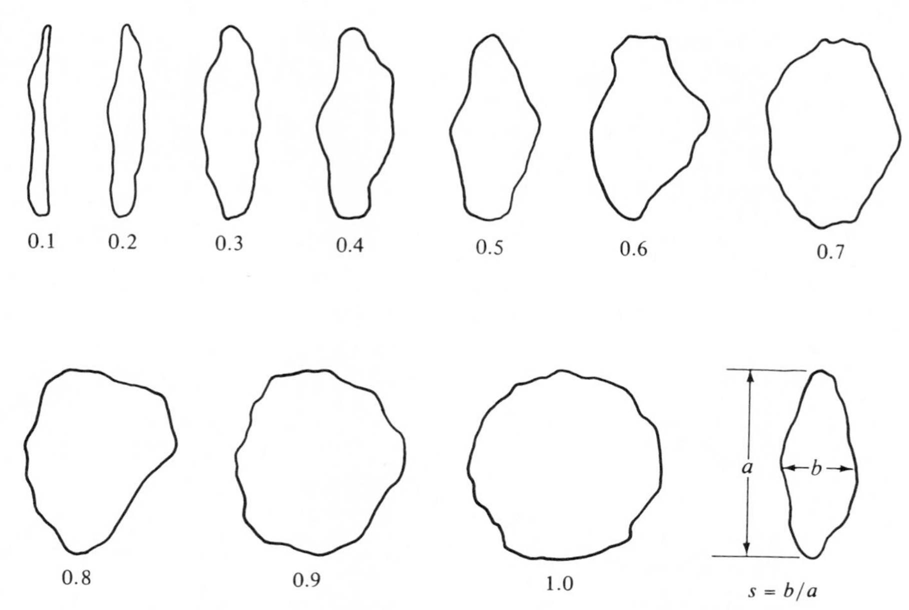

Fig. 7-2. Outline chart for estimating two-dimensional sphericity [modified from Rittenhouse (*10*)].

Particle shape has received relatively little attention from soil engineering researchers, and it is often assumed that higher angularity means better interlocking and higher strength. However, the shape factors that resist shearing also resist compaction, which is a controlled shearing process [Huang *et al.* (*6*)]. As will be discussed later, "interlocking" means only that a soil must increase in volume in order to shear, so it may not relate to particle shape. Micaceous soils are especially resistant to compaction and remain weak and "springy" [McCarthy and Leonard (*9*)].

7.10. PROTECTIVE FILTERS. When percolating or drainage water flows from a body of relatively fine soil into a coarser material, there is danger that the fine soil particles may migrate into the coarse material and eventually clog the pores of the latter. This can be prevented by providing a protective filter between the two bodies of soil. The U.S. Bureau of Reclamation has developed criteria for protective filters which satisfy four main requirements as follows:

1. The filter material should be more pervious than the base material in order that there will be no hydraulic pressure built up to disrupt the filter and adjacent structures.

2. The voids of the in-place filter material must be small enough to prevent base material particles from penetrating the filter and causing clogging and failure of the protective filter system.

3. The layer of the protective filter must be sufficiently thick to provide a good distribution of all particle sizes throughout the filter and also to provide adequate insulation for the base material where frost action is involved.

4. Filter material particles must be prevented from movement into drainage pipes by providing sufficiently small slot openings or perforations, or an additional coarser filter zone if necessary.

A guide for the gradation of materials for a filter or the various filter layers is given below.

For uniform grain-size filters:

$$R_{50} = \frac{50\% \text{ size filter material}}{50\% \text{ size base material}} = 5 \text{ to } 10$$

For graded filters of subrounded particles:

$$R_{50} = \frac{50\% \text{ size filter material}}{50\% \text{ size base material}} = 12 \text{ to } 58\text{, and}$$

$$R_{15} = \frac{15\% \text{ size filter material}}{15\% \text{ size base material}} = 12 \text{ to } 40$$

For graded filters of angular particles:

$$R_{50} = \frac{50\% \text{ size filter material}}{50\% \text{ size base material}} = 9 \text{ to } 30\text{, and}$$

$$R_{15} = \frac{15\% \text{ size filter material}}{15\% \text{ size base material}} = 6 \text{ to } 18$$

7.11. TEXTURE OF SOIL. The sizes of particles in a natural soil mass may differ widely as a result of the mixing effect of geological transporting agencies and variations in the rate of weathering and breakdown of different types of minerals. The distribution of particle sizes and the relative predominance of fine or coarse grains impart to the soil a distinctive appearance and "feel," which is called texture. The feel of a soil is obviously influenced by the shape of the particles as well as their size, since a preponderance of flat and rounded grains would give a soil a less harsh feel than if the grains were sharp and angular. However, as the term texture is used in the United States, it refers wholly to the size characteristics of the soil particles. Thus, when we speak of a "sandy soil," we mean that there is a preponderance of the larger particles; but, in all probability there will be present varying percentages of grains of silt-size, clay-size, and colloidal-size as well, and the term does not give a clue as to the shape of the sand grains or their arrangement. In Europe, the term texture refers to other characteristics than those indicated by grain size alone and includes many of the concepts discussed in connection with the structure of surface soils.

The terms commonly used in the United States to indicate texture of soil, such as loam, sandy loam, and silty clay, are wholly arbitrary in origin; and their general meaning and connotation may vary from place to place and from one organization to another, the significance depending on the needs of the organization and a more or less accidental development of usage of the terms.

Engineers seldom classify soils as loam, silt loam, etc., on the basis of "feel" alone. However, it is useful to recognize that sand is loose and granular; loam will form a cast when squeezed in the hand, but the cast is rather crumbly; clay loam forms a cast and can be kneaded; and clay if moist can be squeezed between the thumb and the finger to form a long, flexible ribbon. Details of these and other textural descriptions are given in the *Soil Survey Manual* (*13*).

7.12. GRAVELLY OR STONY TEXTURE. If soil of any of the textural classes contains sufficient gravel or broken stone to noticeably influence the appearance and feel of the soil—say with 8 to 10% of large particles—it may be described by prefixing the adjective "gravelly" or "stony" before the textual class name, as in the expression "gravelly clay loam" or "stony silt loam."

7.13. DETERMINATION OF TEXTURAL CLASS FROM GRADING. The approximate grading of soils falling in ten commonly used textural classes is shown in Table 7-4.

The textural class to which a soil belongs may be determined readily, if the grading of the soil is known, by means of a triangular chart, either in the form of an equilateral triangle, as shown in Fig. 7-3, or in the form of a right triangle, as shown in Fig. 7-4. Whenever such charts are used, it is important to specify the limits used for particle size grades. On the equilateral-triangle chart, the percent of sand is shown along the left side of the triangle, and the corresponding grid lines extend downward and to the right; the percent of silt-size material is shown along the base of the triangle, and the grid lines extend upward and to the right; the percent of clay-size material is shown along the right side of the triangle, and the grid lines extend horizontally across the chart. The use of this chart can best be described by Example 7-2.

TABLE 7-4. Textural Classification of Soil Based on Grading

Textural Class	*Composition (%)*		
	Sand	*Silt-Size*	*Clay-Size*
Sand	80–100	0–20	0–20
Sandy loam	50–80	0–50	0–20
Loam	30–50	30–50	0–20
Silt loam	0–50	50–100	0–20
Sandy clay loam	50–80	0–30	20–30
Clay loam	20–50	20–50	20–30
Silty clay loam	0–30	50–80	20–30
Sandy clay	55–70	0–15	30–45
Silty clay	0–15	55–70	30–45
Clay	0–55	0–55	30–100

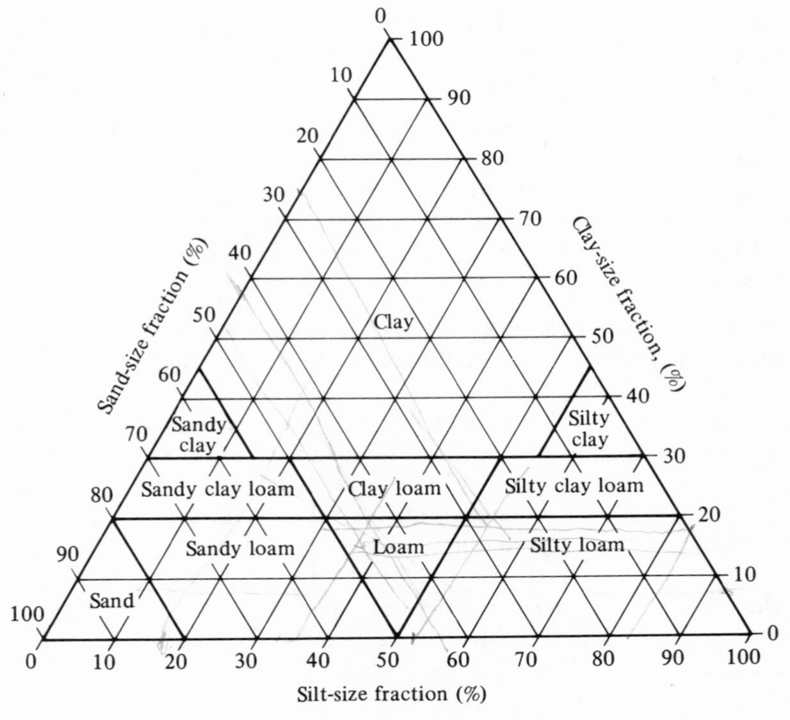

Fig. 7-3. Textural classification chart. Sand-size particles, 2–0.074 mm; silt-size particles, 0.074–0.002 mm; and clay-size particles, less than 0.002 mm.

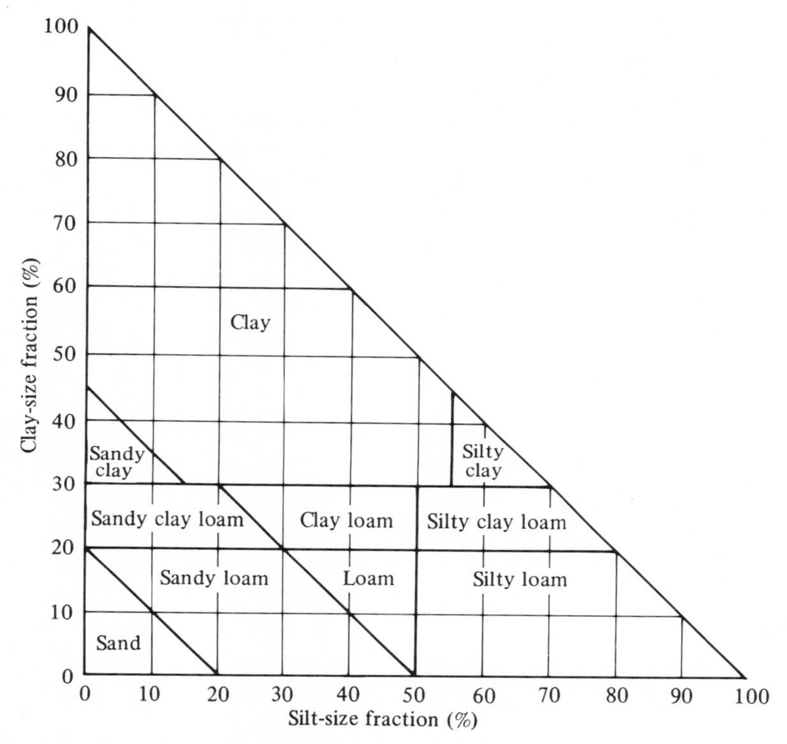

Fig. 7-4. Textural classification chart. Ranges are as defined in Fig. 7-3.

EXAMPLE 7-2. Let it be desired to determine the textural classification of a soil whose grading is as follows: gravel, 4%; sand, 31%; silt-size material, 44%; clay-size material, 21%.

SOLUTION. Enter the diagram in Fig. 7-3 from the sand scale along the left side of the triangle at the value 31. Project a line through this point downward to the right until it intersects a line extending upward to the right from the point corresponding to 44% on the silt scale along the base of the triangle. These two lines intersect within the area labeled "clay loam," which is the textural class of this soil.

The point of intersection of the lines representing the percentages of silt-size and clay-size material and the intersection of the lines for sand and clay-size material may also be determined; and it will be found that these points nearly coincide with the point first determined. If there were no gravel in the soil, the lines representing the three sizes included on the chart would intersect at a common point, since the sum of these three sizes would then be 100%.

When a soil contains sufficient coarse material to be classified as a stony or gravelly soil, its textural class should be determined on the basis of the amounts of sand, silt-size material, and clay-size material expressed as percentages of the total weight of these three sizes only, and not as percentages of the total weight of the soil, including the gravel.

EXAMPLE 7-3. Let it be desired to determine the textural class of a soil whose grading is: gravel, 14%; sand, 21%; silt-size material, 56%; clay-size material, 9%.

SOLUTION. Since this material contains more than 10% of gravel, it will be a "gravelly" soil of some kind. In order to use the triangular chart, it is necessary to multiply the percentages of the sizes smaller than gravel by the ratio 100/86 in order to express these percentages in terms of the weight of the material exclusive of the gravel. Thus, we obtain:

$$\text{Sand} \ldots\ldots\ldots\ldots \quad 21 \times \frac{100}{86} = 24\%$$

$$\text{Silt-size} \ldots\ldots\ldots \quad 56 \times \frac{100}{86} = 65\%$$

$$\text{Clay-size} \ldots\ldots\ldots \quad 9 \times \frac{100}{86} = 11\%$$

When these adjusted percentages are used, the textural class determined from the triangular chart is found to be "silt loam." The complete designation of the textural classification of this soil, therefore, is "gravelly silt loam."

7.14. USE OF RIGHT-TRIANGLE CHART. Since the sum of the percentages of sand, silt-size material, and clay-size material in a soil for which the triangular chart is applicable is approximately 100%, it is possible to determine the textural classification by locating the point of intersection of lines representing any combination of two of these three size classes. This fact suggests that a right-triangle chart can be constructed which will yield the same results as the equilateral-triangle chart. In the right-triangle chart, the percentages of any pair of size separates, such as sand and silt-size material, sand and clay-size material, or silt-size and clay-size materials, are given as the vertical ordinates and the horizontal abscissas, respectively. Such a chart, drawn for the silt-size and clay-size materials, is illustrated in Fig. 7-4. By locating the intersection of the ordinate representing the percentage of clay-size material and the abscissa representing the percentage of silt-size material, the textural class of the soil is readily determined. The right-triangle chart is preferred by some engineers because of the orthogonal arrangement of grid lines and its consequent simplicity.

7.15. OTHER TEXTURAL CHARTS. The textural charts shown in Figs. 7-3 and 7-4 represent the textural classification formerly used by the U.S. Bureau of Public Roads and still widely used by various state highway departments throughout the United States. Other organizations have developed other textural classifications which fulfill the particular needs and desires of those organizations. As a result, the same textural class names may have various quantitative meanings, the proportions of the particle sizes depending on the textural classification system used in the

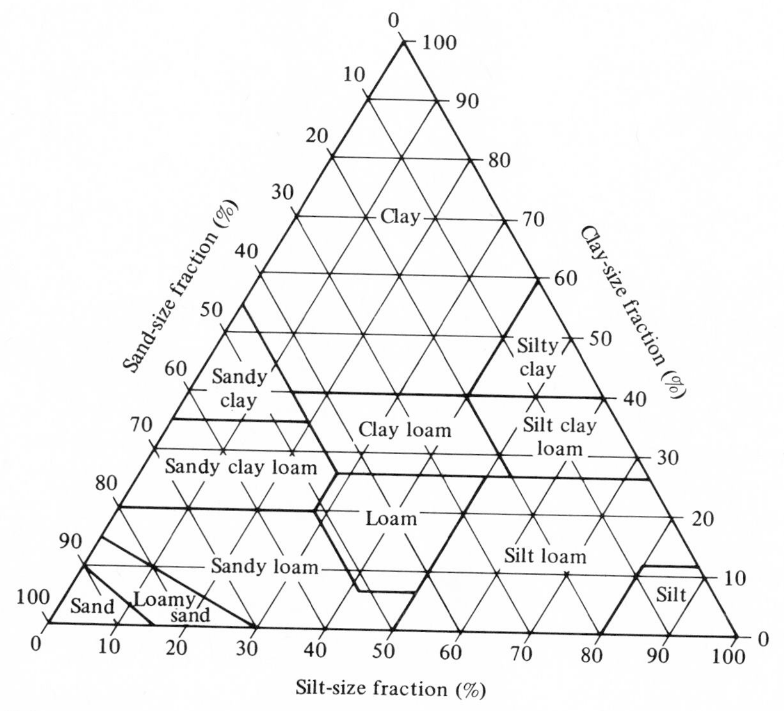

Fig. 7-5. U.S.D.A. textural classification chart. Sand-size particles, 2–0.05 mm; silt-size particles, 0.05–0.002 mm; and clay-size particles, less than 0.002 mm.

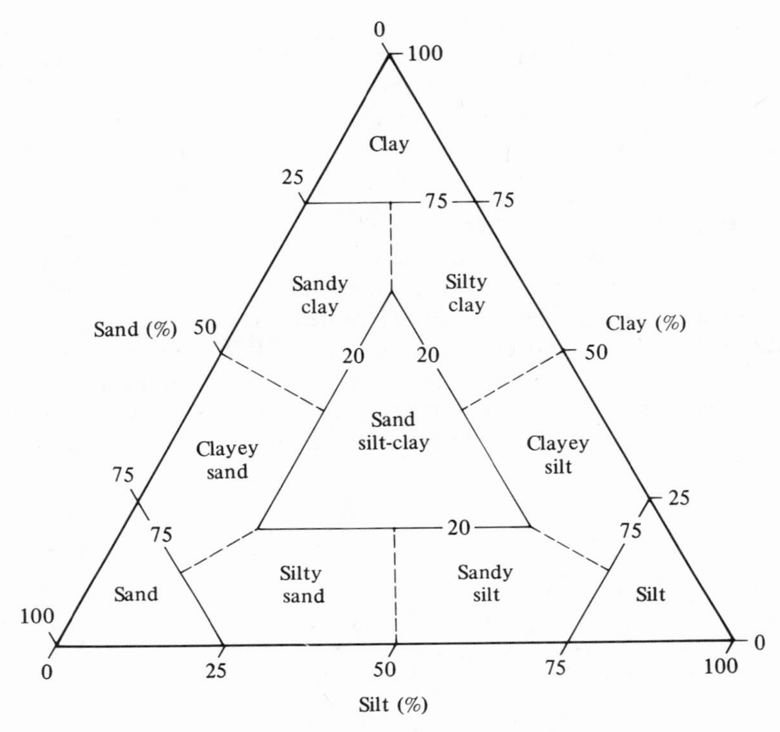

Fig. 7-6. Textural classification chart for sediments. After Shepard (*12*).

determination of the names. It is important, therefore, to specify the basis of classification when the textural name of a soil is given, unless it is certain that the basis is mutually understood by all parties concerned.

The United States Department of Agriculture classifies surface soils by means of the textural chart shown in Fig. 7-5. It is based upon the grain-size classification used by that agency as shown in Table 7-1. This chart has not been widely adopted in engineering practice. Note that the boundaries in Figs. 7-3 to 7-5 were defined to describe texture, or "feel."

A triangular classification system favored by ocean engineers and geologists and more purely descriptive of particle size gradation is shown in Fig. 7-6. In this system any soil containing over 20% each of sand, silt, and clay is called sand-silt-clay. Other intergrades are as shown. Many geologists use size grades based on a geometrical progression: clay is finer than $\frac{1}{256}$ mm (3.9 microns) and silt is finer than $\frac{1}{16}$ mm (62.5 microns). Therefore whenever a textural designation is used, the size grades should be specified.

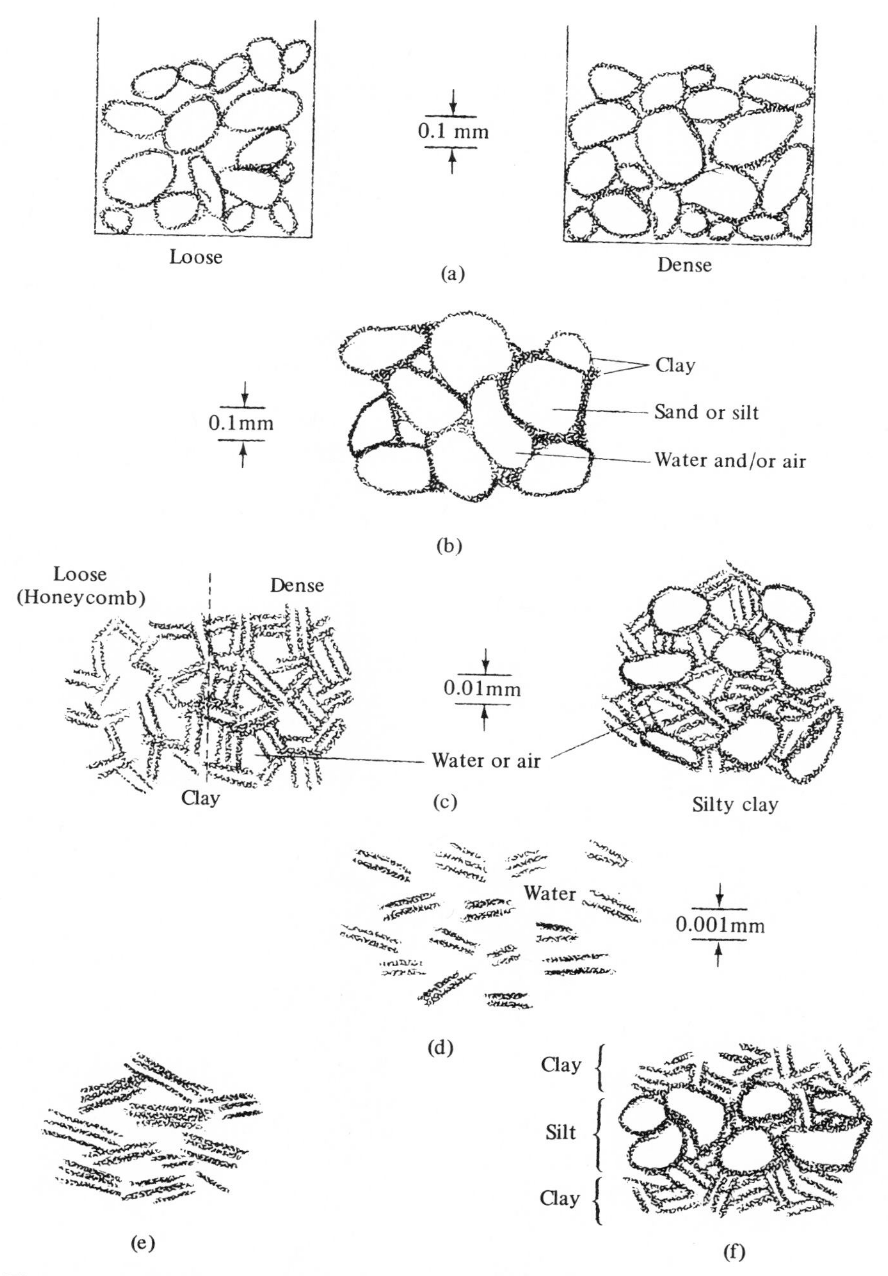

Fig. 7-7. Idealized soil structures. (a) Single-grained structure; (b) wedge-block structure; (c) massive flocculated structure; (d) massive dispersed structure; (e) bedded or fissile; and (f) stratified.

7.16. SOIL MICROSTRUCTURE. Soil microstructure or simply soil structure refers to how the individual mineral and rock grains occur in relation to one another and to interstitial water and air. Structure is difficult to ascertain and imperfectly understood, yet is an extremely important factor in engineering behavior. Soil structure can be changed by manipulation: for example, the purpose of compaction is to strengthen a soil by changing its structure and expelling air, although under certain conditions it may weaken soil by reducing it to a single-grain or dispersed structure. Soil structure also may change spontaneously, as when sliding or liquifaction occurs. Some idealized concepts of soil structure are shown in Fig. 7-7.

7.16.1. Single-Grain Structure. Clean sands and gravels have what is called a single-grain structure; the grains occur as individuals with no particular attachment or liaison with others, but merely touching at points of contacts. There is little or no cementation except for capillary attraction by water, which is transient and disappears on wetting or thorough drying. A single-grain structure may be loose or compact, changing the number of interparticle contacts. Clean silts and sand-silt mixtures such as loams, particularly those with very little clay, also tend to exhibit a single-grain structure, easily pulverizing in the hand, or slaking and disintegrating under water.

7.16.2. Wedge-Block Structure. Usually silts contain sufficient clay to bond the silt grains, and likewise sands and gravels may be cemented by clay into a stronger structure. Microscopic studies show that clay grains tend to occur as oriented coatings or *clay skins* on the larger grains, and concentrate into wedge-shaped rings or bridges at the grain contacts. This has been designated by Grim (5) as a wedge-block structure.

The wedging action accounts for a low tensile strength since the grains are readily plucked apart, whereas the shearing strength may be high because of difficulty in sliding the wedges. The shear strength varies depending on how much clay is available to constitute the wedges: The data for molding sands indicate that shearing strength reaches a maximum and levels off with about 9 to 14% clay depending on the clay mineral, montmorillonite being more effective than illite or kaolinite. If the clay content exceeds about 25% the wedge-block structure disappears, and sand or silt grains become separated and "floating" in a matrix of clay.

7.16.3. Massive Structures. Where sufficient clay is present to form a matrix enclosing coarser grains (normally 25–30% clay) the structure of the clay dictates the structural behavior of the soil. For example, clays which expand on wetting usually contain about 25% or more clay-size material.

Clay particles are by definition fine-grained and therefore have a very large surface area per unit weight or volume. Under most normal soil conditions the clay particles are *flocculated*, or held together by electrostatic attractions arising from electrical surface charges. In very general terms, the flat faces of clay particles tend to be negatively charged, whereas broken bonds at edges of particles give a patchwork of negative and positive charges. The result is that the edge of one particle tends to be attracted to the face of another, resulting in an edge-to-face attraction sometimes described as a "cardhouse" structure. Such a structure is shown in Fig. 7-8(a). As can be seen, practically all particles in the photograph are finer than 2 microns (0.002 mm) and thus fit the prevailing definition of clay-size. Figure 7-8(b) shows the same clay broken parallel to a bedding plane, indicating that edge-to-edge and face-to-face attractions are also common.

The extent and strength of flocculation is very sensitive to the kind

(a)

(b)

Fig. 7-8. Scanning electron micrographs showing soil structures. Photographs courtesy of Major W. Badger and the Iowa State University Soil Research Laboratory. (a) Loess soil, dry density 88 pcf, showing clay-aggregated silt grains and open pores. Index line length is 10 microns. (b) Sedimentary illite-montmorillonite clay. Index line is 1 micron.

and amount of water available and the salts in solution. For example, erosion of a flocculated clay by water usually disperses it by separating and diluting the grains and keeping them agitated. Prolonged settling and concentration of the clay without agitation, as on a lake bottom, may cause weak flocculation, whereas discharging into salt water will cause sudden, strong flocculation and rapid settling of the aggregates. Drying a clay is sometimes an inducement to flocculation, since removal of the water separating clay grains brings strong attractive forces into play. The resulting aggregates may or may not be stable upon rewetting.

7.16.4. Dispersed Massive Structure. When ionic attractive forces between clay particles lessen owing to leaching of soluble salts, or are disrupted by mechanical remolding or erosion by water, the clay may become *dispersed.* In this case clay particles repel one another and are separated by liquid water. Dispersion is used to prepare a soil for particle size analysis, but it is devastating to shear strength. A few natural clays become dispersed by agitation; these are called *quick clays*, and occur along estuaries in Norway and Canada. Usually they are Pleistocene marine clays whose flocculent structure has become unstable as the soluble salt has gradually leached out.

Many clays experience a temporary partial dispersion when they are remolded or reworked by hand or machine. These are termed *sensitive clays* or *thixotropic clays.* The degree of sensitivity may be investigated in the laboratory as follows: First, test an undisturbed cylindrical sample of the soil in unconfined compression, that is, by applying an axial load without a confining pressure. Then, remold the sample without loss of water and pack it into a cylindrical mold of the same size, extrude it, and retest it in unconfined compression. The ratio of the strength of the soil in the undisturbed state to that of the soil in the remolded state is known as the *sensitivity* of clay. The sensitivity of foundation clays usually ranges from 2 to 4, but may be as high as 8 or more. Terms descriptive of sensitivity are shown in Table 7-5.

TABLE 7-5. Sensitivity of Clays[a]

Sensitivity	*Descriptive Term*
<2	Insensitive
2–4	Moderately sensitive
4–8	Sensitive
8–16	Very sensitive
16–32	Slightly quick
32–64	Medium quick
>64	Quick

[a] After Bjerrum, quoted in Leonards (*8*).

7.16.5. Layered Structures. The difference between photographs 7-8(a) and (b) is caused by *preferred orientation* of the clay mineral grains. This could be important, since one would expect to find a difference in compressibility and in shearing strength of this clay depending on orientation of the compressive or shearing stresses. As might be expected, a direction of layering, Fig. 7-8(b), is usually a plane of weakness. Such preferred orientation results from long consolidation under the pressure of overburden. An extreme case of preferred orientation is found in some shales which readily break along bedding planes. Such a shale is said to be "fissile," which means easily split. Clays which are intermediate between massive and fissile are simply called "bedded."

Another type of layered structure results from layer variations in particle size. For example, regularly interlayered silt and clay is common in glacial lakebeds, and is called *varved clay*. When particle size is the major variable causing layering, the soil is said to be *stratified*.

The various soil structures are illustrated in Fig. 7-7. These sketches are highly diagrammatic, since a true section through a group of sand grains as in (a) would not necessarily cut at grain contacts; instead many or most grains would appear to be "floating," the contacts being elsewhere in the third dimension. Similarly the clay structures (b), (c), and (d) are idealized since all particles are shown on edge.

7.17. GROSS STRUCTURE IN THE SOIL PROFILE. The above-mentioned types of soil structure all refer to microstructure, describing the relationship between individual soil particles as seen on a microscopic basis. Agricultural soil scientists also use "structure" to designate the gross tendency of soil to break into lumps or *peds*. Each ped represents a stable aggregation of thousands or millions of soil particles. The tendency to form stable peds is of prime importance in agriculture since plants need air, and is the main reason for "liming" a soil with pulverized calcium carbonate. Pedological structure also has a profound effect on engineering properties of A and B horizons. The structure of the A horizon soils is of relatively minor significance in engineering, common structures being *granular* if the peds are nonporous, or *crumb* if porous.

As discussed in Chapter 2, the structure of the B horizon sometimes appears to result from seasonal wetting and drying causing vertical cracks. One result is a *prismatic* or *columnar* structure (the latter if the tops are rounded), the individual prisms usually being a matter of a few inches across. Intersection by horizontal or inclined cracks gives a *blocky* structure which may be termed *subangular blocky* if the corners are rounded. Such structures tend to be preserved by oriented *clay skins* forming around the peds, preventing strong bonding. A blocky structure may be

TABLE 7-6. Mechanical Analysis of Soils: Percentages Passing Various Sieve Sizes

Soil	Sieve Number and Size of Opening (mm)									Sedimentation Size (mm)						Soil
No.	1 in.	$\frac{3}{4}$ in.	$\frac{3}{8}$ in.	No. 4	No. 10	No. 40	No. 60	No. 100	No. 200					L. L.	P. I.	No.
	26.7	18.8	9.4	4.75	2.00	0.42	0.25	0.149	0.074	0.050	0.005	0.002	0.001			
1	100	90	80	72	67	56	44	34	24	21	11	7	4	29	7	1
2		100	99	98	97	96	91	80	71	63	34	25	18	39	14	2
3					100	99	96	92	80	73	41	31	23	76	21	3
4				100	97	84	66	50	32	24	5	4	4	54	16	4
5					100	96	85	61	34	31	13	10	7	69	7	5
6						100	95	88	80	54	25	14	5	35	10	6
7							100	99	98	95	9	7	6	80	9	7
8						100	97	76	60	45	35	21	10	35	17	8
9					100	95	88	81	65	59	18	11	6	27	5	9
10		100	99	97	93	70	58	56	44	42	24	17	11	41	12	10
11		100	98	92	82	50	42	35	28	25	12	8	5	38	16	11
12	100	93	77	64	48	24	20	16	12	11	8	7	6	13	4	12
13					100	99	84	48	12	8	—	—	—	—	N.P.	13
14							100	99	95	93	68	49	34	86	49	14
15					100	79	60	48	34	30	14	11	9	24	8	15
16				100	94	89	87	85	81	73	39	24	12	59	43	16
17			100	98	94	48	42	33	26	22	13	11	10	47	24	17
18	100	98	97	96	94	91	90	89	86	80	42	27	16	45	17	18
19					100	45	38	30	22	20	10	7	4	16	6	19
20					100	98	96	95	93	89	64	44	29	84	53	20

advantageous for engineering if the blocks are sound and stable, since they greatly increase the ease of excavation, and under low loads behave more like gravel than clay. The individual blocks also may have a high density allowing compaction to a specified overall density with less compactive effort. Nevertheless C horizon material is ordinarily preferred, being less weathered and containing less clay.

PROBLEMS

7.1. Plot a particle size accumulation curve for Soil No. 4, Table 7-6, using four- or five-cycle semilogarithmic paper. (a) Evaluate the effective size and uniformity coefficient. (b) What is the median grain size? (c) Defining clay as <0.002 mm, silt as 0.002–0.074 mm, sand 0.074–2.0 mm, and gravel >2.0 mm, what are the percentages clay, silt, sand, and gravel in Soil No. 4? (d) Classify this soil according to the chart in Fig. 7-3 or 7-4. (e) Classify this soil according to the chart in Fig. 7-6.

7.2. Defining the size grades in Soil No. 4 according to the information in Fig. 7-5, determine its textural classification.

7.3. Plot a particle-size accumulation curve for Soil No. 1, Table 7-6, using five-cycle semilogarithmic paper. (a) Identify the median and mode(s). (b) If there are two modes, what is the approximate percentage of each soil in the mixture? (c) Using size grades defined in Problem 7.1(c), calculate percentages of clay, silt, sand, and gravel. (d) Adjust the grade percentages for gravel, and classify the soil by the chart in Fig. 7-3.

7.4. Calculate the effective size and uniformity coefficient for Soil No. 1, 5, 6, 7, 9. Ans.: Soil No. 1: $D_{10} = 0.0039$ mm, $Cu = 192$.

7.5. Sketch three grain outlines for each of the following two-dimensional sphericities: 0.3, 0.6, 0.9.

7.6. Which of the soils in Table 7-6 is gravelly?

7.7. Classify the soils by inspection of the data in Table 7-6, using Fig. 7-4 but defining the clay-silt break as 0.005 mm.

7.8. Calculate to the nearest percent and by Eq. (7-1), 0.002 mm clay content in Table 7-3.

7.9. Estimate sensitivity of soils with structures (a), (b), and (c) in Fig. 7-7.

7.10. Can an "insensitive" structure lose strength on remolding?

7.11. What distinguishes between a wedge-block and massive structure in a silt-clay soil?

7.12. Draw an idealized triangular diagram hypothesizing a relationship between soil structure and sand, silt, and clay contents. Assume 10% clay is required to remove a soil from single-grained structure.

7.13. Does the chart prepared for Problem 7.12 more closely resemble Fig. 7-5 or 7-6? Is this reasonable?

7.14. Combine soils 1 and 3 in Table 7-6 in such proportions that the resulting mixture will contain 20% of 5-micron clay. Draw the grading curve of the mixture and classify it texturally.

Ans.: 70% soil No. 1 and 30% soil No. 3. Mechanical analysis of combined soil follows:

Size (*mm*)	*Percent Passing*	*Size* (*mm*)	*Percent Passing*
26.7	100	0.25	60
18.8	93	0.149	51
9.4	86	0.074	44
4.75	80	0.050	37
2.00	77	0.005	20
0.42	69	0.001	10

7.15. Combine soils 12 and 14 in Table 7-6 in such proportions that the resulting mixture will contain 25% of 5-micron clay.

REFERENCES

1. American Association of State Highway Officials. *Standard Specifications for Highway Materials*, 10th ed., Part I. Washington, D.C., 1970.
2. American Society of Civil Engineers, Committee on Soil Properties, C. C. Ladd, Chairman. "Standardization of Particle-Size Ranges." *ASCE J. Soil Mechanics and Found. Division* **95** (SM5); 1247–1252 (1969).
3. ASTM Committee D-18. *Procedures for Testing Soils*, 4th ed. Amer. Soc. for Testing and Materials, Philadelphia, Pennsylvania, 1964.
4. Baver, L. D. *Soil Physics*, 3d ed. John Wiley, New York, 1956.
5. Grim, Ralph E. *Applied Clay Mineralogy*. McGraw-Hill, New York, 1962.
6. Huang, Y., Squier, L. R., and Triffo, R. F. "Effect of Geometric Characteristics of Course Aggregates on Compaction Characteristics of Soil-Aggregate Mixtures." *Highway Research Rec.* **22,** 38–47 (1963).
7. Krumbein, W. C., and Pettijohn, F. J. *Manual of Sedimentary Petrography*. Appleton Century-Crofts, New York, 1938.
8. Leonards, G. A. "Engineering Properties of Soils." In Leonards, G. A., Ed., *Foundation Engineering*, pp. 66–240. McGraw-Hill, New York, 1962.
9. McCarthy, D. F., Jr., and Leonard, R. J. "Compaction and Compression Characteristics of Micaceous Fine Sands and Silts." *Highway Research Rec.* **22,** 23–37 (1963).
10. Reclamation, U.S. Bureau of. *Earth Manual*, 1st ed. Washington, D.C., 1963.
11. Rittenhouse, G. "A Visual Method of Estimating Two Dimensional Sphericity." *J. Sedimentary Petrology* **13,** 79–81 (1943).
12. Shepard, Francis P. "Nomenclature Based on Sand-Silt-Clay Ratios." *J. Sedimentary Petrology* **24** (3), 151–158 (1943).
13. Soil Survey Staff. *Soil Survey Manual.* U.S.D.A. Handbook No. 18. Supt. of Documents, U.S. Government Printing Office, Washington, D. C., 1951.

8

Soil Density and Compaction

8.1. IMPORTANCE OF DENSITY. The density of soil is one of the most important of its engineering properties, and many soil-engineering operations are directed toward improving the density characteristics of the material in order to decrease the settlement of embankments and to increase the strength of the soil in bearing and in shearing resistance. Increasing the density of soil also decreases its permeability and reduces the detrimental effects of water absorption and movement. It is necessary for the soil engineer to understand the factors which influence this property, to be cognizant of methods of measuring it, and to be familiar with procedures by which greater density may be obtained.

8.2. DEFINITION OF DENSITY. Density of soil is defined as the weight per unit volume, being expressed in such units as pounds per cubic foot or grams per cubic centimeter. It is synonymous with the terms *unit weight* and *specific weight*. The term density as here defined is widely used in soil-engineering practice, but attention is directed to the fact that such usage is at variance with the meaning of density in certain other fields of science. In physics, density is defined as mass per unit volume instead of weight per unit volume. Since weight is a force equal to the product of mass and the acceleration of gravity, the physics concept of density is weight per unit volume divided by the gravitational constant. In this text, the term density will be used to indicate weight per unit volume, and thus "denseness" of the soil, in accordance with rather common practice in soil engineering; but the relation of this usage to the more scientific meaning of the term should be clearly understood.

The weight of the dry solid particles in a unit volume of soil is called the *dry density;* and the weight of the dry solids plus the moisture contained in the void spaces is called the *wet density.* From these definitions it is apparent that, if a particular mass of soil contains some moisture, the soil may have two numerical values of density. For a given soil in a given state of packing, the dry density is a constant value and is equal to the unit weight of the soil without any moisture in the voids. If the soil contains moisture, the wet density varies between the dry density as a lower limit and the unit weight of the soil at saturation, or when the void spaces are completely filled with water, as an upper limit.

8.3. FACTORS AFFECTING DENSITY. Soil density is dependent mainly on the relative volumes of the solid particles and the void spaces, on the specific gravity of the solids, and, in the case of wet density, on the amount of water in the soil. The fact that soil consists of both solid particles and void spaces is of fundamental importance. This conception of the nature of soil needs to be firmly fixed in the minds of all engineers who deal with this material, since their efforts will often be directed toward attaining a soil structure and particle arrangement having as large a percentage of solids and as small a percentage of void spaces as practicable. A high degree of density is particularly desirable in the case of earth embankments, levees, and dams and also in the case of stabilized-soil road surfaces and their subgrades.

8.4. VOID RATIO AND POROSITY. Void spaces in soil, often called pores or pore spaces, are interspersed in and throughout the soil skeleton or framework of solid particles. An idea of the porous nature of soil and of the relationship between solid matter and void space in a soil mass can be visualized by imagining a cubical container that is 1 ft on each side and is filled with balls 2 in. in diameter, as illustrated in Fig. 8-1. The balls represent the solid particles in a cubic foot of soil, and the spaces between them are the voids.

The relative amount of solid matter in a soil may be expressed conveniently by means of either the *void ratio* or the *porosity.* The void ratio is defined as the ratio of the volume of voids to the volume of solids, without regard to the proportions of liquid, air, or other gases which may occupy the void spaces. The porosity is defined as the ratio of the volume of the voids or pores to the total volume of the soil. Like the void ratio,

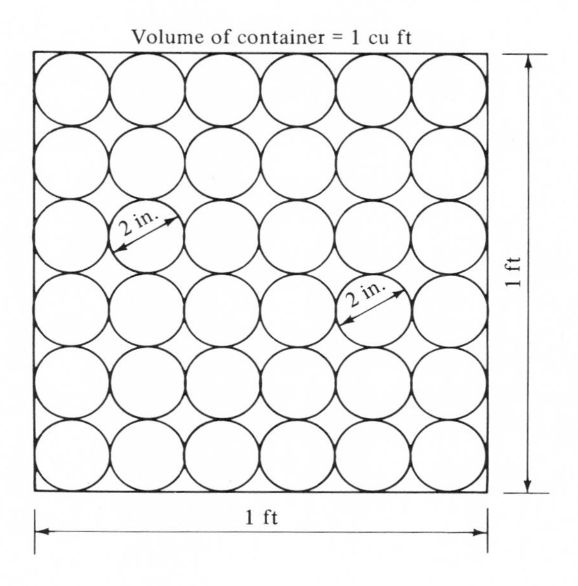

Fig. 8-1

porosity refers to the total amount of void space, without regard to the amount of moisture or air contained in the pores. Porosity, however, is usually expressed as a percentage rather than as an abstract ratio.

The percent of the total volume of the soil which is occupied by air in the voids is called the *percent air voids*. Similarly, the percent of the total soil volume which is occupied by water in the voids is called the *percent water voids*. Obviously, the sum of the percent air voids and the percent water voids is equal to the porosity of the soil. Also, the percent of the total volume of the soil which is occupied by the solid particles is called the *percent solids*. The percent solids plus the porosity equals 100%.

8.5. BASIC PROPERTIES OF SOIL. The relationships defined in the preceding section may be expressed by simple formulas.

Let

V = total volume of a soil mass;
V_s = volume of the solid particles;
V_e = volume of the void spaces;
V_a = volume of the air voids;
V_m = volume of the water voids;
e = void ratio;
n = porosity, in percent;

n_a = percent air voids;
n_m = percent water voids;
n_s = percent solids.

Then, by definition,

$$V = V_s + V_e = V_s + V_a + V_m \quad (8\text{-}1)$$

$$e = \frac{V_e}{V_s} \quad (8\text{-}2)$$

$$n = \frac{V_e}{V} \times 100 \quad (8\text{-}3)$$

Additional formulas for soil properties are listed below. It is suggested that the student derive these equations.

$$V_s = \frac{1}{1 + e} V \quad (8\text{-}4)$$

$$V_e = \frac{e}{1 + e} V \quad (8\text{-}5)$$

$$n_a = \frac{V_a}{V} \times 100 \quad (8\text{-}6)$$

$$n_m = \frac{V_m}{V} \times 100 \quad (8\text{-}7)$$

$$n_s = \frac{V_s}{V} \times 100 \quad (8\text{-}8)$$

$$n_s + n = n_s + n_a + n_m = 100 \quad (8\text{-}9)$$

$$e = \frac{n}{100 - n} \quad (8\text{-}10)$$

$$n = \frac{e}{1 + e} \times 100 \quad (8\text{-}11)$$

EXAMPLE 8-1. Determine (a) the void ratio, (b) the porosity, and (c) the percent solids for the imaginary soil mass represented in Fig. 8-1, where each side of the container is 1 ft in length and it is filled with spheres which are 2 in. in diameter and are arranged as indicated.

SOLUTION. The total number of balls in the container is $6 \times 6 \times 6 = 216$ and the volume of each is

$$\frac{3.14 \times 2^3}{6 \times 1728} = 0.002425 \text{ cu ft}$$

Hence, the volume of the solids is

$$216 \times 0.002425 = 0.524 \text{ cu ft}$$

and the volume of voids is

$$1 - 0.524 = 0.476 \text{ cu ft}$$

(a) By Eq. (8-2), the void ratio is

$$e = \frac{V_e}{V_s} = \frac{0.476}{0.524} = 0.910$$

(b) By Eq. (8-3), the porosity is

$$n = \frac{V_e}{V} \times 100 = \frac{0.476}{1} \times 100 = 47.6\%$$

(c) The percent of solids can be computed by either Eq. (8-8) or Eq. (8-9). Thus,

$$n_s = \frac{0.524}{1} \times 100 = 52.4\%$$

or

$$n_s = 100 - 47.6 = 52.4\%$$

8.6. TRUE SPECIFIC GRAVITY. The specific gravity of any solid or liquid substance is the ratio of the unit weight of the substance to the unit weight of water. It is a measure of, and a means of expressing, the heaviness of a material. The specific gravity of the solid particles of a soil, exclusive of the void spaces, is called the "true" or "real" specific gravity. This property has an important influence on the density of the soil. In the case of the imaginary soil of Fig. 8-1, it is obvious that its density would be much greater if the balls were made of steel than if they were made of wood, since the specific gravity of steel is much greater than that of wood. Also, it is obvious that if half the balls were wood and half were steel the average true specific gravity, and therefore the density of the mass, would be intermediate between the values for all steel and all wood.

The true specific gravity of an actual soil is the weighted average of the specific gravities of all the mineral particles in the soil. About 1000 minerals have been identified by petrologists as the constituents of rocks; and, since soil is derived from rocks, many of these minerals may be found in any one soil. Also, the specific gravities of different minerals vary widely. That of borax is only about 1.7, whereas that of the iron mineral hematite is 5.3 and some rare minerals have much higher values. Nevertheless, the law of averages and the normal preponderance in soils of quartz, with a specific gravity of 2.65, and feldspars, with values of 2.55 to 2.75, narrows the usual range of variation of the true specific gravity of most soils to values between 2.6 and 2.7. Some examples of specific gravities are shown in Table 8-1. The specific gravity of volcanic ash, which is mostly glass, is low because glass has no regular crystal structure with requisite packing of ions.

TABLE 8-1. True Specific Gravities of Selected Soils

Volcanic ash, Kansas	2.32
Kaolinite	2.61
Alluvial montmorillonitic clay	2.65
Platte River sand	2.65
Iowa loess soil	2.70
Micaceous silt, Alaska	2.76
Oxisol (Latosol), Hawaii	3.00

The specific gravity of clays is difficult to measure but may be calculated from crystallographic and chemical data. Specific gravity is usually measured on an oven-dry (110°C) basis, which removes most but not all interlayer water from expansive clays.

EXAMPLE 8-2. The theoretical specific gravity of a dehydrated calcium montmorillonite is 2.70. Assuming G for interlayer water is 1.0, what is the specific gravity: (a) After ordinary oven-drying, d_{001} = 12.4 Å? (b) After air-drying, d_{001} = 15.4 Å? (c) After saturation, d_{001} = 19.6 Å?

SOLUTION. (a) The 12.4 Å mineral consists of 10.1 Å of mineral and 12.4 − 10.1 Å = 2.3 Å of water, giving a total weight per unit area of 10.1 (2.70) + 2.3 (1.0) = 27.3 + 2.3 = 29.6. Dividing by the height gives a specific gravity of 29.6 ÷ 12.4 = 2.39.

(b) 10.1 (2.70) + (15.4 − 10.1)(1.0) = 27.3 + 5.3 = 32.6; G = 32.6 ÷ 15.4 = 2.12.

(c) 10.1 (2.70) + (19.6 − 10.1)(1.0) = 27.3 + 9.5 = 36.8; G = 36.8 ÷ 19.6 = 1.88.

Experimental values are about 0.1 lower, probably because of hydration water around the outside of the clay crystals.

8.7. APPARENT SPECIFIC GRAVITY. Since soil is a porous material, it also has an "apparent" specific gravity. This is defined as the ratio of the dry weight of a unit volume of soil (volume of solids plus volume of voids) to the unit weight of water. It is also known as the volume weight or bulk density of the soil. The unmodified term *specific gravity* is often used to refer to the true specific gravity, but this is not good practice. Unless it is certain that the meaning of the unmodified term will be clearly understood by its connotation or otherwise, it is better practice to use the adjective "true" or "apparent" to indicate clearly which property of the soil is intended.

8.8. MOISTURE CONTENT. The moisture content of soil refers to the total amount of water contained therein, either as free water or capillary water in the soil pores or as adsorbed water films around the solid particles. It is always expressed as a percentage of the weight of the dry particles in the soil. The moisture content is determined by first weighing a representative sample of the soil in its natural or wet state, drying the sample to constant weight in an oven at a temperature between 100° and 110°C, and then weighing the dried sample. The difference between the weights of the sample before and after drying represents the amount of water in the sample; and this weight, computed as a percentage of the weight of the dried sample, is the moisture content of the soil. If a drying oven is not available, the sample may be dried in a pan over an open flame. However, care should be exercised not to overheat the soil, particularly if it is a fine-grained material. Overheating may vitrify some of the particles or otherwise change their character to such an extent that the sample is no longer representative of the original material.

8.9. FORMULAS INVOLVING SPECIFIC GRAVITY AND MOISTURE CONTENT. The definitions in Sections 8.6–8.8 can be expressed conveniently by simple formulas. Again it is suggested that the student derive the equations.

Let

G = true specific gravity;
G_a = apparent specific gravity, or volume weight;
γ_s = unit weight of soil solids (absolute volume);
γ_w = unit weight of water;
w = moisture content, in percent;
γ_d = dry density;
γ' = submerged unit weight, or buoyant unit weight;
γ = wet density (unit weight of soil);
W_d = dry weight of soil;
W_w = wet weight of soil;
W_m = weight of moisture in soil; and

V_a, V_e, V_m, and e have the same meanings as in Section 8.5.

Then

$$G = \frac{\gamma_s}{\gamma_w} \tag{8-12}$$

$$G_a = \frac{\gamma_s V_s}{\gamma_w (V_e + V_s)} \tag{8-13}$$

$$V_s = \frac{W_d}{\gamma_w G} \tag{8-14}$$

$$V_m = \frac{W_m}{\gamma_w} \tag{8-15}$$

$$W_m = W_w - W_d \tag{8-16}$$

$$w = \frac{W_m}{W_d} \times 100 \tag{8-17}$$

$$W_d = \frac{W_w}{1 + (w/100)} \tag{8-18}$$

If a unit volume of soil is considered, $W_d = \gamma_d$ and $W_w = \gamma$. Hence,

$$\gamma_d = \frac{\gamma}{1 + (w/100)} \tag{8-19}$$

In other words, the dry density of a soil is equal to its wet density divided by 1 plus the moisture content expressed as a decimal fraction.

Soil which lies below a free water table is acted upon by an upward force, in accordance with the principle of Archimedes. The magnitude of this upward force is equal to the weight of water displaced by the soil solids. The net unit weight, frequently called the submerged unit weight or buoyant unit weight, is equal to the dry density minus the buoyant force on the solids in a unit volume of soil. It may be expressed by the formula

$$\gamma' = \frac{\gamma_w}{1 + e}(G - 1) \tag{8-20}$$

The saturated unit weight, or weight of a soil mass saturated but not surrounded by water, equals

$$\gamma_t = \gamma' + \gamma_w \tag{8-21}$$

The saturated unit weight may be a little hard to visualize, so one can imagine weighing a bucket of saturated sand with the bucket under water vs. out of the water. The saturated unit weight is used in problems where the standing water level suddenly lowers, leaving the soil still saturated before the water has time to drain out. This condition is referred to as "sudden drawdown," and is discussed more fully in Section 20.5.

Other relationships involving specific gravity, void ratio, and porosity are

$$G_a = G\left(1 - \frac{n}{100}\right) \qquad (8\text{-}22)$$

$$G_a = \frac{G}{1 + e} \qquad (8\text{-}23)$$

$$G = G_a(1 + e) \qquad (8\text{-}24)$$

$$G = \frac{G_a}{1 - (n/100)} \qquad (8\text{-}25)$$

EXAMPLE 8-3. Let it be assumed that the mass of balls represented in Fig. 8-1 contains 25 lb of water interspersed in the void spaces and around the surfaces of the balls. Determine (a) the volume of water voids; (b) the percent water voids; (c) the volume of air voids; (d) the percent air voids.

SOLUTION. (a) By Eq. (8-15), in which the unit of volume is 1 cu ft (the weight of a cubic foot of water will be taken as 62.4 lb),

$$V_m = \frac{W_m}{\gamma_w} = \frac{25}{62.4} = 0.398 \text{ cu ft}$$

(b) By Eq. (8-7),

$$n_m = \frac{0.398}{1} \times 100 = 39.8\%$$

(c) From Eq. (8-1), $V_a = V_e - V_m$. Since V_e was found in Example 8-1 to be 0.476 cu ft,

$$V_a = 0.476 - 0.398 = 0.078 \text{ cu ft}$$

(d) By Eq. (8-6),

$$n_a = \frac{0.078}{1} \times 100 = 7.8\%$$

8.10. DEGREE OF SATURATION. The moisture content of soil may vary from zero if the soil is perfectly dry to a maximum value when the soil is saturated. The degree of saturation, sometimes called the humidity, is the actual amount of moisture in the soil expressed as a percentage of the maximum amount which it would contain if saturated. This percentage may be computed either on a volumetric basis or by weight.

Let S be the degree of saturation, or humidity, in percent.

Then

$$S = \frac{V_m}{V_e} \times 100 \qquad (8\text{-}26)$$

or

$$S = \frac{W_m}{\gamma_w V_e} \times 100 \qquad (8\text{-}27)$$

8.11. DETERMINATION OF DENSITY CHARACTERISTICS FROM BASIC DATA. If the wet density, moisture content, and true specific gravity of a soil are known, its void ratio, porosity, degree of saturation, and percent air voids can be easily determined. The first step is to compute the dry density and the volume of dry solids in a unit volume of soil by applying Eqs. (8-19) and (8-14). By subtracting this volume of dry solids from 1, we can obtain the volume of void spaces; and the void ratio and the porosity can be determined by Eqs. (8-2) and (8-3). The weight of water contained in a unit volume of the soil can then be found by Eq. (8-16); and the degree of saturation can be computed by applying Eq. (8-26). The volume of air voids and percent air voids can be computed by proceeding as just explained in Example 8-3.

EXAMPLE 8-4. The wet density of a soil is 128 pcf, its moisture content is 8%, and its true specific gravity is 2.65. Find (a) the dry density; (b) the void ratio; (c) the porosity; (d) the degree of saturation; (e) the percent air voids.

SOLUTION. (a) By Eq. (8-19),

$$\gamma_d = \frac{\gamma}{1 + (w/100)} = \frac{128}{1 + (8/100)} = 118.5 \text{ pcf}$$

(b) In Eq. (8-14), $W_d = \gamma_d$. Hence, the volume of the solids is

$$V_s = \frac{W_d}{\gamma_w G} = \frac{118.5}{62.4 \times 2.65} = 0.716 \text{ cu ft}$$

and

$$V_e = 1 - 0.716 = 0.284 \text{ cu ft}$$

The void ratio is, by Eq. (8-2),

$$e = \frac{V_e}{V_s} = \frac{0.284}{0.716} = 0.397$$

(c) The porosity is, by Eq. (8-3),

$$n = \frac{V_e}{V} \times 100 = \frac{0.284}{1} \times 100 = 28.4\%$$

(d) By Eq. (8-16), the weight of water in a cubic foot of the soil is

$$W_m = W_w - W_d = 128 - 118.5 = 9.5 \text{ lb}$$

Then, by Eq. (8-27), the degree of saturation is

$$S = \frac{W_m}{\gamma_w V_e} \times 100 = \frac{9.5}{62.4 \times 0.284} \times 100 = 53.5\%$$

(e) The volume of water in 1 cu ft of soil is

$$V_m = \frac{W_m}{\gamma_w} = \frac{9.5}{62.4} = 0.152 \text{ cu ft}$$

The volume of air voids is

$$V_a = V_e - V_m = 0.284 - 0.152 = 0.132 \text{ cu ft}$$

and the percent air voids is

$$n_a = \frac{V_a}{V} \times 100 = \frac{0.132}{1} \times 100 = 13.2\%$$

EXAMPLE 8-5. A block diagram showing phases in 1 cu ft of soil is helpful for the above calculations and avoids the need to memorize so many formulas. As a simple convention, weight quantities can be written to the right of the diagram, unit weights within the diagram, and volume quantities to the left. In the above problem $W_w = 128$ lb and $G = 2.65$. Specific gravities of water and air are assumed to be 1.0 and 0, respectively. Therefore,

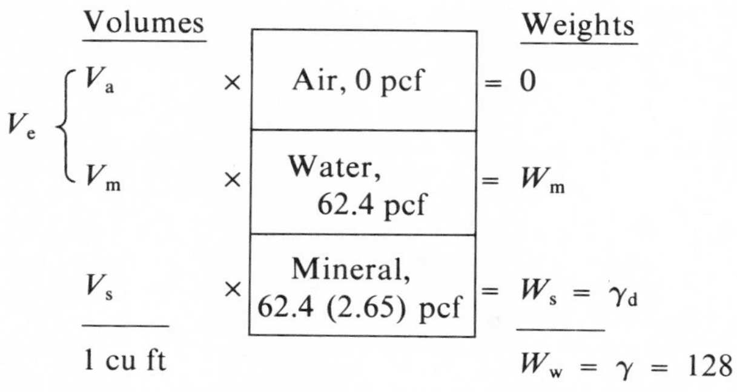

	Volumes			Weights
V_e	V_a	×	Air, 0 pcf	= 0
	V_m	×	Water, 62.4 pcf	= W_m
	V_s	×	Mineral, 62.4 (2.65) pcf	= $W_s = \gamma_d$
	1 cu ft			$W_w = \gamma = 128$

In the above example $\gamma = W_w = 128$ pcf and $W = W_m/W_s = 8\%$. (Note that the definition of moisture content is still needed.) On the right-hand side one sees that: $128 = W_m + W_s$ or $W_m = 128 - W_s$. Substituting,

$$W = \frac{128 - W_s}{W_s} = 0.08$$

$$128 - W_s = 0.08\ W_s$$

$$W_s = 118.5 \text{ lb}$$

Subtracting,

$$W_m = 9.5 \text{ lb}$$

These values are written into the diagram, and respective volumes calculated by dividing by unit weights:

$$V_m = \frac{9.5 \text{ lb}}{62.4 \text{ pcf}} = 0.152 \text{ cf}$$

$$V_s = \frac{118.5 \text{ lb}}{62.4\ (2.65) \text{ pcf}} = 0.716 \text{ cf}$$

By inspection one sees that

$$V_a + V_m + V_s = 1$$

$$V_a = 1 - 0.152 - 0.716 = 0.132 \text{ cu ft}$$

The phase weights and volumes are now all calculated and may be manipulated to obtain desired quantities:

	Volume	Phase	Weight	
V_e	0.132 cf	Air	0 lb	
V_e	0.152 cf	Water	9.5 lb	
	0.716 cf	Mineral	118.5 lb	$= \gamma_d$
	1.0 cf		128 lb	

(a) $\gamma_d = 118.5$ pcf

(b) $e = \dfrac{V_e}{V_s} = \dfrac{0.132 + 0.152}{0.716} = 0.397$

(c) $n = \dfrac{V_e}{V} \times 100 = \dfrac{0.132 + 0.152}{1} \times 100 = 28.4\%$

(d) $S = \dfrac{V_m}{V_e} \times 100 = \dfrac{0.152}{0.132 + 0.152} \times 100 = 53.5\%$

(e) $n_a = \dfrac{V_a}{V} \times 100 = 13.2\%$

EXERCISE. Sketch a block diagram and calculate phase weights and volumes when $\gamma_d = 95$ pcf, $G = 2.70$, and $w = 10\%$.

PARTIAL SOLUTION.

Volume (cf)	Phase	Weight
—	Air	0 lb
0.15	Water	9.5 lb
0.54	Mineral	95 lb
1.0 cf		104.5 lb

8.12. EFFECT OF SIZE OF UNIFORM PARTICLES ON DENSITY. The density of a soil having a certain true specific gravity is inversely related to the void ratio; that is, the greater the density, the smaller the void ratio, and vice versa. Both the void ratio and the density are influenced by the physical state and arrangement of the solid particles in a soil mass. It will be helpful to discuss some of these factors and their effect on density.

First, it can be shown that the density is practically the same whether the soil is composed of large particles or small particles, provided that other characteristics, such as the shape and arrangement of the particles, remain constant. For example, let it be assumed that an imaginary soil is composed of balls which are 1 in. in diameter and which have the same specific gravity and are placed in the same orderly arrangement as the 2-in. balls in Fig. 8-1. The number of 1-in. balls in a cubical container 1ft long in each direction would be 1728, and the volume of each such ball would be 0.000303 cu ft. The total volume of solids would then be $0.000303 \times 1728 = 0.524$ cu ft, which is the same as the volume of the 2-in. balls.

Since the total volume of the soil mass is unchanged in the two cases just considered, the void ratio and the density would be the same. In fact, these properties would remain constant, no matter how small the diameter of the balls might be. This indication is borne out in practice by the fact that a coarse gravel having particles which are fairly uniform in size actually has about the same density as a fine sand or silt of the same character. Also, it is a well-recognized fact that a bushel basket of small potatoes weighs about the same as a basket of large ones.

8.13. EFFECT OF ARRANGEMENT OF PARTICLES ON DENSITY. Although, as indicated in the preceding section, the density of soil is not affected by a difference in size of particles, it is influenced appreciably by their arrangement. Suppose, for example, that the balls composing the imaginary soil of Fig. 8-1 are rearranged by moving each tier of balls a half-diameter to the right, as illustrated in Fig. 8-2. In the new arrangement, the volume of the balls (soil solids) is still 0.524 cu ft, but the total volume of soil (solids plus voids) is reduced to 0.911 cu ft. Hence, the void ratio is reduced to 0.738. Since the specific gravity of the particles was not changed

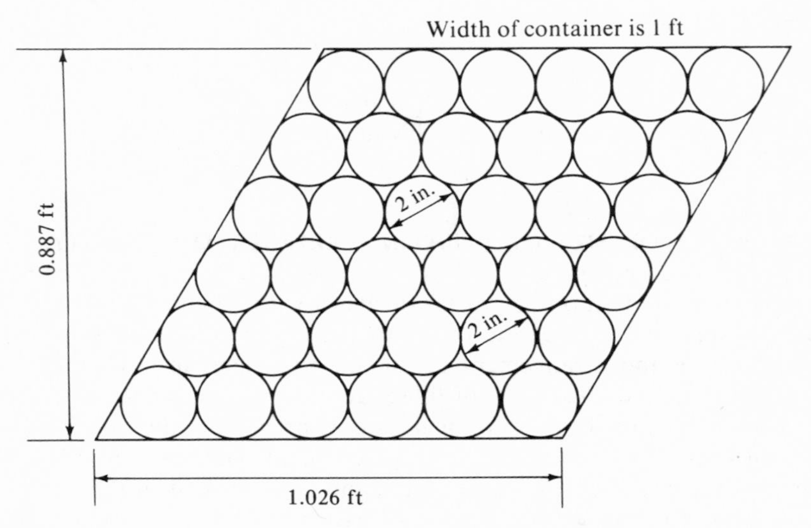

Fig. 8-2

in the process of rearrangement, the density of the imaginary soil in Fig. 8-2 is greater than that of the soil in Fig. 8-1. The density could be still further increased by sliding each tier of balls a half-diameter in the direction normal to the page. This case, termed rhombohedral packing, gives the minimum void ratio for uniform spheres, 0.35.

In practice, the fact that particle arrangement influences soil density is of very great importance. Because of this fact, field compaction operations are given careful consideration. The application of weight and vibratory and kneading action by means of sheep's-foot rollers, smooth rollers, pneumatic-tired rollers, and other types of compaction equipment tends to rearrange the soil particles and to bring them into more intimate contact. This compaction reduces the void ratio or porosity and increases the density of the soil.

8.14. EFFECT OF GRADING OF PARTICLE SIZES ON DENSITY. The density of soil is greater when it consists of particles of various sizes, rather than particles of uniform size. This principle is illustrated by the fact that, in the case of the imaginary soil in Fig. 8-1, smaller balls can be placed in the void spaces between the larger balls as shown in Fig. 8-3. The introduction of the small balls increases the volume of solid material, but does not change the total volume of soil. Therefore, the void ratio is decreased and the density is increased by this process. The greatest possible density will be attained by continuing to put smaller balls in the remaining spaces until particles of microscopic size have been included.

A soil which has a wide range of particle sizes, well distributed from the smallest to the largest in a relationship that is conducive to the attainment of high density when the soil is compacted, is called a *well-graded* soil. The grading curve of a well-graded material, sometimes spoken of as the ideal grading curve, is parabolic in form and may be expressed by the following formula, known as the Talbot formula:

$$p = \left(\frac{d}{D}\right)^x \times 100 \tag{8-28}$$

in which

p = percentage, by weight, smaller than any particle size d;
d = particle size;
D = maximum particle size;
x = an exponent, the value of which depends on the characteristics of the soil.

The best value of x for a particular soil probably depends on the shape of the soil particles to a considerable extent. The limiting values are about 0.25 and 0.40. However, the potential density characteristics of

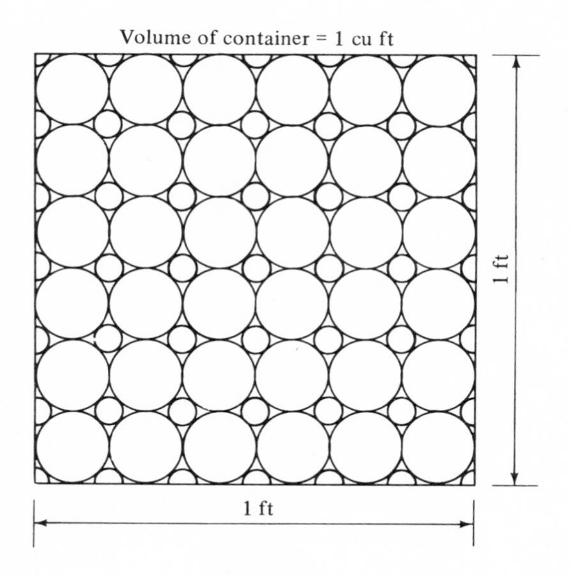

Fig. 8-3

a soil or soil mixture do not vary widely with changes in the exponent x; and a material whose grading curve falls anywhere between the limits defined by taking x as 0.25 and 0.40 is considered to be well graded.

8.15. MAXIMUM POTENTIAL DENSITY. The density of a soil can never be increased sufficiently to equal the density of a solid mass having the same specific gravity as that of the soil particles because, with uneven rounded and flake-shaped particles, there will always be some spaces in the soil mass which do not contain solid matter. The only way such a limiting density could ever be obtained would be for the soil particles to be true cubes or parallelepipeds similar to a child's building blocks. Obviously, this condition never occurs in an actual soil.

The density of a solid mass of material having a specific gravity of 2.65 would be $2.65 \times 62.4 = 165.36$ pcf. The dry density of actual soils in nature is often in the neighborhood of 50 to 60% of this value. However, peats, mucks, tundra, and other highly porous soils may have natural densities as low as 15% of the theoretical maximum limit; while well-graded, dense, stabilized gravel mixtures, on the other hand, may be compacted to dry densities as high as 85 to 90% of the theoretical limit.

It must be kept in mind that the grading characteristics of soil influence only the potential density of the material. Even a well-graded soil may exist in a relatively loose state until it is compacted by some means, either natural or artificial. Natural compaction may result from the weight of overlying material, seismic vibrations, the beating action of rainfall, the puddling action of percolating waters, alternate swell and shrinkage with wetting and drying, and the general influences of time and climate. Artificial compaction may be obtained by the weight, kneading action, and vibration of traffic on soil masses or, more importantly, by the application of energy by means of various kinds of rollers and vibratory equipment at the time soil is placed in an engineering structure.

8.16. RELATIVE DENSITY. One measure of the degree to which a soil has been compacted is the *relative density* or density index, defined as

$$D_r = \frac{e_{max} - e}{e_{max} - e_{min}} \times 100 \tag{8-29}$$

where e is the void ratio; e_{max} is the maximum void ratio, i.e., void ratio of the soil in its loosest state; and e_{min} is the minimum void ratio, i.e., void ratio of the soil in its densest state.

The value of e_{max} is determined by drying and pulverizing the soil to a single-grain structure, pouring it slowly and with minimum disturbance through a funnel into a container of known volume, and then

TABLE 8-2. Some Laboratory Compaction Tests

Test	*A.S.T.M. Designation*[a]	*A.A.S.H.O. Designation*[b]	*Mold Size (cu ft)*	*Compactor*	*Effort*
Standard A.A.S.H.O. density	D 698–70	T99–70	1/30	5.5 lb falling 1 ft	25 blows/layer; 3 layers
Modified A.A.S.H.O. density	D1557–70	T180–70	1/30	10 lb falling 1.5 ft	25 blows/layer; 5 layers
Relative density of cohesionless soils	D 2049–69	—	0.1 or 0.5	60 cps vibrator, 0.025 in. amplitude, pressure 2 psi surcharge	Vibrate 8 min
Harvard Compactor[c]	—	—	1/454	40 lb pushing on 0.5 in. diameter cylinder	Variable: usually >10 tamps/layer; 5 layers

[a] American Society for Testing and Materials Book of Standards, Part 11; issued annually.
[b] American Association of State Highway Officials Standard Specifications, Part II; issued every 5 years.
[c] Special Procedures for Testing Soil and Rock for Engineering Purposes, ASTM STP 479, 1970.

weighing. The soil must be dry to prevent "bulking." The value of e_{min} is determined after compaction with a drop hammer, tamper, or vibrator.

A relative density of 0 indicates no compaction and $e = e_{max}$, whereas a relative density of 1.0 ($D_r = 100\%$) indicates that $e = e_{min}$ and the soil is in its presumed densest state. The use of relative density has been restricted to granular or unstructured soils because of the difficulty of determining e_{max} in clayey soils, although Burmister has suggested that it might be calculated from the liquid limit, or moisture content at which a soil becomes liquid.

Compaction for the determination of e_{min} is usually by use of a vibration table, vibration being very effective for compaction of granular soils. Soil is placed in a cylindrical mold on the vibrating table, and a surcharge weight placed on top to confine the soil and act as an anvil. Tests have shown that the equilibrium density and e_{min} are functions of the acceleration ratio, or ratio of acceleration due to vibrations to that of gravity. They also are affected by the vibration geometry and applied load; therefore the conditions under which the test is performed must be carefully specified and controlled (Table 8-2). A recent study by Youd (*8*) shows that repeated confined shearing will cause relative densities well in excess of 1.0 determined from standard test, confirming the arbitrary nature of e_{min} as presently defined.

8.17. THE PROCTOR RELATIONSHIP. Soil compaction was largely a trial-and-error process until the early 1930s when the Los Angeles County engineer, R. R. Proctor, discovered and described an important relationship between soil density, moisture content, and compactive effort (*6*). Proctor found that by molding a series of specimens with different moisture contents, using the same compactive effort for each specimen, the density on a dry-weight basis would peak out as shown in Fig. 8-4.

It was once believed that water added up to the optimum moisture improved "lubrication" between the soil grains and hence compaction. However, it is now known that nonlayered soil minerals such as quartz have higher sliding friction wet than dry. Instead, moisture contents below optimum cause increased capillarity or negative pore pressure, which pulls grains together and prevents sliding and compaction. This phenomenon also has been referred to as "bulking," from the observation that disturbed wet sand weighs *less* per cubic foot than either dry or saturated sand.

The maximum density shown in Fig. 8-4 is called the Standard Proctor Density. Also, since the American Association of State Highway Officials has adopted this test procedure, it is sometimes referred to as the

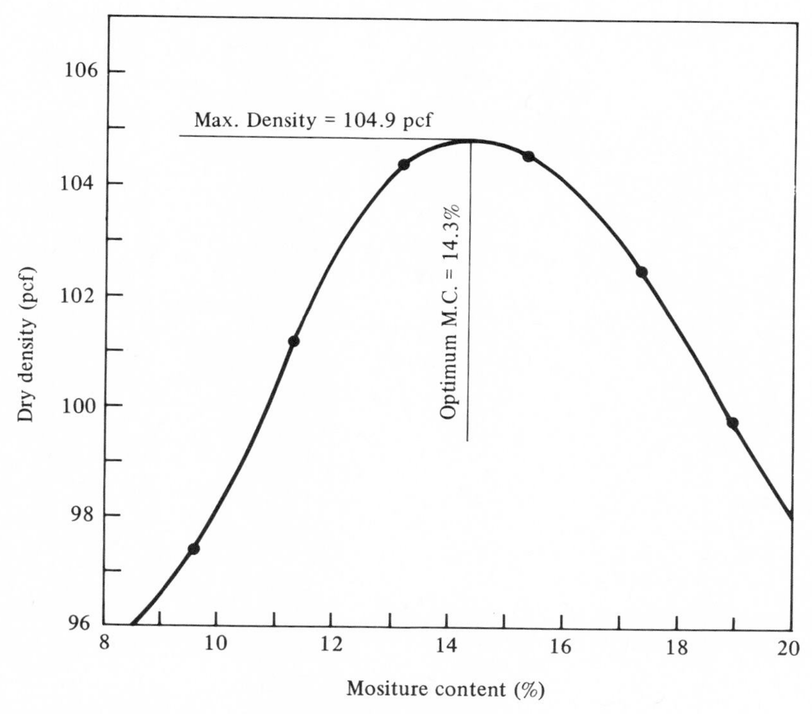

Fig. 8-4. Proctor density curve.

Standard A.A.S.H.O. Density. The moisture content corresponding to this maximum density is called the Standard Optimum Moisture Content.

In the standard Proctor test, the soil is compacted in a cylindrical mold, 4 in. in diameter and about $4\frac{1}{2}$ in. high and having a volume of $\frac{1}{30}$ cu ft. The soil is placed in the mold in three equal layers, and each layer is compacted by 25 blows of a metal tamper weighing $5\frac{1}{2}$ lb and having a striking face 2 in. in diameter. The hammer is allowed to fall freely through a height of 1 ft for each blow. The amount of energy thus applied was established by Proctor as the amount which would give a maximum density in the laboratory test approximately equal to that which it is feasible to obtain in field compaction operations.

8.18. MODIFIED A.A.S.H.O. DENSITY. More recently, the U.S. Army Corps of Engineers felt the need for greater densities of airport-pavement subgrades, embankments, earth dams, etc., than those indicated by the Proctor test, and heavier compacting machinery has been developed to

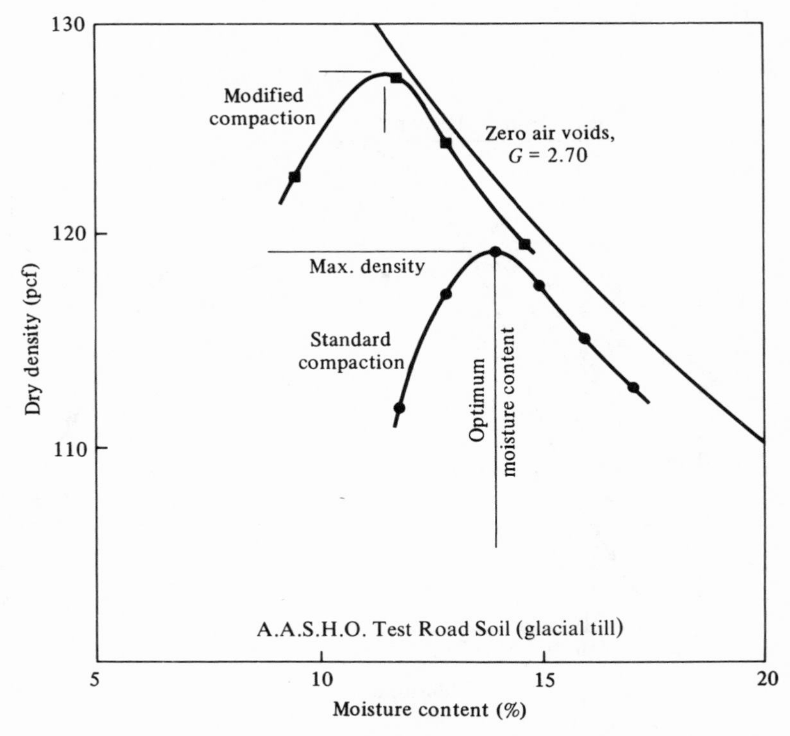

Fig. 8-5. Proctor density curves for a glacial till.

meet this need. Parallel with this development, the Corps has introduced a modification of the standard-density test in which the applied energy is greatly increased. In this modified test the soil is placed in a Proctor mold in five equal layers, and each layer is compacted by 25 blows of a tamper which weighs 10 lb and is allowed to fall freely through a height of 18 in. The maximum density thus indicated is called the Modified A.A.S.H.O. Density (or Modified Proctor Density), and the corresponding moisture content is known as the Modified Optimum Moisture Content.

The maximum density obtained by the modified procedure is greater than that obtained in the standard test, while the modified optimum moisture content is less than the standard. A typical relationship is shown in Fig. 8-5.

8.19. ZERO AIR VOIDS CURVE. In Fig. 8-5 the right-hand limbs of both moisture-density curves roughly parallel a line designated "zero air

TABLE 8-3. Zero-Air-Voids Dry Densities (pcf)

Moisture Content w (%)	*Specific Gravity of Soil Solids, G*					
	2.30	*2.60*	*2.65*	*2.70*	*2.75*	*3.00*
0	143.5	162.2	165.4	168.5	171.6	187.2
10	116.7	128.8	130.7	132.7	134.6	144.0
20	98.3	106.7	108.1	109.4	110.7	117.0
30	84.9	91.1	92.1	93.1	94.0	98.5
40	74.8	79.5	80.3	81.0	81.7	85.1
50	66.8	70.5	71.1	71.7	72.3	75.9
100	43.4	45.1	45.3	45.5	45.8	46.8

voids." This line represents the dry density of the soil if its entire volume is water and solids. Since compaction is a process for expelling air, the moisture-density curves cannot cross this line, although as seen in the figure the higher, modified, compactive effort brings the curve closer to the zero-air-voids line. Since this line represents a theoretical limit on density at any moisture content, its position is often shown on moisture-density plots. The zero-air voids density for any moisture content may be calculated from the formula

$$\gamma_{dt} = \frac{G\gamma_w}{1 + G(w/100)} \tag{8-30}$$

in which γ_{dt} is the dry density at saturation, G the specific gravity of the soil particles, γ_w the unit weight of water, and w the moisture content in percent. Solutions for several values of G and w are shown in Table 8-3.

8.20. OVERCOMPACTION. If a soil contains too much water for a prescribed density, no amount of compactive effort will be sufficient to reach that density unless the soil is dried first. Not only is extra compactive effort wasted, but it sometimes goes to rework and shear the soil, establishing shear planes, dispersing the clay, and causing a large reduction in strength. This is referred to as overcompaction and is avoided except in special circumstances. An example of the effect of overcompaction on strength of a clay is shown in Fig. 8-6, which presents stress-strain curves for two samples of the same soil at the same moisture content and same density, the difference being that the stronger sample was compacted on the dry side of optimum and then soaked in water, whereas the weaker sample was compacted on the wet side of optimum and became dispersed.

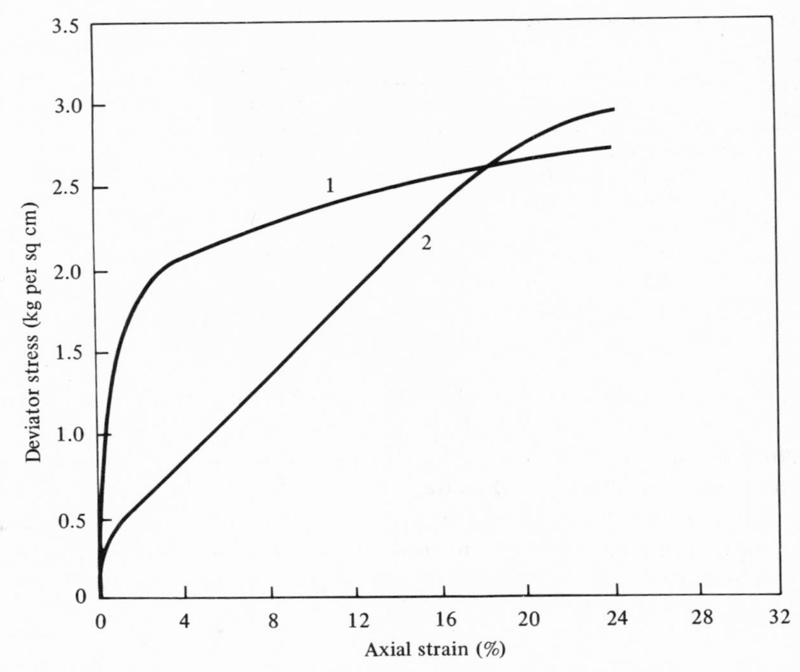

Fig. 8-6. Stress-strain curves for a soil compacted dry of optimum and soaked (curve 1) and then for the same soil compacted wet of optimum (curve 2). After Seed and Chan *(7)*.

8.21. COMPACTION OF EXPANSIVE CLAYS. Artificially compacted expansive clays sometimes do not remain compacted, but expand upon entry of water. Sometimes the action is scarcely detectable because uplift is uniform, but other times it causes major problems, as under buildings or where pavements adjoin existing slabs or bridge abutments. Reducing the compactive effort to give a lower initial density is not an effective control on expansion, since the clay lumps still comprise an expanding skeleton. Furthermore there is a disadvantage in loose compaction: The soil is much more permeable, and therefore more susceptible to daily fluctuations in available water.

Where expansion must be avoided, the clay may be compacted very wet, even though this introduces overcompaction. The proper moisture content to minimize swelling can be determined by immersing compacted laboratory specimens in water and measuring their expansion. A plot of expansion versus initial moisture content will appear as curve A in Fig. 8-7, from which an optimum moisture content for minimum swelling can be selected. In cases where such a high moisture content causes the soil to be

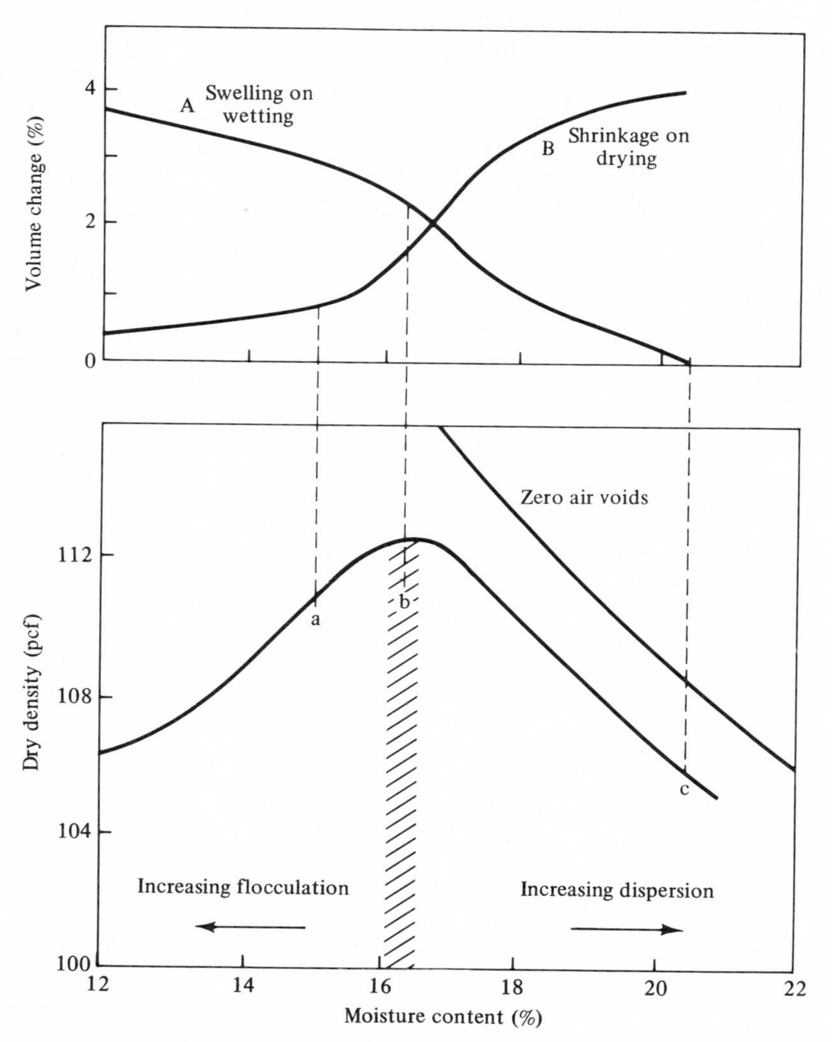

Fig. 8-7. Moisture contents for (a) low shrinkage, (b) maximum density, or (c) zero expansion of an expansive day. Adapted from Seed and Chan (*7*).

too weak—sometimes too weak even to support construction equipment —a common practice is to mix in several percent of hydrated lime, which reacts with the clay minerals, greatly reducing their expansive character and allowing compaction at an optimum moisture content for maximum density.

The other problem of expansive clays, shrinkage, also can be minimized by adjusting the optimum moisture content, as shown in Fig. 8-7, curve B. Shrinkage is less frequently the prime consideration, since moisture tends to accumulate under impervious slabs, causing expansion.

8.22. EFFECT OF SOIL COMPOSITION. The influence of particle size gradation on packing was discussed in general terms in Section 8.14 and is aptly demonstrated by standard A.A.S.H.O. T99 moisture-density curves in Fig. 8-8. Because of the obvious dependence on soil type, a number of investigators have attempted to predict maximum density and optimum moisture content from various other soil properties such as particle gradation, plasticity, or shrinkage test data, or all three. However, none of the formulas yet derived has been found to have a universal application.

An error arising from the standard laboratory test procedures stems from re-use of the same soil over and over again for preparation of density test specimens, adding more water each time the soil is recompacted. This is done for convenience, but causes a shift in the moisture–density curve after the first use, probably caused by a change in structure and perhaps gradation due to degradation of soft particles. An investigation by Nelson and Sowers showed an increase in maximum density of 1 to 8 pcf by re-use of the soil, with relatively little change in optimum moisture content. These and other factors influencing compaction are comprehensively reviewed by Johnson and Sallberg (*4*).

8.23. FIELD DENSITY TESTS. Knowledge of the in-place density of soil is necessary for calculating borrow or cut-and-fill quantities in addition to its use for compaction control. For example, if the natural field density is 100 pcf and the compacted density is 120 pcf, approximately a 20% reduction in volume (loosely termed "shrinkage") will occur when the soil is compacted.

Field density tests may either be direct, by excavating and measuring the volume of a hole and determining the weight of material and moisture content of material taken from the hole, or they may be indirect, by geophysical methods.

In the field density test a hole about 4 in. in diameter and as deep as the thickness of the compacted layer is dug in the soil. All of the material removed is carefully recovered and is immediately weighed. After this

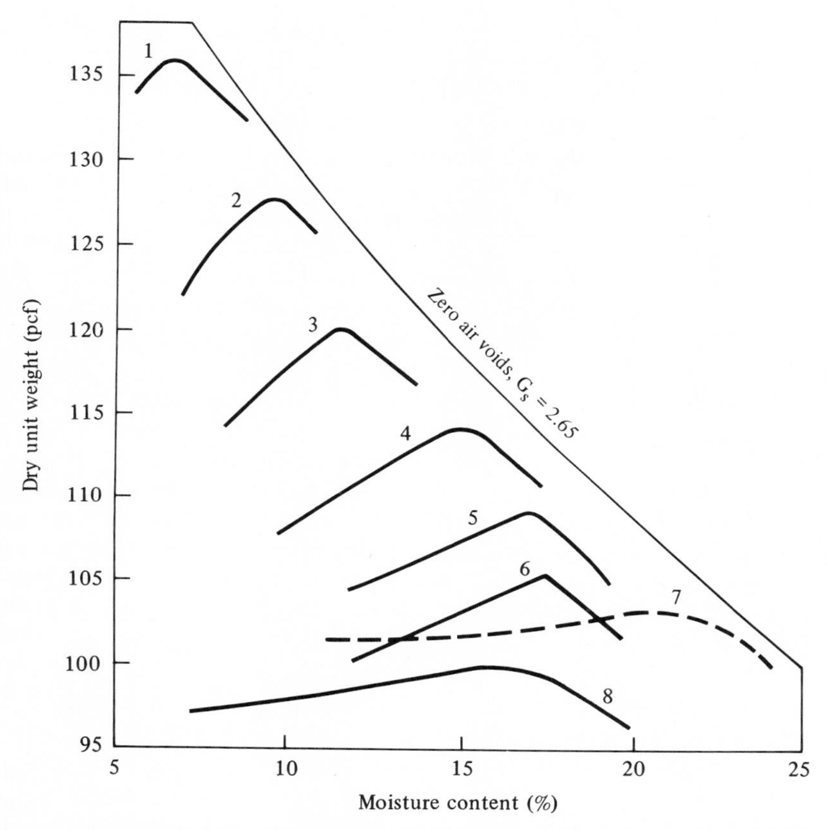

Fig. 8.8. Moisture content vs. dry unit weight relationships for eight soils according to A.A.S.H.O. T99.

Soil Texture and Plasticity Data

No.	*Description*	*Sand*	*Silt*	*Clay*	*LL*	*PI*
1	Well-graded loamy sand	88	10	2	16	NP
2	Well-graded sandy loam	78	15	13	16	NP
3	Medium-graded sandy loam	73	9	18	22	4
4	Lean sandy silty clay	32	33	35	28	9
5	Lean silty clay	5	64	31	36	15
6	Loessial silt	5	85	10	26	2
7	Heavy clay	6	22	72	67	40
8	Poorly graded sand	94	- 6 -			NP

weighing has been made, a representative sample of the removed material is taken for moisture determination. Then the volume of the hole, which is the volume of the soil sample in its undisturbed state before removal, is measured by filling the hole with some substance whose specific gravity is known. The weight of the substance required to fill the hole is converted into volume. The weight of the soil removed from the hole divided by the volume of the hole is the wet density of the soil.

There are three substances in use for measuring the volume of the hole: motor oil; clean, dry, cohesionless sand; and water. If water is used, it must be confined in a thin rubber membrane or balloon which is expanded by water or air pressure into the hole. The volume of the hole is read directly from the lowering of the water level in a calibrated reservoir, termed a "volumeter." Motor oil requires no membrane since it does not penetrate the soil to any appreciable extent if most of the soil voids are filled with water. However, if the soil is unusually dry, it is safer to use either the sand method or the water method. The sand method involves a standard Ottawa sand and a standardized procedure for pouring: the latter is done by utilizing a standardized "sand cone." The amount of sand used is determined by weighing. A weight is subtracted for sand in the cone, and the result is converted to volume. Construction traffic must be halted during the test since vibrations affect packing of the sand.

8.24. NUCLEAR DENSITY TESTS. The most recent trend in shallow field density testing has been toward use of nuclear devices, operating either in drilled holes or more conveniently from the ground surface. In one system the apparatus is mounted in a special truck and trailer and it is possible to continuously monitor and record wet density and moisture content at speeds up to 3 mph. Since the moisture content is recorded in pounds per cubic foot, dry density may be obtained directly by subtraction. Much less expensive are hand-portable units available from a number of manufacturers.

Nuclear density instruments operate on either a direct transmission or backscatter principle. A radioactive source of gamma rays is placed into, against, or close to the surface of the soil. Gamma rays penetrating the soil collide with soil element electrons and are deviated slightly from their path, the energy loss causing a slight shift to a longer wavelength, termed the Compton effect. Repeated collisions cause some gamma rays to return to the surface, where they are detected and counted with a Geiger counter or sodium iodide scintillation counter. Since the amount of such backscatter is proportional to the number of electrons in the soil, the count may be correlated to wet density by utilizing appropriate calibration curves. Calcium and iron in the soil have appreciably larger mass

TABLE 8-4. Radioactive Sources for Density and Moisture Testing

Isotope	*Half-Life (yr)*	*Energy Level (MeV)*
60 Co	5.25	1.17, 1.3
137 Cs	33	0.66, 0.3
226 Ra Be	1620	0.61, 0.35

absorption coefficients for radiation below 0.3 MeV, sufficiently different from other elements to require new calibration curves for iron-rich or calcium-rich soils when measurements utilize a Cs or RaBe source (Table 8-4). On the other hand these sources are sometimes preferred for their longer half-life, requiring less frequent recalibration.

The reported accuracy of nuclear density devices varies rather widely, from a standard deviation of ±1.6 pcf to over ±5 pcf. One problem in evaluating the new density devices is not knowing the variability of the older test methods; for example, the sand-cone and volumeter density tests are probably accurate within about ±2 pcf. One evaluation procedure is to prepare large test blocks with carefully controlled density, and test by different methods. Lower standard deviations have been reported for direct-transmission nuclear instruments, in which the source or counter or both are lowered into holes in the soil; this has been attributed to less dependence on surface irregularities. Similarly, results from surface instruments have been improved by utilizing an air gap between the instrument and the soil, and this technique is now utilized in most instruments. While soil density is measured to a depth of 6 or 8 in., about 50% of the count relates to densities in the upper 2 in., and 75% relates to densities in the upper 3 in. Penetration with the smaller surface instruments may be only one-half of this and varies with soil density and instrument geometry.

Nuclear density devices are potentially as precise as direct measurement methods, are more expensive, and require less time for testing. Nuclear test equipment must be checked periodically for radioactivity leaks, and personnel using the equipment should be provided with film badges. The current maximum allowable dosage is 100 milliroentgens per week.

8.25. FIELD MOISTURE CONTENTS. The direct method of moisture content determination includes weighing a sample, drying in a 110°C oven, and reweighing to determine moisture loss. On field projects this procedure may be speeded with some loss in precision by drying in a frying pan over a gas flame. Another method is to saturate the soil with alcohol, which is miscible with water, and ignite. Wood or denatured alcohol are recommended for this procedure to minimize unauthorized losses.

A more scientific method is the "Speedy Moisture Tester," developed in England. Measured amounts of the wet soil and calcium carbide powder are sealed in a pressure flask and agitated with a steel ball (*1*). Water from the soil reacts with calcium carbide to produce acetylene gas, which increases the pressure in the canister:

$$CaC_2 + 2H_2O \rightarrow Ca(OH)_2 + C_2H_2 \uparrow$$

Moisture content is then read directly from a factory-calibrated pressure gauge. Accuracies of $\pm 0.5\%$ have been reported. However, clay and organic matter decrease the accuracy, and errors as large as 4 to 5% also have been reported, indicating the need for checking by other methods.

Nuclear methods are increasingly being utilized for moisture content determinations and are often built into the same instrument that is used for density determinations. For moisture determinations a fast neutron source is used. Neutrons have approximately the same mass as a hydrogen nucleus or proton, but carry no electrical charge, and therefore do not interact with and are not deflected by electron swarms around the soil atoms. Instead, the neutrons penetrate undeflected until they collide with atomic nuclei having about the same mass, i.e., hydrogen nuclei. Thus the reduction of fast neutrons to slow neutrons mainly relates to the number of hydrogen atoms in soil, and hence soil water. However, other hydrogen atoms, as in organic matter, asphalt, and clay minerals, also affect the determination. Also certain elements such as iron, potassium, and chlorine have a high capture cross section for neutrons and affect the determination. After repeated collisions the velocity of the neutrons is decreased about 10 times such that they can be counted with a boron trifluoride counter or lithium iodide scintillation counter.

The radioactive source of neutrons is usually triggered by gamma radiation from a second source. The RaBe gamma ray source also emits fast neutrons and may be used for both moisture and density determinations, by independently counting slow neutrons and backscattered gamma rays. Americium-beryllium, platinum-beryllium, or other neutron sources may be utilized where the high gamma radiation is not wanted.

Tests have shown the nuclear moisture gauges to give somewhat better accuracy in terms of total weights than their density counterparts, the standard deviations in moisture content being about 1 to 3 pcf. However, this variation can be critical; an excess of 1 or 2 pcf sometimes preventing further compaction. Best accuracy is obtained by finding the count ratio to a reference standard, either a soil block or a pail of water.

Because of the dependence on elemental composition, check tests are needed or the calibrations should be performed on a similar soil.

PROBLEMS

8.1. Convert the following wet densities and moisture contents to dry densities: (a) 120 pcf, 15%; (b) 117 pcf, 17%; (c) 135 pcf, 21%; (d) 101 pcf, 5%; (e) 120 pcf, 12%; (f) 90 pcf, 102%.

8.2. If $G = 2.70$, find V_s, V_e, and the void ratio for each of the above.

8.3. Calculate the saturated and the submerged unit weights of each of the above.

8.4. An undisturbed sample of soil, weighing 11.23 lb, is submerged in kerosene and is found to displace 0.092 cu ft of the liquid. If the sample contains 13.4% moisture and has a true specific gravity of 2.65, calculate the following properties: (a) wet density, (b) dry density, (c) void ratio, (d) porosity, (e) humidity, (f) percent water voids, (g) percent air voids, (h) percent solids.

8.5. If 224,800 cu yd of soil are removed from a highway cut in which the void ratio is 1.22, how many cubic yards of fill having a void ratio of 0.78 could be constructed?

8.6. If the true specific gravity of the soil in Problem 8.5 is 2.70 and it contains 10% moisture, how many tons of soil will have to be moved? How many tons of dry soil will there be? How many tons of water will there be in the soil?

8.7. (a) Explain how a moisture content can exceed 100%. (b) What soil minerals are indicated by moisture contents in excess of 100%? (c) Why is it advantageous to define moisture content on the basis of dry weight rather than total weight?

8.8. (a) Relative density tests on a dry sandy gravel give a maximum and minimum of 128 and 92 pcf, respectively. If $G = 2.65$, find e_{min} and e_{max}. (b) Dry densities of this soil are found to be 122, 116, and 108 pcf. Calculate relative densities.

8.9. Calculate the compactive energy per cubic foot of soil in the standard and modified Proctor tests.

8.10. Calculate the acceleration ratio in ASTM D2049. What is the function of the surcharge load?

8.11. Plot data from the following 20 density tests on probability paper: 102, 101, 100 (4 tests), 99 (2), 98 (3), 98 (2), 94, 93 (2), 92 (3), 91 pcf. (a) Assuming a normal distribution, show $\bar{X}$ and σ and estimate the percent of fill exceeding 90% of maximum density. (b) Explain the large σ. (c) On the basis of these data, how many additional tests would be needed to estimate $\bar{X}$ within ± 0.5 pcf?

8.12. Calculate $\bar{X}$ and σ for data in Problem 8.11 using equations in Chapter 7.

8.13. (a) Calculate the zero-air-voids dry densities in Fig. 8-5. Assume quartz sand with moisture contents of 0, 5, 10, and 15%. (b) What are the percentages of air voids at the two peak densities?

8.14. A proposed earth dam will contain 5,360,000 cu yd of earth. It will be compacted to a void ratio of 0.80. There are three available borrow pits, which are designated as A, B, and C. The void ratio of the soil in each pit and the estimated cost of moving the soil to the dam is as shown in the tabulation. What will be the least earth-moving cost, and which pit will it be most economical to use?

Pit	*Void Ratio*	*Cost of Moving*
A	0.9	28¢ per cu yd
B	2.0	20¢ per cu yd
C	1.6	26¢ per cu yd

REFERENCES

1. Blystone, J. R., A. Pelzner, and G. P. Steffens. "Moisture Content Determination by the Calcium Carbide Gas Pressure Method." *HRB Bull.* **309,** 77–84 (1961).
2. Foster, Charles R. "Reduction in Soil Strength with Increase in Density." *ASCE Trans.* **120,** 803–822 (1955).
3. Hughes, C. S., and M. C. Anday. "Correlation and Conference of Portable Nuclear Density and Moisture Systems." *HRB Record* **177,** 239–279 (1967).
4. Johnson, A. W., and J. R. Sallberg. "Factors Influencing Compaction Test Results." *HRB Bull.* **319,** 148 (1962).
5. Olson, Roy E. "Effective Stress Theory of Soil Compaction." *ASCE J.* **89,** (SM2), 27–45 (1963).
6. Proctor R. R. "Fundamental Principles of Soil Compaction." *Engineering News-Record* (Aug. 31 and Sept. 7, 21, and 28, 1933).
7. Seed, H. B., and C. K. Chan. "Structure and Strength Characteristics of Compacted Clays." *ASCE J.* **85** (SM5:Pt. 1), 87–128 (1959).
8. Youd, T. L. "Maximum Density of Sand by Repeated Straining in Simple Shear." *HRB Record* **374,** 1–6 (1971).

9

Soil Water

9.1. EFFECT OF MOISTURE CONTENT ON SOIL. In connection with the definition of soil given in Chapter 4, it was pointed out that the engineering behavior of this material is greatly influenced by changes in moisture content. This fact is a matter of common knowledge to all who stop to give it consideration. After a rain, clayey soil which may have been dry and hard a short time before, becomes soft and slippery to the extent that one has difficulty walking on it or driving a car over it. Also, bank cave-ins and slides occur most frequently in clayey soil after heavy rainfall or an extended wet period. These evidences clearly indicate that the fine-grained soils are materially weakened, both in bearing capacity and in resistance to shearing stresses, by high moisture content. On the other hand, it is difficult to walk or drive on beach sand when such material is in a dry, loose state; but, after a rain or ebb of the tide, the same sand becomes relatively hard and stable and it is easy to walk on the material.

9.2. HYDROLOGIC CYCLE. Commonly observed phenomena, such as those just mentioned, are indicative of the profound effect of changes in moisture content on the engineering properties of soil, and they clearly dictate how necessary it is for the soil engineer to give extensive consideration to the hydrologic environment to which a soil structure or engineering facility will be subjected. Moisture is continually moving at and near the earth's surface and in the atmosphere surrounding the earth. The

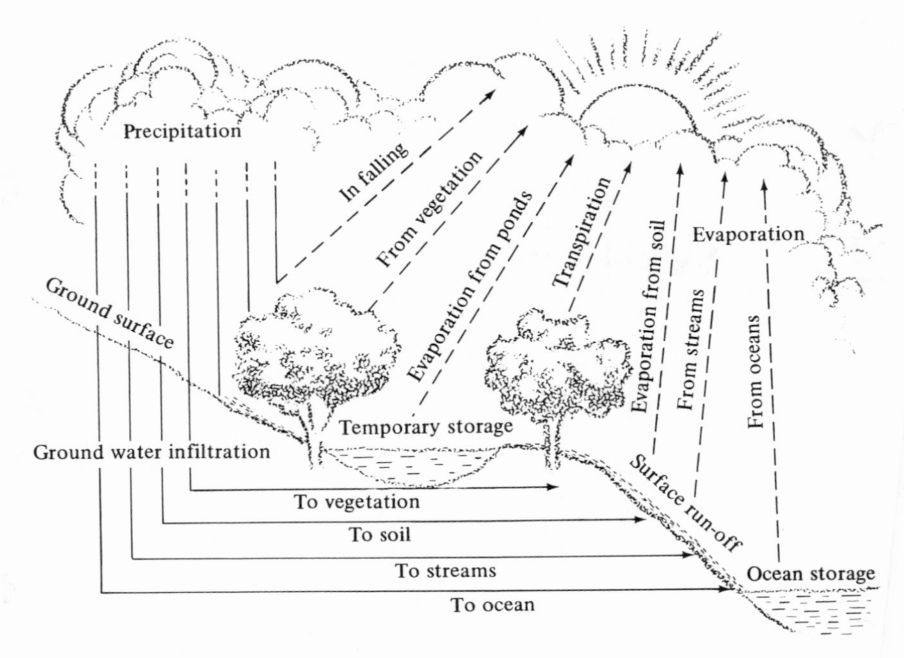

Fig. 9-1. The hydrologic cycle.

role of the soil in the hydrologic cycle, which refers to the disposition of water from the time it is precipitated until it re-enters the atmosphere by evaporation, is illustrated in the diagram in Fig. 9-1. Moisture vapor in the clouds condenses under the influence of temperature changes and falls to the earth as rain, snow, sleet, or hail. A part of this precipitation may not reach the land surface, but may evaporate in the air while falling, may evaporate from the leaves and stems of vegetation, or may evaporate from impervious surfaces such as pavements and roofs of buildings. Most of the precipitation, however, falls on the land. This water is disposed of in three ways. It is evaporated directly from the soil (flyoff), runs off the surface (runoff), or soaks into the soil (cutoff). Runoff water may be temporarily stored in swamps, ponds, lakes, and reservoirs; or it may reach the streams and finally flow to the ocean storage. Runoff is the cause of soil erosion and creats important problems in engineering and in agriculture, but cutoff water is the principal concern of the soil engineer.

9.3. GROUND WATER. Cutoff water, or water which soaks into the soil, percolates downward through the soil pores to replenish the ground water below the surface of saturation. The soil acts as a vast subsurface reservoir from which the swamps, lakes, and streams are fed during the periods between rainstorms when no surface runoff is available. The water stored in the soil is the source of supply for the growth of plants. Without it vegetation would perish. Animal life also draws heavily upon this underground storage water for sustenance, and mankind utilizes it extensively for domestic and industrial purposes. The importance of cutoff water stored in the soil to the existence and well-being of all forms of life on the face of the earth is at once apparent.

The cutoff portion of precipitation also affects many engineering structures adversely by reducing the bearing capacity of the soil and otherwise weakening it. Many problems in soil engineering involve a study of seepage quantities through earth dams, levees, and the sides of irrigation ditches, and a study of flow into underdrains and wells. It is highly important, therefore, that an engineer should have a working knowledge of the principles governing the flow and retention of water in the soil and the effect of water upon the strength and stability of this material.

9.4. LOCATION OF WATER TABLE. When cutoff water enters the soil at the time of precipitation, it tends to migrate downward to a zone of saturation where the soil pores are completely filled with water. Typical distribution of soil water is pictured in Fig. 9-2. The upper surface of the zone of saturation is called the *water table* or the *phreatic surface.* As will be pointed out in Chapter 10, the water table may be defined as the surface at which the pressure in the soil water is equal to atmospheric pressure. For the present purpose it may be thought of simply as the upper surface of the zone of saturation, although there is frequently a zone above the water table where the pores are completely filled with capillary water. The elevation of the water table at any point may be determined by digging a test well or bore hole. The elevation to which water rises in the hole marks the location of the water table, provided sufficient time has elapsed after digging to allow the water in the hole to come to equilibrium with the water in the soil. The time required for equilibrium to develop varies with the ease with which water flows through the soil, being less for highly permeable soils than for impermeable soils. A period of 24 hr will usually suffice to permit equilibrium to be established.

Contrary to a rather widespread misconception among laymen, the water table is not a horizontal surface like the surface of a body of still water. Rather, the water table rises higher under the hilltops and is lower adjacent to lakes, swamps, and streams. Hogentogler describes the water table as a "subdued replica of the surface of the ground," and such a description is very appropriate. Water tends to flow laterally and downward from points where the water table is high to points of lower eleva-

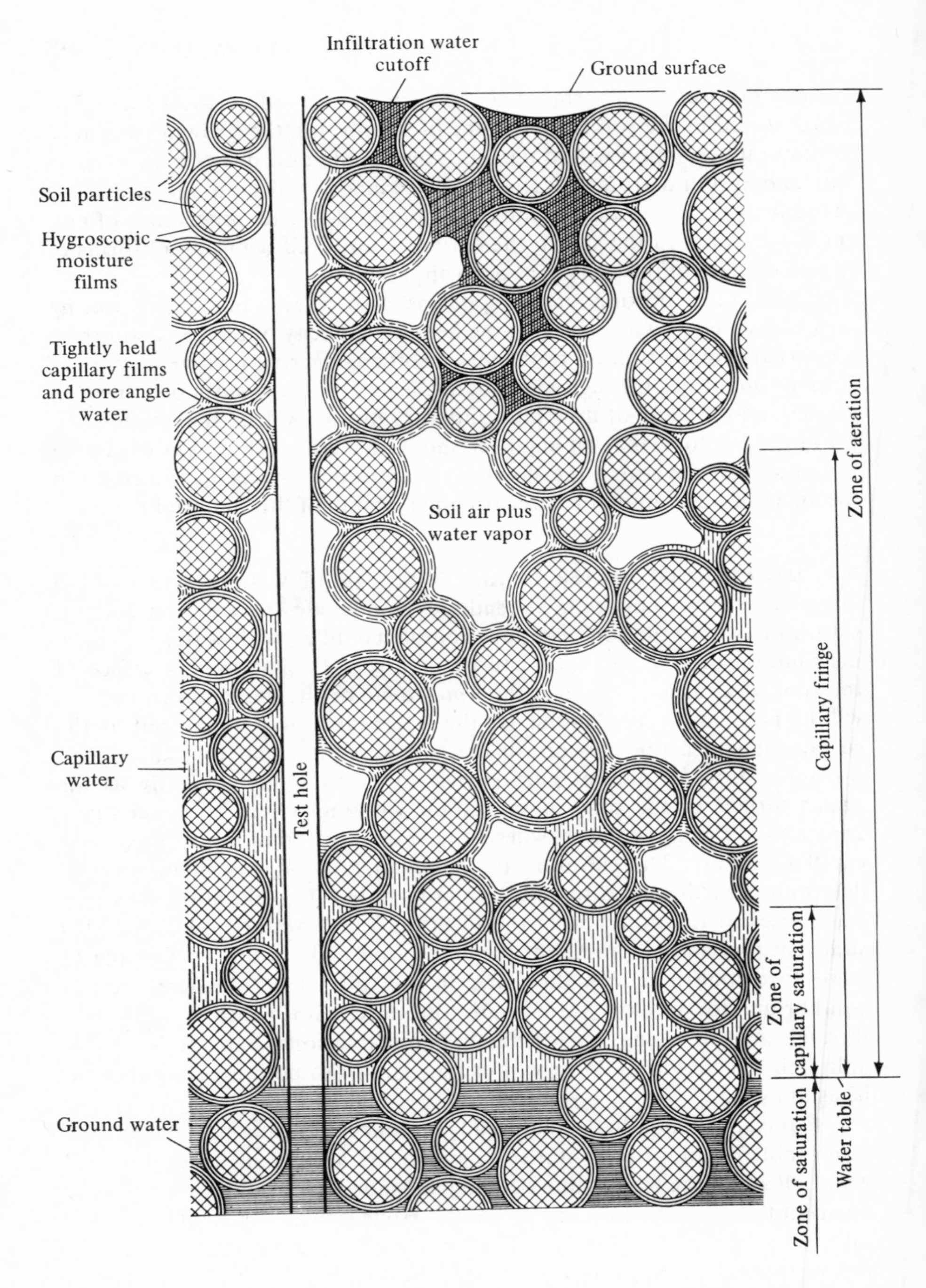

Fig. 9-2. Conventional illustration of soil-water distribution (adapted from Zunker).

tion, and it may emerge as seepage water which feeds the swamps and lakes and is the source of supply for streams during dry weather.

9.5. FLUCTUATIONS IN POSITION OF WATER TABLE. The water table fluctuates up and down, its elevation at any time depending on climatic conditions. During and immediately after a rainstorm, the cutoff water seeps downward and the water table rises closer to the ground surface at all points. Between rainstorms, gravity pulls the water toward the seepage outlets and the water table is lowered, particularly under the hilltops. During an extended period of drought, the water table may be lowered many feet at some locations. For these reasons, the water table does not remain at a fixed elevation at all times of the year, and it may vary over cyclic periods involving many years, its position depending on climatic variations. When the elevation of a water table is measured, therefore, it is desirable to indicate the time of year when the measurement is made and to take into account the normal seasonal variation for the geographical area. For example, in the corn belt of the United States, the water table is nearly always highest in the spring of the year and falls steadily during the summer, because of the high rate of evaporation and the rapid removal of moisture by growing vegetation. A low point is reached in the fall, and then the water table begins rising again toward its springtime peak.

9.6. IRREGULARITIES IN WATER TABLE. A water table may not be continuous in the lateral direction. As represented in Fig. 9-3, a somewhat dish-shaped stratum of impervious soil may exist above a main water table and cause an accumulation of water in the soil above. The upper surface of saturation of this accumulation is a true water table, but it is

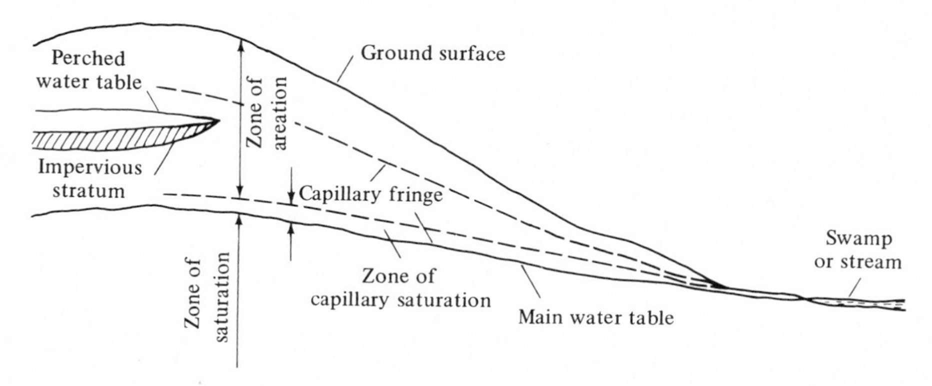

Fig. 9-3. Main and perched water tables.

called a "perched" or "hanging" water table to distinguish it from the main water table located at a lower elevation and extending laterally a much greater distance. Conversely, a main water table may be interrupted by the inclusion of a region of pervious soil surrounded by dikes of impervious material. These divergences from normality are not uncommon, and they indicate the necessity of measuring the water table at many points close together in connection with a soil survey.

9.7. EFFECT OF DRAINAGE SYSTEM ON WATER TABLE. A water table may be lowered artificially by providing an outlet for the ground water, at an elevation below its natural outlet, by means of open ditches or by tile underdrains. It must be realized, however, that the new water table will not be lowered to the elevation of the drains themselves, except right in the vicinity of the drains, because of the natural tendency for a water table to rise as the distance from an outlet or point of release increases. The steepness of a water table in the vicinity of a drain depends on the character of the soil, being relatively flat in a permeable soil and relatively steep in an impermeable soil, as indicated in Fig. 9-4.

For purposes of design of tile spacings, from Dupuit-Fochheimer Theory the ground-water surface between two tile lines may be taken as a portion of an ellipse (*6*):

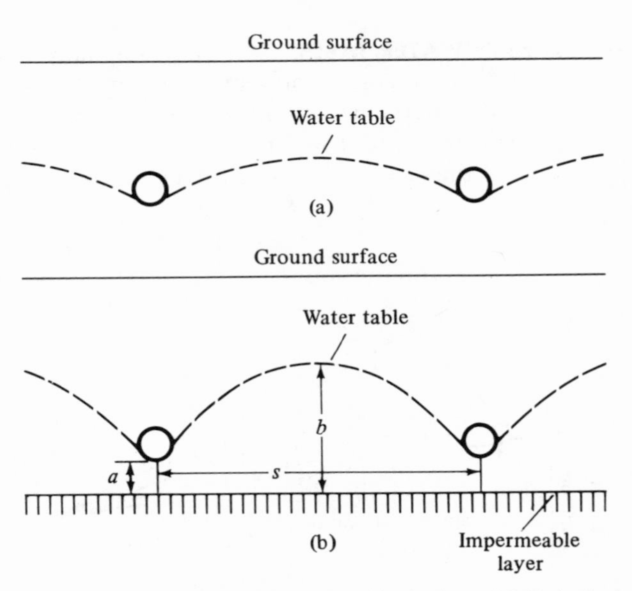

Fig. 9-4. Shape of water table adjacent to tile drains. (a) Relatively permeable soil; (b) less permeable soil.

$$\frac{S^2}{b^2 - a^2} = \frac{4k}{R} \tag{9-1}$$

where S is the tile line spacing, a and b are the depths to an impermeable barrier under the tile and half-way between the tile, respectively, k is the hydraulic conductivity of the soil, and R the infiltrating rainfall. (Fig. 9.4b).

EXAMPLE 9-1. A permeable soil layer 15 ft thick is to be drained so that the water table at its highest point is 5 ft below the ground surface. If $k = 200$ in./yr and $R = 50$ in./yr, suggest two alternate tile depths and spacings.

SOLUTION 1. Let the tile depth be 10 ft. Then $a = 15 - 10 = 5$ ft; $b = 15 - 5 = 10$ ft. Therefore $S^2 = (10^2 - 5^2)\,\dfrac{4(200)}{50} = 1200$. $S = 34.6$ ft.

SOLUTION 2. Let the tile spacing be 50 ft; what tile depth is needed?

$$\frac{(50)^2}{10^2 - a^2} = \frac{4(200)}{50}$$

A solution is not possible, indicating that this spacing is too high.

EXERCISE. Recalculate Solution 1 for a lower permeability soil, $k = 50$ in./yr.
Answer: $S = 17.3$ ft.

9.8. EFFECTS OF RAINSTORMS. The more or less normal regimen just described is interrupted and modified during and after a rainstorm, when the upper layers of soil are saturated. Under these conditions, gravity and capillarity pull the water downward to the zone of saturation and the ground water is thereby replenished. Then the reverse process begins anew. Thus, it is seen that the soil water is continually in motion and it seldom reaches a condition of equilibrium or a steady state, except where the situation may be modified by the construction of engineering works such as impervious pavements or other structures. The effect of these artificial situations will be discussed further in Chapter 10.

9.9. CLASSIFICATION OF CUTOFF WATER. Numerous attempts have been made from time to time by various soil scientists to classify the water in the soil into various categories or kinds, the classification depending on the action of various portions of the water or its degree of availability to plants. One of the simplest, and at the same time the most useful, of such classifications for engineering purposes is that proposed by Briggs, a former soil physicist with the U.S. Department of Agriculture and a former Director of the U.S. Bureau of Standards. Briggs classified the soil water into three categories, as shown in Table 9-1. Each class refers to the portion of the water which is primarily controlled in its movement and retention by a distinct kind of force.

TABLE 9-1. Briggs Classification of Soil Water

Kind of Water	*Primary Controlling Force*
Hygroscopic	Forces of adsorption
Capillary	Forces of capillarity
Gravitational	Force of gravity

Attention is directed to the role of the kinds of primary controlling forces in this classification, and it is emphasized that the water itself is not different in the various categories. For example, capillary water and gravitational water are exactly the same as far as chemical characteristics are concerned; but capillary water moves in the soil and is retained by it primarily in accordance with the laws of surface tension and capillary action rather than the law of gravity. Of course, gravity continues to act upon capillary water, since gravity is always present and active near the earth's surface. However, its influence on this category of soil water is relatively small in comparison with that of the capillary forces. Similarly, gravity is always active with reference to hygroscopic moisture, but its influence on water in this category is insignificant compared with that of the forces of adsorption. Any particular molecule of water in the soil may exist first as a part of one category and then as a part of another, as its physical environment changes and causes it to pass from one category to another.

9.10. ZONE OF AERATION AND CAPILLARY FRINGE. In the region between the water table and the surface of the ground, the pores of the soil contain varying amounts of air and water, and this region is referred to as the *zone of aeration.* Within this zone, water normally is raised for some distance above the water table and held there by the forces of capillarity. This region above the water table within which water exists in the capillary state is called the *capillary fringe.* These regions are indicated in Figs. 9-2 and 9-3.

Within both the zone of saturation and the capillary fringe, the hygroscopic or film moisture tightly adheres to the fine soil particles. Above the capillary fringe, the hygroscopic moisture may be the only water present in the soil. Also, in this upper region the film moisture and capillary water may be converted into water vapor and become a part of the soil air. Surface-air currents pick up this vapor, and thus the soil dries out as more water is raised from the zone of saturation and is converted into vapor and carried away.

9.11. EFFECTS OF COLLOIDAL FRACTION ON SOIL PROPERTIES. The study of hygroscopic moisture in soil, and of the adsorptive forces which primarily control it, is to a very large extent a study of the colloidal fraction of the soil. The term "colloid" is derived from a Greek word meaning "glue-like," since colloidal material, when concentrated, is a mass of gluey, gelatin-like material. The colloidal fraction in soil is largely responsible for its plasticity characteristics in the wet state and for the bonding action which causes a plastic soil to solidify into hard masses or clods when dried. The colloidal fraction in a stabilized soil mixture binds the material together; and, with the internal friction characteristics of the mass, it contributes to the stability of the mixture. The bonding action of a clay admixture in a molding sand is largely caused by the effect of the colloidal fraction of the admixture.

9.12. COLLOIDAL CHARACTERISTICS. The term colloidal size was used in Chapter 7 to indicate the group of particles in a soil which are less than 0.001 mm in diameter. This is a purely conventional use of the word, since some materials may exhibit colloidal characteristics when the particles are larger than 1 micron and others when the particles are smaller. A much more accurate concept of colloidal properties is based on a comparison of gravitational forces acting on the particle to forces from molecular impacts from the suspending medium. The suspending medium ordinarily is water, the water molecules being in constant agitation due to their thermal energy. If the soil particle is small enough to be significantly moved by impacts from water molecules it tends to remain in suspension and is said to be colloidal; if, however, gravity forces dominate and the individual particle settles out of suspension it is not colloidal.

The motion imparted to colloidal particles by molecular impacts may be seen by viewing a suspension under a high-powered light microscope. The movement is referred to as *Brownian movement*. It is mostly vibrational, but also includes random dislocations, since during a very short time interval there will tend to be more impacts acting on one side than on another.

Colloidal particles also carry an electrical surface charge due to surface chemical effects and charge deficiencies within their crystal structure, as discussed in Chapter 2. This charge, if high and consistent enough, causes particles to repel one another and remain dispersed. Addition of an acid or soluble salt to a clay suspension may change the surface charge in such a way as to either increase the charge, improving dispersion, or decrease the charge, allowing particles coming close to one another to become attracted to each other, forming *flocs*. Such electrical attractions are in part due to the nonuniform nature of the charge distribution on a clay particle, causing edge-to-face bonding. Many colloidal particles join together to form a single floc. The floc, being large, is no longer colloidal, and settles out of suspension. Such *flocculation* occurs in nature when muddy river water enters an ocean, the suspended clay rapidly flocculating and settling to form a delta. This explains why salt water deltas tend

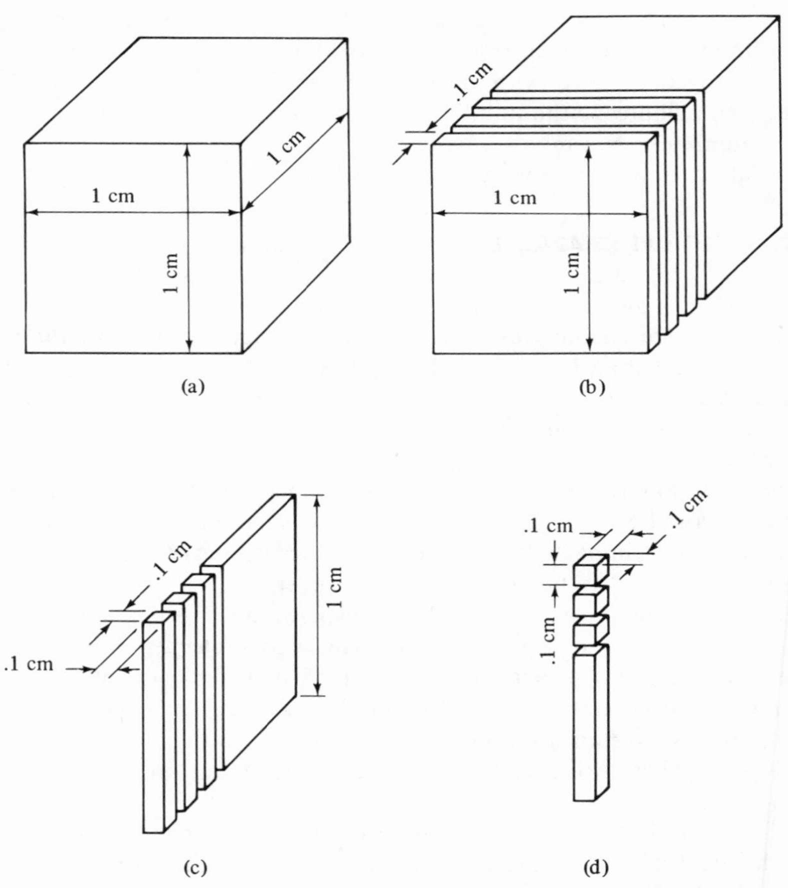

Fig. 9-5. Increases in surface area and edge and corner discontinuities which accompany a decrease in particle size.

to be clayey whereas in fresh water the clays remain suspended and are washed farther out.

9.13. RELATION OF SURFACE AREA AND MASS OF PARTICLES. From the preceding discussion it is seen that colloidal properties are closely associated with the relationship between surface area and mass of a group of particles. It is worthwhile, therefore, to explore this relationship thoroughly in order to understand clearly how these factors vary. Consider a cube of material each side of which is 1 cm long, as indicated in Fig. 9-5(a). The volume of this cube is 1 cc, and the surface area is 6 sq cm. Also the total edge distance is 12 cm, and there are 8 corners on the cube. Now, imagine that this cube is divided into 10 equal slices, as shown in Fig. 9-5(b). The total volume of material remains the same, but all the other factors are increased. The total surface area is now 24 sq cm; the total edge distance is 84 cm; and the total number of corners is increased to 80. Also, imagine that each slice is cut into 10 equal bars, as indicated in Fig. 9-5(c); and each bar is then cut into 10 equal cubes, as indicated in Fig. 9-5(d). Each of these subdivisions greatly increases the surface area, the total length of the exposed edges, and the number of corners associated with the original volume of material, which has remained unchanged. Imagine, further, that each of the smaller cubes is subdivided in the manner just described until exceedingly small cubes are obtained. The increases in the surface area, total length of exposed edges, and number of corner discontinuities become literally enormous, as is shown in Table 9-2.

TABLE 9-2. Increase in Surface Area, Free Edge Length, and Number of Exposed Corners when a 1-cm Cube of Material Is Divided into Equal Cubes of Various Sizes

Property	*Size of Cube (mm)*					
	10	*1*	*0.1*	*0.01*	*0.001*	*0.0001*
Number of cubes	1	10^3	10^6	10^9	10^{12}	10^{15}
Total volume (cc)	1	1	1	1	1	1
Volume of each cube (cc)	1	10^{-3}	10^{-6}	10^{-9}	10^{-12}	10^{-15}
Total surface area (sq cm)	6	60	600	6,000	60,000	600,000
Total free edge length (cm)	12	12×10^2	12×10^4	12×10^6	12×10^8	12×10^{10}
Total number of corners	8	8×10^3	8×10^6	8×10^9	8×10^{12}	8×10^{15}

The table shows that when a 1-cm cube is subdivided into equal cubes 0.0001 mm on a side, which is within the size range of colloidal particles in soil, the volume of each comminuted particle is infinitesimally small and the effect of gravity on its mass becomes negligible. On the other hand, the surface area and other discontinuities such as free edge length and number of corners become relatively very large. Since the surface, edges, and corners of a particle are the seat of certain electrostatic forces, the nature of which will be discussed later, the behavior of the very small particles is primarily controlled by these surface charges, and the material is said to be colloidal in character.

9.14. RELATION OF DIAMETER OF SPHERES TO SURFACE AREA OF MASS. The increase in surface area of a mass of particles as the size of units decreases can be further illustrated by considering a mass of spheres. In Chapter 8 it was shown that the total solid volume of the balls in a cubic-foot container remained constant regardless of their diameter, provided the general arrangement was not changed. In contrast to this fact, it can be shown that the total surface area of such a mass of balls greatly increases as the diameter of the balls decreases. As a typical case, consider the arrangement shown in Fig. 8-1.

Let

L = length of each side of a cubical container;
d = diameter of each sphere;
N = number of spheres in container;
a = surface area of one sphere = πd^2; and
A = total surface area of all spheres.

Then

$$N = \left(\frac{L}{d}\right)^3 \tag{9-2}$$

Also, since the total surface area equals the number of spheres times the area of one sphere,

$$A = \frac{\pi L^3}{d} \tag{9-3}$$

This equation shows that the total surface area is inversely proportional to the diameter of the spheres.

9.15. SPECIFIC SURFACE OF COLLOIDAL MATERIAL. The relationship between the surface area and the mass of a colloidal material is indicated by a quantity called the *specific surface*. This is defined as the surface area per unit volume of the solid particles.

Let s be the specific surface.

Then, for uniform spherical particles of diameter *d*,

$$s = \frac{\pi d^2}{\pi d^3/6} = \frac{6}{d} \quad \text{square units per cubic unit} \tag{9-4}$$

For uniform cubical particles, where *d* is the length of each side of a cube,

$$s = \frac{6d^2}{d^3} = \frac{6}{d} \quad \text{square units per cubic unit} \tag{9-5}$$

Both Eq. (9-3) and Eq. (9-4) show that the specific surface increases as the size of the particles decreases. Various kinds of materials begin to display colloidal characteristics at different values of specific surface, the value for a particular material depending on the chemical composition of the material and on other factors. For material found in soil, colloidal characteristics usually begin to be manifest when the specific surface reaches values from 60,000 to 100,000 sq cm/cc.

9.16. CAUSES OF ELECTRICAL CHARGES. As already mentioned, electrical charges associated with clay mineral surfaces are caused by two factors, the first related to the fact that minerals are composed of ions, or electrically charged atoms, held together in a mass by their electrical charges. For example, table salt is composed of positive sodium ions alternating with and held in place by negative chlorine ions. Such crystals are not molecules (although molecules may form crystals), because whereas molecules are discrete, as in O_2 or CO_2, crystals have no predetermined boundaries. Any edge or surface of an ionic solid, whether crystalline or glassy, therefore is comprised of an array of unsatisfied electrical charges.

The second cause for electrical charges on clay particles was mentioned in Section 4.9, and in Table 4-5. This reason is the clay crystal structure itself, which in many clays gives a net negative charge. If the charge is high, as in the micas, it is balanced in this case by potassium ions more or less permanently adsorbed between the layers, holding them together. If the charge is less, as in montmorillonites, no cations are held strongly enough to bind the structure together against ingress of water; the layers readily separate and the clay expands.

Idealized formulas for a mica and a montmorillonite are shown in Table 9-3. The Roman numeral susperscripts indicate whether cations are in the A tetrahedral layer (IV) or the B octahedral layer (VI). Dotted lines connect ionic substitutions which give rise to the negative charge with positive ions which balance that charge.

The montmorillonite formula shows some other pertinent properties

TABLE 9-3. Idealized Chemical Formulas for Common Clay Minerals

Name	*Formula*	*Negative Charge per Formula Unit*
Kaolinite	$(Si_4)^{IV}(Al_4)^{VI}O_{10}(OH)_8$	0
Muscovite (mica)	$K_2(Si_6Al_2)^{IV}(Al_4)^{VI}O_{20}(OH)_4$	2
Montmorillonite	$(Si_8)^{IV}(Al_{3.33}Mg_{0.67})^{VI}O_{20}(OH)_4 \cdot nH_2O$ ⋮ $Na_{0.67}$	0.67

of this clay mineral: First, the positive ion is shown below the line, since it is exchangeable. (It is shown as a fractional ion merely to show the charge, since fractional ions do not really exist.) Secondly, at the end of the formula is $\cdot nH_2O$, indicating variable amounts of interlayer water, which depend on the relative humidity. If an excess of a calcium solution were added to this clay, calcium ions would exchange for the sodium and would be shown in the formula. As the clay dries out, n changes.

The formulas in Table 9-3 are idealized and not at all representative, because clays in nature are chemically even more complex than shown in these formulas. For example, montmorillonite group minerals often have some Al^{+3} substituting for Si^{+4} in the tetrahedral layer, and Fe^{+2} for Al^{+3} in the octahedral layer. In fact if enough Al is substituted, the chemical composition may be close to that for kaolinite, even though properties of these two minerals are very different. The chemical composition of a soil therefore is not nearly so important as the mineralogical composition.

9.17. DIPOLES. The discussion so far has centered on the electrostatic nature of minerals, their surfaces, and their exchangeable ions. We shall now see that water molecules also exhibit relatively weak electrostatic attractions and repulsions, even though the -2 valence of each oxygen atom is fully balanced by two $+1$ charged hydrogen atoms. The reason for the electrostatic behavior of water is shown in Fig. 9-6: The two hydrogen atoms are not located on opposite sides of the oxygen atom, but are both on the same side, creating a positive end to the molecule. The other end, being locally deficient in positive charge, is negative.

Although the analogy is not strictly accurate, it is convenient to think of a dipolar molecule as being similar to a small magnet, which is positive at one end and negative at the other, as illustrated in Fig. 9-6. When such dipolar molecules enter the vicinity of an electrostatic-force field surrounding an ion, they tend to orient themselves with the positively charged ends toward negative ions and the negatively charged ends toward positive ions, just as a compass needle orients itself. For this

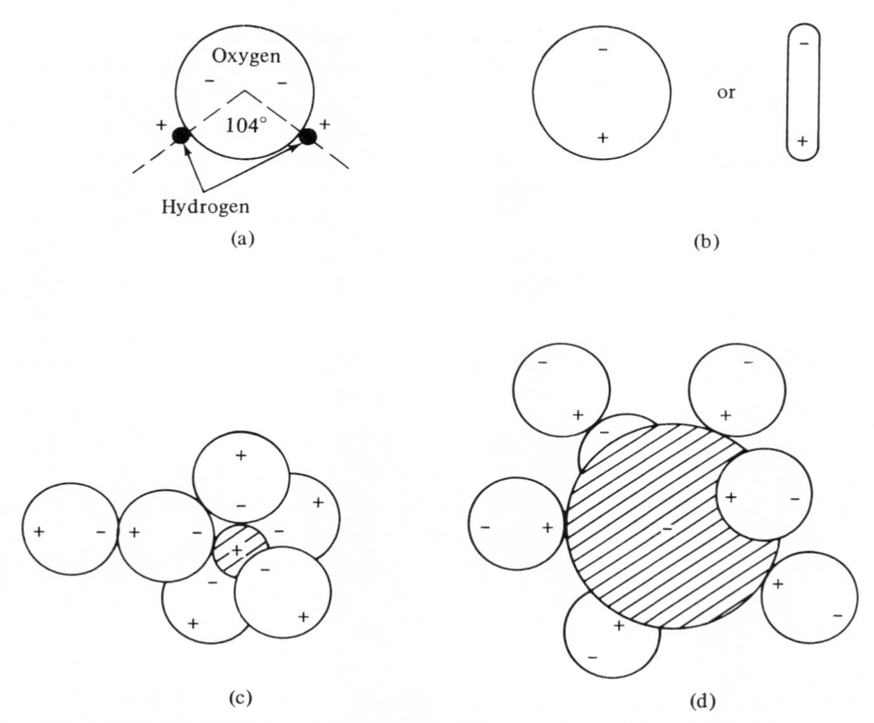

Fig. 9-6. Dipolar character of water molecule and attraction of dipoles to positive and negative ions. (a) Arrangement of atoms in water dipole; (b) symbols for water dipole; (c) clustering of dipoles around a sodium cation; and (d) adsorption of dipoles around a chlorine anion.

reason water may be adsorbed both on the surface of a soil colloid, which is negatively charged, and on the surfaces of its adsorbed ions, which are positively charged.

9.18. HYGROSCOPIC WATER. The water adsorbed on soil colloids and their associated ions is the hygroscopic water. This water is held to the surface of the colloids with very great force, which in extreme cases may be as much as 10,000 atm (142,000 psi). Under these extremely high pressures the water exists in a solidified state. However, the adsorptive force diminishes rapidly as the distance from the surface of the colloid becomes greater, and the total thickness of the adsorbed film may be that of only a few molecular layers.

9.19. HYGROSCOPIC MOISTURE CONTENT OF SOIL. When wet soil is dried in air in the laboratory, moisture is removed by evaporation until the hygroscopic moisture in the soil is in equilibrium with the moisture vapor in the air. The amount of moisture in air-dried soil, expressed as a percentage of the weight of the dry soil, is called the hygroscopic moisture content. This is a useful quantity in connection with certain laboratory tests and procedures, but it must be recognized that the hygroscopic moisture content is a variable quantity for the same soil, its value depending on the temperature and humidity of the laboratory air. Hygroscopic moisture films may be driven off from air-dried soil by heating the material to the temperature of boiling water in an oven until the weight becomes constant. Although water still remains in the soil, even after it is thus dried to constant weight, the amount is small and such soil is considered to be dry.

9.20. HEAT OF WETTING AND HYGROSCOPICITY. When water is brought into contact with oven-dry soil, the first increments of moisture are adsorbed on the colloidal fraction until the adsorptive forces are satisfied. Since heat is required to remove adsorbed moisture films from the colloidal fraction, it follows that heat is produced during the reverse process of film development. This evolved heat is known as the *heat of wetting.*

9.21. ELECTRICAL DRAINAGE OF SOILS. The exchangeable nature of adsorbed cations together with the dipolar nature of water have led to some interesting applications of electrical drainage. This is done where conventional drainage is not possible because of small pore size, low permeability, and the tightly bound nature of the water in the soil. Electrical drainage of clayey soils was first utilized in 1939 by L. Casagrande (*4*) to stabilize sliding soil in a railway cut in Germany. Since then it has been used in numerous applications to stop landslides and dewater foundations under difficult circumstances.

The process of electrical drainage is called *electroosmosis.* Since cations on clay surfaces are exchangeable, they can be moved with d.c. electricity, and since water is adsorbed on the cations [Fig. 9-6(c)], when cations are moved the water moves along. A schematic illustration of electroosmosis is shown in Fig. 9-7. A perforated pipe is used as a cathode; cations reaching the cathode release their water which builds up pressure until it flows out the top of the pipe. The anode, on the other hand, must supply metallic cations; iron waterpipe is usually used and must be replaced periodically.

Figure 9-8 shows why electroosmosis works in clays where conventional drainage does not: With conventional well points the hygroscopic water does not move, whereas it can be moved electrically. Ordinary d.c.

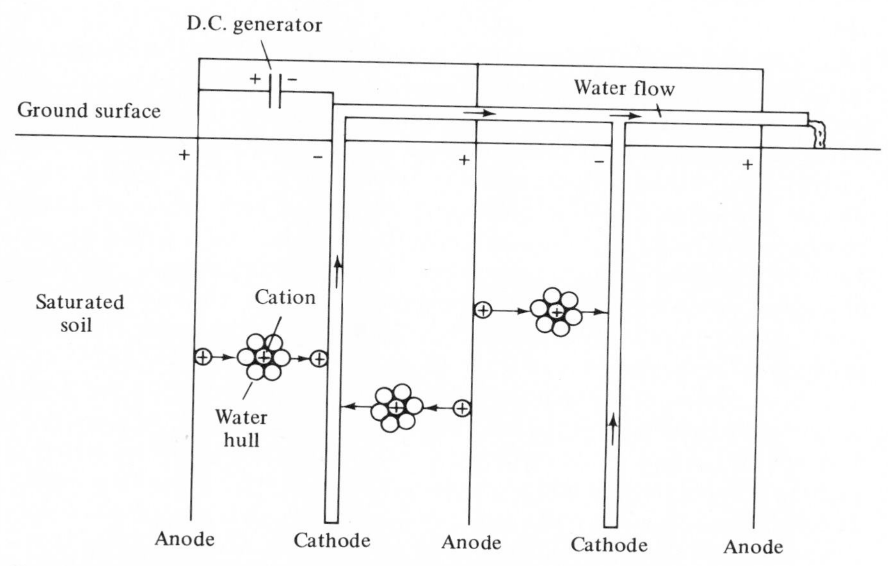

Fig. 9-7. Schematic of electrical drainage of clay soils (no pumps).

welders can be used to supply power, and electrodes are spaced to give a maximum voltage gradient of about 1 in. Because of its higher cost, electrical drainage is not used where conventional or vacuum well points will do the job.

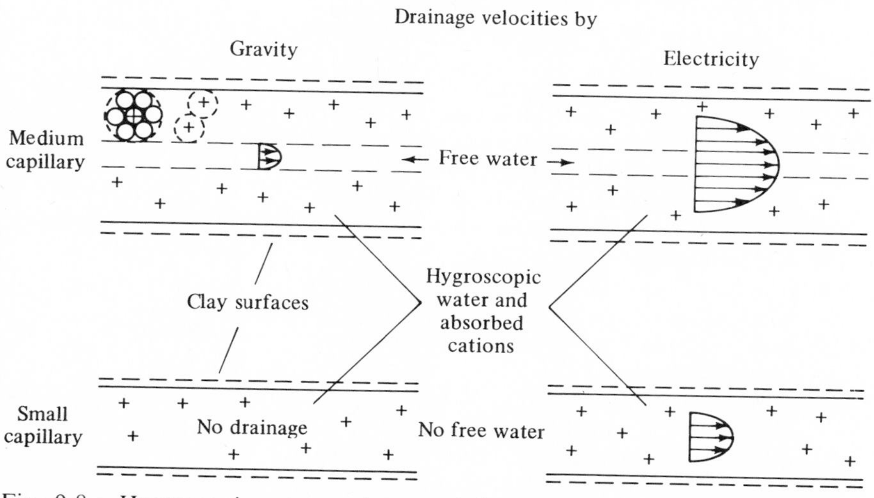

Fig. 9-8. Hygroscopic water can be moved electrically by moving the cations.

9.22. CHARACTERIZATION OF THE SOIL-WATER SYSTEM. Although x-ray diffraction and other elaborate means for analysis are the basic tools for characterization of soils, the relations between character of the soil water and clay content, clay minerals, cation exchange capacity, and identity of adsorbed cations, establish important aspects of readily measurable soil behavior. For example, soils with a high cation exchange capacity necessarily contain many adsorbed cations, each with its water hull. Such soils have a high hygroscopic moisture content and, even more important, when they are wet they are sticky, because of the ionic attractions. Dried, they shrink excessively. Soils with a low cation exchange capacity show the opposite of these effects.

At the present time engineers rely to a very large extent on the Atterberg consistency limits of the fraction of soil passing a No. 40 sieve (0.42 mm) for indications of the surface chemical characteristics of the material, such as plasticity or friability, shrinkage and swell, and bonding power. The Atterberg limits are the *liquid limit*, *plastic limit*, and *plasticity index*. These properties are defined and discussed in detail in Chapter 13. Their relationship to engineering problems, particularly those in the construction of airfield runways and roads, is purely empirical; and considerable judgment and experience are required in their application.

PROBLEMS

9.1. Give two important examples of the influence of water on the strength and stability of soil.

9.2. Define and describe the hydrologic cycle.

9.3. What are the three principal ways in which precipitation is disposed of after falling to the earth?

9.4. Outline the Briggs classification of soil water. Describe each category and state clearly the basis for classification.

9.5. Describe a main water table and a perched water table.

9.6. Distinguish between the zone of saturation, the zone of aeration, the capillary fringe, and the zone of capillary saturation.

9.7. Discuss two main natural processes by which the elevation of a water table may be lowered.

9.8. How may a water table be lowered artificially?

9.9. What causes a water table to rise? Explain in detail.

9.10. After studying Fig. 9-4, discuss the influence of soil permeability on the spacing of tile drains.

9.11. A line of three vertical drill holes spaced 20 ft apart indicates water table depths of 15, 4, and 13 ft, respectively. Suggest several possible explanations, assuming the ground surface is horizontal.

9.12. Rewrite the formula for montmorillonite in Table 9-3 with calcium (Ca^{+2}) as the exchangeable cation.

9.13. The formula for talc (used in talcum powder) is $(Si_8)^{IV}(Mg_6)^{VI}O_{20}(OH)_4$. (a) What is the cation exchange capacity from ionic substitutions in the crystal structure? (b) Would you expect this clay to become sticky in water? Why (not)?

9.14. An idealized formula for vermiculite is

$$Mg_{0.65}{}^{+2}(Si_{6.7}Al_{1.3})^{IV}(Mg_6)^{VI}O_{20}(OH)_4 \cdot 9\,H_2O$$

(a) Compare this to the formula for talc in Problem 9.13(a), and connect a line from the ionic substitution giving rise to a negative charge to the exchangeable ion balancing the charge. (b) What is the negative charge per formula unit? (c) Comparing this negative charge with those in Table 9-3, would you expect vermiculite to behave more like mica, montmorillonite, or an intermediate between these two?

9.15. The cation exchange capacity is usually expressed in milliequivalents (mEq) per 100 gm of soil. (a) Calculate the formula weight of kaolinite from the formula given in Table 9-3.

Answer: $4(28) + 4(27) + 10(16) + 8(16) + 8(2) = 524.$

(b) Calculate the formula weight for the montmorillonite in Table 9-3, omitting the water. Approximate atomic weights are Si = 28, Al = 27, Mg = 24, Na = 23, O = 16, H = 1. (c) The cation exchange capacity being 0.67 equivalent weight = 670 mEq, express the c.e.c. in mEq/100 gm.

9.16. The measured cation exchange capacity of kaolinite is usually 3 to 15 mEq/100 gm. (a) How much of this is due to negative charge arising from ionic substitutions? (b) How would you expect the c.e.c. of kaolinite to be affected by grinding? (c) Will electrical drainage work in a kaolinite soil? Why (not)?

REFERENCES

1. Baver, L. D. *Soil Physics*, 3rd ed. John Wiley, New York, 1956.
2. Bjerrum, L., J. Moum, and O. Eide. "Application of Electro-osmosis to a Foundation Problem in a Norwegian Quick Clay." *Geotechnique* **17,** 214–235 (1967).
3. Briggs, L. J. *The Mechanics of Soil Moisture.* U.S.D.A., Bureau of Soils, Bulletin 10, 1897.
4. Casagrande, L. "Review of Past and Current Work on Electro-osmotic Stabilization of Soils." Harvard Soil Mech. Ser. No. 45, rev. 1957.
5. Grim, R. E. *Clay Mineralogy.* McGraw-Hill, New York, 1953.
6. Kirkham, Don and W. L. Powers. *Advanced Soil Physics,* 91. Wiley-Interscience, New York, 1972.
7. Van Olphen, H. *An Introduction to Clay Colloid Chemistry.* Interscience, New York, 1963.

10

Capillary Water

10.1. CAPILLARY FORCES. The movement and retention of water in the capillary fringe above a ground water table is similar in many respects to the rise and retention of water in a capillary tube, although there also are important differences between the two cases. Because of the similarity with this commonly observed capillary-tube phenomenon, it is convenient and appropriate to introduce the subject of capillary water in the soil by reviewing the action of water in a capillary tube and studying the forces which control this action. Such forces, which are usually referred to as capillary forces, are dependent on the surface tension of the water (or other liquid), on the pressures in the water in relation to atmospheric pressure, and on the size and conformation of the tube in which the water rises or, in the case of soil, on the size and conformation of the soil pores.

10.2. SURFACE TENSION. The surface tension of a liquid is a property which is associated with the resultant force at the air-liquid surface caused by the fact that molecular attractions on molecules at the surface of the liquid differ from the molecular attractions on the molecules in the interior of the mass of liquid. All molecules, when in close contact with each other, are held together by molecular attraction and the amount of this attractive force depends on the kinds of molecules involved. For example, the amount of force with which two molecules of water will at-

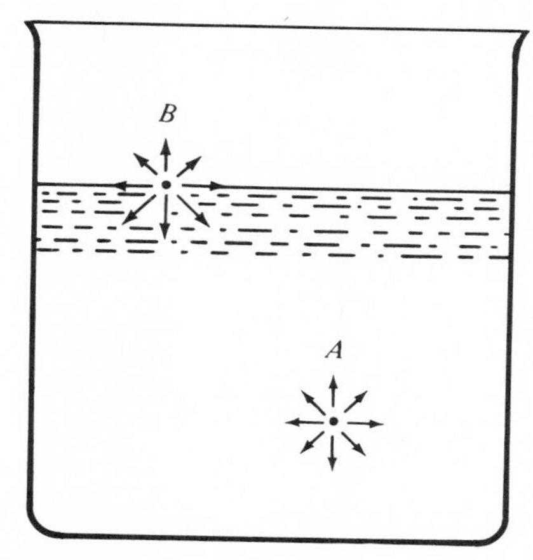

Fig. 10-1. Surface tension.

tract each other will differ from the amount of force developed by a molecule of water and a molecule of air. Thus, as indicated in Fig. 10-1, the set of forces acting on a particle in the interior of a mass of water, as at point A, differs from the set of forces acting on a particle at B in the air-water surface. This different force situation on the surface particles and the interior particles causes a resultant force which is directed inward. As a result the liquid surface tends to contract and to occupy the least possible area consistent with the volume of the mass and the shape of a vessel in which the water is placed. The force required to balance this tendency to contract is called surface tension. It constitutes an important physical property of a liquid, and is expressed as a force per unit width of surface. Typical units are dynes per centimeter.

Surface tension forces act in such a manner that the surface of the liquid behaves as though it were covered with a tightly stretched membrane or skin. Therefore, surface tension is sometimes defined as the tensile stress in such a membrane. This membrane analogy is a very appropriate and useful concept in helping to explain surface-tension phenomena. In fact, many people who observe conditions at an air-liquid surface have been led to believe that such a membrane actually exists; but this opinion is false.

10.3. LEVEL OF WATER IN CAPILLARY TUBE. When a small glass tube is inserted into a vessel of water which is open to the atmosphere, water will rise in the tube and come to rest at some definite height above the free water surface, as indicated in Fig. 10-2. The height of rise can be expressed by a formula derived by equating the force which holds the water

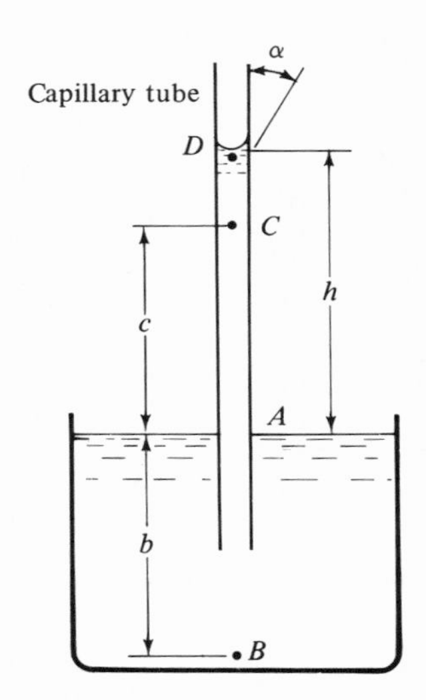

Fig. 10-2. Pressure and tension (negative pressure) in water.

up in the tube to the weight of the column of water above the free water surface. Thus

$$2\pi r T \cos \alpha = \pi r^2 h d g$$

from which

$$h = \frac{2T}{rdg} \cos \alpha \tag{10-1}$$

in which

r = radius of a circular capillary tube, in centimeters;
T = surface tension of liquid, in dynes per centimeter;
α = angle of contact between the meniscus and the wall of the tube;
h = height of rise of the liquid, in centimeters;
d = density of the liquid, in grams (mass) per cubic centimeter;
g = acceleration of gravity = 980 cm/sec/sec.

In this equation, the contact angle is dependent on the affinity between the liquid and the material of which the capillary tube is made. This affinity is sometimes referred to as the degree of wetting, or the wetability, of the solid and the fluid; and the contact angle may be called the wetting angle. For example, pure water wets clean glass perfectly and the angle of contact equals zero for these materials. In other words, the meniscus is tangent to the side of the tube at their line of contact. On the other hand, if a glass tube is dirty or covered with a film of grime, water does not wet the surface completely and the angle of contact will be some finite value whose cosine is less than unity. In accordance with Eq. (10-1), the height of rise of water in such a tube will be less than the height of rise in a clean tube.

For some combinations of liquid and solid, the lack of affinity between the materials may be so great that the angle of contact will be greater than 90°. In such a case cos α will have a negative value and the height of rise will be negative; that is, the liquid surface inside the tube is at equilibrium at an elevation below the free liquid surface outside the tube. This condition exists in the case of mercury and glass, it being well known that mercury is depressed when a glass tube is inserted in a vessel of this liquid.

The importance of the affinity or degree-of-wetting concept in relation to soil lies in the fact that all minerals are not wetted to the same extent by water. Therefore, the capillary characteristics of soil are influenced by the mineralogical character of the soil grains which comprise the walls of the soil pores or capillaries.

10.4. COMPRESSIVE STRESS DUE TO CAPILLARY ACTION. Another matter of interest in the capillary-tube phenomenon is the fact that the rise of the liquid causes a compressive stress in the walls of the tube. In Fig. 10-2, for example, the column of capillary water is held up by the tube; and, by virtue of this fact, the walls of the tube are subjected to a stress equal to the weight of the column. It is as though a man were to hang by his hands inside a chimney. The chimney supports the man, and the reaction to his weight causes a compressive stress in the walls of the chimney. Similarly, in the case of soil, the capillary water in the soil pores sets up compressive stresses in the soil skeleton which are directed inward and which contribute to the strength and stability of the soil mass. This capillary-induced strength is temporary and ephemeral in character and may disappear entirely if the soils becomes saturated, since saturation eliminates the capillary menisci.

10.5. CAPILLARY PRESSURE. When water rises to equilibrium in a capillary tube, the pressures existing in the water and in the adjacent atmosphere are of interest. It is customary in capillary-water considerations to take the free water surface as a datum of zero pressure and to call pressures greater than atmospheric positive. Then pressures less than atmospheric are negative; and water acted upon by a negative pressure is said to be under tension. The word tension in this case indicates a pressure deficiency in relation to atmospheric pressure. In soils the pressure deficiency also is referred to as negative pore pressure.

Consider the conditions illustrated in Fig. 10-2. The free water surface at A is acted upon by atmospheric pressure. Also, since the capillary tube is open at the upper end, the upper surface of the meniscus is at atmospheric pressure. The pressure at any point B below the free water surface is equal to the weight of a column of water of unit cross section extending from the level of B to the water surface. On the other hand, a negative pressure or tension exists at any point within the tube above the water surface. Thus, the tension at any point C is equal to the weight of a unit column of water extending from C to the free water surface. From these considerations it is evident that a tension exists in the water at the under side of the meniscus, as at D. The numerical value of this tension is equal to the product of the unit weight of water and the height or rise, or $\gamma_w h$, and it represents the difference between the pressures on the outside and on the inside surfaces of the meniscus. The curvature of the meniscus is directly related to this pressure difference or tension.

10.6. TENSION ACROSS SPHERICAL MENISCUS. When a thin flexible membrane is subjected to different pressures on its two sides, equilibrium requires that the membrane assume a curved shape and that it be subjected to a tensile stress. Since a capillary meniscus acts in all respects like a thin stretched membrane, this general law of physics may be utilized to determine the relationship between the curvature of a meniscus, the surface tension of the capillary liquid, and the pressure difference across the meniscus, that is, the tension in the liquid inside the meniscus.

In Fig. 10-3 is represented a half-section of a meniscus in a circular capillary tube. The pressure on the outside (concave) surface is atmospheric pressure, which is represented by p_a; that on the inside is some lesser value, which is represented by p_1. The pressure deficiency across the meniscus is the tension in the capillary water and is represented by p. Then

$$p = p_a - p_1$$

Also, let T be the surface tension of the liquid and let R be the radius of the meniscus. Equating the vertical components of the forces, we obtain

$$2\pi R^2 p \int_0^\theta \sin\phi \cos\phi d\phi = 2\pi R \sin^2\theta \, T \qquad (10\text{-}2)$$

from which

$$p = \frac{2T}{R} \qquad (10\text{-}3)$$

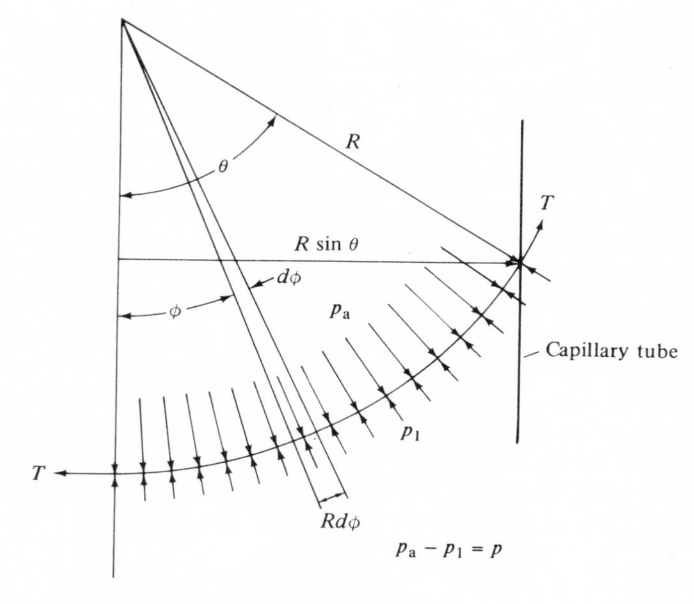

Fig. 10-3. Cross section of capillary meniscus.

Equation (10-3) represents the relationship between the tension or pressure deficiency in capillary water, the surface tension of the liquid, and the curvature of the meniscus in a circular capillary tube in which the meniscus is a segment of a sphere having the same curvature at all points.

10.7. TENSION ACROSS MENISCUS OF ANY SHAPE. In a tube that is not circular, and particularly in a soil capillary, the meniscus is not spherical in shape, but it may be warped or saddle-shaped surface having an infinite variety of curvatures. The equation for soil-water tension in such a case is

$$p = T\left(\frac{1}{R_1} + \frac{1}{R_2}\right) \tag{10-4}$$

in which R_1 and R_2 are radii of curvature of a warped surface in two orthogonal principal planes and the expression $[(1/R_1) + (1/R_2)]$ is called the total curvature of the surface. Thus, it may be said that the tension or negative pore pressure in capillary water is equal to the surface tension of the water times the total curvature of the meniscus or air-water interface.

10.8. ATTRACTION OF SOIL FOR WATER. Capillary water in soil seeks to adjust itself to the state of pressure equilibrium represented by Eq. (10-4).

By virtue of this fact, the soil is said to have an attraction for water. This capillary attraction of the soil for water is exerted in all directions. When water rises above a water table, the capillary attraction draws it up in opposition to the pull of gravity. When cutoff water soaks into the ground and migrates downward toward a water table, gravity and capillarity combine to cause the downward flow. Capillary water may also move horizontally in soil. This movement is not influenced by gravity, however, because the force of gravity does not have a horizontal component.

10.9. CAPILLARY POTENTIAL. The attraction which soil has for capillary water may be expressed quantitatively by a stress property called capillary potential. It is defined as the work required to pull a unit mass of water away from a unit mass of soil and represents the security or tenacity with which the soil holds capillary water. Gardner (*3*) has demonstrated that the capillary potential at any point in a soil mass is numerically equal to the tension in the soil water at the point, but is of the opposite sign. Thus, if ψ represents the capillary potential,

$$\psi = -p = -\gamma_w h = -T\left(\frac{1}{R_1} + \frac{1}{R_2}\right) \tag{10-5}$$

It is seen that capillary potential is always a negative quantity and can never have a value greater than zero, since the pressure in capillary water is always less than atmospheric pressure. It is a true potential, and flow of moisture through an unsaturated soil is similar in many respects to the flow of electricity through a wire or the flow of heat through a wall, the flow in each case being from regions of high potential to regions of lower potential. Since capillary potential is always a negative quantity, the phenomenon of flow of capillary water in soil is analogous to the flow of heat in a conductor when all the temperatures involved are below zero. *Capillary potential will be referred to in the algebraic sense; that is, the potential increases when it changes from −10 to −5 to 0.* Also it is pointed out that a low capillary potential is equal in magnitude but opposite in sign to a high value of tension in the soil water. This is readily visualized if it is recalled that the moisture in a drier soil is under relatively high tension and moisture flow is toward the drier soil, other factors being the same. Therefore, the capillary potential is low. On the other hand, in a wet soil the tension in the water is low and the capillary potential is high. In a saturated soil, the capillary potential is zero, this being the highest possible value of that property.

10.10. METHODS OF EXPRESSING CAPILLARY POTENTIAL. Capillary potential may be expressed in any units which are appropriate for indicating pressure, such as psi (pounds per square inch) or gm per sq cm. Also, it may be convenient in certain cases to express the potential in terms of the height of a column of liquid, such as feet of water, centimeters of water, or inches of mercury. Since the possible values of capillary potential in soil vary over a wide range, Schofield (*7*) has suggested the use of the term pF to designate the value of the potential. The pF of a soil is defined as the common logarithm of the capillary potential expressed in centimeters of water. Thus, a pF of 2.40 represents a capillary potential of −251 cm of water, or −3.57 psi. It is customary to neglect the negative sign when pF values are used.

10.11. FLOW OF MOISTURE IN SOIL. In the case of moist soil, the attraction of the soil for additional moisture and the earth's gravitational pull provide the principal potentials which influence moisture movement; and, wherever these potentials are unbalanced, flow will take place. Wherever potentials are balanced, the moisture seeks to attain a state of static equilibrium. Ordinarily, the capillary potential is less in dry soils than in wet ones. This is not always true for soils which are texturally different, however, and moisture may flow from a relatively dry sandy soil into a wetter clay if the characteristics of the two materials are such that there is a potential gradient toward the wet soil. In other words, the criterion for unsaturated moisture flow from one soil region to another is the relative potential of the two regions, and not their relative moisture contents. Also, when capillary potentials are balanced with the gravitational potential, the capillary soil moisture will approach a state of static equilibrium with a water table below; and no upward or downward flow will take place. For the case of a texturally homogeneous soil with capillary water in static equilibrium, the soil will be saturated at the water table and the moisture content will decrease as the distance above the water table increases.

10.12. WATER IN CAPILLARY FRINGE. An analogy of static equilibrium of capillary water in soil with water in capillary tubes is illustrated in Fig. 10-4. Here a closed water tank *A* (with access to atmospheric pressure) is fitted with capillary tubes *a*, *b*, and *c*, having different diameters as indicated; and *B* is a column of soil supported by a cylindrical sieve *C* and a porous plate *D* which are readily permeable to water. When the tank *A* is filled with water, the liquid will rise by capillarity in the tubes and in the soil column. The capillary water in the soil column will adjust itself to a static equilibrium in which the total curvature of the air–water interfaces or menisci of the wedge-shaped masses of water in the soil pores

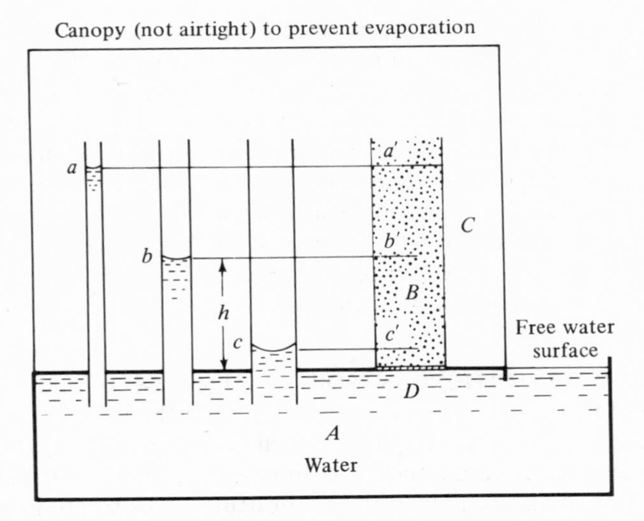

Fig. 10-4. Capillary tube and soil-moisture analogy [after Richards (*5*)].

will be the same as the curvature of the menisci in the capillary tubes at equal heights above the free water surface.

A water table in the soil is a free water surface, being the locus of all points in the soil water at which the pressure is atmospheric. It follows, therefore, that the water in the capillary fringe seeks to adjust itself to an equilibrium situation similar to that indicated in Fig. 10-4. However, moisture in a field soil seldom reaches a state of equilibrium, because of the relatively slow rate at which capillary water moves in the soil and because of changing weather conditions. It is well known that, when a capillary tube is inserted into a vessel of water, the liquid rises in the tube almost immediately; but, in the soil, considerable time is required for a similar movement to take place. Consequently, when rain falls on the surface, that portion of the precipitation which soaks into the soil causes a wave of wetness which migrates slowly downward. Before an equilibrium distribution of moisture is established, however, evaporation may dry out the soil in the upper few inches near the surface to such an extent that the capillary potential in this region is greatly reduced. Under these conditions a potential gradient will develop in the upward direction from the wave of wetness, and some of the moisture that is on its way downward will be caused to reverse its direction of flow and move upward in opposition to the pull of gravity. For this reason, and also because of other effects, such as plant transpiration and variations in barometric pressure and soil temperature, the moisture in the unsaturated soil above a water table is kept in motion by repeated rains and interspersed periods of evaporation.

10.13. EFFECT OF IMPERVIOUS COVER. When an impervious pavement or a building is constructed, rain is kept out of the soil and evaporation and transpiration are prevented. Also the pavement or building serves as an insulator which reduces the range and the rapidity of change of temperature in the underlying soil. These artificial conditions created by engineering construction are conducive to the establishment of equilibrium with the water table below. As a result, moisture tends to accumulate in a pavement subgrade after the pavement is laid. Observations show that such accumulation occurs slowly and that 3 to 5 years may elapse before the maximum accumulation is reached. Even subgrades under pavements in arid regions have been known to become very wet, and the soil may lose a substantial part of its bearing capacity by accumulation of capillary moisture from a water table below. It is desirable, therefore, to study the capillary characteristics of a subgrade soil and to know the probable maximum elevation of the water table under a pavement.

10.14. FACTORS AFFECTING RELATIONSHIP BETWEEN CAPILLARY POTENTIAL AND MOISTURE CONTENT. It was indicated in the consideration of Fig. 10-4 that the equilibrium value of the capillary potential in a soil column at a given height above a free water surface is a constant, which is independent of the grain size, state of packing, temperature, and angle of contact (degree of wetting), of the soil and also of the amount of dissolved salts in the soil water. The percentage of moisture in the soil at this height, however, is dependent on all these factors. It will be of value to discuss the effects of these various properties of the soil and the soil water upon the relationship between capillary potential and moisture content. The capillary potential is dependent on the surface tension of the soil water and on the radii of curvature of the air-water surfaces of the tiny wedges of water which exist between the soil grains, in accordance with Eq. (10-4). It would be possible to compute the capillary potential at any point in moist soil if the surface tension and the radii of curvature of the water wedges could be measured. Obviously, such measurements are impossible. Nevertheless, a consideration of these relationships serves to point out some qualitative effects of the characteristics of the soil and soil water upon capillary potential.

10.15. EFFECT OF TEMPERATURE. The surface tension of water is an inverse function of temperature. Hence, a decrease in temperature increases the surface tension and decreases the capillary potential. In other words, cooling the soil increases its attraction for water. In all probability this relationship accounts, in part at least, for the fact that the moisture content in soil subgrades in the Northern Hemisphere is usually greater in the late winter or early spring when the soil is at its coolest temperature than it is in the late summer and fall when the soil is warmest.

10.16. EFFECT OF DISSOLVED SALTS. An increase in the amount of dissolved salts in the soil water increases its surface tension and thereby lowers the capillary potential of the soil. Theoretically, therefore, subgrades in coastal areas or in regions where the soil has high alkaline content should attract more capillary water than do similar subgrades in freshwater areas. There is little factual evidence to support this theory, and it is probable that the effect of dissolved salts in the soil water is relatively minor.

10.17. EFFECT OF MOISTURE CONTENT. Moisture content, size of soil grains, angle of contact, and state of packing affect the value of capillary potential because of their influence on the radii of curvature of the water-wedge surfaces. As is indicated in Fig. 10-5, if the amount of water collected between two soil grains is decreased, the water will recede further into the interstices between the grains and the curvature of the air-water interface will increase (radius will decrease). This change, according to Eq. (10-4), causes a decrease in capillary potential and indicates a greater attraction for water. It is, of course, a well-known fact that a moderately wet soil has less attraction for additional moisture than a drier soil, just as a moist blotter has less attraction for a liquid than a dry blotter. Parts of the attractive forces within the moist blotter have been satisfied by the previous moistening.

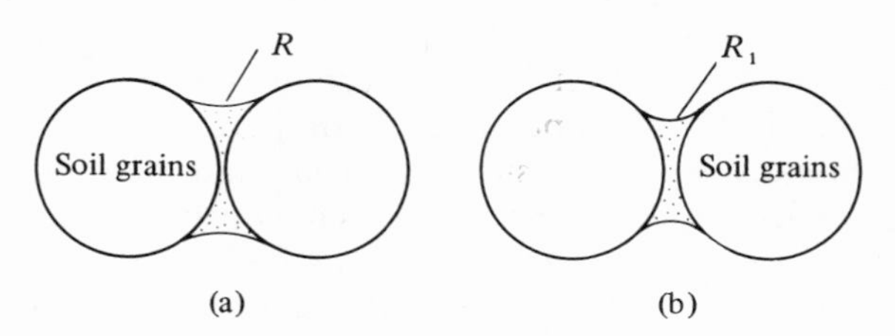

Fig. 10-5. Effect of moisture content upon curvature of air-water interface. (a) Wet soil; (b) dry soil.

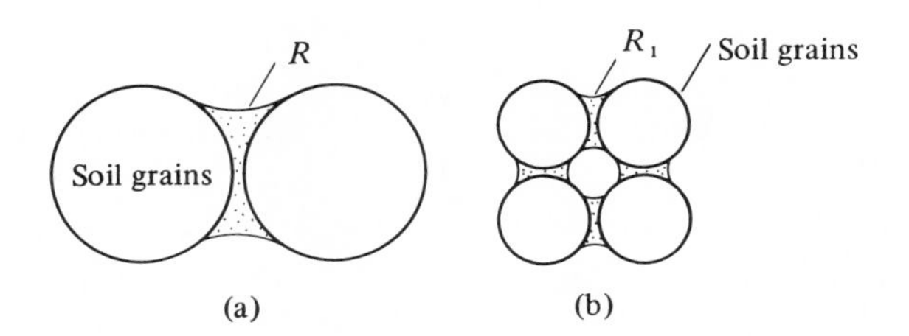

Fig. 10-6. Effect of particle size upon curvature of air-water interface. (a) Coarse-grained soil; (b) fine-grained soil.

10.18. EFFECT OF GRAIN SIZE. If equal weights of a fine soil and a coarse soil have the same moisture content, the fine soil will have more surface area and more points of proximity and contact between soil particles. Less water will be collected at each of the contact points, and the curvature of the surfaces will be greater, as indicated in Fig. 10-6. There is, therefore, a lower capillary potential in fine-grained soils and a greater attraction for moisture. This fact is also a matter of common observation.

10.19. EFFECT OF STATE OF PACKING OF SOIL. If two particles of moist soil are pushed together, the curvature of the menisci will be decreased, as indicated in Fig. 10-7. When an isolated mass of relatively dry soil is sufficiently compressed, it will become saturated; that is, all the pores will become filled with water, although the moisture content expressed as a percentage of dry weight will remain unchanged. During the process of compressing the soil, the curvature of the air-water interfaces gradually decreases; and, finally, no curvature remains and the capillary potential increases from a low negative value to zero.

In the foregoing discussion an isolated mass of soil was considered to have been made more dense by compression or reduction in volume. If densification of the soil is obtained by increasing the amount of solid matter in a given volume, the result will differ greatly from that just

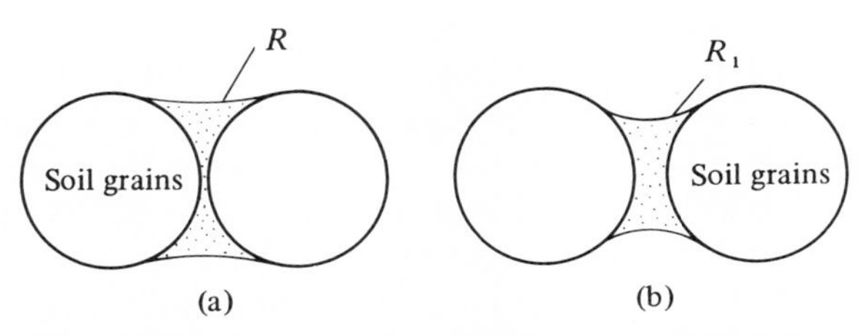

Fig. 10-7. Influence of state of packing of soil on curvature of air-water interface. (a) Closely-packed soil; (b) loosely-packed soil.

described. In this case, the number of points of contact between soil grains will be increased, and the water wedges will be much smaller and will have greater curvature. Such changes tend to reduce the capillary potential of the soil.

These facts are extremely important in connection with studies on subgrade-soil moisture, and they dictate the necessity of measuring the capillary potential of a soil in its actual state of density and structural arrangement, if quantitatively useful values of this potential are to be determined. Tests should therefore be made on the soil *in situ,* or on undisturbed samples in the laboratory, or on laboratory samples compacted to the density of the material in the finished subgrade.

10.20. EFFECT OF ANGLE OF CONTACT. If the mineralogical composition of the soil is such that the soil grains are not completely wetted by water, the angle of contact between the menisci and the soil particles will be greater than zero. An increase in the angle of contact will tend to decrease the curvature of the menisci, as shown as Fig. 10-8, and thereby increase the capillary potential of the soil at a given water content. A soil with an angle of contact greater than zero will have less attraction for water than one in which the particles are completely wetted.

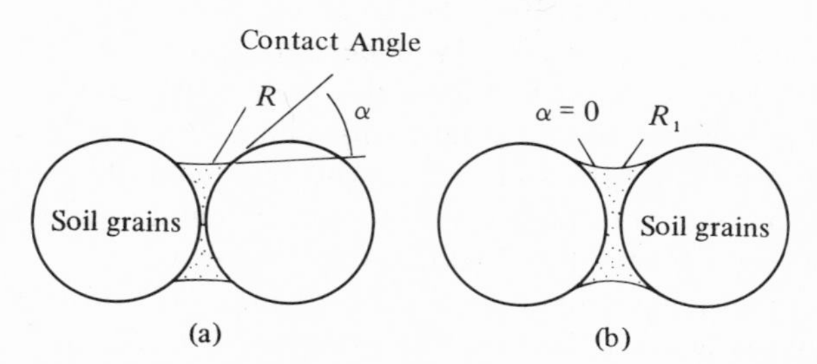

Fig. 10-8. Influence of wettability of soil grains on curvature of air-water interface. (a) Low wettability; (b) high wettability.

10.21. SORPTION CURVES. Of all the many factors which influence capillary potential, the moisture content of the soil is the most significant and useful. Curves showing the relationship between moisture content and capillary potential, when other factors are held constant, are known as sorption curves. It has been demonstrated that a sorption curve for a given soil, in a given state of packing and at constant temperature, is a continuous curve from saturation to oven dryness. Also, it has been shown that there is a decided hysteresis effect in the relationship between capillary potential and moisture content (see Fig. 10-16), the extent of this effect depending on whether the soil is wetting up or drying out. Such hysteresis is much more pronounced in fine-grained soils than in coarser soils, and the tension due to it in a very fine soil may amount to several feet of water, the tension being greater (capillary potential being less) for a given moisture content when the soil is draining than when it is wetting up. A sorption curve obtained when a soil is draining is often called a *desorption curve*; when the soil is wetting up, it is called an *absorption curve*.

10.22. CAUSE OF HYSTERESIS. The reasons for the hysteresis phenomenon just mentioned are not entirely understood, but it probably may be attributed to the fact that a suction is required to make air penetrate into a porous medium, whereas water will penetrate more readily. During the process of draining a soil, air enters and water is withdrawn—first from the larger pores and then from successively smaller pores as drainage progresses. A greater suction is required to empty small pores. Consequently, at a given soil moisture tension, fine-textured soils will retain more water than coarser soils.

The maximum tension that can be developed at any point in field soil as downward drainage progresses, expressed in equivalent height of water column, is equal to the elevation of the point above the water table. That is to say, the maximum tension which can exist at an elevation of 2 ft above a water table is equal to the pressure corresponding to a water column 2 ft high, which is 124.8 psf or 0.87 psi. If the soil pores at this elevation are sufficiently small, the suction required to pull air into them may be greater than this value. In this case the pores would remain filled with water, even though the pressure in the water is less than atmospheric. The region above a water table within which the pores are completely filled with water under tension is called the zone of capillary saturation (see Fig. 9-2). Such a region extends to a considerably greater height in fine soils than in coarse ones and is much more in evidence above a falling water table than above a water table which is rising.

10.23. TYPICAL SORPTION CURVES. Several typical sorption curves for soils of different textural characteristics are shown in Fig. 10-9. Curves constructed from the same data but with the moisture content expressed as a percentage of the saturation value are shown in Fig. 10-10. In these diagrams, the capillary potentials are expressed in negative feet of water and they are numerically equal to the soil-water tensions.

To obtain a better conception of the significance of these curves, let it be assumed that long tubes were filled with these soils at the temperature and state of packing used in these determinations, that their lower

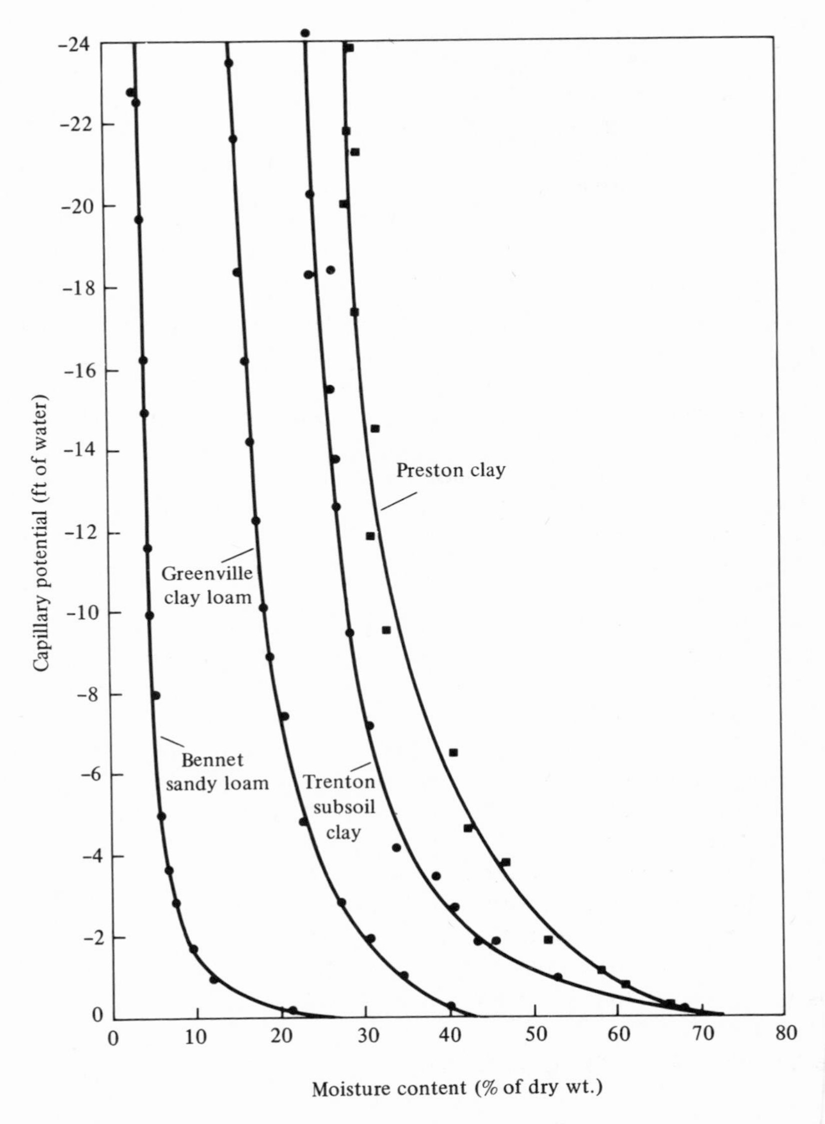

Fig. 10-9. Desorption curves for four soil types (data by Richards).

ends were dipped in free water, and that sufficient time was allowed for the water to come to static equilibrium. Then the distances above the water level would correspond numerically to the capillary potentials; and the equilibrium moisture percentages at any height for the different soils could be taken from the curves. Stated in another way, if the soils used

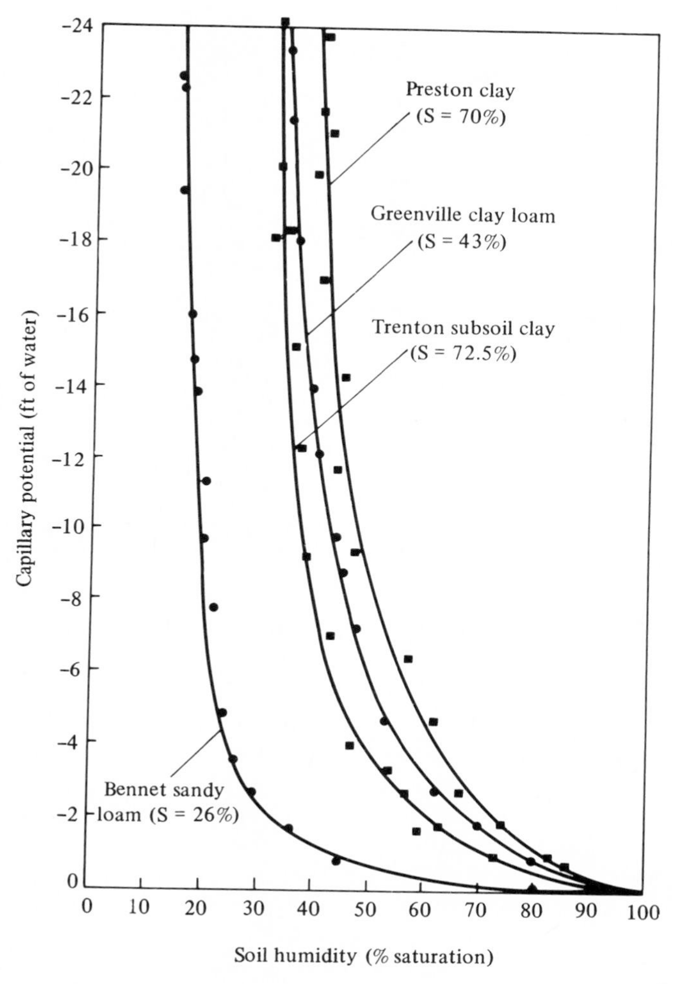

Fig. 10-10. Desorption curves for four soil types (data same as for Fig. 10-9).

in these tests and the conditions under which the tests were conducted were representative of soils in actual pavement subgrades, the equilibrium moisture content at any height above a free water table could be obtained from the curves in the manner illustrated in Fig. 10-11.

Sorption curves of this kind may be very useful in the study of moisture conditions in pavement subgrades. It is pointed out that an actual subgrade may consist of several strata of different textured soils between the base of a pavement and the water table. In such a case the sorption curve would be a composite curve of the various soils encountered.

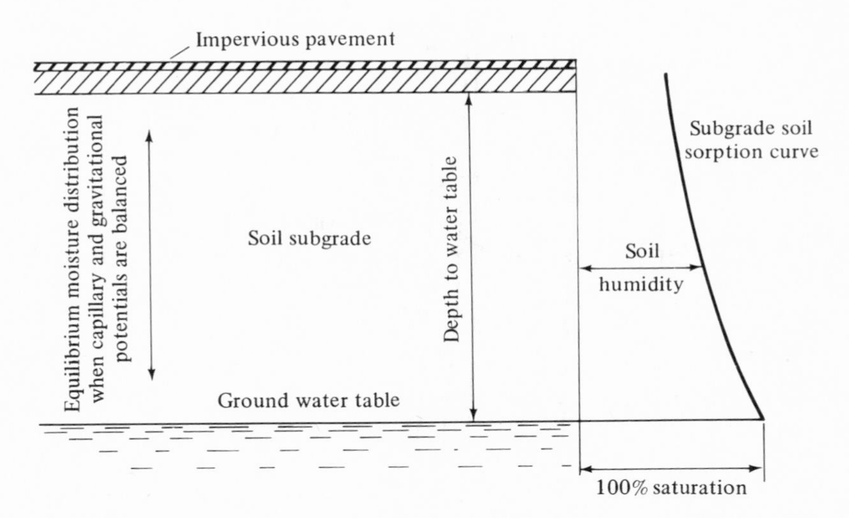

Fig. 10-11. Subgrade soil moisture distribution at static equilibrium with a water table.

10.24. SORPTION CURVES FOR STRATIFIED SOIL. A demonstration of this principle is indicated in Figs. 10-12 and 10-13 wherein actual capillary moisture contents of two texturally and geologically different soils are plotted in comparison with the composite sorption curves of the two materials arranged in two stratified relationships. First the sorption curve for each soil was determined by means of tensiometers with the results shown in Fig. 10-14. Then the soils were placed in glass tubes in stratified arrangement and the tubes were mounted with their lower ends immersed in a vessel of water which was open to the atmosphere. After sufficient time had elapsed for the capillary water to attain an equilibrium distribution, the moisture contents at 5 cm intervals were determined. These experiments were conducted at room temperature and barometric pressure. It is probable that if they had been conducted under isothermal and isobaric conditions, the correspondence between actual and theoretical moisture contents would have been even closer.

These experiments also demonstrated the wide difference between the capillary conductivity of the two soils. As indicated in Fig. 10-15, it took approximately three times as long for the wetting front to rise 105 cm in the tube with the glacial soil in contact with the free water surface as in the tube with the loess soil at the bottom.

10.25. DISCUSSION OF SORPTION CURVES. Attention is called to the fact that the slopes of typical soil sorption curves are very steep in the

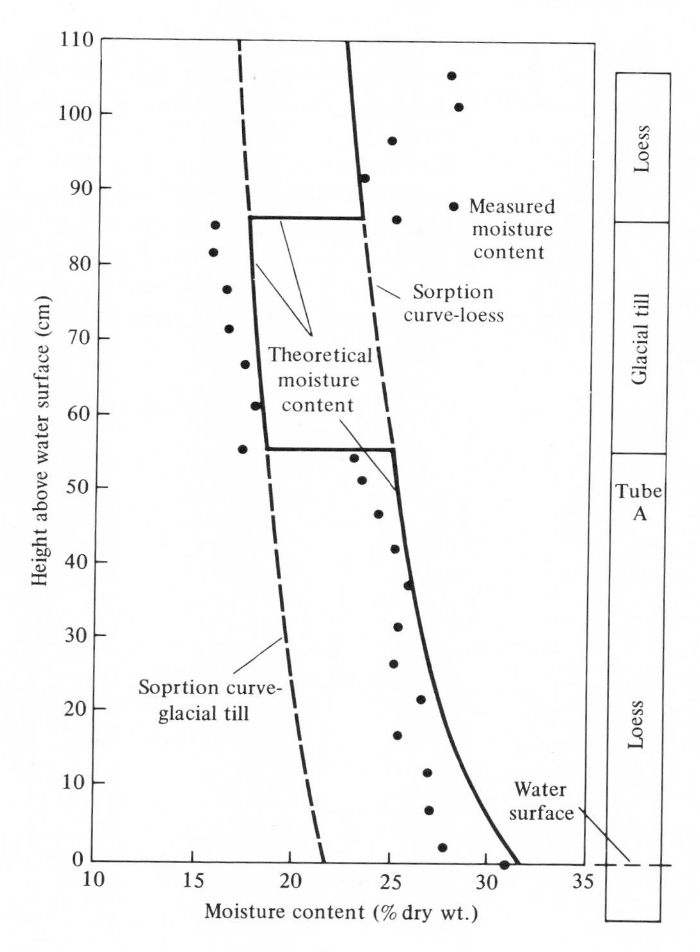

Fig. 10-12

regions of high moisture tension. This fact indicates that there is a very gradual reduction in the equilibrium moisture content as the distance above a water table increases. It therefore may be concluded that fairly high moisture contents may eventually develop in fine-grained subgrades beneath impervious pavements, even though the distance to the water table is relatively great. For example, in semiarid regions where normal evaporation rates are high, the soil in its natural state may be very dry. Nevertheless, when the soil is covered with an impervious pavement, subgrade moisture contents may increase markedly as equilibrium develops, especially in fine-grained, heavy-textured soils. Kersten (*4*) has observed that some pavement subgrades in regions where the annual precipitation was as low as 14 in. had accumulated moisture contents in excess of the plastic limit of the soils. This observation seems to be compatible with the principle just noted.

Two additional sorption curves are presented in Fig. 10-16 to illustrate the character of the hysteresis loops which develop when soils are alternately drained and wetted. The numbers on these curves show the chronology of the points of observation at which the moisture content was allowed to reach equilibrium for various applied tensions. These curves show the characteristically wider hysteresis loops which develop in

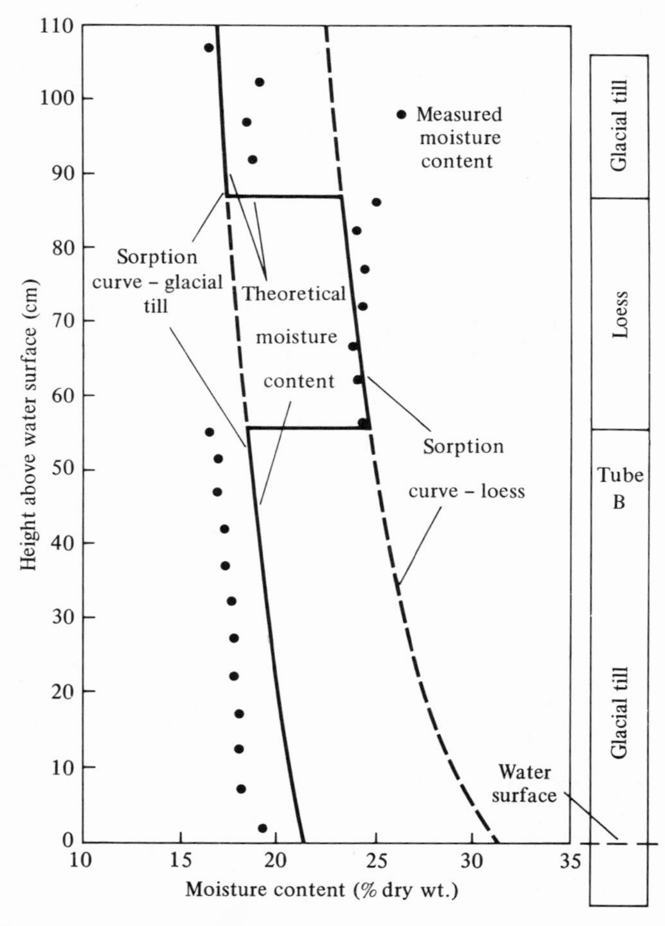

Fig. 10-13

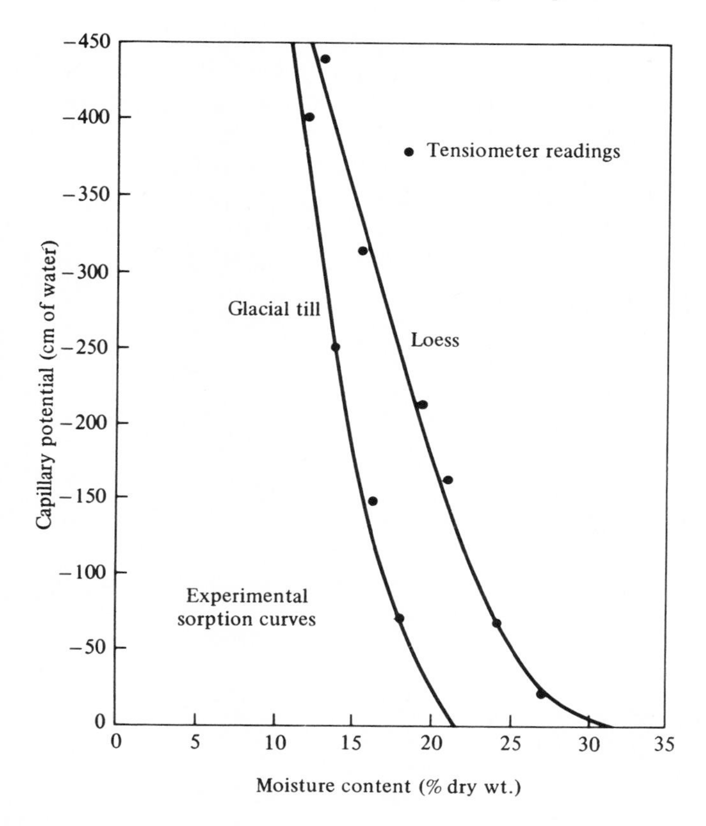

Fig. 10-14

the case of a fine-grained soil as compared to a coarser soil. They give an indication of the hysteresis in moisture content when a water table rises and falls in a range of about 16 ft. The loops would be correspondingly smaller for fluctuations of lesser amounts.

In the discussion accompanying the curves in Fig. 10-16 Richards (*5*) pointed out that considerable time was required for the attainment of moisture equilibrium after a change in the applied water tensions, even though the maximum distance of travel for water in the samples was only 3 in. Equilibrium moisture adjustments corresponding to tension increases (drying) over a range from 0 to $\frac{1}{2}$ atm took place in 1 to 3 days, whereas 4 to 20 days were required when tensions were decreased (wetting). Air-dry loam soil required as long as 120 days to wet up to equilibrium when water was supplied at tensions as low as 5 ft of water.

When we are considering moisture movements in unsaturated soils and the time for establishing equilibrium, it appears to be necessary to distinguish between drying and wetting processes, the latter taking place at much slower rates under corresponding moisture conditions. Richards' experience in this regard is in harmony with the data collected by Kersten (*4*), which indicate that several years may be required for a subgrade under a pavement to reach a fairly constant terminal moisture content. Here again, as in practically all matters involving the flow of moisture through soils, there was a wide difference in the characteristics exhibited by fine-grained and coarse soils.

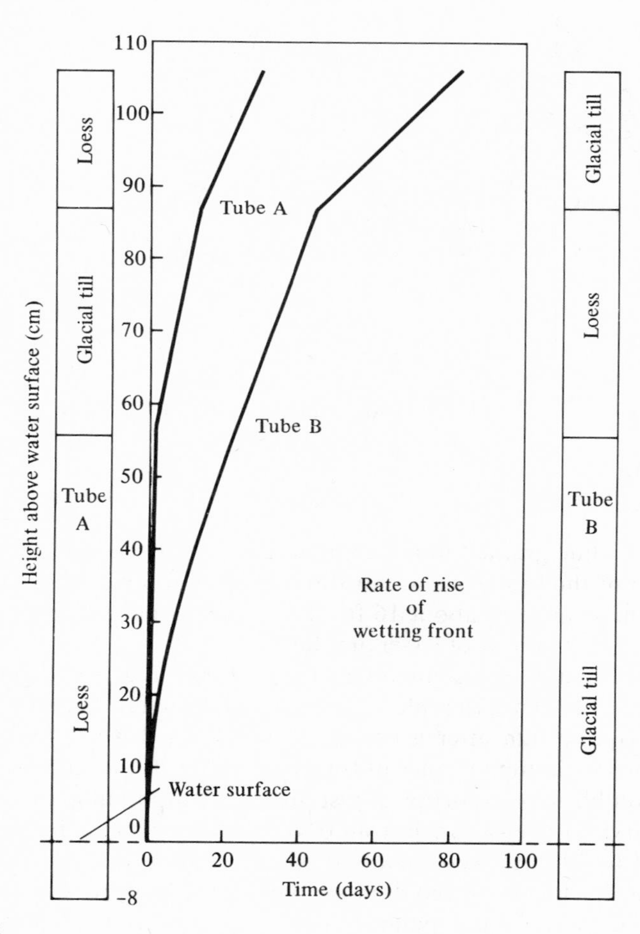

Fig. 10-15

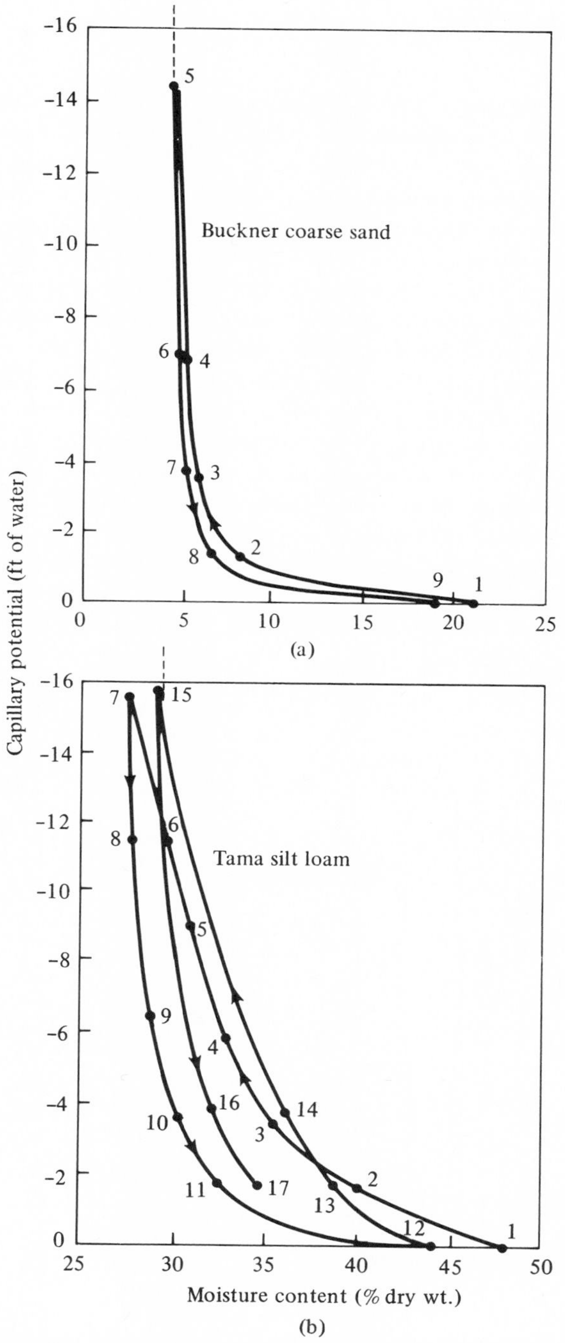

Fig. 10-16. Soil sorption curves showing hysteresis loops (Richards). The dashed lines show the moisture equivalents.

10.26. MEASUREMENT OF CAPILLARY POTENTIAL BY TENSIOMETER. Several methods of measuring the capillary potential of a soil at varying moisture contents have been developed. Two of these will be described here, namely, measurements by means of tensiometers and measurements by means of porous plates and membranes. The essential features of a soil tensiometer are shown in Fig. 10-17. It consists of a porous ceramic cup *A*, sealed onto a glass tube which is connected to a mercury manometer *B*. The porous cup and the manometer tube above the mercury are filled with water, and a rubber stopper is inserted at *C*. Then the cup is imbedded in the soil, the operator making certain that there is intimate contact between the cup and the soil. The soil must be kept covered at the top to prevent evaporation of soil moisture, but the cover should be perforated in a few places to maintain atmospheric pressure in the soil air.

The soil, which is at some moisture content less than saturation, tends to draw water through the porous walls of the cup. As a result, a negative pressure is produced in the cup water and the mercury rises in the closed leg of the manometer. This process continues until the soil water and the cup water have the same pressure deficiency or tension. The pressure deficiency, which is numerically equal to the capillary

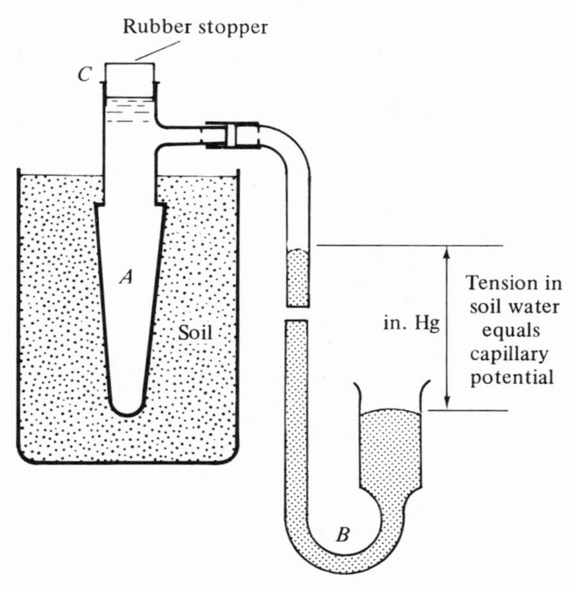

Fig. 10-17. Soil tensiometer (Richards).

potential, may be determined directly by reading the difference in elevation of the mercury in the manometer legs.

The same process should be repeated on the same sample at different moisture contents or on different samples of the same soil compacted to the same state of density, but containing different amounts of moisture. A sorption curve may then be drawn for the soil. The end point, or the point of zero capillary potential, will always be at the saturation moisture content of the soil. Instruments based upon this same principle have been developed for measuring the capillary potential of soils *in situ* and are described in Ref. (5). A vacuum gauge may be substituted for the mercury manometer. For extended observations over a period of time, a self-recording vacuum gauge is convenient.

The tensiometer provides data on the wet end of the energy-moisture relationship, and the maximum measurable tension cannot exceed 1 atm. When the tension in the soil water exceeds this value, air enters the porous cup and the instrument is no longer operative. Nevertheless this technique offers many possibilities for studying soil-moisture changes within its range of applicability, which embraces practically all situations of interest in engineering.

10.27. MEASUREMENT OF CAPILLARY POTENTIAL BY USE OF A MEMBRANE. Techniques involving the use of porous plates and porous membrane have been devised for measuring moisture tension relationships in

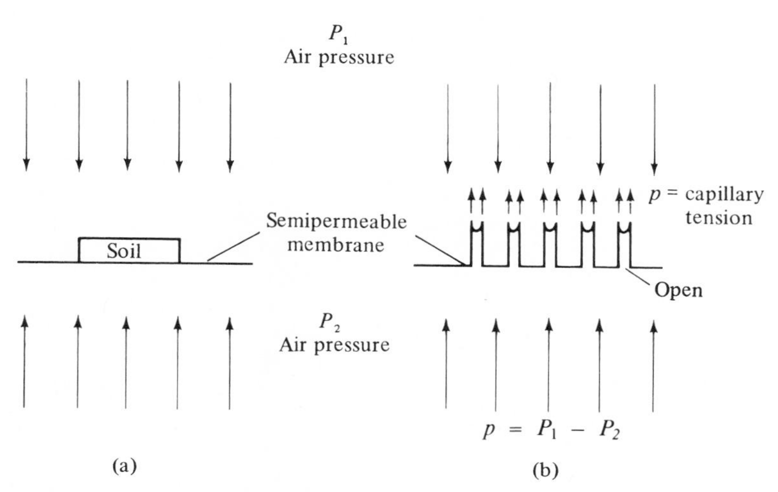

Fig. 10-18. Membrane method for measuring capillary tension. (a) Physical arrangement; (b) schematic showing pressures involved.

the range from 1 to 15 atm (*1*). Soil is placed in contact with a porous plate or membrane in which pores are so fine they do not allow passage of air but will allow flow of water in or out of the soil. Air pressure is applied to the soil, driving capillary water through the membrane until an equilibrium is reached whereby air pressure against the soil is balanced by capillary forces in the soil plus air pressure on the other side of the membrane (Fig. 10-18). The capillary tension at equilibrium thus equals the difference between the two air pressures.

PROBLEMS

10.1. (a) Calculate the height of rise of pure water in a clean glass tube whose inside diameter is 0.15 mm, using $T = 75$ dynes/cm. (b) What is the height of rise if the tube is sufficiently dirty to produce a wetting angle of 15°?

10.2. If the outside diameter of the tube in Problem 10.1 is 0.35 mm, what is the compressive stress in the walls of the tube in each of the two cases considered in that problem?

10.3. Two soils have the sorption characteristics shown in the accompanying table. Draw two sets of sorption curves for these soils, using capillary potential expressed in pF and in centimeters of water.

pF	*Moisture Content (%)*	
	Soil A	*Soil B*
1.0	47	32
1.5	33	17
2.0	22	12
2.5	17	9
3.0	13	7
3.5	10	5

10.4. Assume that two masses of soils A and B in Problem 10.3 are at the same elevation above a water table and are in intimate contact with each other. If soil A contains 17% of moisture and B contains 12%, will capillary water flow from A to B or from B to A, or will it remain static?

10.5. A subgrade under an impervious pavement has the sorption characteristics of soil A in Problem 10.3. What will be the accumulated moisture content of the soil 1 ft below the bottom of the pavement, if the water table is constant at an elevation 5 ft below the bottom of the pavement?

10.6. Repeat Problem 10.5, using soil B in Problem 10.3.

10.7. What is meant by the zone of capillary saturation? Is the pressure in the soil water within this zone greater than, less than, or equal to atmospheric pressure?

10.8. In a certain locality the water table rises steadily from October to April and then falls from May through September. Will the zone of capillary saturation extend a greater or less distance above the water table in July than in March?

10.9. The water in a capillary tube extends 3 ft above a free water surface. Determine the pressure in the water in this capillary tube: (a) at the free water surface; (b) 1 ft above the free water surface; (c) 2 ft above it; (d) just inside the meniscus; (e) on the outside of the meniscus.

REFERENCES

1. Black, C. A. *Methods of Soil Analysis, Part 1.* American Society of Agronomy, Madison, Wisconsin, 1965.
2. Edlefsen, N. E., and Alfred B. C. Anderson. "Thermodynamics of Soil Moisture." *Hilgardia* **15,** No. 2. University of California, Berkeley, California, 1943.
3. Gardner, Willard. "The Capillary Potential and Its Relation to Soil Moisture Constants." *Soil Science* **10,** 103–126 (1920).
4. Kersten, M. S. "Survey of Subgrade Moisture Conditions." *Proc. Highway Research Board* **24,** 497–512 (1944).
5. Richards, L. A. "Uptake and Retention of Water by Soil as Determined by Distance to a Water Table." *J. Amer. Soc. Agron.* **33,** 778–786 (1941).
6. Russell, M. B., and M. G. Spangler. "The Energy Concept of Soil Moisture and the Mechanics of Unsaturated Flow." *Proc. Highway Research Board* **21,** 435–470 (1941).
7. Schofield, R. K. "The pF of the Water in Soil." *Transactions 3rd. Intern. Cong. Soil Sci.* **2,** 37–48 (1935).
8. Spangler, M. G. "Distribution of Capillary Moisture at Equilibrium in Stratified Soil." *Highway Research Board*, Special Report No. 2 (1952).

11

Gravitational Water and Seepage

11.1. NATURE OF GRAVITATIONAL FLOW IN SOIL. The flow of gravitational water in soil is caused by the action of gravity which tends to pull the water downward to a lower elevation. It is similar in many respects to the free flow of water in a conduit or an open channel in that it is attributable to the gravitational pull which acts to overcome certain resistances to movement or flow of the water. Such resistances are due mainly to friction or drag along the surfaces of contact between the water and the conduit in free flow and to friction and viscous drag along the sidewalls of the pore spaces in the case of flow through soils. In hydraulics, gravity flow of water may be either laminar or turbulent in character, its nature depending on the velocity of flow and on the size, shape, and smoothness of the sides of the conduit or channel. In the study of gravitational flow in soils, we are primarily interested in the laminar type of flow, since the velocity of ground water rarely, if ever, becomes high enough to produce turbulence in the sense in which it is used here.

11.2. CHARACTERISTICS OF LAMINAR FLOW. Laminar flow is said to exist when all particles of water move in parallel paths and the lines of flow are not braided or intertwined as the water moves forward. The quantity of water flowing past a fixed point in a stated period of time is equal to the cross-sectional area of the water multiplied by the average velocity of flow. This relationship may be expressed by the formula

$$Q = Av \qquad (11\text{-}1)$$

in which Q is the volume of flow per unit of time, such as cubic feet per day or cubic centimeters per minute; A is the cross-sectional area of flowing water, in square feet or square centimeters; and v is the velocity of flow, in feet per day or centimeters per minute.

11.3. HYDRAULIC GRADIENT. The driving force which causes water to flow may be represented by a quantity known as the *hydraulic gradient*. This is defined as the drop in head divided by the distance in which the drop occurs. It may be expressed by the relation

$$i = \frac{h}{d} \qquad (11\text{-}2)$$

in which i is the hydraulic gradient; h is the drop in head; and d is the distance in which the drop occurs.

For example, if an open channel is 5000 ft long and drops 50 ft in that distance, the hydraulic gradient is 50/5000 or 0.01.

11.4. DARCY'S LAW. The general relationship between hydraulic gradient and the character and velocity of flow is indicated in the diagram of Fig. 11-1. As the hydraulic gradient is increased through zones I and II, the flow remains laminar and the velocity increases in linear proportion to the gradient. At the boundary between zones II and III the flow breaks from laminar to turbulent and the proportional relationship between velocity and gradient no longer prevails.

Under decreasing hydraulic gradient, the flow remains turbulent through zones III and II and does not resume the laminar characteristic until the boundary between zones II and I is reached. Here the relationship between velocity and gradient again becomes linear and coincides with that for an increasing gradient.

The gravitational flow of water in soil is represented by the curve in zone I of Fig. 11-1 and we may write the equation

$$v = ki \qquad (11\text{-}3)$$

in which k is a proportionality constant.

By substituting this expression for v in Eq. (11-1), we obtain the relation

$$Q = Aki \qquad (11\text{-}4)$$

This relationship is general and may be applied to any situation in

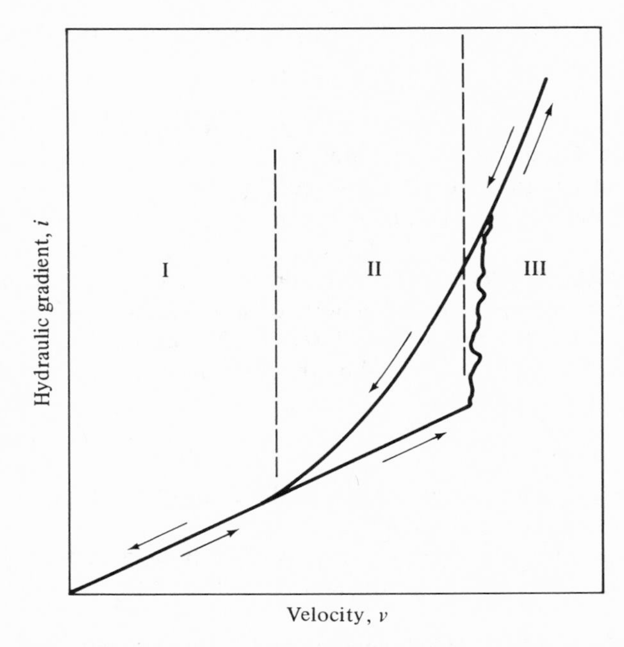

Fig. 11-1. Relationship between hydraulic gradient, velocity, and type of flow.

which the flow is laminar in character. It is an expression of Darcy's law applied to the flow of gravitational water through soil. A more general statement of the Darcy law is necessary in connection with the flow of other fluids through other types of porous media, but the foregoing statement is sufficient for the purpose of this discussion.

Let us assume that we have a conduit in which a mass of soil is placed in such a manner that all of the water flowing through the conduit must flow through the soil, as illustrated in Fig. 11-2. Since practically all the resistance to flow in this case is caused by the mass of soil, the value of the proportionality constant in Eq. (11-4) depends on the characteristics of the soil which influence the flow of water through its pores. The equation indicates that the quantity of water flowing through a given cross-sectional area of soil is equal to a constant multiplied by the hydraulic gradient.

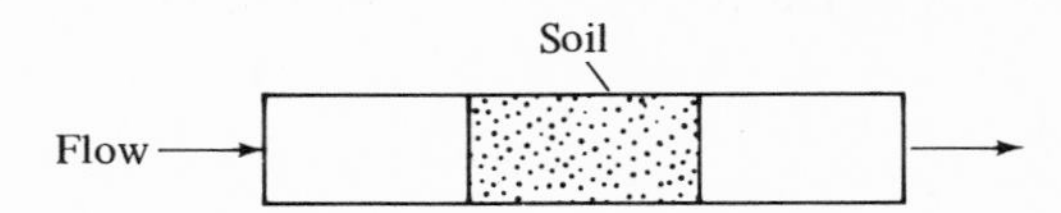

Fig. 11-2. Hypothetical flow through soil.

11.5. COEFFICIENT OF PERMEABILITY. The constant k in Eq. (11-4) is known as the *coefficient of permeability* or, more recently, as the *coefficient of hydraulic conductivity*. It constitutes an important property of soil, and its value depends largely on the size of the void spaces, which in turn depends on the size, shape, and state of packing of the soil grains. A clayey soil with very fine grains will have a very much lower permeability coefficient than will a sand with relatively coarse grains, even though the void ratio and the density of the two soils may be nearly the same. The reason is the greater resistance offered by the very much smaller pores or flow channels in the fine-grained soil through which the water must pass as it flows under the influence of a hydraulic gradient. From this standpoint, we may say that the coefficient of permeability is independent of the void ratio or density when we are comparing soils of different textural characteristics. On the other hand, when we consider the same soil in two different states of density, the permeability is dependent on the void ratio, since the soil grains are brought into closer contact by the process of compaction and densification. The pore spaces are reduced in size, and resistance to flow is increased.

11.6. VELOCITY OF APPROACH OF WATER. Attention is directed to the fact that, in the application of the Darcy law and Eq. (11-4), the cross-sectional area A is the area of the soil including both solids and void spaces. Obviously, the water cannot flow through the solids, but must pass only through the void spaces. Therefore, the velocity ki in Eq. (11-4) is a factitious velocity at which the water would have to flow through the whole area A in order to yield the quantity of water Q which actually passes through the soil. This factitious velocity is referred to as the "velocity of approach" or the "superficial velocity" of the water just before entering, or after leaving, the soil mass.

A dimensional analysis of Eq. (11-4) indicates that the coefficient of permeability k has the dimensions of a velocity, that is, a distance divided by time. Therefore, permeability is sometimes defined as "the superficial velocity of water flowing through soil under unit hydraulic gradient."

11.7. COEFFICIENT OF PERCOLATION. If the actual velocity of flow through the pores of the soil is considered, then the corresponding area which must be used in writing the flow equation is the area of the pore spaces cut by a typical cross section of the soil. The Darcy equation for this case is

$$Q = A_v k_p i \tag{11-5}$$

in which A_v is the area of pore spaces in a soil cross section; and k_p is a proportionality constant.

The product $k_p i$ in this case is equal to the average actual velocity of the water through the soil pores. Since the area of the pores in any cross section will always be less than the total area, it is obvious that this actual velocity will always be greater than the velocity of approach. The proportionality constant k_p is called the *coefficient of percolation,* and it always has a greater value than the coefficient of permeability for any given soil.

11.8. RELATION BETWEEN COEFFICIENTS OF PERMEABILITY AND PERCOLATION. The distinction between the two flow coefficients should be clearly understood by the student. The coefficient of percolation refers to the average actual velocity of water flowing through the actual pore area of the soil; whereas, the coefficient of permeability refers to a factitious velocity of flow through the total area of solids plus pore spaces, as pointed out in Section 11.6. Since, as a rule, the total area of soil is more conveniently determined in gravitational flow problems, the permeability coefficient is used more often than the percolation coefficient.

The area of the pore spaces in a typical cross section of soil is equal to the total area multiplied by the porosity. It therefore follows that the coefficient of permeability of the soil is equal to the coefficient of percolation multiplied by the porosity. Thus,

$$A_v = nA \tag{11-6}$$

By substituting this value of A_v in Eq. (11-5) and setting the result thus obtained equal to the expressions for Q given by Eq. (11-4), we get

$$Aki = nAk_p i \tag{11-7}$$

from which

$$k = nk_p \tag{11-8}$$

11.9. APPLICATIONS OF PERMEABILITY CHARACTERISTICS OF SOIL. There are numerous types of problems in connection with engineering projects which require knowledge of the permeability characteristics of the soil involved, such as computations of seepage through earth dams and levees and losses from irrigation ditches. Estimates of pumpage-capacity requirements for unwatering cofferdams or excavations below a water table are familiar examples of such problems. The spacing and depth of underdrains for lowering the water table under a road or runway in order to improve subgrade stability or for draining waterlogged agricultural land is another type of problem in which the permeability of the soil is of paramount importance. Also, the rate of settlement of a structure resting on a soil foundation is a function of the rate at which water moves through and out of the foundation soil.

11.10. MEASUREMENT OF PERMEABILITY CHARACTERISTICS OF SOIL. Several methods of measuring permeability characteristics of soils are available. Some methods involve laboratory procedures on disturbed or undisturbed samples, and others are adapted to determination of the permeability of the soil in place below a water table. Each of these procedures has advantages which are important in different types of problems; and the method which is most feasible and appropriate for the particular problem in hand should be chosen. For example, in studying the seepage through a rolled earth dam, it would be appropriate to make a laboratory type of test on a sample of the soil to be used which would be compacted to the same density as in the prototype structure. On the other hand, a field test of the soil in place would be more appropriate in the case of studies relating to the unwatering of an excavation. In every case, the objective should be to determine the permeability of the soil in its natural or normal operating condition or to do so as nearly as is possible. Furthermore, soils in nature are frequently nonisotropic with respect to flow; that is, the coefficient of permeability in the vertical direction may differ considerably from that in the horizontal direction. If this condition exists, it may be necessary to measure the permeability in both directions.

11.11. TEST WITH CONSTANT-HEAD PERMEAMETER. A laboratory test which is particularly adapted to determination of the coefficient of perme-

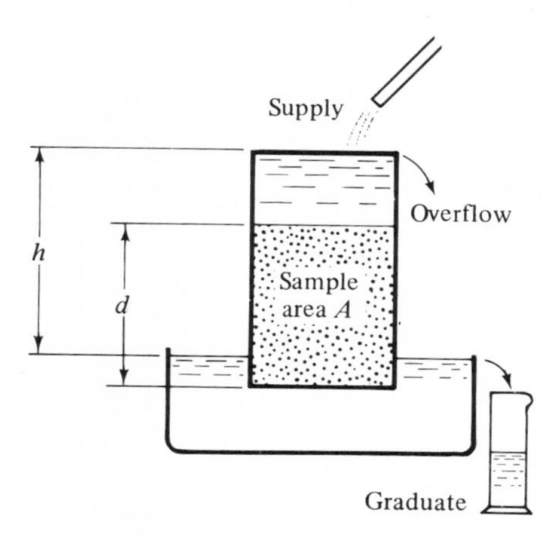

Fig. 11-3. Constant-head permeameter.

ability of relatively coarse-grained soils is one in which the hydraulic gradient is held constant throughout the testing period. A typical arrangement of apparatus for this test, called a constant-head permeameter, is shown in Fig. 11-3.

In the conduct of the test, all the water passing through the soil sample in a measured period of time is collected, and the quantity is measured. This quantity of water and the appropriate dimensions of the apparatus and the soil sample are substituted in Eq. (11-4), and a value of the permeability coefficient is obtained.

EXAMPLE 11-1. A soil sample in a constant-head permeameter is 6 in. in diameter and 8 in. long. The vertical distance from headwater to tailwater is 11 in. In a test run, 766 lb of water passes through the sample in 4 hr 15 min. Determine the coefficient of permeability.

SOLUTION: From this test, $h = 11$ in. and $d = 8$ in.; and

$$i = \frac{h}{d} = 1.375$$

Also, $A = 28.27$ sq in. $= 0.196$ sq ft and

$$Q = \frac{766}{62.4 \times 255} = 0.048 \text{ cfm}$$

Substituting these values in Eq. (11-4), we obtain

$$0.048 = 0.196 \times k \times 1.375$$

from which

$$k = 0.178 \text{ fpm}$$

In computing the value of the permeability coefficient from data obtained in a test of this type, as in all permeability problems, it is important to keep the computations dimensionally correct. A relatively easy and sure way to do this is to decide in advance the units in which the coefficient of permeability is desired. Then reduce the values of Q and A to those units before making the computation. In the preceding example, Q and A were reduced to feet and minutes and the resulting value of k was expressed in feet per minute. Since the hydraulic gradient is a dimensionless quantity, the units of h and d are not important, provided the same units are used for both distances.

11.12. TEST WITH FALLING-HEAD PERMEAMETER. Another laboratory test, which is more appropriate in the case of fine-grained soils, is called a

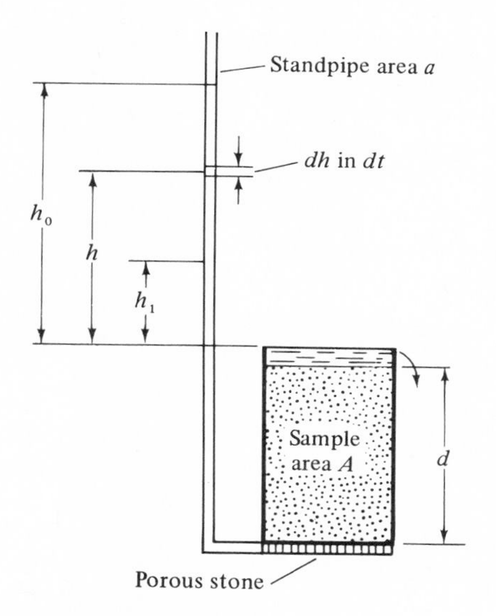

Fig. 11-4. Falling-head permeameter.

variable-head permeameter or *falling-head permeameter.* A typical arrangement of apparatus for this test is shown in Fig. 11-4. In the conduct of the test, the water passing through the soil sample causes water in the standpipe to drop from h_0 to h_1 in a measured period of time t_1. The head on the sample at any time t between the start and finish of the test is h; and, in any increment of time dt, there is a decrease in head equal to dh. From these facts, the following relationships may be written:†

$$k \frac{h}{d} A = -a \frac{dh}{dt} \tag{11-9}$$

Then

$$k \frac{A}{d} \int_0^{t_1} dt = -a \int_{h_0}^{h_1} \frac{dh}{h} \tag{11-10}$$

from which

$$k = \frac{ad}{At_1} \log_e \frac{h_0}{h_1} \tag{11-11}$$

EXAMPLE 11-2. A sample of clay soil, having a cross-sectional area

†The minus sign in Eq. (11-9) is appropriate because the head decreases with elapsed time.

of 78.5 sq cm and a height of 5 cm, is placed in a falling-head permeameter in which the area of the standpipe is 0.53 sq cm. In a test run, the head on the sample drops from 80 cm to 38 cm in 1 hr 24 min 18 sec. What is the coefficient of permeability of the soil?

SOLUTION: From this test, $a = 0.53$ sq cm; $t_1 = 1$ hr 24 min 18 sec = 84.3 min; $h_0 = 80$ cm; $d = 5$ cm; and $A = 78.5$ sq cm.

Substitution in Eq. (11-11) gives†

$$k = \frac{0.53 \times 5}{78.5 \times 84.3} \log_e \frac{80}{38} = 0.000299 \text{ cm/min}$$

11.13. EFFECT OF AIR IN PORES. The permeability of the soil sample in either of the two laboratory tests just described may be affected appreciably by pocketed bubbles of air in the soil pores. Attempts should be made to eliminate entrapped air from the sample by passing water through it for a considerable period of time before a test run is made. Also, since difficulty may be encountered if dissolved air is released from the permeating water and trapped in the pores as the water passes through the soil, it is advisable to use air-free or distilled water as the permeate. Furthermore, since water tends to absorb air as it cools and to release dissolved air as it warms up, the temperature of the permeating water should preferably be somewhat higher than that of the soil sample. This precaution not only will prevent air from being released in the soil, but may assist in removing entrapped air in the pores since the water will be cooled as it passes through the soil and will have a tendency to absorb air.

11.14. EFFECT OF VISCOSITY OF WATER. The coefficient of permeability is primarily influenced by the size and shape, or tortuousness, of the soil pores and by the roughness of the mineral particles of the soil. However, it is also affected by the viscosity of the permeating water. Since the viscosity of water is a function of its temperature, it may be advisable in some cases to correct the laboratory-measured permeability coefficient for temperature difference between that of the laboratory water and that of the water which will flow through the prototype structure. For example, laboratory measurements of permeability may be made at a room temperature of say 80°F, whereas it is known that the temperature of the seepage water through the prototype structure will be in the neighborhood of 50°F. The coefficient determined in the laboratory may be too high in this case because the viscosity of 80° water is less than of 50° water.

A correction factor for the permeability coefficient with water at

†$\log_e N = 2.303 \log_{10} N$.

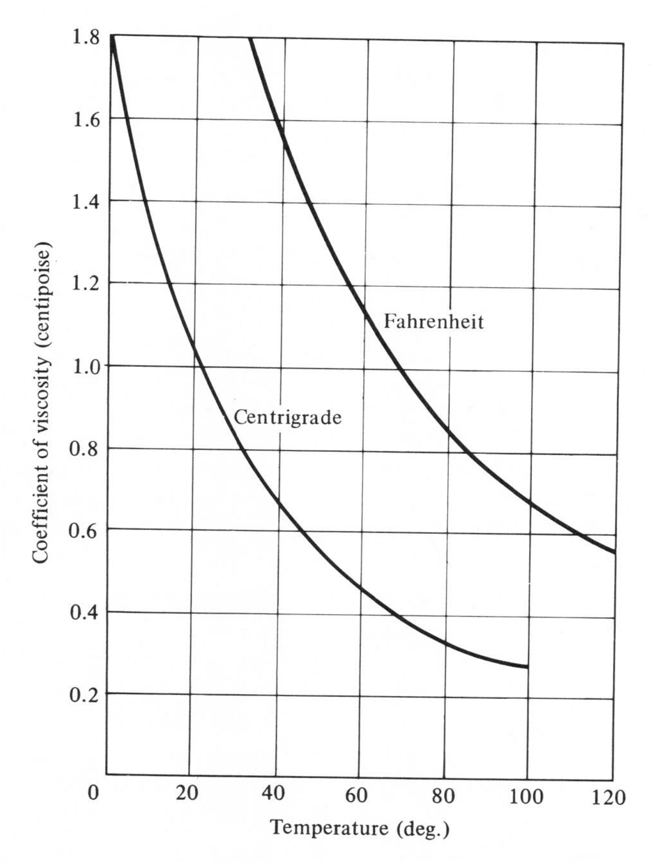

Fig. 11-5. Correction factor to permeability with water at 20.20°C.

various temperatures may be determined on the basis of the relationship between temperature and the coefficient of viscosity for water. The unit of the coefficient of viscosity in the metric system is the dyne-second per square centimeter and is called the *poise*. The coefficient of viscosity of water at 20.20°C (68.36°F) is 0.01 poise or 1 centipoise. The curves in Fig. 11-5 show the values of this coefficient, in centipoises, for a range of temperatures on both the centigrade and Fahrenheit scales.

Since the coefficient of permeability is inversely proportional to the viscosity of the permeating water and directly proportional to its temperature, the coefficient of viscosity can be used as a correction factor by which the permeability determined at one temperature can be reduced to that at the base temperature of 20.20°C.

For the temperature situation mentioned in the first paragraph of this section, the following relationship may be stated:

$$k_0 = 0.86\,k_{80}$$

and

$$k_0 = 1.31\,k_{50}$$

where k_0 is the permeability at 20.20°C (68.36°F). Then,

$$1.31\,k_{50} = 0.86\,k_{80}$$

and

$$k_{50} = \frac{0.86}{1.31} \times k_{80} = 0.656\,k_{80}$$

Thus, if the coefficient of permeability of a soil as measured in the laboratory with water at 80°F is 0.000299 cm/min, then the permeability of the same soil with water at 50°F will be 0.656 × 0.000299 = 0.000196 cm/min.

11.15. ADVANTAGES OF FIELD TESTS. Determinations of the coefficient of permeability in the field by pumping tests or by observations of the rate of infiltration into a test hole are often more reliable than the results of laboratory tests. This statement is particularly true in situations involving flow through relatively sandy soils, because it is difficult to obtain a sample of such soil in the undisturbed state and field conditions cannot be easily duplicated for laboratory testing. Field determinations are particularly appropriate in connection with investigations of seepage into excavations or through the foundation material under dams, in underdrainage problems, or in any other situation involving ground-water flow through natural soil. They are distinctly advantageous in many cases because the results represent the permeability characteristics of the soil in place with a minimum of disturbance and because the "sample" involved is large and the effects of local variations in the soil are averaged out to a considerable extent in the testing process.

11.16. PUMPING TEST. In the pumping test for permeability, a test well having a perforated casing is sunk through the water-bearing soil to an underlying impervious stratum; or, if no impervious stratum is present, the well is sunk to a considerable depth below the water table. The perforations in the well casing must be of sufficient size and must be close enough to one another to permit the ground water to flow into the well as fast as it reaches the casing. Observation wells are put down at various radial distances from the test well, the number of such wells and the radial distances depending on the uniformity of the local situation and the extent to which the investigations are carried out. If these wells need to be cased, the casings should be perforated or equipped with a screen sand point at the lower end.

With the test well and the observation wells installed, initial elevations of the ground water table are recorded. Then pumping is started at the test well at a known uniform rate of discharge and is continued until a steady state of flow into the well is established. This steady state will be indicated when the water level in the test well and the water-table elevations at the observation wells become constant. The drop in elevation of the water table at each of the various observation wells, the radial distances out to these wells, and the rate of discharge from the test well provide the necessary data for computing the coefficient of permeability of the soil within the zone of influence of the test well.

11.17. COMPUTATION OF COEFFICIENT. When a steady state of flow into the well has been established, the ground-water surface is lowered within the zone of influence of the test well and assumes the shape of an inverted hyperbolic cone. A radial section through the well and two observation wells is shown in Fig. 11-6. The flow toward the test well is considered to be radial at the two observation wells, and the hydraulic gradient may be expressed as dh/dr. The area of soil through which the

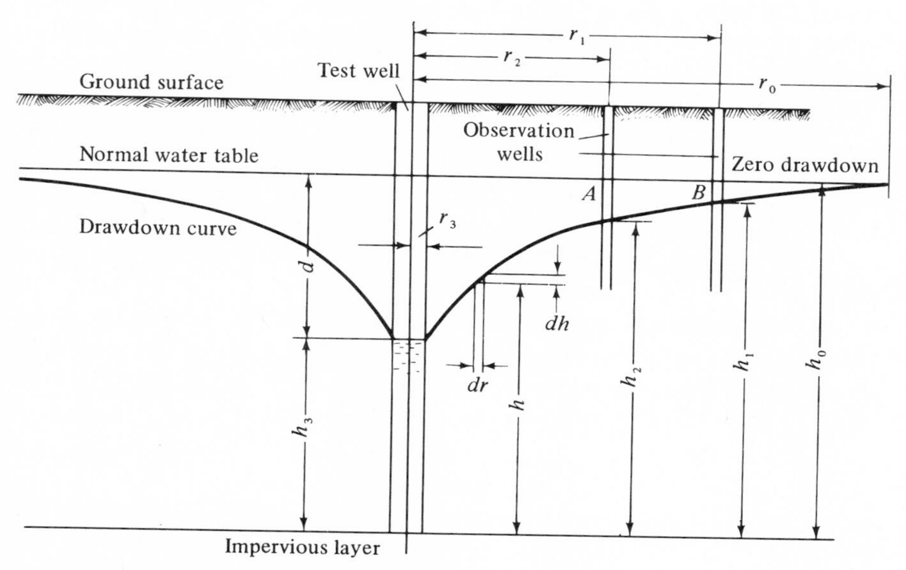

Fig. 11-6. Section through test well and observation wells.

water flows toward the test well is equal to that of a cylindrical surface of radius r and height h, or $2\pi rh$. Substituting these values of the hydraulic gradient and the area in Eq. (11-4), we obtain

$$Q = 2\pi rhk \frac{dh}{dr} \tag{11-12}$$

Integrating between the limits of r and h for the two observation wells, we get

$$\int_{r_2}^{r_1} \frac{dr}{r} = \frac{2\pi k}{Q} \int_{h_2}^{h_1} h\,dh \tag{11-13}$$

from which

$$k = \frac{Q \log_e \frac{r_1}{r_2}}{\pi(h_1^2 - h_2^2)} \tag{11-14}$$

EXAMPLE 11-3. Let it be assumed that a steady state of flow is reached under a uniform discharge of 125 gpm from the test well in Fig. 11-6. Also, assume that the elevations of the water table above an impervious layer (or above the bottom of the test well in case no impervious layer exists) at the observation wells A and B are 34.1 ft and 34.7 ft, respectively; and these wells are located 60 ft and 100 ft, respectively, from the center of the test well. Determine the coefficient of permeability of the soil.

SOLUTION. From the observed data, Q = 125 gpm = 16.67 cfm; h_1 = 34.7 ft; h_2 = 34.1 ft; r_1 = 100 ft; and r_2 = 60 ft.

Substitution of these values in Eq. (11-14) gives

$$k = 0.066 \text{ fpm} = 95 \text{ ft/day}$$

11.18. APPROXIMATE METHOD. If the point A in Fig. 11-6 is taken on the circumference of the test well and the point B is on the circle of zero drawdown of the water table, then the quantity $(h_1^2 - h_2^2)$ in Eq. (11-14) is equal to $(h_0^2 - h_3^2)$. For this case, that equation reduces to

$$k = \frac{Q \log_e \frac{r_0}{r_3}}{\pi(h_0^2 - h_3^2)} \tag{11-15}$$

in which

r_3 = radius of the test well;
r_0 = maximum radius of influence, or radius of circle of zero drawdown;
d = drawdown in the test well at steady flow;
h_0 = distance from water table to impervious layer; and
h_3 = $h_0 - d$.

Since the radius of the circle of zero drawdown will always be several hundred times the radius of the test well, the value of $\log_e r_0/r_3$ will vary over only a relatively narrow range. Because of this fact, Eq. (11-15) may be used to determine an approximate value of k by assuming a value of r_0. It is then unnecessary to observe the drawdown in observation wells.

EXAMPLE 11-4. Let it be assumed that the radius of the test well of Fig. 11-6 is 12 in., the drawdown produced by a steady pumping discharge of 125 gpm is 25.1 ft, and the distance from the impervious layer to the normal water table is 35.2 ft. Assume also that the radius r_0 of zero drawdown is 500 ft. Determine the coefficient of permeability.

SOLUTION. In Eq. (11-15), $Q = 125$ gpm $= 16.67$ cfm; $d = 25.1$ ft; $r_3 = 1$ ft; $r_0 = 500$ ft; $h_0 = 35.2$ ft; and $h_3 = 35.2 - 25.1 = 10.1$ ft. In this case,

$$k = 0.029 \text{ fpm} = 41.9 \text{ ft/day}$$

If r_0 is assumed to be 1000 ft and the observed values are as given in Example 11-4, the value of k is found to be 46.5 ft/day or only 11% greater than the value obtained with r_0 was assumed to be 500 ft. Since the permeability coefficient can seldom be determined with a high degree of precision, even by the most elaborate methods, the approximate method just described will suffice in many cases.

11.19. NONEQUILIBRIUM PUMPING TEST. The pumping test described in Sections 11.16–11.18 is based upon the assumption of a completely homogeneous soil region within the zone of influence of the test well. If barriers to uniform flow exist within this zone, such as dikes, faults, scarps, underground streams, etc., a more sophisticated procedure should be employed, such as Theis' nonequilibrium pumping test and equation, described in Ref. 4.

11.20. TUBE METHOD. A relatively simple procedure, called the "tube method," may be used for measuring the permeability of soil *in situ* below a shallow water table. In this method, which has been described by Frevert and Kirkham (*2*), a tube of known diameter is placed tightly in a hole of the same size to a known depth below a water table, as shown in Fig. 11-7. Then the water in the tube is pumped out to some known elevation below the water table and above the bottom of the tube; and water from the surrounding soil is allowed to flow into the tube through the bottom. The rise of the water level in a measured period of time is observed, and the permeability is computed by means of the following formula:

$$k = \frac{A \log_e (h_0/h_1)}{Et_1} \tag{11-16}$$

in which

A = area of tube;
h_0 = distance from water table to water level in tube at beginning of test;
h_1 = distance from water table to water level in tube at end of test;
t_1 = elapsed time within which distance from water table to water level decreases from h_0 to h_1; and
E = the E-factor, which is a coefficient.

The E-factor is a function of the diameter of the tube, of the depth of the tube below the water table, and of the shape of the soil surface at the bottom of the tube. This factor also takes into account the pattern of flow of the ground water toward the bottom entrance to the tube. It always has the dimension of a length. Frevert and Kirkham used an electric analog to determine values of the E-factor for tubes of various diameters, for various depth-diameter ratios, and for a horizontal soil surface at the bottom of the tube. These values in inch units are given in Table 11-1.

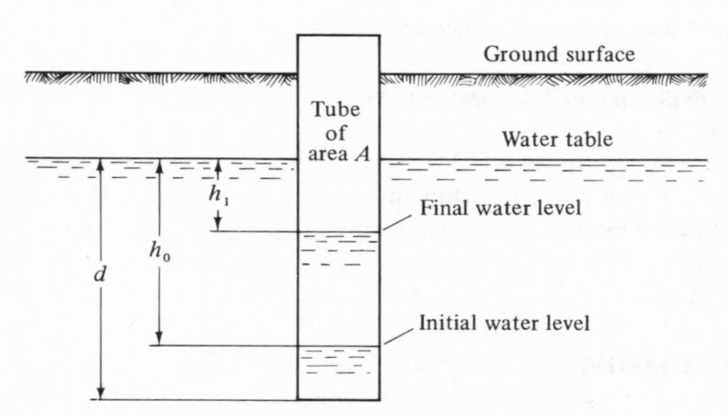

Fig. 11-7. Tube method of determining coefficient of permeability.

TABLE 11-1. Values[a] of *E*-Factor

Depth / *Diameter*	*Diameter of Tube (in.)*						
	1	*2*	*3*	*4*	*5*	*6*	*8*
1	...	...	...	...	...	15.6	20.9
2	...	...	...	...	13.1	15.5	20.8
3	...	...	...	10.3	13.0	15.5	20.7
4	...	...	7.7	10.3	12.9	15.4	20.5
5	...	...	7.7	10.2	12.9	15.3	20.4
6	...	5.1	7.6	10.2	12.8	15.2	20.3
7	...	5.1	7.6	10.1	12.7	15.2	20.2
8	...	5.1	7.5	10.1	12.7	15.1	20.1
10	...	5.0	7.5	9.9	12.5	14.9	...
12	2.5	5.0	7.4	9.8	12.4	...	...
15	2.4	4.9	7.2	9.7	...	...	...
25	2.3	4.6	6.8	...	...	...	...
40	2.1	4.0	...	...	...	...	...
60	1.9	...	...	...	...	...	...
100	1.5	...	...	...	...	...	...

[a] Values are in inch units.

EXAMPLE 11-5. Let it be assumed that the tube in Fig. 11-7 has a radius of 3 in. and is driven to a depth of 24 in. below a water table. After the water in the tube has been pumped out until the surface is 18 in. below the water table, the water in the tube rises to an elevation 6 in. below the water table in 34.5 min. Determine the coefficient of permeability.

SOLUTION. In this case, $A = 28.27$ sq in.; $d = 24$ in.; $d/2r = 4$; $h_0 = 18$ in.; $h_1 = 6$ in.; and $t_1 = 34.5$ min. Also, from Table 11-1, the *E*-factor is 15.4. Substitution of these values in Eq. (11-16) gives

$$k = \frac{28.27 \times \log_e 3}{15.4 \times 34.5} = 0.584 \text{ in./min}$$

11.21. PRECAUTIONS IN USE OF TUBE METHOD. In carrying out the tube method of measuring permeability, it is essential that the tube fit tightly in the hole in the soil in order to prevent channeling of the water down the outside of the tube. Also, it is desirable that the soil at the bottom of the hole be as nearly undisturbed as possible. If the soil is badly puddled at this surface, the pores may be sealed over to such an extent that the accuracy of the results is questionable. When the soil is of a type which puddles readily, it may be advisable to pump the tube out several times and allow it to refill in order to flush out the soil pores near the surface at the bottom of the hole. Tubes of any diameter may be used in this method. However, there are practical limits to which tubes of the larger diameters can be driven, and they are most useful in shallow-depth determinations. Also, since the flow of water into the tube is upward, this method of measurement tends to accentuate the coefficient of permeability in the vertical direction in nonisotropic soils.

11.22. PIEZOMETER METHOD. For measurement of permeability at greater depths, a thin-walled electrical conduit having an inside diameter of 1 in. may be used. This method is called the "piezometer method." The pipe is driven a short distance into the soil, a soil auger with a diameter of $\frac{15}{16}$ in. is bored through the pipe and into the soil to a depth of about 4 in. below the bottom, and the soil is removed. The pipe is then driven into the soil 4 in., and the augering is repeated. This process is continued until the bottom of the pipe has reached the desired depth, there being finally a space 4 in. deep below the bottom of the pipe, as illustrated in Fig. 11-8. This method of driving the pipe prevents compaction of the soil "sample" to be tested.

After water has been pumped from the tube several times to flush out the soil pores in the cavity at the lower end, the rise of the water in the tube and the time interval corresponding to the rise are measured. These data may be utilized in Eq. (11-16) to determine the permeability coefficient of the soil in the region of the cavity. The *E*-factor for this method of measurement is dependent on the diameter of the tube and

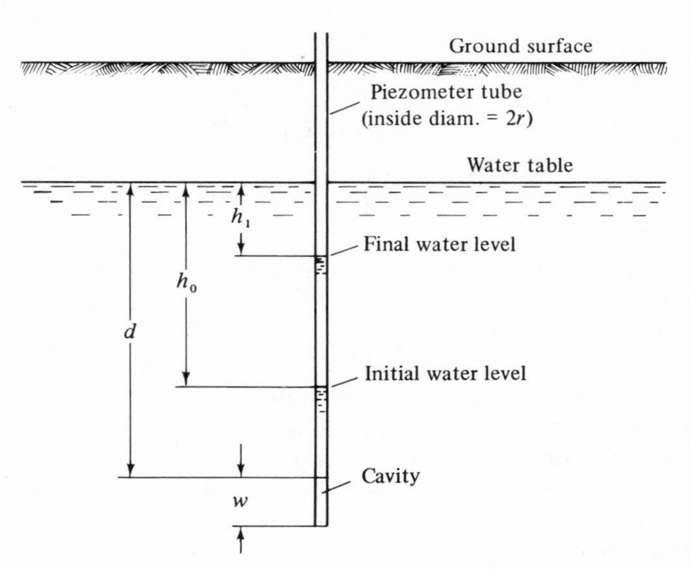

Fig. 11-8. Piezometer method for determining coefficient of permeability.

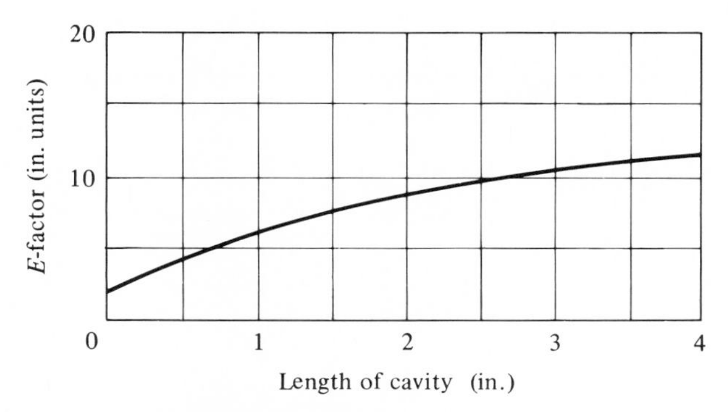

Fig. 11-9. Values of *E*-factor for cavities 1 in. in diameter.

cavity and on the length of the cavity below the bottom of the tube. Theoretically, it is also dependent on the depth of the cavity below the water table and on the distance from the cavity to an impermeable layer; but, if these dimensions are large in relation to the length of the cavity, their effect is negligible. Values of the *E*-factor which have been determined by an electrical analogy are given in Fig. 11-9 for cavities 1 in. in diameter and from 0 to 4 in. long.

The elevation of the water level in the piezometer tube may be measured by means of a well-insulated and waterproofed copper wire which projects slightly from the insulation at one end. Connect the other end of the wire and the metal tube to a hearing-aid battery and an earphone. Then lower the projecting end of the wire into the tube until a click sounds in the earphone to indicate that the point of the wire touches the water surface. Its elevation can be measured.

The flow of water into the cavity is mainly horizontal. Therefore, in a nonisotropic soil, the piezometer method tends to accentuate the coefficient of permeability in the horizontal direction. Also, the volume of the "sample" through which water moves to reach the cavity is relatively small. These facts indicate that this method is adaptable to the measurement of horizontal permeability in various strata of stratified soils.

11.23. USE OF AUGER HOLE. van Bavel and Kirkham (*5*) have suggested a method of measuring soil permeability which would appear to be the ultimate in simplicity and practicability for shallow-depth determinations. It consists merely of boring an auger hole to some measured depth below a water table, as shown in Fig. 11-10, and observing the rate of rise of water in the hole. In this method, as in those described previously, it is advisable to pump water from the hole several times and allow the hole to refill, in order to flush out the soil pores at its sides. If

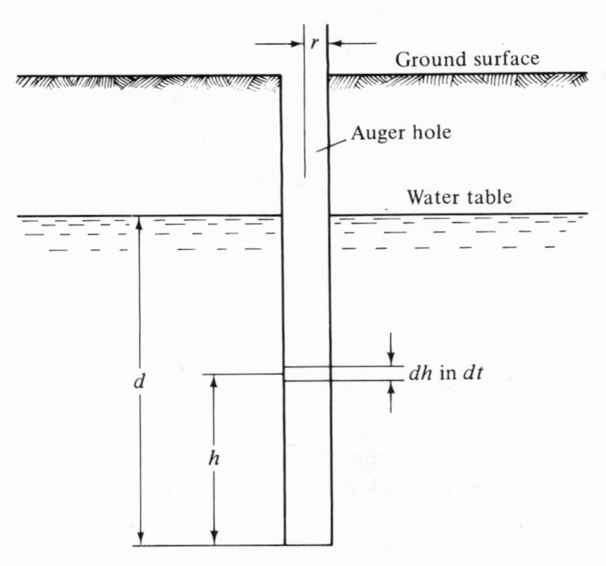

Fig. 11-10. Auger-hole method for determining coefficient of permeability.

the auger hole extends completely through a pervious stratum to an impervious layer, the flow situation is subject to exact mathematical analysis. However, the results can be used to obtain a good approximation, even in the absence of an impervious layer, if the ratio of the depth of the auger hole to its diameter is large. The formula for determining the permeability coefficient by this procedure is

$$k = 0.617 \frac{r}{Sd} \cdot \frac{dh}{dt} \qquad (11\text{-}17)$$

in which

h = depth of water in hole at the time dh/dt is determined;
dh/dt = rate of rise of water level in hole at depth h;
r = radius of hole;
d = depth of hole below water table; and
S = a coefficient which is dependent on the ratios h/d and r/d.

Values of the coefficient S are given in Fig. 11-11.

A suggested procedure for measuring the coefficient of permeability by the auger-hole method is as follows: Bore a hole, say 4 in. in diameter, down to an impermeable layer if one exists; if there is no impermeable layer, bore the hole to a depth below the water table equal to at least 10 times the diameter of the hole. Allow the water in the hole to rise to a constant level, which is considered to be the elevation of the water table. Measure the depth of the hole below the water table. Pump the water out of the hole and allow it to refill two or three times—or more if the soil puddles readily. Then pump the water out of the hole again and observe the rate of rise of the water level at each of several elevations. This may be done by measuring the rise in water level in a short period of time, say 1 min, and dividing the increment of rise by the increment of time. A stop watch, a hook gage, and a flashlight will be convenient equipment for these measurements. The radius of the hole being known, select from Fig. 11-11 the value of S for each value of h at which the rate of rise was

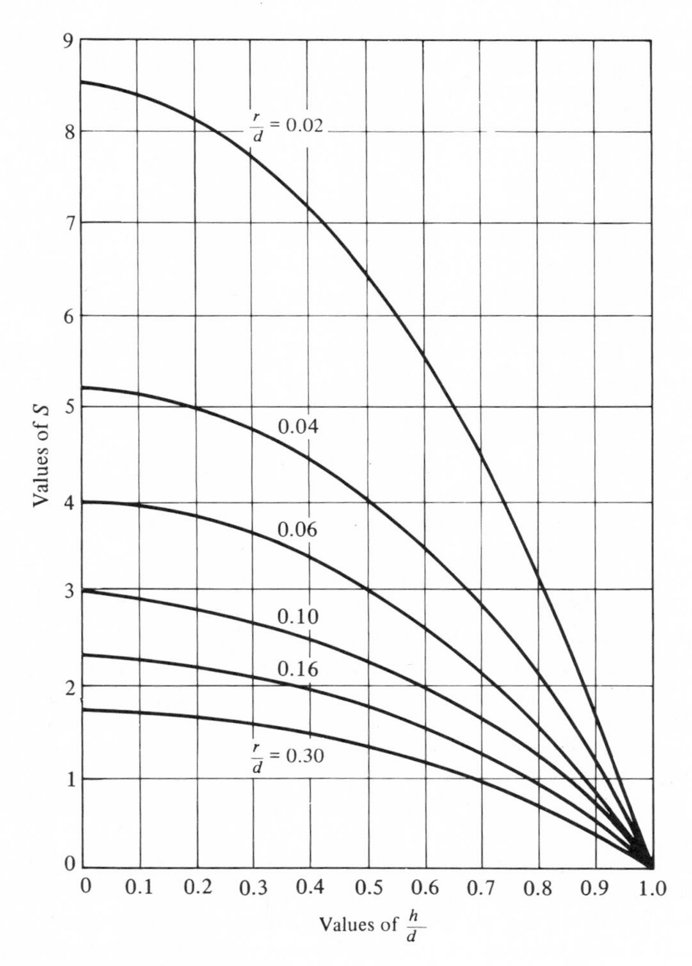

Fig. 11-11. Values of S in Eq. (11-17).

measured. Substitution of values in Eq. (11-17) will give values of the coefficient of permeability of the soil in the vicinity of the hole.

In this method the flow into the auger hole is almost wholly horizontal. It is, therefore, especially adapted to measurement of the horizontal permeability coefficient in nonisotropic soils. If the soil is stratified, the flow into the auger hole is from all the strata through which the hole is driven. The resulting coefficient is a composite value representing all these strata, and individual values cannot be isolated.

More recently, Boast and Kirkham (*1*) have published a more exact mathematical method for solving the problem of flow to a finite auger hole partly penetrating a water-saturated porous stratum. This method permits measurements of the coefficient of permeability in a wider range of sizes and depths of auger holes than previous methods and is applicable regardless of the proximity or depth of an impermeable layer.

PROBLEMS

11.1. Draw a graph of velocity versus hydraulic gradient in a soil for which the proportionality constant equals 0.21.

11.2. The fall of a certain stream is 5 ft/mile. What is the hydraulic gradient?

11.3. State the Darcy law of flow of water through soil. Define the coefficient of permeability and the coefficient of percolation.

11.4. Explain the term "velocity of approach," as used in connection with the Darcy law.

11.5. The coefficient of permeability of a soil is 0.25 cm/min, and it has a porosity of 0.60. What is the coefficient of percolation?

11.6. Some soils are said to be nonisotropic with respect to flow of gravitational water. What does this mean?

11.7. Describe and draw sketches of apparatus for two types of laboratory tests for measuring the permeability characteristics of soil.

11.8. A constant-head permeameter is set up with a soil specimen 8 in. long and 3.0 in. in diameter. The vertical distance between the head-water and tail-water surfaces is 12 in. In a test run, 361.5 lb of water passes through the sample in a period of 18 hr 20 min. What is the coefficient of permeability of the soil?

11.9. A falling-head permeameter is set up with a soil sample 6 in. long and 2 in. in diameter. The area of the standpipe is 0.33 sq in. In a test run, the water in the standpipe falls from an elevation of 48 in. above the tail water to 21.2 in. in 7 hr 26 min 32 sec. What is the coefficient of permeability of the soil?

11.10. The coefficient of permeability of a soil, as determined in the laboratory at a temperature of 76°F, is 0.00035 cm/min. What is the coefficient of permeability of the same soil when the permeating water is at 40°F?

11.11. A test well is installed in permeable sand to a depth of 30 ft below the water table, and observation wells are put down at distances of (a) 50 ft and (b) 100 ft from the test well. When pumping at the rate of 255 gpm reaches a steady state, the water table at (a) is lowered 3.8 ft and at (b) 3.1 ft. Compute the coefficient of permeability of the sand.

11.12. A test well 30 in. in diameter is sunk, through a water-bearing sand, down to an impervious layer which is 24 ft below the normal water table. Steady pumping at the rate of 200 gpm causes the water in the test well to be lowered 9 ft 4 in. What is the approximate value of the coefficient of permeability of the sand?

11.13. A thin-walled tube having an inside diameter of 4.5 in. is driven into the soil to a depth of 36 in. below the water table. After the soil is excavated from inside the tube, the water is pumped out to an elevation 28 in. below water table. The water then rises to within 6 in. of the water table in 4 hr 36 min. What is the coefficient of permeability of the soil?

11.14. A 1 in. diameter piezometer tube is driven to a depth of 36 in. below the water table and the soil is augered from inside the tube as it is driven. Then a cavity is augered to a depth of 2 in. below the bottom of the tube. The time required for water to rise from 30 in. to 8 in. below the water table is 5 hr 16.5 min. Calculate the coefficient of permeability of the soil.

11.15. A hole 6 in. in diameter is bored in the soil to a depth 75 in. below the water table. After the hole has been pumped out several times and allowed to refill, it is pumped out again and the rate of rise of water at an elevation 40 in. below the water table is observed to be 0.75 in. in 1 min. What is the coefficient of permeability of the soil?

REFERENCES

1. Boast, C. W., and Don Kirkham. "Auger Hole Seepage Theory." *Proc. Soil Science Soc. Amer.* **35,** 365 (1971).
2. Frevert, Richard K., and Don Kirkham. "A Field Method for Measuring the Permeability of Soil Below a Water Table." *Proc. Highway Research Board* **48,** 433–442 (1948).
3. Taylor, Donald W. *Fundamentals of Soil Mechanics.* John Wiley, New York, 1948.
4. Todd, David K. *Ground Water Hydrology.* John Wiley, New York, 1959.
5. van Bavel, C. H. M., and Don Kirkham. "Field Measurement of Soil Permeability Using Auger Holes." *Proc. Soil Science Soc. Amer.* **13** (1948).

12

Flow Nets and Seepage Forces

12.1. NATURE OF FLOW NET. The study of gravitational flow and the computation of seepage quantities through the ground or through such soil structures as earth dams and levees is greatly facilitated by the use of a graphical device known as a flow net. A flow net consists of two groups or families of curves which bear a fixed relationship to each other and which represent the pattern of flow and the dissipation of the head or driving force that causes the flow. The curves of one of these groups are called *equipotential lines.* An equipotential line passes through points of equal head and may be thought of as a potential contour. The curves of the other group are known as *flow lines* or *stream lines.* Each flow line represents the path which a particle of water follows as it travels from point to point in a soil mass.

The space between two adjacent equipotential lines is known as an *equipotential space* and represents a definite increment of drop in head. The space between two adjacent flow lines is called a *flow path.* When a true flow net has been constructed for any problem of gravitational flow and the coefficient of permeability of the soil is known, the quantity of seepage water flowing through the mass of soil involved can be easily and quickly computed. A flow net exists for any situation involving gravitational flow in a saturated soil mass, but the applications of flow nets are usually limited to two-dimensional or plane flow.

Flow nets are based upon certain definite mathematical relationships, but a presentation of these relationships is beyond the scope of this text. The following discussion will be limited to certain general characteristics of flow nets and the manner in which they are used in seepage problems.

12.2. CHARACTERISTICS OF FLOW LINES. A simple case of rectilinear seepage, which may be used to illustrate the basic characteristics of a flow net, is that of a constant-head permeameter in which water flows through a mass of soil of rectangular cross section 1 unit wide (length normal to

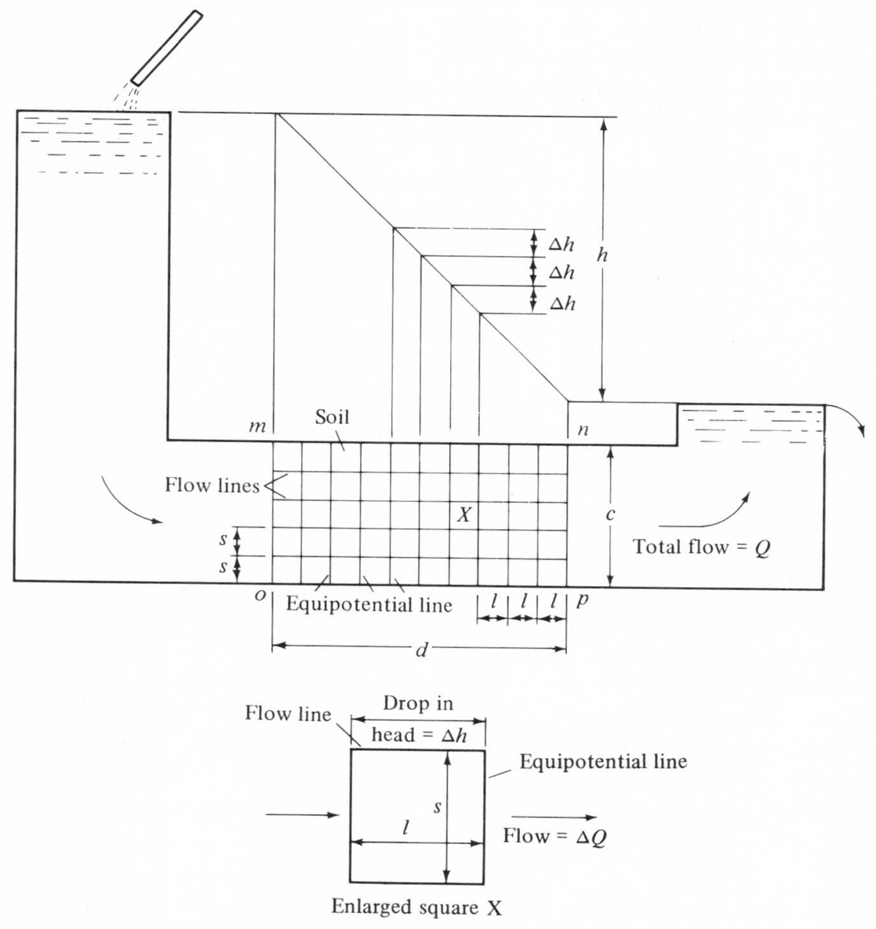

Fig. 12-1. Simple flow net.

the page), *c* units high, and *d* units long. A longitudinal cross section of such a mass of soil is shown in Fig. 12-1. In this very simple case, all the flow lines are parallel to each other and are shown as a series of lines drawn parallel to the longitudinal axis of the soil mass. Actually, there are a very large number of flow lines through the soil, since each particle of water entering the soil on the longitudinal section shown traces a flow line. For practical reasons, however, only a small number of lines are drawn, this number being arbitrarily chosen by the person drawing the flow net.

Although the number of flow lines drawn is purely arbitrary, the lines of a true flow net must be spaced in such a way that the quantity of water flowing in each flow path will be the same. In the simple case under consideration, this requirement results in flow paths of the same width because of the rectilinear nature of the seepage; but in an actual flow problem the flow paths will not be of the same width, as will be shown in later illustrations.

12.3. CHARACTERISTICS OF EQUIPOTENTIAL LINES. The equipotential lines in a flow net are oriented at right angles to the flow lines; and it is a requirement of a true flow net that the geometrical figures formed by the two families of curves must be essentially squares. In the simple case under consideration in Fig. 12-1, the figures are true squares, but this ideal condition results from the rectilinear nature of the flow. In a more practical case involving curvilinear flow, the figures cannot be true squares. However, they must have right angles at the corners and the two median dimensions of each figure must be equal. These requirements are associated with a further requirement that each equipotential space must represent an equal drop in head.

12.4. BOUNDARY LINES OF FLOW NET. In any flow problem, the boundary characteristics of both the flow lines and the equipotential lines must be known, or it must be possible to estimate the positions of these lines, before a flow net can be drawn. In Fig. 12-1 the lines *mn* and *op* are the boundary flow lines, while *mo* and *np* are the boundary equipotential lines. Line *mo* represents the surface of the soil at which the water enters, and the head at this surface is a maximum and equal to *h*. Line *np* represents the surface at which the water leaves the soil. Since the total head *h* has been used up in forcing the water through the soil to *np*, the head at this surface is zero.

12.5. MEASUREMENT OF HEAD. The head at any point in a soil mass may be determined by inserting a piezometer tube or standpipe to the point and observing the height to which water rises in the tube. For the conditions represented in Fig. 12-1, the water in two piezometer tubes terminating at any points on the line *mo* will rise to the same height h above the tail water. Hence, *mo* is an equipotential line. Likewise, the water in two tubes terminating at any points in line *np* will rise to the tail-water level; the potential at this line is taken equal to zero. Piezometer tubes placed at points on any intermediate equipotential line will show a water level somewhere between the elevation of the tail water and that of the head water.

12.6. DETERMINATION OF FLOW FROM FLOW NET. The total quantity of water Q flowing through a unit width of a soil mass is equal to the sum of the quantities in all the flow paths of the flow net. But it is a basic requirement of a flow net that every flow path must transmit the same quantity of water. Therefore, the quantity in each flow path, which will be designated as ΔQ, must be equal to the total flow divided by the number of flow paths. Likewise, the total head h is the sum of the drops in head in all the equipotential spaces of the flow net; and the drop in head in each such space, which will be designated as Δh, must be equal to the total head divided by the number of equipotential spaces.

In the case of the net shown in Fig. 12-1, the flow through any single square, as X, is ΔQ and the drop in head from the upstream face of this square to the downstream face is Δh. Applying the Darcy law, $Q = kiA$, to this arbitrarily selected square and remembering that the cross-sectional area through which flow takes place is equal to the height s of the square times unity, we may write

$$\Delta Q = \frac{k\,\Delta h s}{l} \tag{12-1}$$

Since the figures are squares, $s/l = 1$ and

$$\Delta Q = k\,\Delta h \tag{12-2}$$

If the complete flow net has N spaces between equipotential lines, then $N\,\Delta h = h$ or

$$\Delta h = \frac{h}{N} \tag{12-3}$$

The flow through any square, and thus through any flow path, is

$$\Delta Q = k\,\frac{h}{N} \tag{12-4}$$

Also, if there are F flow paths in the net,

$$\Delta Q = \frac{Q}{F} \tag{12-5}$$

Then the total flow through all the flow paths is

$$Q = kh\,\frac{F}{N} \tag{12-6}$$

According to Eq. (12-6), the total quantity of water that will seep through a unit width (length normal to the page) of a soil structure can be found as follows: Draw a true flow net on a cross section of the structure; and multiply together the coefficient of permeability, the total head, and the ratio of the number of flow paths to the number of equipotential spaces. To find the total seepage through the entire structure, it is only necessary to multiply the flow through a unit width by the total width of the structure.

The validity of the flow-net procedure for computing seepage quantities may be demonstrated by comparing the result obtained by use of Eq. (12-6) with the result obtained by use of Eq. (11-4) when both equations are applied to the conditions represented in Fig. 12-1, as in the following example.

EXAMPLE 12-1. Assume that the soil mass for which the flow net is shown in Fig. 12-1 is 1 ft wide, 1 ft high, and 2 ft long; that the head is 4 ft; and that $k = 0.15$ ft/hr. Determine the total seepage: (a) by Eq. (12-6) and (b) by Darcy's law or Eq. (11-4).

SOLUTION. (a) In this case, the distance d in Fig. 12-1 is 2 ft and the distance c is 1 ft. In the figure, there are 5 flow paths and 10 equipotential spaces; hence, the ratio of the number of flow paths to the number of equipotential spaces is $\frac{1}{2}$. By Eq. (12-6),

$$Q = 0.15 \times 4 \times \frac{5}{10} = 0.3 \text{ cu ft/hr}$$

(b) By Eq. (11-4),

$$Q = kiA = 0.15 \times \frac{4}{2} \times 1 = 0.3 \text{ cu ft/hr}$$

12.7. TYPICAL FLOW NETS. Of course, for the simple case of rectilinear flow indicated in Fig. 12-1, the use of a flow net would be superfluous, since the Darcy equation can be applied directly. However, in an actual seepage problem the flow is practically always curvilinear and the use of a flow net greatly simplifies the solution of such a problem. Flow

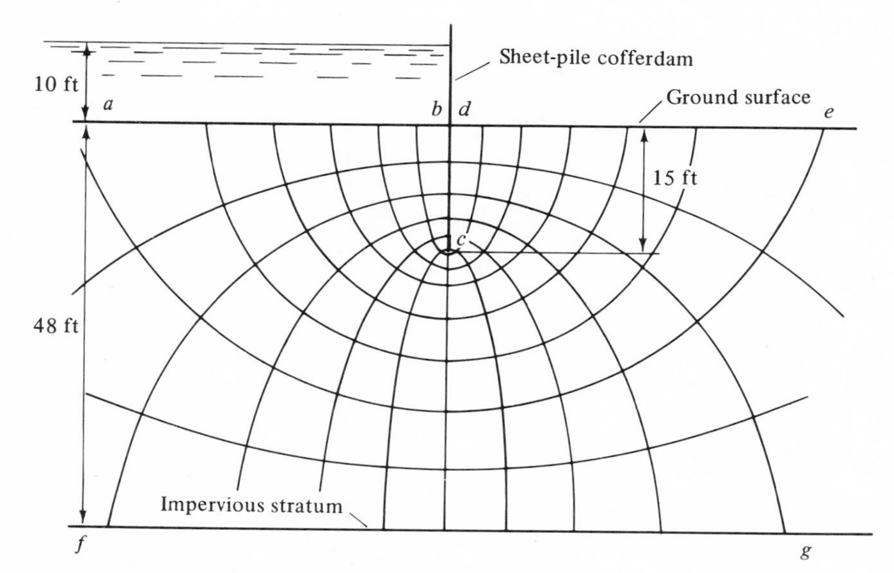

Fig. 12-2. Flow net for sheet-pile cofferdam.

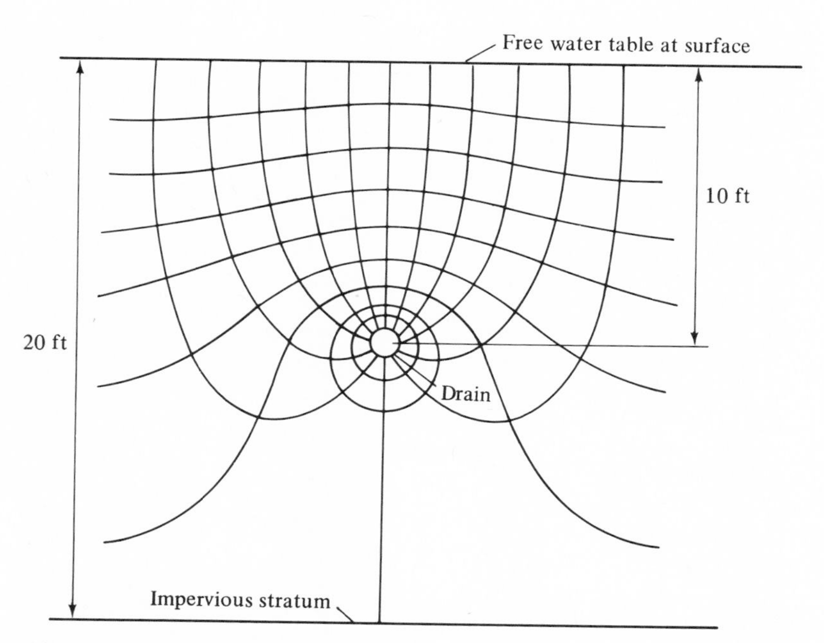

Fig. 12-3. Flow net for drain.

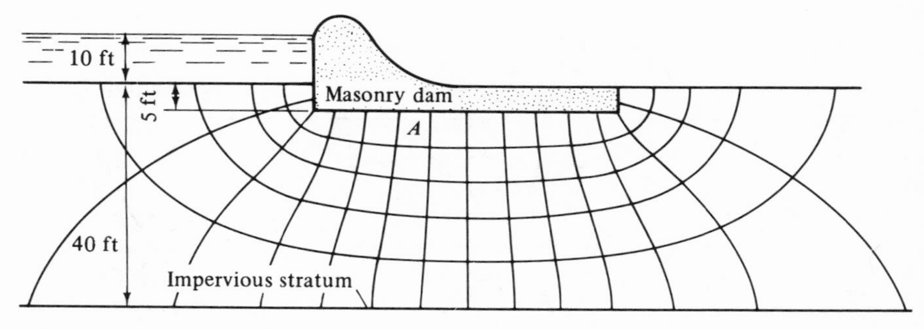

Fig. 12-4. Flow net for masonry dam.

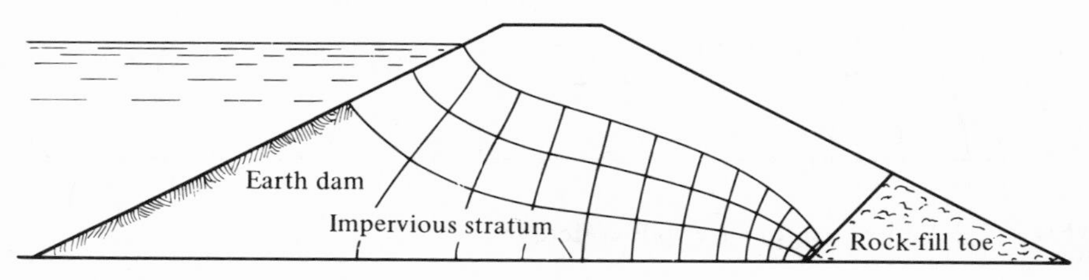

Fig. 12-5. Flow net for earth dam with rock-fill toe.

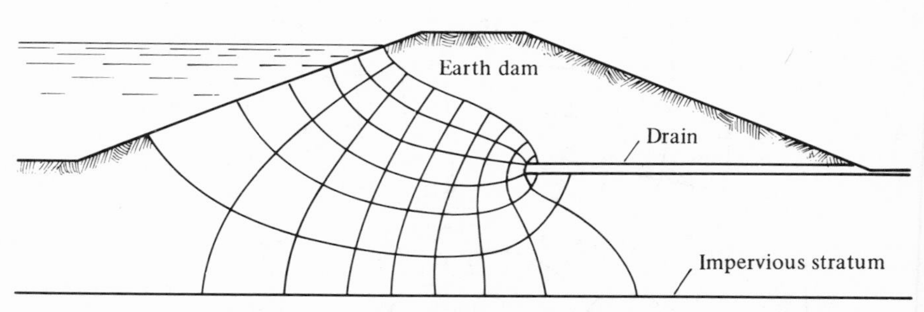

Fig. 12-6. Flow net for earth dam with downstream drain.

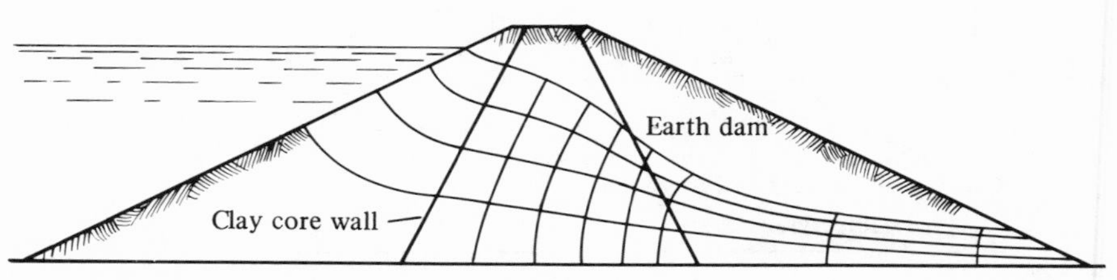

Fig. 12-7. Flow net for earth dam with clay core wall.

nets for a number of different types of structures involving seepage problems are shown in Figs. 12-2 to 12-7. For practice the student should identify the flow lines, the equipotential lines, the boundary conditions, and the other characteristics of each of these flow nets; and should compute the value of the ratio F/N in order to gain a measure of familiarity with this graphical device for solving seepage problems.

12.8. SUMMARY OF FUNDAMENTAL PROPERTIES OF FLOW NETS. The foregoing discussion of flow nets may be summarized by stating five fundamental properties of flow nets, as follows:

1. Every flow path of a net must transmit the same quantity of seepage.
2. The potential difference between any pair of adjacent equipotential lines must be the same as that between any other pair.
3. Flow lines must intersect equipotential lines at right angles.
4. All geometrical figures formed by the flow lines and the equipotential lines must be essentially squares.
5. At any point in the net, the spacing of lines is inversely proportional to the hydraulic gradient and to the seepage velocity.

12.9. PROCEDURES FOR DRAWING FLOW NETS. There are a number of procedures available by which an engineer may draw a flow net for the solution of a seepage problem. Among these are:

1. Mathematical solution
2. Electrical analogy
3. Sand models
4. Trial sketching

Each of the first three methods requires specialized knowledge and may require extensive apparatus. It is probable that the method of trial sketching is the best method available to the beginner in this field. By strict adherence to the rules given in Section 12.8 and by the exercise of patience and a certain amount of intuition, an engineer can—after having acquired some experience—sketch a flow net with a fair degree of accuracy. The precision of this method is probably comparable with the precision with which it is possible to determine the permeability characteristics of the soil involved. Many large engineering organizations rely principally on trial sketching for obtaining flow nets, and then use the electrical-analogy method or the sand-model method to check the salient features of the sketched flow net.

12.10. STEPS IN TRIAL-SKETCHING METHOD. In the method of trial sketching, the first step is to draw to a convenient scale a cross section of the soil structure at a point along the axis normal to the direction of seepage flow. Next, it is necessary to establish the two boundary flow lines and the two boundary equipotential lines. Then, by trial, sketch in the intermediate flow lines and equipotential lines, adhering rigidly to right-angle intersections and essentially square figures. Locate the unknown lines as well as possible, and continue sketching until an inconsistency develops. Each inconsistency will indicate the direction and magnitude of change for a second trial. Successive trials must be made until the net is reasonably consistent throughout.

Do not use too many flow lines. Four, five, or six will generally suffice. Keep in mind that every transition in a homogeneous soil is smooth and either elliptical or parabolic in shape. The size of the squares in each flow path will change gradually. The beginner in this field should study flow nets constructed by experts and should try to develop a sense of fitness and intuition to assist in trial sketching.

12.11. LOCATION OF BOUNDARY LINES FOR EARTH DAM. The boundary lines of a flow net sometimes coincide with certain physical boundaries of the cross section of the structure or mass of soil. They can then be located by inspection. In other cases, one of the flow boundaries may be a line of saturation within the soil mass. A typical example of such a situation is an earth dam with water impounded on one side. Seepage water passing through the dam will saturate the lower portion, while the upper portion remains relatively dry. The seepage flow occurs below the upper limit of saturation. This limit, which is sometimes called the *phreatic surface*, is a true ground water table and marks the locus of all points in the soil mass where the pressure in the soil water is equal to atmospheric pressure.

A. Casagrande (*2*) has given certain empirical rules for estimating the location and shape of the phreatic surface or line of saturation on a cross section of a dam. In general, the line of saturation is parabolic in shape, but it deviates from a parabola at the upstream or entrance face of the dam and at the exit face; the amount and character of the deviations depend on the details of these portions of the dam. In Fig. 12-8 is represented a cross section of a trapezoidal dam of homogeneous soil. For simplicity, it is assumed that this dam rests on an impervious foundation and that all seepage water therefore flows through the dam. Then the line of contact with the foundation is one boundary flow line. The upstream or entrance face of the dam represents one equipotential boundary. At the downstream toe of the dam there is provided a horizontal drainage layer which extends some distance back under the structure and acts as a collecting gallery for the seepage water. The line of contact be-

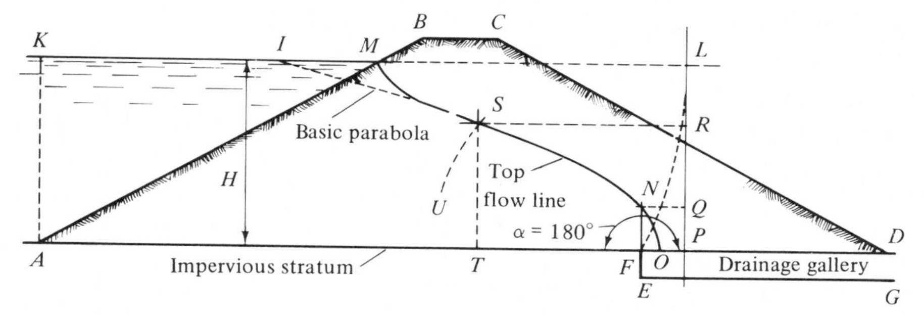

Fig. 12-8. Boundary lines for dam with drainage gallery at downstream toe.

tween the dam and this drainage element is the outlet face and represents the downstream equipotential boundary of the flow net.

12.12. CONSTRUCTION OF BASIC PARABOLA. The basic parabola from which the upper flow line in Fig. 12-8 is developed has its focus at point F, which is at the intersection of the lower flow line and the downstream equipotential boundary. A point I on the parabola near the upstream face of the dam is taken on the water surface of the reservoir at a distance back of the dam equal to 0.3 times the horizontal distance from the intersection of the water line and the dam to the upstream toe of the dam; thus, $MI = 0.3\,MK$. With the focus and one point on the parabola established, the basic parabola may be drawn.

In order to clarify this construction, step-by-step instructions for locating the basic parabola will be given for the dam in Fig. 12-8. After the cross section $ABCD$ of the dam has been drawn to scale and the positions of the reservoir water line KM and the drainage gallery FEG have been established, locate the point K vertically above the point A, and lay off the distance MI equal to $0.3\,MK$. This point I is on the basic parabola and F is its focus. Now, remembering that a parabola is the locus of all points which are equidistant from a point and a line, the directrix of the parabola may be located as follows: Draw the arc FL with I as the center and IF as the radius; extend the water line KM to intersect this arc at L; and draw the vertical line LP, which is the directrix. Next, locate the vertex O of the parabola by bisecting the distance FP. Also, locate the point N by laying off FN parallel to the directrix and equal to FP. We now have three points on the parabola, namely, O, N, and I. If an additional point on the parabola is desired, as at S, proceed in the following manner: In the vicinity of S, draw a line ST that is parallel to the directrix; then, with the point F as the center and the distance PT ($=RS$) as the radius, draw an arc SU which intersects the line ST at the required point S. The parabola $ONSI$ may now be drawn.

12.13. LOCATION OF UPPER LIMIT OF SATURATION. The upper end of the line of saturation in Fig. 12-8 will obviously be at point M. This point may be connected to the basic parabola by drawing by eye an easy transition curve which must intersect the upstream face of the dam at right angles, since AM is an equipotential line and the saturation line is a flow line. For the particular type of downstream drainage arrangement shown in Fig. 12-8, the basic parabola does not need modification at the lower end. Therefore, the line of saturation for this case is the

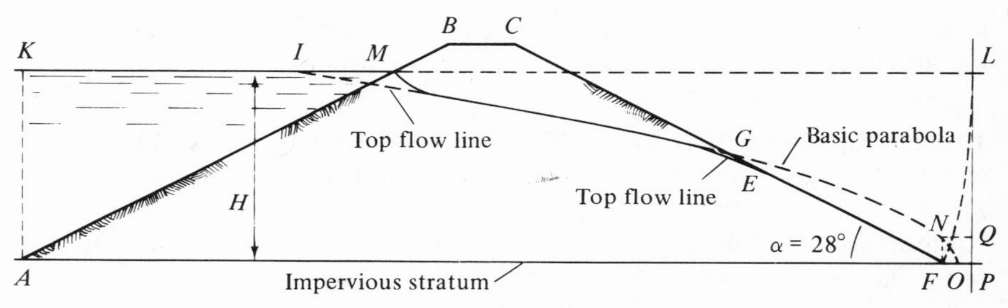

Fig. 12-9. Boundary lines for dam without special drainage at downstream toe (for use of Fig. 12-10, $a = GE$ and $b = GF$).

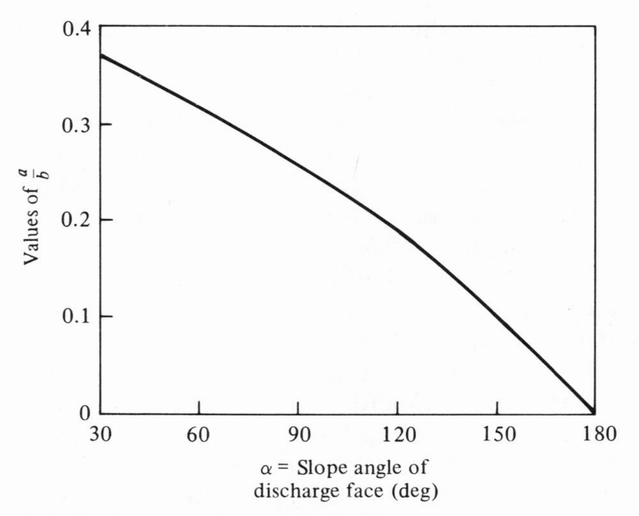

Fig. 12-10. Relation between slope of discharge face and ratio a/b.

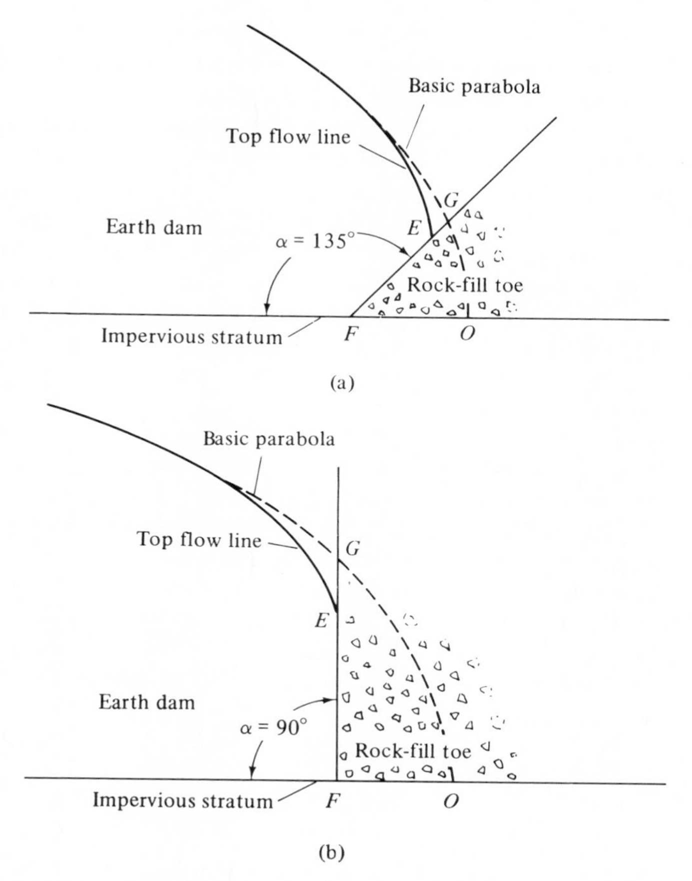

Fig. 12-11. Lower end of saturation line (for use of Fig. 12-10, $a = GE$ and $b = GF$).

line $MSNO$. This line having been located, all four boundary lines are known and sketching of the flow net can begin.

For other types of downstream drainage, the procedure for locating the basic parabola and the transition section at the upper end are the same as outlined for the conditions in Fig. 12-8, but the lower end of the parabola has to be modified to fit the particular type of drainage situation involved. For example, suppose a dam similar to that shown in Fig. 12-8 is not provided with any kind of drainage structure or collecting gallery, as shown in Fig. 12-9. In this case, the line of saturation would be relatively high and seepage water would break out on the downstream slope of the dam. The focus of the basic parabola is at the point F, which is at the intersection of the lower boundary flow line and the outlet face of the dam through which the seepage water emerges. Since this parabola extends beyond the limits of the dam, it is obvious that the actual line of saturation must lie somewhere below the parabola. The distance from the basic parabola to the point E at which the seepage water breaks out is a function of the angle α, which is the clockwise angle between the base of the dam and the discharge face. The ratio of the distance GE, denoted by a, to the distance GF, denoted by b, is given in Fig. 12-10 for various values of α from 30° to 180°. After the break-out point E is located, a transition curve from this point to the basic parabola can be sketched by eye, and the upper limit of saturation can be completed. The lower ends of the lines of saturation for two types of drainage galleries are shown in Fig. 12-11.

12.14. EFFECT OF CORE WALL IN DAM. When seepage water passes from a mass of soil having a certain permeability coefficient to another mass having a different permeability coefficient, the flow lines are deflected at the boundary between the two masses, as indicated in Fig. 12-12, much the same as a beam of light is deflected when it enters a body of water. This fact may be of considerable importance when seepage through certain types of structures is being considered, as, for example, an earth dam having a relatively impermeable core. When the situation repre-

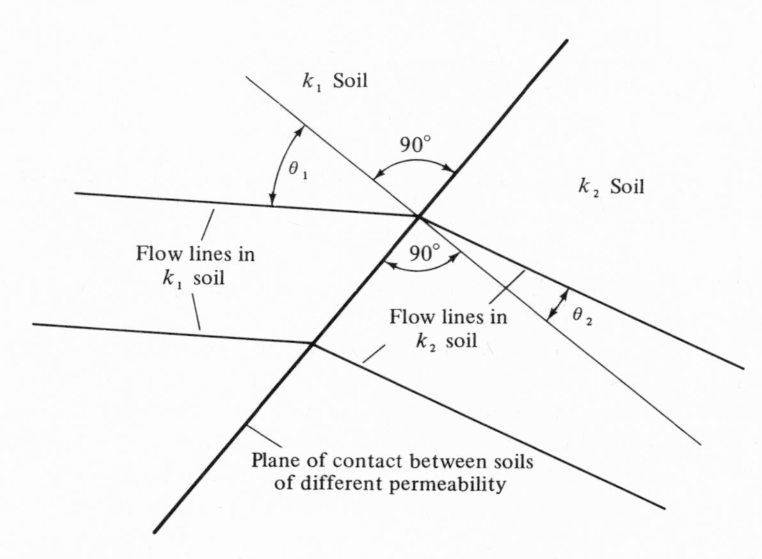

Fig. 12-12. Deflection of flow lines at surface of contact between soils of different permeability.

sented in Fig. 12-12 exists, the flow lines are deflected an amount which is proportional to the values of the permeability coefficients, in accordance with the following relation:

$$\frac{\tan \theta_1}{\tan \theta_2} = \frac{k_1}{k_2} \tag{12-7}$$

in which k_1 and k_2 are coefficients of permeability of the soils; θ_1 is the angle between flow line and normal to line of contact of soils, on side of k_1 soil; and θ_2 is the angle between flow line and normal to line of contact of soils, on side of k_2 soil.

A flow net for an earth dam having a dense clay core wall is shown in Fig. 12-7.

12.15. FLOW NET FOR NONISOTROPIC SOIL. All of the foregoing discussion of flow nets has referred to gravitational flow through isotropic soils, that is, soils in which the coefficient of permeability is the same in all directions. Not infrequently, soil deposits are more permeable in the horizontal direction than in the vertical direction, or vice versa. If this situation exists, it may be necessary to transform the scale of the cross section of the soil mass before starting to sketch a flow net. When the horizontal permeability is the greater, this transformation is made by dividing the horizontal dimensions of the cross section by the quantity $\sqrt{k_h/k_v}$, where k_h is the coefficient of permeability in the horizontal direction and k_v is the coefficient of permeability in the vertical direction.

After the scale transformation has been made, a flow net is drawn in the normal manner on the transformed section, adhering to the rules of right-angle intersections and square figures, as shown in Fig. 12-13(a). The final flow net is then obtained by retransforming the cross section, including the flow net, back to the natural scale, as in Fig. 12-13(b). The flow lines and equipotential lines of the flow net thus obtained will not intersect at right angles; nor will the figures be squares. The value of the coefficient of permeability to be used when Eq. (12-6) is to be applied to nonisotropic soil is

$$k' = \sqrt{k_v k_h} \tag{12-8}$$

12.16. EFFECT OF FORCE OF SEEPAGE WATER ON STABILITY OF SOIL. Everyone is familiar with the fact that flowing water has the power to move solid material. A common effect of this power is the rolling and tumbling of sand and pebbles along the bottom of a stream. The size of the particles which are moved is dependent on the velocity of the water; and this velocity, in turn, is dependent on the hydraulic gradient. In somewhat the same manner, seepage water flowing through soil exerts a force on the soil mass in the direction of flow; and this force is proportional to the hydraulic gradient. In some cases, this force is sufficient to cause movement of the soil. Whether the soil actually moves or not, seepage forces exist in all cases of gravitational flow through soil, and they may need to be investigated to determine their effect upon the stability of the soil mass.

As an illustration of the nature of seepage forces, consider the vessel shown in Fig. 12-14. Here a container of cross-sectional area A and height d is filled with spherical balls representing soil. In Fig. 12-14(a), the head water and the tail water are at the same elevation and there is no flow through the voids of the soil. The downward force at the base of the soil mass is equal to the weight of the saturated soil, which is

$$W = (V_s G + V_e)\gamma_w \tag{12-9}$$

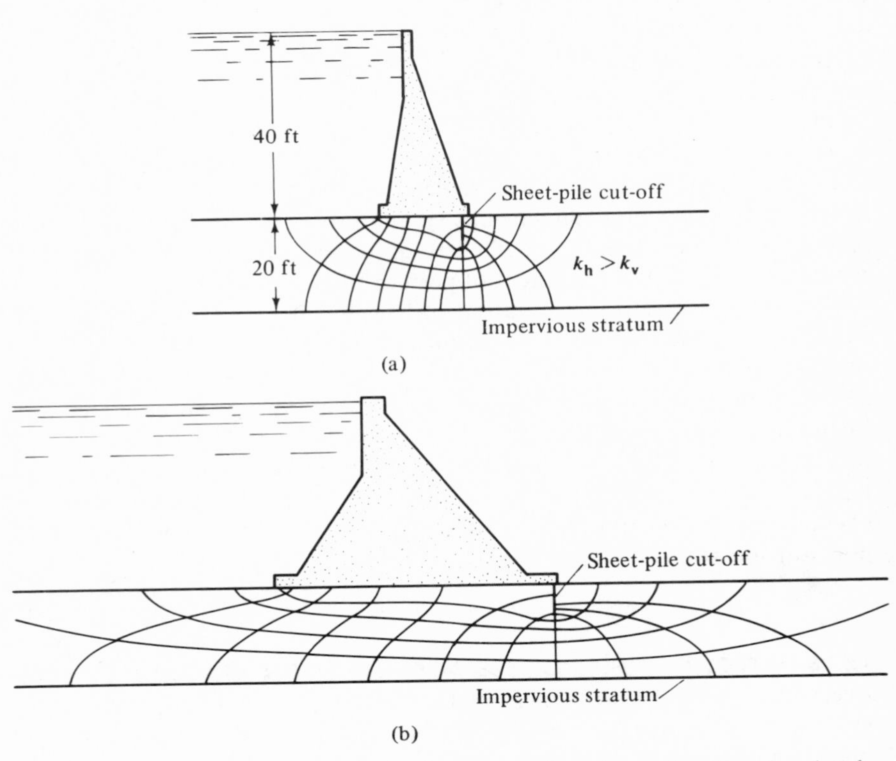

Fig. 12-13. Flow net for nonisotropic soil. (a) Transformed section for sketching flow net; (b) retransformed section (flow net as sketched in (a) elongated to take into account greater permeability in the horizontal direction).

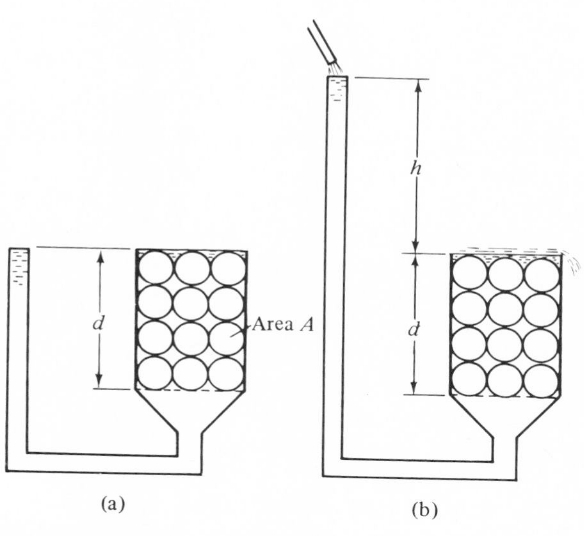

Fig. 12-14. Effect of seepage force on soil.

If the elevation of the head water is raised and maintained at a distance h above the level of the tail water, as shown in Fig. 12-14(b), water flows upward through the soil and the velocity of flow creates an upward force. The magnitude of the upward force at the base of the soil mass is equal to the product of the height of the water column above the base, the unit weight of water, and the area of the cross section of the soil. This product is

$$S = (h + d)A\gamma_w \tag{12-10}$$

12.17. BOILING ACTION IN QUICKSAND. As long as the upward force S in Eq. (12-10) is less than the downward force W in Eq. (12-9), their resultant is downward and the soil in the container is stable. But, if S is greater than W, the whole soil mass may be pushed upward and out of the container. When S and W are equal, the soil acts as though it were weightless and becomes highly unstable. A condition known as "boiling" develops, and the soil is described as "quick" or "quicksand." The hydraulic gradient at which a soil becomes quick is called the critical gradient, and an expression for its value in terms of soil properties can be determined by equating the upward and downward forces indicated by Eqs. (12-9) and (12-10). Thus,

$$(h + d)A\gamma_w = (V_sG + V_e)\gamma_w \tag{12-11}$$

or

$$\frac{h}{d} = i_c = \frac{G - 1}{1 + e} \qquad (12\text{-}12)$$

in which i_c is the critical gradient; G is the true specific gravity of soil; and e is the void ratio of soil.

It is apparent from Eq. (12-12) that the critical gradient is dependent on the true specific gravity of the soil and on its void ratio or density. Attention is directed to the fact that the critical gradient is not dependent on the size of the soil particles and that a coarse gravel therefore would become quick under the same hydraulic gradient as a fine sand. However, it is pointed out that, in accordance with the Darcy law, a considerably greater quantity of water is required to maintain flow under the critical gradient in the case of coarse gravel having a high coefficient of permeability than in the case of the fine sand.

12.18. TYPICAL QUICKSAND CONDITIONS. Examples of quicksand conditions are rather numerous in nature and in connection with engineering construction operations. Natural quicksand may exist in the bed of a lake or river at points of emergence of water flowing upward under artesian pressure. These points of emergence constitute underwater springs, and a quick condition is produced when the hydraulic gradient is equal to or greater than the critical value. Since the critical gradient can develop more readily in fine sand than in coarser material, quicksand conditions are usually associated with fine sand. This is the background for the practice in the masonry trades of referring to fine sand consisting of rounded particles as "quicksand." However, it is pointed out that the true meaning of quicksand refers to a condition of instability resulting from upward flow of water and not to the size or kind of sand involved. It is entirely possible that quicksand conditions may exist in a certain area in the spring of the year when artesian pressures and the quantity of water flowing from underwater springs are greatest; and that at other times of the year such conditions may not exist. Yet the soil involved has not changed in character.

Quicksand or boiling of the soil may develop in excavations for footings of buildings or bridges when water tends to flow upward into unwatered areas in caissons or cofferdams. Sand boils may develop on the downstream side of an earth dam after the reservoir is filled and high seepage pressures are built up in the soil beneath the dam. Such boils occur frequently on the land side of levees along the lower Mississippi River during flood stage. A common method of combating these river boils is to build an enclosing wall of sandbags around the area and to allow the water to pond within the enclosure. This stored water builds up a back pressure, which reduces the hydraulic gradient and quiets the soil in the boil area.

A quick condition in a submerged mass of sand may be triggered by vibrations caused by earthquakes, pile driving, or in rare cases, by highway, railway, or airplane traffic. A case is known where a relatively moderate earthquake (between III and IV on the Richter scale) caused a severe quick condition in connection with a structure resting on sand at a distance of 40 mi from the epicenter.

12.19. PREVENTION OF SAND BOILS. Sand boils on the downstream side of an earth dam may be prevented by installing relief wells in the area where a flow net indicates the seepage water may escape to the surface. These wells extend into the foundation soil and provide a means of release for high pressures in the soil water, thus reducing the gradient to a safe value below the critical gradient. Water rises in the wells and is allowed to flow away under control to outlets where it will do no damage.

Another method of preventing boiling or quicksand conditions is to cover the area with a blanket of heavy, coarse-grained material which will add weight to the soil but which is permeable enough to permit seepage water to pass through it without appreciable resistance. Such a blanket of material is called a protective filter. It must be relatively very permeable, but not sufficiently coarse to permit the blanketed material to be washed into the voids of the filter by the seepage water. In order to meet these requirements, a protective filter of uniform material should be as coarse as possible, but the particle size should not exceed about 10 times the diameter of a uniform foundation soil.

12.20. RESULTANT BODY FORCE ON SOIL MASS. Seepage forces exist in every soil through which gravitational water is flowing, even though the gradient is less than critical and there is no evidence of boiling or other movement of the soil. Such forces act in the direction of flow and must be combined vectorially with the weight of the soil in order to obtain the resultant body forces acting on the soil mass. For the conditions in Fig. 12-14, the force required to move the water through the soil is that produced by the head h, and is equal to $h\gamma_w A$. If the soil is texturally homogeneous, this force is uniformly distributed throughout the volume of the mass, since the whole mass of soil is effective in retarding the flow and dissipating the head. The force per unit volume may then be expressed by the relation

$$s = \frac{h\gamma_w A}{Ad} = i\gamma_w \tag{12-13}$$

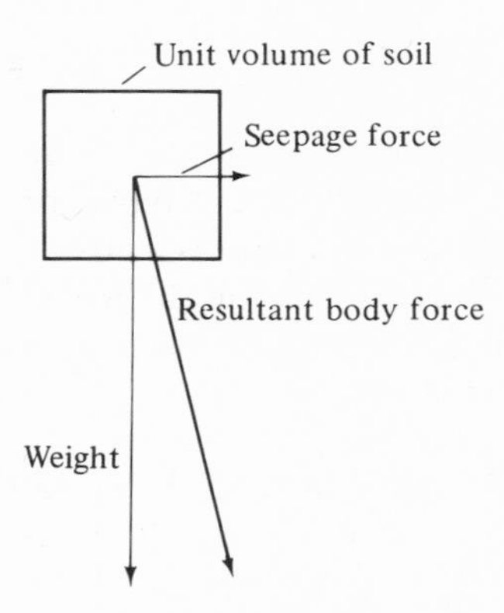

Fig. 12-15. Resultant body force on soil mass.

By way of illustration, suppose that a mass of saturated soil weighs 120 pcf and that it is subjected to horizontal seepage under a hydraulic gradient of 0.25. The seepage force on this soil is $0.25 \times 62.4 = 16.4$ pcf. The resultant body force acting on this mass of soil is the resultant of the unit weight acting downward and the unit seepage force acting horizontally, as shown in Fig. 12-15.

12.21. EFFECTIVE STRESS AND NEUTRAL STRESS. The unit stress within a mass of soil at any point below a water table is equal to the sum of two components, which are known as *effective stress* and *neutral stress*. It is necessary in many engineering problems to distinguish clearly between these two components.

Effective stress is defined as the total force on a cross section of a soil mass which is transmitted from grain to grain of the soil divided by the area of the cross section, including both solid particles and void spaces. It is sometimes referred to as intergranular stress.

Neutral stress is defined as the unit stress carried by the water in the soil pores in a cross section. It is sometimes called the pore water pressure.

An idea of the nature of these types of stresses may be gained by imagining a column of soil of unit cross section, as shown in Fig. 12-16. The total unit stress on a section B–B, which is a horizontal plane at a distance h below the water table A–A, is equal to the weight of all material above B–B divided by the area of the column. This total unit stress is the sum of the stress transmitted from grain to grain of the soil (the effective stress or intergranular stress) and the stress in the water at B–B (the neutral stress or pore water pressure). In this illustration the neutral stress is simply the hydrostatic pressure in the water due to its position below a water table, or $\gamma_w h$.

There are two methods by which the effective stress at *B–B* may be determined: (a) Compute the total weight of all material—soil solids plus water—above this plane, and then subtract the neutral stress from the total unit stress. (b) Compute the weight of material above the water table *A–A*, and add to it the buoyant weight of the soil solids between *A–A* and *B–B*. Since these solid particles are submerged below a free

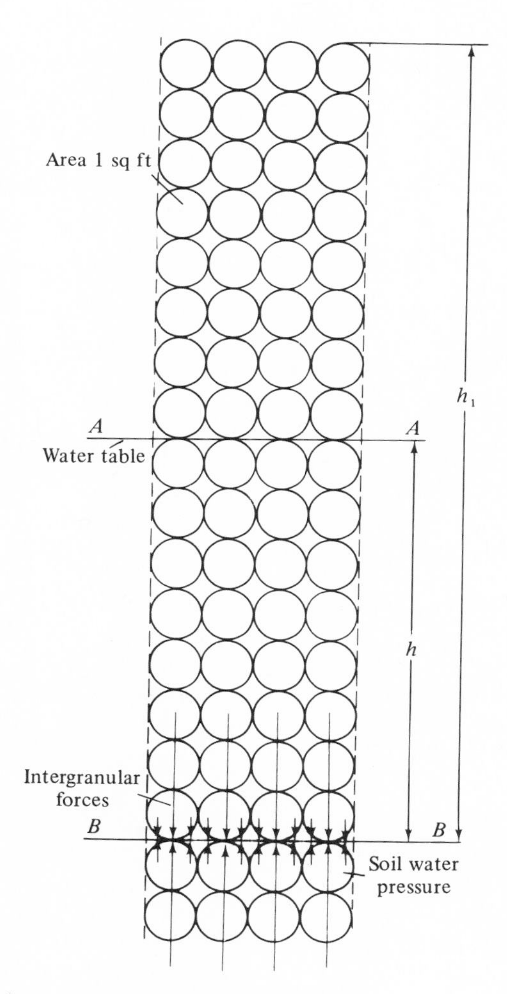

Fig. 12-16. Effective stress and neutral stress.

water surface, they are buoyed up by a force equal to the weight of water which they displace, in accordance with Archimedes' principle; therefore, the force transmitted from grain to grain is reduced by the amount of this buoyant force.

EXAMPLE 12-2. Assume that the dry density of the soil shown in Fig. 12-16 is 90 pcf; the wet density (unit weight) of the soil above *A–A* is 108 pcf; the true specific gravity is 2.75; $h_1 = 4$ ft; and $h = 2$ ft. Determine the effective stress and the neutral stress on plane *B–B* by the two methods just outlined.

SOLUTION. Method (a): In this case the total weight of soil above *A–A* is $2 \times 108 = 216$ lb. The weight of the solids between *A–A* and *B–B* is $2 \times 90 = 180$ lb. The volume of void space between *A–A* and *B–B*, by Eq. (8-1) and (8-14), is 0.952 cu ft, and the weight of water in the void space is $0.952 \times 62.4 = 59.4$ lb. Then the total weight of all material above *B–B* is

$$216 + 180 + 59.4 = 455.4 \text{ lb}$$

Since the cross-sectional area of this column is 1 sq ft, the total stress on *B–B* is 455.4 psf.

The hydrostatic pressure at *B–B* is

$$2 \times 62.4 = 124.8 \text{ psf}$$

This is also the neutral stress at *B–B*. Therefore, the effective stress is

$$455.4 - 124.8 = 330.6 \text{ psf}$$

Method (b): The weight of the soil above *A–A* is

$$2 \times 108 = 216 \text{ lb}$$

The buoyant weight of the soil between *A–A* and *B–B* is equal to the weight of the solid particles minus the weight of water which they displace. By Eq. (8-14),

$$V_s = \frac{2 \times 90}{62.4 \times 2.75} = 1.048 \text{ cu ft}$$

The weight of water displaced is

$$1.048 \times 62.4 = 65.4 \text{ lb}$$

The buoyant weight of solids between *A–A* and *B–B* is

$$180 - 65.4 = 114.6 \text{ lb}$$

Then the effective stress at *B–B* is

$$216 + 114.6 = 330.6 \text{ psf}$$

The neutral stress, as in method (a), is

$$2 \times 62.4 = 124.8 \text{ psf}$$

The principles of effective stress and total stress will again be discussed in connection with soil shearing strength in Chapter 19, since frictional strength of a soil relates to how hard the individual particles bear against one another, i.e., effective stress, rather than total stress.

12.22. UPLIFT PRESSURE. When free water is in contact with a structure, such as a bridge pier or a masonry dam, uplift pressures are exerted against the base of the structure. If the water is static, the uplift pressure is equal to the hydrostatic pressure, in accordance with the principle of Archimedes. If seepage water is flowing beneath the structure, the uplift pressure at any point can be estimated from the appropriate flow net.

The total head at any point in flowing water is equal to the sum of the velocity head, the pressure head, and the elevation head. In practically all cases of flow through soil, velocities are so small that velocity heads are negligible, and we may consider that total head is equal to pressure head plus elevation head. Elevation head is the vertical distance of a point from a datum plane, which is usually chosen as the elevation of the tail water. Pressure head is equal to the water pressure at any point divided by the unit weight of water.

The total head at any point on the base of a structure can be determined from the equipotential line that intersects the base at the point. Then the difference between the total head and the elevation head is the

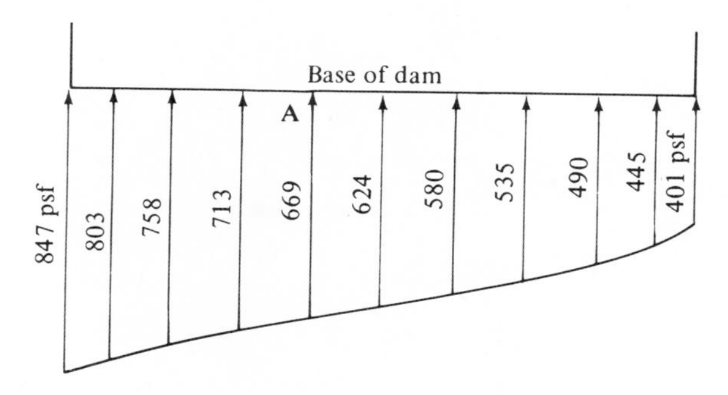

Fig. 12-17. Uplift pressure on masonry dam shown in Fig. 12-4.

pressure head, which may be multiplied by the unit weight of water to obtain the uplift pressure.

Consider the masonry dam and flow net shown in Fig. 12-4. The total head loss h in this case is 10 ft; and since there are 14 equipotential spaces in the net, the head loss represented by each space is $0.0714h$. Therefore the head loss at point A is $6 \times 0.0714h$ or 4.28 ft, and the total head at this point is 5.72 ft. Also, the elevation head at the point is -5.0 ft. Therefore the pressure head is $5.72 - (-5.0) = 10.72$ ft, and the uplift pressure is $10.72 \times 62.4 = 669$ psf. The distribution of uplift pressure across the base of this structure is shown in Fig. 12-17.

PROBLEMS

12.1. What is a flow net and what is its purpose?

12.2. Define flow line; equipotential line; flow path; and equipotential space.

12.3. State five fundamental properties of a flow net.

12.4. Identify the boundary flow lines and equipotential lines in Fig. 12-2.

12.5. Assume that the coefficient of permeability of the soil in Fig. 12-2 is 0.00042 cm/min. Determine the seepage per 100 ft of length of the sheet-pile cofferdam.

12.6. The pervious foundation soil beneath the masonry dam in Fig. 12-4 has a coefficient of permeability of 0.14 ft/day. Determine the seepage per 100-ft length of the dam.

12.7. The pervious foundation soil in Fig. 12-13 has a coefficient of permeability of 0.004 fpm in the vertical direction and 0.02 fpm in the horizontal direction. Compute the seepage per linear foot of dam.

12.8. Construct a flow net and estimate the seepage per linear foot of the dam in Fig. 12-18.

12.9. What is the critical gradient associated with a soil having a true specific gravity of 2.65 and a void ratio of 0.85?

12.10. Determine the hydraulic gradient of seepage water at point d in Fig. 12-2. If the true specific gravity of the soil inside the cofferdam is 2.62 and the

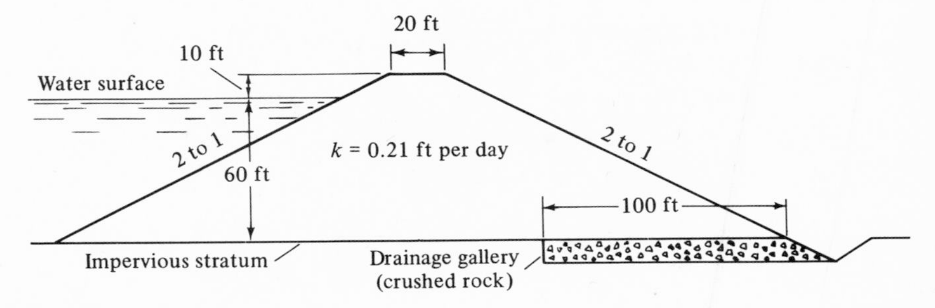

Fig. 12-18

void ratio is 1.14, will sand boils develop? What is the factor of safety against boiling at this point?

12.11. What is the escape hydraulic gradient at point *e* in Fig. 12-2?

12.12. What is the hydraulic gradient at a point 10 ft below the bottom of the sheet piling in Fig. 12-2?

12.13. The pervious soil in Fig. 12-2 has a true specific gravity of 2.73 and a void ratio of 0.9. Determine the hydraulic gradient at points 17 and 24 ft from the cofferdam on the dry side. What is the factor of safety against boiling at each of these points?

12.14. Construct a flow net and estimate the seepage per linear foot of the dam in Fig. 12-19.

12.15. Calculate the effective stress at a plane 16 ft below the surface of a soil mass in which the water table is 5 ft below the surface. Assume that the dry density of the soil is 90 pcf, the moisture content above the water table is 15%, and the true specific gravity is 2.70.

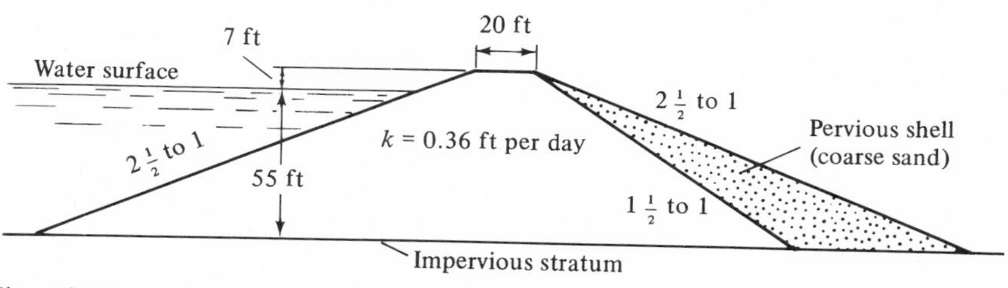

Fig. 12-19

REFERENCES

1. Bureau of Reclamation, U.S. Dept. of Interior, *Design of Small Darms.* U.S. Government Printing Office, Washington, D.C., 1965.
2. Casagrande, A. "Seepage through Dams." *J. New England Waterworks Assoc.*, June 1937.
3. Gregg, L. E. "Typical Flow Nets for the Solution of Probems in Ground Water Flow and Seepage." *Proc. Purdue Conference on Soil Mechanics and Its Applications*, Purdue University, Lafayette, Indiana, July 1940.
4. Taylor, Donald W. *Fundamentals of Soil Mechanics.* John Wiley, New York, 1948.

13

Soil-Water Consistency

13.1. SOIL-WATER MIXTURES. When certain soil-water relationships and properties were discussed in previous chapters, the soil was considered to be either in a natural state or in some condition normally encountered in engineering practice. If soil and water are mixed together and manipulated or stirred, changes in the consistency of the mixture at various moisture contents, and changes in volume of the mixture in relation to changes in moisture, reveal very important characteristics of the soil, even though the condition of the soil-water mixture is not the same as that encountered in practice. These characteristics of consistency and shrinkage or swell have been correlated with engineering behavior of soil to a considerable extent; and such relationships, together with the mechanical analysis or grain-size distribution, provide a basis for the identification and engineering classification of a soil and for judgment as to its suitability for certain engineering uses.

A mixture of soil and water may exist in any one of several states of matter, the actual state in any case depending on the amount of water in relation to the amount of soil and, to some extent, on the amount of manipulation or stirring of the mixture. As the moisture content decreases, these states of matter may range from that of a true liquid at very high moisture contents, down through that of a viscous liquid, a plastic solid, and a semisolid, to that of a solid—in the case of a cohesive soil.

In order to visualize these changes in state, imagine a small volume of soil mixed with a relatively large volume of water. Such a mixture would have all the properties of a true liquid. It would have definite buoyant properties and would be as truly a liquid as water itself, although its unit weight would be greater than that of water because of the suspended soil particles.

13.2. SOIL IN THE LIQUID STATE. A true liquid is an ideal Newtonian fluid in that it approaches zero shearing resistance as the rate of shear becomes very slow. As the shearing rate increases, as by stirring, the resistance to shearing (or stirring) increases in linear proportion, the relation between the two being the *viscosity*.

Temperature and dissolved ions and compounds strongly influence viscosity by affecting molecular attractions within the liquid. For ex-

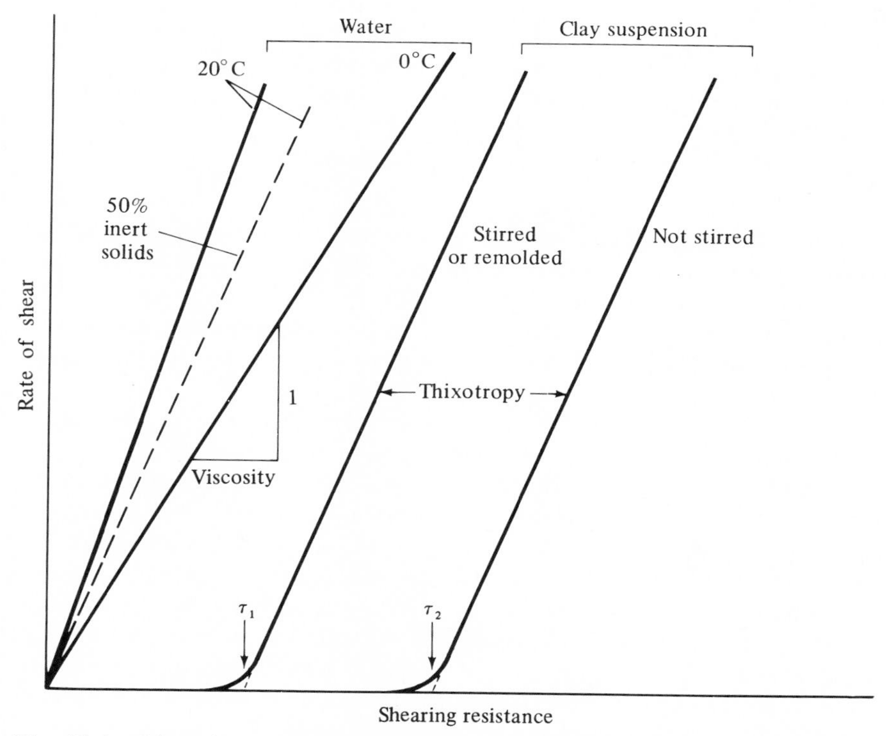

Fig. 13-1. Viscosity of Newtonian and non-Newtonian fluids. τ_1 and τ_2 are yield stresses.

ample, the viscosity of water at 0°C is 1.8 times that at 20°C (Fig. 13-1). By comparison, suspended inert soil particles have relatively little effect, working only to displace liquid and thereby increase the shearing rate in liquid between the particles. The theoretical equation describing this effect is one of the early contributions of Einstein. Suspended electrically repulsive particles, for example, clay particles dispersed in dilute suspensions, do not greatly influence viscosity.

On the other hand as the water content decreases, the clay particles tend to interact with one another through water dipoles which link together like weak magnets. Although such bonds are very weak and are continually being made and broken owing to thermal energy and Brownian movement, they greatly influence viscous behavior. In fact, the clay suspensions no longer behave like true liquids because a stress called the *yield stress* must be applied to start shearing. This type of behavior is shown by the τ_1 curve in Fig. 13-1, τ_1 being the yield stress. Many clay suspensions gain a higher yield stress upon setting, because of the time needed to optimize the skeleton structure; this is shown by τ_2, and is described as *thixotropy*. Soil engineers are mainly concerned with clays below such a yield stress, and in particular, with how to keep them that way. The loss of strength through disturbance has previously been described as *sensitivity* and can be very important wherever dynamic stresses are applied, whether by pile driving, blasting, or earthquakes.

13.3. LIQUID LIMIT. The moisture content above which a soil readily becomes a liquid upon stirring is called the *liquid limit*. However, because of the imperfect dividing line between solid and liquid states of a soil, a standardized agitation procedure must be used. Skempton (*4*) found that soil at the liquid limit has a shearing strength (τ_1 in Fig. 13-1) of about 0.1 psi.

The liquid limit of a soil is determined in the laboratory by the following standardized procedure: A small quantity of air-dry powdered soil passing the No. 40 sieve is mixed with water to a paste consistency. It is then placed in a spherical-shaped brass cup and the surface is struck off with a spatula so that the maximum thickness is 1 cm. The soil pat is next divided into two segments by means of a grooving tool of standard shape and dimensions. The brass cup is mounted in such a way that, by turning a crank, it can be raised and allowed to fall sharply onto a hard-rubber block or base. The shock produced by this fall causes the adjacent sides of the divided soil pat to flow together. The wetter the mixture, the fewer shocks or blows will be required to cause the groove to close; the drier the mixture, the greater will be the number of blows. The moisture content at which 25 blows cause the groove to close is defined as the liquid limit by this laboratory procedure.†

In liquid-limit tests, it is customary to determine the number of blows required to close the groove in the soil pat at three or more moisture contents, some above the liquid limit and some below it. Then the numbers of blows are plotted against the moisture contents, and a curve is drawn through the plotted points. This curve is called the *flow curve*, and the moisture content at which it intersects the 25-blow abscissa is considered to be the liquid limit.

The flow curve is logarithmic in character. Therefore it is most conveniently plotted as a straight line on semilogarithmic graph paper, with moisture contents as ordinates and numbers of blows as abscissae on the logarithmic scale. The slope of this line is called the *flow index*, which is denoted by F_i. Its numerical value is the difference in moisture content intercepted by the flow curve in one cycle of the logarithmic scale of number of blows.

13.4. PLASTIC LIMIT. When the moisture content of a soil–water mixture is reduced below the liquid limit, the material becomes more strongly plastic in character. In this state external force is required to cause it to deform, and its bearing power begins to increase appreciably. It loses its stickiness and can be molded in the hands into a ball or any other shape. The minimum moisture content at which the mixture acts as a plastic solid is called the *plastic limit*, which is denoted by PL.

The plastic limit of a soil is determined in the laboratory by a standardized procedure, as follows: A small quantity of the soil–water mixture is rolled out with the palm of the hand on a frosted glass plate or on a mildly absorbent surface until a thread or worm of soil is formed. When the thread is rolled to a diameter of $\frac{1}{8}$ in. it is balled up and rolled out again, the mixture gradually losing moisture in the process. Finally the sample dries out to the extent that it becomes brittle and will no longer hold together in a continuous thread. The moisture content at which the thread breaks up into short pieces in this rolling process is considered to be the plastic limit.‡

13.5. PLASTICITY INDEX. One of the most important of the properties of soil which are revealed by changes in consistency of a soil–water mix-

†See A.A.S.H.O. Specification T-89-68 or A.S.T.M. Specification D 423-66 for details of the liquid-limit test.

‡See A.A.S.H.O. Specification T-90-70 or A.S.T.M. Specification D 424-59 for details of the plastic-limit test.

ture is the *plasticity index*, denoted by PI. It is defined as the numerical difference between the liquid limit and the plastic limit of the soil. It represents the range of moisture content within which the soil exhibits the properties of a plastic solid. Also, it is a measure of the cohesive properties of the soil and indicates the degree of surface chemical activity and the bonding properties of the fine clay and colloidal fraction of the material.

The plasticity index is an empirical indicator of the suitability of the clay fraction of a binder material in a stabilized soil mixture. Experience has indicated that a sand-clay mixture or clay-gravel mixture in which the plasticity index of the binder fraction is too high tends to soften in wet weather. A pavement constructed of such material develops ruts under traffic and may shift and shove to develop a washboard surface and other evidences of instability. When a pavement of this type is used without a bituminous wearing surface, a high plasticity index indicates that the surface will become slippery in wet weather. On the other hand, if the plasticity index is too low or the mixture is nonplastic in character, it will become friable in dry weather, ravel at the edges, and abrade severely under traffic. Such a pavement becomes very dusty in service, and much of the binder soil may gradually be blown away in dry seasons of the year. The plasticity index, in conjunction with the mechanical analysis, provides the basis for several of the engineering classifications of soil.

A soil may exhibit a plasticity index equal to zero; that is, the numerical values of the plastic limit and the liquid limit may be the same, and their difference may be zero. Such a soil is definitely plastic in character, although the range of moisture content within which it exhibits the properties of a plastic solid is so small that this range cannot be measured by the laboratory tests for liquid limit and plastic limit. Soils having a zero plasticity index should not be confused with nonplastic soils, which are discussed in Section 13.10.

Soils having finite values of plasticity index may vary in toughness. A gumbo clay is usually very tough, while a glacial clay is only moderately so. This property of soil is expressed numerically by the *toughness index*, denoted by T_i, which is equal to the plasticity index PI divided by the flow index F_i. Thus,

$$T_i = \frac{PI}{F_i} \tag{13-1}$$

The *sticky limit*, denoted by T_w, is the lowest water content at which soil adheres to metal tools. The sticky limit is determined by gradually reducing the water content of a clay-soil pat until it is possible to clean a nickel-plated spatula by drawing the spatula over the surface of the pat.

13.6. LIQUIDITY INDEX OR RELATIVE CONSISTENCY. A natural deposit of clay lying below a water table usually is completely saturated, and its moisture content may lie within the plastic range, that is, between the liquid limit and the plastic limit. The relative softness of such a clay is indicated by the nearness of the natural moisture content to the liquid limit. Softness is expressed numerically by the *liquidity index* or *relative consistency*. This index is denoted by L_i, and is equal to the difference between the natural moisture content m and the plastic limit PL divided by the plasticity index PI. Thus,

$$L_i = \frac{m - \mathrm{PL}}{\mathrm{PI}} = \frac{m - \mathrm{PL}}{\mathrm{LL} - \mathrm{PL}} \tag{13-2}$$

If the natural soil is at the liquid limit the liquidity index is 1.0; if at the plastic limit, 0.0.

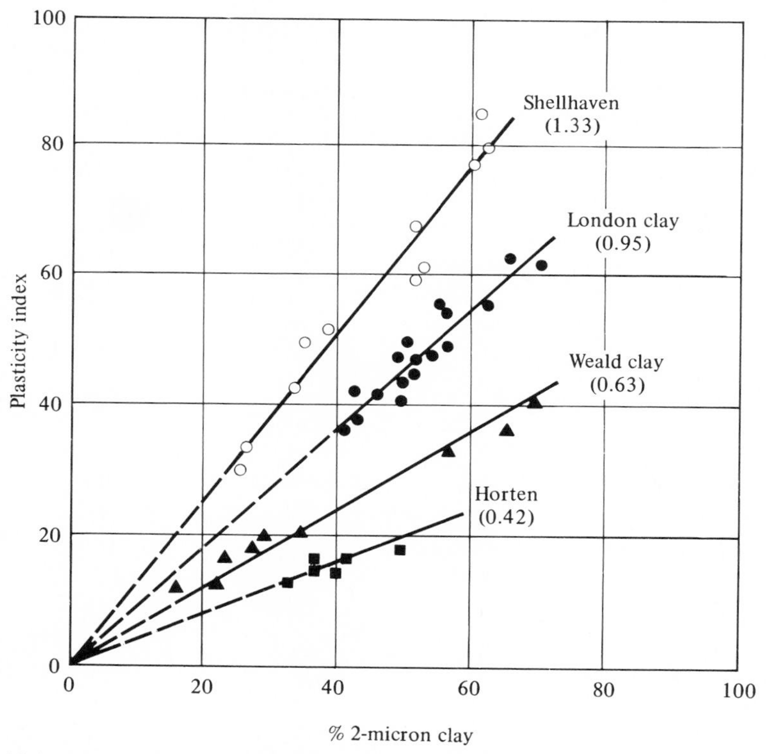

Fig. 13-2. Activity index of four soils [after Skempton (*4*)].

13.7. ACTIVITY INDEX AND CLAY MINERALOGY. The liquid limit, plastic limit, and plasticity index are affected by both the amount of clay present in a soil and the kind of clay mineral. The activity index (AI) was defined by Skempton (*4*) to describe plasticity of the clay-size fraction of soils. As might be expected, the activity index correlates fairly well with the clay mineralogy, and is a rough guide to the kind of clay minerals. Skempton defined the AI as the PI divided by the 0.002 mm clay content. Thus

$$AI = \frac{PI}{\%\ 2\text{-micron clay}} \qquad (13\text{-}3)$$

In general the liquid limit is directly proportional to the clay content, whereas the plastic limit is directly proportional above about 45% clay and inversely proportional below [Davidson and Handy (*2*); Ballard (*1*)]. Since the activity index depends on the difference between these, which is the PI, it therefore should relate to clay minerals only in clay-rich soils (Fig. 13-2). Representative activity indices are in Table 13-1. Clay mineral mixtures and interlayers have intermediate activities. Data in the table also show how activity might be modified by the use of chemical additives to change the adsorbed cation.

TABLE 13-1. Activity Indices of Selected Clay Minerals[a]

Na^{+}-montmorillonite	3–7
Ca^{+2}-montmorillonite	1.2–1.3
Illite	0.3–0.6
Kaolinite (poorly crystallized)	0.3–0.4
Kaolinite (well crystallized)	<0.1

[a]Data from Grim (*3*), Table 5-1.

13.8. SHRINKAGE LIMIT. At moisture contents below the plastic limit, a soil–water mixture becomes semisolid in character and its bearing values increase very markedly. Finally, although the pore spaces in the soil have become very small and the inward capillary pressures are correspondingly high, the soil solids can no longer be forced closer together by these forces and the mass reaches its minimum volume in the drying process. The moisture content of the soil at which further reduction in moisture is not accompanied by a reduction in volume is called the *shrinkage limit*.†

Below the shrinkage limit, a cohesive soil has only the properties of a

†See A.A.S.H.O. Specification T-92-68 or A.S.T.M. Specification D 427-61 for details of the shrinkage tests.

solid, and it develops its maximum bearing capacity and strength, the values depending on the bonding power of the fine clay and colloidal matter in the soil. A soil containing relatively small quantities of inactive colloidal-sized material may be very weak and friable in this state, while one which contains a large amount of chemically active colloids may develop very high strength and hardness at low moisture contents. The transition from the semisolid state to the solid state is also accompanied by a noticeable change from a darker to a lighter shade of color due to the entry of air.

13.9. SHRINKAGE DIAGRAM. Figure 13-3 shows the volume changes occurring during drying of a soil suspension or slurry. At higher water contents water is a continuous phase separating the soil grains, so drying 1 gm of water results in 1 cc volume change. Ideally at the shrinkage limit the soil grains suddenly come into contact and prevent further volume change, but as might be expected this point is not always sharply defined, especially in clays. It is obtained by intersection of the 45° volume change line with the horizontal V_{SL} line in Fig. 13-3. If the soil grains were to fit perfectly together with no voids, shrinkage would continue along the dotted line to a point representing the specific gravity of the soil solids, $G = W_S/V_S$. By use of the relationships in Fig. 13-3 a number of interrelationships can be derived.

By definition the shrinkage limit is

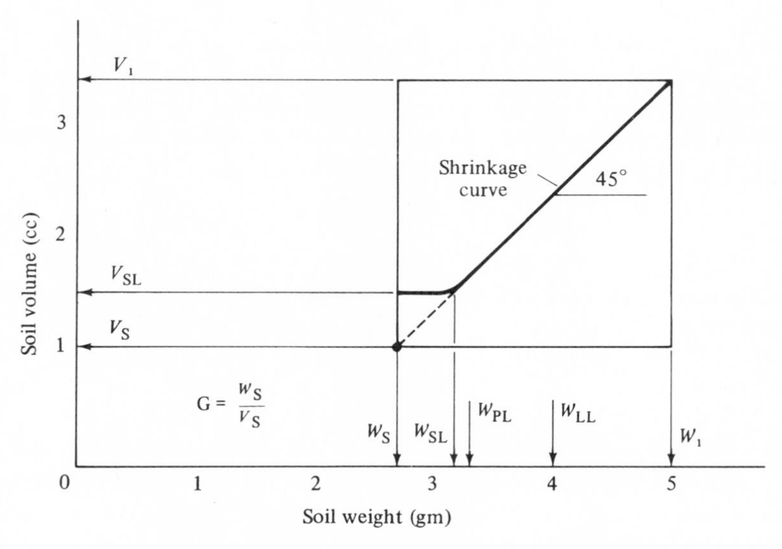

Fig. 13-3. Shrinkage diagram for a loessial soil containing 30% 2-micron montmorillonitic clay.

$$SL = \frac{W_{SL} - W_S}{W_S} \times 100 \tag{13-4}$$

in which W_{SL} is the soil weight at the shrinkage limit and W_S is the dry weight of the soil. In the laboratory shrinkage limit test W_S is measured, and in addition the following are measured: W_1 and V_1, weight and volume, respectively, of soil at any arbitrary moisture content above the shrinkage limit; and V_{SL}, volume of the soil at the shrinkage limit. W_1 and V_1 are found by simply filling a small dish of known volume with wet soil and weighing; W_S and V_{SL} are measured after oven drying the soil in the dish—W_S by weighing the dried soil pat and V_{SL} by immersing it in mercury to determine its volume (A.S.T.M. Designation: D 427-61).

Since W_{SL} in Eq. (13-4) is not measured directly, it is evaluated from W_1, V_1, and V_{SL} by assuming that the shrinkage amount, $V_1 - V_{SL}$, is due to the loss of water which has a specfic gravity of 1.0. Referring to Fig. 13-3,

$$W_1 - W_{SL} = (V_1 - V_{SL}) \times 1 \text{ gm/cc}$$

$$W_{SL} = W_1 - V_1 + V_{SL}$$

Substituting in Eq. (13-4),

$$SL = \frac{W_1 - V_1 + V_{SL} - W_S}{W_S} \times 100$$

$$= \left[\frac{W_1 - W_S}{W_S} - \frac{V_1 - V_{SL}}{W_S}\right] 100 \tag{13-5}$$

The first term in the brackets is the moisture content of the wet soil.

The *shrinkage ratio* was originally defined in terms of slope of the volume change curve, equaling the volume change in percent of V_{SL} divided by the corresponding change in moisture content; that is, if SR is the shrinkage ratio and the other symbols are as above,

$$SR = \frac{V_1 - V_{SL}}{V_{SL}} \div \frac{W_1 - W_{SL}}{W_S}$$

However, since as shown in Fig. 13-3, the two numerators are equal, they cancel and

$$SR = \frac{W_S}{V_{SL}} \tag{13-6}$$

which is the dry density of the soil at the shrinkage limit. The "shrinkage ratio" therefore is a "shrinkage density." The shrinkage ratio and shrinkage limit can be used to calculate percent volume change or volumetric shrinkage, defined as

$$\mathrm{VC} = \frac{V_1 - V_{SL}}{V_{SL}} \times 100$$

This may be written

$$\mathrm{VC} = \frac{V_1 - V_S - (V_{SL} - V_S)}{V_{SL}} \times 100$$

$$= \frac{W_1 - W_S - (W_{SL} - W_S)}{V_{SL}} \times 100$$

Substituting from Eqs. (13-6) and (13-4),

$$\mathrm{VC} = \frac{W_1 - W_S - (W_{SL} - W_S)}{W_S} \times \mathrm{SR} \times 100$$

$$= \left(100\,\frac{W_1 - W_S}{W_S} - \mathrm{SL}\right)\mathrm{SR} \tag{13-7}$$

The first term in the brackets is the moisture content of the wet soil.

13.10. NONPLASTIC SOIL. Noncohesive soils, such as sands which are relatively free from clayey material, undergo the same general pattern of volume change with reduction in moisture content as that just described. However, a noncohesive soil does not solidify upon drying out; nor does it pass through the plastic solid state. Rather, it changes abruptly from the viscous liquid state to a dry incoherent granular material which does not form into clods or other solid forms. It is not possible to roll a material of this kind into a thread as small as $\frac{1}{8}$ in. in diameter. Therefore, a plastic limit cannot be determined. Such a soil is said to be nonplastic and in test reports it is designated as NP.

13.11. SLAKING OF DRY SOIL. The shrinkage process described in the preceding sections, in which a soil-water mixture undergoes various changes in state from that of a true liquid to that of a solid, is not a reversible process. If a dry clod or pat of soil is placed in water, the adsorptive power of the fine soil particles tends to cause films of hygroscopic moisture to form around the particles. As a result the bonds between the particles are broken and the soil mass disintegrates. This process is known as slaking.

13.12. SLAKING CHARACTERISTICS OF SOIL IN PLASTIC STATE. In contrast to the slaking action of dry soil masses upon contact with water, soil which is wet enough to be in the plastic state will resist slaking to a very great extent. When the soil is in this state, the adsorptive attraction of the fine particles has been satisfied by the formation of moisture films and the pore spaces between the grains are well filled with capillary water. Therefore, there is little opportunity for additional water to enter the soil, and the expansive forces which produce slaking do not develop to any appreciable extent. This contrast between the slaking characteristics of dry and wet masses of soil has a very important practical significance in connection with soil-stabilization work in highway or airport construction.

When a granular soil material is to be stabilized for a highway or runway pavement, clay soil is frequently added to bind the mixture into a hard coherent mass. It is important that the clay binder be thoroughly dispersed or distributed throughout the granular material in order to produce a homogeneous pavement material of uniform strength. If the binder clay is added in a wet state, it cannot be mixed thoroughly with the granular material, and it will remain in chunks or "clay balls" which weaken the stabilized mixture. On the other hand, if the clay is reasonably dry when mixed with the sand and gravel, it will slake more readily when water is added in connection with compaction operations; and good dispersion of the binder material can be obtained.

13.13. BULKING. When cohesive soil in a powdered state is dried in an oven and then brought into contact with moisture in either the liquid or vapor form, the adsorptive power of the fine soil particles causes the moisture to condense into thin hygroscopic films on the particles. This phenomenon causes the soil to undergo a considerable increase in volume, since the films serve to increase the size of the soil particles. Much of this volume increase may subsequently disappear, however, if more water is gradually added to the soil. A similar increase in volume develops when a small quantity of water, say 6–8%, is added to sand. Such an increase in volume is called "bulking," and may amount to as much as 25%. Flooding the sand with additional water causes much of the bulking to disappear.

13.14. SHRINKAGE PROPERTIES OF SOIL. The ratio of the shrinkage limit to the liquid limit of a soil indicates the general shrinkage properties of the soil. The smaller the shrinkage limit as compared with the liquid limit, the greater will be the tendency for the soil to undergo detrimental volume changes with changes in moisture content. The ratio of the liquid limit to the content of 5-micron clay discloses the relative surface chemical activity of the clay fraction. A liquid limit approximately equal to the clay content indicates the presence of inactive fine particles capable of forming a paste with the addition of a relatively small amount of water. The more the liquid limit exceeds the clay content, the greater is the likelihood that undesirable gluey colloidal material is present in the soil.

13.15. CENTRIFUGE MOISTURE EQUIVALENT. The centrifuge moisture equivalent† is a property which indicates the water-holding characteristics of a soil when it is drying out or draining. In the test for this property a small quantity of dry powdered soil passing a No. 40 sieve is placed in a porcelain cup with a perforated bottom which is covered by a piece of filter paper. The soil is saturated with water by allowing the perforated cup and its contents to stand in shallow water for several hours. Then the cup is placed in a centrifuge and whirled for 1 hr at an angular velocity which will cause the sample to be subjected to a centrifugal force equal to 1000 times the force of gravity. The moisture content of the soil after this treatment is defined as the *centrifuge moisture equivalent* (CME).†

The CME represents a true equilibrium condition between the capillary potential of the soil sample and the centrifugal force applied in the test procedure. In the standard test it represents a pF of 2.70. However, since the sample is compressed by the centrifugal force which expels a portion of its moisture, the soil as tested is not representative of the soil in nature or in an engineering structure. The test, therefore, gives only qualitative results for comparison with the CME of other soils and cannot be considered a true quantitative value of capillary potential of the soil.

In the case of a fine-grained clayey soil, the rate at which the water is whirled out of the sample is relatively slow. Centrifugal compression of the soil may occur before all of the removable water is expelled, with the result that part of the water is squeezed to the surface of the sample and remains there throughout the test. Such a soil is said to be "waterlogged," and this fact should be reported in connection with the results of the CME test.

13.16. FIELD MOISTURE EQUIVALENT. The field moisture equivalent‡ of a soil indicates the water-holding characteristics when the soil is wetted. In the test for determining this property, a quantity of dry soil powder is mixed with water to form a plastic paste. A sample of the paste is placed in a porcelain evaporating dish, its surface is smoothed off with a spatula, and a drop of water is applied to the smooth surface. If the water is immediately absorbed, the sample is mixed and smoothed off again, and another drop of water is added. This procedure is repeated again and again until the drop of water is not immediately absorbed but remains as a shiny spot on the surface for a period of 30 sec. The moisture content at which this shiny spot develops is defined as the *field moisture equivalent* (FME).‡ Like the CME, it is a qualitative indicator property which must

†See A.S.T.M. Specification D 425-69 for details of the centrifuge-moisture-equivalent test.

‡See A.A.S.H.O. Specification T-93-68 for details of the field-moisture-equivalent test.

be correlated with soil performance in a particular geographical area in order to have definite meaning.

PROBLEMS

13.1. Define liquid limit; plastic limit; plasticity index; and activity index.

13.2. In the liquid-limit test, four trials give the following data relative to moisture content and number of blows to close the groove in the soil pat.

Number of blows	Moisture content (%)
45	29
31	35
21	41
14	48

Plot the flow curve. Determine the liquid limit and the flow index of the soil.

13.3. If the plastic limit of the soil in Problem 13.2 is 13%, what is the plasticity index?

13.4. If the soil in Problem 13.2 contains 30% of 2-micron clay, what is the activity index?

13.5. What is the toughness index of the soil in Problem 13.2?

13.6. The liquid limit of a soil is 59 and the plastic limit is 23. If the soil in its natural state lies below a water table and the saturation moisture content is 46%, what is its liquidity index or relative consistency?

13.7. The liquidity index of a soil is 1.1, the liquid limit is 69, and the saturated natural moisture content is 73%. What are the plastic limit and plasticity index?

13.8. Define shrinkage limit; shrinkage ratio.

13.9. The volume of the dish used in a shrinkage limit test is measured and found to be 20.0 cc and the volume of the oven-dry soil pat is 14.4 cc. The weights of the wet and dry soil are 41.0 and 30.5 gm, respectively. Calculate the shrinkage limit and shrinkage ratio.

13.10. The shrinkage limit of a soil is 24% and the shrinkage ratio is 2.2 Calculate the percent volume expansion when the moisture content increases from 16 to 29%.

13.11. Complete the following derivation to obtain soil specific gravity G in terms of the shrinkage limit and shrinkage ratio.

$$G = \frac{W_S}{V_S}$$

$$\frac{1}{G} = \frac{V_S}{W_S} = \frac{V_{SL}}{W_S} - \frac{V_{SL} - V_S}{W_S}$$

13.12. Describe a nonplastic soil and explain how this characteristic is determined in the laboratory.

13.13. Distinguish clearly between a nonplastic soil and one which has a PI equal to zero.

13.14. Explain why it is necessary for fine-grained binder soil to be reasonably dry when it is added to a coarse-grained soil to form a stabilized soil mixture.

13.15. If the PI of a stabilized-soil pavement is too high, what adverse characteristics are likely to develop under service conditions? What may happen if the PI is too low?

13.16. Define sensitivity in terms of τ_1 and τ_2 in Fig. 13-1 (Reference: Chapter 7).

13.17. Is soil containing water in excess of the liquid limit necessarily a liquid? Explain.

13.18. (a) Calculate G, LL, PL, PI, SL, and AI for the soil in Fig. 13-3, taking W_S as 2.7 gm. (b) Is the AI indicative of the observed clay mineral? (c) What are the approximate wet and dry densities of the soil in Fig. 13-3 at the liquid limit, in gm/cc and pcf? (d) What is the dry density in gm/cc at the shrinkage limit? (e) What is the latter called?

REFERENCES

1. Ballard, G. E. H. "The Plastic Limit as a Binary Packing Phenomenon." *ASTM Materials Res. and Standards*, pp. 366–374 (1964).
2. Davidson, D. T., and R. L. Handy. "Studies of the Clay Fraction of Southwestern Iowa Loess." *Proc. 2nd National Clay Conf.*, pp. 190–208 (1953).
3. Grim, Ralph E. *Applied Clay Mineralogy*. McGraw-Hill, New York, 1962.
4. Skempton, A. W. "The Colloidal Activity of Clays," *Proc. 3rd Intern. Conf. Soil Mech. and Foundation Eng.*, Vol. 1, p. 57. Switzerland, 1953.

14

Engineering Soil Classification

14.1. ORIGINAL SYSTEM OF BUREAU OF PUBLIC ROADS. In earlier chapters of this text, several systems of soil classifications were presented, such as the geological classification, the pedological classification, the grain-size classification, and the textural classification. Each of these provides a language by means of which one person's knowledge of the general characteristics of a soil can be conveyed to another person or group in a brief and concise manner, without the necessity of entering into lengthy descriptions and detailed analyses. For example, if a soil is described as "drift," it is immediately identified by any person familiar with the classification system as a material which was transported and deposited in its present location by glacial action in the geological past. If it is further described as "silty clay," this means that it contains not more than 15% of sand, 55–70% of silt-size particles, and 30–45% of clay-size particles. Additional characteristics can be described by similar classifying terms.

None of the above classifications, however, is based primarily upon characteristics which influence the engineering behavior of soil. In order to meet the need for a language to convey information of this kind, a number of engineering organizations have devised classification systems. Several of these have proved to be very useful and are extensively employed in engineering circles. The oldest of the engineering classification systems, and one which is still widely used in engineering work, was developed by the United States Bureau of Public Roads and introduced in 1928. In this system all soils are divided into eight major groups, which are designated by the symbols A-1, A-2, A-3, etc.

In the first three groups, or those designated A-1, A-2, and A-3, are the coarse-grained soils, such as gravel, sand, and mixtures of sand and gravel. In the next two groups, or A-4 and A-5, are the predominantly silty soils. The A-6 and A-7 groups include the clayey soils, while the A-8 group refers to peat and muck soils. Thus, it is seen that the group designations are numerical indicators of the quality of soils. Generally speaking, the lower the number, the better is the soil from an engineering standpoint. In the original Public Roads system of classification the determination of the group to which a soil belonged was based upon a number of soil properties in addition to the mechanical analysis, such as plasticity index, shrinkage properties, and the centrifuge-moisture and field-moisture equivalents.

14.2. A.A.S.H.O. SYSTEM. As the usefulness of the Public Roads system of soil classification became apparent, it was adopted by engineering organizations on an ever widening scale. However, many of these organizations found that it was necessary to modify the system somewhat to meet their needs; and quite a few subgroup classifications came into rather widespread use. For example, a soil which exhibited the properties of both the A-5 group and the A-7 group was frequently designated as an A-7-5 soil. After this system had been in use about 15 years, the Highway Research Board of the National Research Council appointed a committee of engineers to review the experience with the system and to recommend its modification in the light of this experience.

The report of this committee constituted a revision of the original classification system. This revision was adopted by the American Association of State Highway Officials and is now known as the A.A.S.H.O. systme. In it a number of subgroup designations were recommended and the number of physical properties of a soil upon which its classification is based was reduced to three, namely, the mechanical analysis, the liquid limit, and the plasticity index. The process of determining the group or subgroup to which a soil belongs was greatly simplified by the preparation of the tabular chart shown in Table 14-1. The procedure for classifying a soil whose mechanical analysis, liquid limit, and plasticity index are known is as follows: Begin at the left-hand column of the chart and see if all these known properties of the soil comply with the limiting values specified in the column. If they do not, move to the next column to the right; and continue across the chart until the proper column is reached. The first column in which the soil properties fit the specified limits indi-

TABLE 14-1. Classification[a] of Soils and Soil-Aggregate Mixtures (With suggested subgroups)

General Classification	Granular Materials (35% or less passing No. 200)							Silt-Clay Materials (More than 35% passing No. 200)			
Group Classification	A-1		A-3	A-2				A-4	A-5	A-6	A-7
	A-1-a	A-1-b		A-2-4	A-2-5	A-2-6	A-2-7				A-7-5, A-7-6
Sieve analysis, percent passing:											
No. 10	50 max.										
No. 40	30 max.	50 max.	51 min.								
No. 200	15 max.	25 max.	10 max.	35 max.	35 max.	35 max.	35 max.	36 min.	36 min.	36 min.	36 min.
Characteristics of fraction passing No. 40:											
Liquid limit				40 max.	41 min.	40 max.	41 min.	40 max.	41 min.	40 max.	41 min.[b]
Plasticity index	6 max.		N.P.	10 max.	10 max.	11 min.	11 min.	10 max.	10 max.	11 min.	11 min.
Usual types of significant constituent materials	Stone fragments, gravel and sand		Fine sand	Silty or clayey gravel and sand				Silty soils		Clayey soils	
General rating as subgrade	Excellent to good							Fair to poor			

[a]Classification procedure: With required test data available, proceed from left to right on above chart and correct group will be found by the process of elimination. The first group from the left into which the test data will fit is the correct classification.

[b]Plasticity index of A-7-5 subgroup is equal to or less than LL minus 30. Plasticity index of A-7-6 subgroup is greater than LL minus 30 (see Fig. 14-1).

cates the group or subgroup to which the soil belongs. Group A-3 is placed before Group A-2 in the table to permit its use in this manner. This arrangement does not indicate that A-3 soils are "better" than A-2 soils.

The ranges of the liquid limit and the plasticity index for groups A-4, A-5, A-6, and A-7 are shown in Fig. 14-1.

EXAMPLE 14-1. If a soil contains 65% of material passing a No. 200 sieve, has a liquid limit of 48, and has a plasticity index of 17, to what group does it belong?

SOLUTION: Since more than 35% of the soil material passes the No. 200 sieve, it is a silt-clay material and the process of determining its classi-

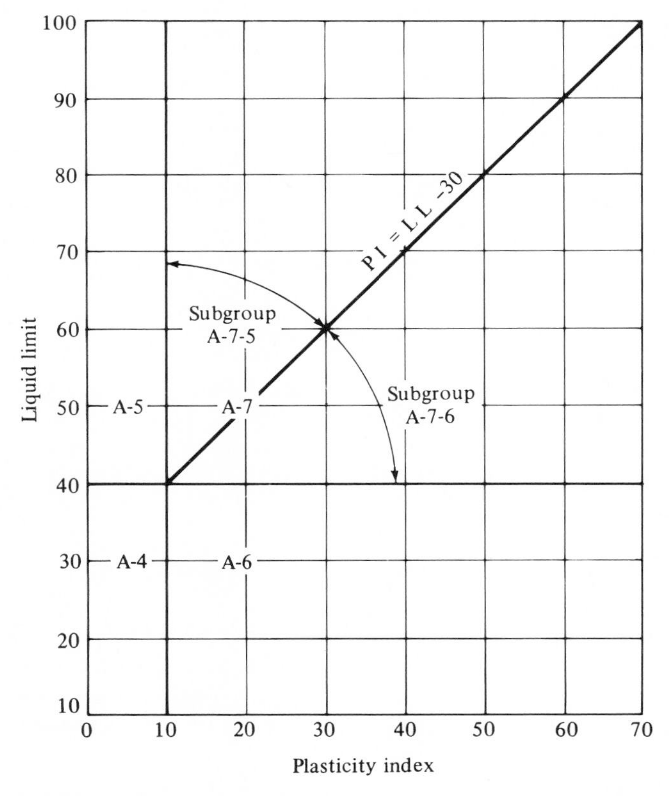

Fig. 14-1. Ranges of liquid limit and plasticity index for groups A-4, A-5, A-6, and A-7.

fication can begin by examining the specified limits for Group A-4. The maximum value of the liquid limit in this column is 40, and the liquid limit of the soil being classified is 48. Therefore, it cannot be an A-4 soil. By proceeding to the right and comparing the actual soil properties with the values given in the various columns, it is found that the soil fits the specifications in the last column and it therefore belongs to Group A-7.

If it is desired to further classify this soil into a subgroup, subtract 30 from the liquid limit of the soil. The difference in this example is 18. Since the plasticity index is less than this difference, the soil is in the A-7-5 subgroup. If the plasticity index had been greater than the liquid limit minus 30, the material would have been an A-7-6 soil.

14.3. DESCRIPTION OF GROUPS AND SUBGROUPS

14.3.1. A.A.S.H.O. Definitions of Gravel, Sand, and Silt-Clay

Gravel—Material passing sieve with 3-in. square openings and retained on the No. 10 sieve.

Coarse sand—Material passing the No. 10 sieve and retained on the No. 40 sieve.

Fine sand—Material passing the No. 40 sieve and retained on the No. 200 sieve.

Silt-clay (combined silt and clay)—Material passing the No. 200 sieve.

Boulders (retained on 3-in. sieve) should be excluded from the portion of the sample to which the classification is applied, but the percentage of such material, if any, in the sample should be recorded.

The term "silty" is applied to fine material having plasticity index of 10 or less and the term "clayey" is applied to fine material having plasticity index of 11 or greater.

14.3.2. Group A-1. The typical material of this group is a well-graded mixture of stone fragments or gravel, coarse sand, fine sand, and a nonplastic or feebly plastic soil binder. This group also includes stone fragments, gravel, coarse sand, volcanic cinders, etc., without soil binder.

Subgroup A-1-a includes those materials consisting predominantly of stone fragments or gravel, either with or without a well-graded binder of fine material.

Subgroup A-1-b includes those materials consisting predominantly of coarse sand either with or without a well-graded soil binder.

14.3.3. Group A-3. The typical material of this group is fine beach sand or fine desert blow sand without silty or clayey fines or with a very small amount of nonplastic silt. The group includes also stream-deposited mixtures of poorly graded fine sand and limited amounts of coarse sand and gravel.

14.3.4. Group A-2. This group includes a wide variety of "granular" materials which are at the border line between the materials falling in Groups A-1 and A-3 and the silt-clay materials of Groups A-4 through A-7. It includes any material not more than 35% of which passes a No. 200 sieve and which cannot be classified as Group A-1 or Group A-3 because of having fines content or plasticity, or both, in excess of the limitations for those groups.

Subgroups A-2-4 and A-2-5 include various granular materials not more than 35% of which passes a No. 200 sieve and containing a minus No. 40 portion having the characteristics of the A-4 and A-5 groups. These subgroups include such materials as gravel and coarse sand with silt content or plasticity index in excess of the limitations of Group A-1, and fine sand with nonplastic silt content in excess of the limitations of Group A-3.

Subgroups A-2-6 and A-2-7 include materials similar to those described under subgroups A-2-4 and A-2-5 except that the fine portion contains plastic clay having the characteristics of the A-6 group or the A-7 group. The approximate combined effect of plasticity indexes greatly in excess of 10 and percentages passing the No. 200 sieve considerably in excess of 15 is reflected by group-index values of 0 to 4.

14.3.5. Group A-4. The typical material of this group is a nonplastic or moderately plastic silty soil 75% or more of which usually passes the No. 200 sieve. The group includes also mixtures of fine silty soil and up to 64% of sand and gravel retained on the No. 200 sieve.

14.3.6. Group A-5. The typical material of this group is similar to that described under Group A-4, but it is usually of diatomaceous or micaceous character and may be highly elastic, as indicated by the high liquid limit.

14.3.7. Group A-6. The typical material of this group is a plastic clay soil 75% or more of which usually passes the No. 200 sieve. The group includes also mixtures of fine clayey soil and up to 64% of sand and gravel retained on the No. 200 sieve. Materials of this group usually have high volume change between wet and dry states.

14.3.8. Group A-7. The typical material of this group is similar to that described under Group A-6, but it has the high liquid limits characteristic of the A-5 group and may be elastic as well as subject to high volume change.

Subgroup A-7-5 includes those materials which have moderate plasticity indexes in relation to liquid limit and which may be highly elastic as well as subject to considerable volume change.

Subgroup A-7-6 includes those materials which have high plasticity indexes in relation to liquid limit and which are subject to extremely high volume change.

14.3.9. Group A-8 of Original Classification. The typical material of this group is a peat or muck soil ordinarily found in obviously unstable, swampy areas. It is characterized by low density, high compressibility, high water content, and high organic-matter content. Attention is directed to the fact that the classification of soils in this group is based largely upon the character and environment of their field occurrence, rather than upon laboratory tests of the material. As a matter of fact, A-8 soils usually show the laboratory-determined properties of an A-7 soil, but are properly classified as group A-8 because of the manner of their occurrence.

14.4. GROUP INDEX. A new feature incorporated in the A.A.S.H.O. system is the group index. By means of this index, a soil can be rated within its group or subgroup. The group index is a number which is dependent upon the percentage of the soil passing the No. 200 sieve, its liquid limit, and its plasticity index. This index is computed by the following empirical formula:

$$\text{Group index} = (F\text{-}35)[0.2 + 0.005(\text{LL-}40)] + 0.01(F\text{-}15)(\text{PI-}10) \qquad (14\text{-}1)$$

in which F is the percentage passing No. 200 sieve, expressed as a whole number. This percentage is based only on the material passing the 3-in. sieve; LL is the liquid limit, and PI is the plasticity index as previously defined. When the calculated group index is negative, the group index is reported as zero (0). The group index is expressed to the nearest whole number and is written in parentheses after the group or subgroup designation. A group index should be given for each soil, even if the numerical value is zero, in order to indicate that the classification has been determined by the A.A.S.H.O. system rather than the original Public Roads system. The solution of Eq. (14-1) is made easy and rapid by the chart shown in Fig. 14-2.

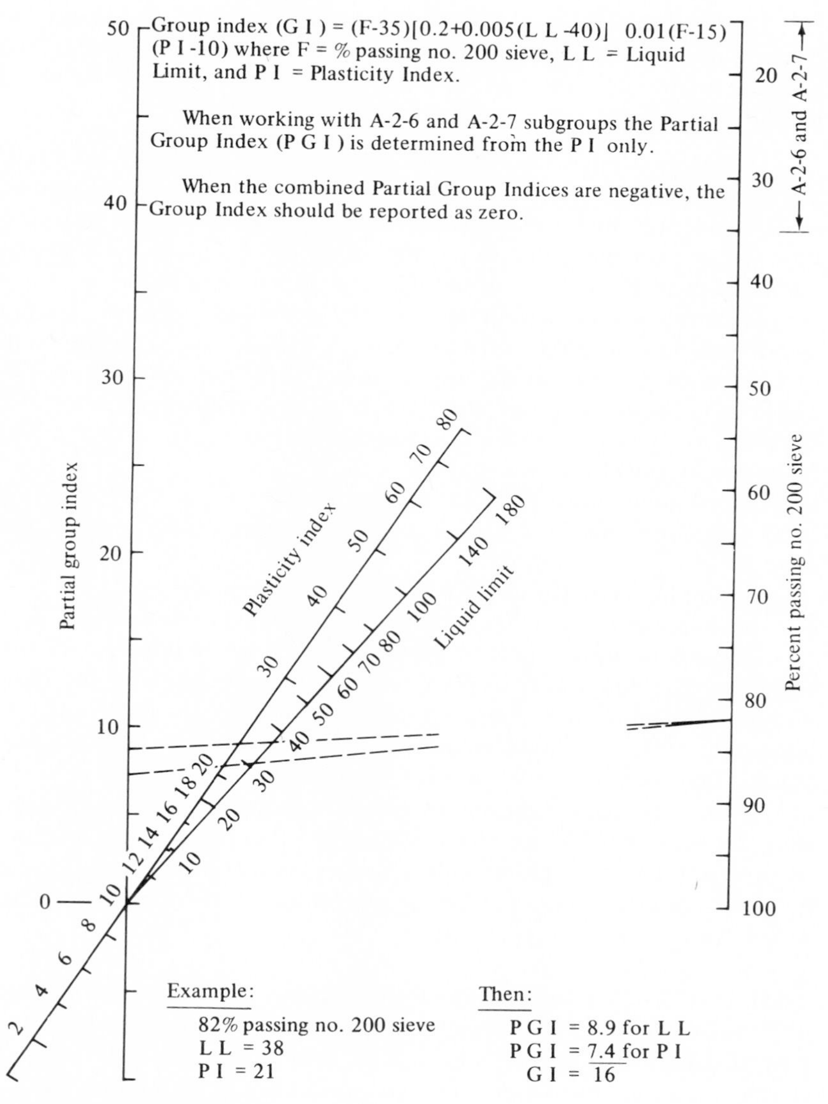

Fig. 14-2. Group index chart.

14.4.1. Basis for Group Index Formula. The empirical group index formula devised for approximate within-group evaluation of the "clayey granular materials" and the "silt-clay materials" is based on the following assumptions:

Materials falling within Groups A-1-a, A-1-b, A-2-4, A-2-5, and A-3 are satisfactory as subgrade when properly drained and compacted under moderate thickness of pavement (base and/or surface course) of a type suitable for the traffic to be carried, or can be made satisfactory by additions of small amounts of natural or artificial binders.

Materials falling within the "clayey granular" Groups A-2-6 and A-2-7 and the "silt-clay" Groups A-4, A-5, A-6, and A-7 will range in quality as subgrade from the approximate equivalent of the good A-2-4 and A-2-5 subgrades to fair and poor subgrades requiring a layer of subbase material or an increased thickness of base course in order to furnish adequate support for traffic loads.

The assumed critical minimum percentage passing the No. 200 sieve is 35 neglecting plasticity, and 15 as affected by plasticity indexes greater than 10.

Liquid limits of 40 and above are assumed to be critical.

Plasticity indexes of 10 and above are assumed to be critical.

There is no upper limit of group index value obtained by use of the formula. The adopted critical values of percentage passing the No. 200 sieve, liquid limit, and plasticity index, are based on an evaluation of subgrade, subbase, and base course materials by several highway organizations that use the tests involved in this classification system.

Under average conditions of good drainage and thorough compaction, the supporting value of a material as subgrade may be assumed as an inverse ratio to its group index, that is, a group index of 0 indicates a "good" subgrade material and group index of 20 or greater indicates a "very poor" subgrade material.

14.5. LIMITATIONS OF SOIL CLASSIFICATION. It is not good practice to rely too heavily upon the classification of a soil for actual design information, because many factors which enter into design are not indicated by the classification of soil alone. Design characteristics should be based upon tests of the individual soils involved, such as tests for shearing strength, permeability, bearing strength, and shrinkage and swell. Of course, as experience is accumulated in a given geographical region, it is reasonable to translate and codify that experience in terms of one or several soil classifications. For example, in Iowa, experience has indicated that glacial clay of the A-6 group makes a better subgrade for both rigid and flexible pavements than a loess of the A-4 group, and this experience is properly translated into design through the medium of the soil classification. However, in other regions, the reverse might very well be true because of different climatic and geophysical conditions.

14.6. DESIRABLE SUBGRADE SOILS. In spite of the precaution urged in the preceding paragraph, there are a number of general characteristics associated with the soil groups of the A.A.S.H.O. system which are helpful in visualizing the usefulness of a soil in engineering works when its classification is known. For instance, because of the granular character of soils of the A-1, A-2, and A-3 groups, they may be expected to have relatively high values of internal friction and permeability and in general to be satisfactory as subgrades for moderate combined thicknesses of base and surface courses. The better materials of the other groups may also be satisfactory as subgrades for moderate combined thicknesses of base and surface courses.

14.7. DESIRABLE SUBBASE MATERIALS. While it is usually desirable to include in specifications for individual projects definite requirements for selected subbase materials, the group symbols A-1-a, A-1-b, A-2-4, A-2-5, and A-3 will serve to identify those materials most suitable as bank-run blanket courses over inferior silt and clay soils. Where thick subbase layers are required over the poorest clay soils, materials of only fair quality, as those of group A-4 or group A-7, may be used in the lower portions of such layers if this use will not introduce danger of frost damage due to unfavorable climatic conditions and inadequate drainage. The group-index values serve as a general guide to the depth below the surface at which it is safe to use a particular material as a subbase layer. In order to obtain maximum benefit from selected materials, placing them in the thinnest practicable layers with thorough compaction is essential.

Materials of the A-1, A-2, and A-3 groups may be so devoid of fines as to require addition of fine materials in order to form a firm subgrade or subbase on which to place a base course or pavement. Also, when a material of the A-1 or A-2 group is used as a subbase over a clay subgrade, attention is directed to the need for a substantial sand content in the material or a blanket of sandy soil directly on the subgrade to prevent intrusion of clay into the subbase.

14.8. MATERIALS FOR BASE COURSES. Materials of the A-1 group, particularly those of the A-1-a subgroup, include friable coarse granular soils which are suitable for granular base courses or which can be made suitable by processing. A natural material may be encountered in construction which falls in this group and would be satisfactory as a base course for a thin bituminous surfacing without processing. While this group indicates those materials which may be suitable or can be made suitable for use as granular base courses, comparison with properties of materials known to be adequate for base courses should be the basis of selection for that purpose.

14.9. MATERIALS FOR EMBANKMENTS. In general, materials evaluated as best for subgrades will also form the best embankments with a minimum of construction and maintenance difficulties. In the case of inferior materials, special tests and investigations may be necessary to determine safe designs for high embankments.

14.10. FROST-PRONE MATERIALS. The groupings of this classification system may be valuable in identifying materials likely to be weakened by frost action, although the general subject is one involving climatic and drainage conditions as well as materials. Those materials falling in Groups A-4 and A-5 are of the types which are usually frost-prone. Materials of the A-2, A-6, and A-7 groups may also be frost-prone, but to a lesser extent than those of the A-4 and A-5 groups.

14.11. UNIFIED CLASSIFICATION SYSTEM. Another engineering soil classification which is widely used is the Unified Classification. This system is an outgrowth of the Airfield Classification developed by Professor A. Casagrande. It was adopted jointly by the Corps of Engineers and the Bureau of Reclamation in 1952.

The Unified System incorporates the textural characteristics of the soil into the engineering classification and utilizes the grain-size classification shown in Table 7-1. The essentials of the system are given in Table 14-2. All soils are classified into fifteen groups, each group being designated by two letters. These letters are abbreviations of certain soil characteristics, as follows:

G—Gravel
S—Sand
M—Nonplastic or low plasticity fines
C—Plastic fines
Pt—Peat, humus, swamp soils
O—Organic
W—Well graded
P—Poorly graded
L—Low liquid limit
H—High liquid limit

14.11.1. GW and SW Groups. These groups comprise well-graded gravelly and sandy soils which contain less than 5% of nonplastic fines passing the No. 200 sieve. Fines which are present must not noticeably change the strength characteristics of the coarse-grained fraction and must not interfere with its free-draining characteristic. In areas subject to frost action, the material should not contain more than about 3% of soil grains smaller than 0.02 mm in size.

14.11.2. GP and SP Groups. These groups are poorly graded gravels and sands containing less than 5% of nonplastic fines. They may consist of uniform gravels, uniform sands, or nonuniform mixtures of very coarse material and very fine sand with intermediate sizes lacking. Materials of this latter type are sometimes referred to as skip-graded, gap-graded, or step-graded.

14.11.3. GM and SM Groups. In general, these groups include gravels or sands which contain more than 12% of fines having little or no plasticity. The plasticity index and liquid limit of a soil in either of these groups plot below the "A" line on the plasticity chart in Table 14-2. Gradation is not important, and both well-graded and poorly graded materials are included. Some sands and gravels in these groups may have a binder composed of natural cementing agents, so proportioned that the mixture shows negligible swelling or shrinkage. Thus the dry strength is provided by a small amount of soil binder or by cementation of calcareous materials or iron oxide. The fine fraction of noncemented materials may be composed of silts or rock-flour types having little or no plasticity, and the mixture will exhibit no dry strength.

14.11.4. GC and SC Groups. These groups comprise gravelly or sandy soils with more than 12% of fines which exhibit either low or high plasticity. The plasticity index and liquid limit of a soil in either of these groups plot above the "A" line on the plasticity chart in Table 14-2. Gradation of these materials is not important. The plasticity of the binder fraction has more influence on the behavior of the soils than does variation in gradation. The fine fraction is generally composed of clays.

14.11.5. ML and MH Groups. These groups include the predominantly silty materials and micaceous or diatomaceous soils. An arbitrary division between the two groups has been established where the liquid limit is 50. Soils in these groups are sandy silts, clayey silts, or inorganic silts with relatively low plasticity. Also included are loessial soils and rock flours. Micaceous and diatomaceous soils generally fall within the MH group but may extend into the ML group when their liquid limit is less than 50. The same is true for certain types of kaolin clays and some illite clays having relatively low plasticity.

14.11.6. CL and CH Groups. The CL and CH groups embrace clays with low and high liquid limits, respectively. They are primarily inorganic

TABLE 14-2. Unified Soil Classification (Including Identification Description)

Major Divisions			Group Symbols	Typical Names	Field Identification Procedures (Excluding particles larger than 3 in. and basing fractions on estimated weights)		
1	2		3	4	5		
Coarse-grained Soils More than half of material is *larger* than No. 200 sieve size.	Gravels More than half of coarse fraction is larger than No. 4 sieve size.	Clean Gravels (Little or no fines)	GW	Well-graded gravels, gravel-sand mixtures, little or no fines.	Wide range in grain sizes and substantial amounts of all intermediate particle sizes.		
			GP	Poorly graded gravels or gravel-sand mixtures, little or no fines.	Predominantly one size or a range of sizes with some intermediate sizes missing.		
		Gravels with Fines (Appreciable amount of fines)	GM	Silty gravels, gravel-and-silt mixture.	Nonplastic fines or fines with low plasticity (for identification procedures see ML below).		
			GC	Clayey gravels, gravel-sand-clay mixtures.	Plastic fines (for identification procedures see CL below).		
	Sands More than half of coarse fraction is smaller than No. 4 sieve size. (For visual classification, the 1/4-in. size may be used as equivalent to the No. 4 sieve size)	Clean Sands (Little or no fines)	SW	Well-graded sands, gravelly sands, little or no fines.	Wide range in grain size and substantial amounts of all intermediate particle sizes.		
			SP	Poorly graded sands or gravelly sands, little or no fines.	Predominantly one size or a range of sizes with some intermediate sizes missing.		
		Sands with Fines (Appreciable amount of fines)	SM	Silty sands, sand-silt mixtures.	Nonplastic fines or fines with low plasticity (for identification procedures see ML below).		
			SC	Clayey sands, sand-clay mixtures.	Plastic fines (for identification procedures see CL below).		
Fine-grained Soils More than half of material is *smaller* than No. 200 sieve size.					Identification Procedures on Fraction Smaller than No. 40 Sieve Size		
					Dry Strength (Crushing characteristics)	Dilatancy (Reaction to shaking)	Toughness (Consistency near PL)
	Silts and Clays	Liquid limit is less than 50	ML	Inorganic silts and very fine sands, rock flour, silty or clayey fine sands or clayey silts with slight plasticity.	None to slight	Quick to slow	None
			CL	Inorganic clays of low to medium plasticity, gravelly clays, sandy clays, silty clays, lean clays.	Medium to high	None to very slow	Medium
			OL	Organic silts and organic silty clays of low plasticity.	Slight to medium	Slow	Slight
	Silts and Clays	Liquid limit is greater than 50	MH	Inorganic silts, miscaceous or diatomaceous fine sandy or silty soils, elastic silts.	Slight to medium	Slow to none	Slight to medium
			CH	Inorganic clays of high plasticity, fat clays.	High to very high	None	High
			OH	Organic clays of medium to high plasticity, organic silts.	Medium to high	None to very slow	Slight to medium
Highly Organic Soils			Pt	Peat and other high organic soils.	Readily identified by color, odor, spongy feel and frequently by fibrous texture.		

The No. 200 sieve size is about the smallest particle visible to the naked eye.

(1) Boundary classification: Soils possessing characteristics of two groups are designated by combinations of group symbols. For example GW-GC, well-graded gravel-sand mixture with clay binder.

(2) All sieve sizes on this chart are U.S. standard.

FIELD IDENTIFICATION PROCEDURES FOR FINE-GRAINED SOILS OR FRACTIONS

These procedures are to be performed on the minus No. 40 sieve size particles, approximately 1/64 in. For field classification purposes, screening is not intended, simply remove by hand the coarse particles that interfere with the tests.

Dilatancy (reaction to shaking)

After removing particles larger than No. 40 sieve size, prepare a pat of moist soil with a volume of about one-half cubic inch. Add enough water if necessary to make the soil soft but not sticky.

Place the pat in the open palm of one hand and shake horizontally, striking vigorously against the other hand several times. A positive reaction consists of the appearance of water on the surface of the pat which changes to a livery consistency and becomes glossy. When the sample is squeezed between the fingers, the water and gloss disappear from the surface, the pat stiffens, and finally it cracks or crumbles. The rapidity of appearance of water during shaking and of its disappearance during squeezing assist in identifying the character of the fines in a soil.

Very fine clean sands give the quickest and most distinct reaction whereas a plastic clay has no reaction. Inorganic silts, such as a typical rock flour, show a moderately quick reaction.

Dry Strength (crushing characteristics)

After removing particles larger than No. 40 sieve size,

Adopted by Corps of Engineers and Bureau of Reclamation, January 1952

Information Required for Describing Soils	Laboratory Classification Criteria		
6	7		
For undisturbed soils add information on stratification, degree of compactness, cementation, moisture conditions, and drainage characteristics.	Use grain-size curve in identifying the fractions as given under field identification. Determine percentages of gravel and sand from grain-size curve. Depending on percentage of fines (fraction smaller than No. 200 sieve size) coarse-grained soils are classified as follows: Less than 5% GW, GP, SW, SP; More than 12% GM, GC, SM, SC; 5% to 12% *Borderline* cases requiring use of dual symbols.	$C_u = \frac{D_{60}}{D_{10}}$ Greater than 4 $C_c = \frac{(D_{30})^2}{D_{10} \times D_{60}}$ Between 1 and 3	
		Not meeting all gradation requirements for GW	
Give typical name; indicate approximate percentages of sand and gravel, maximum size; angularity, surface condition, and hardness of the coarse grains; local or geologic name and other pertinent descriptive information; and symbol in parentheses.		Atterberg limits below "A" line or PI less than 4	Above "A" line with PI between 4 and 7 are *borderline* cases requiring use of dual symbols.
		Atterberg limits above "A" line with PI greater than 7	
		$C_u = \frac{D_{60}}{D_{10}}$ Greater than 6 $C_c = \frac{(D_{30})^2}{D_{10} \times D_{60}}$ Between 1 and 3	
Example: *Silty sand*, gravelly; about 20% hard, angular gravel particles 1/2-in. maximum size; rounded and subangular sand grains, coarse to fine; about 15% nonplastic fines with low dry strength; well compacted and moist in place; alluvial sand; (SM).		Not meeting all gradation requirements for SW	
		Atterberg limits above "A" line or PI less than 4	Limits plotting in hatched zone with PI between 4 and 7 are *borderline* cases requiring use of dual symbols.
		Atterberg limits above "A" line with PI greater than 7	
For undisturbed soils add information on structure, stratification, consistency in undisturbed and remolded states, moisture and drainage conditions. Give typical name; indicate degree and character of plasticity; amount and maximum size of coarse grains; color in wet condition; odor, if any; local or geologic name and other pertinent descriptive information; and symbol in parentheses. Example: *Clayey silt*, brown; slightly plastic; small percentage of fine sand; numerous vertical root holes; firm and dry in place; loess; (ML).		Plasticity chart (see below) For laboratory classification of fine-grained soils	

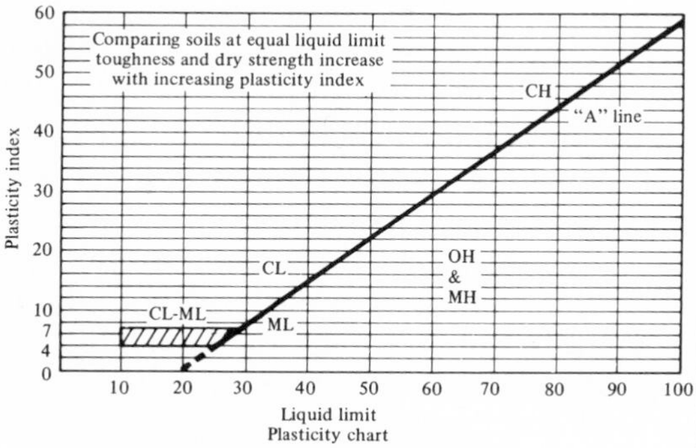

Plasticity chart

For laboratory classification of fine-grained soils

mold a pat of soil to the consistency of putty, adding water if necessary. Allow the pat to dry completely by oven, sun, or air-drying, and then test its strength by breaking and crumbling between the fingers. This strength is a measure of the character and quantity of the colloidal fraction contained in the soil. The dry strength increases with increasing plasticity.

High dry strength is characteristic for clays of the CH group. A typical inorganic silt possesses only very slight dry strength. Silty fine sands and silts have about the same slight dry strength, but can be distinguished by the feel when powdering the dried specimen. Fine sand feels gritty whereas a typical silt has the smooth feel of flour.

Toughness (consistency near plastic limit)

After particles larger than the No. 40 sieve size are removed, a specimen of soil about one-half inch cube in size is molded to the consistency of putty. If too dry, water must be added and if sticky, the specimen should be spread out in a thin layer and allowed to lose some moisture by evaporation. Then the specimen is rolled out by hand on a smooth surface or between the palms into a thread about one-eighth inch in diameter. The thread is then folded and re-rolled repeatedly. During this manipulation the moisture content is gradually reduced and the specimen stiffens, finally loses its plasticity, and crumbles when the plastic limit is reached.

After the thread crumbles, the pieces should be lumped together and a slight kneading action continued until the lump crumbles.

The tougher the thread near the plastic limit and the stiffer the lump when it finally crumbles, the more potent is the colloidal clay fraction in the soil. Weakness of the thread at the plastic limit and quick loss of coherence of the lump below the plastic limit indicate either inorganic clay of low plasticity, or materials such as kaolin-type clays and organic clays which occur below the "A" line.

Highly organic clays have a very weak and spongy feel at the plastic limit.

clays. Low-plasticity clays are classified as CL and are usually lean clays, sandy clays, or silty clays. The medium-plasticity and high-plasticity clays are classified as CH. These include the fat clays, gumbo clays, certain volcanic clays, and bentonite. The glacial clays of the northern United States cover a wide band in the CL and CH groups.

14.11.7. OL and OH Groups. The soils in these groups are characterized by the presence of organic matter, including organic silts and clays. They have a plasticity range which corresponds with the ML and MH groups.

14.11.8. Pt Group. Highly organic soils which are very compressible and have undesirable construction characteristics are classified in one group with the symbol Pt. Peat, humus, and swamp soils with a highly organic texture are typical of the group. Particles of leaves, grass, branches of bushes, or other fibrous vegetable matter are common components of these soils.

14.11.9. Borderline Classifications. Soils in the GW, SW, GP, and SP groups are nonplastic materials having less than 5% passing the No. 200 sieve, while GM, SM, GC, and SC soils have more than 12% passing the No. 200 sieve. When these coarse-grained materials contain between 5 and 12% of fines, they are classed as borderline and are designated by a dual symbol, such as GW–GM. Similarly, coarse-grained soils which have less than 5% passing the No. 200 sieve, but which are not free draining or in which the fine fraction exhibits plasticity, are also classed as borderline and given a dual symbol. Still another type of borderline classification occurs when the liquid limit of a fine-grained soil is less than 29 and the plasticity index lies in the range from 4 to 7. These limits are indicated by the shaded area on the plasticity chart in Table 14-2. When the liquid limit and plasticity index of a soil plot within the shaded area, a double symbol, such ML–CL, is used to designate the soil.

14.11.10. "Silty" and "Clayey." In Chapter 7 the terms silt and clay were used to indicate certain particle-size ranges of soil material. In the Unified System these terms are used to describe soils whose Atterberg limits plot below and above the "A" line on the plasticity chart in Table 14-2, and the adjectives "silty" and "clayey" may be used to describe soils whose limits plot close to the "A" line. For example, a clay soil for which the liquid limit is 40 and the plasticity index is 16 may be called a silty clay.

14.12. FIELD IDENTIFICATION. Final classification of a soil will require laboratory tests to determine the critical properties, but a tentative field classification may often be of value. For a coarse-grained material, a dry sample is spread on a flat surface to determine gradation, grain size and shape, and mineral composition. Considerable skill and experience are required to visually differentiate between a well-graded soil and a poorly-graded soil.

From the standpoint of durability, pebbles and sand grains of sound rock, including weathered granite rocks and quartzite, are easily identified. Unsound weathered material may be recognized because of its discolorations and the relative ease with which the grains can be crushed. Fragments of shaly rock may render a coarse-grained soil unsuitable for certain purposes, since alternate wetting and drying may cause them to disintegrate partially or completely. This characteristic can be identified by submerging thoroughly dried particles in water for at least 24 hr and then testing for the difference between the final strength and the original strength.

14.13. FINE FRACTION. An important criterion for classification of coarse-grained soils is the amount of material passing the No. 200 sieve. This amount can be approximated in the field by decanting a sample until the water is clear and estimating the amount of material thus removed. Another method is to place a sample of soil in a test tube, fill the tube with water and shake the contents thoroughly, and then allow the material to settle. Particles retained on a No. 200 sieve will settle out of suspension in about 20 to 30 secs, whereas finer particles will take a longer time. An estimate of the relative amounts of coarse and fine material can be made on the basis of the relative volumes of the coarse and fine portions of the sediment.

14.14. FINE-GRAINED SOILS. The principal procedures for field identification of fine-grained soils are the tests for dilatancy (reaction to shaking), plasticity characteristics, and dry strength, which are performed on the fraction passing the No. 40 sieve. In addition, observations of color and odor are of value, particularly for organic soils. If a No. 40 sieve is not available, removal of the fraction retained on this sieve may be partially accomplished by hand picking. No doubt some plus No. 40 particles will remain in the soil after hand separation, but they probably will have only a minor effect on the field tests.

14.14.1. Dilatancy. For the dilatancy test, enough water is added to about $\frac{1}{2}$ cu. in. of the minus 40 fraction of a soil to make it soft but not sticky. The pat of soil is shaken horizontally in the open palm of one hand, which is struck vigorously against the other hand several times. A fine-grained soil that is nonplastic or one that exhibits very low plasticity will become livery and will show free water on the surface while being shaken. Squeezing the pat with the fingers will cause the water to disappear from the surface and the sample to stiffen and finally crumble under increasing pressure, like a brittle material. Shaking the pat will cause it to flow together and water to again appear on the surface. A distinction should be made between a rapid reaction, a slow reaction, or no reaction to the shaking test, the rating depending on the speed with which the pat changes its consistency and the water on the surface appears or disappears. Rapid reaction is typical of nonplastic, uniform fine sand, of silty sand (SP or SM), of inorganic silt (ML), particularly the rock-flour type, and of diatomaceous earth (MH).

The reaction becomes somewhat more sluggish as the uniformity of gradation decreases and the plasticity increases, up to a certain degree. Even a slight amount of colloidal clay will impart some plasticity to the soil and will materially slow the reaction to the shaking test. Soils which react in this manner are somewhat plastic inorganic and organic silts (ML or OL), very lean clays (CL), and some kaolin-type clays (ML or MH). Extremely slow reaction or no reaction to the shaking test is characteristic of all typical clays (CL or CH) and of highly plastic organic clays (OH).

14.14.2. Plasticity Characteristics. Field estimates of the plasticity characteristic of fine-grained soils and the binder fraction of coarse-grained soils may be made by rolling a small sample of minus 40 material between the palms of the hand in a manner similar to the standard plastic-limit test. The sample should be fairly wet, but not sticky. As it is rolled into $\frac{1}{8}$-in. threads and folded and rerolled, the stiffness of the threads and the toughness of the lumps after crumbling should be observed. The higher the soil above the "A" line on the plasticity chart (CL or CH), the stiffer are the threads as the water content approaches the plastic limit, and the tougher are the lumps after crumbling and remolding. Soils slightly above the "A" line (CL or CH) form a medium-tough thread which can be rolled easily as the plastic limit is approached;. but when the soil is kneaded below the plastic limit, it crumbles readily. Soils below the "A" line (ML, MH, OL, or OH) form a weak thread, and with the exception of an OH soil, such a soil cannot be lumped into a coherent mass below the plastic limit. Plastic soils containing organic material or much mica form threads which are very soft and spongy near the plastic limit.

In general, the binder fraction of a coarse-grained soil with silty fines

(GM or SM) will exhibit plasticity characteristics similar to those of ML soils. The binder fraction of a coarse-grained soil with clayey fines (GC or SC) will be similar to CL soils.

14.14.3. Dry Strength. To determine dry-strength characteristics, a pat of minus 40 soil is moistened and molded to the consistency of putty, and is then allowed to dry in an oven or in the sun and air. When dry the pat should be crumbled between the fingers. Soils with slight dry strength (ML or MH) crumble readily with very little finger pressure. Also, organic silts and lean organic clays of low plasticity (OL) and very fine sandy soils (SM) have slight dry strength. Most clays of the CL group and some OH soils, as well as the binder fraction of gravelly and sandy clays (GC or SC), have medium dry strength and require considerable finger pressure to crumble the sample. Most CH clays and some organic clays (OH) having high liquid limits and located near the "A" line have high dry strength, and the test pat can be broken with the fingers but cannot be crumbled.

14.14.4. Color and Odor. Color identification of soil has been discussed in Chapter 2. In connection with field classification by the Unified System, certain dark or drab shades of gray or brown, including almost-black colors, are indicative of fine-grained soils containing organic colloidal matter (OL or OH). In contrast, brighter colors, including medium and light gray, olive green, brown, red, yellow, and white, are generally associated with inorganic soils.

An organic soil (OL or OH) usually has a distinctive odor, which can be used as an aid in its identification. This odor is especially apparent from a fresh sample. It gradually diminishes on exposure of the soil to air, but it can be revived by heating a wet sample.

14.15. FEDERAL AVIATION ADMINISTRATION CLASSIFICATION. Another engineering soil classification system is that employed by the Federal Aviation Administration. It is based upon three test properties of soils, namely, the mechanical analysis, the liquid limit, and the plasticity index. In addition, certain geophysical characteristics of the soil environment, such as drainage and degree of frost susceptibility, are utilized in conjunction with the classification groups to assign to a soil a rating symbol which indicates its suitability as a pavement-foundation material. The soil groups of this system are shown in Table 14-3, and the subgrade rating of each group under various conditions of drainage and frost susceptibility are shown in Table 14-4.

TABLE 14-3. Classification of Soils for Airport Pavement Construction

Soil Group		Mechanical Analysis				Liquid Limit	Plasticity Index
			Material Finer than No. 10 Sieve (%)				
Type	*No.*	*Material Retained on No. 10 Sieve*[a] *(%)*	*Coarse Sand, Passing No. 10; Retained on No. 40*	*Fine Sand, Passing No. 40 Retained on No. 200*	*Combined Silt and Clay; Passing No. 200*		
Granular	E-1	0–45	40+	60–	15–	25–	6–
	E-2	0–45	15+	85–	25–	25–	6–
	E-3	0–45	—	—	25–	25–	6–
	E-4	0–45	—	—	35–	35–	10–
	E-5	0–55	—	—	45–	40–	15–
Fine grained	E-6	0–55	—	—	45+	40–	10–
	E-7	0–55	—	—	45+	50–	10–30
	E-8	0–55	—	—	45+	60–	15–40
	E-9	0–55	—	—	45+	40+	30–
	E-10	0–55	—	—	45+	70–	20–50
	E-11	0–55	—	—	45+	80–	30+
	E-12	0–55	—	—	45+	80+	—
	E-13	Muck and peat—field examination					

[a]If percentage of material retained on the No. 10 sieve exceeds that shown, the classification may be raised, provided such material is sound and fairly well graded.

TABLE 14-4. Airport Paving Subgrade Classification

Soil Group	Subgrade Class[a]		
	Good Drainage	Poor Drainage	
	(No Frost or Frost)	No Frost	Frost
E-1	Fa or Ra	Fa or Ra	F1 or Ra
E-2	Fa or Ra	F1 or Ra	F2 or Rb
E-3	F1 or Ra	F2 or Rb	F3 or Rb
E-4	F1 or Ra	F2 or Rb	F4 or Rb
E-5	— — — — — —	F3 or Rb	F5 or Rb
E-6	— — — — — —	F4 or Rc	F6 or Rc
E-7	— — — — — —	F5 or Rc	F7 or Rc
E-8	— — — — — —	F6 or Rc	F8 or Rd
E-9	— — — — — —	F7 or Rd	F9 or Rd
E-10	— — — — — —	F8 or Rd	F10 or Rd
E-11	— — — — — —	F9 or Re	F10 or Re
E-12	— — — — — —	F10 or Re	F10 or Re
E-13		Not suitable for subgrade	

[a]Note: Subgrades classed as Fa for flexible pavements and Ra for rigid pavements furnish adequate subgrade support without the addition of subbase material. The soil's value as a subgrade material decreases as the number increases. See Ref. 2 for subbase thickness.

14.15.1. Group E-1. Included are well-graded, coarse, granular soils that are stable even under poor drainage conditions and are not generally subject to detrimental frost heave. Soils of this group may conform to well-graded sands and gravels with little or no fines. If frost is a factor, the soil should be checked to determine the percentage of the material less than 0.02 mm in diameter.

14.15.2. Group E-2. Similar to Group E-1 but having less coarse sand, the group may contain greater percentages of silt and clay. Soils of this group may become unstable when poorly drained as well as being subject to frost heave to a limited extent.

14.15.3. Groups E-3 and E-4. Included are the fine, sandy soils of inferior grading. They may consist of fine cohesionless sand or sand-clay types with a fair-to-good quality of binder. They are less stable than Group E-2 soils under adverse conditions of drainage and frost action.

14.15.4. Group E-5. Here all poorly graded soils having more than 35% but less than 45% of silt and clay combined are included. This group also includes all soils with less than 45% of silt and clay but which have plasticity indices of 10 to 15. These soils are susceptible to frost action.

14.15.5. Group E-6. Consisting of the silts and sandy silts having zero-to-low plasticity, these soils are friable and quite stable when dry or at low moisture contents. They lose stability and become very spongy when wet and for this reason are difficult to compact unless the moisture content is carefully controlled. Capillary rise in the soils of this group is very rapid; and they, more than soils of any other group, are subject to detrimental frost heave.

14.15.6. Group E-7. The silty clay, sand clay, clayey sands, and clayey silts comprise this group. They range from friable to hard consistency when dry and are plastic when wet. These soils are stiff and dense when compacted at the proper moisture content. Variations in moisture are apt to produce a detrimental volume change. Capillary forces acting in the soil are strong, but the rate of capillary rise is relatively slow and frost heave, while detrimental, is not as severe as in the E-6 soils.

14.15.7. Group E-8. These soils are similar to the E-7 soils but the higher liquid limits indicate a greater degree of compressibility, expansion, shrinkage, and lower stability under adverse moisture conditions.

14.15.8. Group E-9. Consisting of the silts and clays containing micaceous and diatomaceous materials, these soils are highly elastic and very difficult to compact. They have low stability in both the wet and dry state and are subject to frost heave.

14.15.9. Group E-10. Included are the silty clay and clay soils that form hard clods when dry and are very plastic when wet. They are very compressible, possess the properties of expansion, shrinkage, and elasticity to a high degree and are subject to frost heave. Soils of this group are more difficult to compact than those of the E-7 or E-8 groups and require careful control of moisture to produce a dense, stable fill.

14.15.10. Group E-11. These soils are similar to those of the E-10 group but have higher liquid limits. This group includes all soils with liquid limits between 70 and 80 and plasticity indices over 30.

14.15.11. Group E-12. All soils having liquid limits over 80 regardless of their plasticity indices belong here. They may be highly plastic clays that are extremely unstable in the presence of moisture, or they may be very elastic soils containing mica, diatoms, or organic matter in excessive amounts. Whatever the cause of their instability, they will require the maximum in corrective measures.

14.15.12. Group E-13. This group encompasses organic swamp soils such as muck and peat which are recognized by examination in the field. In their natural state, they are characterized by very low stability, very low density, and very high moisture content.

14.16. SPECIAL CONDITIONS AFFECTING FINE-GRAINED SOILS. A soil may possibly contain certain constituents that will give test results which would place it, according to Table 14-3, in more than one group. This could happen with soils containing mica, diatoms, or a large proportion of colloidal material. Such overlapping can be avoided by the use of Fig. 14-3 in conjunction with Table 14-3, with the exception of E-5 soils which should be classified strictly according to Table 14-3.

Soils with plasticity indices higher than those corresponding to the maximum liquid limit of the particular group are not of common occurrence. When encountered, they are placed on the higher numbered group as shown in Fig. 14-3. This is justified by the fact that, for

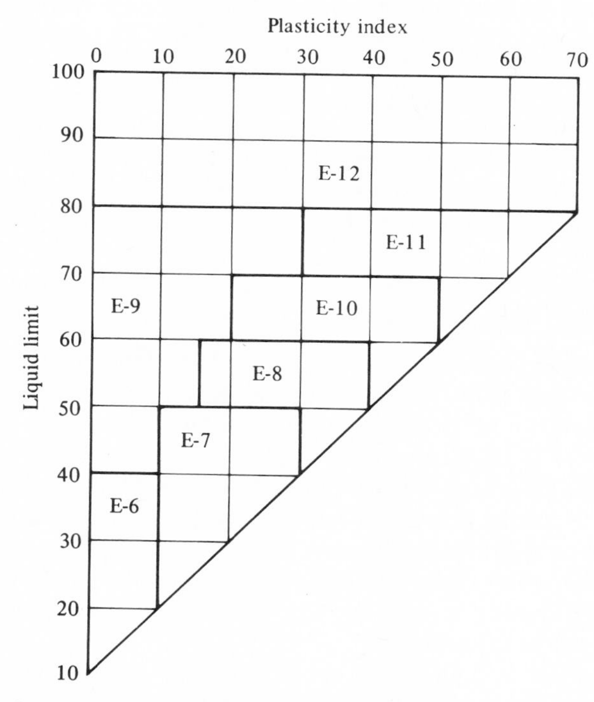

Fig. 14-3. Classification chart for fine-grained soils.

equal liquid limits, the higher the plasticity index, the lower the plastic limit at which a slight increase in moisture causes the soil to rapidly lose stability.

14.17. COARSE MATERIAL RETAINED ON NO. 10 SIEVE. Only that portion of the sample passing the No. 10 sieve is considered in the above-described classification. Obviously, the presence of material retained on the No. 10 sieve should serve to improve the overall stability of the soil. For this reason, upgrading the soil from 1 to 2 classes is permitted when the percentage of the total sample retained on the No. 10 sieve exceeds 45% for soils of the E-1 to E-4 groups and 55% for the others. This applies when the coarse fraction consists of reasonably sound material which is fairly well graded from the maximum size down to the No. 10 sieve size. Stones or rock fragments scattered through a soil should not be considered of sufficient benefit to warrant upgrading.

14.18. SUBGRADE CLASSIFICATION. For each soil group there are corresponding subgrade classes. These classes are based on the performance of the particular soil as a subgrade for rigid or flexible pavements under different conditions of drainage and frost. The subgrade class is determined from the results of soil tests and the information obtained by means of the soil survey and a study of climatological and topographical data. The subgrade classes and their relationship to the soil groups are shown in Table 14-4. The prefixes "R" and "F" indicate subgrade classes for rigid and flexible pavements, respectively. These subgrade classes determine the total pavement thickness for a given aircraft load.

14.19. DRAINAGE. Good and poor drainage in this classification system refer to the subsurface soil drainage.

Poor drainage denotes soil that cannot be drained because of its composition or because of the conditions at the site. Soils primarily composed of silts and clays for all practical purposes are impervious and as long as a water source is available the soils' natural affinity for moisture will render these materials unstable. These fine grain soils cannot be drained and are classified as having poor drainage as indicated in Table 14-4. A granular soil that would drain and remain stable except for conditions at the site such as high water table, flat terrain, or impervious strata, should also be designated as poor drainage. In some cases this condition may be corrected by the use of subdrains.

Good drainage is defined as a condition where the internal soil drainage characteristics are such that the material can and does remain well drained, resulting in a stable subgrade material under all conditions.

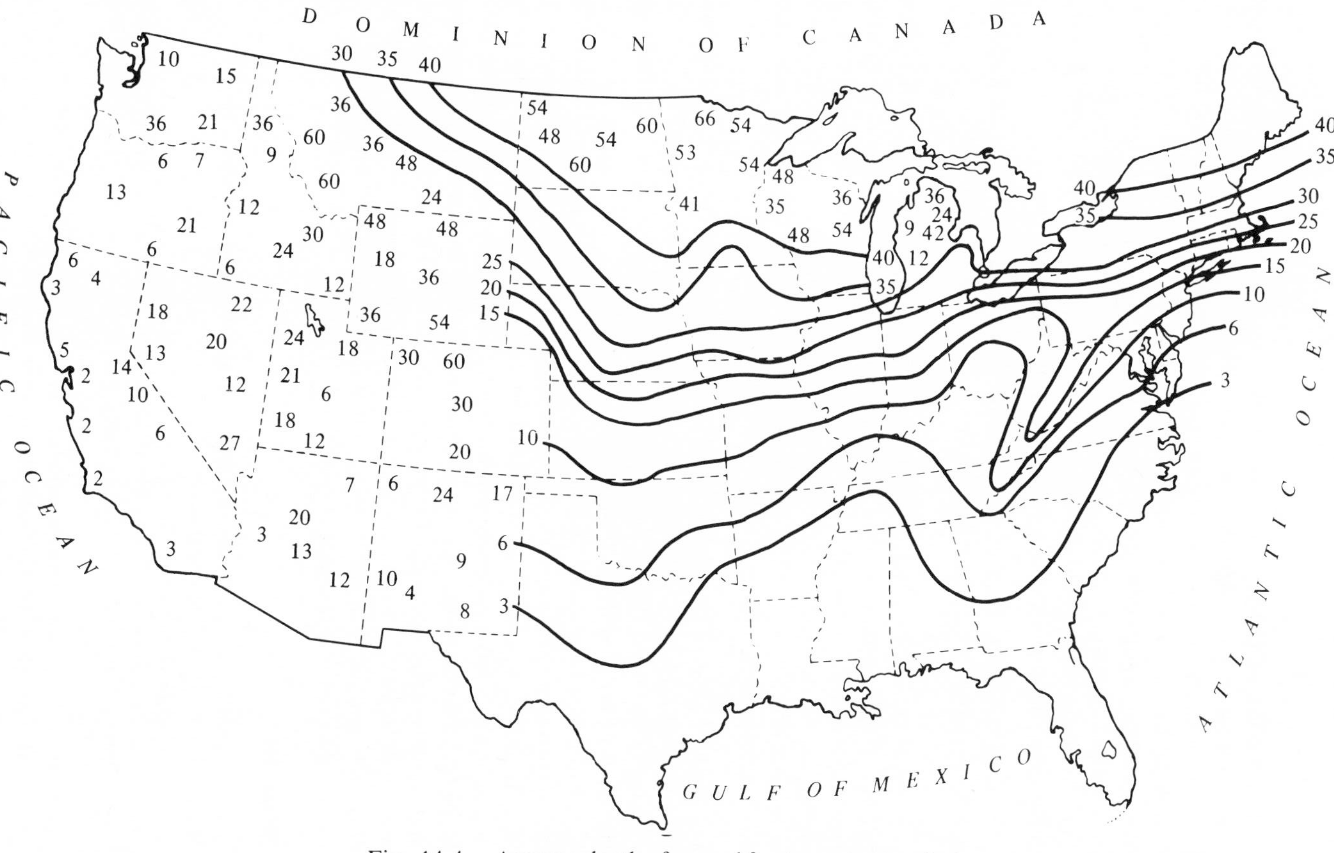

Fig. 14-4. Average depth of annual frost penetration (in.).

14.20. FROST ACTION. There is a tendency to overlook the detrimental effect of frost in pavement design. The effects of frost are widely known; however, experience shows that all too often pavements are damaged or destroyed by frost that was not properly taken into account in the design. Most inorganic soils containing 3% or more of grains finer than 0.02 mm in diameter by weight are frost susceptible for pavement design purposes. The subgrade soil should be classified either as "No Frost" or "Frost" depending on one of the two following conditions:

No frost should be used in the design when the average frost penetration anticipated is less than the thickness of the pavement section.

Frost should be used when the anticipated average frost penetration exceeds the pavement sections. The design should consider including non-frost susceptible material below the required subbase to minimize or eliminate the detrimental frost effect on the subgrade. The extent of the subgrade protection needed depends on the soil and the surface and sub-surface environment at the site.

Figure 14-4 shows the average annual frost penetration throughout the continental United States. It is included primarily as a guide. Actual depth of frost penetration should be determined for each particular site on the basis of reliable local information.

Subgrade treatment should be considered in the pavement design if one or more of the following conditions exist: poor drainage, adverse surface drainage, or frost. A stabilized or modified subgrade will to some degree make a hard-to-work soil more workable, provide a working platform for construction, act as a moisture barrier between untreated soil and the pavement section, and be frost resistant. The agent for the treatment will depend on the soil and site conditions. Lime is used for most clay, silt, and silt-clay soils, while portland cement and bituminous materials are readily adaptable to some soils.

PROBLEMS

14.1. Name three widely used engineering systems of soil classification.

14.2. What is the significance of the group index in connection with the A.A.S.H.O. system of classification?

14.3. State the broad general character of soils included in groups A-1, A-2, and A-3 of the A.A.S.H.O. system; groups A-4 and A-5; groups A-6 and A-7; group A-8.

14.4. Give the major characteristics of soils included in the GW, GC, GP, and GF groups of the Unified Classification system; the SW, SC, SP, and SF groups; the ML, CL, and OL groups; the MH, CH, and OH groups.

14.5. Classify soils No. 1, 2, and 3 in Table 7-6 according to the A.A.S.H.O. system, the Unified system, and the F.A.A. system.
14.6. Classify soils No. 4, 5, and 6 in Table 7-6 according to the A.A.S.H.O., Unified, and F.A.A. systems.
14.7. Classify soils No. 7, 8, and 9 in Table 7-6 according to the A.A.S.H.O., Unified, and F.A.A. systems.
14.8. Classify soils No. 10, 11, and 12 in Table 7-6 according to the A.A.S.H.O., Unified and F.A.A. systems.
14.9. A soil is classified as SC in the Unified system. Write a complete description of this soil.
14.10. What are the principal differences between two soils which are classified as A-4 and A-5 in the A.A.S.H.O. system?

REFERENCES

1. American Association of State Highway Officials. *Standard Specifications for Highway Materials and Methods of Sampling and Testing,* Part I, A.A.S.H.O. Designation: M 145–66, 10th ed. Washington, D.C., 1971.
2. Federal Aviation Administration, Department of Transportation. *Airport Paving.* Washington, D.C., 1967.
3. Hogentogler, C. A., and Charles Terzaghi. "Interrelationship of Load, Road and Subgrade." *Public Roads* **10,** No. 3 (May 1929).
4. U.S. Army Waterways Experiment Station. "The Unified Soil Classification System." Tech. Mem. No. 3-357, Office of the Chief of Engineers, U.S. Army. April 1960, reprinted May 1967.

15

Frost Action in Soil

15.1. EFFECTS OF FROST ACTION ON PAVEMENTS. The damage to highway and airport pavements caused by lowered temperatures and frost action in soil subgrades constitutes one of the most difficult problems with which a highway or airport engineer must cope. Pavements are frequently broken up or severely damaged as the subgrades freeze in winter and thaw out in the spring, and repairs and maintenance costs may severely strain the engineer's budget for operations. Pavements of the flexible type, such as clay-bound sand and gravel, sand-clay, soil-cement, and soil-bitumen, are the most vulnerable to damage of this kind. However, rigid type concrete slabs or bituminous surfaces on concrete bases also may be caused to heave and crack. Such cracking results in deterioration of the riding qualities of these high type rigid pavements and a marked increase in maintenance costs, as well as a reduction in service life.

In addition to the physical damage suffered by flexible pavements and the high cost of repairs and maintenance caused by frost action under adverse conditions, the economic and social loss to the users of the highway or airport thus affected may be very great. In the spring of the year, when frozen subsoils are thawing out, they may become extremely unstable. In order to protect the pavement during this period, it is frequently necessary to place an embargo on the facility to restrict the weight of vehicles which may use it. In some severely affected areas, it may be necessary to close the highway or airport to all traffic for a period of time to allow the subgrade to recover its stability. The length of time for which partial or complete traffic embargoes must be enforced depends on the severity of the frost damage and on weather conditions which affect the rate of thawing and drying out of the subgrade. In some regions, even school buses may have to be denied use of the highways during the critical thawing period in the early spring.

15.2. PHASES OF FROST DAMAGE. Frost damage to pavements and their subgrades occurs in two distinct phases. First the pavements *heave*

Fig. 15-1. Soft, unstable, supersaturated soil beneath a stabilized gravel surface in a frost-boil area.

vertically during the winter season when air temperatures are below freezing for sustained periods of time. As the frost table penetrates the subgrade, layers of ice may form in the soil in areas where conditions are favorable to their growth. The amount of the pavement heave is approximately equal to the thickness of these ice layers, and in some regions may be many inches or even several feet. This heaving of the pavement, of itself, might not be seriously detrimental if it were uniform in amount; but, since soils and other conditions which influence the growth of ice layers are never uniform, heaved pavements present a rough riding surface and rigid pavements may be cracked and broken by the unequal heaving.

The second phase of frost damage occurs toward the end of the winter season or in early spring when thawing sets in. The frozen subgrade thaws both from the top and from the bottom. Thus, for a period of time during the spring thaw, there is a layer of frozen soil in the subgrade which gradually diminishes in thickness as thawing progresses. The thawed soil between the pavement and this frozen layer contains an excess amount of moisture resulting from the melting of the ice which it contained when frozen. Since the frozen layer of soil is impervious to the water, the thawed soil above it cannot drain; and this thawed soil loses practically all of its bearing power because of its supersaturated condition. In Fig. 15-1 is illustrated soft unstable soil under a stabilized gravel surface. If the pavement is of a flexible or nonrigid type and traffic is allowed to use the road or runway while the subgrade is in this condition, the wheels of vehicles easily break through the pavement and churn up the soft soupy layers of soil immediately below. This condition is known as a *frost boil* and is a direct consequence of frost heave when heavy traffic is

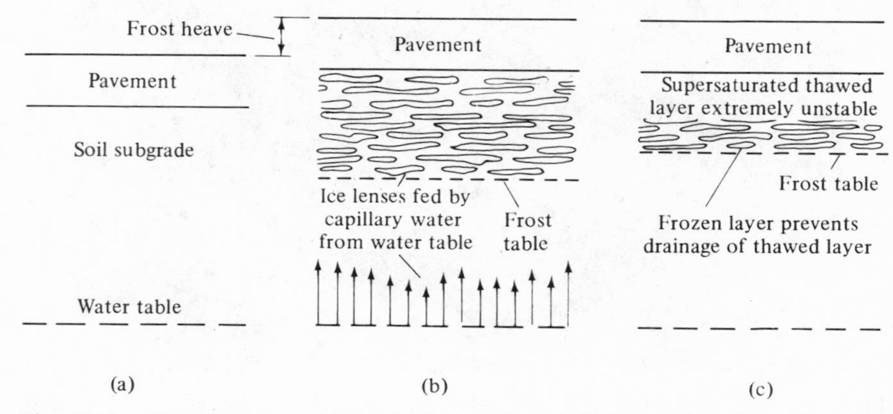

Fig. 15-2. Frost heave and frost boil in soil. (a) Normal situation; (b) frost heave (winter); (c) frost boil (early spring).

permitted to use heaved flexible pavements during the thawing period. The sequence of events in connection with the development of frost heave and subsequent frost boil is illustrated in Fig. 15-2.

15.3. DURATION OF FROST BOILS. Since the soft unstable condition of the subgrade in frost-boil areas is caused by the presence of a frozen layer of soil below, which cuts off drainage of the upper layer of subgrade, the length of time for which a frost-boil condition lasts depends on the time required for the frozen layer to thaw completely. If the weather warms up rapidly, the layer may thaw out in a relatively few days. But, if the thaw is gradual and freezing weather and thawing weather are interspersed over a long period of time, frost boils may persist for several weeks; and it may be necessary to continue a traffic embargo for a long period in order to protect the pavement. If a warm rain occurs during the frost-boil season, the rain water may soak down into the soil and hasten the thawing of the frozen cut-off layer in the subgrade. When that layer is completely thawed, the supersaturated layer above will drain rather quickly by gravity. Such conditions may result in the anomalous situation of a rainstorm causing a road to dry out more quickly, sometimes overnight.

There is no practical artificial method of curing a frost boil after it has developed. Efforts toward thawing the frozen cut-off layer by artificial means, such as injecting steam into the subgrade, have been effective over limited areas, but the procedure is far too expensive to be used extensively. Highway engineers have resorted to planking the roads over frost-boil areas in some instances, but this too is an expensive procedure and few maintenance budgets can stand the strain of annually placing and removing planks over such areas. Also it is obvious that planked roads are a hindrance to the free movement of traffic. The best defense against frost boils is to construct the road in such a manner that frost heave will be prevented or greatly inhibited. This requires a thorough knowledge of the mechanics of frost heave and methods of identifying the parts of a road or runway where heave is likely to occur.

15.4. CAUSE OF FROST HEAVE. In the early days of highway engineering it was assumed that road surfaces heaved simply because of the expansion upon freezing of the moisture contained in the pores of the soil. However, since water expands only about 9% as it freezes, a simple computation indicates that heaves of only a fraction of an inch can be attributed to this cause in ordinary mineral soils, whereas heaves of many inches are very commonly observed. More recently, field observations and laboratory studies have indicated definitely that surface heave is due to the growth of ice layers or lenses in the subgrade; and that efforts to eliminate frost boils and the spring break-up must be directed toward creating conditions which will prevent the growth of the ice.

15.5. FACTORS AFFECTING GROWTH OF ICE LAYERS. There are many factors which contribute to the growth of ice layers in soil subgrades, and the extent to which heave develops depends on the combined influence of all these factors. The principal factors referred to are the climate or weather, the nature of the soil with reference to its capillary characteristics, and the existence and elevation of a ground water table in the soil. Other relatively minor factors are the insulating properties of the pavement and the subgrade, the heat reflective character of the pavement surface, and the presence of dissolved salts in the soil water.

Because there are so many factors involved, the development of frost heave and frost boils is highly unpredictable. A certain road or the roads in a certain county may suffer very little damage one year, whereas during the next year damage may be widespread because of an unfavorable combination of weather and ground-water conditions. In this connection, it is of interest to point out that the practice of snow removal from roads and runways, while very desirable from the standpoint of traffic movement, contributes to the growth of ice layers in the soil beneath the pavement since snow is a good insulator. The speed and depth of frost penetration are much greater in areas from which snow is removed than in adjacent areas where it is left in place.

15.6. BASIC CONDITIONS FOR GROWTH OF ICE LAYERS. The ice layers, or lenses, which cause highway and airport surfaces to heave, grow in the soil in essentially the following manner. First, as the frost line penetrates downward, small volumes of water in the soil pores freeze. This action, in effect, dries the soil in the region of the ice formation, because the frozen water is no longer available to satisfy the attraction of the soil for capillary water. Since drying a soil decreases its capillary potential, water from below—where the potential is greater—tends to flow upward toward this newly created region of low potential. This water also freezes when it reaches the frost line. As it freezes, the newly formed ice crystals become attached to the lower side of the originally minute ice particles, causing them to increase in thickness in much the same way as ice on a pond or lake thickens in the downward direction. Thus the heave-producing ice layers grow, being fed by capillary water from below, until a state of equilibrium is reached and the growth stops.

If the frost line continues to penetrate into the soil, another ice layer may grow below the first one in a similar manner; and so the freezing may proceed as long as the frost line continues to penetrate further into the soil. The sum of the thicknesses of all the ice layers which form is the amount by which the surface heaves. From this brief description, it is evident that three basic conditions must exist in order for ice layers to form and grow in thickness. These are: (1) freezing temperatures in the soil; (2) a reservoir of ground water sufficiently close to the frost line to feed the growing ice layers; (3) soil material having favorable characteristics for rapid movement of capillary water upward from the water table.

15.7. LOWERING WATER TABLE. Obviously an engineer cannot do anything to prevent the first of the basic conditions accompanying frost heave. If a road or runway is constructed in a climate where freezing temperatures occur in winter, in all probability the soil beneath the pavement will freeze unless the period of lowered temperature is very short. There are, however, several principles of design which can be applied to combat the second and third basic conditions. These principles are illustrated in Fig. 15-3. If a soil survey or other sources of information indicate that ground water and soil conditions may be conducive to the development of frost heave, every effort should be made to lower the ground water table in relation to the grade of the road or runway. As indicated in Fig. 15-3(a), this may be accomplished by the installation of tile drains or open side ditches, provided suitable outlets are available and provided also that the subgrade soil is drainable; or it may be accomplished by raising the grade line in relation to the water table. Whatever may be the means employed for producing the condition, the distance from the bottom of a pavement

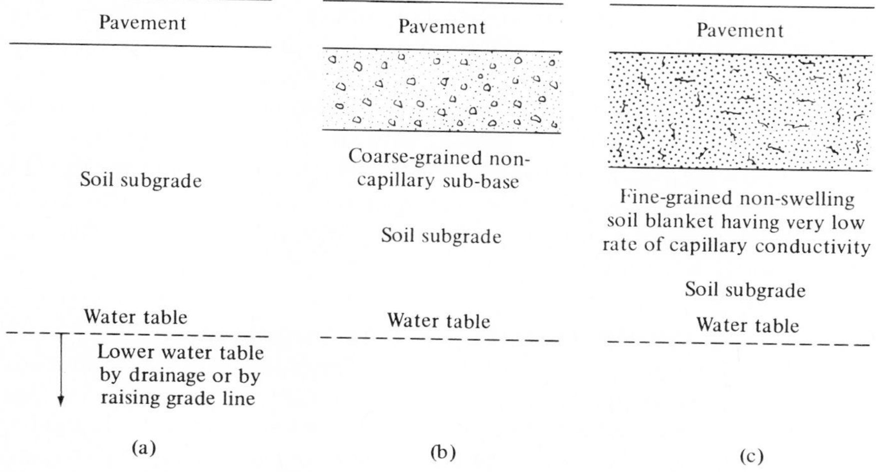

Fig. 15-3. Three methods of combating frost heave.

to the highest probable elevation of the water table should not be less than 6 ft; and distances greater than this are very desirable if they can be obtained at reasonable cost. Since the growth of heave-producing ice lenses is made possible by the capillary flow of water upward from the water table, their rate of growth will be reduced as the verticle distance through which the water must flow is increased.

15.8. USE OF COARSE-GRAINED SUBBASE. A second principle of design which may be employed to combat frost heave is represented in Fig. 15-3(b). The treatment is to under-cut the subgrade in areas where the natural soil is conducive to high capillary flow and to replace the soil removed with coarse-grained material having a low value of capillary conductivity. This layer is often referred to as a subbase and its purpose is to interrupt the upward flow of capillary water and thereby limit the growth of ice layers. Subbases are also used for the purpose of distributing wheel-load pressures over wide areas to reduce the unit pressure on the subgrade, as will be discussed in Chapter 16. To be most effective in reducing frost heave, a subbase should be as coarse-grained as possible and poorly graded; that is, it should be free from fines which might fill the voids and provide capillary channels around the large particles. Crushed stone is an ideal material for cutting off the capillary flow of water.

A subbase employed mainly for the purpose of inhibiting growth of ice layers should be as thick as is economically possible with materials at hand, but it need not extend below the maximum probable depth of frost penetration. In the Corn Belt of the United States, for example, frost rarely penetrates the soil to a depth greater than 4 ft and this would represent a maximum effective thickness of subbase as a barrier against ice growth. However, in many instances it is not feasible to obtain this maximum. The rule in such cases is to make the subbase as thick as possible. A thickness of 6 in. will help; but 12 in. would be better, 24 in. would be very much better, and so on. Not infrequently, when roads are being reconstructed, the only available source of subbase material may be the old road metal, that is, the gravel or crushed stone with which it was previously surfaced. In such cases it is worthwhile to salvage this material and to stockpile it for use in frost-prone areas of the road.

15.9. USE OF CUT-OFF BLANKET. A third principle of design against frost heave is to provide a capillary cut-off blanket of fairly well graded, nonswelling, clayey material under the pavement, as indicated in Fig. 15-3(c). Because of its high clay content, the capillary forces tending to lift water from the water table may be large. However, by reason of this same fine-grained characteristic, the capillary conductivity or the rate at which capillary water moves through the soil may be very low, with the result that only a relatively small amount of water may be drawn up to feed the growing ice layers during the period when freezing temperatures prevail. In other words, although a subbase of this kind may exert large forces tending to draw excess water into the subgrade, the rate at which it moves is so slow that the winter may be over before the ice layers can grow suffiently thick to produce detrimental heaving.

15.10. OBJECTIONS TO SILTY SOILS. The poorest soils from the standpoint of frost damage—soils which are frost prone—are those which are sufficiently fine-grained to exert moderately high capillary forces tending to lift water above a water table and, at the same time, are sufficiently coarse-grained to have a high rate of capillary conductivity. Soils of this kind will transmit relatively large quantities of water in a relatively short time to feed growing ice lenses. Uniformly graded soils which contain more than 10% of particles smaller than 0.02 mm and fairly well-graded soils containing more than 3% of this size should be looked upon with suspicion whenever encountered in a soil survey.

15.11. PERMAFROST. Another matter of engineering interest in frost action is the presence of *permafrost*, or permanently frozen ground, in arctic and subarctic regions. This permanently frozen ground underlies vast areas of the earth's surface. It is estimated that nearly one-fifth of the total land area of the world is thus affected. Permafrost is most widespread in the northern hemisphere around the shores of the Arctic Ocean. It extends as far south as the 50th parallel, although the southern boundary of the area is very irregular, as indicated in Fig. 15-4. There are also extensive areas of permafrost in the Antarctic Region. It is believed that permafrost originated during the refrigeration of the earth's surface at the beginning of the Pleistocene or Ice Age, perhaps a million years ago.

The thickness of permafrost varies widely from a few feet to several hundred feet. In general, permafrost is thicker along the coast of the Arctic Ocean and diminishes in thickness toward the south. However, the thickness varies with local climatic, topographic, and soil conditions, as well as with latitude. At one locality in northern Russia, the permafrost was penetrated to a depth of 750 ft and its thermal gradient indicated that it probably extended downward nearly 400 ft farther. Above the permafrost is a layer of ground that thaws in the summer and freezes again in the winter. This layer represents the seasonally frozen ground and is called the *active layer*. The permanently unfrozen ground that is below, adjacent to, or included within, the permafrost is called *talik*.

The permafrost below the active layer may be continuous or discon-

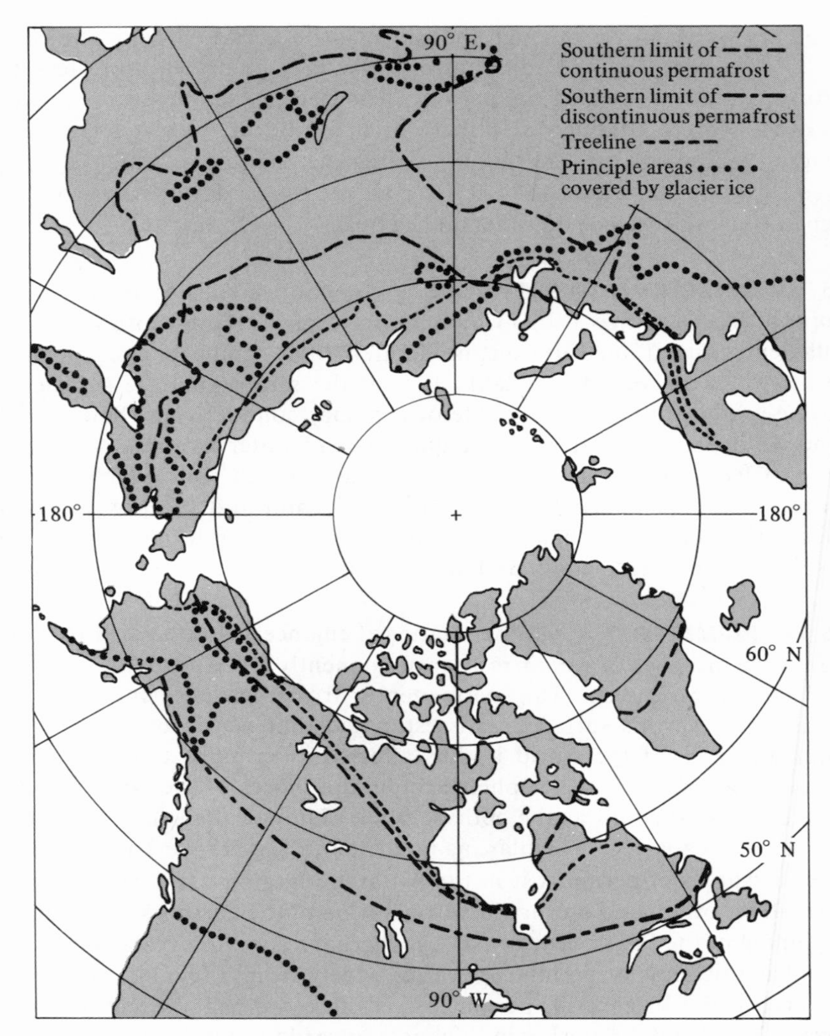

Fig. 15-4. Northern hemisphere permafrost distribution [after R. J. E. Brown (*1*)].

tinuous. In the continuous zone, as the name implies, the frozen soil extends uninterruptedly downward to the talik underlying the permafrost layer. In the discontinuous zone, unfrozen layers of talik may be tortuously interspersed through the permafrost. As indicated in Fig. 15-4, the discontinuous zone extends farther south than the continuous zone in the northern hemisphere.

15.12. FACTORS AFFECTING SEASONAL THAWING. The rate and depth of seasonal thawing are markedly influenced by the type of exposure, by the hydrology, and by vegetation, snow cover, and soil type. An insulating cover of snow or vegetation will cause a shallower thaw, as contrasted with a deeper thaw in places that are bare or receive more sunshine. These conditions may vary considerably within short distances, with the result that the bottom of the thawed ground is a rather uneven surface. By analogy with ground-water terminology, this surface is called the *frost table*.

With the downward progress of the seasonal thaw, the frost table moves progressively lower until it reaches an underlying talik or merges with the upper surface of the permafrost. This upper surface of the permafrost is called the permafrost table. It does not necessarily coincide with the frost table. Like the frost table, however, its irregularities are determined by differences in the insulating cover of vegetation, in the thermal conductivity of the ground, in the geographic position and character of exposure, and in the hydrology of the ground water.

15.13. PERMAFROST LAND FORMS. A permafrost region may be recognized, both on the surface and from the air, by the occurrence of a number of distinctive land forms, such as pingos, palsas, polygons, ice wedges, drunken forests, thermokarst landscape, and beaded streams.

A pingo is a large mound perhaps 100–150 ft in diameter and 25–30 ft high, raised by frost action above the permafrost.

A palsa is similar to a pingo, but consists mainly of peat soil and occurs most frequently near the southern fringe of the discontinuous permafrost.

Polygonal ground refers to polygon-shaped soil patterns produced on level ground by thermal contraction cracks in the upper permafrost layer. The cracks form a hexagonal or rectangular pattern and are filled from the surface by melt water which freezes to form vertical *ice wedges*, or wedge-shaped veins of ground ice. Ice wedges may also develop in areas other than polygonal ground.

A drunken forest is a forested region in which trees lean crazily out of plumb owing to thawing of the permafrost and gross distortion of the land surface resulting therefrom.

Thermokarst landscape is marked by sink holes caused by melting of large volumes of ground ice.

A beaded stream has a queer angular meander pattern which results from following along the tops of polygonal ice wedges. It is characterized by frequent enlargements caused by melting of large blocks of ground ice, which give the appearance of beads on a string when viewed from the air.

15.14. EFFECT OF PERMAFROST ON ENGINEERING STRUCTURES. The existence of permafrost in a region greatly influences the performance of engineering installations and presents problems in design and construction which are not encountered in the more temperate zones. In planning an engineering project, from the simplest type of building to an extensive highway, airfield, or public-utility installation, it is necessary to determine in advance the nature of seasonal changes in the active layer and the permafrost table, the depth of permafrost, the existence of talik above or within the permafrost, and the hydrologic features of ground water within the active layer.

Fluctuations of the permafrost are influenced by the thermal regime to which it is subjected as a result of all the influential factors of its environment. If this thermal regime is changed as a result of engineering operations, then the permafrost will change and the influence of this change on the stability and function of the engineering installation may be very great. It is necessary, therefore, to plan the engineering project in such a way that either the thermal regime will not be disturbed or the influence of the project on the thermal regime and the permafrost can be predicted and

Fig. 15-5. Air space beneath building preserved the thermal regime, permafrost remained frozen, and settlement of the structure was negligible. This illustrates the passive method of design.

Fig. 15-6. Thermal regime was disturbed, permafrost melted from heat of building and the structure settled grossly. Note relatively little settlement of the unheated porch.

the effect of these changes on the engineering work can be provided for in design. Planning in the first way is called the passive method of design; and planning in the second way is the active method.

When any construction work is to be done in a permafrost area, it is very important to consider the time of year during which the work will be done. For instance, if the fill shown in Fig. 15-7 were constructed in the

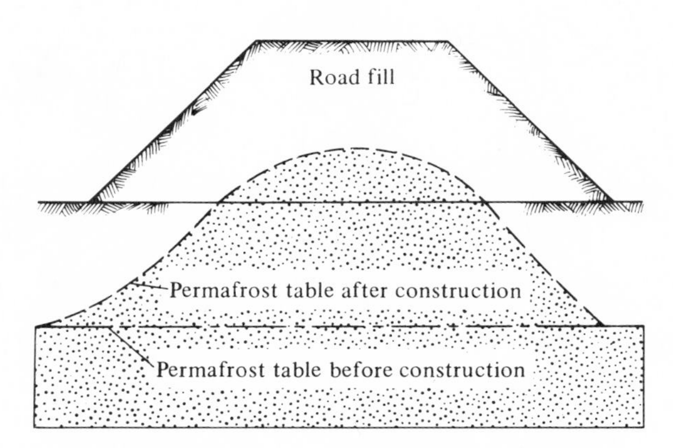

Fig. 15-7. Effect of road fill on permafrost table (after Bobkov).

late summer or early fall, it would contain a considerable quantity of stored heat. This heat would probably lower the permafrost table at first, and this table would later tend to become stabilized at a higher elevation, as shown in the illustration. The rise of permafrost under the fill not only will cause the embankment to heave and distort, but may also act as a barrier to the flow of ground water in the active layer. As a result, artesian pressures may be introduced in the ground water, causing swamp areas and springs to develop in places where they did not exist before. Water flowing from springs of this kind frequently spreads over large areas, possibly engulfing parts of highways and other facilities, and then freezes. This phenomenon is known as icing. Its prevention in permafrost areas is an important part of engineering planning. Block diagrams, illustrating road-fill construction across a swamp area to inhibit icing conditions, are shown in Fig. 15-8.

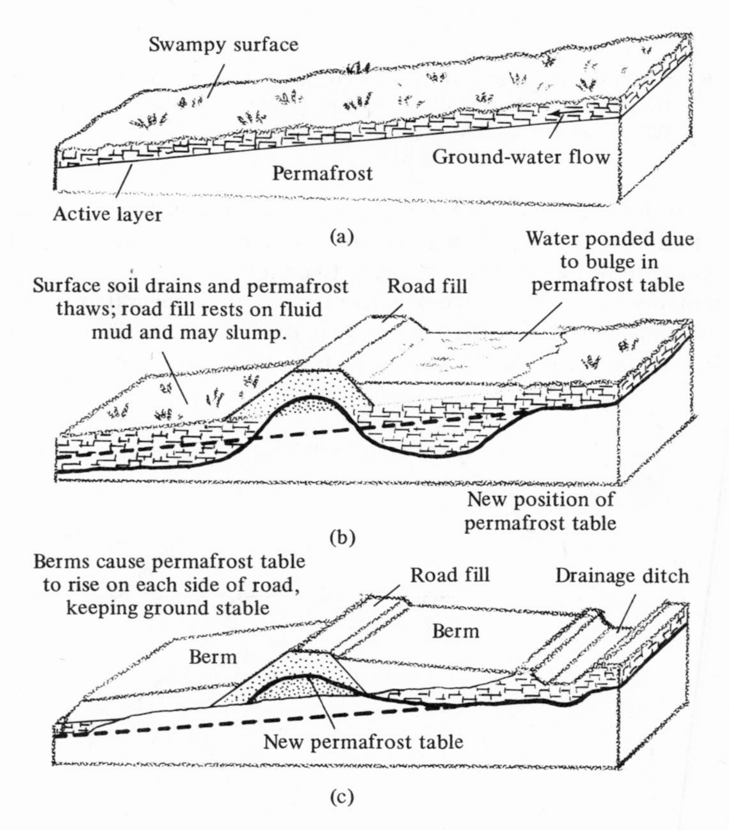

Fig. 15-8. Conditions inhibiting icing [after Muller (*4*)].

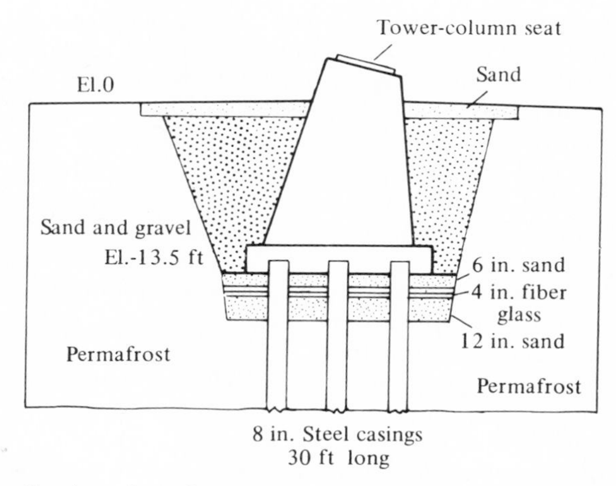

Fig. 15-9. Application of passive method (from *Engineering News-Record*).

15.15. APPLICATION OF PASSIVE METHOD. An interesting application of the principle of maintaining the thermal regime of permafrost—the passive method—in order to prevent settlement of a tower structure is shown in Fig. 15-9. The structure shown is one of four rhombic pedestals constructed by the United States Navy to support a steel tower in a permafrost region. After excavation was carried through the active layer and well into the permafrost, piles consisting of steel casings 8 in. in diameter and 30 ft long were set by drilling into the frozen material and permitting them to adfreeze to the permafrost. They were then filled with clean dry sand to within 5 ft of the top and with concrete for the rest of the length. The pedestal is supported on a slab footing 14 ft square and 18 in. thick. Before the footing was poured, however, there were placed on top of the permafrost insulating layers consisting of 12 in. of clean dry sand, 4 in. of fiber glass, and 6 in. of sand. The footing and the pedestal were then constructed in a normal manner. Backfill around the pedestal consisted of sand and gravel. By this method of design and construction, the permafrost below the footing is kept from thawing by the layers of insulation material, and settlement of the tower structure has been negligible.

PROBLEMS

15.1. Name and describe two phases of frost damage to highway and airport pavements. In what order do they occur chronologically?

15.2. At what stage of frost action in the subgrade may it be necessary to impose a traffic embargo on a road or runway? How long should the embargo last?

15.3. Explain how a warm rain may cause a pavement subgrade to dry out.

15.4. The dry density of a soil subgrade is 110 pcf and it contains 18% of moisture. If the soil freezes to a depth of 2.5 ft below a pavement and all the volume change is in a vertical direction, how much will the pavement heave? Is this sufficient to cause serious damage?

15.5. Name, in the order of importance, at least six factors which influence the growth of heave-producing ice lenses in a pavement subgrade.

15.6. Describe, in detail, the growth of ice lenses in a pavement subgrade.

15.7. Name three basic conditions which must exist before ice lenses will grow in a soil subgrade.

15.8. Describe three principles which may be employed in pavement design to inhibit the growth of heave-producing ice lenses. Explain how each method accomplishes its purpose.

15.9. Examine the soils listed in Table 7-6 and list them in the order of frost proneness.

15.10. What is permafrost and where does it occur?

15.11. Describe the active and passive methods of design of engineering structures in the permafrost region.

15.12. What is meant by "icing" in permafrost areas? Describe its cause.

15.13. Name and describe four distinctive land forms which are typical of a permafrost region and by which such a region can be identified.

REFERENCES

1. Brown, R. J. E. "Comparison of Permafrost Conditions in Canada and the USSR." Technical Paper No. 255, Division of Building Research, National Research Council of Canada, Ottawa, Aug. 1967.
2. Johnson, A. W. "Frost Action in Roads and Airfields." *Highway Research Board*, Special Report No. 1, 1952.
3. Jumikis, Alfreds R. *Thermal Soil Mechanics.* Rutgers University Press, New Brunswick, N.J., 1966.
4. Muller, S. W. *Permafrost or Permanently Frozen Ground and Related Engineering Structures.* J. W. Edwards, Inc., Ann Arbor, Mich., 1947.
5. Roberts, P. W., and F. A. F. Cooke. "Arctic Tower Foundations Frozen into Permafrost." *Engineering News-Record* **144,** No. 6, Feb. 9, 1950.
6. Taber, Stephen. "Freezing and Thawing of Soils as Factors in the Destruction of Road Pavements." *Public Roads* **11,** No. 6, 1930.
7. Winn, H. F. "Frost Action in Highway Subgrades and Bases." *Proc. Purdue Conference on Soil Mechanics and Its Applications.* Purdue University, Lafayette, Ind., 1940.

16

Granular Soil Stabilization

16.1. DEFINITION. The term soil stabilization in its broadest sense refers to any process, natural or artificial, by which soil is made stronger and more resistant to deformation and displacement under applied loads. Some observers go so far as to consider the simple drying-out of soil after a rainstorm as a form of soil stabilization, and it is true that, in general, a dried soil is stronger and more stable than a wet soil. As the term is used in this chapter, however, stabilization has a much more restricted meaning. Here, it refers to the processes of design, mixing, placing, and compaction by which granular soil materials are rendered strong and resistant to displacement to such an extent that they will serve as usable road surfaces or pavements in all kinds of weather.

16.2. SIMPLE EXAMPLES OF SOIL STABILIZATION. Although modern soil stabilization, as applied in highway and airport construction, is an art of relatively recent origin, the fundamental principles upon which it is based have been known and practiced for a long time. As an illustration, consider a path in a garden or a road on the farm. If the path or road becomes muddy in wet weather, it is customary for the housewife or farmer to apply some kind o. .eadily available granular material, such as ashes, cinders, sand, gravel, or crushed rock. Under the influence of traffic, these coarse-grained materials become mixed with the natural soil and these mixtures are compacted as the traffic continues. Applications of the granular materials may be repeated from time to time until the path or road becomes usable under all weather conditions. On the other hand, if the natural soil is very loose and sandy, small additions of fine-grained clayey soil over a period of time will produce a satisfactory path or road.

These two simple and common-place procedures involve all the underlying principles of granular soil stabilization. By these means, surfacings of appreciable thickness, consisting of natural soil and granular material in the right proportions and sufficiently compacted to produce the desired firmness or stability, have been imposed on the underlying soil. These may be called "cut-and-try" methods of soil stabilization.

16.3. ADVANTAGES OF TESTS AND SPECIFICATIONS. While the cut-and-try methods of soil stabilization always produce considerable improvement in the character of the surface, they may not approach the maximum improvement possible with the materials at hand; nor do they provide a basis for judgment in regard to the relative merits of various materials which may be available. The value of soil tests and specifications developed for governing the materials and processes in soil stabilization lies in the fact that by their use the stability obtainable by cut-and-try methods over long periods of time can be secured with confidence during construction or shortly thereafter. These tests and specifications also provide a definite basis for comparison of available materials and facilitate the selection of those which are most suitable and economical.

16.4. PARTS OF PAVEMENT STRUCTURE. A stabilized-soil pavement structure may consist of two, three, or four distinct layers or elements. As shown in Fig. 16-1, these elements are: (1) the wearing surface at the top, (2) the base course, (3) the subbase course, and (4) the subgrade at

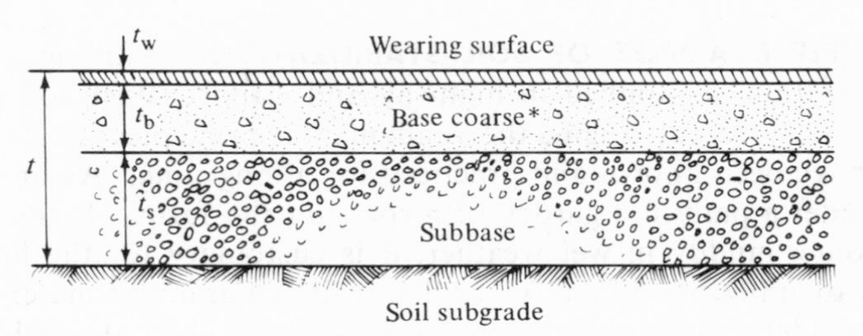

Fig. 16-1. Elements of a flexible pavement structure (*called a surface course when wearing surface is omitted).

the bottom. The wearing surface may consist of a bituminous layer ranging from a light surface treatment about $\frac{3}{4}$ in. thick to a dense-graded, hot-mix asphaltic concrete wearing course 2 to 3 in. or more in thickness. The function of the wearing surface is to provide a smooth, dustless, impervious cover over the pavement and to protect the base course from excessive wetting and drying and traffic abrasion. It is frequently omitted on a road carrying only light traffic when construction funds are limited. When a base course of a flexible pavement is constructed without a wearing surface, it is frequently referred to as a surface course.

16.4.1. Base Course. The base course is the main structural element of the pavement. It must have adequate strength to withstand the stresses produced by the wheels of vehicles. The weight of a wheel at rest or in uniform motion is transmitted through the pavement into the foundation, and causes stresses in the various elements which tend to produce vertical deformation or settlement of the loaded part of the pavement. Also, when the speed of a vehicle is being accelerated, there is produced at the pavement surface a tangential shearing force which tends to push the pavement backward; and, when a vehicle is decelerated by braking, a similar effect is produced in the opposite direction. Furthermore, wheel loads applied through pneumatic tires cause surface shearing forces directed inward toward the center of the tire contact at right angles to the line of travel of the wheel, and rolling wheels produce vibratory or dynamic stresses which tend to destroy the bond between a thin bituminous wearing surface and the base course.

16.4.2. Subbase. A subbase may be provided as a part of the pavement foundation for one or both of two principal reasons. It may be necessary in order to inhibit the growth of heave-producing ice layers in the frost zone, as discussed in Chapter 15, or it may be employed to distribute the wheel-load pressures more widely over the subgrade soil, thus reducing the unit pressures on the soil and the amount of its vertical deformation under the influence of the wheel load. Since many subgrade soils which are susceptible to frost heave are also low in bearing capacity or resistance to deformation under load, a subbase may readily fulfill both of these functions. The heavier the wheel loads and the greater the deformation characteristics of the subgrade soil, the thicker the subbase course must be made. The determination of thickness of surface, base, and subbase courses for various traffic and subgrade conditions constitutes the structural design of a flexible pavement.

16.4.3. Subgrade. The subgrade is the ultimate foundation of the pavement structure. It may be soil in its natural state or filled materials.

Essentially the pavement foundation must satisfy two requirements. It must be able to carry the wheel-load pressures to which it is subjected without excessive deformation; and it must be of such character that the danger of detrimental volume changes due to environmental variations and moisture fluctuations will be eliminated or held to a minimum.

16.5. MATERIALS FOR SURFACE COURSES AND BASE COURSES. A clear understanding of the nature and function of each element of the stabilized-soil pavement structure is necessary for two reasons. First, the material requirements of the several elements are different. For a base course or a surface course, only materials which conform to more or less rigid specification requirements are suitable; these requirements for a base course that is protected by a wearing surface are different from those for a surface course that is subjected directly to traffic and weather. It is also necessary to distinguish between the various elements from the viewpoint of thickness required to withstand wheel loads of different weights.

Many materials with characteristics suitable for surface courses are not suitable for base courses which are protected with bituminous wearing surfaces. Both surface and base courses must have sufficient inherent stability to support the superimposed loads without detrimental deformation. A surface course also must withstand the abrasive action of traffic, should shed a large proportion of the rain which falls upon the surface without becoming muddy and slippery, and should have sufficient capillarity to permit replacement of moisture lost by evaporation. A base course, in contrast, is protected by a cover which takes the abrasion, sheds rain, and prevents evaporation.

It has been observed repeatedly that granular mixtures which served excellently as surfaces have failed when later covered by an impervious wearing surface. On the other hand, mixtures that were too porous to give the best service as surface courses have made excellent base courses when covered by a bituminous wearing surface. The reason for this behavior is as follows. The presence of clay is required in a surface course for the retention against evaporation of the amount of water necessary for stability. If, however, this same amount of clay is present in a base course where evaporation from the surface is prevented, it may swell and soften and total failure of the pavement may result. As a rule, less clay should be used in bases than is needed in surface courses. The amount of clay in the mixture will be controlled by the character and quantity of the binder soil used.

16.5.1. Coarse and Fine Aggregates. Materials used in granular stabilized roads are commonly classified as coarse aggregate and fine aggregate. Coarse aggregate refers to all material retained on a No. 10 (2 mm) sieve. It should consist of hard durable particles or fragments of stone, gravel, or slag, or a combination of these. Materials that break up when alternately frozen and thawed, or when alternately wetted and dried, are not suitable for stabilized road construction. The coarse aggregate should have a percentage of wear by the Los Angeles abrasion test of not more than 50. This latter requirement may sometimes be waived in the case of some locally available materials which have a high percentage of wear, but which are known by experience to produce satisfactory stabilized roads.

Fine aggregate refers to all material passing the No. 10 sieve, including the fine mineral particles passing the No. 200 sieve. It should consist of natural or crushed sand combined with silt-clay materials or stone dust. The portion of the fine aggregate passing the No. 40 sieve is called the binder soil.

The composite mixture of coarse and fine aggregates should be free from vegetable matter and lumps or balls of clay.

16.5.2. Important Characteristics of Soil-aggregate Mixture. In a stable soil mixture the granular soil fraction is assumed to furnish strength, internal friction, and hardness, and otherwise to function much the same as aggregate in a portland-cement-concrete mixture. The silt functions as a filler to help seat the granular particles, and the clay acts as a cementing medium to bind all the fractions into a strong, durable mass. The principal characteristics of the total soil-aggregate mixture which are important in soil stabilization practice are: (1) the grading of the particles, and (2) the binding properties of the soil fraction passing the No. 40 sieve, as indicated by the Atterberg limits of this fraction.

For use in granular soil stabilization processes, the grading of the particles is determined by sieve analyses, as discussed in Chapter 7; and the binding properties of the binder-soil fraction are determined by the tests for liquid limit, plastic limit, and plasticity index, as discussed in Chapter 13.

16.5.3. Specifications for Materials. The working principles employed in granular soil stabilization with reference to the grading of the particles and the binding properties of the binder-soil fraction are at the present time wholly empirical. Knowledge of the application of physical chemistry to this subject has not progressed to the point where a stabilized soil mixture can be designed on a rational basis, although the binding action of the binder-soil fraction is definitely a physical-chemical phenomenon. For this reason local practice and experience is the best

TABLE 16-1. State Highway Specifications for Plasticity Index (PI) and Liquid Limit (LL) for Highway Construction[a]

State	Base Courses	PI	LL	Surface Courses	PI	LL	Year
Alabama	Granular soil, soil aggregate,	6	25	Soil or aggregate	6	26	1964
	stabilized clay,	6	26				
	Calcium chloride						
Alaska							1965
	Subbase	6	—				
Arizona	Select material or aggregate	5	—				1965
Arkansas	Gravel or crushed stone	6	25				1966
California	*R* value and sand equivalent						1969
Colorado	Aggregate base course						1967
	Class 1, 2, and 3	6	35				
	Class 4, 5, 6, and 7	6	25				
Connecticut	Rolled gravel	N.P.[b]	—	Rolled gravel	N.P.[b]		1963
Delaware							1965
District of Columbia	Soils base courses	6	25				1963
	Subbase	10	40				
Florida	Ocala limerock base	6	35				1966
	Ocala limerock stabilized base	10	35				
	Miami limerock base	6	35				
	Miami limerock stabilized base	6	35				
	Sand-clay base	6	25				
	Stabilized base (Florida)	10	40				
	Stabilized base (LBR)	10	40				
Georgia	Soil base	9(3)[c]	25				1966
	Soil aggregate	6(2)[c]	25				
Hawaii	Waterbound macadam base	6	—				1969
	Subbase	10–15	—				
Idaho	Sand equivalent						1967
Illinois	Aggregate-gravel	2–6	—	Aggregate-gravel	2–9	—	1968
	Crushed stone or gravel	0–4	—				
Indiana	Compacted aggregate	5	25	Compacted aggregate	5	25	1969
	Slag	5	35	Slag	5	35	

Iowa	Rolled stone-soil aggregate	6	25	Soil aggregate	6	25	1964
	Graded stone base	4	25	Stabilized surface course	5–12	35	
Kansas	Aggregate base or subbase						1966
	AB-1 (2 in. max)	1–6	25				
	AB-2 (1 in. max)	1–6	25				
	AB-3 (limestone)	2–8	30				
	AB-4 (chert)	3–8	25				
Kentucky							1965
Louisiana	Sand-gravel-clay	6	25	Sand-gravel-clay	6	25	1966
	Cement stabilized	12	35	Cement stabilized	12	35	
	Clam and reef shell	6	25	Clam and reef shell	6	25	
	Reef shell and sand	6	25	Reef shell and sand	6	25	
	Clam shell and sand	6	25	Clam shell and sand	6	25	
	Clam shell, reef shell, and sand	6	25	Clam shell, reef shell, and sand	6	25	
Maine							1965
Maryland	Bank run gravel	9	30	Bank run gravel	9	30	1968
	Gravel (SBII)	3	30	Gravel (SBII)	3	30	
Massachusetts							1965
Michigan	Mineral aggregates	5	—	Bituminous stabilized	0–5	25	1967
				Chemical stabilized	4–7	25	
				Surfacing	6–12	35	
Minnesota	Binder soil	—	45	Binder soil	—	45	1969
Mississippi	Base aggregates						1967
	Group A	6	25				
	Group B	8	25				
	Group C	10	30				
	Group D	15	35				
	Group E	6–15	35				
Missouri	Aggregate–Type 1	6	—				1968
	Crushed stone or chat–Type 2	0–6	—				
	Sand and gravel–Type 2	2–6	—				
	Sand and gravel–Type 3	2–8	—				
Montana	Crushed Type A	6	25	Selected	6	30	1966
	Crushed Type B	6	25	Sand	0	25	
				Crushed Type A	6	25	
				Crushed Type B	NP-9	35	
Nebraska	Soil aggregate	6	25				1965
	Granular foundation course	6	—				

TABLE 16-1. Continued.

State	Base Courses	PI	LL	Surface Courses	PI	LL	Year
Nevada	Aggregate base		35				1968
	Percent Passing No. 200 — Max. PI						
	0.1–3.0 — 15						
	3.1–4.0 — 12						
	4.1–5.0 — 9						
	5.1–8.0 — 6						
	8.1–11.0 — 4						
	11.1–15.0 — 3						
New Hampshire							1969
New Jersey							1961
New Mexico	Base	6	25				1963
	Subbase	6	35				
New York	Base course	0	—	Stabilized gravel surface cover (include CaCl and NaCl)	3–8	30	1962
North Carolina	Soil base course (A, B, and C)	6	25				1965
	Coarse aggregate base course	6	30				
	Stabilized aggregate base	6	30				
North Dakota	Base aggregates			Surface aggregates			1965
	Class 3, 4, 5	6[d]	25[d]	Class 3, 4, 5	6[d]	25[d]	
	Class 10	2–6[d]	25[d]	Class 10	2–6[d]	25[d]	
	Class 11	10 min.	45	Class 11	10 min.	45	
	Class 12, 13		25[d]	Class 12, 13		25[d]	
Ohio	Aggregate base	6	25				1969
	Subbase	6	30				
Oklahoma	Aggregate base	6	25	Traffic bound surface course	8–18	35	1967
	Subbase (I and II)	10	30				
	Subbase (IIIA)	15	35				
	Subbase (IIIB)	12	35				
	Subbase (IIIC)	10	30				
	Subbase (IV)	6	25				
	Sand cushion	8	35				
	Caliche	10	35				
	Cement treated	9	35				

Oregon	Percent Passing No. 40						1964
	0.1–3.0	20	40				
	3.1–4.0	15	40				
	4.1–5.0	12	37				
	5.1–8.0	6	33				
	8.1–10.0	6	30				
	10.1–15.0	4	27				
	15.1–20.0	3	24				
	20.1–25.0	2	21				
	Over 25	0	21				
Pennsylvania	Base layer (DG)	6	25				1967
	Aggregate lime-Pozzolan	6	25				
	Sand bituminous	10	30				
	Soil-lime-Pozzolan	6	25				
Rhode Island							1965
South Carolina	Soil aggregate base course	10	25				1964
South Dakota	Subbase			Gravel surfacing	4–15	—	1969
	A	10	—	Service gravel	0–15	—	
	B	10	—				
	C	10	25				
	D	6	25				
	E	3	25				
	Gravel cushion	4	25				
	Aggregate base course						
	1	6	25				
	1A	6	25				
	2	6	25				
	2A	6	—				
Tennessee	Granular subgrade treatment	6	25	Mineral aggregate surfacing	6	25	1968
	Mineral aggregate base	6	25				
	Aggregate cement base	10	35				
Texas	Flexible base (caliche)	12	40				1962
	Flexible base (shell and sand)	10	35				
	Flexible base (bank run gravel)	12	35				
	Flexible base (processed gravel)	12	35				
	Flexible base (iron ore)	12	35				
	Flexible base (class 1)	10	40				
Utah	Base course	6	25	Gravel surface course	6	25	1968

TABLE 16-1. Continued.

State	*Base Courses*	*PI*	*LL*	*Surface Courses*	*PI*	*LL*	*Year*
Vermont	Soil cement base course	NP	40				1964
Virginia	Crusher run aggregate	3	25				1966
	Road stabilized aggregate	6	25				
	Select materials Type I	6	25				
	Select materials Type II	12	40				
	Select materials Type III	9	30				
	Subbase course[e]	6	25				
	Aggregate base course Type I	3	25				
	Aggregate base course Type II	6	25				
Washington	(Sand equivalent)						1963
West Virginia	Aggregate for base and subbase	6	25				Dec. 1968
Wisconsin							1969
Wyoming	Subbase (or if N.P. LL may be 30)	6	25	Asphaltic treatment surfacing	6	—	1967
	Base (or if N.P. LL may be 30)	6	25				
Federal Highway Project FP-61	Aggregate bases	6	25	Aggregate surfacing	4–9	35	1961

[a]Compiled by John W. Guinnee, Engineer of Soils, Geology and Foundations of the Highway Research Board, from data on file with the Bureau of Public Roads.

[b]On the minus 100 portion.

[c]Calculated PI = (actual PI × percent −40)/100 when percent −40 exceeds 30. Calculated PI cannot exceed number in parentheses.

[d]The maximum allowable PI shall be 6.0 except where the PI computed by the formula (10 − Percent of Material Passing No. 40 Sieve)/10 results in a higher PI. Then the higher PI shall be the maximum allowable. The PI and LL shall be reported to nearest whole number. Where a PI greater than 6.0 is permitted the maximum liquid limit shall be increased by the amount the computed PI exceeds 6.0.

[e]Except for size 19 where the maximum PI is 3.

guide upon which to rely for the choise of materials. Table 16-1 is a summary of a number of state highway specifications for plasticity index and liquid limit of materials suitable for highway construction. Since these specifications are based upon the experience of the various states with stabilized soil roads, the table is a valuable guide to an engineer working in this field. However, local experience in geographical areas smaller than an individual state should govern, when it is available, and is in conflict with the values given in the table.

The American Association of State Highway Officials and the American Society for Testing and Materials have written specifications† for materials and mixtures suitable for subbases, base courses, and surface courses. These specifications are essentially the same, differing only in the amount of fine material passing the No. 200 sieve. They recognize the following two general types of mixtures: Type I, which is a relatively coarse-grained mixture with a considerable amount of material up to 1 in. and 2 in. maximum size; and Type II, which is predominantly sand and clay either with or without some larger fragments up to 1 in. maximum size.

16.5.4. Recommended Gradings. Six recommended gradings designated A, B, C, D, E, and F are given in Table 16-2 and are shown graphically in Figs. 16-2 and 16-7, inclusive. In any grading, the ratio of the fraction passing the No. 200 sieve to that passing the No. 40 sieve should not exceed two-thirds. This ratio is frequently referred to as the dust ratio. Any mixture whose grading curve falls within the shaded area on one of the diagrams and which has a dust ratio less than two-thirds is considered to conform to that particular grading requirement.

Subbase and base-course mixtures should conform to one of the six recommended gradings. However, where local experience has shown that lower percentages passing the No. 200 sieve are required to prevent damage to subbase and base courses by frost action, such lower percentages should be specified.

Surface-course materials should conform to one of the four finer gradings, that is, to grading C, D, E, or F. However, when it is planned that the soil-aggregate surface course is to be maintained for several years without a bituminous surface treatment or other superimposed impervious surfacing, grading C, D, or E with a minimum of 8% passing the No. 200 sieve should be specified instead of the minimums shown in Table 16-2 for such a grading.

†A.A.S.H.O. Standard Specifications for Materials for Soil-Aggregate Sub-Base, Base, and Surface Courses, M 147–65.
A.S.T.M. Standard Specifications for Materials for Soil-Aggregate Sub-Base, Base, and Surface Courses, D 1241–68.

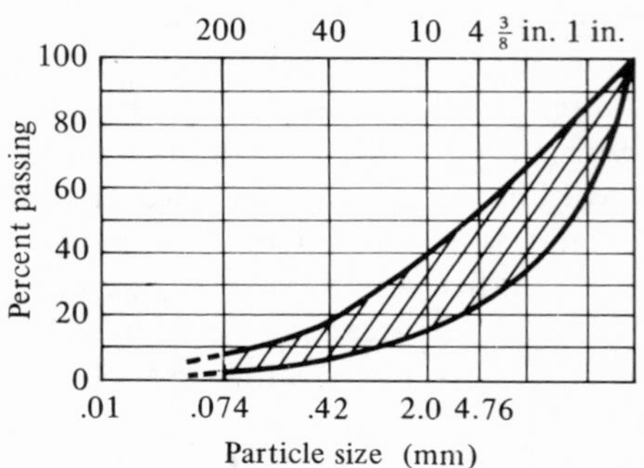

Fig. 16-2. Specification band for grading A. Note: Solid lines indicate grading bands according to A.A.S.H.O. M 147-65T; dotted lines are for A.S.T.M. D 1241-64T in Figs. 16-3, 16-5, 16-6, and 16-7.

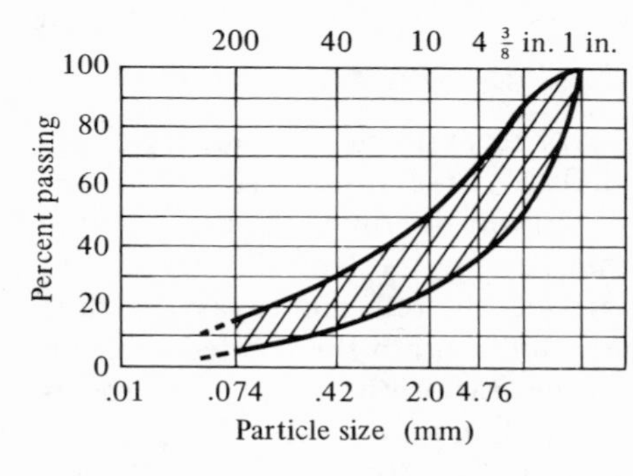

Fig. 16-3. Specification band for grading B.

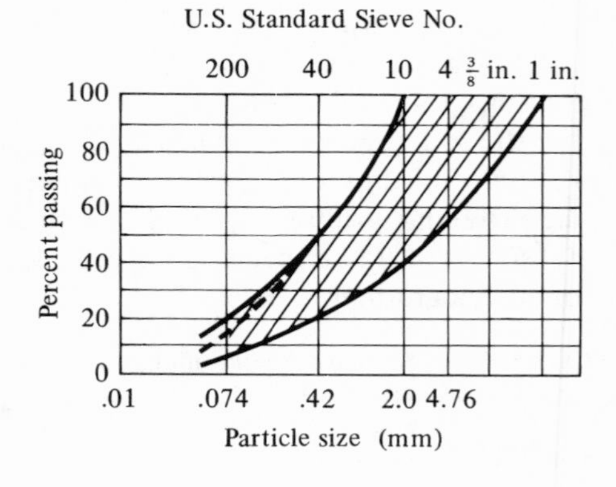

Fig. 16-4. Specification band for grading C.

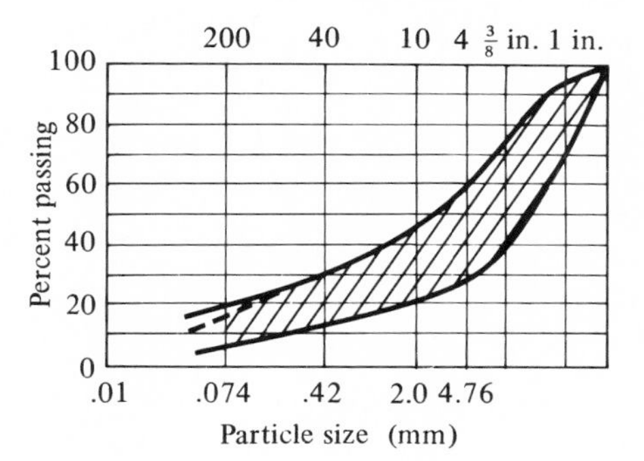

Fig. 16-5. Specification band for grading D.

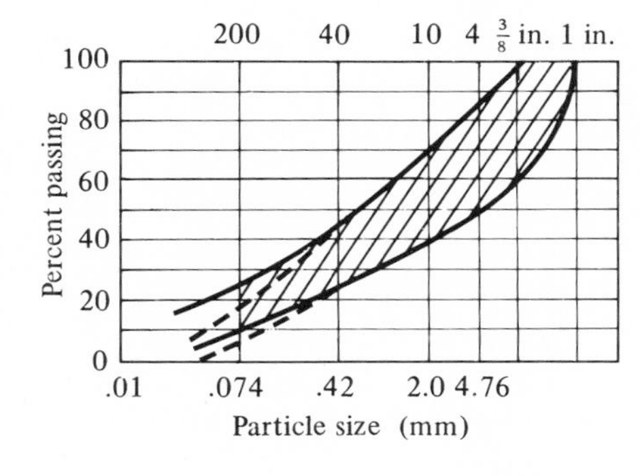

Fig. 16-6. Specification band for grading E.

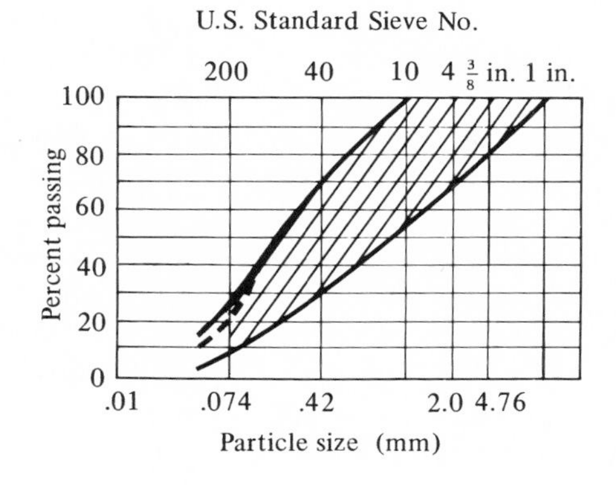

Fig. 16-7. Specification band for grading F.

TABLE 16-2. Gradation Requirements[a] for Soil-Aggregate Materials

Sieve Size (Square Openings)	Percentage, by Weight Passing Square-Mesh Sieves					
	Type I				Type II	
	Grading A	Grading B	Grading C	Grading D	Grading E	Grading F
2 in.	100	100	—	—	—	—
1 in.	—	75–95	100	100	100	100
$\frac{3}{8}$ in.	30–65	40–75	50–85	60–100	—	—
No. 4	25–55	30–60	35–65	50–85	55–100	70–100
No. 10	15–40	20–45	25–50	40–70	40–100	55–100
No. 40	8–20	15–30	15–30	25–45	20–50	30–70
No. 200	2–8	5–15 (5–20)[b]	5–15 (5–15)[b]	8–15 (5–20)[b]	6–15 (6–20)[b]	8–15 (8–25)[b]

[a]A.S.T.M. D1241-68 and A.A.S.H.O. M147-65 (except as noted).

[b]Figures in parentheses indicate percent passing No. 200 sieve according to A.A.S.H.O. D147-65T.

16.5.5. Recommended Binding Properties. The binding properties of the binder-soil fraction passing the No. 40 sieve are indicated by the Atterberg limits of this fraction. For Type I and Type II mixtures the liquid limit should not exceed 25 and the plasticity index should not exceed 6, except that when a mixture with grading C, D, or E is to be maintained for several years as a surface course without an impervious surfacing, the liquid limit should not exceed 35 and the plasticity index should fall in the range from 4 to 9.

16.5.6. Use of Materials Not Meeting Foregoing Specifications. The purpose of the foregoing recommendations is to produce well-graded dense surface courses and well-graded base courses, which have enough stability to remain intact under normal service conditions and at the same time to be free from the possibility of softening under extreme climatic and unfavorable moisture conditions. However, all road and runway locations are not subjected to unfavorable climatic, ground-water, and drainage conditions. Also, many local materials have superior qualities which are not reflected by the grading and plasticity tests.

Experience has indicated that materials which conform to the foregoing recommended requirements will produce the desired result. Nevertheless, many materials which fall outside the recommended specification limits have been found suitable under certain local conditions. Granular soil stabilization practice is not a matter of writing a specification and then obtaining material in conformity with that specification from any distance and at any cost. The problem is rather to make satisfactory use of materials immediately at hand. A great variety of materials may be used in stabilization processes; and on any particular job the specifications must be based upon the characteristics of the available local materials, as indicated by the appropriate A.S.T.M. and A.A.S.H.O. tests on the materials.

16.5.7. Local Materials. Among the local materials which either meet the requirements for granular-type base or surface courses or can be made to meet them by the addition of relatively small amounts of fine or coarse material are: topsoil (as found in certain localities in the southeastern part of the United States), sand-clay, sand-clay-gravel, sand-gravel, crusher-run quarry products, blast-furnace slag, lime rock, caliche, chert, volcanic cinders, and burnt shale. Again it is emphasized that the best guide to satisfactory practice is local experience with locally available materials.

16.6. SIMPLE FIELD TESTS. There a number of simple field tests which can be performed on a soil-aggregate mixture to determine its suitability as a base course under emergency conditions when laboratory equipment is not available. A sample of well-graded moist material, with particles coarser than about $\frac{1}{4}$ in. excluded, will be satisfactory if, when it is squeezed in the hand, the following characteristics are noted: (a) The soil is extremely gritty. (b) It can be formed into definite shapes that retain their forms even when the material dries out. (c) When the clay alone adheres to the hands, there is only enough to discolor them slightly. (d) When enough soil adheres to the hands to discolor them appreciably, it will consist of both sand and clay rather than clay alone. (e) When the moist sample is patted in the palm of the hand, it will compact into a dense cake that cannot be penetrated readily with a blunt stick the size of a lead pencil. (f) If the moist sample is compacted into a container and allowed to dry, it will show little or no shrinkage.

Grittiness of the sample indicates sufficient granular material. Development of some strength on drying indicates sufficient binder soil. Resistance to the penetration of the pencil or stick, even when the sample is thoroughly wetted, indicates desirable interlocking of the grains and a sufficient amount of capillary force. Too much sand would cause the sample to fall apart when dried. Too much clay would leave the hand muddy after the moist sample was squeezed, would cause the moist sample after being patted to offer little resistance to the penetration of the stick, and would cause excessive shrinkage of the sample on drying.

16.7. SELECTING MATERIALS FOR MIXTURES. Unless the native material in place on a road or runway is a naturally stable mixture, it is necessary to combine two or more materials from different sources in order to obtain the desired result for the purpose at hand, whether it be surface course, base course, or subbase. There are four conditions which may be encountered, although the processes of designing are essentially the same. These conditions are: (a) adding aggregate to the native soil in place; (b) adding soil or soil fines to native material not sufficiently cohesive; (c) proportioning mixtures of soil and aggregate for plant mixing; and (d) adding noncohesive material of the proper composition to a surface course which is to be revamped for use as a base.

The first part of the problem consists of finding a near-by source of material which can be combined with the material in place on the graded road or with another material to produce a mixture which will be within the specified limits for liquid limit and plasticity index of the minus No. 40 sieve fraction and, in so far as practicable, within the specified grading limits. If the county in which the project is located has been mapped for soil classification by the U.S. Soil Conservation Service, the location of sources of binder soil may often be greatly facilitated by consulting the soil map.

16.8. PROPORTIONING OF MATERIALS. After the materials have been selected, it will then be necessary to determine the relative proportions in which they are to be mixed. The solution may be approached in three ways. One is to combine the materials so that the plasticity index of the mixture may be expected to be within the allowable range. Then make up a combined sample and test it to see if the PI is as expected; if the liquid limit is within the allowable limits; and if the grading is reasonably close to the specified limits. Another method is to combine the materials so that the grading is within the specified limits and then to test a sample of the mixture to see if the LL and PI are satisfactory. The third method is simply to make up trial mixes until a satisfactory mixture is secured. All three procedures are really trial-and-error methods, but the first two have the advantage of offering guidance which may shorten the work. Experienced engineers may have no difficulty in using the third method and arriving at a satisfactory mix on the first or second try, but less experienced men will find it worth while to employ one or the other of the first two methods. These methods may best be described by illustrative examples.

16.8.1. Proportioning to Obtain Desired Plasticity Index. For the first method of proportioning, assume that a rather coarse material A in place on a roadway is to be stabilized for a surface course by the addition of binder soil B. The characteristics of the materials to be combined are shown in columns 2 and 3 of Table 16-3, and the specification limits are shown in column 4. It is apparent that adding more binder to the coarse

TABLE 16-3

Sieve Size (1)	Percent Passing: Road Material A (2)	Binder Material B (3)	Specification (4)	Final Mixture (5)
1 in.	100	—	100	100
$\frac{3}{4}$ in.	90	—	85–100	91.5
$\frac{3}{8}$ in.	75	—	65–100	78.5
No. 4	60	—	55–85	65.6
No. 10	40	100	40–70	48.4
No. 40	28	99	25–45	37.9
No. 200	6	70	10–25[a]	15.0
Liquid limit	23	28	<35	
Plasticity index	0	14	4 to 9	5

[a]The fraction passing the No. 200 sieve should be less than two-thirds of the fraction passing the No. 40 sieve (dust ratio less than $\frac{2}{3}$).

material will increase the PI; hence, in order to keep the amount of binder to be hauled and added as low as possible, it will be desirable to work toward a PI for the mixture near the lower limit. Assume that the desired PI is 5.

The Michigan State Highway Department has devised a formula for the solution of this problem which is

$$P = KRM \tag{16-1}$$

in which

P = pounds of binder soil to add to each 100 lb of road material;
K = a conversion factor which depends on the PI of the binder, the PI of the road material, and the PI desired for the mixture;

$$R = 1 + \frac{\text{percent of binder soil retained on No. 40 sieve}}{\text{percent of binder soil passing No. 40 sieve}};$$

M = percentage of road material passing No. 40 sieve.

Values of K are given in Table 16-4. In this problem $K = 0.56$. Also, $R = 1 + (1/99) = 1.01$, and $M = 28$. Then

$$P = 0.56 \times 1.01 \times 28 = 15.83 \text{ lb}$$

Hence, it is necessary to add approximately 16 lb of binder soil to each 100 lb of coarse road material; and the stabilized mixture will consist of 86% of road material and 14% of binder soil.

Applying these factors to the percentages of road material and binder soil passing each sieve, then adding the results, the sieve analysis of the

TABLE 16-4. Values of Conversion Factor, K, for Use in Eq. (16-1)

PI of Binder Soil	Desired Plasticity of Mix																											
	PI = 7							PI = 6						PI = 5					PI = 4				PI = 3			PI = 2		PI = 1
	PI (Road Material)[a]							PI						PI					PI				PI			PI		PI
	0	1	2	3	4	5	6	0	1	2	3	4	5	0	1	2	3	4	0	1	2	3	0	1	2	0	1	0
10	2.33	2.00	1.67	1.33	1.00	0.67	0.33	1.50	1.25	1.00	0.75	0.50	0.25	1.00	0.80	0.60	0.40	0.20	0.67	0.50	0.33	0.17	0.43	0.29	0.14	0.25	0.13	0.11
11	1.75	1.50	1.25	1.00	0.75	0.50	0.25	1.20	1.00	0.80	0.60	0.40	0.20	0.84	0.67	0.50	0.33	0.17	0.57	0.43	0.29	0.14	0.38	0.25	0.13	0.22	0.11	0.10
12	1.40	1.20	1.00	0.80	0.60	0.40	0.20	1.00	0.84	0.67	0.50	0.33	0.17	0.72	0.57	0.43	0.29	0.14	0.50	0.38	0.25	0.13	0.33	0.22	0.11	0.20	0.10	0.09
13	1.17	1.00	0.84	0.67	0.50	0.33	0.17	0.86	0.72	0.57	0.43	0.29	0.14	0.63	0.50	0.38	0.25	0.13	0.44	0.33	0.22	0.11	0.30	0.20	0.10	0.18	0.09	0.08
14	1.00	0.86	0.72	0.57	0.43	0.29	0.14	0.75	0.63	0.50	0.38	0.25	0.13	0.56	0.44	0.33	0.22	0.11	0.40	0.30	0.20	0.10	0.27	0.18	0.09	0.17	0.08	0.08
15	0.88	0.75	0.63	0.50	0.38	0.25	0.13	0.67	0.56	0.44	0.33	0.22	0.11	0.50	0.40	0.30	0.20	0.10	0.36	0.27	0.18	0.09	0.25	0.17	0.08	0.15	0.08	0.07
16	0.78	0.67	0.56	0.44	0.33	0.22	0.11	0.60	0.50	0.40	0.30	0.20	0.10	0.46	0.36	0.27	0.18	0.09	0.33	0.25	0.17	0.08	0.23	0.15	0.08	0.14	0.07	0.07
17	0.70	0.60	0.50	0.40	0.30	0.20	0.10	0.55	0.46	0.36	0.27	0.18	0.09	0.42	0.33	0.25	0.17	0.08	0.31	0.23	0.15	0.08	0.21	0.14	0.07	0.13	0.07	0.06
18	0.64	0.55	0.46	0.36	0.27	0.18	0.09	0.50	0.42	0.33	0.25	0.17	0.08	0.39	0.31	0.23	0.15	0.08	0.29	0.21	0.14	0.07	0.20	0.13	0.07	0.13	0.06	0.06
19	0.58	0.50	0.42	0.33	0.25	0.17	0.08	0.46	0.39	0.31	0.23	0.15	0.08	0.36	0.29	0.21	0.14	0.07	0.27	0.20	0.13	0.07	0.19	0.13	0.06	0.12	0.06	0.06
20	0.54	0.46	0.39	0.31	0.23	0.15	0.08	0.43	0.36	0.29	0.21	0.14	0.07	0.33	0.27	0.20	0.13	0.07	0.25	0.19	0.13	0.06	0.18	0.12	0.06	0.11	0.06	0.05
21	0.50	0.43	0.36	0.29	0.21	0.14	0.07	0.40	0.33	0.27	0.20	0.13	0.07	0.31	0.25	0.19	0.13	0.06	0.24	0.18	0.12	0.06	0.17	0.11	0.06	0.11	0.05	0.05
22	0.47	0.40	0.33	0.27	0.20	0.13	0.07	0.38	0.31	0.25	0.19	0.13	0.06	0.29	0.24	0.18	0.12	0.06	0.22	0.17	0.11	0.06	0.16	0.11	0.05	0.10	0.05	0.05
23	0.44	0.38	0.31	0.25	0.19	0.13	0.06	0.35	0.29	0.24	0.18	0.12	0.06	0.28	0.22	0.17	0.11	0.06	0.21	0.16	0.11	0.05	0.15	0.10	0.05	0.10	0.05	0.05
24	0.41	0.35	0.29	0.24	0.18	0.12	0.06	0.33	0.28	0.22	0.17	0.12	0.06	0.27	0.21	0.16	0.11	0.05	0.20	0.15	0.10	0.05	0.14	0.10	0.05	0.09	0.05	0.04
25	0.39	0.33	0.28	0.22	0.17	0.11	0.06	0.32	0.27	0.21	0.16	0.11	0.05	0.25	0.20	0.15	0.10	0.05	0.19	0.14	0.10	0.05	0.14	0.09	0.05	0.09	0.04	0.04
26	0.37	0.32	0.27	0.21	0.16	0.11	0.05	0.30	0.25	0.20	0.15	0.10	0.05	0.24	0.19	0.14	0.10	0.05	0.18	0.14	0.09	0.05	0.13	0.09	0.04	0.08	0.04	0.04
27	0.35	0.30	0.25	0.20	0.15	0.10	0.05	0.29	0.24	0.19	0.14	0.10	0.05	0.23	0.18	0.14	0.09	0.05	0.17	0.13	0.09	0.04	0.13	0.08	0.04	0.08	0.04	0.04
28	0.33	0.29	0.24	0.19	0.14	0.10	0.05	0.27	0.23	0.18	0.14	0.09	0.05	0.22	0.17	0.13	0.09	0.04	0.17	0.13	0.08	0.04	0.12	0.08	0.04	0.08	0.04	0.04
29	0.32	0.27	0.23	0.18	0.14	0.09	0.05	0.26	0.22	0.17	0.13	0.09	0.04	0.21	0.17	0.13	0.08	0.04	0.16	0.12	0.08	0.04	0.12	0.08	0.04	0.07	0.04	0.04
30	0.31	0.26	0.22	0.17	0.13	0.09	0.04	0.25	0.21	0.17	0.13	0.08	0.04	0.20	0.16	0.12	0.08	0.04	0.15	0.12	0.08	0.04	0.11	0.07	0.04	0.07	0.04	0.04
31	0.29	0.25	0.21	0.17	0.13	0.08	0.04	0.24	0.20	0.16	0.12	0.08	0.04	0.19	0.15	0.12	0.08	0.04	0.15	0.11	0.07	0.04	0.11	0.07	0.04	0.07	0.04	0.03
32	0.28	0.24	0.20	0.16	0.12	0.08	0.04	0.23	0.19	0.15	0.12	0.08	0.04	0.19	0.15	0.11	0.07	0.04	0.14	0.11	0.07	0.04	0.10	0.07	0.04	0.07	0.03	0.03
33	0.27	0.23	0.19	0.15	0.12	0.08	0.04	0.22	0.19	0.15	0.11	0.07	0.04	0.18	0.14	0.11	0.07	0.04	0.14	0.10	0.07	0.04	0.10	0.07	0.03	0.07	0.03	0.03
34	0.26	0.22	0.19	0.15	0.11	0.07	0.04	0.21	0.18	0.14	0.11	0.07	0.04	0.17	0.14	0.10	0.07	0.04	0.13	0.10	0.07	0.03	0.10	0.07	0.03	0.06	0.03	0.03
35	0.25	0.21	0.18	0.14	0.11	0.07	0.04	0.21	0.17	0.14	0.10	0.07	0.04	0.17	0.13	0.10	0.07	0.03	0.13	0.10	0.07	0.03	0.09	0.06	0.03	0.06	0.03	0.03
36	0.24	0.21	0.17	0.14	0.10	0.07	0.04	0.20	0.17	0.13	0.10	0.07	0.03	0.16	0.13	0.10	0.07	0.03	0.13	0.09	0.06	0.03	0.09	0.06	0.03	0.06	0.03	0.03
37	0.23	0.20	0.17	0.13	0.10	0.07	0.03	0.19	0.16	0.13	0.10	0.07	0.03	0.16	0.13	0.09	0.06	0.03	0.12	0.09	0.06	0.03	0.09	0.06	0.03	0.06	0.03	0.03
38	0.23	0.19	0.16	0.13	0.10	0.07	0.03	0.19	0.16	0.13	0.09	0.06	0.03	0.15	0.12	0.09	0.06	0.03	0.12	0.09	0.06	0.03	0.09	0.06	0.03	0.06	0.03	0.03
39	0.22	0.19	0.16	0.13	0.09	0.06	0.03	0.18	0.15	0.12	0.09	0.06	0.03	0.15	0.12	0.09	0.06	0.03	0.11	0.09	0.06	0.03	0.08	0.06	0.03	0.05	0.03	0.03
40	0.21	0.18	0.15	0.12	0.09	0.06	0.03	0.18	0.15	0.12	0.09	0.06	0.03	0.14	0.11	0.09	0.06	0.03	0.11	0.08	0.06	0.03	0.08	0.05	0.03	0.05	0.03	0.03

[a] This horizontal column indicates the plasticity index of the raod material throughout.

mixture is found to be as shown in column 5 of Table 16-3. It is seen that the calculated sieve analysis of the mixture falls within the specification limits. Also the dust ratio is less than $\frac{2}{3}$, and that requirement of the specification is satisfied. It only remains to make up a trial sample in accordance with the calculated results and to test it for sieve analysis, LL and PI. If the variations from the specification limits are larger than those which should be allowed, it will be necessary to try again with new assumptions based upon this experience, and possibly to repeat the procedure until a mixture in reasonable compliance with the specifications is found.

16.8.2. Proportioning to Obtain Desired Grading. The second method of combining materials is on the basis of their sieve analysis, and again a method devised by the Michigan State Highway Department will be described. The proportions of two given materials A and B required to approximate a specified grading of the mixture of the two may be calculated from the summations of the differences between the desired grading and the gradings of materials A and B, respectively, provided one of the two materials is finer and the other is coarser than the desired grading. The differences between the percentages passing each sieve for the specified grading and each of the two materials are calculated and added without respect to sign. The ratio of the two summations indicates the proportions desired.

Whenever the sieve analyses of the materials to be blended are such that both gradings in whole or in part are either finer or coarser than the desired combination grading, the ratio computed by this method may be in error. In those cases of unbalanced or nonuniform sieve analyses, it may be necessary to raise or lower the ratio to secure a more satisfactory result. It is best to satisfy the grading requirements of the finer sieve fractions and to neglect those of the coarser fractions whenever mixtures must be used which will not meet the desired grading requirements on all sieves.

The process just outlined may be best described by means of an illustrative example, as follows. Table 16-5 gives the gradings of two materials A and B, the specified grading, and the desired grading of the mixture. The grading which is desired is one that falls within the specification limits and is therefore assumed to be satisfactory. In the case shown, the desired grading is taken as the average of the specification limits, except that the percentage passing the No. 200 sieve is taken as somewhat less than the average in order to hold down the amount of binder.

From Table 16-5 the summations of the differences between the desired grading and materials A and B are 22 for A and 80 for B. According

TABLE 16-5

Sieve Size	*Percent Passing*					
	Material A	*Material B*	*Specifi-cation*	*Desired Mixture E*	*Difference E—A*	*Difference E—B*
$\frac{3}{4}$ in.	98	100	100	100	2	0
$\frac{3}{8}$ in.	65	98	60–85	73	8	−25
No. 10	35	79	40–50	45	10	−34
No. 40	25	15	20–35	27	2	12
No. 200	10	1	10–18	10	0	9
				Totals	22	80[a]

[a]Numerical sum without regard to sign.

to this method, the ratio of material A to material B in the final mix is to be the same as the ratio of the summations of the differences. Therefore,

$$\frac{\mathrm{A}}{\mathrm{B}} = \frac{22}{80}$$

or B = 3.64 A. That is, the proportions of these two materials for the final mixture will be 1 part of B to 3.64 parts of A; and the mixture will contain 100/4.64 = 21.6% of B and 3.64 × 21.6 or 78.4% of A. When these percentages are applied to the sieve analyses of materials A and B, the sieve analysis of the final mixture is found to be as shown in Table 16-6. These values are as close to those of the desired mix as may be expected.

After the proportions of the materials in the mixture have been computed, a combined sample should be prepared and tested for compliance with the specifications.

16.9. PREPARATION OF OLD SURFACING. In many cases it is planned to place a bituminous wearing surface on a stabilized road that has been

TABLE 16-6

Sieve	*0.78A*	*0.22B*	*Percent Passing*	
			Final Mix	*Desired Mix (Table 16-5)*
$\frac{3}{4}$ in.	76.4	22.0	99	100
$\frac{3}{8}$ in.	50.7	21.6	72	73
No. 10	27.3	17.4	45	45
No. 40	19.5	3.3	23	27
No. 200	7.8	0.2	8	10

in use for some time as a surface course. Unless the surface course was originally laid with this eventuality in mind, and hence consists of a mix suitable for either purpose, it will usually be found that the old surface course is not satisfactory for a base. In some cases of this kind it may be desirable to scarify the old surface course, add granular material that will bring the combined mixture into compliance with base-course requirements, then reconstruct the course according to the proper procedure for a base course. In other cases it may be desirable to place a base course of proper thickness over the old surface.

16.10. COMPACTION OF SOIL. Compaction of the soil-aggregate mixture is a vital part of the stabilization process. To obtain the desired result it is necessary that the surface, base, or foundation be left in as compact a mass as possible. A granular stabilized road must be properly compacted before it is covered with a bituminous top. This compaction requires a judicious use of water and strict control of the rolling operations. The amount of water which will produce adequate compaction varies with the character of the soil, as discussed in Chapter 8. After compaction, there will be greater assurance of satisfactory results if a seasoning period is allowed, until about 40% of the moisture has evaporated, before the bituminous wearing surface is applied.

16.11. COMBINING THREE MATERIALS. It is sometimes necessary to combine three separate materials to produce a stabilized soil mixture for use as a base course or a surface course. The proportions of the materials required for the final mixture may be determined by means of a triangular chart like that shown in Fig. 16-8. This method of proportioning was developed by the Indiana State Highway Commission. Again, an illustrative example will be used to describe the method of procedure.

Suppose that it is desired to combine a predominantly gravel material with sand and a silty-clay binder soil to produce a mixture which complies with the grading and plasticity-index specifications in column 5 of Table 16-7. The three materials to be combined are designated as A, B, and C. Their mechanical analyses are also shown in this table, and the percentages of their constituent parts are as given in Table 16-8. The terms gravel, sand, and silt-clay in this connection refer only to size of particles. Actually the materials may be crushed rock, screenings and stone dust, or other similar materials.

The next step is to plot points *A*, *B*, and *C* on the triangular chart in Fig. 16-8 to represent each of the three materials. Also, plot the specified grading characteristics on the same chart. Since the specification permits a range of variation of the various size constituents, plotting will

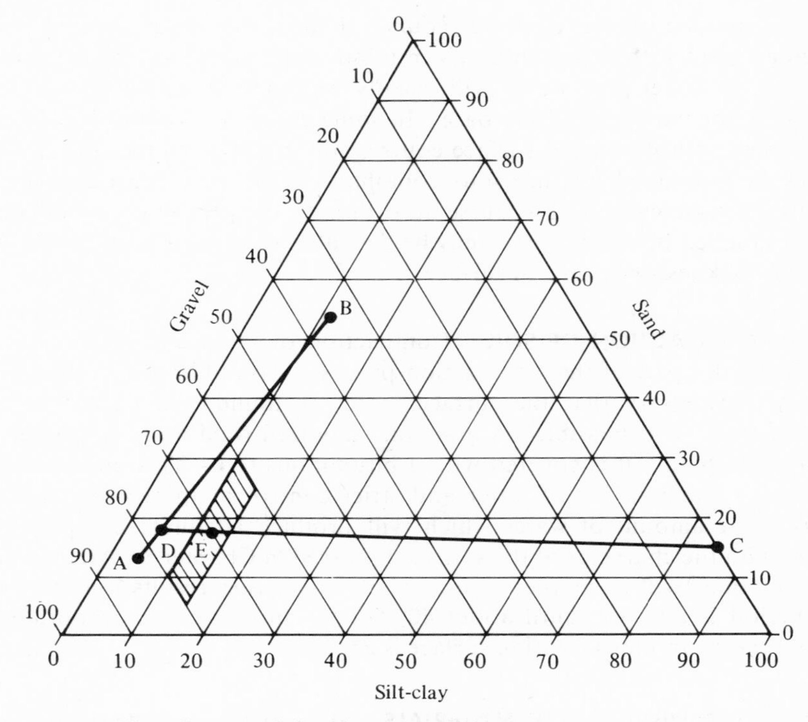

Fig. 16-8. Chart for combining three materials.

TABLE 16-7

Sieve Size (1)	Material A (2)	Material B (3)	Material C (4)	Specification of Mixture (5)	
	Percent Passing				
$1\frac{1}{2}$ in.	100	—	—	100	Gravel
1 in.	98	—	—	85–100	60–80
$\frac{3}{4}$ in.	76	—	—	65–95	
$\frac{1}{2}$ in.	48	—	—	30–80	Sand
No. 4	27	100	—	25–60	30–5
No. 10	18	64	100	20–40	
No. 40	10	28	97	15–30	Silt-clay
No. 200	4	11	85	10–15	10–15
PI	0	0	13.3	4–9, incl.	

TABLE 16-8.

Material	*Size Fraction (%)*		
	Gravel	*Sand*	*Silt-clay*
A	82	14	4
B	36	53	11
C		15	85

yield an area on the chart which can be cross-hatched to clearly define its limits. Select a point E at about the center of the hatched area. Draw a line AB. Then draw a line from C through the point E until it intersects AB at D. Using any convenient scale, measure and tabulate the lengths of various lines as follows:

Line	Length	Line	Length
AB	10.9	DC	19.7
DB	9.7	EC	17.8
AD	1.2	DE	1.9

The proportions of the materials A, B, and C in the final mixture may be obtained from the following formulas:

$$\text{A} = \frac{EC}{DC} \times \frac{DB}{AB} = \frac{17.8}{19.7} \times \frac{9.7}{10.9} = 80.4\%$$

$$\text{B} = \frac{EC}{DC} \times \frac{AD}{AB} = \frac{17.8}{19.7} \times \frac{1.2}{10.9} = 10.0\%$$

$$\text{C} = \frac{DE}{DC} = \frac{1.9}{19.7} = \qquad 9.6\%$$

$$100.0\%$$

By applying these percentages to the gradation of the original materials, the gradation of the final mixture is obtained as illustrated in Table 16-9.

TABLE 16-9.

Sieve Size	*A × 0.804*	*B × 0.10*	*C × 0.096*	*Final Mix (Percent Passing)*
$1\frac{1}{2}$ in.	80.4	10.0	9.6	100
1 in.	78.8	10.0	9.6	98
$\frac{3}{4}$ in.	61.1	10.0	9.6	81
$\frac{1}{2}$ in.	38.6	10.0	9.6	58
No. 4	21.7	10.0	9.6	41
No. 10	14.5	6.4	9.6	31
No. 40	8.0	2.8	9.3	20
No. 200	3.2	1.1	8.2	12

It is very desirable to estimate the plasticity index of the final mixture in order to judge the suitability of the binder soil under consideration. An approximate estimate of this property can be calculated by the formula

$$\text{PI} = \frac{XS_1P_1 + YS_2P_2 + ZS_3P_3}{XS_1 + YS_2 + ZS_3} \tag{16-2}$$

in which

X = percent of material A in the mixture;
Y = percent of material B in the mixture;
Z = percent of material C in the mixture;
S_1 = percent of material A passing No. 40 sieve;
S_2 = percent of material B passing No. 40 sieve;
S_3 = percent of material C passing No. 40 sieve;
P_1 = PI of material A;
P_2 = PI of material B;
P_3 = PI of material C.

If P_1 or P_2 is zero, substitute the value 1 for this computation. Thus, if $P_1 = 0$, $P_2 = 0$, and $P_3 = 13.3$, as in Table 16-7, the estimated PI of the mixture is

$$\text{PI} = \frac{(80.4 \times 10 \times 1) + (10.0 \times 28 \times 1) + (9.6 \times 97 \times 13.3)}{(80.4 \times 10) + (10.0 \times 28) + (9.6 \times 97)} = 6.7$$

A trial mixture must now be made, and the actual grading and the P.I. of the mix should be determined. If either the grading or the P.I. does not conform to the specifications, the proportions of the materials should be adjusted and a new trial mixture should be made and tested.

16.12. CHEMICAL SOIL STABILIZATION. The last three decades have seen a rapid increase in the use of chemicals to stabilize soils, in particular by mixing the soil with portland cement, hydrated lime, or bitumen. Such additives may impart a substantial improvement and permanence compared to granular stabilization. Chemical stabilization also is widely used to improve the properties of soils that do not meet gradation requirements discussed above. The details of chemical stabilization are beyond the scope of this text; Table 16-10 indicates some of the more common stabilizers and their applicability to different soils.

Hydrated lime, $Ca(OH)_2$, in particular is finding uses in everyday construction, since a small percentage of lime will effect a dramatic change in wet, plastic clays by causing an immediate increase in the plastic limit, changing the soil from plastic to solid and making it appear to "dry up"

TABLE 16-10. Some Chemical Soil Stabilizers

Chemical	*Preferred Soil*	*Chemical Reactions*	*Uses*
Hydrated lime $Ca(OH)_2$	Clayey soils	1. Flocculation, decreasing PI 2. Slow pozzolanic cementation	1. "Drying up" wet clays for subbases or platforms for compaction 2. Road base courses 3. Landslides
Portland cement	Gravelly, sandy, and silty soils	1. Flocculation, decreasing PI 2. Rapid cementation	1. Road base courses 2. Riprap on earth dams 3. Canal linings
Lime + cement	Clays	1. Same as each alone	1. Clay base courses where higher strength is needed
Asphalt	Gravelly, sandy, and silty soils	1. Coats and waterproofs 2. Sticks together grains and aggregates	1. Base courses 2. Surface spray (road oil) for dust palliation
Lignins (Lignosulfonates)	Densely graded gravelly, sandy, and silty soils	1. Disperses, increasing PI and improving compaction 2. Cements upon drying 3. Water soluble; surface drainage required	1. Surface courses for secondary roads 2. Surface spray for dust palliation 3. Periodic maintenance required
Sodium chloride NaCl	Same as for lignins	1. Lubricates, improving compaction 2. Cements upon drying 3. Reduces freezing point to −6°F 4. Water soluble; surface drainage required	1. Surface courses for secondary roads 2. Partial dust palliation 3. Reduction of frost boils 4. Periodic maintenance required; usually $CaCl_2$ surface treatment
Calcium chloride $CaCl_2$	Same as for lignins	1. Lubricates and increases surface tension, increasing density as much as 15% on drying 2. Retains moisture at relative humidity above about 25% 3. Reduces freezing point to −60°F 4. Water soluble; surface drainage required	1. Surface courses for secondary roads 2. Dust palliation 3. Prevention of frost boils 4. Periodic maintenance required

without any real change in the moisture content. Field uses therefore range from the bag or two of lime thrown into a troublesome mud hole and worked in with a crawler tractor or road grader, to stabilization with road mixers of entire building sites where improvement in trafficability can result in a large saving in hours and dollars during construction. Lime also has been introduced in drill holes to stabilize weak clay subgrades and permanently halt landslides, as discussed later in Section 20.14.

Another very common chemical additive is portland cement, which mixed with soil and compacted at optimum moisture content hardens the soil, converting it into "soil-cement." Soil-cement is used mainly for base courses and as riprap on earth dams. Portland cement liberates about one-fourth of its weight as hydrated lime, and thus tends to give in addition the same benefits as lime stabilization. Clayey soils may deplete the system of lime; in this case lime alone or lime plus cement may be used.

Bituminous stabilization is frequently used to upgrade stabilized roads, providing a waterproof base course for later surfacing. As in the case of lime and cement, the bitumen is mixed with the soil, which is then compacted and covered with an asphaltic surface course. "Road oil," which includes asphalt cutbacks or road tars sprayed on the surface of granular stabilized roads, is a minimal treatment requiring annual maintenance and reapplication.

Lignins are waste products from the paper industry, produced in a quantity approximately equal to the production of paper. They may be mixed in or sprayed on granular stabilized roads, reducing maintenance and dusting by a factor up to 10. Chlorides are used similarly, but may be more objectionable as water pollutants. The maximum life of such treatments on unsurfaced roads without annual reapplication is about 3 to 5 years.

PROBLEMS

16.1. Define soil stabilization as practiced in connection with highway and airport work.

16.2. Name and describe the elements of a flexible or stabilized soil pavement.

16.3. What are the two principal functions of a subbase or ballast course of a flexible pavement?

16.4. Describe the various forces applied to a highway or runway pavement by traffic wheel loads.

16.5. Explain why a base course which has served satisfactorily as a surface course may not be satisfactory if it is later covered with a bituminous wearing surface.

16.6. Explain the functions of the granular fraction, the silt-size fraction, and the clay-size fraction in a stabilized granular soil mixture.

16.7. What are the two most important characteristics of a stabilized granular soil mixture?

16.8. Distinguish between Type I and Type II surface-course and base-course materials as specified by A.A.S.H.O. and A.S.T.M.

16.9. Define the dust ratio of a stabilized granular soil mixture.

16.10. Give the A.A.S.H.O. and A.S.T.M. specification requirements for PI, LL, and dust ratio for surface courses; for base courses.

16.11. List six field tests which will give a general indication of whether or not a granular soil mixture will give satisfactory service as a stabilized base course.

16.12. A coarse-grained road material is to be stabilized by adding fine-grained binder soil. Twenty-four percent of the road material and 85% of the binder soil passes a No. 40 sieve. The PI of the road material is 0, and that of the binder soil is 13. How much binder soil will be required for each 100 lb of road material to produce a mixture having a PI of 6?

16.13. The grading characteristics of the materials in Problem 16.12 are shown in the accompanying tabulation. Compute the grading of the final mixture of these materials according to the proportions determined in Problem 16.12.

Particle Size	*Percent Passing*	
	Road Material	*Binder Soil*
1 in.	100	—
$\frac{3}{4}$ in.	96	—
$\frac{3}{8}$ in.	90	—
No. 4	61	—
No. 10	32	100
No. 40	21	85
No. 200	3	70

16.14. To which types and gradings does the trial mixture of Problem 16.13 conform?

16.15. What is the dust ratio of the trial mixture of Problem 16.13?

16.16. If the mixture of Problem 16.13 is to be maintained for several years as a surface course, what values of LL and PI are recommended?

16.17. Three materials have the percentages of gravel, sand, and silt-clay shown in the accompanying table. Draw a triaxial chart similar to that in Fig. 16-8 and compute the percentage of each material required to produce a mixture which meets the specification given in the table. Show the final mixture in tabular form.

Material	*PI*	*Size Fraction (%)*			*Specification Limits (%)*
		Gravel	*Sand*	*Silt-clay*	
A	0	92	8	—	Gravel 60–75
B	2	30	64	6	Sand 5–30
C	12	—	5	95	Silt-clay 10–20

16.18. The three materials in Problem 16.17 have the percentages passing the No. 40 sieve shown in the tabulation. Estimate the plasticity index of the final mixture obtained in Problem 16.17.

Material	*Percent Passing*
A	3
B	21
C	98

16.19. Suggest one or more possible chemical stabilizers for the following soils:

Soil	Use
SP	Road base
SC	Road base
ML	Riprap
CH	Building site
GM	Road surface

16.20. (a) Calculate activity indices of the following soils before and after lime treatment. [Data adopted from Hilt and Davidson, *Highway Research Board Bull.* **262,** 20–32 (1960).]

Soil	*Major Clay Mineral*	*Percent 0.002 mm Clay*	*Soil PI*	*Soil + 3% $Ca(OH)_2$ PI*
Gumbotil	Montmorillonite	67	50	29
Loess	Montmorillonite	33	32	19
Glacial till	Illite	44	23	7
Residual soil	Kaolinite	30	26	19

(b) Do the untreated clay activities agree with the ranges in Table 13.1? Explain. (c) Based on the reduction in activity by lime treatment, what is the significance of the kind of clay minerals?

16.21. The soils in Problem 16.20 with 6% lime were compacted at standard A.A.S.H.O. density and gave the following unconfined compressive strengths after 27 days moist curing and 1 day soaking in water. (The strength with no lime was zero.)

Soil	q_u (psi)
Gumbotil	160
Glacial till	90
Residual soil	130

(a) Based on the increase in unconfined compressive strength, what is the significance of clay minerals? (b) Little or no strength increase results from addition of more than 6% lime to the above clay soils. What else could be added to further improve these soils? Should this be in addition to or in place of lime?

REFERENCES

1. Highway Research Board. "Granular Stabilized Roads." Wartime Road Problems No. 5, 1943.
2. Ho, Clara, and R. L. Handy. "Electrokinetic Properties of Lime-Treated Bentonites." *Proc. 12th National Conference on Clays and Clay Min.*, pp. 267–280. Pergamon Press, New York, 1964.
3. O'Flaherty, C. A. *Highways.* Edward Arnold Ltd., London, 1967.
4. Spangler, M. G. "Wheel Load Stress Distribution through Flexible Type Pavements." *Proc. Highway Research Board* **21** (1941).
5. Wang, J. W. H., and R. L. Handy. "Role of MgO in Soil-Lime Stabilization." Highway Research Board Special Report 90, pp. 475–492. 1966.
6. Woods, K. B., Ed. *Highway Engineering Handbook.* McGraw-Hill, New York, 1960.
7. Yoder, E. J. *Principles of Pavement Design.* John Wiley, New York, 1959.

17

Stress Distribution in Soil

17.1. TYPES OF PROBLEMS. Many problems in structural and foundation engineering require a study of the transmission and distribution of stresses in large and extensive masses of soil. Some examples of problems of this kind are: wheel loads transmitted through embankments to culverts or through the backfilling material to sewers; foundation pressures transmitted to soil strata at various elevations below footings; wheel loads or pressures from isolated footings transmitted to retaining walls. In problems of these kinds the lines of action of the stresses produced by the loads at the surface of the soil are not determined by the geometry of the structure, as in the case of superstructure beams, columns, and trusses. Rather, the stresses are transmitted in all downward and lateral directions.

Adding to stresses from loads superimposed on soils are stresses originating from weight of the soil itself. These stresses, called *body stresses* or *geostatic stresses* were discussed in Section 12.21. Both applied stresses and body stresses play important roles in settlement predictions, discussed in the next chapter.

17.2. STRESSES BY BOUSSINESQ SOLUTION. The study of distribution of applied stresses in soils is greatly facilitated by a theoretical approach which is known as the Boussinesq (pronounced Boo-si-nesk′) solution for stresses in a semi-infinite elastic medium due to a point load applied at its surface. The term "semi-infinite medium," also called a "half space," refers to mass on one side of a theoretically infinite plane. The solution obtained by the French elastician assumes an infinite half space to be filled with an ideally elastic, homogeneous, isotropic mass of material. Although soil is known to be nonelastic, nonhomogeneous, and nonisotropic, experimental measurements of stress distribution indicate that the classical Boussinesq solution, when properly applied, serves as a reasonably good guide for the determination of stresses in the undersoil. Problems frequently arise, however, in which the environmental conditions may be considerably different from those for which the solution was obtained, and some modifications of the Boussinesq formulas must then be made. Therefore, a soil engineer must understand the assumed conditions upon which the formulas are based, in order to be able to identify those problems to which they are directly applicable and those in which some modifications of the classical formulas are necessary, even though he may not be skilled in the advanced mathematical procedures by which the solution was obtained.

The orthogonal axes X–X' and Y–Y' in Fig. 17-1(a) lie in a horizontal

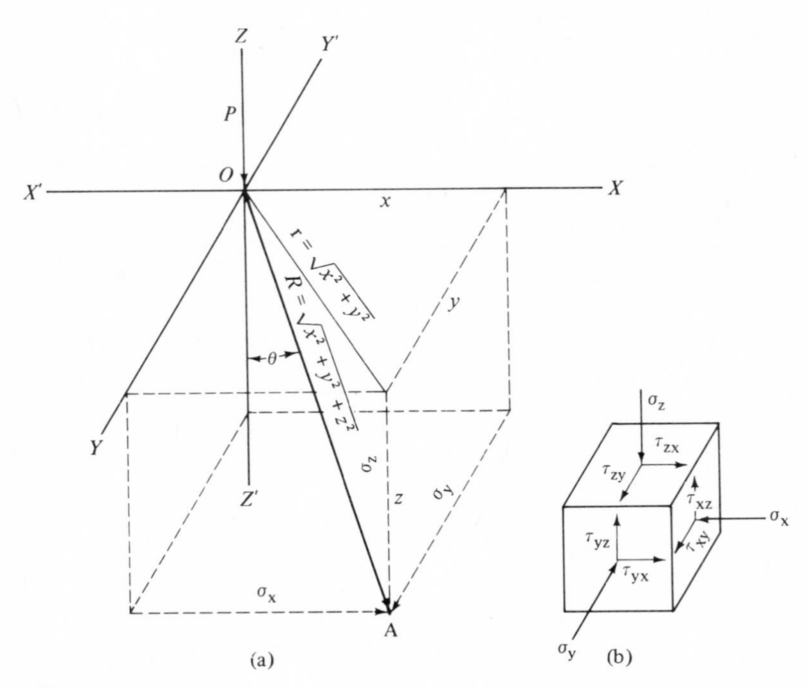

Fig. 17-1

plane which is the upper boundary of a semi-infinite mass. The axis Z–Z' extends vertically downward into the mass and passes through the origin of coordinates, which is the point of application of the vertical load P. Then the location of any point A in the soil mass is indicated by the coordinate distances x, y, and z.

The unit stresses acting at point A on planes normal to the coordinate axes are shown in Fig. 17-1(b). They consist of normal stresses, denoted by σ, and shearing stresses, denoted by τ. The subscripts of σ indicate the orientation of the lines of action of the several normal stresses. Thus, σ_x is a stress parallel to the x axis through the point A. The first letter of each double subscript of τ indicates the plane in which the shearing stress acts, and the second letter indicates the orientation of its line of action. Thus, τ_{zx} is a shearing stress acting in a plane normal to the z axis and on a line of action parallel to the x axis; while τ_{zy} is a shearing stress acting in the same plane but on a line of action parallel to the y axis. The stresses on only three faces of a minute cube of soil at point A are shown in Fig. 17-1(b). Since this cube is in equilibrium, the stresses on the other three faces will have the same magnitudes as those shown but they will act in the opposite directions.

The Boussinesq equations for the normal stresses are as follows:

$$\sigma_z = \frac{3P}{2\pi}\frac{z^3}{R^5} \tag{17-1}$$

or

$$\sigma_z = \frac{3P\cos^5\theta}{2\pi z^2} \tag{17-1a}$$

$$\sigma_x = \frac{P}{2\pi}\left[\frac{3x^2z}{R^5} - (1 - 2\nu)\left(\frac{x^2 - y^2}{Rr^2(R + z)} + \frac{y^2z}{R^3r^2}\right)\right] \tag{17-2}$$

$$\sigma_y = \frac{P}{2\pi}\left[\frac{3y^2z}{R^5} - (1 - 2\nu)\left(\frac{y^2 - x^2}{Rr^2(R + z)} + \frac{x^2z}{R^3r^2}\right)\right] \tag{17-3}$$

in which ν is Poisson's ratio; and all other symbols have the meanings indicated in Fig. 17-1.

17.3. POISSON'S RATIO. Poisson's ratio is defined as the ratio of the unit strain in a material in a direction normal to an applied stress to the unit strain parallel to the applied stress. It is an inherent property of elastic materials and has an important influence on the relative volumes of a material in the unstressed and stressed states. The normal range of values of Poisson's ratio for all kinds of engineering materials is from 0 to 0.5. When a vertical cylindrical specimen of material for which Poisson's ratio is zero is loaded axially, the cylinder will undergo strain in the direction of the load, but its dimensions in the lateral direction will not change. Therefore, the volume of the cylinder is reduced in the stressed state. Cork is a common material for which Poisson's ratio is approximately equal to zero. In contrast, if the material of a vertical cylinder has a finite value of Poisson's ratio, it will undergo strain in a lateral direction as well as the vertical direction under an axial load. If the value of this ratio is approximately the maximum of 0.5, the increase in volume due to lateral expansion will be just equal to the decrease in volume due to vertical compression. In other words, the volume of the cylinder remains constant when it is loaded. A material for which Poisson's ratio is 0.5 is called an incompressible material because of this constant-volume characteristic.

Poisson's ratio for a soil is a highly tenuous property and one which it is very difficult to determine. The few experimenters who have tried to measure this property have obtained widely scattered results; and, in general, they have not been satisfied with their determinations. However, as a broad generalization, it can be said that the order of magnitude of Poisson's ratio for soil is closer to the upper limit of 0.5 than it is to the lower limit of zero. Furthermore, examination of Eqs. (17-2) and (17-3) indicates that, if $\nu = 0.5$, these equations are very much simplified. Therefore, Eqs. (17-2) and (17-3) are frequently written in the simplified form

$$\sigma_x = \frac{3P}{2\pi} \frac{x^2 z}{R^5} \tag{17-4}$$

$$\sigma_y = \frac{3P}{2\pi} \frac{y^2 z}{R^5} \tag{17-5}$$

17.4. ELASTIC CRITERIA. In the theory of elasticity, there are three fundamental criteria or conditions which must be complied with before the Boussinesq solution can be considered applicable. First, the stresses must be in equilibrium; this means that the total vertical pressure on any horizontal plane in the under-soil must equal the applied load at the surface. Second, the strains in the material must be compatible; that is, there must not be any separation or cracks in the material. Third, the boundary conditions of the actual case must be reasonably close to those assumed in the mathematical case.

The first of these criteria is always complied with and can be considered axiomatic in any stress-distribution problem. Strictly speaking, the second or strain-compatibility criterion is not complied with in the case of a soil mass. In the Boussinesq stress distribution there is a shallow zone, near the surface and adjacent to the load, where the material is stressed in tension; whereas, in all other regions the material is in compression. Since soil is not elastic and not isotropic and since it has little or no strength in tension, the theoretical stress situation is modified by these facts. However, the difference between actual and theoretical stress distribution due to lack of strain compatibility is appreciable only at relatively shallow depths below the soil surface.

The third or boundary condition criterion may not be complied with where the soil mass is of limited extent in one or more directions and is in contact with a surface that is highly strain-resistant, such as ledge rock at a fairly shallow depth or a relatively unyielding retaining wall. A soil mass confined in a large box or bin is another example where the boundary conditions may not be compatible with the Boussinesq assumptions.

17.5. LAYERED SOIL MASSES. Another situation sometimes encountered in practice which may depart from the ideal conditions represented by the Boussinesq equations is the case where the soil mass consists of relatively shallow layers of material which vary widely in stiffness. An example of this condition is a flexible-type pavement resting on a soil subgrade. Here the stiffness of the various elements of the pavement—wearing surface, base course, and subbase—is usually very great in comparison with that of the subgrade, and the combined soil mass through which wheel loads are transmitted and distributed is definitely nonhomogeneous and nonisotropic.

Theoretical analyses of this problem were developed by Burmister (*1*) in 1943 and some experimental studies of wheel-load stresses transmitted through stabilized gravel and sand-clay flexible pavements to soil subgrades were reported by Spangler and Ustrud (*11*) in 1940.

17.6. MODIFIED BOUSSINESQ EQUATIONS. The distribution of vertical stresses on a horizontal plane in the under-soil due to a concentrated surface load, when the soil mass or its boundary conditions do not comply with the conditions assumed in the Boussinesq solution, is indicated by a general equation, as follows:

$$\sigma_z = \frac{(n-2)P}{2\pi}\frac{z^{n-2}}{R^n} \tag{17-6}$$

or

$$\sigma_z = \frac{(n-2)P}{2\pi}\frac{\cos^n\theta}{z^2} \tag{17-6a}$$

in which n is a parameter which is sometimes called the "concentration factor" or "dispersion factor." These modified equations were developed by Griffith (*4*) in 1929 and by Froelich (*3*) in 1932.

Stresses indicated by Eq. (17-6) or Eq. (17-6a) are in equilibrium with the applied load, but their distribution depends on the value of n. It may be noted that, when $n = 5$, Eqs. (17-6) and (17-6a) are identical with Eqs. (17-1) and (17-1a). Values of the concentration factor for specific cases must be determined by experimental evidence or from experience with problems of a similar nature.

17.7. DISTRIBUTION OF PRESSURE. The distribution of the vertical pressure on any horizontal plane in the soil beneath a concentrated load is represented by a bell-shaped surface. The maximum ordinate of this stress surface is at the vertical axis directly beneath the load, and the stress decreases in all directions outward from this axis. Theoretically, the stress is zero at infinity; but for practical purposes it may be considered to reach the zero value at a relatively small finite distance. The maximum pressure ordinate is relatively high at shallow depths, and it decreases as the depth increases. In other words, the bell-shaped surface flattens out with increasing depth.

If the stress surface is plotted for each of a series of horizontal planes at various depths and if points of equal stress on the various planes are

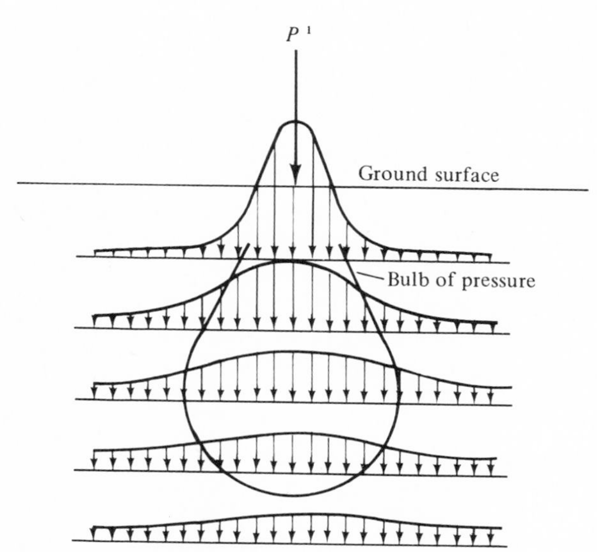

Fig. 17-2. Bulb of pressure or iso-stress surface.

connected, a surface of revolution having a bulk shape is developed. Such a surface, which is often called the bulb of pressure, is illustrated in Fig. 17-2. The pressure at each point on the pressure bulb is the same. Pressures at points inside the bulb are greater than that at a point on the bulb surface; and pressures at points outside the bulb are smaller than the pressure at points through which the bulb surface passes. Any number of bulbs of pressure may be drawn for any applied load, since a different one corresponds to each arbitrarily chosen value of pressure. Note that the σ_z pressure bulb, or isostress contour, such as in Fig. 17-2 is for vertical stresses which are not principal stresses except directly under the load axis.

17.8. TOTAL PRESSURE ON FINITE AREA IN UNDERSOIL. Problems frequently arise in engineering practice in which it is desirable to estimate the total pressure on a finite area in the undersoil due to a concentrated load applied at the surface. A typical example of this kind of problem is the computation of the total load produced on a section of an underground conduit, such as a culvert or sewer, by a truck wheel applied at the ground surface. Although an underground structure of this kind constitutes a relatively rigid inclusion in the soil mass which would appear to seriously violate the tenets of the Boussinesq theory, nevertheless, extensive experimental evidence indicates that there is reasonably close correlation between calculated and actual loads. In the case of a rectangular conduit, loads are calculated on the horizontal plane of the top of the structure. For a circular or arch-shaped conduit, experimental evidence indicates that it is valid to calculate loads on the projection of the structure onto the horizontal plane through its top.

If it is desired to determine the total load on a rectangular area due to a concentrated load applied to the surface directly over the center of the area, the Boussinesq formula for load, i.e., Eq. (17-1), may be integrated over one quadrant of the area in question. Multiplying this quadrant load by 4 gives the total load on the area. Holl (5) completed this integration for the load on a rectangular area with dimensions A and B and having one corner directly below the origin of coordinates and at a vertical distance H from the origin, as illustrated in Fig. 17-3. He obtained the following formula:

$$\frac{\sum \sigma_z}{P} = 0.25 - \frac{1}{2\pi}\left[\left(\sin^{-1} H \sqrt{\frac{A^2 + B^2 + H^2}{(A^2 + H^2)(B^2 + H^2)}}\right) - \frac{ABH}{\sqrt{(A^2 + B^2 + H^2)}}\left(\frac{1}{A^2 + H^2} + \frac{1}{B^2 + H^2}\right)\right] \qquad (17\text{-}7)$$

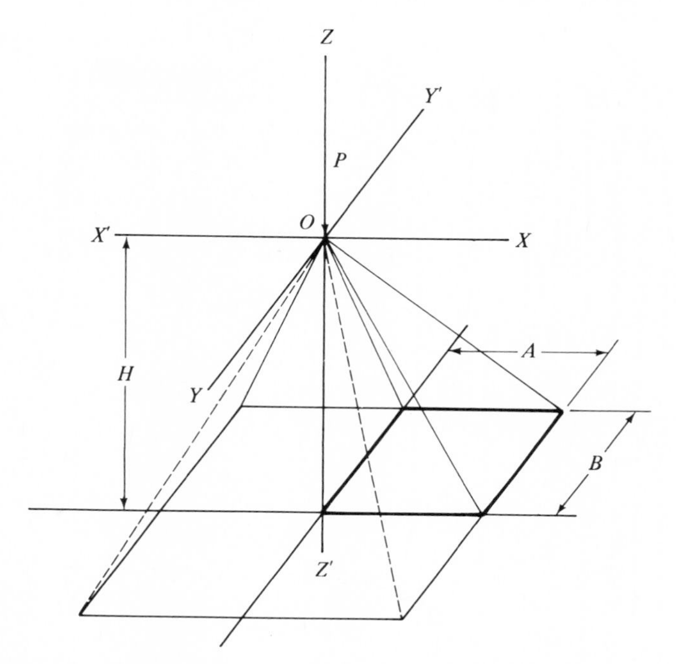

Fig. 17-3. Position of area in Eq. (17-7).

in which $\sum \sigma_z$ is the total load on rectangular area A by B at depth H, in pounds; and P is the point load applied at origin, in pounds.

The value of the right-hand member in Eq. (17-7) for known values of A, B, and H is called the influence coefficient. Determination of this coefficient is made easy by the use of the diagram in Fig. 17-4. The curved lines in that diagram indicate the values of the coefficient. To use the chart it is only necessary to locate the point having the known coordinates $m = A/H$ and $n = B/H$ and to determine the corresponding coefficient from the position of that point with respect to the curves. Multiplying this coefficient by the applied load P gives the total load on the area having the dimensions A and B; and multiplying this product by 4 gives the load on the total area which has the dimensions $2A$ and $2B$ and is symmetrical with respect to the applied load.

Values of the influence coefficients are given also in Table 17-1, which may be used if more precise results are desired.

EXAMPLE 17-1. Determine the total load on an area 4 ft wide and 6 ft long at a depth of 4 ft, due to a 10,000-lb wheel load at the surface directly above the center of the area.

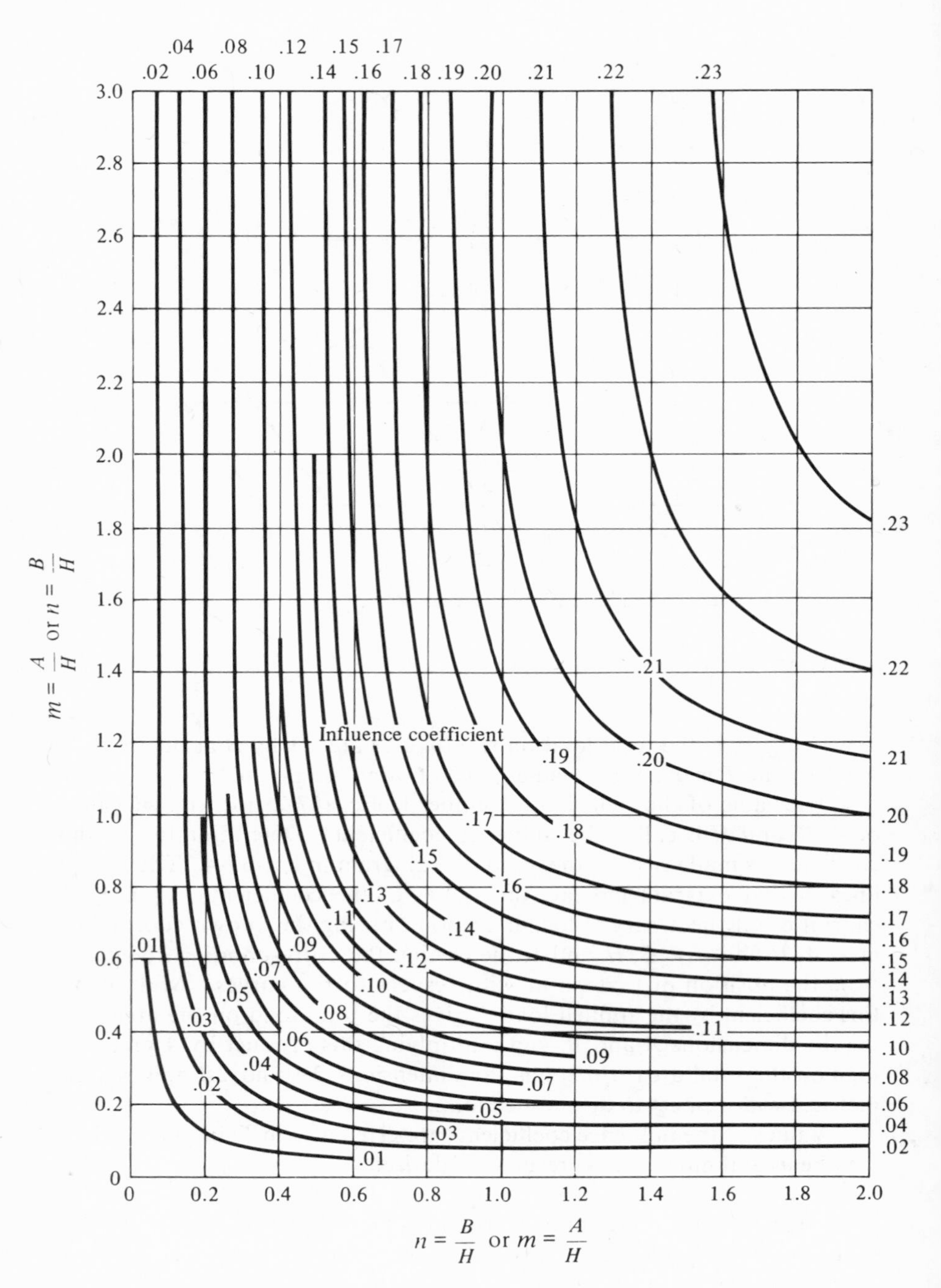

Fig. 17-4. Chart for Eqs. (17-7) and (17-8).

TABLE 17-1. Influence Coefficients for Rectangular Areas[a]

$m = A/H$ or $n = B/H$	$n = B/H$ or $m = A/H$								
	0.1	*0.2*	*0.3*	*0.4*	*0.5*	*0.6*	*0.7*	*0.8*	*0.9*
0.1	0.005	0.009	0.013	0.017	0.020	0.022	0.024	0.026	0.027
0.2	0.009	0.018	0.026	0.033	0.039	0.043	0.047	0.050	0.053
0.3	0.013	0.026	0.037	0.047	0.056	0.063	0.069	0.073	0.077
0.4	0.017	0.033	0.047	0.060	0.071	0.080	0.087	0.093	0.098
0.5	0.020	0.039	0.056	0.071	0.084	0.095	0.103	0.110	0.116
0.6	0.022	0.043	0.063	0.080	0.095	0.107	0.117	0.125	0.131
0.7	0.024	0.047	0.069	0.087	0.103	0.117	0.128	0.137	0.144
0.8	0.026	0.050	0.073	0.093	0.110	0.125	0.137	0.146	0.154
0.9	0.027	0.053	0.077	0.098	0.116	0.131	0.144	0.154	0.162
1.0	0.028	0.055	0.079	0.101	0.120	0.136	0.149	0.160	0.168
1.2	0.029	0.057	0.083	0.106	0.126	0.143	0.157	0.168	0.178
1.5	0.030	0.059	0.086	0.110	0.131	0.149	0.164	0.176	0.186
2.0	0.031	0.061	0.089	0.113	0.135	0.153	0.169	0.181	0.192
2.5	0.031	0.062	0.090	0.115	0.137	0.155	0.170	0.183	0.194
3.0	0.032	0.062	0.090	0.115	0.137	0.156	0.171	0.184	0.195
5.0	0.032	0.062	0.090	0.115	0.137	0.156	0.172	0.185	0.196
10.0	0.032	0.062	0.090	0.115	0.137	0.156	0.172	0.185	0.196
∞	0.032	0.062	0.090	0.115	0.137	0.156	0.172	0.185	0.196

	1.0	*1.2*	*1.5*	*2.0*	*2.5*	*3.0*	*5.0*	*10.0*	∞
0.1	0.028	0.029	0.030	0.031	0.031	0.032	0.032	0.032	0.032
0.2	0.055	0.057	0.059	0.061	0.062	0.062	0.062	0.062	0.062
0.3	0.079	0.083	0.086	0.089	0.090	0.090	0.090	0.090	0.090
0.4	0.101	0.106	0.110	0.113	0.115	0.115	0.115	0.115	0.115
0.5	0.120	0.126	0.131	0.135	0.137	0.137	0.137	0.137	0.137
0.6	0.136	0.143	0.149	0.153	0.155	0.156	0.156	0.156	0.156
0.7	0.149	0.157	0.164	0.169	0.170	0.171	0.172	0.172	0.172
0.8	0.160	0.168	0.176	0.181	0.183	0.184	0.185	0.185	0.185
0.9	0.168	0.178	0.186	0.192	0.194	0.195	0.196	0.196	0.196
1.0	0.175	0.185	0.193	0.200	0.202	0.203	0.204	0.205	0.205
1.2	0.185	0.196	0.205	0.212	0.215	0.216	0.217	0.218	0.218
1.5	0.193	0.205	0.215	0.223	0.226	0.228	0.229	0.230	0.230
2.0	0.200	0.212	0.223	0.232	0.236	0.238	0.239	0.240	0.240
2.5	0.202	0.215	0.226	0.236	0.240	0.242	0.244	0.244	0.244
3.0	0.203	0.216	0.228	0.238	0.242	0.244	0.246	0.247	0.247
5.0	0.204	0.217	0.229	0.239	0.244	0.246	0.249	0.249	0.249
10.0	0.205	0.218	0.230	0.240	0.244	0.247	0.249	0.250	0.250
∞	0.205	0.218	0.230	0.240	0.244	0.247	0.249	0.250	0.250

[a] After Newmark (*9*).

SOLUTION. First divide the area into four quadrants such that the common corner of each rectangular quadrant is directly under the load, as indicated in Fig. 17-3. Since $A = 2$, $B = 3$, and $H = 4$, then $m = 0.5$ and $n = 0.75$. From Table 17-1 the value of the coefficient corresponding to the point having these coordinates is 0.107. The total load on the area under consideration is, therefore,

$$0.107 \times 10{,}000 \times 4 = 4280 \text{ lb}$$

17.9. UNIT PRESSURE AT POINT IN UNDERSOIL. Another problem of frequent occurrence is that of determining the unit pressure in the soil at a point some depth below a uniformly distributed load applied over a rectangular area at or near the soil surface. An example of this type of problem is the calculation of the pressures at various points in the undersoil for the purpose of estimating the probable settlement of a structure such as a bridge pier or building foundation resting on the soil. Newmark (*9*) solved this problem by integrating the Boussinesq formula for the case of loads uniformly distributed over a rectangular area. The resulting formula for the unit pressure at a point in the undersoil at depth H directly under one corner of a loaded rectangle of dimensions A and B, as indicated in Fig. 17-5, is

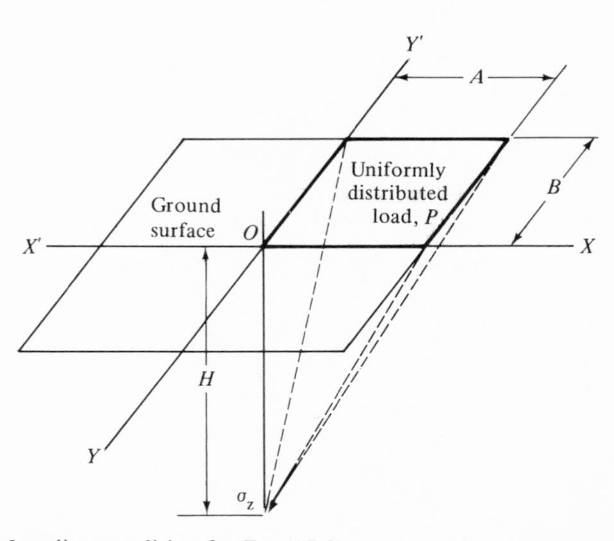

Fig. 17-5. Loading condition for Eq. (17-8).

$$\frac{\sigma_z}{p} = \frac{1}{4\pi}\left[\frac{2ABH\sqrt{(A^2 + B^2 + H^2)}}{H^2(A^2 + B^2 + H^2) + A^2B^2} \cdot \frac{A^2 + B^2 + 2H^2}{A^2 + B^2 + H^2} + \left(\sin^{-1}\frac{2ABH\sqrt{A^2 + B^2 + H^2}}{H^2(A^2 + B^2 + H^2) + A^2B^2}\right)\right] \qquad (17\text{-}8)$$

in which σ_z is the unit pressure at depth H; and p is the unit load applied over rectangular area A by B.

Again, the value of the right-hand member in Eq. (17-8) for known values of A, B, and H, as in Eq. (17-7), is called the influence coefficient.

Although Holl's problem and Newmark's problem were directed toward different objectives, they both involved the integration of the Boussinesq equation over a rectangular area and the right-hand member of Eq. (17-8) has the same value as that of Equation (17-7). Therefore, the solution of Eq. (17-8) can also be obtained by the use of the diagram in Fig. 17-4 or Table 17-1 in a manner similar to that outlined in Section 17.8.

EXAMPLE 17-2. Calculate the unit pressure at a depth of 20 ft under the center of a 20 ft by 40 ft rectangular area which carries a load of 1600 tons, or 2 tons per sq ft.

SOLUTION. The applied unit pressure is 2 tons per sq ft. With the origin of coordinates directly over the center and the area divided into four quadrants, the values of A, B, and H are 10 ft, 20 ft, and 20 ft, respectively. Then, $m = 0.5$ and $n = 1.0$. From Table 17-1 the influence coefficient corresponding to these values is 0.120. For an applied unit pressure of 2 tons per sq ft, the value of σ_z is 0.24 ton per sq ft. This is the unit pressure in the undersoil at a depth of 20 ft under a corner of one quadrant of the loaded area. Since all four quadrants have a common corner at the center, the unit pressure due to the total area load is $4 \times 0.24 = 0.96$ ton per sq ft.

17.10. PRESSURE UNDER ECCENTRIC LOADING. Equation (17-7) and Fig. 17-4 or Table 17-1 may also be used to solve problems in which the concentrated surface load is not over the center of the area on which the total load is desired, as illustrated in Fig. 17-6. The general procedure in a situation of this kind is to extend the area in question in such a manner as to form a series of rectangles having a common corner directly under the load. Then determine the influence coefficient for the largest rectangle which includes the original area; and subtract the influence coefficients for the smaller rectangles lying outside the original area. The result is the influence coefficient for the desired area. For example, in Fig. 17-6 let the

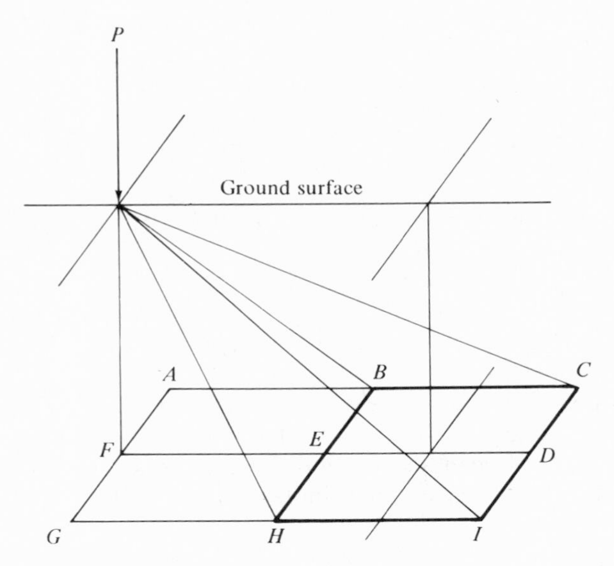

Fig. 17-6. Eccentric load on area.

rectangle *BCIH* be the area on which the total load due to the surface load *P* is desired. In accordance with the procedure just outlined, extend this area to form rectangles *ACDF* and *DFGI*. Since these rectangles have a common corner under the load, the influence coefficient for them may be determined. Next, subtract the influence coefficient for the rectangles *ABEF* and *FEHG*. Multiplying the resulting coefficient by the load *P* gives the total load on the area *BCIH*.

In a similar manner, Eq. (17-8) and Fig. 17-4 or Table 17-1 may be used to determine the unit pressures at points in the soil which are not under the center of the area over which a uniform load is applied. This type of problem is illustrated in Fig. 17-7. The influence coefficient for the unit pressure under point *O* due to the uniform load on the area *BCFE* may be obtained from the coefficients for various rectangles, as follows:

$$\text{Rectangle } BCFE = ACIO - ABHO - DFIO + DEHO$$

Note that in subtracting the coefficients for the rectangles *ABHO* and *DFIO*, the influence of the small rectangle *DEHO* is subtracted twice. Therefore, it is necessary to add the coefficient for this small rectangle to obtain the final result.

There are six situations involving the position of an applied concentrated load relative to a rectangular load receiving area in the undersoil, or the position of a point beneath a uniformly distributed load at which it

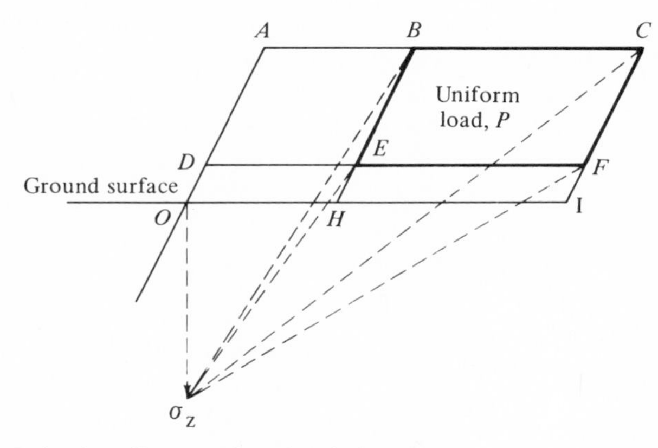

Fig. 17-7. Point in soil eccentric to loaded area.

is desired to determine the unit stress. These problems are solved by applying Eq. (17-7) or (17-8) to a series of rectangles and algebraically combining the results to obtain the net result for the area in question, as indicated above. The six situations and the rectangles to be combined are shown in Fig. 17-8.

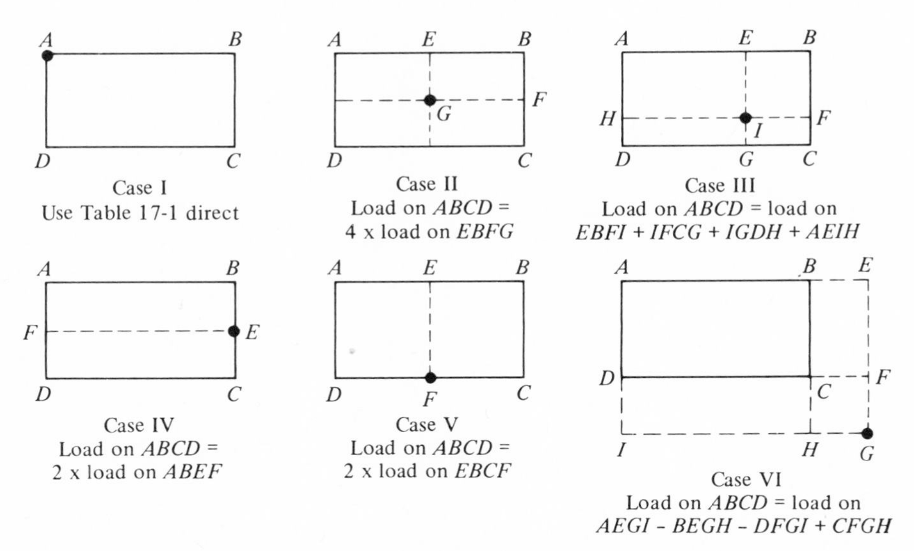

Fig. 17-8. Relationship between area and position of load or position of unit pressure in application of Eqs. (17-7) and (17-8).

17.11. NEWMARK INFLUENCE CHART. When the pressure at a point in the undersoil due to the load on an irregularly shaped footing or a group of spread footings is desired, it may be advantageous to use an influence chart, which is a graphical representation of Newmark's integration of the Boussinesq equation. Such a chart drawn for an influence value of 0.005 is shown in Fig. 17-9.

To use the chart, draw an outline of the footing or footings on a piece of transparent paper to such a scale that the distance *XY* on the chart

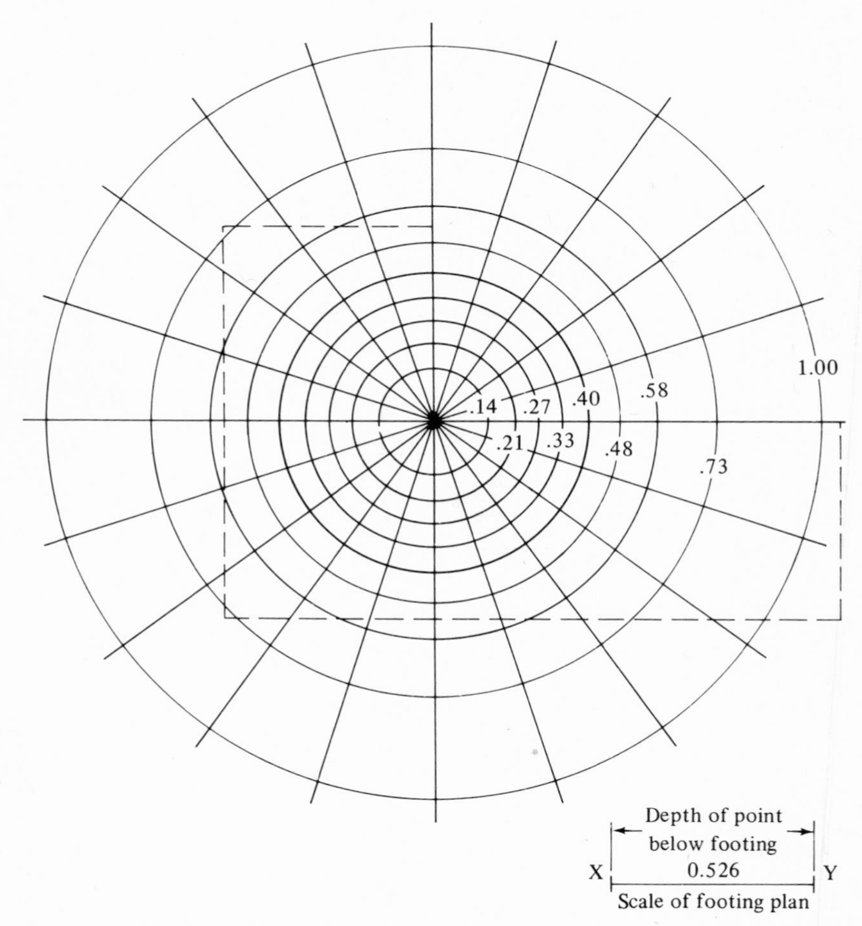

Fig. 17-9. Newmark influence chart for computation of vertical pressure in undersoil (influence value is 0.005).

equals the depth from the base of the footing to the point in question. The point on the footing plan beneath which the pressure is desired is then placed over the center of the chart. In this position, each rectangular area bounded by arcs of adjacent circles and radial lines represents a pressure equal to the influence coefficient for the chart, as 0.005 for the chart in Fig. 17-9, and this value multiplied by the unit pressure at the base of the footing gives the unit pressure at the point in question. Therefore, it is only necessary to count the rectangular areas embraced by the footing plan and multiply this number by the pressure corresponding to one area to determine the total unit pressure at the point caused by the applied load. If the pressure at a point at a different depth is desired, it is necessary to redraw the footing plan to a scale such that the distance XY on the chart is equal to this different depth and proceed as before.

The numbers on the various circles of the chart in Fig. 17-9 indicate the relative values of the radii of the circles and are given to enable anyone to redraw the chart to a larger scale if desired. The scale distance XY must also be increased in the same ratio as the radii of the circles.

EXAMPLE 17-3. It is required to determine the soil pressure at a depth of 35 ft beneath the interior corner A of an L-shaped building shown below. The load exerted by the structure is 1500 psf.

SOLUTION. Draw an outline of the building to such scale that the distance X–Y in Fig. 17-9 is equal to 35 ft. Then superimpose the outline on the Newmark chart with the point A directly over the center of the chart. The number of "squares" covered by the building outline is 110.

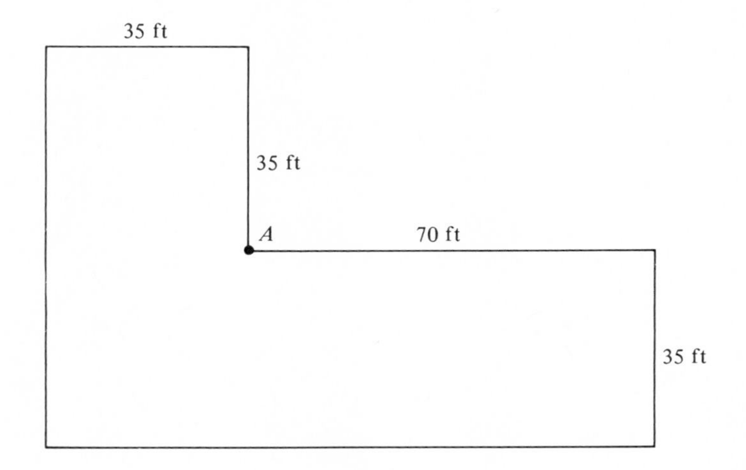

This number multiplied by the influence value of 0.005, and by the foundation pressure of 1500 psf, gives 825 psf, which is the desired pressure.

NOTE. It is suggested that the student also solve this problem by the method in Section 17-9, using the chart in Fig. 17-4 or Table 17-1.

17.12. PRESSURE DUE TO UNIFORM LOAD ON CIRCULAR AREA. Influence coefficients for determining unit pressures at various depths in the soil beneath a uniformly loaded circular area of radius a are given in Table 17-2. In this table, H is the depth of the point and r is the radial horizontal distance to the point from the center of the circle. This problem may arise in connection with settlement studies of structures on circular foundations, such as gasoline tanks, grain elevators, and storage bins. As in the case of a rectangular footing area, the influence coefficient obtained from Table 17-2 should be multiplied by the uniform load on the circular area to obtain the unit pressure in the undersoil.

EXAMPLE 17-4. Compute the unit pressures at points o, m, and n in Fig. 17-10 at a depth of 25 ft directly under points O, M, and N due to a footing load of 1500 psf applied uniformly over a circular area 50 ft in diameter.

SOLUTION. Since a = 25 ft and H = 25 ft, the ratio H/a = 1.0. The horizontal distance from the center of the loaded area to the point

TABLE 17-2. Influence Coefficients for Points under Uniformly Loaded Circular Area

	r/a									
H/a (1)	0 (2)	0.25 (3)	0.50 (4)	1.0 (5)	1.5 (6)	2.0 (7)	2.5 (8)	3.0 (9)	3.5 (10)	4.0 (11)
0.25	0.986	0.983	0.964	0.460	0.015	0.002	0.000	0.000	0.000	0.000
0.50	.911	.895	.840	.418	.060	.010	.003	.000	.000	.000
0.75	.784	.762	.691	.374	.105	.025	.010	.002	.000	.000
1.00	.646	.625	.560	.335	.125	.043	.016	.007	.003	.000
1.25	.524	.508	.455	.295	.135	.057	.023	.010	.005	.001
1.50	.424	.413	.374	.256	.137	.064	.029	.013	.007	.002
1.75	.346	.336	.309	.223	.135	.071	.037	.018	.009	.004
2.00	.284	.277	.258	.194	.127	.073	.041	.022	.012	.006
2.5	.200	.196	.186	.150	.109	.073	.044	.028	.017	.011
3.0	.146	.143	.137	.117	.091	.066	.045	.031	.022	.015
4.0	.087	.086	.083	.076	.061	.052	.041	.031	.024	.018
5.0	.057	.057	.056	.052	.045	.039	.033	.027	.022	.018
7.0	.030	.030	.029	.028	.026	.024	.021	.019	.016	.015
10.0	.015	.015	.014	.014	.013	.013	.013	.012	.012	.011

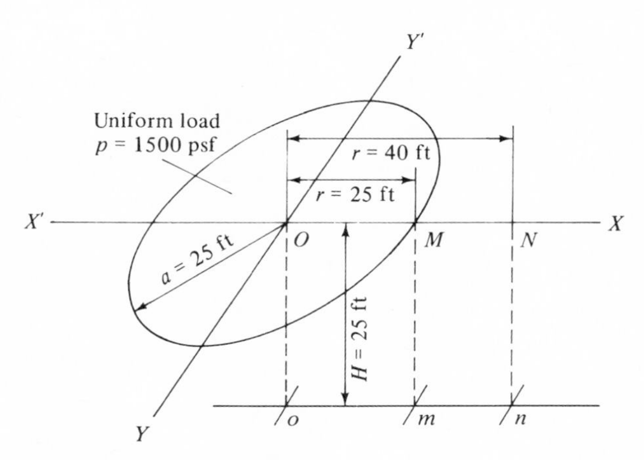

Fig. 17-10. Conditions in Example 17-4.

O is 0. Therefore, the ratio $r/a = 0$ for point O, and the influence coefficient is found from Table 17-2 in column 2 opposite $H/a = 1.0$ in column 1. It is equal to 0.646. The required unit pressure is $1500 \times 0.646 = 969$ psf.

For a point below M, the ratio

$$\frac{r}{a} = \frac{25}{25} = 1.0$$

and the coefficient, which is found in column 5, is 0.335. The corresponding unit pressure is $1500 \times 0.335 = 502$ psf.

For a point below N,

$$\frac{r}{a} = \frac{40}{25} = 1.6$$

The desired coefficient, found by interpolation between columns 6 and 7, is 0.109. Therefore, the pressure at n is $1500 \times 0.109 = 163$ psf.

17.13. APPROXIMATE ESTIMATE OF STRESS DISTRIBUTION. Not infrequently, certain approximate methods of estimating vertical pressures in the undersoil are sufficiently accurate for a particular purpose, such as a preliminary estimate of settlement. One such method is to assume that the applied load spreads laterally and downward in such a manner that the stress on a subsurface horizontal plane is uniform over an area which is defined by planes descending from the edges of the applied load on an angle of 30 deg with the vertical, as indicated in Fig. 17-11.

Another method is to consider each dimension of the stressed area

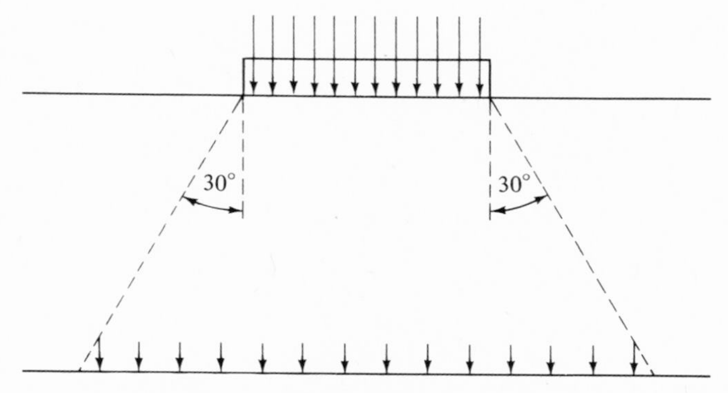

Fig. 17-11. Approximate stress distribution on a subsurface plane.

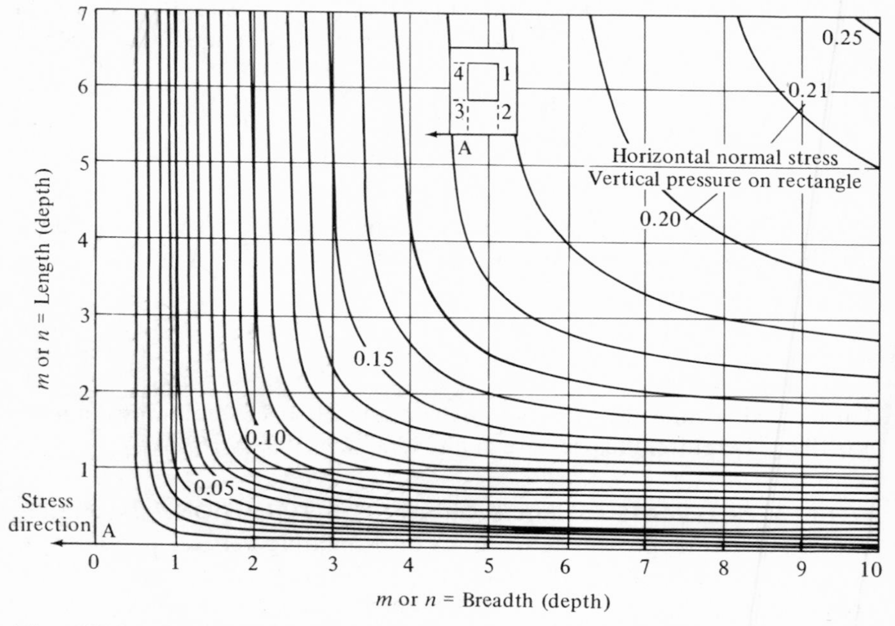

Fig. 17-12. Graph of horizontal normal stress under corner of rectangle loaded with unit pressure. From Barber (*13*). The stress determined is below point *A* when *A* is not below corner of rectangle. Locate rectangle to scale; add stress ratios at 1 and 3, and subtract stress ratios at 2 and 4.

to be larger than the corresponding dimension of the loaded area by an amount equal to the depth of the subsurface area. Thus, if a load is applied on a rectangle with dimensions A and B, the stress on the soil at depth H is considered to be uniformly distributed on an area with dimensions $(A + H)$ and $(B + H)$.

17.14. LATERAL STRESSES UNDER VERTICAL APPLIED PRESSURE. Although most settlement problems in soil engineering deal with vertical stress, the lateral stress can be important in settlement and in retaining wall problems. Lateral stresses from a point surface load are indicated by Eqs. (17-2) through (17-5). As in the case of the vertical stress formula, the equations for lateral stresses have been integrated for line and strip loads and for area loads, by Holl (*5*) and by Mickle (*8*), respectively. These equations are presented in Section 22.18 below, in connection with stresses on retaining walls.

Figure 17-12 shows a chart prepared by Barber (*13*) for horizontal stresses from an area surface load on a semi-infinite mass with Poisson's ratio equal to 0.5. Note that the isoinfluence lines are not symmetrical with respect to the X and Y axes, the stress direction being to the left, indicated by the small arrow.

EXAMPLE 17-5. Calculate the horizontal unit pressure in the X and Y directions under the center of the loaded area as in Example 17-2.

SOLUTION. As in Example 17-2, the area is divided into four quadrants with $m = 0.5$ and $n = 1.0$ where m corresponds to the short side. From Fig. 17-12 the influence coefficients corresponding to these values are 0.01 and 0.025 for planes parallel to the long and short sides, respectively. For a surcharge load of 2 tons per sq ft., $\sigma_x = 0.02$ and $\sigma_y =$ 0.05 tons per sq ft. Since all four quadrants have a common corner at the center, the unit pressures from the total area load are $\sigma_x = 0.08$ and $\sigma_y = 0.2$ tons per sq ft on planes parallel to the long and short sides, respectively.

PROBLEMS

17.1. For a concentrated surface load of 10,000 lb, compute and plot the vertical stresses along a diametral line in a horizontal plane in the undersoil at each of the following depths: 2, 4, 6, and 10 ft.

17.2. A concentrated load of 15,000 lb is applied at the soil surface. Compute the unit vertical pressure at the following points in the undersoil.

Point	*Coordinates (ft)*		
	x	y	z
A	4	6	3
B	10	0	5
C	0	4	10
D	3	3	6
E	7	7	7
F	0	0	6

17.3. Compute and plot the vertical stresses at several points along the vertical axis beneath a concentrated load of 1000 lb applied at the surface.

17.4. Compute the horizontal stress in the direction parallel to the x-axis, due to a concentrated surface load of 8000 lb, at a point whose coordinates are $x = 4$ ft, $y = 5$ ft, and $z = 6$ ft. Assume Poisson's ratio of the soil to be 0, 0.25, and 0.5.

17.5. Using concentration factors of $n = 4$ and $n = 6$, compute and plot the vertical stresses along a diametral line in a horizontal plane in the undersoil at a depth of 5 ft due to a concentrated surface load of 4000 lb.

17.6. Determine the total load on an area 6 ft by 8 ft whose center is 4 ft below a truck-wheel load of 10,000 lb applied at the ground surface.

17.7. Using the data of Problem 17.6, determine the total load on the area if the wheel load is directly above the center of the 8-ft side.

17.8. Determine the total load on an area 2 ft by 3 ft at a depth of 4 ft, due to an axle load of 18,000 lb applied at the soil surface. One wheel of the axle is directly over the center of the area. The wheels are spaced 6 ft, center to center, and the axle is oriented parallel to the 3-ft sides of the area.

17.9. Determine the unit pressure at a point 20 ft beneath the center of an area, 20 ft by 40 ft, on which a load of 2400 tons is uniformly distributed.

17.10. Using the data of Problem 17.9, determine the unit pressure at a point beneath one corner of the area.

17.11. Using the data of Problem 17.9, determine the unit pressure at a point beneath a point on the surface located outside the loaded area at a distance 10 ft each way from one corner.

17.12. A system of spread footings for a building is shown in Fig. 17-13. All footings exert a pressure of 6 kips per sq ft on the underlying soil. Using Newmark's chart in Fig. 17-9, estimate the pressure at a depth of 16 ft beneath footings A-1, A-3, and B-3.

Footing No.	*Size (ft)*
A-1, A-5, D-1, D-5	5 × 5
A-2, A-3, A-4, B-1, B-5, C-1, C-5, D-2, D-3, D-4	6 × 6
B-2, B-3, B-4, C-2, C-3, C-4	8 × 8

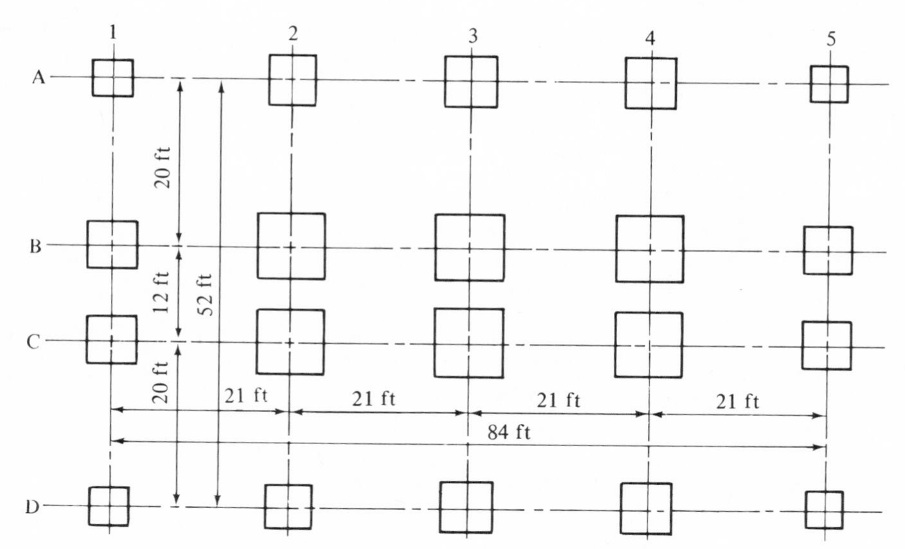

Fig. 17-13. Problem 17-12.

17.13. A pressure of 5000 psf is applied over a circular area 20 ft in diameter. Determine the unit pressure at a depth of 30 ft in the undersoil (a) beneath the center of the area, (b) beneath the perimeter of the loaded area, and (c) beneath a point 10 ft beyond the loaded area.

REFERENCES

1. Burmister, Donald M. "The Theory of Stresses in Layered Systems and Applications to Design of Airport Runways." *Highway Research Board Proc.* **23,** 126 (1943).
2. Cummings, A. E. "Distribution of Stresses under a Foundation." *Trans. Amer. Soc. C.E.* (1936).
3. Froelich, O. K. "Drukverdeeling in Bouwground." *Der Ingenieur* (April 15, 1932).
4. Griffith, John H. "The Pressures under Substructures." *Engineering and Contracting* **68,** 113–119 (1929).
5. Holl, D. L. "Plane-Strain Distribution of Stress in Elastic Media." *Iowa Eng. Exp. Sta. Bull.* Iowa State College, Ames, Iowa **148,** 55 (1941).
6. Jurgenson, L. "The Application of Theories of Elasticity and Plasticity to Foundation Problems." Boston Soc. Civ. Eng., Contr. to Soil Mech. 1925–1940, 148–183.
7. Marston, Anson. "The Theory of External Loads on Closed Conduits in the Light of the Latest Experiments." Bull. **96,** Iowa Engineering Experiment Station, Ames, Iowa, 1930.

8. Mickle, J. L. "Lateral Pressures on Retaining Walls Caused by Uniformly Distributed Surface Loads." Unpublished M.S. Thesis. Iowa State Univ. Libr., Ames, Iowa, 1955.
9. Newmark, Nathan M. "Simplified Computation of Vertical Pressures in Elastic Foundations." Circular No. 24, Engineering Experiment Station, University of Illinois, 1935.
10. Scott, R. F. *Principles of Soil Mechanics.* Addison-Wesley, Reading, Massachusetts, 1963.
11. Spangler, M. G., and H. O. Ustrud. "Wheel Load Stress Distribution through Flexible Type Pavements." *Highway Research Board Proc.* **20,** 235 (1940). (See also Vols. 21 and 22.)
12. Spangler, M. G., and Richard L. Hennessy. "A Method of Computing Live Loads Transmitted to Underground Conduits." *Highway Research Board Proc.* **26,** (1946).
13. Spangler, M. G., and Mickle, J. L. "Lateral Pressures on Retaining Walls due to Backfill Surface Loads." *Highway Research Board Bull.* **141,** 1–15 (1956). Discussion by E. S. Barber, follows on pp. 15–18.

18

Consolidation and Settlement of Structures

18.1. DEFORMATION OF MATERIALS. Every material undergoes a certain amount of strain when a stress is applied. A steel rod lengthens when it is subjected to tensile stress, and a concrete column shortens when a compressive load is applied to it. If a material is truly elastic and obeys Hooke's law, each equal increment of applied stress causes a proportionate increase in strain; and, when the stress is removed, the body regains its initial dimensions. A material may be partially elastic in the sense that equal increments of stress cause equal increments of strain but, when the stress is removed, the body does not regain its initial dimensions. In other words, a part of the strain caused by the applied load is not recoverable, and the volume of the body has been permanently changed. Some materials undergo strains which are not proportional to applied stress increments, and they do not recover all or even a major proportion of the strain when the load or a portion of the load is removed. As a generalization it may be said that soil falls in the last category. This is especially true in the case of a foundation soil beneath a heavy structure where the greatest bulk of the load, which is the weight of the structure itself, is applied only once.

18.2. SETTLEMENT OF STRUCTURE ON SOIL. Probably 90% or more of the structures which engineers build rest upon soil, and the load which a structure imposes causes the soil to undergo compressive strains which are responsible for settlement of the structure. It is necessary, therefore, to study the stress-strain characteristics of the foundation soil in order to understand the settlement behavior of a structure and to predict and make provision for the settlements which may occur during its life. Settlement is not necessarily an adverse characteristic of a structure, provided it is uniform throughout and does not reach excessive proportions. But, if the settlements are unequal, that is, if one corner or one end goes down more than the rest of the structure, serious consequences may result.

If the structure is a building, unequal settlements may cause plaster to crack badly, door and window frames to bind, brick and masonry work to loosen, and floors to crack and fault. As a result there will be a general increase in maintenance costs and rapid deterioration of the value of the building. In an extreme case, unequal settlements may impair the structural integrity of the framework and cause the building to be condemned. If the structure is a tall smokestack, monument, or church spire, unequal settlements of the foundation may cause the structure to lean in an unsightly manner. In an extreme case, it may lean far enough to become dangerously unstable. Unequal settlements of the piers and abutments of a bridge may produce vertical misalignment of the grade line, thus hindering the flow of traffic over the bridge; or, in the case of a continuous structure, excessive settlement of one pier in relation to others may cause serious overstress to some members.

18.3. CAUSES OF SETTLEMENT. Settlement of a structure resting on soil may be caused by two distinct kinds of action within the foundation soil. In one case the load imposed may cause shearing stresses to develop within the soil mass which are greater than the shearing strength of the material. When this occurs, the soil fails by sliding downward and laterally, and the structure settles and perhaps tips out of vertical alignment. This action results from a distinct failure of the soil and is discussed in Chapter 23 on bearing capacity.

In the other case, a structure settles by virtue of the compressive stress and the accompanying strain which are developed in the soil as a result of the load imposed upon it. This strain is a normal phenomenon and in no sense is to be thought of as a failure of the soil, although it may cause the structure to fail if the settlements are excessive and nonuniform. The reduction in volume of a soil mass resulting from the application of a foundation load and the accompanying compressive stress and strain are called consolidation. This type of action is the subject of this chapter.

18.4. TERZAGHI'S THEORY. A theory of consolidation of soil has been proposed by Terzaghi, and it constitutes one of his greatest contributions to the science of soil mechanics. By the application of this theory, it is possible to make a reasonable estimate of the probable magnitude and time-rate of settlement of a structure resting on soil. In the development of the theory, numerous simplifying assumptions were employed which, together with the normal heterogeneity of natural soil masses, make it impossible to predict the settlement of a structure with the same degree of precision that is usual in some other types of structural problems such as the compression of a steel column in a tall building. Nevertheless, a settlement analysis is very important in the case of a large and costly structure and will provide the designer with a valuable estimate of the order of magnitude and the rate at which settlements may be expected to develop.

18.5. TWO TYPES OF COMPRESSION STRAIN. The compression strain of a foundation soil is attributable to two distinct phenomena within the soil, although the resulting strains overlap each other as the settlement develops and it is impossible to detect with certainty when one type of strain ends and the other begins. The first of these types of strain results from reduction in volume of the soil mass and is accompanied by outflow of water from the soil pores. It is generally referred to as primary consolidation. The second type of strain is the result of plastic deformation of the mass under constant load and is called secondary consolidation. The Terzaghi theory deals only with primary consolidation, which is normally the greater of the two and occurs earlier in the life of a structure. There is no rational method available at the present time for computing the secondary consolidation. It must be estimated wholly on the basis of observation, experience, and judgment. The Terzaghi theory also is based upon the assumption that the outflow of water from the soil pores occurs only in the vertical direction. It is sometimes referred to as one-dimensional consolidation.

18.6. PRIMARY CONSOLIDATION. The primary compression strain or consolidation of a foundation soil mass is the result of a decrease in the volume of the void spaces in the soil. There may be some slight compression of the soil grains at their points of contact due to the intergranular pressure and there may be a very slight amount of compression of the water in the soil voids, but both of these sources of reduction in volume are negligible. For this reason it is convenient to express the stress-strain relationship for soil in consolidation studies in terms of void ratio and unit pressure instead of unit stress and unit strain used in the case of most other engineering materials.

18.7. VOID RATIO AT ANY PRESSURE DURING COMPRESSION. An idealized void ratio-pressure diagram for soil is shown in Fig. 18-1, in which unit pressures are plotted as abscissae and void ratios are plotted

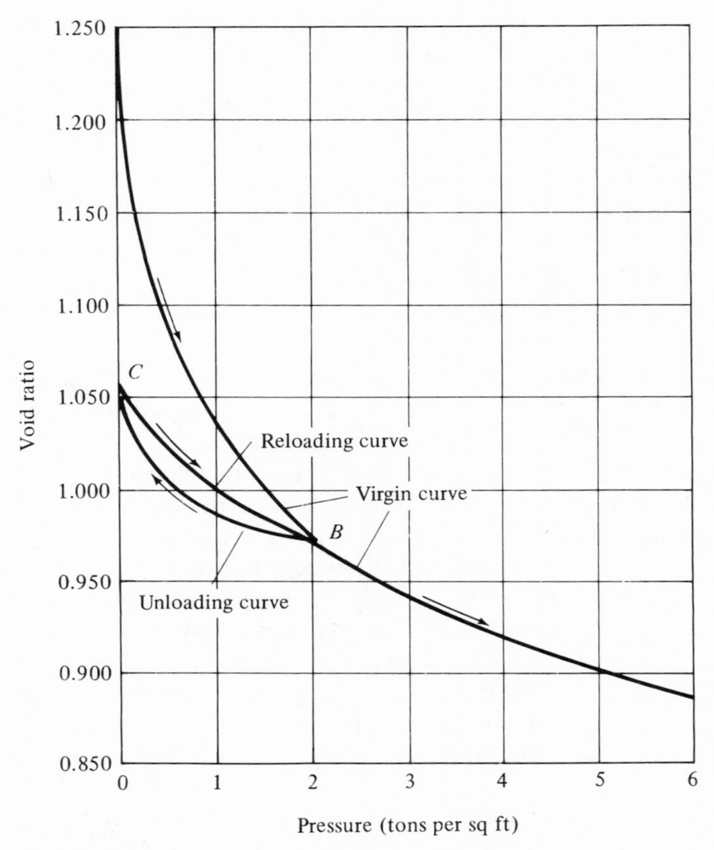

Fig. 18-1. Idealized void ratio-pressure diagram.

as ordinates. Since the volume of a soil mass decreases as pressure is increased, the void ratios are plotted in a descending scale. Some important characteristics of this curve, which indicate the general nature of the relationship between compression strains in soil and applied pressure, will be discussed. First, it will be noticed that the void ratio decreases at a relatively rapid rate as early increments of pressure are applied, but the rate of decrease diminishes as the pressure increases. The curve is roughly logarithmic in form; that is, the void ratio decreases in proportion to the logarithm of the pressure. An empirical equation for the curve can be written

$$e = e_0 - C_C \log \frac{p}{p_0} \tag{18-1}$$

in which

e = void ratio at pressure p;
p = any applied pressure;
e_0 = known void ratio at pressure p_0;
p_0 = pressure at which void ratio is known; and
C_C = compression index, which defines the relationship between applied pressure and void ratio.

Thus, if the void ratio e_0 at pressure p_0 and the compression index of a soil are known, the void ratio e at any other pressure p can be determined by Eq. (18-1). If the void ratio e_1 at unit pressure is known, Eq. (18-1) may be simplified to

$$e = e_1 - C_C \log p \tag{18-2}$$

For example, suppose that the void ratio of a soil at 1 ton/sq ft is 0.838 and the compression index of the soil is 0.049. Then, by Eq. (18-2), the void ratio at 4 tons per sq ft is

$$e = 0.838 - 0.049 \times \log 4 = 0.809$$

18.8. VOID RATIO AT ANY PRESSURE DURING EXPANSION. Now suppose that the soil represented by the void ratio-pressure diagram in Fig. 18-1 is loaded to some pressure such as indicated at point B, and the load is then removed. Since the soil is not elastic, it will not return to the same void ratio which it originally had at zero pressure before the initial loading. Rather, it will expand to some void ratio which is considerably less than the initial value and the difference between the two void ratios at zero pressure represents the permanent change in volume caused by the loading cycle from zero to point B and back to zero again. The unloading curve from B to C also approximates a logarithmic curve and may be represented by the following formula:

$$e = e_0 - C_E \log \frac{p}{p_0} \tag{18-3}$$

in which C_E is the expansion index for soil.

Equation (18-3) may be simplified to

$$e = e_1 - C_E \log p \tag{18-4}$$

in which e_1 is the void ratio at unit pressure during expansion of the soil as it is unloaded.

18.9. EFFECT OF RELOADING. If the soil is reloaded, the void ratio will again decrease. The void ratio-pressure curve for the reloading will closely follow the unloading curve, but will usually lie a little above it, as indicated in Fig. 18-1. As the reapplied load approaches the maximum value applied initially, the reloading curve reverses in curvature and continues as an extension of the initial void ratio-pressure curve.

18.10. VIRGIN CURVE. The initial curve, on either side of the point of unloading and reloading, is sometimes called the virgin curve. Since the void ratio-pressure relationship for soil is approximately logarithmic in character, the virgin curve will approximate a straight line when plotted on semilogarithmic graph paper with the pressure on the logarithmic scale, as shown in Fig. 18-2. The slope of the curve is negative on this type of plot and is equal numerically to the compression index C_C.

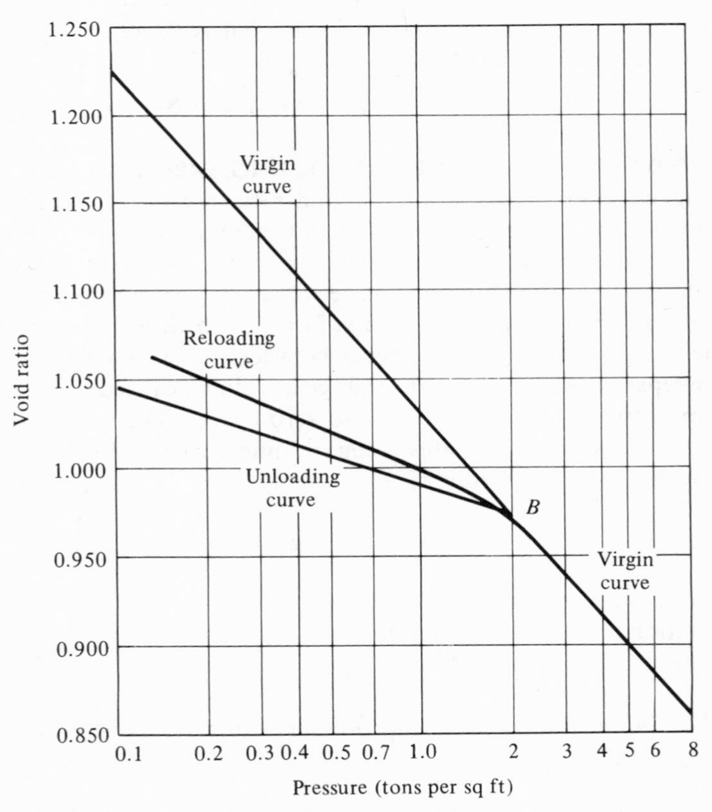

Fig. 18-2. Semilogarithmic form of idealized void ratio-pressure curve.

By transposing in Eq. (18-1), we obtain

$$C_C = \frac{e_0 - e}{\log \dfrac{p}{p_0}} \tag{18-5}$$

Also, since the logarithm of 10 is 1.0, it follows that when $p = 10p_0$

$$C_C = e_0 - e \tag{18-6}$$

Study of Eqs. (18-5) and (18-6) indicates that the compression index may be quickly determined from a void ratio–pressure diagram by subtracting the void ratio at $10p_0$ from the void ratio at p_0. Thus, in Fig. 18-2, the void ratio at 5.0 tsf is 0.900 and that at 0.5 tsf is 1.090. Therefore $C_C = 1.090 - 0.900 = 0.190$.

18.11. APPROXIMATE VALUE OF COMPRESSION INDEX. There appears to be an approximate relationship between the liquid limit (LL) of a clay soil and the compression index. Skempton (*9*) has demonstrated that this relationship can be expressed by the empirical formula

$$C_C = 0.009(\text{LL} - 10) \tag{18-7}$$

Eq. (18-7) is of great practical value, since it enables a designer to make an approximate estimate of the settlement of a foundation on clay without carrying out expensive and time-consuming consolidation tests.

18.12. DEVIATION OF IDEALIZED CURVE. The idealized curves in Figs. 18-1 and 18-2 indicate that the void ratio-pressure relationship for soil is logarithmic in character at all ranges of the applied load. Actually, however, the curve for practically any soil deviates from the logarithmic shape at low values of pressure, as typically indicated in Fig. 18-3. This deviation from the idealized curve may be caused by prior loading of the soil during its geological history, such as loading resulting from the weight of glacial ice or from the weight of soil overburden which has since been eroded away. Or the deviation may be caused simply by the fact that, in taking an undisturbed sample of soil, it is inevitable that the pressure of the overburden is released. Whatever may be the cause of the deviation, the early portion of an actual void ratio–pressure diagram represents a reloading operation and corresponds to the reloading portion of the idealized curve in Fig. 18-1.

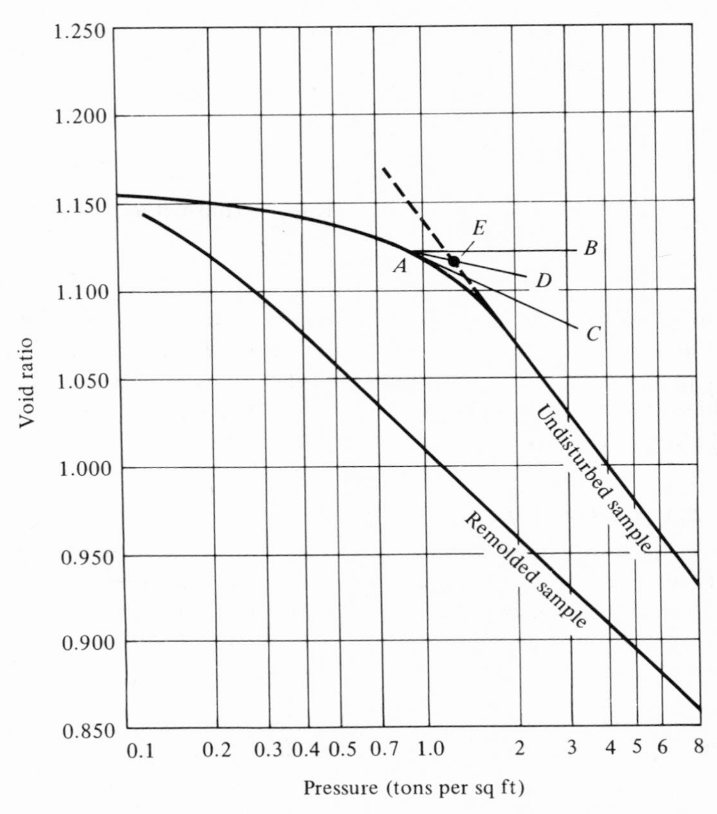

Fig. 18-3. Casagrande method of determining preconsolidation pressure.

18.13. PRECONSOLIDATION LOAD. The probable load to which a soil stratum may have been subjected in the geological past is frequently referred to as the preconsolidation load or the precompression load. In some foundation studies it is desirable to know the approximate value of this load. Casagrande (*2*) has suggested the following method for estimating the preconsolidation load. On a semilogarithmic plot of a void ratio–pressure diagram, as in Fig. 18-3, the first step is to prolong the straight-line or virgin portion of the diagram backward. Next select point *A* at the point of maximum curvature or shortest radius on the curved portion of the diagram. Through this point, draw the horizontal line *AB* and also the line *AC* tangent to the curve. Then bisect the angle between these two lines. The point of intersection of the bisector *AD* and the prolonged straight-line portion of the diagram represents the approximate preconsolidation load; or, in other words, the intersection point E represents the maximum load to which the soil was subjected during its geological history or prior to the removal of the soil sample for testing.

18.14. NORMALLY CONSOLIDATED AND PRECONSOLIDATED CLAYS. If the effective overburden pressure on the in-place prototype soil stratum from which an undisturbed sample was taken is approximately the same as the preconsolidation load, the soil is said to be *normally consolidated.* If the preconsolidation load is substantially greater than the effective overburden pressure, the soil is said to be *preconsolidated.* The effective overburden pressure can be estimated from knowledge of the depth of the stratum below the surface, the densities of the overlying strata, and the elevation of the water table, as outlined in Section 12.21. Frequently, it is possible to determine whether or not a clay stratum is normally loaded or preloaded by a geological study of the site.

18.14.1. Underconsolidated and Collapsible Soils. A few soils occur in locations where the existing overburden pressure exceeds that indicated by the normal consolidation curve. Such soils are said to be *underconsolidated.* Clays in geologically recent deposits as in deltas, etc., may be underconsolidated because the pore water has not had time or opportunity to drain away; in this case underconsolidation can be recognized either from in-place pore water pressure measurements, or from comparison of the existing void ratio to that obtained from a consolidation curve at a pressure equal to the weight of the overburden.

Deltaic sands are sometimes underconsolidated because of their quiet mode of deposition and a loose structure sustained by their grain-to-grain static coefficient of sliding friction. Vibrations from machinery or earthquakes may trigger a very rapid normal consolidation of such soils, whereby the loss of grain-to-grain contact and the presence of excess pore water leads to *liquification* or a complete loss of strength. The recognition of underconsolidated sands from consolidation tests is difficult or impossible because of the difficulty in acquiring undisturbed samples. A correlation to standard penetration test blow counts is discussed in Section 19.9.

A third class of underconsolidated soils is silts, usually loess, deposited and held in a loose structure by cohesive forces at grain contacts. Upon wetting, the cohesive forces may reduce sufficiently that the soil structure is unstable, and collapses. The existence of such underconsolidated collapsible soils can be detected by loading a consolidometer with an undisturbed sample of soil carefully preserved at its natural moisture content. The sample is loaded to the calculated overburden pressure, and then saturated with water. Collapsibility is ascertained from the change in height upon saturation by water. A sudden decrease in volume of 5% or more is not uncommon.

A more convenient and rather reliable criterion to detect collapsible loess was proposed by Gibbs and Bara (*4*) on the basis of bulk density and liquid limit data: If the bulk density is such that upon saturation with water the moisture content exceeds the liquid limit (indicating that upon disturbance the soil would be liquid) the soil is identified as collapsible.

EXAMPLE 18-1. The *in situ* dry density of a loess soil is 85 pcf and the liquid limit is 29. Assuming $G = 2.70$, predict whether or not the soil is collapsible.

SOLUTION. The volume of soil solids in 1 cu ft is $V_s = 85 \div 2.7(62.4) = 0.50$ cu ft. Hence $V_e = 0.50$ cu ft. The moisture content upon saturation is therefore

$$W = \frac{0.50(62.4)}{85} \times 100 = 37\%$$

Since this exceeds the liquid limit, the soil is most likely collapsible.

Data on Iowa loess show a rough correlation of collapsibility to clay content: Iowa loess containing less than 30% 0.005 mm clay has a 50% likelihood of being collapsible, whereas with less than 20% 0.005 mm clay the likelihood of being collapsible exceeds 80% (*5*).

18.15. EFFECT OF REMOLDING. Up to this point, the discussion of void ratio–pressure or stress–strain relationships has applied only to soils in an undisturbed state. Information relating to undisturbed soil is, of course, of greatest interest in settlement analysis, because compressible strata beneath structures, which permit them to settle, are practically always in a natural undisturbed state. However, consolidation tests on remolded samples of the soil may provide additional information of interest and value.

If the sample is remolded at the same moisture content and to the same density as in the undisturbed state, the void ratio–pressure diagram will usually be flatter and the compression index will be less than is indicated by the curve for the undisturbed state, as shown typically in Fig. 18-3. When an undisturbed sample is recovered, even by the most careful methods, it has probably been reworked or remolded to some unknown extent. Therefore, in all probability, a curve which most nearly represents the actual in-place soil would be somewhat steeper than that obtained from testing the undisturbed sample. This condition suggests that values of the compression index and other void ratio–pressure relationships should be evaluated on the high or conservative side of available data rather than on the average.

18.16. INFLUENCE OF SENSITIVITY. The void ratio–pressure diagram for a highly sensitive clay (see Section 7.16.4) is usually distorted from the idealized semilogarithmic shape shown in Figs. 18-1 and 18-2. This type of distortion is illustrated in Fig. 18-4. Obviously, a single value of the compression index is not applicable in such a situation. When a clay of this kind is encountered, the value of the void ratio for a specific value of pressure must be taken from the curve rather than calculated by a formula involving the compression index.

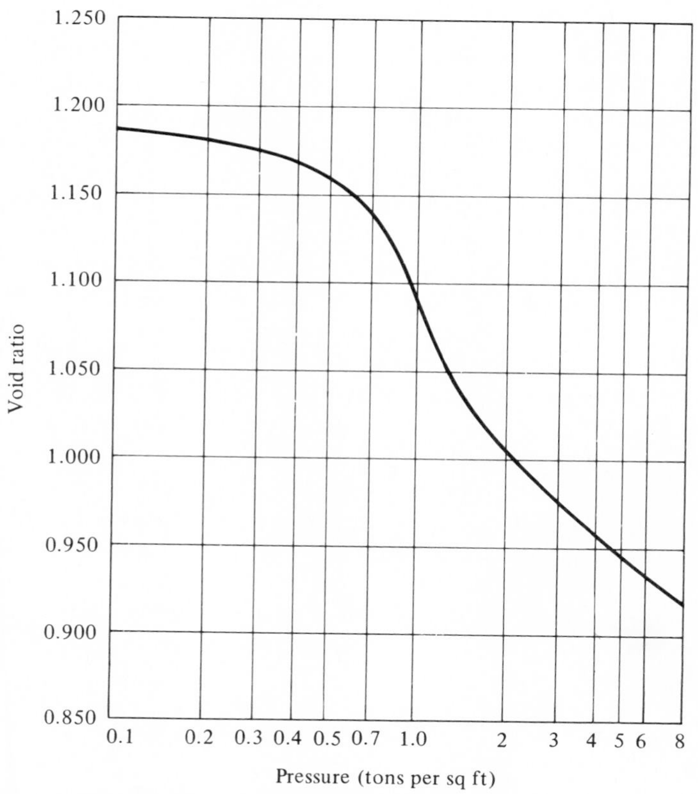

Fig. 18-4. Typical distortion of $e - \log p$, curve for sensitive clay.

18.17. INFLUENCE OF TIME ON STRAIN. An additional characteristic of the void ratio-pressure relationship of soil, which is not indicated by a void ratio–pressure diagram but which is of very great importance in the study of settlement of structures, is the influence of time upon the development of compression strains. Unlike steel and concrete, wherein strains develop practically instantaneously with the application of stress, the strain in soil develops over a period of time after an increment of consolidating pressure is applied. The primary reason for this time lag is the fact that some of the water contained in the voids of foundation soils has to be squeezed out before the volume of the voids can decrease. The rate of outflow of this pore water depends on the permeability of the soil.

In relatively coarse-grained soils the pore water can escape rapidly and the time lag between application of pressure and the development of strain is relatively small. A structure founded on a soil of this character will usually attain its maximum settlement early, and very little further settlement will occur after the structure is completed. On the other hand, if a structure is founded on a fine-grained clayey soil or if a stratum of such soil is present at some depth beneath, the outflow of water from the voids due to the pressure imposed by the structure will be very slow because of a relatively small coefficient of permeability. Therefore, the settlement of the structure will develop at a very slow rate and may require tens or even hundreds of years to be completed, the period depending on the permeability of the soil, the thickness of the layer, and whether the pore water can escape from both the top and bottom of the clay layer or only from one or the other of these surfaces.

18.18. TIME–COMPRESSION CURVE. In order to analyze the probable settlement behavior of a proposed structure, it is necessary to know those characteristics of the soil which influence the total settlement and those which influence the time rate at which the settlement will develop. The void ratio–pressure diagram gives the information required for the first of these objectives. For the second, a time–compression curve for each increment of pressure applied to the soil must be obtained, and the coefficient of permeability or its equivalent must be determined. Two time–compression curves, for a coarse-grained soil and a fine-grained soil, respectively, are shown in Fig. 18-5.

According to Fig. 18-2, it is seen that a pressure increment from 3 to 4 tsf causes a change in void ratio from 0.942 to 0.916. This change in void ratio represents the total consolidation of the soil under this particular load increment. If the material is a coarse-grained, highly permeable, sandy soil, the total void ratio change will occur in a very short

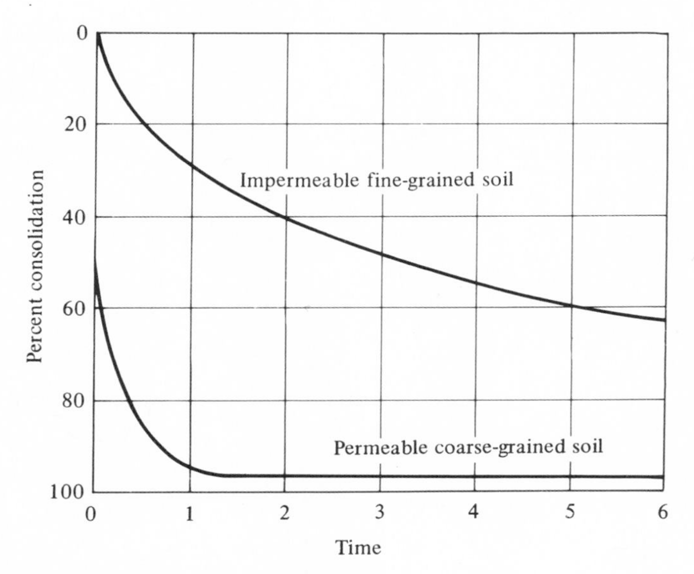

Fig. 18-5. Time-compression curves.

time, as shown in Fig. 18-5. If it is a fine-grained and impermeable soil, considerable time is required for the total change to develop. These curves are drawn to represent the time effects in a small laboratory sample on the order of 1 in. thick, and the time is expressed in minutes. The time for complete consolidation of a compressible layer in the field is many times as great; but it can be computed from the laboratory and field data, as will be shown later.

18.19. COMPARISON OF SOIL AND SPRING. An excellent mechanical analogy illustrating the principles of consolidation is shown in Fig. 18-6. In (a) a coil spring 12 in. high is represented. A load of 20 lb applied to the spring causes it to compress to a height of 7 in., as indicated in (b). Now imagine that this same spring is placed in a cylinder filled with water and fitted at the top with a frictionless water-tight piston and a petcock, as in (c). This assembly is analogous to a saturated clay stratum in nature. It is subjected only to its own weight and that of the overburden above it. The spring represents the soil skeleton, and the water in the cylinder represents the pore water in the soil.

Now suppose that a load of 20 lb is placed on top of the piston, as shown in (d). So long as the petcock remains closed, the spring will not compress because the water is incompressible. Instead, the load causes an

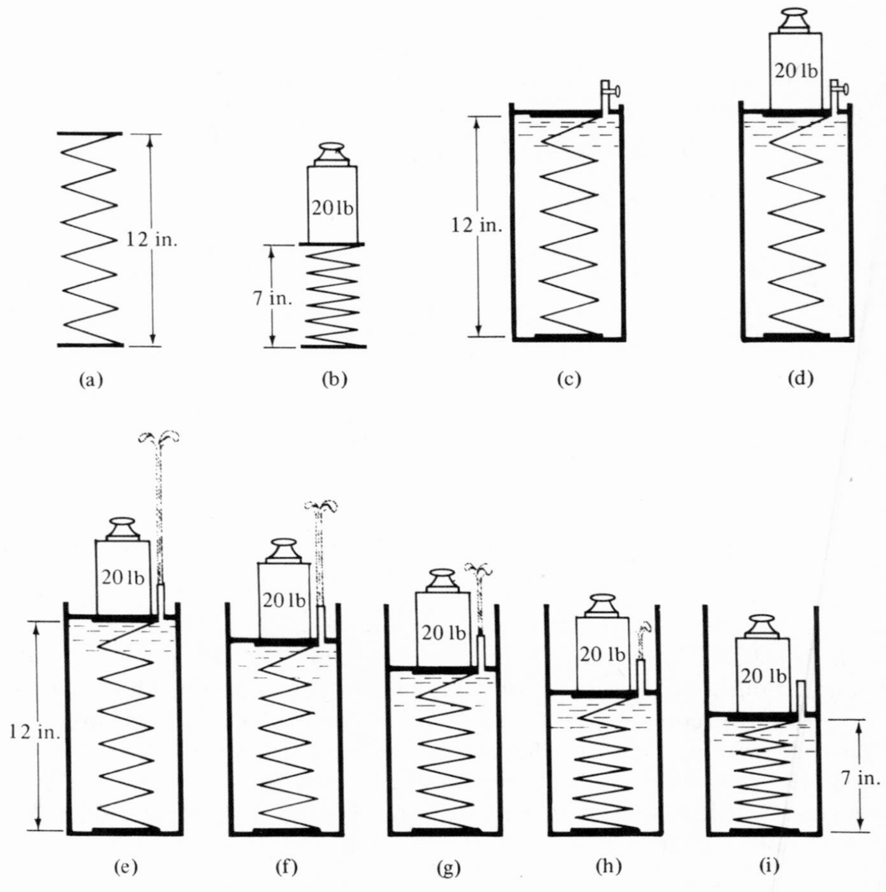

Fig. 18-6. Analogy of soil and a spring. Reproduced by permission from *Fundamentals of Soil Mechanics* by D. W. Taylor, published by John Wiley & Sons, Inc., 1948.

increase in pressure in the water. This increased pressure is called the hydrostatic excess pressure. In an actual situation the hydrostatic excess pressure is the increase in pore-water pressure over and above the normal hydrostatic pressure due to position below a water table. The hydrostatic excess pressure is a maximum at the time at which a foundation load is applied. It gradually decreases to zero as consolidation progresses to completion.

With the petcock closed so that no water can escape from the cylinder, the situation in (d) is analogous to a structure resting on soil having a zero coefficient of permeability. Of course, no such soil exists in nature, although many foundation clays retard the flow of water to an exceedingly small rate. Now suppose that the petcock is opened, as in (e). Immediately, water will flow outward under the influence of the hydrostatic excess pressure, the spring will begin to compress, and the applied load will gradually be transferred from the water in the cylinder to the spring, as indicated in (f), (g), and (h). The rate at which this transfer occurs is dependent on the size of opening in the petcock or, in an actual situation, on the coefficient of permeability of the soil. Finally, as shown in (i), the hydrostatic excess pressure is completely dissipated, the water ceases to flow out through the petcock, and all of the applied load is carried by the spring in a compressed state similar to that indicated in (b). Consolidation is now complete, or 100% under the applied load increment of 20 lb and no further settlement may be expected. Of course, if an additional 20-lb increment of load were now added, the cycle just described would be repeated and further settlement would develop.

18.20. ONE-DIMENSIONAL CONSOLIDATION TEST. The consolidation characteristics of a foundation soil, both those relating to the amount of compression or change in void ratio caused by an applied load and those relating to the time rate at which the compression strains develop, are determined in the laboratory by a consolidation test. A diagrammatic sketch of a consolidation machine or consolidometer is shown in Fig. 18-7. In a consolidation test a small representative sample of undisturbed soil is carefully trimmed and fitted into a rigid metal or plastic ring several inches in diameter and 1 to $1\frac{1}{2}$ in. high. The soil sample is mounted on a porous stone base, and a similar stone is placed on top to permit water which is squeezed out of the sample to escape freely at the top and bottom. Prior to loading, the height of the sample should be accurately measured to the nearest 0.0001 in. Also, a micrometer dial graduated in ten-thousandths of an inch is mounted in such a manner that the vertical strains in the sample can be measured as loads are applied.

The consolidation-test apparatus is designed to permit the sample to be submerged in water during the test, to simulate the position below a water table of the prototype soil layer from which the test sample was taken. Also, in earlier days, the apparatus was fitted with a vertical glass tube connected with the base which served as the standpipe of a falling head permeameter to measure the coefficient of permeability of the sample. At present, the influence of permeability on the rate of consolidation of the sample and the coefficient of consolidation are determined directly by empirical methods described in Sections 18.27 and 18.28.

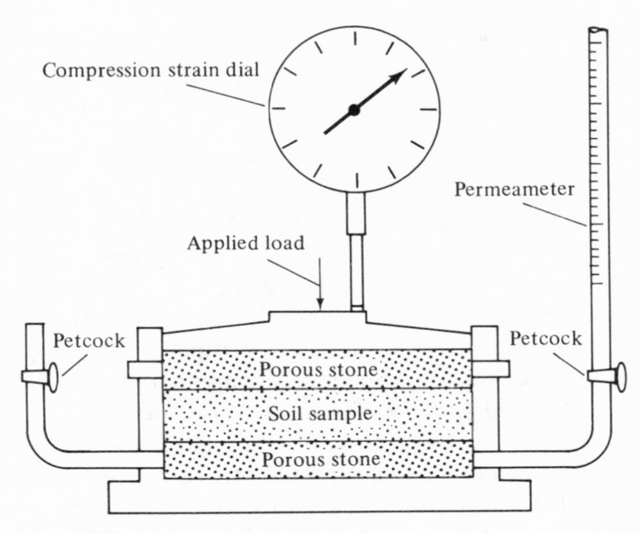

Fig. 18-7. Consolidometer.

The procedure for conducting a consolidation test is as follows. With no load on the sample and with the petcock to the permeameter standpipe closed, record the zero-load reading of the compression-strain dial. Then apply a suitable increment of load, say 0.1 tsf to the sample; and read the compression dial at various intervals of time. Readings should be taken frequently at first, but may be less frequent as compression under the load increment progresses. When movement of the vertical dial indicates that the sample has virtually reached its maximum compression under the applied load, another increment of load is applied to the sample, and the time-rate of compression strain under this new increment is observed. This cycle of loading and measuring the time-rate of strain is repeated until the total applied load exceeds that to which the prototype soil will be subjected by a proposed structure. A procedure for the one-dimensional consolidation test was recently adopted as a standard by A.S.T.M. (A.S.T.M. Designation D2435–70).

One difference between the consolidation test as normally performed and the actual field situation is the hydrostatic pore water pressure. That is, although load increments may create the same *excess* pore water pressure which then dissipates as the pore water seeps out, the pressure from hydrostatic head is much less in the laboratory than in the field. A *back-pressure consolidometer* may be used to more closely simulate the field conditions, the main advantage being to prevent exsolution and blocking of capillaries by gases from the pore water.

18.21. DETERMINATION OF VOID RATIO AND OTHER SOIL CHARACTERISTICS. After the consolidation test is completed, the sample should be removed from the apparatus, and its dry weight and true specific gravity should be determined. Then, since the diameter of the sample and its initial height are known, the initial void ratio and the void ratio at equilibrium under each load increment can be computed. These equilibrium void ratios may be plotted against the applied loads to give the void ratio–pressure diagram of the soil. If the diagram follows the typical logarithmic form, the value of the compression index can be determined empirically from the data by means of Eq. (18-5). Also it may be helpful to plot a void ratio–coefficient of permeability curve, if this property of the soil is measured. However, an empirical method of determining the effect of the coefficient of permeability without actually measuring it is available. This method is discussed in Section 18.27.

18.22. CONSOLIDATING PRESSURE. It is necessary, in the application of the consolidation theory to a foundation settlement problem, to distinguish clearly between consolidating pressure and nonconsolidating pressure. Consolidating pressure may be defined as pressure which causes a decrease in void ratio of a compressible layer of soil below the initial void ratio which existed prior to the beginning of engineering operations at the site of a structure. To illustrate, let it be assumed that a foundation exerts a pressure of 3 tons/sq ft at its plane of contact with the soil. If the footing rests directly on the surface of the soil, all of this pressure is available to cause a reduction in void ratio of the soil below it and the full value of 3 tons/sq ft is consolidating pressure. If, on the other hand, the bottom of the footing is at some elevation below the ground surface, a certain amount of excavation is required before the foundation is constructed. This excavation is an unloading process and the soil below the footing level will expand, that is, the void ratio will increase, somewhat in the manner indicated by the expansion portion of the curve in Fig. 18-1.

During construction of the foundation, the early increments of weight of the structure merely act to replace the weight of the excavated material. This constitutes a reloading of the soil and the void ratio decreases to the initial value before excavation was started. Any additional foundation pressure, over and above that required to replace the weight of excavated material, causes a reduction in void ratio below the initial value and is considered to be consolidating pressure. If it is assumed that the bottom of the footing considered in the preceding paragraph is at an elevation 10 ft below the ground surface, instead of at the ground surface, and that the excavated soil weighs 100 lb/cu ft, then the pressure release at this elevation is 0.5 ton/sq ft. Therefore the first 0.5 ton/sq ft of applied foundation pressure acts to replace the pressure removed during excavation, and only the remaining 2.5 tons/sq ft is effective in consolidating the soil below the footing.

18.23. TOTAL COMPRESSION OF A SOIL LAYER. The total compression of a soil layer under the influence of a consolidating load either in a testing machine or in a natural stratum can be computed from the void ratio-pressure relationship determined in a consolidation test by

$$s = d\left[\frac{e_1 - e_2}{1 + e_1}\right] \tag{18-8}$$

in which

s = total compression of soil layer over long period of time;
d = thickness of compressible soil layer;
e_1 = initial void ratio of soil layer before consolidating load is applied;
e_2 = final void ratio of soil layer after consolidating load has been applied for a long time.

To derive this expression, consider the column of soil of unit cross-sectional area, shown in Fig. 18-8. This column is representative of a compressible layer of soil of thickness d under an initial unit pressure p_1 in (a) and under a final pressure p_2 in (b). The consolidating pressure due to the application of a foundation load is, therefore, $p_2 - p_1$. The thickness of the layer will decrease to d_2 under the increased pressure, and the

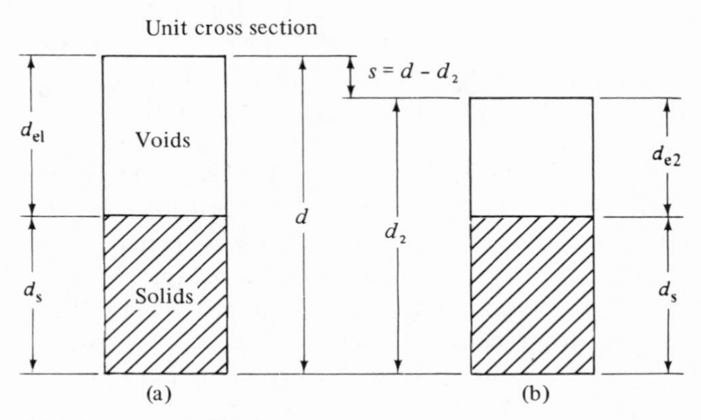

Fig. 18-8. Settlement of soil column.

total settlement is $d - d_2 = s$. The soil column in the figure is conventionalized by showing the total volume of voids concentrated at the top and the total volume of solids at the bottom. Since the cross-sectional area is unity and since all of the compression of the soil is accounted for by a decrease in the volume of the voids, the volume of the solids may be represented by the height d_s and the volume of the voids may be represented by the height d_{e1} under the initial pressure and by d_{e2} under the final pressure. Then,

$$d = d_{e_1} + d_s \quad \text{and} \quad d_2 = d_{e_2} + d_s$$

Also,

$$e_1 = \frac{d_{e_1}}{d_s} = \frac{d - d_s}{d_s} = \frac{d}{d_s} - 1$$

and

$$e_2 = \frac{d_{e_2}}{d_s} = \frac{d_2 - d_s}{d_s} = \frac{d_2}{d_s} - 1$$

$$e_1 - e_2 = \frac{d - d_2}{d_s}$$

Since

$$s = d - d_2 \qquad \text{and} \qquad d_s = \frac{d}{1 + e_1}$$

Eq. (18-8) follows:

$$s = d\left[\frac{e_1 - e_2}{1 + e_1}\right]$$

18.24. TIME-RATE OF SETTLEMENT. Equation (18-8) gives the total compression of the soil layer or the total settlement of the structure due to the compression, but it tells nothing about the time-rate at which the settlement will occur. In order to study this phase of the settlement problem, Terzaghi employed mathematical procedures similar to those involved in the study of one-dimensional flow of heat through a body. A discussion of the mathematical treatment is beyond the scope of this text. According to this concept, the time-rate of compression of a clay stratum depends mainly on the following four factors:

1. Thickness of the clay layer;
2. Number of drainage faces;

3. Permeability of the soil;
4. Magnitude of the consolidating pressure acting on the clay layer.

It will be noted that the first two of these factors influence the distance through which the water in the soil voids must travel in order to escape and permit the volume of the voids to decrease. The third factor controls the rate at which the water can escape, and the fourth influences the hydrostatic excess pressure which causes the outflow of water.

The basic equation for fraction of total settlement vs. time may be written as

$$q = 1 - \frac{8}{\pi^2}\left(\epsilon^{-N} + \frac{1}{9}\epsilon^{-9N} + \frac{1}{25}\epsilon^{-25N} + \cdots\right) \tag{18-9}$$

where q = fraction of total settlement

$$N = \frac{\pi^2}{4}T \tag{18-10}$$

and

$$T = \frac{ct}{h^2} \tag{18-11}$$

In Eq. (18-9), ϵ is the base of the system of natural logarithms, or 2.718.... In Eq. (18-11),

T = a time factor;
t = elapsed time in which q fraction of total settlement occurs;
h = maximum distance through which pore water must flow in order to escape from the clay layer;
c = coefficient of consolidation.

The coefficient of consolidation is a quantity analogous to the coefficient of diffusivity in the one-dimensional heat-flow problem. It can be determined by the formula

$$c = \frac{k(1 + e_1)}{\gamma_w a} \tag{18-12}$$

in which

k = coefficient of permeability;
e_1 = initial void ratio under an increment of consolidating pressure;
γ_w = unit weight of water;
a = coefficient of compressibility.

The coefficient of compressibility a is the slope of the void ratio–

pressure curve at the value e_{avg}, which is equal to $(e_1 + e_2)/2$. This coefficient may be determined by

$$a = \frac{\Delta e}{\Delta p} = \frac{e_1 - e_2}{p_2 - p_1} \tag{18-13}$$

Throughout the settlement analysis it is necessary to be consistent with respect to the units of measurement of the various quantitites involved. A convenient system of units is to express time in years, distances in feet, and weights or pressures in tons. If these units are used, the various quantitites will be expressed as follows:

k will be in feet per year;
p_1 and p_2 will be in tons per square foot;
γ_w will be equal to 0.0312 ton per cubic foot;
c will be in square feet per year.

Then,

$$c = \frac{k(1 + e_1)}{0.0312a} \tag{18-14}$$

Most soil-testing laboratories habitually express the coefficient of consolidation in square centimeters per second and express the coefficient of permeability in centimeters per second. The following factors may be used to convert these coefficients to the units suggested above:

Multiply square centimeters per second by 33,967 to obtain square feet per year.

Multiply square centimeters per minute by 566.1 to obtain square feet per year.

Multiply centimeters per second by 1,035,330 to obtain feet per year.

18.25. FACTORS AFFECTING *T*. It will be noted in Eqs. (18-9) and (18-10) that T is the only variable upon which q depends. Also, transposing in Eq. (18-11) gives

$$t = \frac{Th^2}{c} \tag{18-15}$$

Therefore, if we know the value of the time factor T that corresponds to a selected value of q, together with the coefficient of consolidation and the maximum distance through which pore water must flow in order to escape from the clay layer, then the elapsed time t for q percent of the total settlement s to develop may be calculated by means of Eq. (18-15). The numerical value of T for any particular percentage of consolidation q depends on the character of the distribution of the consolidating pressure throughout the vertical thickness of the clay layer relative to the direction of outflow of the pore water during consolidation. A number of variations in this distribution are possible, and a value of T for each of them has been determined by several investigators. However, the influence of these variations on T is relatively minor, and in view of the fact that many simplifying assumptions are employed in the theory of consolidation, they may be ignored without materially influencing the accuracy of the settlement analysis. Values of T corresponding to various values of q are given in Table No. 18-1.

TABLE 18-1. Values of *T* for Various Values of *q*

Fraction of Total Settlement q	*Value of Time Factor* T
0.00	0.000
0.05	0.003
0.10	0.008
0.15	0.018
0.20	0.031
0.25	0.049
0.30	0.071
0.35	0.097
0.40	0.126
0.45	0.165
0.50	0.197
0.55	0.240
0.60	0.287
0.65	0.342
0.70	0.403
0.75	0.478
0.80	0.567
0.85	0.684
0.90	0.848
1.00	∞

18.26. NUMBER OF DRAINAGE FACES. The most important condition which influences the time-rate of settlement of a structure overlying a stratum of compressible clay is the number of drainage faces. If the pore water in the clay can flow freely from both the upper and lower faces of the stratum, the maximum distance through which a particle of water must flow in order to escape is one-half the thickness of the layer. This condition is referred to as double drainage. On the other hand, if the clay stratum is underlain by an impermeable material, such as ledge rock, the maximum distance through which a particle must flow is equal to the full thickness of the layer. This condition is called single drainage. It is obvious that the pore water can escape more rapidly, and therefore that settlements will develop at a faster rate, in the case of double drainage.

If the thickness of the compressible stratum is designated as d, then $h = d/2$ for double-drainage conditions and $h = d$ for single-drainage conditions. When these values are substituted in Eq. (18-15), the results are:

for double drainage, $$t = \frac{T(d/2)^2}{c} \tag{18-16}$$

for single drainage, $$t = \frac{Td^2}{c} \tag{18-17}$$

Thus, it is seen that settlements will develop four times as fast for double drainage as for single drainage.

18.27. DETERMINATION OF COEFFICIENT OF CONSOLIDATION. The coefficient of permeability of a soil has an important influence on the time-rate of settlement of a structure resting on compressible soil. It enters into the computation of settlements as a factor in the expression for coefficient of consolidation, as indicated in Eq. (18-12). The measurement of the coefficient of permeability during the conduct of a consolidation test is very time consuming. In order to eliminate the necessity for making permeability measurements, Taylor (*12*) has devised an empirical method of determining the coefficient of consolidation directly from the compression-dial readings taken at various time intervals during compression of the soil sample under the several increments of load applied during a consolidation test. It is called the "square root of time fitting method." It is simple and speedy; and, although it is empirical in character, it compares favorably in accuracy with the whole process of estimating settlements by the consolidation theory.

In the square root of time fitting method the compression-dial readings for each increment of load in a consolidation test are plotted as ordinates against the square root of the corresponding time intervals, as shown in Fig. 18-9. It will usually be found that the early portion of this curve is essentially a straight line. Draw a straight line through this early portion of the curve, such as OA. Then draw a second straight line, as OB, which at each point has an abscissa 1.15 times that of OA. The intersection of this second line and the laboratory curve for $\sqrt{t}$ is assumed to be the point at which 90% of primary consolidation has occurred. Its time value is determined and is designated as t_{90}. According to Eq. (18-9) and Table 18-1, the theoretical value of the time factor T at 90% con-

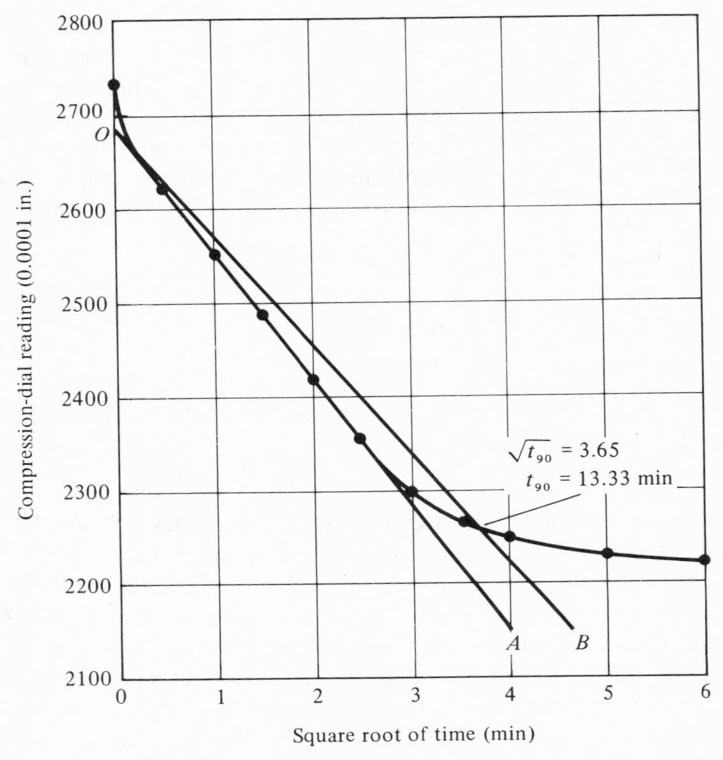

Fig. 18-9. Taylor's square root of time fitting method.

solidation is 0.848. Therefore, Eq. (18-11) may be written

$$c = \frac{0.848\ h^2}{t_{90}} \tag{18-18}$$

The coefficient of consolidation may be determined by substituting the empirical value of t_{90} in Eq. (18-18). A value of c should be determined for each increment of load applied in the consolidation test, and the average value for the range of consolidating pressure to which the compressible layer is subjected should be used in the settlement calculations.

18.28. LOGARITHM OF TIME-FITTING METHOD. A similar empirical procedure has been developed by A. Casagrande in which the compression dial readings are plotted against the logarithm of time. Typically, such a plot yields a curve which has two distinct straight-line portions, one in the early time phase of loading and another in the later phase.

These tangent portions are prolonged until they intersect. The time ordinate at this intersection is considered to be the logarithm of time at 100% consolidation. Then a point of zero consolidation is determined after which the 50% point and its time may be located. From Table 18-1, the time factor T for 50% is 0.197 which may be substituted in Eq. (18-11) to obtain the coefficient of consolidation. Complete descriptions of both the $\sqrt{t}$ and log t fitting methods may be found in Refs. (*12*) and (*13*).

18.29. DATA FOR APPLICATION OF CONSOLIDATION THEORY. In order to summarize and illustrate the principles of the consolidation theory which have been presented in this chapter, a complete example of settlement analysis will be given. Assume that a bridge pier is to be constructed as shown in Fig. 18-10. The base of the pier is 20 ft by 60 ft and it exerts a total load of 4800 tons which is assumed to be uniformly distributed

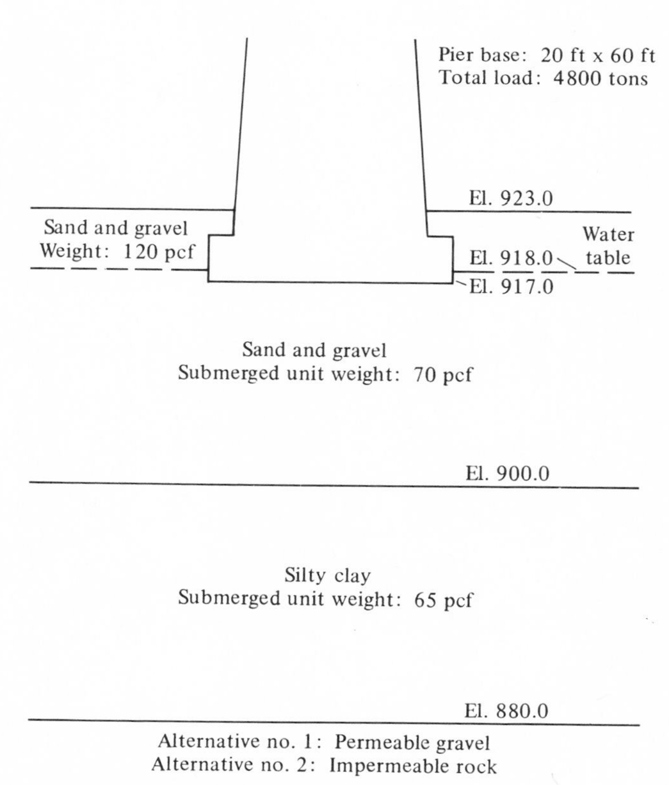

Fig. 18-10. Conditions for application of consolidation theory.

TABLE 18-2. Consolidation Test on Undisturbed Silty Clay Soil[a]

Elapsed Time (min) t	$\sqrt{t}$	*Load Increments (tsf) and Compression Dial Readings (0.0001 in.)*													
		0.1–0.2		*0.2–0.5*		*0.5–1*		*1–2*		*2–4*		*4–8*		*8–16*	
		Dial	*Diff.*	*Dial*	*Diff.*	*Dial*	*Diff.*	*Dial*	*Diff.*	*Dial*	*Diff.*	*Dial*	*Diff.*	*Dial*	*Diff.*
0	0	4896	0	4850	0	4761	0	4478	0	3793	0	2954	0	2332	0
0.25	0.5	4892	4	4842	8	4742	19	4400	78	3723	70	2897	57	2280	52
1	1	4889	7	4836	14	4728	33	4343	135	3672	121	2855	99	2241	91
2.25	1.5	4886	10	4830	20	4714	47	4296	182	3621	172	2814	140	2203	129
4	2	4883	13	4825	25	4700	61	4230	248	3570	223	2773	181	2165	167
6.25	2.5	4880	16	4819	31	4686	75	4173	305	3518	275	2732	222	2127	205
9	3	4877	19	4813	37	4673	88	4116	362	3468	325	2692	262	2090	242
12.25	3.5	4874	22	4807	43	4659	102	4059	419	3417	376	2651	303	2052	280
16	4	4871	25	4801	49	4645	116	4011	467	3374	419	2610	344	2014	318
25	5	4867	29	4794	56	4627	134	3946	532	3308	485	2562	392	1967	365
36	6	4864	32	4788	62	4614	147	3900	578	3259	534	2525	429	1936	396
49	7	4863	33	4785	65	4607	154	3880	598	3228	565	2503	451	1916	416
64	8	4862	34	4783	67	4603	158	3872	606	3207	586	2494	460	1908	424
100	10	4861	35	4781	69	4597	164	3860	618	3171	622	2473	481	1889	443
225	—	4858	38	4777	73	4587	174	3848	630	3124	669	2443	511	1861	471
400	—	4853	43	4767	83	4548	213	3821	657	3034	759	2366	588	1820	512
1440	—	4850	46	4761	89	4478	283	3793	685	2954	839	2332	622	1762	570

[a]True specific gravity = 2.68; dry weight = 410.5 gm = 0.905 lb; area of sample = 81.076 sq cm = 12.566 sq in.; thickness of sample = 1.2500 in. with dial reading at 0.3147.

over the area of the pier base. The profile of the soil beneath the pier indicates that practically all the anticipated settlement of the structure will result from compression of the silty-clay layer between elevations 900.0 and 880.0. Undisturbed samples of this soil are obtained; and laboratory consolidation tests give the data shown in Table 18-2.

After completion of the consolidation test, it is necessary to determine the dry weight and the true specific gravity of the sample and also its area and its thickness at some arbitrary dial reading. These data are also shown at the bottom of Table 18-2.

18.29.1 Computations for Void Ratios and Coefficients of Consolidation. From the data in Table 18-2, it is possible to calculate the results shown in Table 18-3. The void ratio for the initial pressure of each increment is determined first and entered in column 2. The void ratio-pressure diagram in Fig. 18-11 is then drawn from these void ratios. Also, the compression index of the virgin curve can be evaluated by means of Eq. (18-5) or Eq. (18-6). In this particular case, the e-log p curve is somewhat distorted, and the virgin curve is not a straight line throughout the range of applied pressures. It may be more appropriate, therefore, not to use a compression index, but to evaluate void ratios directly from the curve. The difference between settlements estimated by these two procedures is insignificant.

TABLE 18-3. Void Ratios and Coefficients of Consolidation Computed from Consolidation Test Data

Load Increment (tsf)	*Void Ratio at Initial Load*	*Average Height of Sample, d (in.)*	*Maximum Flow Distance, h = d/2*	*Square Root of 90% Consolidation Time, $\sqrt{t_{90}}$ (min)*	t_{90}	*Coefficient of Consolidation, c (sq ft/yr)*
(1)	(2)	(3)	(4)	(5)	(6)	(7)
0.1 to 0.2	0.915	1.4226	0.7113	5.75	33.1	47.3
0.2 to 0.5	0.909	1.4159	0.7080	5.9	34.8	44.6
0.5 to 1	0.897	1.3972	0.6986	5.9	34.8	43.4
1 to 2	0.859	1.3489	0.6744	5.6	31.4	44.8
2 to 4	0.767	1.2727	0.6364	5.65	31.9	39.4
4 to 8	0.654	1.1996	0.5998	5.75	33.1	33.4
8 to 16	0.570	1.1400	0.5700	5.7	32.5	30.9
16	0.494					

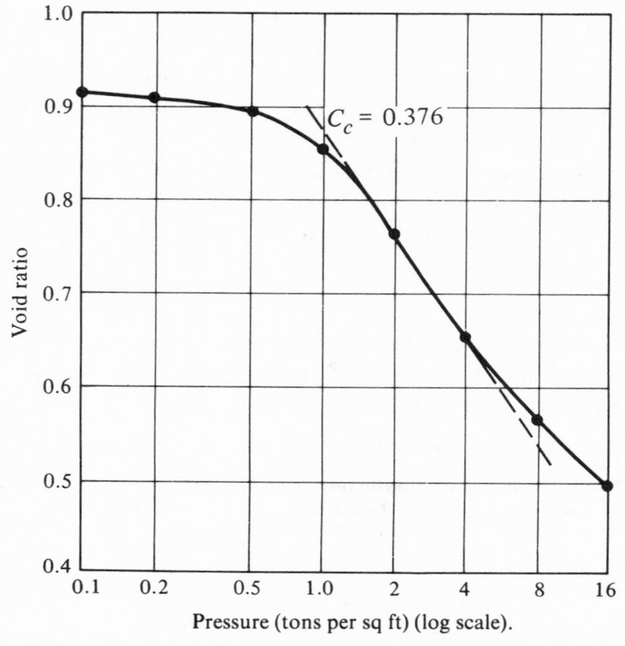

Fig. 18-11. Void ratio-pressure curve for data in Table 18-2.

18.29.2 Determination of Coefficients of Consolidation. Next the coefficient of consolidation is determined for each load increment by the square root of time fitting method, as outlined in Section 18-27. In Fig. 18-12 are shown the square root of time-compression curves for increments of load from 1 to 2 and from 2 to 4 tsf. The 90% consolidation time for each load increment is determined from the proper curve. Since the compression dial readings are in inches and the time intervals are measured in minutes, these units should be converted to feet and years when substituting the values of t_{90} and h^2 in Eq. (18-18) in order to obtain the coefficient of consolidation in square feet per year. (Any other units of time and distance may be used, if desired, provided that the units are consistent throughout.)

18.29.3 Computations for Total Settlement. The initial pressure on the mid-plane of the compressible layer is the weight of the soil overburden above this plane before any excavation or construction operations begin at the site. Since only the intergranular pressure is effective in pro-

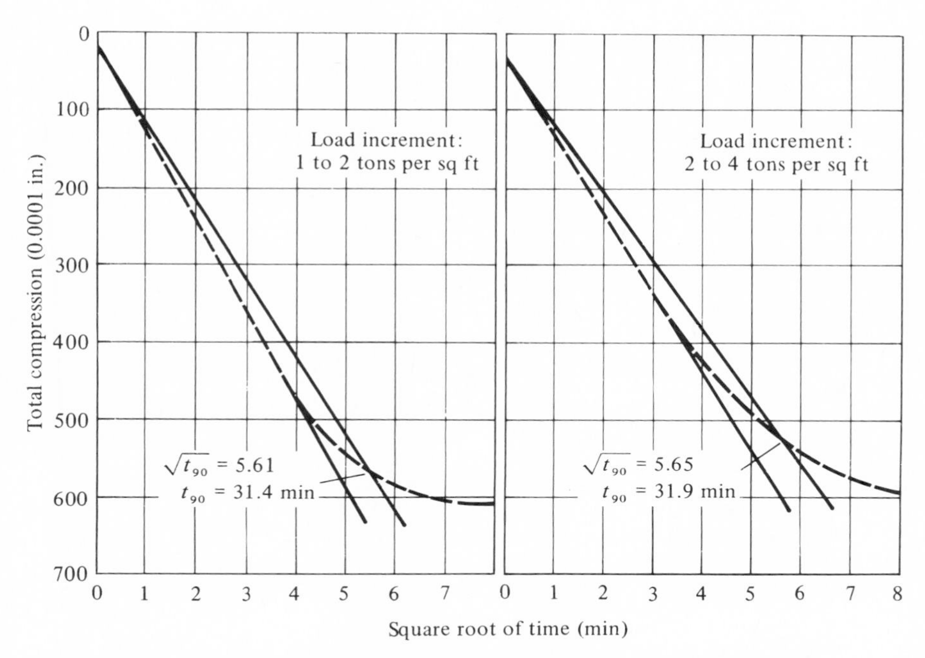

Fig. 18-12. Square root of time-compression curves for data in Table 18-2.

ducing compression strains, the submerged unit weight of the overburden soil lying below the water table is used in computing the initial pressure.

The computations for determining the initial pressure p_1 are

$$\begin{aligned} 5 \times 120 &= 600 \text{ psf} \\ 18 \times 70 &= 1260 \\ 10 \times 65 &= \underline{650} \\ p_1 &= 2510 \text{ psf} = 1.255 \text{ tsf} \end{aligned}$$

From the void ratio–pressure diagram, the initial void ratio corresponding to the pressure p_1 is $e_1 = 0.837$.

To find the final pressure p_2 corresponding to an applied load of 4 tsf, the first step is to determine the weight of excavation

$$\begin{aligned} 5 \times 120 &= 600 \text{ psf} \\ 1 \times 70 &= \underline{70} \\ \text{Weight of excavation} &= 670 \text{ psf} = 0.335 \text{ tsf} \end{aligned}$$

The net consolidating pressure is equal to the footing pressure minus the weight of the excavation, or $4 - 0.335 = 3.665$ tsf. This consolidating pressure is applied uniformly over the area of the pier footing. The pressure increment produced at the mid-plane of the compressible layer directly under the center of the pier by the applied consolidating pressure can be computed according to the Boussinesq theory by means of the diagram in Fig. 17-4 or Table 17-1. The depth from the bottom of the footing to the mid-plane is 27 ft, and the dimensions of one quadrant of the footing are 10 ft and 30 ft. Therefore, $m = 10/27 = 0.37$ and $n = 30/27 = 1.11$. The influence coefficient corresponding to these values is 0.096. Then the consolidating pressure at the mid-plane under the center of the pier is

$$4 \times 0.096 \times 3.665 = 1.408 \text{ tons/sq ft}$$

The final pressure at the mid-plane equals the initial pressure plus the consolidating pressure, or

$$p_2 = 1.255 + 1.408 = 2.663 \text{ ton/sq ft}$$

The final void ratio corresponding to the final pressure p_2 is $e_2 = 0.730$. Knowing e_1 and e_2 and the thickness d of the compressible layer, we can compute the total settlement s by applying Eq. (18-8). Thus,

$$s = 20 \times \frac{0.837 - 0.730}{1 + 0.837} = 1.165 \text{ ft}$$

18.29.4. Time-Settlement Relationship for Alternative No. 1. Next the time-settlement analysis will be made for alternative No. 1. In this situation, since the compressible clay layer is sandwiched between two permeable layers, it is free to drain in both the upward and downward directions and the maximum distance through which pore water has to travel in order to escape is one-half the thickness of the layer. Therefore, $h = d/2$ and Eq. (18-16) applies. Also, since the pressures involved in this analysis are between 1 and 4 tons/sq ft, the coefficient of consolidation to be used is the average of the values obtained for the pressure increments from 1 to 2 and from 2 to 4 tsf. Therefore, from Table 18-3, $c = (44.8 + 39.4)/2 = 42.1$ sq ft/yr. A typical computation for this case follows: For $q = 10\%$, $s_{10} = 0.117$ ft, $h = 10$ ft, $T = 0.008$, and $c = 42.1$ sq ft/yr, Eq. (18-16) gives

$$t_{10} = \frac{0.008 \times 10^2}{42.1} = 0.019 \text{ yr}$$

This indicates that the pier will settle approximately 0.117 ft in 0.019 yr. Similar computations for other values of q may be made. The results are tabulated in Table 18-4 and are shown graphically by the lower time-settlement curve in Fig. 18-13.

TABLE 18-4. Computed Time-Settlement Relationship for Alternatives No. 1 and No. 2

Values of q (%)	*Time Factor*	*Settlement (ft)*	*Alternative No. 1 Time (yr)*	*Alternative No. 2 Time (yr)*
0	0	0	0	0
10	0.008	0.117	0.019	0.076
20	0.031	0.233	0.074	0.30
30	0.071	0.350	0.17	0.68
40	0.126	0.466	0.30	1.20
50	0.197	0.583	0.47	1.87
60	0.287	0.699	0.68	2.73
70	0.403	0.816	0.96	3.83
80	0.567	0.932	1.35	5.39
90	0.848	1.049	2.01	8.06
100	∞	1.165	∞	∞

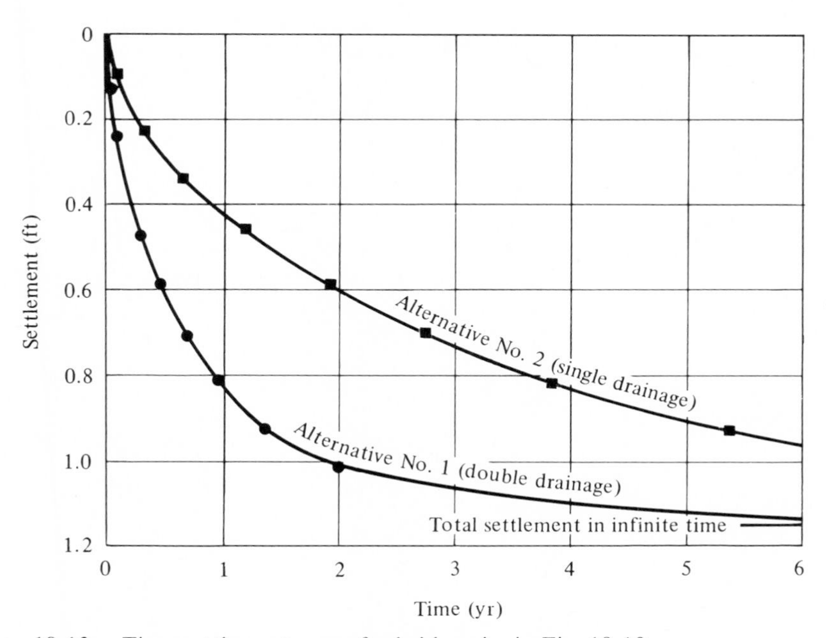

Fig. 18-13. Time-settlement curve for bridge pier in Fig. 18-10.

18.29.5. Computations for Alternative No. 2. In the case of alternative No. 2, the clay layer can drain only in the upward direction, and the maximum distance through which the pore water must travel to escape is the full thickness of the layer; that is, $h = d$. Then, by Eq. (18-17),

$$t_{10} = \frac{0.008 \times 20^2}{42.1} = 0.076 \text{ yr}$$

Again, similar computations are made for other values of q. The results are tabulated in Table 18-4 and are shown by the upper time-settlement curve in Fig. 18-13.

18.30. TIME OF CONSTRUCTION. The foregoing discussion and illustrations are based upon the assumption that consolidating pressure is applied to a soil stratum instantaneously. In practice, this is never the case. Time is required to build a structure and the load is applied gradually, or perhaps intermittently over the construction period, which in some instances, may extend over several years. This period of gradual load application does not influence the estimated total settlement over a

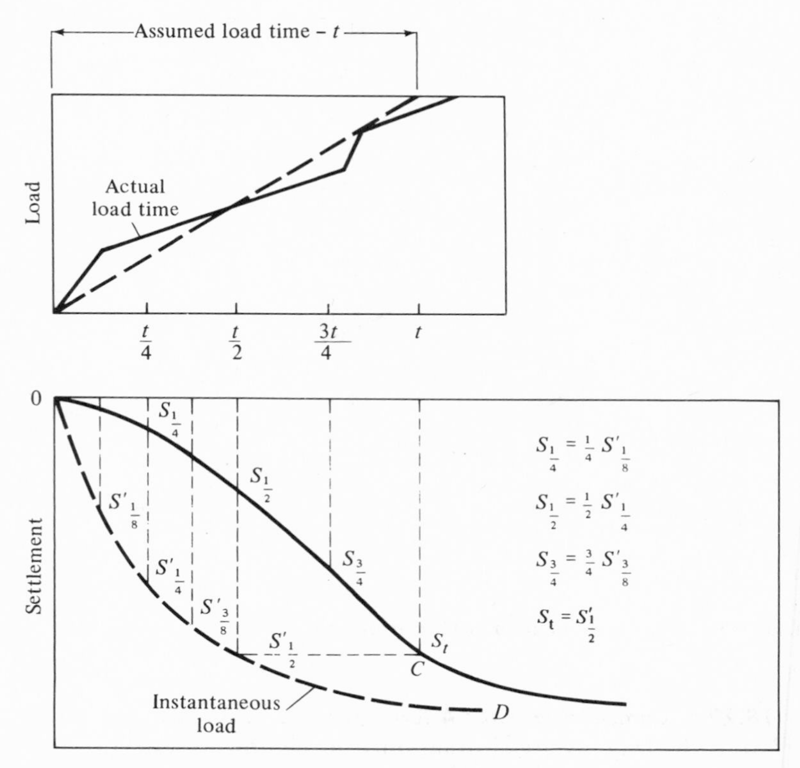

Fig. 18-14. Illustration of procedure for determining time-settlement curve during construction period.

long period of time, but it does influence the rate at which the settlement develops, especially in the early stages of settlement.

An approximate method of estimating settlements during the construction period has been outlined by Taylor (*12*) and is based upon the assumption of a straight-line increase in consolidating pressure from beginning to end of the construction period. At any elapsed time, expressed as a fraction of total time, the settlement is considered to be this fraction multiplied by the settlement at one-half the elapsed time on the curve of time vs. settlement under instantaneous loading.

The procedure is to first calculate the instantaneous load-settlement curve, as *OD* in Fig. 18-14. Then according to the above assumption, the settlement at the end of construction on curve *OC* will be equal to the settlement under instantaneous loading at one-half the total construction time. At any intermediate time, say three-fourths of the total construction period, the settlement equals this fraction times the instantaneous settlement at three-eighths of the total time. This procedure may be repeated to obtain as many points on the settlement curve during construction as may be desired.

The settlement curve beyond completion of construction is assumed to be the same as the instantaneous curve, offset to the right by one-half the construction period.

18.31. ACTUAL AND CALCULATED SETTLEMENTS. There have been sufficient comparisons made between actual settlements and those estimated by the one dimensional consolidation theory to establish its general correctness and reliability, even though many circumstances such as nonhomogeneity of the soil, anisotropy of its permeability, and drainage characteristics, to name a few, may contribute to a divergence between

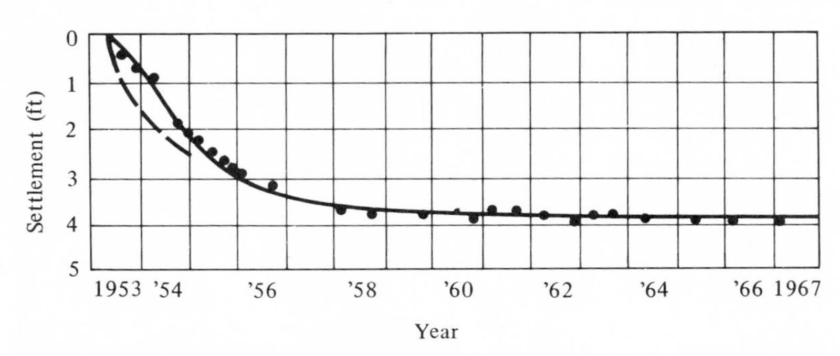

Fig. 18-15. Measured and calculated foundation settlements at Gage No. 33, Ft. Randall Dam. Key: Solid line, calculated settlement; dashed line, instantaneous load; circles, measured settlements.

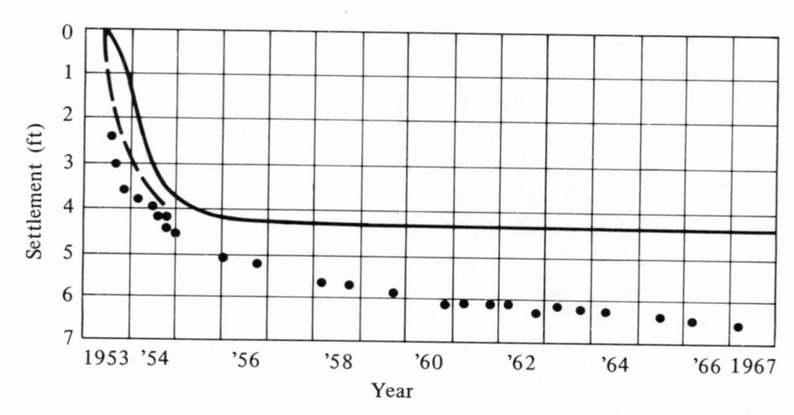

Fig. 18-16. Measured and calculated foundation settlements at Gage No. 36, Ft. Randall Dam. Key: same as in Fig. 18-15.

actual and estimated settlements. On the whole, however, carefully conducted settlement analyses will indicate the order of magnitude of settlement to be anticipated.

A comparison between actual and calculated settlements was made at twelve locations within the Ft. Randall Dam on the Missouri River in South Dakota (*10, 11*). The best and the poorest correspondence between calculated and actual settlements among the twelve stations studied are shown in Figs. 18-15 and 18-16.

18.32. LATERAL STRESSES DURING CONSOLIDATION. The one-dimensional consolidation test has been criticized because it does not allow lateral deformation to occur, and because of shearing stresses around the perimeter of the specimen. The latter may be minimized by use of a "floating" ring to confine the soil, a Teflon ring liner, or coating the ring with silicone grease. Lateral confinement during the test is consistent with soil conditions in the field at relatively shallow depths under large loaded areas, where lateral movement of the soil is precluded by extent of the stress field. Lateral stresses developed during the consolidation test also are of interest since they simulate those developed in naturally consolidating soils where lateral movement is prevented by extent of the deposit.

The lateral pressure during consolidation may be measured by using a special consolidometer or axial loading device whereby the sample is sealed in a rubber membrane and confined on the sides by oil or water pressure. A metal ring and strain gauge arrangement are placed around the sample to sense when the loaded sample starts to bulge at the sides, and lateral oil pressure is increased just enough to prevent bulging. The oil pressure then equals the lateral pressure developed by the soil. The ratio of lateral to vertical effective stress is designated K_0, the *coefficient of earth pressure at rest:* (see Chapter 22)

$$K_0 = \frac{\sigma'_h}{\sigma'_v} \tag{18-19}$$

where σ'_h is horizontal stress and σ'_v vertical stress.

In normally consolidated soils K_0 is usually 0.4 to 0.5 for granular soils, and up to 0.7 for clays. The development of lateral stress and therefore the value of K_0 are resisted by sliding friction between grains of soil, which tends to be greater for sand than for clay particles. An empirical relationship between K_0 and plasticity index, based on data from Brooker and Ireland (*1*), is

$$K_0 = 0.4 + 0.0065 \text{ PI} \lesssim 0.7 \tag{18-20}$$

where PI designates the plasticity index. A widely used and accepted equation by Jaky is

$$K_0 = 1 - \sin \phi \tag{18-21}$$

where ϕ is the soil friction angle, discussed in Chapter 19.

Equations (18-20) and (18-21) apply to normally consolidated clays. The relation to K_0 in overconsolidated clays can be visualized from their stress history: Let us say a clay is normally consolidated under vertical effective stress of 1000 psf. If $K_0 = 0.5$, the horizontal effective stress will be 500 psf. The vertical load is then partially removed: σ_v decreases, but σ_h tends to remain essentially the same. For example if σ_v is reduced to 500 psf and σ_h remains the same, $K_0 = 500 \div 500 = 1.0$. Or if σ_v were reduced to 200 psf, K_0 would be 2.5. Eventually if σ_v were reduced low enough, σ_h should induce a reverse sliding of the soil grains, partially relieving lateral stress. Thus K_0 for overconsolidated soils may reach the reciprocal of the values mentioned above or even exceed these values in clays, because of improved bonding in overconsolidated soils.

The above analysis applies only to soils overconsolidated by vertically applied external loads; those overconsolidated by drying shrinkage will have a reduced K_0, until vertical shrinkage cracks render $K_0 = 0$.

18.33. STRESS PATH METHOD FOR PREDICTION OF SETTLEMENT. A more precise method for predicting total settlement under small loaded areas is the stress path method devised by Lambe (*6*), consisting of subjecting a soil sample to the vertical and lateral stresses existing and anticipated in the field. The apparatus for doing this is the same as is used for triaxial shear testing and is shown in Fig. 19-14 in the next chapter. For such triaxial loading the soil sample is covered with a rubber membrane and placed in a pressure chamber in order to apply all-around pressure which represents σ_h. A vertical load is then applied to simulate σ_v. Drainage of pore water is allowed through the base plate.

First the soil sample is in effect "put back" into its field stress load situation by applying σ_v equal to that calculated from overburden pressure and σ_h from σ_v times an assumed K_0. The sample is then loaded with additional vertical and horizontal stresses predicted from elastic theory by methods discussed in Chapter 17, and vertical deformations are measured. Several samples from several depths may be tested and deformations summed, or a single sample from an "average" depth equal to one-half to three-fourths the width of the loaded area may be used as an approximation.

Unfortunately in this field situation not only are lateral stresses different from those in the usual consolidation test, the rate of settlement involves three-dimensional rather than one-dimensional drainage of the pore water. This problem has not been thoroughly resolved, but qualitatively we can say that settlement should occur faster than would be predicted by one-dimensional consolidation theory.

A very common load condition, a narrow strip load, is not covered by either the consolidation test or the triaxial stress-path method. In this case σ_h is not constant but depends on direction: In the direction parallel to the load strip σ_h will be $K_0\sigma_v$, as in the consolidation test, and perpendicular to this it will be in accord with elastic criteria used in the stress path method. As for a small area load, the conventional consolidation test should tend to underpredict total settlement and overpredict the time required.

PROBLEMS

18.1. What is the difference between primary consolidation and secondary consolidation of a foundation soil?

18.2. The void ratio of a soil is 0.798 at a pressure of 1 tsf and the compression index is 0.063. Draw a void ratio–pressure diagram for this soil on both a semilogarithmic scale and a natural scale.

18.3. Plot the following void ratio–pressure data on a semilogarithmic scale and estimate the preconsolidation load of the soil.

18.4. The data in Problem 18.3 is for a soil sample taken at a depth 20 ft below the surface and 15 ft below the water table. Assume the soil above the water table weighs 110 pcf and the buoyant unit weight below water is 66 pcf. Is this soil normally loaded or preloaded?

18.5. Determine the compression index of the soil in Problem 18.3.

18.6. A clayey soil has a liquid limit of 28. What is the approximate value of its compression index?

Data for Problem 18.3

Load Increment (tons/sq ft)	*Void Ratio*
0.1	1.072
0.2	1.063
0.5	1.044
1.0	1.013
2.0	0.941
4.0	0.834
8.0	0.725
16.0	0.616

18.7. Explain the meaning of the term nonconsolidating pressure, and illustrate with a numerical example.

18.8. A compressible soil layer is 28 ft thick and its initial void ratio is 1.046. Tests and computations indicate that the final void ratio after construction of a building will be 0.981. What will be the probable total settlement of the building over a long period of time?

18.9. In a consolidation test, the void ratio of the soil decreases from 1.239 to 1.110 when the pressure is increased from 2 to 4 tons/sq ft. If the coefficient of permeability of the soil at this increment of pressure is 8.4×10^{-8} cm/sec, determine the coefficient of consolidation, in square feet per year.

18.10. Using the data for the increment of load from 4 to 8 tsf in Table 18-2, determine the coefficient of consolidation of the soil, in square inches per minute and in square feet per year, by Taylor's square root of time fitting method.

18.11. In connection with the development of Eq. (18-18) it is stated that the theoretical value of the time factor T at 90% consolidation is 0.848. Verify this by solving Eq. (18-9).

18.12. A foundation 20 ft by 40 ft in plan is to be constructed at a site where the geological profile is as shown in Fig. 18-17. The foundation load is 3 tsf and the unit weights of various soils are as shown in the sketch. Consolidation tests of the compressible clay indicate that it has a void ratio of 1.680 at 1 ton/sq ft and a compression index of 0.69. The coefficient of consolidation is 8.3×10^{-4} cm^2/sec. Calculate the total settlement, and plot the rate of settlement at the center of the foundation, in inches, vs. years.

18.13. Assume that the material beneath the compressible clay in Problem 18.12 is impervious rock instead of gravel. Calculate and plot the rate of settlement versus time for this condition, and compare the results with those of Problem 18.12.

18.14. Assume that the total time of construction of the foundation in Problem 18.12 is 1 yr and that the load is applied uniformly over this period of time. Calculate and plot the settlement vs. time for this condition.

18.15. A foundation 40 ft square to be built on 20 ft of interbedded sand and clay soil over permeable bedrock. The soil is normally consolidated, its unit weight is 120 pcf, and the plasticity index is 20. The water table is at the ground surface and the foundation will exert a vertical stress of 4 tsf. (a)

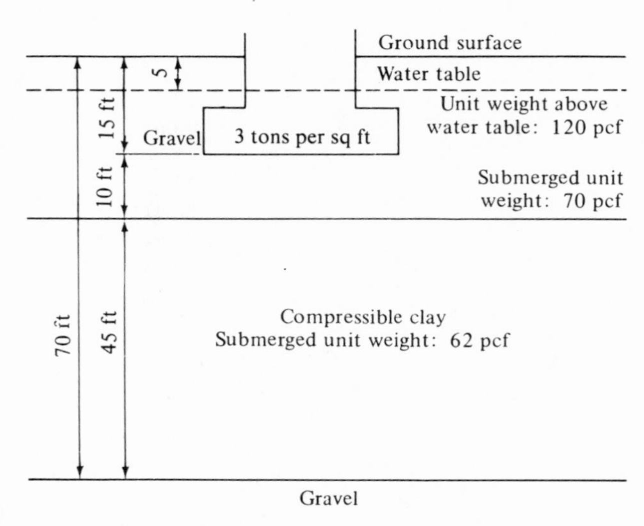

Fig. 18-17. Conditions for Problem 18.12.

Calculate the initial and final stresses which should be applied to a sample from the middle of the clay layer in order to predict total settlement by the stress path method. (Use the elastic solutions of Fig. 17-4 and 17-12 to calculate final stresses.) (b) Repeat (a), but assume the clay layer is dewatered by well points for a period prior to construction, then the water table is allowed to rise to its former level as the tank is being built. Would there be any advantage in this procedure? (c) A test is conducted as outlined in part (a) and the sample fails by shearing. What do you conclude relative to the proposed load at this site?

18.16. A building resting on linear footings on clay appears stable for 20 yr after construction, and then appears to settle and becomes severely cracked. (a) Can this be explained by consolidation theory, and if so, how? (b) Suggest other possible causes of settlement. (c) Suggest another explanation for the cracking, related to soil mineralogy and moisture content. (d) What data could be obtained to prove or disprove each possible cause listed above?

18.17. What remedies could be applied to each possible cause in Problem 18.16?

REFERENCES

1. Brooker, E. W., and H. O. Ireland. "Earth Pressures at Rest Related to Stress History" *Canadian Geotech. J.* **11**, 1 (1965).
2. Casagrande, A. "The Determination of the Pre-Consolidation Load and Its Practical Significance." *Proc. Intern. Conference on Soil Mechanics and Foundation Engineering,* Vol. 3, 60. Harvard University, Cambridge, Massachusetts, 1936.

3. Crawford, Carl B. "Interpretation of the Consolidation Test." *ASCE J. Soil Mechanics and Foundation Eng.* **90** (SM5), 87–102 (1964).
4. Gibbs, H. J. and J. P. Bara. "Stability Problems of Collapsible Soils." *ASCE J. Soil Mechanics and Foundation Eng.* **93** (SM4), 577–594 (1967).
5. Handy, R. L. "Collapsible Loess in Iowa." *Proc. Soil Sci. Soc. Amer.* (1973, in press).
6. Lambe, T. W. "Stress Path Method." *ASCE J. Soil Mechanics and Foundation Eng.* **93** (SM6), 309–331 (1967).
7. Lowe, John, III, P. F. Zaccheo, and H. S. Feldman. "Consolidation Testing with Back Pressure." *ASCE J. Soil Mechanics and Foundation Eng.* **90** (SM5), 69–86 (1964).
8. Palmer, L. A., and E. S. Barber. "The Theory of Soil Consolidation and Testing of Foundation Soils." *Public Roads* **18**, No. 1 (March 1937).
9. Skempton, A. W. "Notes on the Compressibility of Clays." *Quart. J. Geol. Soc. London* **100,** 119–135 (1944).
10. Spangler, M. G., and A. L. Griebling. "Foundation Settlements in the Ft. Randall Dam Embankment." *Highway Research Board,* Bull. **173,** 1958.
11. Spangler, M. G. "Further Measurements of Foundation Settlements in the Ft. Randall Dam Embankment." *Highway Res. Board Rec.* **223,** 60–62 (1968).
12. Taylor, Donald W. "Research on Consolidation of Clays." Department of Civil and Sanitary Engineering, Massachusetts Institute of Technology Serial 82 (August 1942).
13. Taylor, Donald W. *Fundamentals of Soil Mechanics.* John Wiley, New York, 1948.
14. Terzaghi, Charles. "Principles of Final Soil Classification." *Public Roads* **8,** No. 3 (May 1927).

19

Shearing Resistance and Strength

19.1. IMPORTANCE OF SHEARING STRENGTH OF SOIL. One of the most important engineering properties of soil is its shearing strength or its ability to resist sliding along internal surfaces within a mass. Shearing strength is the property which enables soil to maintain equilibrium on a sloping surface, such as a natural hillside, the backslope of a highway or railway cut, or the sloping sides of an embankment, levee, or earth dam. This strength materially influences the bearing capacity of a foundation soil and the lateral pressure which a soil backfill exerts against a retaining wall, bulkhead, or other type of restraining structure. Shearing stresses in the backfill over an underground conduit, such as a sewer or a culvert, exert a surprisingly great influence upon the load to which a structure of this kind is subjected in service. There is hardly a problem in the field of soil engineering which does not involve the shear properties of the soil in some manner.

19.2. BASIC PRINCIPLES RELATING TO FRICTION BETWEEN SOLID BODIES. In order to introduce the subject of shearing resistance and shearing strength, it will be helpful to review some of the basic principles of friction between solid bodies; since shear in soils is similar in many respects to this widely observed phenomenon, although there are important differences between the two subjects also. Imagine a brick resting on a horizontal ta-

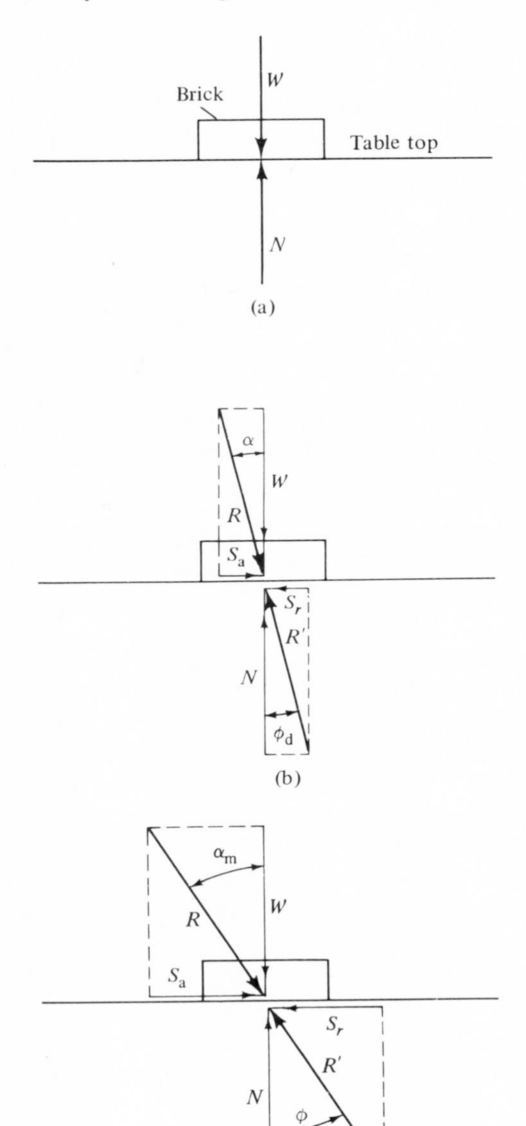

Fig. 19-1. Friction on horizontal surface.

ble top, as shown in Fig. 19-1(a). The brick is in equilibrium under its own weight W and the equal and opposite reaction N provided by the table. Now, suppose that, as indicated in Fig. 19-1(b), a horizontal force S_a is applied to the brick near the plane of contact between the brick and the table. If this force is relatively small, the brick will remain at rest and the applied horizontal force will be balanced by an equal and opposite force S_r in the plane of contact. This resisting force is developed as a result of the roughness characteristics of the bottom of the brick and the table surface.

If the applied horizontal force is gradually increased, the resisting force will likewise increase, always being equal in magnitude and opposite in direction to the applied force. There is a limit, however, to the amount of resistance which can be developed at the plane of contact; and, when the applied force equals or exceeds the maximum possible resistance, equilibrium will be destroyed and the brick will move along the table top. This movement or slippage is a shear failure. The applied horizontal force is a shearing force and the developed force is friction or shearing resistance. The maximum shearing resistance which the materials are capable of developing is called the shearing strength.

19.3. FRICTION ANGLE. When a shearing force is applied to the brick in Fig. 19-1, the resultant R of the shearing force and the weight acts at an angle with the line of action of the force representing the weight, which in this case is normal to the shear plane. This angle, designated as α in Fig. 19-1(b), is called the *obliquity of the resultant* or simply the *obliquity angle.* When the shearing force is increased to a value just equal to the shearing strength, that is, when sliding or failure is impending, the obliquity angle reaches its maximum value and is designated α_m, as shown in Fig. 19-1(c). The forces that are applied normal and tangential to the shear plane are related to each other in accordance with the following equations:

$$\tan \alpha = \frac{S_a}{W} \tag{19-1}$$

$$S_a = W \tan \alpha \tag{19-2}$$

In a similar manner, the reaction N to the weight of the brick, which also acts normal to the shear plane, may be combined with the shearing resistance to obtain a resultant R' which makes an angle ϕ_d with the normal. This angle ϕ_d is called the *developed friction angle;* and it is equal to the obliquity angle α, since the reaction is equal to the weight and the shearing resistance is equal to the applied shearing force. Attention is directed to the fact that the angle ϕ_d depends on the magnitude of the applied shearing force, as long as this force is not sufficient to cause shear failure. However, when the applied shearing force is large enough to cause failure, the shearing resistance has reached its maximum possible value for the particular materials involved. The angle ϕ_d reaches its maximum value at failure and this maximum value is designated as ϕ, as shown in Fig. 19-1(c). This limiting angle ϕ is called the *friction angle* and constitutes a physical property of the materials, which in this illustrative case are the brick and the table top.

If shear on an interior plane in a mass of soil is considered, the angle ϕ is a property of the soil, and the value of tan ϕ is called the *coefficient of internal friction* of the soil. This coefficient is here denoted by μ. The value of tan ϕ is equal to the shearing strength of the soil divided by the reaction normal to the shear plane. Also it is equal to the shearing stress at failure divided by the applied weight force normal to the shear plane. Thus,

$$\tan \phi = \mu = \frac{S_r}{N} = \frac{S_a}{W} = \tan \alpha_m \qquad (19\text{-}3)$$

For example, if the friction angle ϕ of a cohesionless soil is given as 30°, it means that the soil is capable of providing sufficient shearing resistance to maintain equilibrium as long as the applied shearing force produces an angle ϕ_d less than 30°. When the applied shearing force causes ϕ_d to become equal to or greater than 30°, shear failure will result.

19.4. FRICTION ALONG INCLINED SURFACE. Another illustration of friction between solids is shown in Fig. 19-2. Suppose that a brick is placed on a table top which is hinged at one end. As long as the table remains horizontal, the situation will be as shown in Fig. 19-1. But, if the table top is tipped upward through an angle *i*, the force relationships shown in Fig. 19-2(a) develop. In this case the component S_a of the weight that is parallel to the sloped surface tends to pull the brick down hill and is, therefore, a shearing force. As long as the angle *i* is relatively small, the brick will remain in equilibrium because a shearing resistance S_r is developed which balances the shearing force. When the angle *i* is just large enough to cause the brick to slide, the shearing resistance equals the shearing strength of the material and the forces are related as shown in Fig. 19-2(b). Then,

$$N = W \cos \phi \qquad (19\text{-}4)$$

$$S_r = W \sin \phi \qquad (19\text{-}5)$$

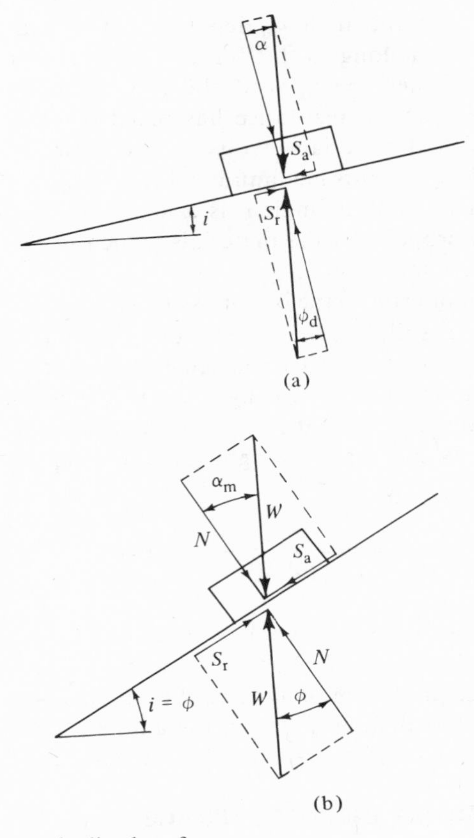

Fig. 19-2. Friction on inclined surface.

$$\tan\,\phi = \mu = \frac{S_r}{N} = \frac{W \sin\phi}{W \cos\phi} \tag{19-6}$$

19.5. SHEARING STRENGTH RELATED TO FRICTION. The preceding discussion indicates that shearing stress and strength attributable to friction can exist only under the following conditions. First, there must be a force normal to the plane on which shear is being considered; second, the material must exhibit friction characteristics, that is, it must have a finite coefficient or angle of friction. These conditions may be illustrated by considering two sheets of sandpaper with their sanded surfaces in contact. These may be very easily caused to slide over each other when no normal force is applied. When a normal force is applied, the resistance to sliding or the shearing strength increases in direct proportion to the normal force.

If a quantity of clean, dry, cohesionless sand is poured out on a level surface, it will come to rest in a cone-shaped heap. The reason the sand can be piled up in this manner is because it has the property of internal friction. It has a definite value of the angle ϕ and is in equilibrium at a definite *angle of repose.* In contrast to the sand, water has no shearing strength, and its angle ϕ is zero. It flows freely down hill under the influence of gravity because of this fact. Furthermore, if a quantity of water is poured out on a level surface, it will spread out in a very thin layer and cannot be heaped up. Its angle of repose is zero.

19.6. ADHESION THEORY FOR SLIDING FRICTION. Although proposed by Terzaghi in 1925 and others even earlier, the adhesion theory for sliding friction had few adherents until it was elaborated and demonstrated by Bowden and Tabor (*3*). This theory suggests that even smooth surfaces in contact actually touch each other at only a few tiny areas. Friction is caused by chemical or adsorptive bonding at these points of contact (Fig. 19-3). The higher the normal stress the larger the contact areas become, and hence the higher the resistance to sliding. When normal stress is re-

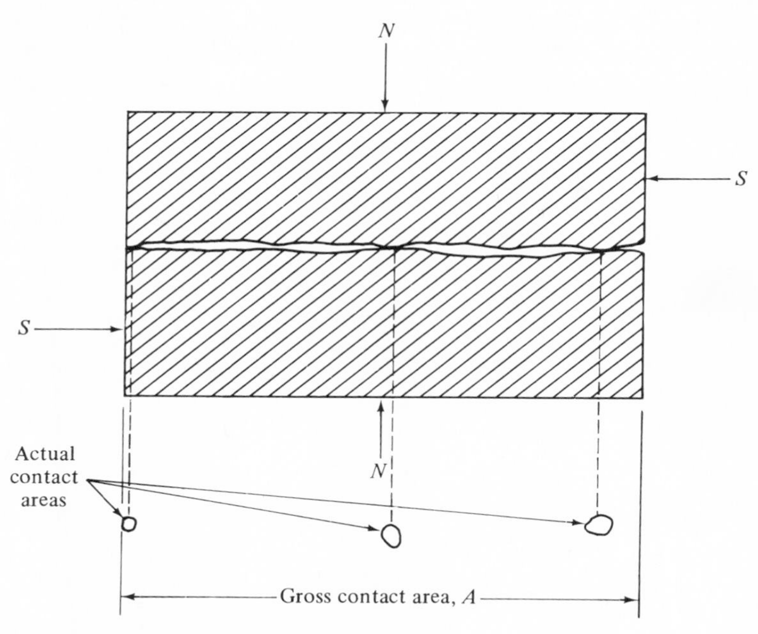

Fig. 19-3. Adhesion theory of friction.

duced, the contact areas of hard materials reduce owing to elastic rebound. However, softer materials such as clays may suffer permanent deformation and retain much of their adhesional strength as *cohesion.* This is discussed in Section 19-11.

19.7. SHEARING STRENGTH OF GRANULAR SOIL. The shearing strength of a granular soil, such as a clean sand, a sand-gravel mixture, and to some extent a friable silt, is closely analogous to the frictional resistance of solids in contact. The relationship between the normal stress acting on a plane in the soil and its shearing strength can be expressed by an equation which is compatible with Eq. (19-3). In terms of stresses,

$$\tau_f = \sigma \tan \phi \tag{19-7}$$

in which

τ_f = the shearing stress at failure, or the shearing strength;

σ = normal stress; and

ϕ = friction angle.

This relationship is illustrated in Fig. 19-4.

The friction angle ϕ and therefore the coefficient of internal friction of a granular soil is primarily influenced by density or grain packing. In dense soils, as in rubbing together two sheets of sandpaper, ϕ may far exceed the angle of sliding friction of two mineral surfaces. The reason in

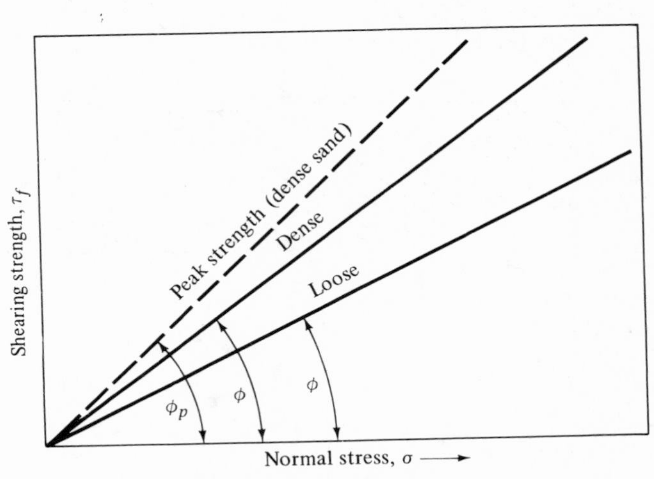

Fig. 19-4. Shearing strength vs. normal stress for cohesionless sand.

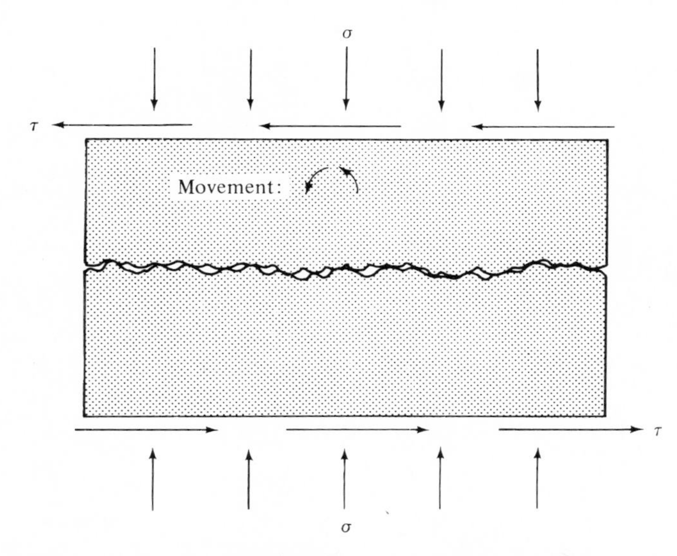

Fig. 19-5. Shearing of dense granular soil requires "up-and-over" movements (small arrows) because of interlocking of grains in the shear zone.

both cases is that the grains *interlock*, much like the teeth of two highly irregular gears, and they must be lifted over one another in order for sliding to occur (Fig. 19-5). The lifting must be done against the normal stress σ, and the force required therefore is proportional to σ, just as sliding friction is proportional to σ. In soils, ϕ therefore is designated the angle of *internal* friction, because it represents the sum of sliding friction plus interlocking.

Whereas sliding friction is more or less constant in a given soil, interlocking is greatly influenced by gradation and densification. In order to understand the influence of gradation, one must examine interlocking in terms of volume change energy. A tightly packed granular soil must increase in volume in order to shear. This behavior is called *dilatancy*. From a consideration of work done during volume change, Taylor (*15*) first proposed a method to separate sliding and interlocking components of internal friction. A more precise equation was derived by independent analyses by Newland and Allely (*11*) and by Rowe *et al.* (*12*):

$$\phi = \phi_s + \theta \tag{19-8}$$

where

$$\tan\theta = \frac{\delta V}{\delta \Delta}, \tag{19-9}$$

in which

ϕ = angle of internal friction,
ϕ_s = angle of sliding friction,
θ = effect of interlocking,
$\delta V/\delta\Delta$ = unit volume change per unit shearing deformation, equal to slope of the height-displacement curve.

This equation applies in the case of the direct shear test, discussed in Section 19.19.

From Eqs. (19-8) and (19-9), the greater the rate of increase in volume as a soil shears, the higher the shearing strength. This is shown diagramatically in Fig. 19-6. The optimum gradations discussed in Chap-

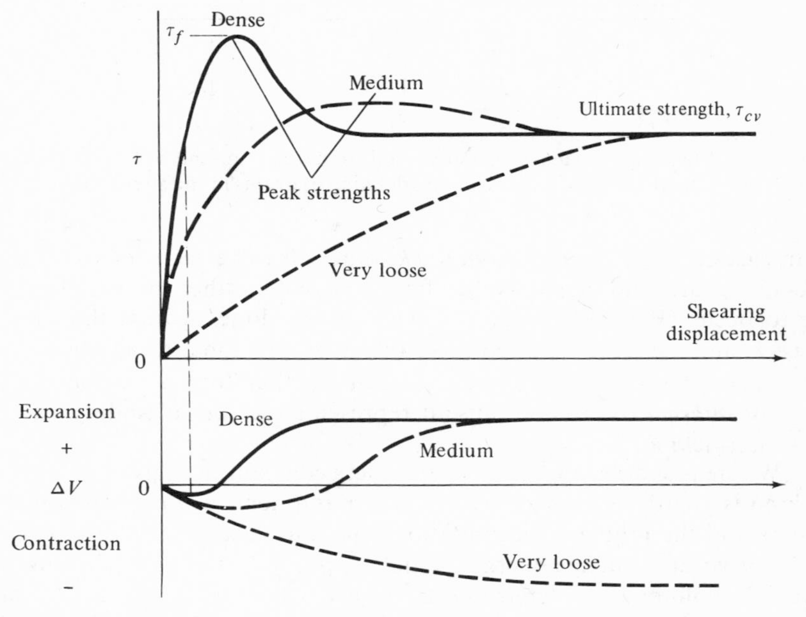

Fig. 19-6. Ideal shearing stress–volume change–displacement relationships for granular soils.

ter 16 therefore give the densest packing of solid particles, which prevents dislodged particles from falling into open pores without inducing volume change. A highly angular or platy particle shape may detract from strength by not allowing such dense packing during compaction.

After the soil volume has increased throughout the shearing zone, no further dilatancy occurs, and as shown in Fig. 19-6, the strength decreases from the *peak strength* to the *ultimate or residual strength,* which may be lower.

Very loose soils may undergo *negative dilatancy* or decrease in volume during shearing, subtracting from the sliding frictional strength. These also eventually reach a stable volume whereby no further volume change occurs.

The void ratio whereby volume remains the same during shearing is called the *critical void ratio.* It is important to recognize that whether a soil is densifying or dilating, the critical void ratio will be reached only in the shearing zone, although it is often calculated from the initial density and total volume change data. Since the residual strength represents strength at constant volume, i. e., when there is neither an increase nor a decrease in volume, it is sometimes designated τ_{cv}.

19.8. SATURATED GRANULAR SOIL. The friction angle of a saturated sand is only slightly less than that of a dry sand having the same relative density. However, if the sand is submerged below a water table, the buoyant effect of the water on the normal stress on a plane must be taken into account. The normal stress σ in Eq. (19-7) should be the effective or intergranular stress. According to Section 12.21, this stress is equal to the total stress at a point minus the neutral stress or pore-water pressure. Therefore

$$\tau_f = (\sigma - u) \tan \phi' \tag{19-10}$$

in which

τ_f = shearing strength,
σ = normal stress,
u = pore-water pressure, and
ϕ' = friction angle, on an effective stress basis.

Therefore, the shearing strength of a sand is reduced when it is submerged, even though the friction angle is not affected appreciably.

The neutral stress in submerged sand is usually equal to the hydrostatic pore-water pressure due to position below a water table. However, if shearing stresses cause the voids in sand to decrease in volume, the neutral stress may be increased by hydrostatic excess pressure. Any such changes in neutral stress influence the effective stress, and therefore the shearing strength, of the material.

19.9. LIQUEFACTION OF SAND. If a mass of sand in a saturated condition has a void ratio that is greater than the critical ratio and it is subjected to a suddenly applied shearing stress, as from an earthquake, heavy blasting, pile driving, or other dynamic forces, the sand tends to decrease in volume. As a result, the pore water is subjected to a suddenly applied hydrostatic excess pressure, and a portion of the weight of overlying ma-

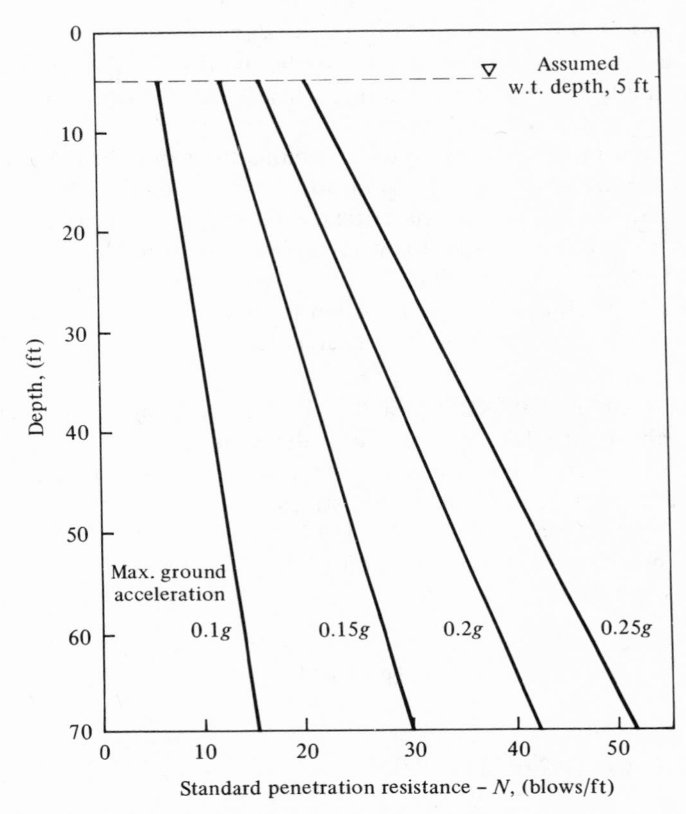

Fig. 19-7. Standard penetration values for which liquifaction is unlikely to occur. After Seed and Idriss (*13*).

terial is transferred from intergranular pressure to pore water pressure, increasing u in Eq. (19-10). This transfer of pressure causes a sudden decrease in shearing strength. If it is reduced to a value below the applied shearing stress, the mass will fail in shear. Such a failure occurs very suddenly, and the whole mass appears to flow laterally as if it were a liquid. This type of failure is sometimes referred to as liquefaction.

Widespread disastrous consequences of liquefaction are very apparent after earthquakes, which cause sudden bearing capacity failures, landslides, and quicksand "boils" in affected areas. Recent severe earthquakes in Japan, Alaska, and California and the likelihood that such quakes will continue to occur have led to intensive study of the factors causing liquefaction. Earthquakes induce cyclical horizontal shearing stresses which increase with depth into soil due to the increasing overburden soil mass. However, repetitive laboratory shear tests show that the stress level required to cause liquefaction also increases with depth, the result being that liquefaction probably first occurs in a discrete shallow depth zone. Seed and Idriss (*13*) recently summarized liquefaction potential of sands in relation to relative density, and expressed the results in terms of standard penetration test blow count and depth, Fig. 19-7. Ground accelerations in moderate to severe earthquakes commonly reach 0.1 to 0.15 *g* (*17*). *N* values (blow count, see Section 6.3) above those indicated in the graph mean the sand should be safe from liquefaction under various accelerations. If *N* is less than about one-half of the indicated values, liquefaction becomes extremely likely; if more than one-half but less than the full value, it may or may not occur, depending on the soil and earthquake magnitude; i.e., number of significant stress cycles.

19.10. SHEARING STRENGTH RELATED TO COHESION. Some soils have a finite shearing strength even when they are not subjected to external forces normal to a shear plane. Furthermore, when soils of this kind are subjected to normal forces, the shearing strength is not increased. These are called cohesive soils; and their shearing strength, which is independent of normal pressure, is called cohesion or no-load shearing strength. Cohesion may be illustrated by considering two sheets of fly-paper with their sticky sides in contact. Considerable force is required to slide one sheet over the other, even though no normal pressure is applied. The shearing resistance in this case is due to cohesion between the sticky surfaces.

19.11. SHEARING STRENGTH OF COHESIVE CLAY. When a saturated cohesive clay is subjected to normal and shearing stresses under conditions in which consolidation takes place, that is, when the volume of voids decreases under the influence of normal and shearing stresses and water is squeezed out of the pores, the shearing strength increases with normal stress somewhat according to the relationship shown in Fig. 19-8. The curve of shearing strength vs. normal stress is roughly a straight line which makes an angle with the horizontal much the same as the similar curve for sand in Fig. 19-4. Some authorities refer to this angle as the friction angle of the clay, but obviously it is not of the same nature as the friction angle of a granular soil or of solids in contact, as was discussed in previous sections.

If one measures the shearing strength of clays which have a previous history of consolidation, i.e., preconsolidated clays, a particularly interesting relationship is found which gives an important clue to the nature of cohesion in clays. Instead of a single line as shown in Fig. 19-8, one ob-

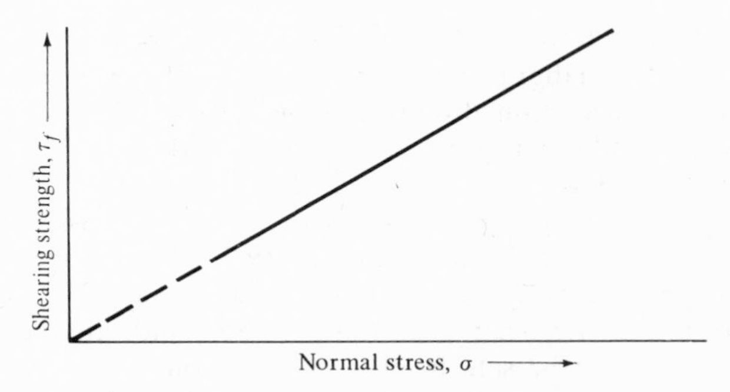

Fig. 19-8. Shearing strength vs. normal stress for consolidating cohesive clay ($u = 0$).

tains a "dog-leg" intersecting at the preconsolidating stress on the shear plane as in Fig. 19-9. The intersection of the first line with the ordinate is called the *cohesion,* or more precisely the cohesive shear strength. The higher the preconsolidating pressure, the higher the line labeled ϕ_1, and the higher the cohesion.

The slope ϕ_1 of this line therefore represents the degree of relaxation of shearing strength after removal of the preconsolidation pressure. In sands no strength is retained so ϕ_1 is steep and equal to ϕ_2, and c is zero.

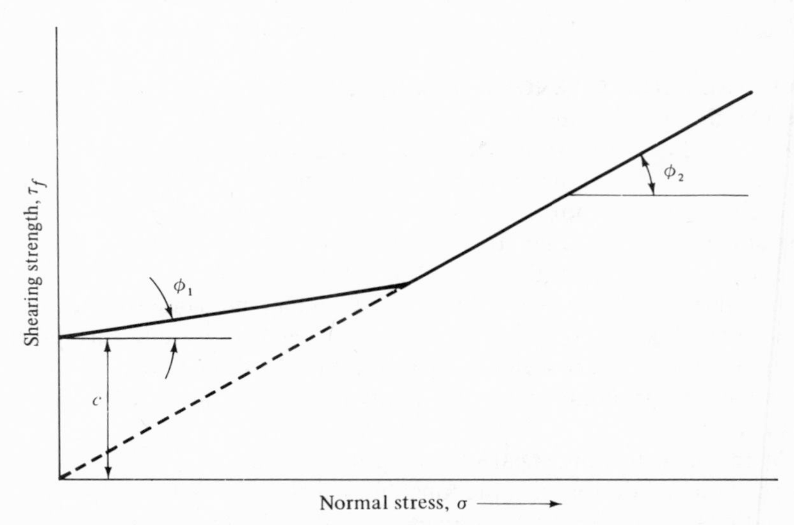

Fig. 19-9. Drained shearing strength vs. normal stress for preconsolidated clay.

In clays considerable strength may be retained, ϕ_1 is low, and c may be high. Once the external consolidation pressure is removed, the only way it's effect can be "saved" is by tensile stresses within the soil, discussed in the next section. A preconsolidation effect therefore can develop indirectly without external application of stress if tension develops in a soil, as when it dries out. Clay may become heavily overconsolidated and attain a high cohesion, merely by a process of drying.

19.12. NEGATIVE PORE WATER PRESSURE OR INTRINSIC PRESSURE.

The first (ϕ_1) shear envelope in Fig. 19-9 may be extrapolated back to zero shearing stress to give the tensile stress, as shown in Fig. 19-10. This is the capillary tension or (without the negative sign) the capillary potential, as discussed in Section 10.9. If clay is present in a soil, very high negative pressures become possible since the water is held by the clay. Such negative pressures are very difficult to measure because unbound liquid water in a water column as shown in Fig. 10-17 vaporizes when the absolute pressure approaches zero. Although methods discussed in Section 10.27 to measure high negative pressures have been used by agronomists and engineers for many years, it is only recently that efforts have been made to measure them during shear testing [Gibbs and Coffey (*8*)]. Such tests confirm the concept of initial stress shown in Fig. 19-9.

19.13. COULOMB EQUATION FOR SHEARING STRENGTH.

In conventional engineering practice only the first, ϕ_1, line in Fig. 19-9 is used. It more often happens that tests are run at only three or four values of σ, and an "average" line or least-squares fit is used. This is appropriate if the σ values used in the tests cover the range which will be experienced in the field. In construction it is best not to exceed the preconsolidation stress of

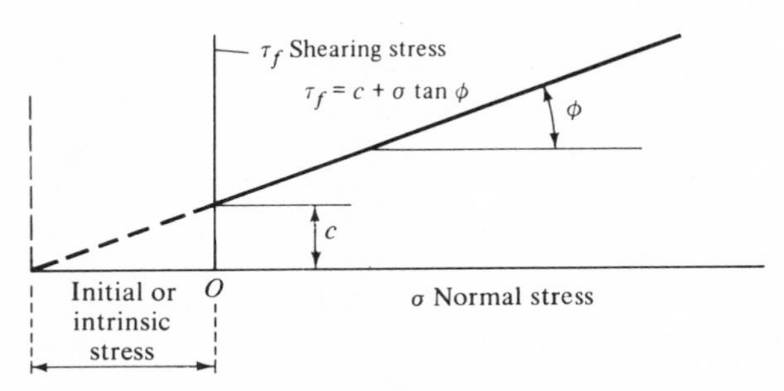

Fig. 19-10. Graph of Coulomb formula for shearing strength of soil.

clays very far because the ϕ_2 line in Fig. 19-9 depends on complete drainage: If drainage does not occur, part of the stress will be taken by pore water pressure, decreasing strength as shown in Eq. (19-10). Even if drainage does occur there may be excessive consolidation and settlement.

The equation for shearing strength as a linear function of normal stress is called the Coulomb equation because it was first proposed by a French military engineer who later became a famous scientist, C. A. Coulomb, in 1773:

$$\tau_f = c + \sigma \tan \phi \qquad (19\text{-}11)$$

in which

τ_f = shearing strength;
c = cohesion;
σ = normal stress on the shear plane; and
ϕ = friction angle.

Equation (19-11) is one of the most widely used equations in soil engineering, since it approximately describes the shearing strength of any soil under drained conditions.

19.14. UNDRAINED SHEARING STRENGTH. Equation (19-11) may be rewritten to include pore water pressure:

$$\tau_f = c' + (\sigma - u) \tan \phi' \qquad (19\text{-}12)$$

in which u is the pore-water pressure and the primes (′) indicate that c and ϕ are expressed on an effective stress basis. The reason for this distinction is shown in Fig. 19-11.

Let us assume that a test is performed with a normal stress at A. The soil initially is not saturated, so at A insignificant pore pressure develops, and the shearing strength at A is the same as the drained strength. A test is then performed with a higher normal stress, at B, but somewhere between A and B the soil consolidates enough to become saturated, such that part of the increase in σ is taken by pore water pressure. The shearing strength at B is higher, but not so high as if the soil were fully drained. A line drawn through A and B therefore gives values of c and ϕ which are based not on effective stresses but on *total stresses*. Note that ϕ is lower and c is higher than ϕ' and c'. Also the degree of difference depends on the amount of pore pressure developed.

As an extreme example that is not uncommon, suppose that additional normal stress is applied to reach a value C in Fig. 19-11, but all of the addition goes to pore water pressure. The line connecting B and C gives a higher cohesion, but $\phi = 0$. Because of the dependence of the un-

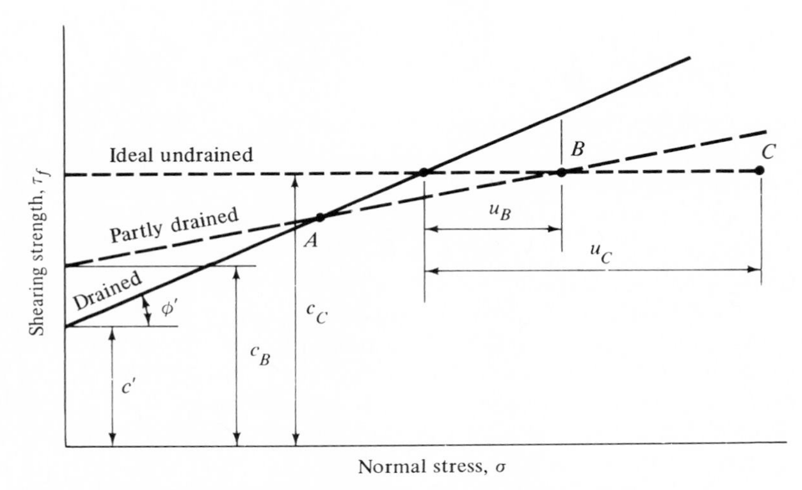

Fig. 19-11. Effect of pore water pressure u on undrained c_u and ϕ_u.

drained total stress c and ϕ values on pore pressure, which may or may not be evaluated, results from undrained tests are often designated c_u and ϕ_u.

19.15. NEGATIVE PORE WATER PRESSURE IN SANDS. Shear tests of moist sands give a linear Coulomb relationship as in Fig. 19-10, but the cohesion is small and disappears if the sand is either very wet or dry. Cohesion which disappears upon saturation is not much to count on, and is called *apparent cohesion.* The cause is capillary tension or negative pore water pressure, illustrated in Figs. 10-5 through 10-8. As shown by Fig. 10-5 and Eq. (10-4), as the soil dries and outward curvature of the moisture films decreases, tension within the water increases, and the grains are more strongly pulled together. However when the negative pressure approaches 1 atm, the moisture vaporizes and the sand dries out and becomes cohesionless. The capillary forces joining such particles were first analyzed by Haines in 1925 (7).

Negative pore pressure also occurs even in saturated sands or other dense soils such as preconsolidated clays, if the grains or peds are interlocked and must dilate in order to shear. In order to dilate, water or air must be pulled in to occupy the expanding pore spaces. Since u is negative instead of positive as in Fig. 19-11, undrained tests of such soils can give an anomalously high value for ϕ which could be dangerous if not corrected for pore pressure. Such dilatancy is readily observed in walking on beach sand at the water's edge; sand reaching incipient shear around the footprint will gain strength and appear to partially dry out.

19.16. COMPARISON OF DRAINED AND UNDRAINED SHEAR DATA. It has already been implied that cohesion and internal friction values derived from undrained or partially drained tests agree with data from drained tests if allowance is made for either positive or negative pore pressure. Fortunately this often is true, but not always. The exceptions are loose, unconsolidated soils such as very loose subaqueous sandy and silty sediments and very sensitive or quick clays. In a drained laboratory test these soils behave as in Fig. 19-8, gradually gaining strength during shearing as a result of consolidation and excessive volume change. In an undrained test there is no opportunity to consolidate, the normal stress being quickly converted to pore water pressure, and strength is less. The best procedure therefore is to select the test condition that most nearly simulates that anticipated in the field. For example, to anticipate conditions of rapid loading, as for a saturated highway subgrade soil under traffic, an undrained test may be preferred. For slow loading conditions on freely draining soils, a drained test may be preferred. To anticipate partly drained conditions an undrained test is performed and pore pressures are measured. To be useful, pore pressures in the field must be predicted, by methods discussed in a later section.

19.17. EFFECT OF SOIL STRUCTURE. As indicated in some detail in Sections 19.7, 19.11, and 19.14, the shearing strength of soil closely relates to its microstructure, which affects interlocking, intrinsic pressure, and pore water pressure. Grosser aspects of structure, as in fissured or blocky clays, also have a major influence, particularly at low normal stresses. This is shown in Fig. 19-12. At low normal stresses the soil peds or lumps may act as densely packed aggregate, giving a medium to high friction angle ϕ, and zero or very low cohesion. At higher normal stresses the granular strength increases but since it cannot exceed the strength of the unfissured clay, the failure envelope follows that for the unfissured clay, ϕ_2. Often the transition is somewhat gradual, giving a curved envelope sometimes likened to a parabola. Precisely the same kind of behavior is observed in rock mechanics but at much higher normal stresses.

The behavior illustrated in Fig. 19-12 has been observed in laboratory and field shear tests where failure can follow the weakest planes. Recent studies of landslides in Canadian quick clays confirm initial failure similar to that of a granular material, whereas failure would not have occurred had the strength been as indicated by tests at higher normal stresses [Eden and Mitchell (*6*)].

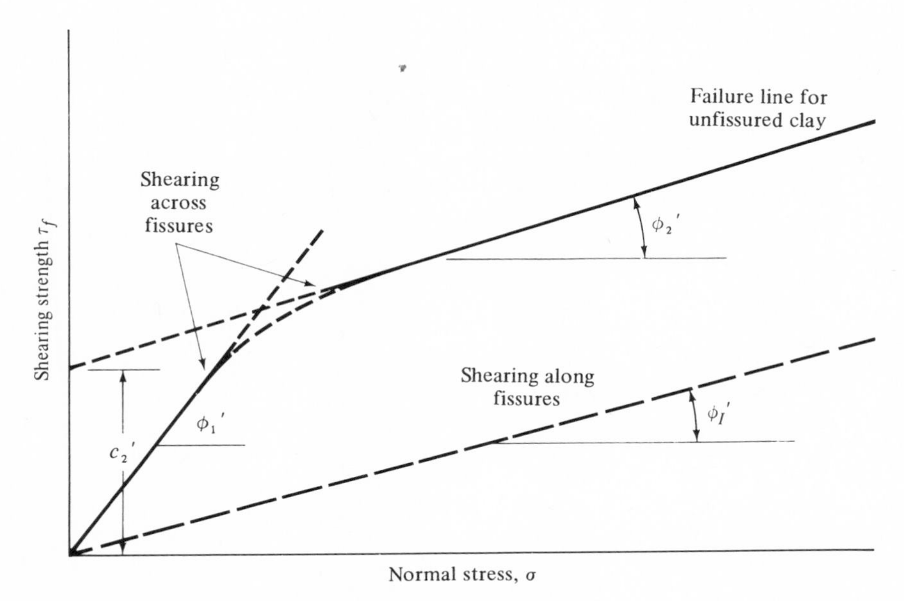

Fig. 19-12. Shear diagrams for fissured clay.

Another possibility occurs when the direction of fissuring closely coincides with the direction of shearing (Fig. 19-12). This strength can be simulated in laboratory tests by continuing shear deformations until the strength drops off because of formation of a shear plane and stabilizes at a *residual strength*. Such tests show that cohesion is zero or very low, and the friction angle ϕ_r' is equal to or a few degrees lower than ϕ' for the unfissured clay [Skempton and Petley (*14*)]

19.18. EFFECT OF RATE OF SHEARING. Except for earthquake and similar dynamic loadings, laboratory and field shear strength tests proceed at a much higher rate than occurs under natural conditions. Initial movements in a landslide may be only a fraction of an inch per day until the sum of soil strengths along the shear surface starts to decrease, when the movement becomes swift and spectacular. In contrast to the variable rate of natural shear failures, laboratory tests usually involve a fixed rate of shearing.

One influence of a rapid laboratory shear rate is to increase pore pressure; however this generally is more than compensated for by the small laboratory sample size, which allows drainage to occur much faster than in the field (see Sections 18.25 and 18.26).

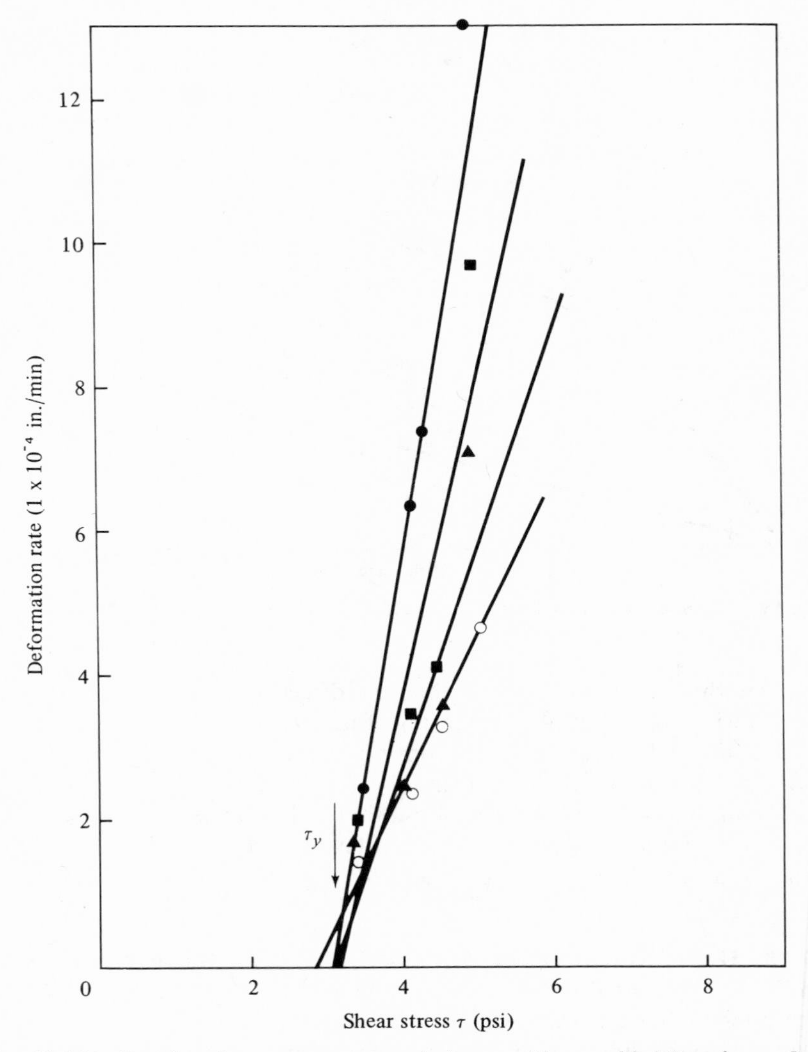

Fig. 19-13. Results of creep tests on sandy loam with w = 20.4% and σ = 5 psi using the bore-hole shear device. Although the creep rate decreases with time, the yield stress is constant. Key to time from start: solid circle, 20 min; square, 30 min; triangle, 40 min; open circle, 50 min. From Lohnes *et al.* (*10*).

A second influence of shear rate is as it pertains to viscous behavior of soils. This is important only with clayey soils. The subject of viscosity was introduced in Section 13.2, and reference should be made to Fig. 13-1, showing viscosity and yield stress. The clay represented in Fig. 13-1 is near the liquid limit; as the water content is reduced, the viscosity and yield stress both greatly increase.

Specialized very slow tests proceding at several different rates may be used to evaluate soil viscous behavior or *creep*. Values of τ_f obtained from creep tests may be plotted against shearing rate and extrapolated to zero rate to give a yield stress, shown in Fig. 19-13. The relation of yield stress to normal stress has been hypothesized but little data are available. Tests by Lohnes *et al.* (*10*) on a porous, actively creeping soil gave data similar to the residual shear strength line in Fig. 19-12, i.e., ϕ is about the same as in more rapid tests but c is zero. This is consistent with the concept of movement along pre-existing planes. In other cases where the soil was not actively creeping, c' was essentially unchanged.

19.19. DIRECT SHEAR TEST FOR DETERMINING SHEARING STRENGTH OF SOIL. At present only two laboratory methods are commonly used for determining shearing strength of soil, namely, the direct shear test and the triaxial compression test. In the direct shear test, a sample of soil in the form of either a low cylinder or a rectangular solid is placed in a testing machine in such a manner that a portion of the mass can be caused to slide in relation to the rest of the specimen. The plane through the sample on which sliding motion is produced is the shear plane. The specimen and the applied forces are usually oriented in such a manner that the shear plane is horizontal, although it may be vertical or at any other angle. A typical laboratory apparatus for the direct shear test is shown diagrammatically in Fig. 19-14. Since drainage of pore water is allowed, this is a drained test.

After the specimen has been carefully trimmed and fitted in the shear machine, a known load is applied in the direction normal to the shear plane. Then a shearing force is applied parallel to the shear plane. The normal load is held constant throughout the test, but the shearing force is increased gradually or in increments. Most tests are strain-controlled—that is, one part of the shear box is driven at a constant rate relative to the other by a screw mechanism, and the shearing resistance is monitored. The shearing force gradually increases and reaches a peak value, after which it decreases to a stable ultimate or residual value, as shown in Fig. 19-6.

Dividing the normal load and the maximum applied shearing force by the cross-sectional area of the specimen at the shear plane gives the

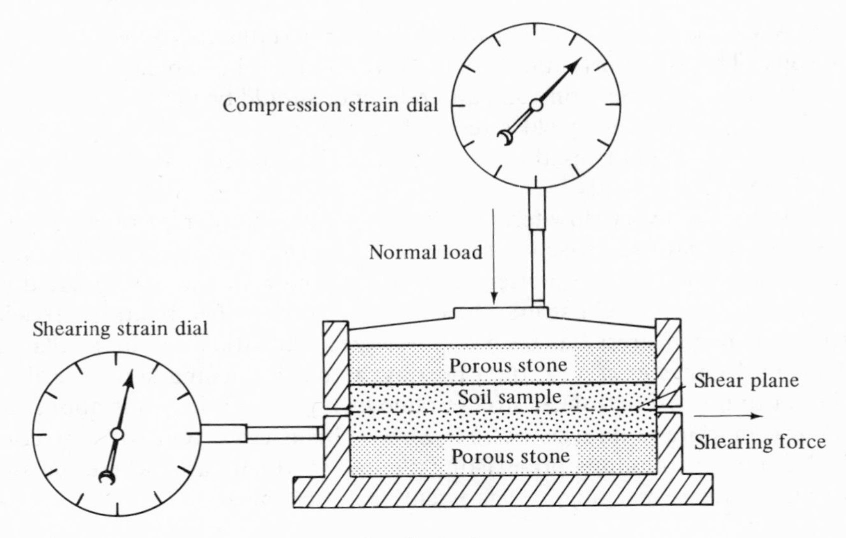

Fig. 19-14. Laboratory apparatus for direct shear test.

unit normal pressure and the shear stress at failure of the sample. These values may be plotted on a shear diagram like that in Fig. 19-6. The results of a single test establish one point on the graph representing the Coulomb formula for shearing strength. In order to obtain sufficient

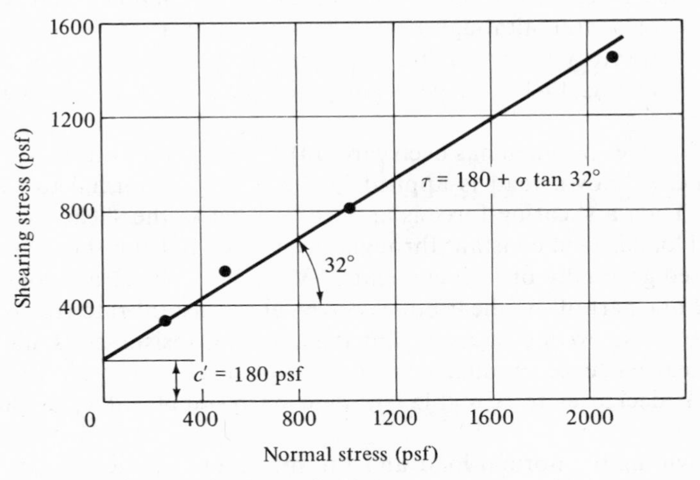

Fig. 19-15. Shear diagram for Example 19-1.

points to draw the Coulomb graph, additional tests must be performed on other specimens which are exact duplicates of the first. The procedure in these additional tests is the same as in the first, except that a different normal stress is applied each time. As a generalization, the plotted points of normal and shearing stresses at failure of the various specimens will approximate a straight line, although there will be deviations due to normal testing errors and some lack of uniformity of the specimens. Also for reasons already discussed the relationship between normal stress and shearing strength may be a curved line. Nevertheless, it is common practice to draw the best straight line through the test points to establish the Coulomb graph and equation for the soil. In this way, values of the friction angle and cohesion can be determined.

EXAMPLE 19-1. Assume that the data in the accompanying tabulation were obtained in tests on four identical samples of a nonsaturated glacial drift. Determine the cohesion and the angle of friction of the soil.

Specimen No.	*Stresses at Failure (psf)*	
	Normal	*Shearing*
1	255	345
2	505	550
3	1010	820
4	2080	1450

SOLUTION. The data are plotted on the shear diagram in Fig. 19-15. A straight line through the plotted points makes an angle of 32° with the horizontal, and the intercept on the vertical axis represents 180 psf. Therefore, the shearing strength properties of the soil are designated as $c' = 180$ psf and $\phi' = 32°$.

19.20. TRIAXIAL COMPRESSION TEST. In the triaxial compression test, a cylinder of soil is subjected to lateral pressure uniformly distributed around the cylindrical surface in a testing device such as that shown in Fig. 19-16. Then an axial load is applied to the sample until it fails. Although the applied loads, axial and lateral, are compressive in character, the sample fails by shear on internal surfaces; and it is possible to determine the shearing strength of the soil from the applied loads at failure.

Ordinarily, it is necessary to test two or more cylinders of soil at different lateral pressures in order to interpret the results in terms of shearing strength, although there are exceptions to this requirement as dis-

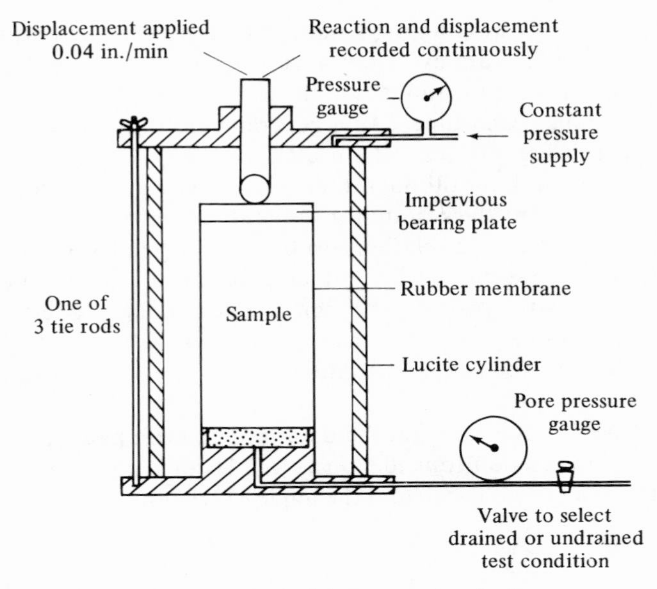

Fig. 19-16. Schematic diagram of triaxial compression device.

cussed in Section 19.26. In order to interpret the results of a triaxial compression test, it is necessary to have a clear understanding of the stress relationships which exist at a point in the interior of a stressed body. The following discussion of the stress situation at a point is generally applicable to any engineering material and appears in many textbooks on mechanics. It is given here for ready reference since it is used extensively in studies of the shearing strength of soil.

19.21. PRINCIPAL PLANES AND PRINCIPAL STRESSES. Through every point in a stressed body there are three planes at right angles to each other which are unique as compared to all other planes passing through the point, because they are subjected only to normal stresses with no accompanying shearing stresses acting in the planes. These three planes are called principal planes, and the normal stresses acting on these planes are principal stresses. Ordinarily the three principal stresses at a point differ in magnitude; and they may be designated as the maximum principal stress, intermediate principal stress, and minimum principal stress, respectively. Principal stresses at a point are important because, once they are evaluated, the stresses on any other plane through the point can be determined. Also they represent the maximum and minimum normal stresses at the point.

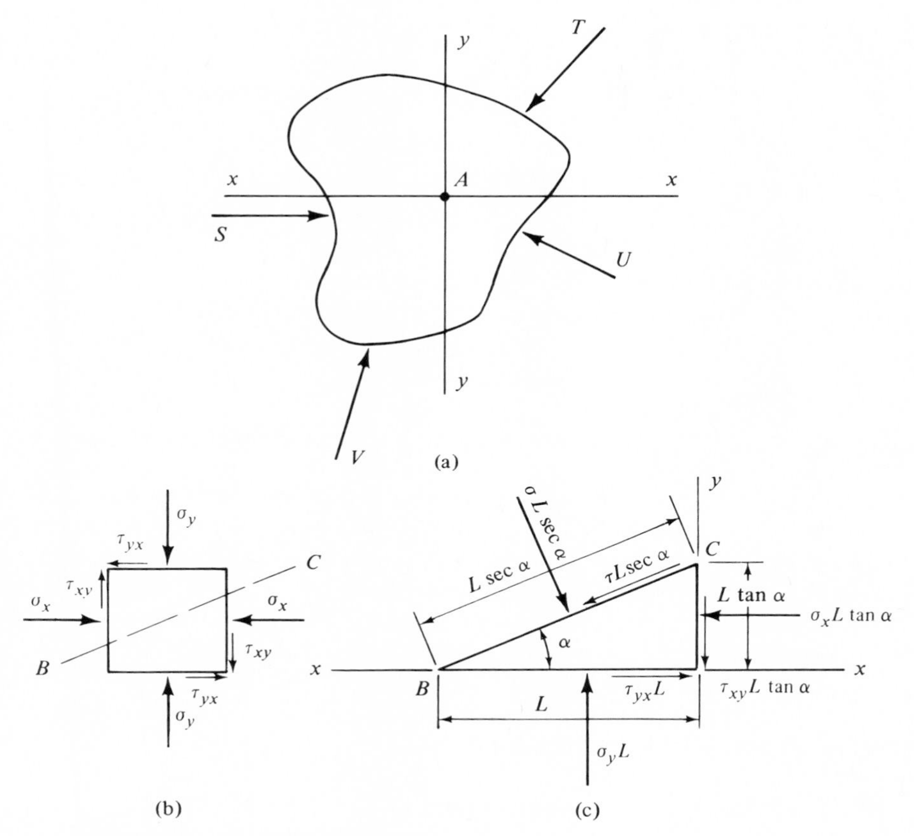

Fig. 19-17. Stresses at a point in a body.

A two-dimensional demonstration of the existence and the orientation of principal or no-shear planes follows. Let the diagram in Fig. 19-17(a) represent a body acted upon by a system of forces, such as S, T, U, and V, whose magnitudes, lines of action, and directions along these lines are known. Then the unit stresses acting at any point A and parallel to the arbitrarily chosen axes x and y can be determined by the equations of equilibrium. These stresses on an element are shown in Fig. 19-17(b). Now pass any plane BC through the element, making the angle α with the x axis. The normal stress and the shearing stress acting on this plane are shown in Fig. 19-17(c). The unit stresses in compression and in shear are designated as σ and τ, respectively. Considering the portion of the element lying below the plane BC as a free body, letting L be unity, and noting that $\tau_{xy} = \tau_{yx}$, we may express the normal and shearing stresses by the following formulas:

$$\sigma = \sigma_x \cos^2 \alpha + \sigma_y \sin^2 \alpha + 2\tau_{xy} \sin \alpha \cos \alpha \qquad (19\text{-}13)$$

and

$$\tau = \frac{\sigma_x - \sigma_y}{2} \sin 2\alpha - \tau_{xy} \cos 2\alpha \qquad (19\text{-}14)$$

By definition, a principal plane is one on which the shearing stress is equal to zero. Therefore, when τ is made equal to zero in Eq. (19-14), the orientation of the principal planes is defined by the relationship

$$\tan 2\alpha = \frac{2\tau_{xy}}{\sigma_x - \sigma_y} \qquad (19\text{-}15)$$

Equation (19-15) indicates that there are two principal planes through the point A and that they are at right angles to each other. Also, by differentiating Eq. (19-13) with respect to α and equating the derivative to zero, it is possible to determine the orientation of planes on which the normal stresses σ are maximum and minimum. This orientation coincides with Eq. (19-15). Therefore, it follows that the principal planes are also planes on which the normal stresses are maximum and minimum.

Similarly, by more complicated procedures, it can be shown that in a three-dimensional situation there is a third principal plane through the point A, and that this plane lies at right angles to each of the other two principal planes, or in the plane of the paper.

19.22. NORMAL AND SHEARING STRESSES ON A PLANE. If a pair of coordinate planes are chosen which are coincident with the maximum and minimum principal planes, certain stress relationships which are of interest may be derived. In Fig. 19-18(a) let $ACDE$ represent a small cube of material located at a point within a body and oriented with side AC in the maximum principal plane and side CD in the minimum principal plane. Then the normal stress on face AC or on face ED is the maximum principal stress σ_1; and the normal stress on face CD or on face AE is the minimum principal stress σ_3. The stress on the face of the cube in the plane of the paper is the intermediate principal stress σ_2. However, in the application of this analysis to the determination of shearing strength from the data of a triaxial compression test, it is not necessary to consider the intermediate principal stress. Therefore, σ_2 will be dropped from further consideration as a simplifying measure in this discussion of the stress situation at a point, although the fact that it exists should be clearly understood.

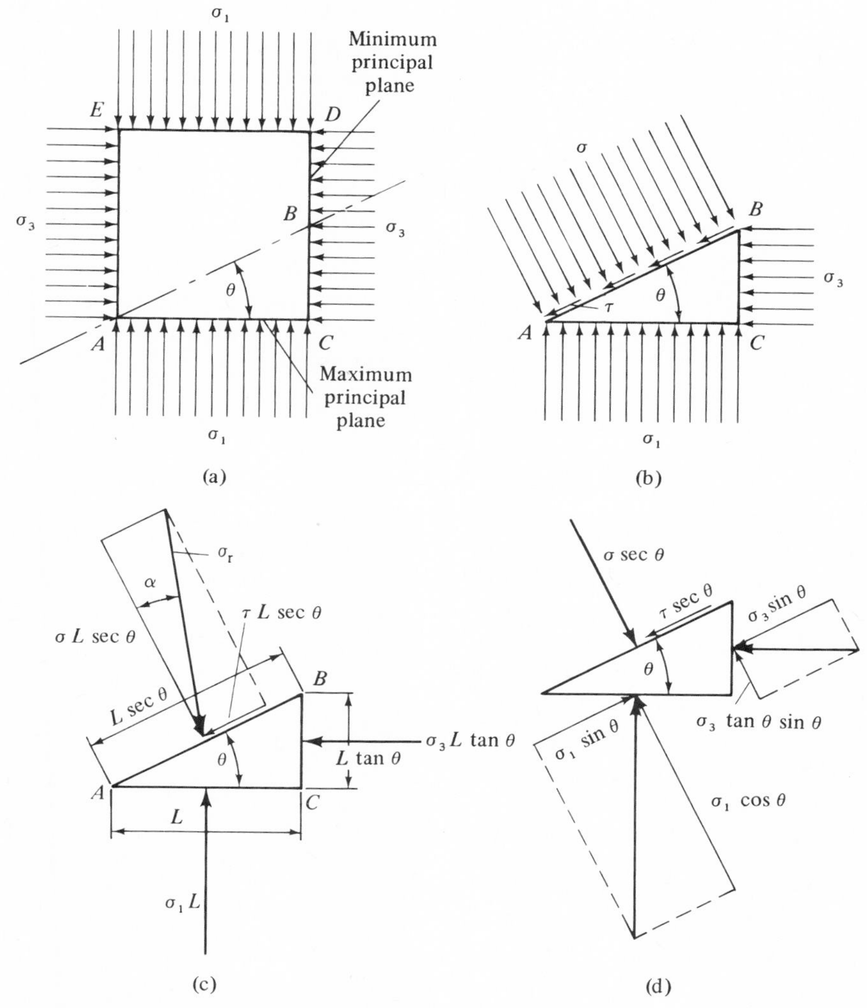

Fig. 19-18. Relation of stresses at a point.

Let plane AB be passed through the cube at an angle θ with the maximum principal plane, and consider a free-body diagram of the portion of the cube below this plane, as in the diagram in Fig. 19-18(b). The unit stress on face AC is σ_1, and that on face BC is σ_3. On the plane AB there is a normal stress σ; and, since AB is not a principal plane, there is also a shearing stress τ. Let the distance AC be denoted by L. Then $AB = L \sec \theta$ and $BC = L \tan \theta$. As shown in Fig. 19-18(c) the total stresses on the sides of the free body are: on AB, $\sigma L \sec\theta$ and $\tau L \sec\theta$; on AC, $\sigma_1 L$; on BC, $\sigma_3 L \tan\theta$. If L is equal to unity and if the total stresses on AC and BC are resolved into components normal and parallel to the plane AB, the forces shown in Fig. 19-18(d) are obtained. Summing up the normal and parallel forces separately, we get

$$\sigma = \sigma_1 \cos^2\theta + \sigma_3 \sin^2\theta \tag{19-16}$$

$$\tau = (\sigma_1 - \sigma_3)\sin\theta\cos\theta \tag{19-17}$$

Equations (19-16) and (19-17) indicate that the normal and shearing stresses on a plane can be expressed in terms of the maximum and minimum principal stresses and the angle that the plane makes with the maximum principal plane. Another fact to be noted in Fig. 19-18(c) is that the resultant σ_r of the normal and shearing stresses on the plane AB makes an angle α with the normal to that plane. This angle is the obliquity of the resultant.

19.23. MOHR DIAGRAM FOR STRESSES. Elimination of the angle θ from Eqs. (19-16) and (19-17) leads to an expression which relates the variables τ and σ in terms of the principal stresses σ_1 and σ_3. This expression is

$$\tau^2 + \left[\sigma - \left(\frac{\sigma_1 + \sigma_3}{2}\right)\right]^2 = \left(\frac{\sigma_1 - \sigma_3}{2}\right)^2 \tag{19-18}$$

Equation (19-18) is the equation of a circle whose center has the coordinates $\tau = 0$ and $\sigma = [(\sigma_1 + \sigma_3)/2]$, and whose radius is $[(\sigma_1 - \sigma_3)/2]$. A plot of this equation, as shown in Fig. 19-19, is known as a Mohr diagram. It provides a convenient graphical method for determining the normal and shearing stresses on any plane through a point in a stressed body when the principal stresses are known. To construct a Mohr diagram, lay off distance OA equal to σ_3 and distance OB equal to σ_1. Then draw a semicircle through the points A and B having a diameter equal to $\sigma_1 - \sigma_3$. Next draw a line through A making the angle θ with the horizontal axis and cutting the semicircle at point C. The abscissa of C is equal to the normal stress σ, and the ordinate of C is the shearing stress τ. The length of line OC represents the resultant stress σ_r, and the angle between OC and the horizontal is the obliquity angle α.

In the Mohr diagram the angle θ represents the orientation of any plane through a point, just as it does in Fig. 19-18. If $\theta = 0°$, the plane which it defines is coincident with the maximum principal plane and the shearing stress is zero. As the angle θ is increased from zero, the shearing stress τ increases to its maximum value $(\sigma_1 - \sigma_3)/2$ when $\theta = 45°$ and then

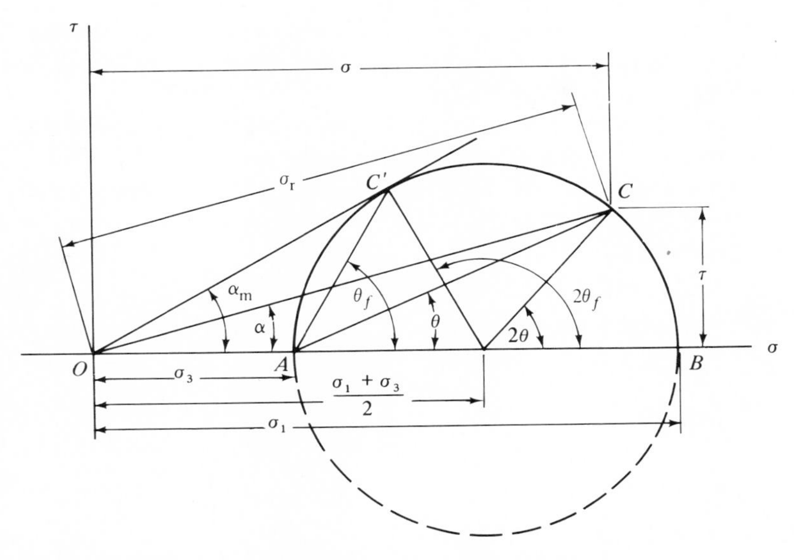

Fig. 19-19. Mohr diagram for normal and shearing stresses.

decreases to zero when $\theta = 90°$, that is, when the defined plane is coincident with the minimum principal plane.

Since in Fig. 19-19 the θ lines showing orientations of planes with stresses indicated by C, C', etc., all pass through point A, this point is referred to as an *origin of planes*. Point A in Fig. 19-19 is the origin of planes if σ_1 is vertical and σ_3 is horizontal, as in the triaxial test and in many field situations. However, in many other instances the major and minor principal stresses will not be oriented vertically and horizontally, respectively. In such cases the origin of planes may be found as follows:

1. Plot the major and minor principal stresses and draw the Mohr circle.
2. Draw a line through either principal stress point *parallel to the plane on which it acts.*
3. The intersection of this line with the Mohr circle is the origin of planes, O_p.
4. The stresses on any other plane can be found by drawing a line parallel to such a plane through O_p, and reading the intersection with the Mohr circle.

EXAMPLE 19-2. A compressive stress of 100 psi acts in a direction inclined 30° from vertical. What are the stresses on horizontal and vertical planes?

SOLUTION 1. Assuming that 100 psi is a major principal stress and the minor principal stress is zero, a Mohr circle is drawn as shown in Fig. 19-20.

2. A line is drawn through (0,100) parallel to the plane on which the 100 psi acts, i.e., inclined 30° from horizontal.

3. The intersection of this line with the Mohr circle is designated O_p. Vertical and horizontal lines are drawn through O_p to give stresses on vertical and horizontal planes.

ALTERNATE SOLUTION. The same procedure is used for step 1. However, for step 2 the other principal stress (O) is used, and a line is

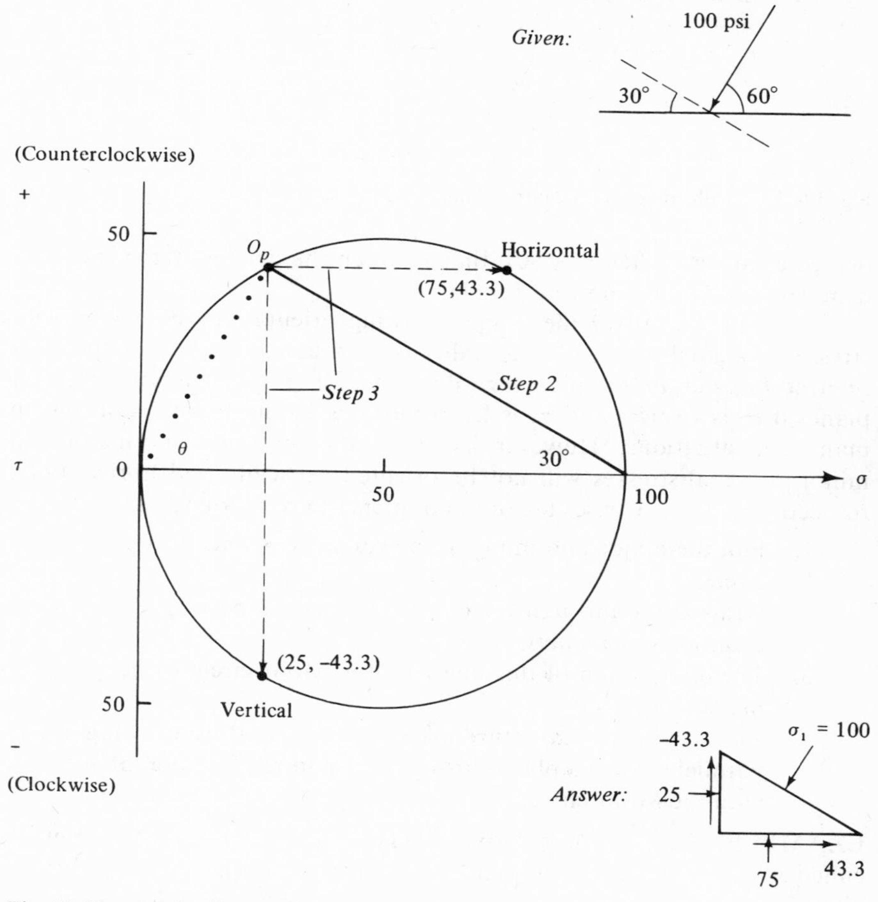

Fig. 19-20. Mohr diagram for Example 19-2.

drawn through (O, O) inclined 30° from vertical. O_p is the same, and the solution proceeds as before.

The meaning of a negative value for τ can be seen by reference to Fig. 19-6. Here τ was assumed to be positive, and acted in a counterclockwise direction. With this convention if τ is negative, it acts in a clockwise direction. A negative τ acts on the vertical plane in the figure for Example 19-2, equal to but opposite in sign from the τ on the horizontal plane, required for static equilibrium.

The origin of planes procedure also works to find the direction of principal stresses from other stresses, so long as data are sufficient to draw a Mohr circle. For example, if one were given the stresses on the vertical and horizontal planes in Figure 19-20, only one Mohr circle could be constructed with its center on the σ axis and passing through these two plotted points. O_p would be found by a line through either H or V, as before, and lines from O_p to the principal stress points define directions of the principal planes.

19.24. MOHR THEORY OF FAILURE. Various theories relative to the stress situation in engineering materials at the time of failure are available in the engineering literature. Each of these theories appears to explain satisfactorily the actions of certain kinds of material at the time they fail, but no one of them is applicable to all materials. The failure of a soil mass is more nearly in accordance with the tenets of the Mohr theory of failure than with those of any other theory, and the interpretation of the data of the triaxial compression test and many field problems depends to a large extent on this fact. According to the Mohr theory, a material fails along the plane and at the time at which the angle between the resultant of the normal and shearing stresses and the normal stress is a maximum; that is, when the combination of normal and shearing stresses produces the maximum obliquity angle α_m.

In the Mohr diagram in Fig. 19-19, it is seen that the optimum stress combination which fulfills the criterion just stated is that represented by the point C'; and the orientation of the failure plane is represented by the line AC' which makes an angle θ_f with the maximum principal plane. Since, according to the Mohr theory, the tangent line OC' represents the stress situation at failure, the maximum obliquity α_m is equal to the friction angle ϕ, just as indicated in the case of the brick sliding on a table top in Section 19.3. The value of the obliquity angle α can never exceed $\alpha_m = \phi$ without the occurrence of failure. Therefore, the tangent line OC' is a locus of points representing failure. It is called the *Mohr envelope* or *line of rupture*.

Normal and shearing stresses which yield plotted points below the envelope represent stress situations which do not produce failure. Points above the envelope cannot exist, since the material will fail before such combinations of stresses can develop. In the case of cohesionless soils, the envelope passes through the origin representing applied stresses, as shown in Fig. 19-18. For cohesive soils the envelope passes through the origin of total stress (initial plus applied), and the ordinate for an applied stress equal to zero represents the value of shearing strength which is the cohesion of the soil (see Fig. 19-18). Thus, the Mohr envelope constitutes a shear diagram and is a graph of the Coulomb equation for shearing stress. The principal objective of a triaxial compression test is to establish the Mohr envelope for the soil being tested. The cohesion and the angle of friction can be determined from this envelope.

19.25. PROCEDURE IN TRIAXIAL COMPRESSION TEST. In the triaxial compression test, a cylindrical specimen is first encased in a thin rubber membrane and subjected to fluid pressure around the cylindrical surface, as indicated in Fig. 19-16. Then an axial load is applied and its magnitude is increased until the specimen fails in shear. The applied axial load produces normal vertical stress on each horizontal plane in the specimen, while the lateral pressure produces normal horizontal stress on any two vertical planes at right angles to each other. These vertical and horizontal stresses fulfill the requirements of principal stresses, since there are no components of applied loads to produce shearing stresses in the horizontal plane or either of any two orthogonal vertical planes. The vertical stress is the maximum principal stress. The horizontal stress on any vertical plane is the minimum principal stress, and that on the vertical plane normal to the minimum principal plane is the intermediate principal stress. The minimum and intermediate principal stresses are equal in this case. However, since the intermediate principal stress does not influence the failure of the material, according to the Mohr theory, it need not be considered.

When one cylindrical sample has been loaded to failure, a Mohr circle may be drawn having a diameter equal to $\sigma_1 - \sigma_3$, or the difference between the vertical and horizontal failure stresses. There is one point on this circle which represents the combination of normal and shearing stress at which the soil failed. This point lies on the Mohr envelope. However, unless it is definitely known that the soil is completely cohesionless, in which case the Mohr envelope passes through the origin of applied stresses, there is no clue as to the location of the failure point. Therefore, it is necessary to conduct another test on a duplicate sample, a different lateral pressure being applied in this test. The applied stresses at failure in this second test may be designated σ_{1a} and σ_{3a} and a second Mohr circle may be drawn with the diameter $\sigma_{1a} - \sigma_{3a}$. Since σ_{3a} is different from σ_3, then σ_{1a} and σ_1 will also be different and the second circle will not coincide with the first. Again, there is one point on the second circle which represents the failure stresses and lies on the Mohr envelope. These two failure points can be located by drawing a common tangent to the circles. This tangent is the Mohr envelop for the soil. It is also the graph of the Coulomb equation for shearing strength of the soil.

The intercept of the tangent line on the vertical axis through the origin of applied stresses is the cohesion c of the soil, and the angle that the tangent makes with the horizontal is the friction angle ϕ of the soil. Also, the intercept of the tangent line on the horizontal axis is the origin of total stresses, and the distance between the two origins represents the intrinsic stress in the material. Since two points establish a straight line, only two tests are theoretically necessary to determine c and ϕ for the soil. However, as is always true of results based upon experimental data, the greater the number of points, the greater will be the accuracy. Therefore, it is advisable to conduct as many triaxial compression tests on duplicate samples as practicable.

EXAMPLE 19-3. The data in the accompanying tabulation were obtained in triaxial compression tests of four identical soil samples. Determine the cohesion and the friction angle of the soil.

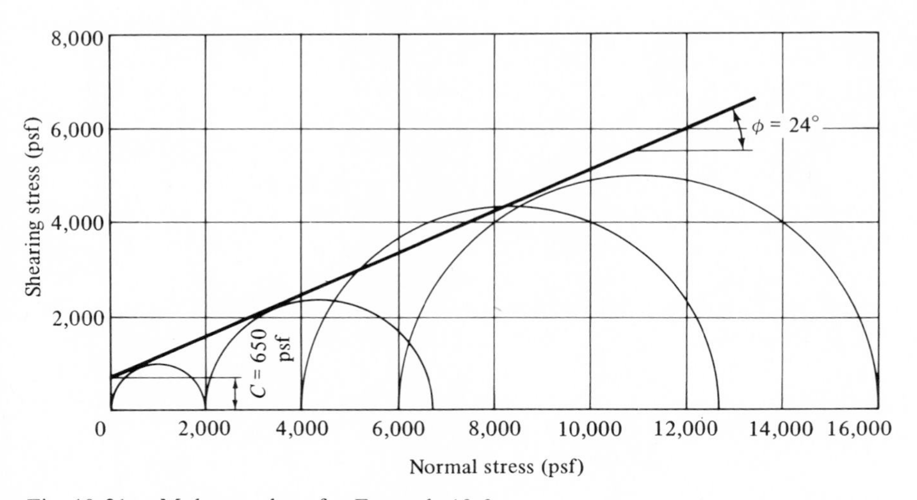

Fig. 19-21. Mohr envelope for Example 19-3.

SOLUTION. As shown in Fig. 19-21, draw four Mohr circles, each having a diameter equal to the difference between the axial-load failure stress and the lateral-pressure failure stress for one test. Then draw the Mohr envelope of rupture as nearly as possible tangent to all four circles. This line is a graph for the shearing strength of the soil. The cohesion c of the soil, which is the intercept of the envelope on the vertical axis representing zero normal stress, is 650 psf. The friction angle of the soil, which is the angle that the envelope makes with the horizontal axis, is 24°.

	Failure stresses (psf)	
Specimen No.	*Lateral* σ_3, σ_{3a}, *etc.*	*Axial* σ_1, σ_{1a}, *etc.*
1	0	1,990
2	2000	6,750
3	4000	12,700
4	6000	16,000

19.26. TYPES OF FAILURE IN TRIAXIAL COMPRESSION TEST. It was shown in Section 19.24 that the chord of a Mohr circle representing the stress situation at failure, extending from the abscissa σ_3 on the horizontal diameter to the point of tangency of the Mohr envelope, is oriented so as to be parallel to the failure plane in the material. The angle between this chord and the horizontal axis represents the angle between the failure plane and the maximum principal plane, which is a horizontal plane in the triaxial test specimen. Therefore, when a specimen fails in such a manner that the shear-failure plane can be identified, the angle which this plane makes with the horizontal should be measured and compared with the angle θ_f on the Mohr diagram as a check on the results.

The failure of a specimen of soil used in the triaxial compression test may occur in a number of different ways. The stresses in the specimen are distributed quite uniformly around the cylindrical axis. If the specimen is uniform in strength throughout, failure may occur on a very large number of closely spaced planes making an angle θ_f with the horizontal. A failure of this kind is evidenced by a uniform bulging of the sample, as shown by the photograph in Fig. 19-22(a). This is a sample of Marshall silt loam, a loess soil from western Iowa. When the sample was removed from the test, it contained a considerable amount of moisture. The failure planes were not visible in the wet condition; but, when the sample was dried in air, they became easily visible. Another type of failure is shown in Fig. 19-22(b). This is a sample of Clarion sandy clay loam, a glacial soil from north central Iowa. In this case the failure occurred along a single plane

Fig. 19-22. Two types of failure of triaxial compression specimens.

which was plainly visible at the time of failure and the angle which the plane makes with the horizontal was readily measured. It is apparent that the strength of the soil was a minimum along this plane. Many specimens fail by a combination of the phenomena just described.

19.27. VARIATIONS IN TYPES OF SHEAR TESTS. The preceding descriptions of shear testing are intended to cover only the basic principles upon which the tests are founded. A number of variations in the test procedures are introduced in various laboratories, being based on the kind of soil and the type of problem involved and in some cases on the judgment and preference of the engineers in charge of the work. Some of these variant procedures will be mentioned briefly here. A complete discussion is beyond the scope of this text.

In one type of variation, shear tests are conducted at a fairly rapid rate; that is, the normal load is applied, and then the shearing force is applied immediately and without drainage. In this procedure, which is known as the *quick-shear* test or the *undrained* test, the hydrostatic excess pore water pressure created by the application of both the confining and axial stresses does not have time or opportunity to be dissipated by water being squeezed out of the sample. This is frequently referred to as the UU (unconsolidated, undrained) test.

In another type of shear test, called the *slow-shear* test or the *drained* test, considerable time is allowed after application of the confining stress to permit the sample to become consolidated before the axial stress is applied. Also considerable time is allowed during application of axial stress in order to permit dissipation of pore water pressures, giving what is called the CD (consolidated, drained) test.

A third type of test is called the *quick-consolidated* test or the *consolidated-undrained* test. The confining stress is applied and the sample is allowed to consolidate; then the axial stress is applied at a fairly rapid rate, resulting in a CU (consolidated, undrained) test.

An engineer charged with the responsibility of properly determining the shearing strength of soil for a specific problem should study the prototype conditions thoroughly and then select the type of test procedure which seems most appropriate.

For example, the slow or CD test is appropriate under many conditions where the anticipated loading in the field will be slow enough that the drainage of excess pore water from the soil will be essentially complete. Typical examples are stage construction earth dams, structures on stratified soils, etc. At the other extreme, the quick or UU test is most applicable for structures built rapidly on thick, impermeable clay, or for impact or earthquake loading of any saturated soil.

An important characteristic of quick, UU tests of saturated soils is that there is virtually zero volume change. The tendency toward volume change during consolidation and shear induces positive and/or negative pore pressures, weakening or strengthening the soil, respectively. Another important feature is that since the confining stress is transferred to the pore water, an increase in σ_3 increases σ by the same amount, in effect translating the Mohr circle to the right without changing its size. The tangent failure envelope therefore gives $\phi = 0$. This greatly simplifies analyses where the undrained condition may be reasonably assumed to exist in the field. In clays the quick or UU test may be approximated by a rapidly performed unconfined compression test or by a vane shear test.

Undrained strengths may be converted to an effective stress basis by subtracting pore pressures. This closely approximates the drained strength of clayey soils, but gives less than the peak drained strength of loose sandy or silty soils where the undrained zero volume change condition does not allow full mobilization of grain-to-grain resistance to shear.

The consolidated, undrained CU test is a widely used compromise between the UU and CD tests, not requiring so long to perform as a slow CD test, while giving a more meaningful value for ϕ than the quick UU test. Results expressed on an effective stress basis may be used to closely predict drained shear strengths, or the shearing strength with any anticipated field pore pressure. This therefore is probably the most popular type of shear test, even though the testing conditions do not simulate real field conditions.

19.28. STRESS PATHS. The various types of triaxial tests as well as various anticipated field loading situations can be shown on a shear diagram by successive Mohr circles. For example, in the usual drained triaxial test σ_1 and σ_3 are applied all around the sample, then σ_1 is gradually increased while the other stresses are held constant, until the sample fails. The Mohr circles showing this loading are indicated in Fig. 19-23(a). Since such diagrams quickly become cluttered, each circle may be represented by a point of maximum shearing stress, and a line connecting successive points is designated a *stress path of maximum shearing stress*. That is, only maximum ordinates are shown, as in Fig. 19-23(b). This is sometimes referred to as a p-q diagram where the abscissa $p = (\sigma_1 + \sigma_3)/2$ and $q = (\sigma_1 - \sigma_3)/2$.

Stress paths can be shown for several different tests, as in Fig. 19-24, and may be used to define a failure line designated the K_f line. The relationship between the slope of the K_f line, α, and friction angle, ϕ, may be found from triangles ABC and ABD in Fig. 19-24.

$$BC = BD$$

$$\frac{BC}{AB} = \tan\alpha = \frac{BD}{AB} = \sin\phi$$

$$\tan\alpha = \sin\phi \tag{19-19}$$

Similarly it can be shown that cohesion $c = a \sec\phi$ where a is the K_f line intercept on the shear axis.

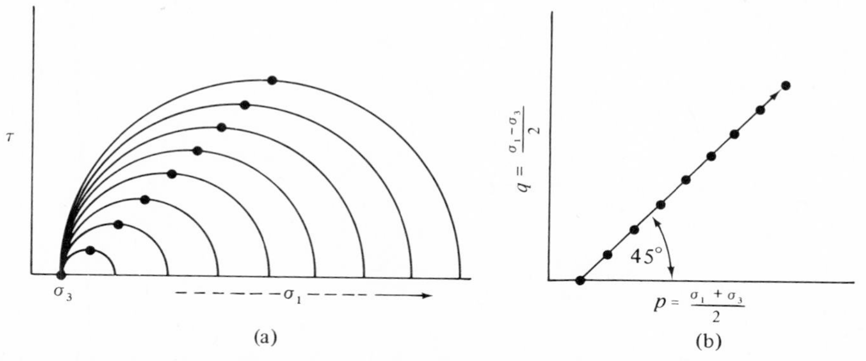

Fig. 19-23. Stress paths during a drained triaxial test. (a) Mohr circle representation; (b) stress path representation.

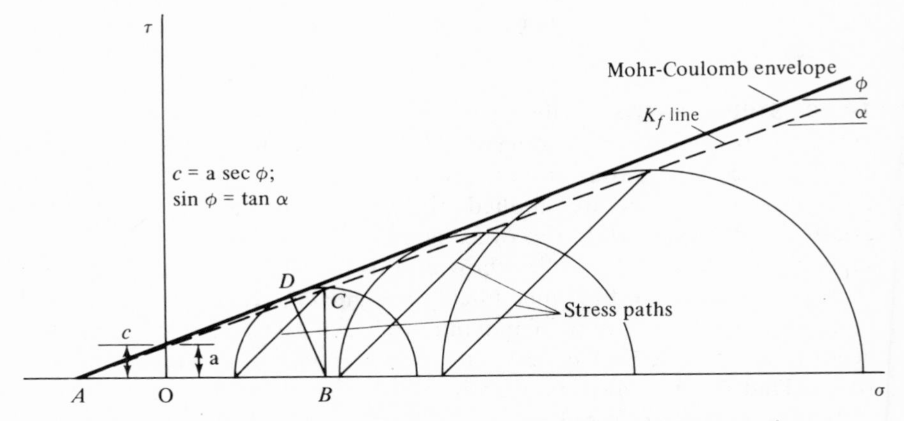

Fig. 19-24. Relation between the stress path K_f line and the failure envelope.

Stress path points are particularly useful for plotting triaxial shear test data and fitting a line, since it is easier to draw an average or "best-fit" line for several points than it is to draw a "best-fit" tangent for several circles. A best-fit line through stress path points may also be obtained statistically by the "method of least squares," found in elementary statistics texts and readily programmed on a small computer.

19.29. SATURATED UNDRAINED CONDITIONS: PORE PRESSURE PARAMETERS.

In an undrained or consolidated-undrained test the direction of the stress path is not 45° because pore pressure u subtracts from both σ_1 and σ_3 equally. This is shown in Fig. 19-25. In order to plot the effective stress path,

$$p' = \frac{\sigma_1' + \sigma_3'}{2} = \frac{(\sigma_1 - u) + (\sigma_3 - u)}{2} = \frac{\sigma_1 + \sigma_3}{2} - u$$

and

$$q' = \frac{\sigma_1' - \sigma_3'}{2} = \frac{(\sigma_1 - u) - (\sigma_3 - u)}{2} = \frac{\sigma_1 - \sigma_3}{2} = q$$

Note that parameter q is the same either on a total or an effective stress basis. This is reasonable since q represents shearing stress, which can not be carried by pore water.

As an example, assume the effective consolidation stress $\sigma_3' = 3$ psi. Initially $\sigma_1' = \sigma_3'$, and then as σ_1 is increased from 3 to 5, the pore pressure increases the same amount, or 2 psi. Then $p' = (5 + 3)/2 - 2 = 2$, and $q' = q = (5 - 3)/2 = 1$. This point is shown by a circle in Fig. 19.25.

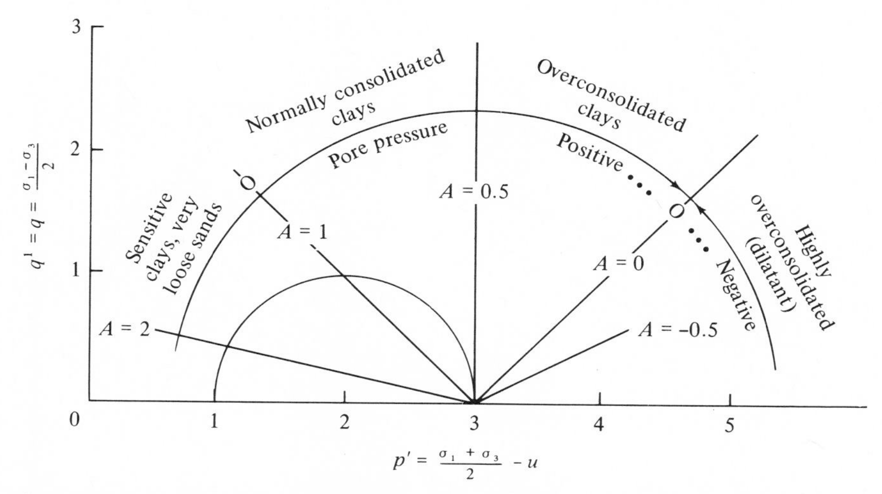

Fig. 19-25. Ideal effective stress paths for undrained triaxial tests, σ_3 constant.

Note that the resulting stress path instead of being 45° to the right is 45° to the left.

The change in pore pressure per unit change in σ_1 was defined by Skempton as *pore pressure parameter A*. In the example discussed above, $A = 1$. Other stress paths for other values of A also are shown in Fig. 19-25. Parameter A may exceed 1.0 in very loose sands or sensitive clays, where the structure collapses upon application of σ_1. In normally consolidated clays it is usually 0.5 to 1.0; in overconsolidated clays and dense fine granular materials it is 0.5 to −0.5, negative values indicating negative pore pressures from volume expansion on shearing. $A = 0$ coincides with a drained test.

Parameter A does not remain constant even during a single test, and the stress path for a constant lateral pressure test typically is curved as shown in Fig. 19.26. The value for A may be determined at any point by measuring Δu from the 45° ($A = 0$) line, and $\Delta\sigma_1/2$ as shown. Then by definition

$$A = \frac{\Delta u}{\Delta \sigma_1} \tag{19-20}$$

For field prediction purposes A therefore should be determined near the anticipated loading conditions. Because of the importance of pore pressure on stability calculations and the need to predict it before it happens, much effort has been expended in recent years devising new measurement

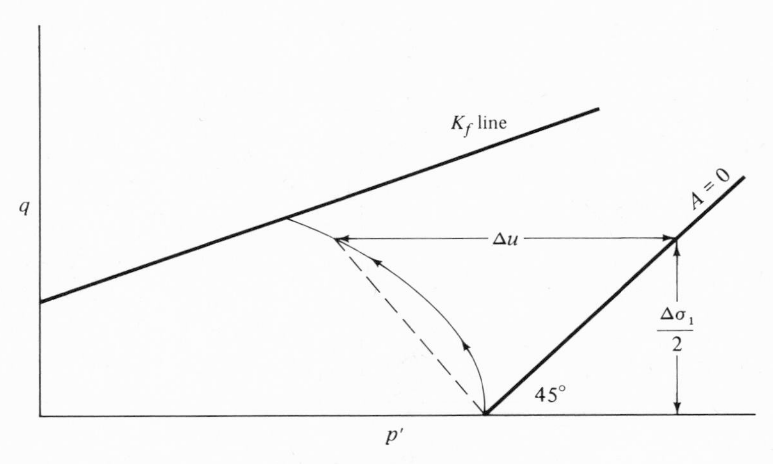

Fig. 19-26. Determination of pore pressure parameter A.

techniques. If σ_3 also changes a more general equation for A is

$$A = \frac{\Delta u - \Delta\sigma_3}{\Delta\sigma_1 - \Delta\sigma_3}. \tag{19-21}$$

or

$$\Delta u = A(\Delta\sigma_1 - \Delta\sigma_3) + \Delta\sigma_3 \tag{19-21a}$$

EXAMPLE 19-4. Tests of a saturated 10 ft thick clay layer under a prospective building site indicate $c' = 3.5$ psi and $\phi' = 20°$ on an effective stress basis. By methods given in Chapter 17 we predict that a surface load will increase the in-situ stresses σ_3 by 0.18 psi and σ_1 by 8.2 psi. (a) If in this stress range $A = 0.2$, what excess pore pressure will develop under conditions of rapid loading? (b) If the *in situ* stresses are $\sigma_1 = 5$ psi and $\sigma_3 = 2.5$ psi, plot the stress path.

SOLUTION. (a) From Eq. (19.21a),

$$\Delta u = 0.2(8.2 - 0.2) + 0.2 = 1.8 \text{ psi}$$

(b) The *in situ* stresses are plotted as Mohr circle A in Fig. 19-27. The final stresses with zero pore pressure are plotted at C. Subtracting Δu = 1.8 psi from the final σ_1 and σ_3 gives circle B. The approximate stress path is therefore AB upon loading followed by BC if there is drainage. At B the Mohr circle is very close to the failure envelope, indicating a dangerous condition.

19.30. PORE PRESSURES IN UNSATURATED SOILS. Loading of soil containing even a small percentage of air greatly diminishes the changes in

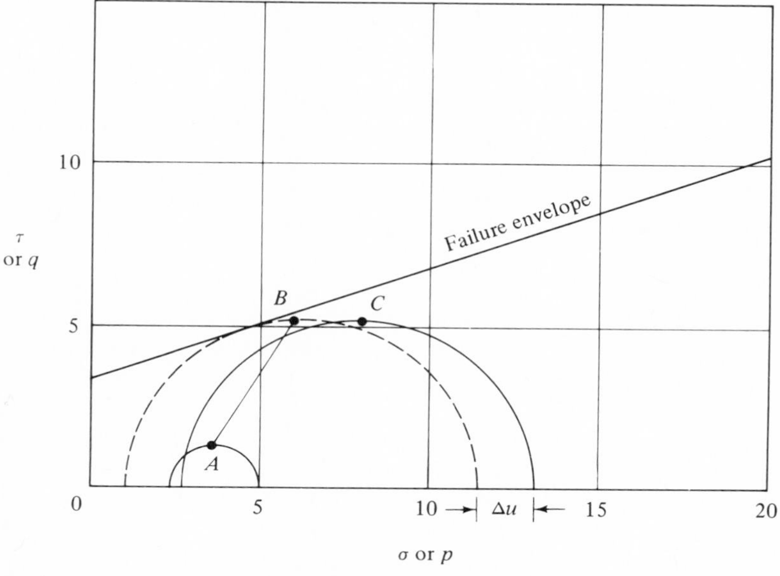

Fig. 19-27. Mohr circles and stress path for Example 19-4.

pore pressure since air is readily compressible. If the soil is obtained from below the water table, it is therefore very important that it be kept saturated or resaturated prior to performing an undrained triaxial test. Pore air and pore water pressures may be independently measured in a triaxial test and their combined effect on shear strength evaluated, but methods are not clearly defined and are beyond the scope of this text.

19.31. PORE PRESSURES RELATIVE TO EQUILIBRIUM PORE PRESSURE. When evaluating pore pressures from conventional triaxial tests the pore pressure is assumed to be zero prior to application of external load; or for evaluation of parameter A it is assumed to be zero prior to application of σ_1 in excess of σ_3. As discussed in Section 19.12 high negative pore pressures exist in clay soils prior to application of any load because of clay-water attractions. Similarly as discussed in Section 12.21 pore pressures below a water table are higher than atmospheric due to hydrostatic head. The pore pressures and pore pressure parameter discussed in Sections 19.29 and 19.30 are therefore relative, intended to describe only what happens to pore pressure under various conditions of loading. To these must be added the hydrostatic head, or allowance made for hydrostatic head by using the buoyant unit weight, as shown in Example 12-2. Allowance ordinarily is not made for intrinsic pressure, since this is taken into account as cohesion.

19.32. UNCONFINED COMPRESSIVE STRENGTH. One of the simplest and most commonly performed tests of cohesive soils is the unconfined compressive strength test, whereby a cylinder or prism of the soil is loaded axially without any lateral confinement. In wet, fine-grained soils if the test is performed quickly enough it simulates an undrained triaxial test with $\sigma_2 = \sigma_3 = 0$. Even if the soil is not held completely saturated, which is difficult in this type of test, the pore air squeezes out more readily than pore water, and the soil may become saturated during the test.

Shear diagrams for the unconfined compression tests with different degrees of drainage are shown in Fig. 19-24. Pore water pressures are not measured, and regardless of the degree of drainage or value of A in a particular test, the unconfined compressive strength, designated q_u, should be the same if $\phi = 0$. Furthermore the undrained cohesion is $c_u = \frac{1}{2}q_u$. Shearing strength therefore may be assumed to be one-half the unconfined compressive strength regardless of normal stress, for design of small structures on cohesive soils. Except at very low normal stresses this procedure is on the safe side, as can be seen by comparing c_u from the $\phi = 0$ line with actual strengths shown along the dotted line. It is not recommended for large structures since it is overly conservative, and of course it is unusable on sands, where $q_u = 0$.

Because of its simplicity the unconfined compressive strength test is

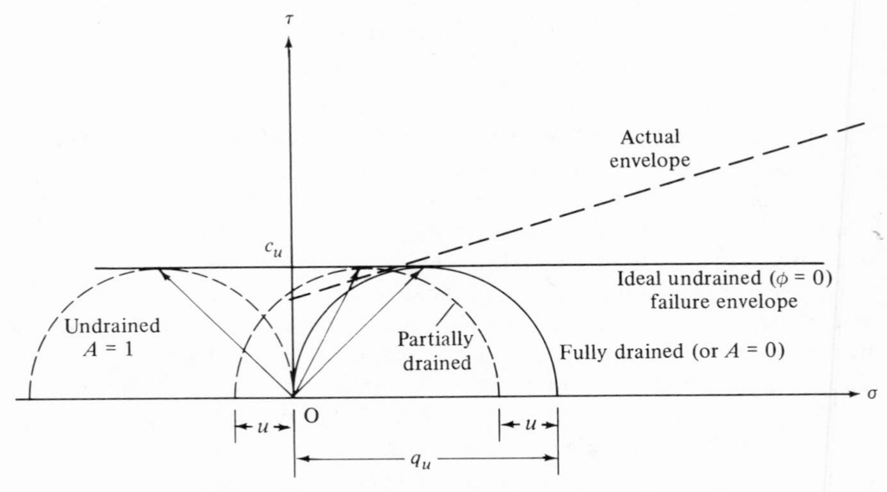

Fig. 19-28. Shear diagrams for the unconfined compression test.

frequently performed on unsaturated or permeable samples whereby $\phi > 0$ and $c < q_u/2$. If the friction angle is approximately known, c may be evaluated from

$$c = \frac{q_u}{2} \frac{1 - \sin \phi}{\cos \phi} \tag{19-22}$$

or

$$c = \frac{q_u}{2} \tan \left(45 - \frac{\phi}{2}\right) \tag{19-22a}$$

The student may develop these equations as follows: Sketch a Mohr circle tangent to the failure envelope in Fig. 19-9, with $\sigma_3 = 0$ and $\sigma_1 = q_u$. Draw a perpendicular from the center of the circle to the envelope, and reduce the equation for $\sin \phi = q_u/2 \div (q_u/2 + c \cot \phi)$. The identity of equations (19-22) and (19-22a) can be shown from $\tan \frac{1}{2} (90 - \phi) = (\sin 90 - \sin \phi)/(\cos 90 + \cos \phi)$.

It should be emphasized that while the use of the above equations may become a necessary expedient, a far more reliable procedure is to perform tests which do separably ascertain c and ϕ.

EXAMPLE 19-5. Unconfined compression tests performed on unsaturated glacial till give an average q_u of 12 psi. If the friction angle is nominally 25°, estimate c.

SOLUTION. From equation 19-22b, $c = \frac{12}{2} \tan \left(45 - \frac{25}{2}\right)$

$= 3.8$ psi.

19.33. OTHER LABORATORY TESTS. The direct shear test (Fig. 19-14) has been criticized because shearing ostensibly occurs only on one surface, and actually occurs in a lens-shaped zone shown in Fig. 19-29(a). The

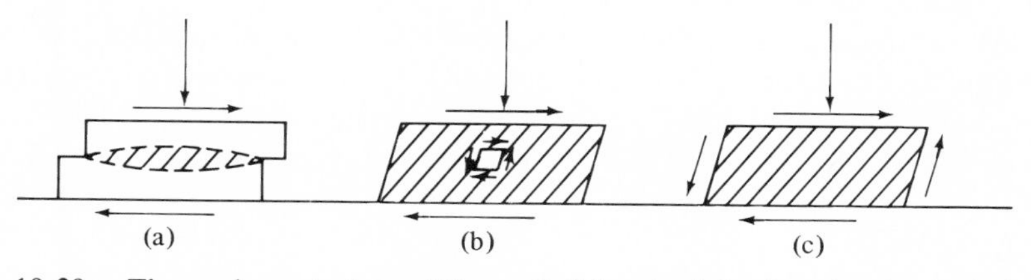

Fig. 19-29. Three shear test conditions. Soil involved in shearing is shaded. (a) Direct shear; (b) simple shear; (c) pure shear.

shearing unit strain, or shearing movement per unit thickness of the shear zone, therefore cannot be evaluated. The conventional triaxial test is criticized because the intermediate principal stress σ_2 equals the minor principal stress σ_3, whereas in field situations it would usually be larger.

A *simple shear* test developed by the Norwegian Geotechnical Institute is an ingenious attempt to improve the direct shear test by enclosing a cylindrical sample in a rubber membrane reinforced by wire rings. As the end plates are moved as in the direct shear test, the sample distorts as shown in Fig. 19-29(b). Another simple shear holder devised by Roscoe at Cambridge University encloses a rectangular specimen in a rubber membrane and confines it in a hinged box.

Even though simple shear represents an improvement over direct shear, the shearing stresses still are not uniformly distributed in the sample. The reason can be seen by comparing Fig. 19-29(b) with the condition of pure shear shown in Fig. 19-29(c). Pure shear exists near the center of a simple shear sample but not at the edges, giving a nonuniform stress distribution.

Many field situations approximate a linear distribution of loads, e.g., linear footings, retaining walls, landslides, etc. Under these conditions the soil is confined by adjacent soil along the axis of loading and must deform in *plane strain* (not plain strain). Plane strain conditions may be imposed in the laboratory by use of a simple shear device, or by confining a rectangular soil specimen between two glass plates a set distance apart, and loading the exposed ends and sides of the specimen in compression. In this case σ_1 and σ_3 are controlled, and σ_2 adjusts to an intermediate value. The effect of confinement along the intermediate stress axis is most pronounced for dense sands, sometimes increasing ϕ by several degrees. This might be expected from dilatancy theory since the sand now is free to expand only in one direction. While plane strain tests are perhaps the most realistic of the test methods, they are not yet practical for routine testing. Furthermore results from triaxial tests are on the safe side.

19.34. SHEAR TESTS OF SOIL *IN SITU*. The advantages of shear-testing of soils *in situ* are readily apparent, since this procedure eliminates the necessity of obtaining undisturbed samples of the material, transporting them to the laboratory, and setting up and operating the laboratory testing equipment. A device for measuring, the shearing strength of medium to soft clay *in situ* is the shear vane, discussed in Section 6.4. The Standard Penetration Test (Sections 6.3 and 6.6) may be used to evaluate internal friction in granular soils, since ϕ is related to relative density. Since this is primarily used to evaluate bearing capacity in foundation en-

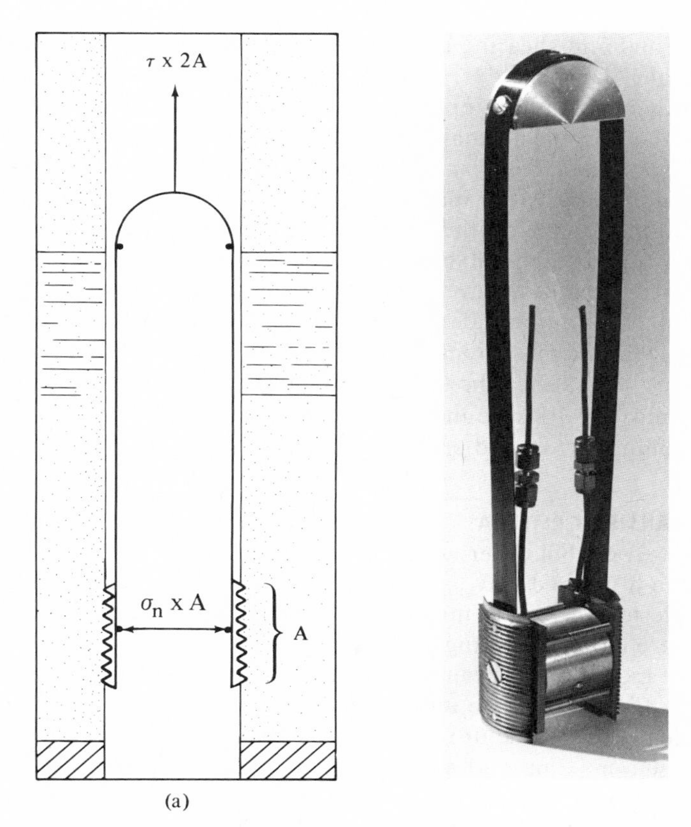

Fig. 19-30. Diagram and photograph of the Iowa bore-hole direct-shear test device.

gineering, the relationship between ϕ and N (blow count) is presented in Fig. 23-13.

A recently developed apparatus for *in situ* testing is the Iowa Bore-Hole Direct-Shear Test Device, shown in Fig. 19-30. It consists of two curved contact plates with V-shaped teeth and grooves cut circumferentially in the outside surfaces. In operation, a hole is bored to the desired depth at which shear strength is to be measured. Then the plate assembly is lowered into the hole and the plates are pushed laterally into intimate contact with the soil by fluid pressure, which provides a measure of the normal force between the plates and the soil. Next the device is pulled axially in the hole until shear failure develops. The magnitude of the axial load is a measure of the shearing stress on the plates.

These normal and shearing stresses provide the coordinates of one point on the Mohr diagram.

Another cycle of the operation is carried out by increasing the fluid pressure and thus the normal force. Then the device is again pulled axially until the soil fails. The results of this cycle provide the coordinates of a second point on the Mohr diagram, and the cycle may be repeated to obtain additional points on the envelope. As many as ten such repetitions have been found to be reliable in the case of a high friction soil, whereas two or three repetitions are about all which can be conducted in a cohesive soil. If additional data points in clay are desired, the device may be rotated 90° and measurements made on the other two faces of the hole at the same elevation. The test is gaining use in engineering practice for slope stability analysis, foundations of structures where settlement is not the controlling factor, and prediction of skin friction on pile.

19.35. FAILURE CRITERIA. Engineers usually think of failure in terms of maximum stress, but other criteria may be preferable for some uses. For example in a fissured clay as shown in Fig. 19-12, or in weathered soils susceptible to creep, one may wish to omit cohesion, utilizing only the residual strength friction angle in stability calculations [Bjerrum (*2*)].

Little mention has been made of stress level as a function of strain (shown in Fig. 19-6). The ultimate strength of a loose sand may not be attainable before a structure fails due to excessive deformations involved; in this case a maximum allowable unit strain may be specified. For example if the sand were 10 ft thick and the allowable settlement is 0.1 ft, one could evaluate the stress giving 0.01 axial unit strain in the triaxial test.

Most soils fit neither of the above two categories, but show load and volume change response similar to curves for medium and dense soils in Fig. 19-6. In this case where part of the peak strength relates to interlocking one may prefer not to depend on this strain-dependent strength, but to use the ultimate strength instead. For example let us say a potential shear surface cuts through both a thin, dense sand layer and a thick clay layer, the two materials having stress-strain responses of the upper two curves in Fig. 19-6. As a load is applied and displacement occurs, the sand strength passes the peak and reaches its ultimate strength before the clay reaches its peak. Here ultimate strength would be a safer criterion, and Mohr envelopes, c, and ϕ values would be determined from those values.

Figure 19-6 also shows that most soils coming under load will first decrease in volume owing to densification or elastic deformation or both, and then increase because of reduction in interlocking. It was suggested by Holtz in 1947 that the initial increase in volume therefore signals initiation of failure, even though the soil has not yet reached a maximum shearing resistance. The minimum volume point also is the point where pore pressure reaches a maximum, although there may be some lag in tests due to permeability affects. The strength at minimum volume (or maximum pore pressure) is pertinent and may be used where initiation of failure is not allowable, as in road bases, etc., subjected to repeated loads. Repeated loading to stresses above the minimum volume strength results in rapid disintegration and failure.

19.36. SHEARING STRENGTH DETERMINED FROM FIELD OBSERVATIONS. There are situations in which it is virtually impossible to determine the shearing strength of soil by sampling and testing procedures. For example, if a mass of clayey soil is highly fissured and slickensided, the strength of the mass will be controlled by the strength along these partings and discontinuities. In some instances, no amount of sampling can adequately yield laboratory specimens which are representative of the overall strength of such a mass. It may be possible in a case of this kind to obtain an estimate of strength from observations of natural phenomena, such as the angle of repose of slopes which have become stabilized with time. This procedure was followed in connection with the estimate of required slopes of cuts in various soils in connection with planning for a sea-level canal at Panama (*1*).

19.37. CONSOLIDATION STRESS PATH. This and the preceding chapter both deal with deformations of soil under loads, and it is instructive to relate the two kinds of behavior. In particular one may ask, does consolidation involve shearing?

As discussed in Section 18.36, consolidation must involve at least minor shear displacements in order that lateral pressure can develop as a consequence of axial loading. Furthermore the ratio of lateral to vertical stress during one-dimensional consolidation was found to equal a constant K_0 because of resistance from sliding friction between mineral grains in the soil. The stress path during consolidation follows a straight line called the K_0 line. The K_0 line passes through the origin and has a slope

$$\begin{aligned} \tan \alpha_{K_0} &= \frac{q}{p} = \frac{(\sigma_1 - \sigma_3)/2}{(\sigma_1 + \sigma_3)/2} \\ &= \frac{1 - \sigma_3/\sigma_1}{1 + \sigma_3/\sigma_1} \\ &= \frac{1 - K_0}{1 + K_0} \end{aligned} \tag{19-23}$$

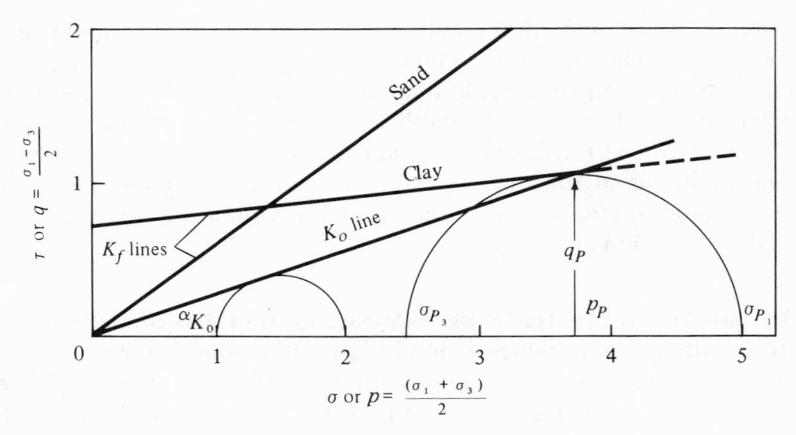

Fig. 19-31. K_0 line for normal consolidation.

As can be seen in Fig. 19-31, the K_0 line is below the K_f failure line for a dense sand or a preconsolidated clay. Since sliding friction is believed to be involved in development of K_0, let us combine Eqs. (19-19) and (19-23):

$$\sin \phi_s \approx \tan \alpha_{K_0} = \frac{1 - K_0}{1 + K_0} \tag{19-24}$$

where ϕ_s is the angle of sliding friction. Substituting the lowest common consolidation value of 0.4 for K_0 gives $\phi_s = 25°$, which is approximately the angle of sliding friction for wet quartz on quartz. Substituting the highest value for K_0, 0.7, $\phi_s = 10°$, approximately the angle of sliding friction for wet micas. Normally consolidated soils therefore should and do have K_0 value between these limits, depending on mineral composition of grain surfaces, and it is not surprising that K_0 has been found to relate to the plasticity index [Eq. (18–20)].

The K_0 line now gains importance because if failure conditions are reached below this line they indicate a higher-than-critical void ratio. For example, sand with $\phi < 25°$ is above the critical void ratio, and readily shears internally when compressive stress is applied. The K_0 line therefore is sometimes referred to as a *critical state line.* Note that the critical void ratio is not constant, because the higher the confining pressure p the higher the allowable shearing stress q and the lower the critical void ratio.

In the case of a preconsolidated clay, intersection of K_0 and K_f lines should be indicative of the preconsolidation pressure (Fig. 19-9), above which there will be internal shearing and the K_f line may follow the K_0

relationship. The vertical and horizontal stresses at the preconsolidation pressure p_p are from Fig. 19-31.

$$\sigma_p = p_p \pm q_p \qquad (19\text{-}25)$$

19.38. STRESS PATHS FOR COMPACTION. Compaction is a fundamental concern in soil engineering, yet relatively little has been done on the mechanics of compaction. Compaction necessarily involves sliding of soil grains over one another, but obviously must avoid inducing a general shear failure with accompanying dilatancy and expansion to the critical void ratio. Thus the three-dimensional confining stress $p = (\sigma_1 + \sigma_2 + \sigma_3)/3$ must be increased without allowing the maximum shearing stress $q = (\sigma_1 - \sigma_3)/2$ to exceed that indicated by the K_f line. During initial stages of compaction, soil will start at or above the critical void

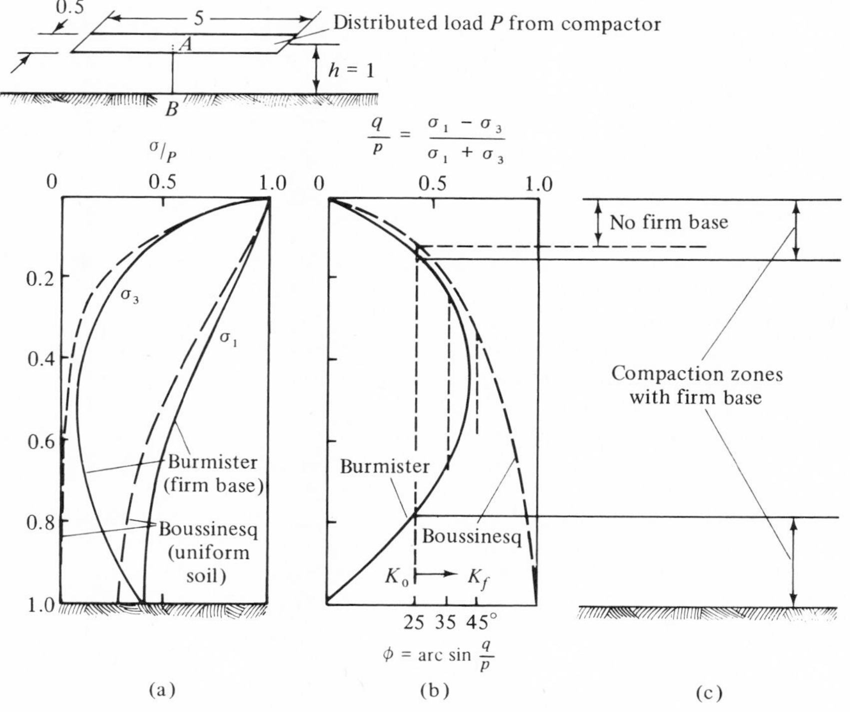

Fig. 19-32. (a) Distribution of vertical and horizontal compressive stresses in a soil layer being compacted with and without a firm base; (b) q/p stress ratios; (c) zones of compaction on the first pass.

ratio, so the K_f and K_0 lines will coincide, and the maximum allowable q will be low. Therefore considerable unproductive shearing may occur during the first pass of a compactor, but on successive passes the K_f line will be higher, and higher shearing stresses can be tolerated without dilatancy. It is important that the K_f line be defined on the basis of stresses at minimum volume rather than peak strength, since the latter involves dilatant shearing.

One problem in compaction is that lateral or σ_3 stresses induced in soil under a roller dissipate more rapidly with depth than do the vertical or σ_1 stresses. This is pictured in Fig. 19-32, showing σ_3 and σ_1, according to Boussinesq elastic theory. Lateral confinement is provided by compacting on a firm base of already compacted soil. The influence of such a base is shown by Burmister elastic theory, also illustrated in Fig. 19-32, and demonstrates the importance of compacting in thin lifts. The compaction zones shown are only for the first pass of the compactor; successive passes will cause compaction of the middle zone as the lower zone becomes part of the firm base.

The function of the base for compaction can be demonstrated by pressing a lump of moist soil between the palm of the hand and either a smooth glass plate or a rough surface such as sandpaper. Pressed against a smooth surface the soil will shear and spread out without compacting, whereas against a rough surface it will compact into a firm mass.

PROBLEMS

19.1. A 50-lb weight resting on a concrete floor is acted upon by a horizontal force of 9.7 lb. What is the value of the obliquity angle? What is the friction angle that is developed?

19.2. If a horizontal force of 30 lb is required to move the weight in Problem 19.1 what are the coefficient of friction and the friction angle between the weight and the floor?

19.3. Suppose that the weight in Problems 19.1 and 19.2 rests on a concrete ramp which makes an angle of 30° with the horizontal. Will the weight slide down the ramp or not?

19.4. Explain fully the nature of the friction component and the cohesion component of the shearing strength of soil.

19.5. Three trials in a direct shear test yield the tabulated results. Draw the shear

Specimen No.	*Failure Stresses (psf)*	
	Normal	*Shearing*
1	300	605
2	600	745
3	1050	950

diagram, and determine the cohesion and the angle of friction of the soil. Also determine the internal initial stress in the soil.

19.6. The maximum and minimum principal stresses at a point in a stressed body are 1000 psf and 300 psf, respectively. Determine the normal and shearing stresses on a plane that passes through the point and makes an angle of 25° with the maximum principal plane. Also determine the resultant of the normal and shearing stresses and find the obliquity of the resultant.

19.7. What is the maximum shearing stress at the point under consideration in Problem 19.6? What normal stress, resultant stress, and obliquity angle are associated with this maximum shearing stress?

19.8. An unconfined compression test is conducted on a cylinder of very fine-grained cohesive soil. The stress at failure is 640 psf. What is the shearing strength of this soil? Draw the Mohr diagram and the failure envelope.

19.9. A triaxial compression test is conducted on three idential cylinders of soil, with the tabulated results. Draw the Mohr diagram, and determine the cohesion and the angle of friction of the soil. Also determine the applied normal stress, the shearing stress, and the resultant stress on the failure plane of each specimen at the time of failure.

Specimen No.	*Failure Stresses (psf)*	
	Lateral	*Axial*
1	1000	4800
2	2000	7600
3	3000	9800

19.10. When an unconfined compression test is conducted on a cylinder of soil, it fails under an axial stress of 2200 psf. The failure plane makes an angle of 50° with the horizontal. Draw the Mohr diagram, and determine the cohesion and the angle of friction of this soil sample.

19.11. In a series of triaxial compression tests it is indicated that the cohesion of the soil is 257 psf and the angle of friction is 22°. If the applied lateral pressure in one of the tests was 400 psf, at what axial stress did the specimen fail?

19.12. Develop an expression for preconsolidation principal stresses in Fig. 19-9 in terms of τ_p at the intersection and ϕ_2. (Hint: Draw a Mohr circle tangent to ϕ_2 at the intersection.)

19.13. Tests on a preconsolidated glacial till gave τ_p = 10.5 psi, ϕ_2 = 19.5°, and P.I. = 11.1. (a) Estimate K_0 for a normally consolidated condition taking ϕ_2 as equal to the angle of sliding friction. (b) Compare K_0 from (a) with K_0 estimated from the P.I. and from the Jaky equation. (c) Calculate σ_p from the equation in Problem (19.12) indicating preconsolidation pressure and horizontal stress. (d) If the soil density is 105 pcf, to what depth will horizontal stress exceed vertical stress, assuming there has been no lateral relief?

19.14. The test of Problem 19.9 gave the following measured pore pressures at the respective failure stresses. Draw the *p-q* diagram on an effective stress basis, draw and label the K_f line, and find α, c' and ϕ.

Specimen No.	*u(psi)*
1	3.1
2	5.0
3	9.6

19.15. In Fig. 19-25 show that the $A = 0.5$ line is vertical.

19.16. A CU triaxial test on clay gives the following data with σ_3 maintained constant at 10 psi. Plot the total and effective stress paths. Draw the $A = 0$ and evaluate parameter A at failure. Qualitatively characterize the amount of preconsolidation relative to the applied σ_3.

Pore Pressures under Various Axial Stresses					
σ_1 (psi)	10	14	18	22	24.5 (failure)
u (psi)	2	4.5	7	10	12

19.17. In consolidometer tests of saturated clays, increments of pore pressure developed immediately after loading approximately equal the increments of load. Assuming that $\Delta\sigma_3 = K_0 \, \Delta\sigma_1$, evaluate pore pressure parameter A from Eq. (19-21). Does this value appear reasonable?

19.18. (a) Discuss the role of pore pressure in compaction. (b) Give two reasons for scarifying the surface of a compacted clay layer prior to the addition of more material.

19.19. The average standard penetration test blow count in sand at a depth of 30 ft is 10. Is this sand likely to liquify (a) if the ground acceleration is less than 0.1 g; (b) if the ground acceleration exceeds 0.15 g?

19.20. A water tower resting on sand for 20 yr suddenly tilts and collapses. Suggest two possible explanations, and how one could go about proving or disproving each possibility.

REFERENCES

1. Bishop, A. W., and D. J. Henkel. *The Measurement of Soil Properties in the Triaxial Test*, 2d ed. Edward Arnold, London, 1962.
2. Bjerrum, L. "Progressive Failure in Slopes of Overconsolidated Plastic Clay and Clay Shales." *ASCE J. Soil Mechanics and Foundation Eng.* **93** (SM5), 3–49 (1967).
3. Bowden, F. P., and D. Tabor. *The Friction and Lubrication of Solids*, Parts 1 (1950) and 2 (1964). Clarendon Press, Oxford.
4. Burmister, D. W. "Stress and Displacement Characteristics of a Two-Layer Rigid Base Soil System: Influence Diagrams and Practical Applications." *Highway Research Board Proc.* **35,** 773–814 (1956).

5. Duncan, J. M., and P. Dunlop. "Behavior of Soils in Simple Shear Tests." *Proc. 7th Intern. Conf. Soil Mechanics and Foundation Eng.* **1,** 101–109 (1969).
6. Eden, W. J., and R. J. Mitchell. "The Mechanics of Landslides in Leda Clay." *Can. Geot. J.* **11** (3), 285–296 (1970).
7. Fisher, R. A. "On the Capillary Forces in an Ideal Soil; Correction of Formulae by W. B. Haines." *J. Agricultural Sci.* **10** (1926).
8. Gibbs, H. J., and C. T. Coffey. "Techniques for Pore Pressure Measurements and Shear Testing of Soil." *Proc. 7th Intern. Conf. Soil Mechanics and Foundation Eng.* **1,** 151–157 (1969).
9. Henkel, D. J. "Geotechnical Considerations of Lateral Stresses." *ASCE Spec. Conf. Lateral Stresses and Earth-Retaining Struct.* **1,** 49 (1970).
10. Lohnes, R. A., A. Millan and R. L. Handy. "*In-situ* Measurement of Soil Creep." *ASCE J. Soil Mechanics and Foundation Eng.* **98** (SM1), 143–147 (1972).
11. Newland, P. L., and G. H. Allely. "Volume Changes in Drained Triaxial Tests on Granular Materials." *Geotechnique* **7,** 17–34 (1957).
12. Rowe, P. W., L. Borden, and I. K. Lee. "Energy Components during the Triaxial Cell and Direct Shear Tests." *Geotechnique* **14,** 247–261 (1964).
13. Seed, H. B., and I. M. Idriss. "Simplified Procedure for Evaluating Soil Liquification Potential." *ASCE J. Soil Mechanics and Foundation Eng.* **97** (SM9), 1249–1273 (1971).
14. Skempton, A. W., and D. J. Petley. "The Strength along Structural Discontinuities in Stiff Clays." *Proc. Geotech. Conf. Oslo* **2,** 29–46 (1967).
15. Taylor, D. W. *Fundamentals of Soil Mechanics.* John Wiley, New York, 1948.
16. Terzaghi, Karl. *Erdbaumechanik.* Franz Deuticke, Vienna, 1925.
17. U.S. Geological Survey. "The San Fernando, California, Earthquake of February 9, 1971." *Geol. Surv. Prof. Paper 733.* U.S. Government Printing Office, Washington, D.C., 1971.

20

Stability of Slopes

20.1. CONSIDERATIONS IN ANALYSIS OF STABILITY OF SLOPES. Every mass of soil which is bounded by a sloping surface is subjected to shearing stresses on nearly all its internal surfaces because of the gravitational force which tends to pull the upper portions of the mass downward toward a more nearly level surface. If the shearing strength of the soil is at all times greater than the stress on the most severely stressed internal surface, the slope will remain stable. On the other hand, if the strength at any time should become less than the stress, the soil will slump or slide down the slope until a position is reached such that the stress is reduced to a value less than the strength. An analysis of the stability of a slope consists of two parts: (1) the determination of the most severely stressed internal surface and the magnitude of the shearing stress to which it is subjected; (2) the determination of the shearing strength along this surface.

Soil structures which have sloping surfaces whose stability may need to be analyzed include highway and railway embankments, levees, earth dams, highway and railway cut slopes, canals, the sides of excavations for foundations, and trenches. The value of the shearing stress to which any of these masses of soil is subjected depends on the unit weight of the material and on the geometry of the slope. The shearing strength which can be mobilized to resist the stresses depends on the character of the soil and its condition with respect to moisture content and density. In general, the shearing strength of the soil can be determined by laboratory tests, although it is difficult in many cases to predict the weakest condition which may develop within the lifetime of the structure. As an example, a 50-ft railway cut in southern Iowa had been in existence for about 45 yr without mishap or maintenance except for routine ditch cleanout occasioned by small amounts of surface erosion. However, in the spring of 1947, a long sustained period of rainfall and generally wet conditions so weakened the soil that slides developed on both side slopes and covered the track with 8 to 12 ft of soil. This difficulty of predicting future soil strength requires the exercise of judgment in slope analysis and the application of a factor of safety which is appropriate to the particular problem.

20.2. FORCES ACTING ON A SMALL ELEMENT OF A SLOPE OF INFINITE EXTENT. As an introduction to methods of slope analysis, the problem of a slope of infinite extent will be presented here. This problem has limited practical application because slopes extending to infinity do not exist in nature; but it serves to introduce some fundamental concepts in a relatively simple and straightforward manner. Imagine an infinite slope, as shown in Fig. 20-1 at *AB*, and assume that the soil is cohesionless and homogeneous throughout. Then the stresses acting on any vertical plane in the soil are the same as those on any other vertical plane. Also, the stress at any point on a plane *CD* which is parallel to the surface will be the same as at every other point on this plane.

Now consider a vertical slice of material *abcd* having a unit dimension in the direction normal to the page. The forces acting on this slice are its weight *W*; a vertical reaction *R* on the base of the slice; and two lateral forces *F*, acting on the sides. Since the slice is in equilibrium, the weight and the reaction are equal in magnitude and opposite in direction; and they have a common line of action which passes through the center of the base *cd*. Also the lateral forces must be equal and opposite, and their line of action must be parallel to the sloped surface. Otherwise, there would be a residual moment with respect to the point *E*, and the slice would not be in equilibrium. Thus, it is seen that all the forces acting on the slice are either vertical or parallel to the sloped surface.

Now consider a small element of soil at any depth having vertical sides and its top and bottom planes parallel to the slope, as the element in Fig. 19-1(b). By applying the same line of reasoning as that adopted in the preceding paragraph, it can be shown that the forces acting on this small volume will consist of vertical stresses *P* on the top and bottom

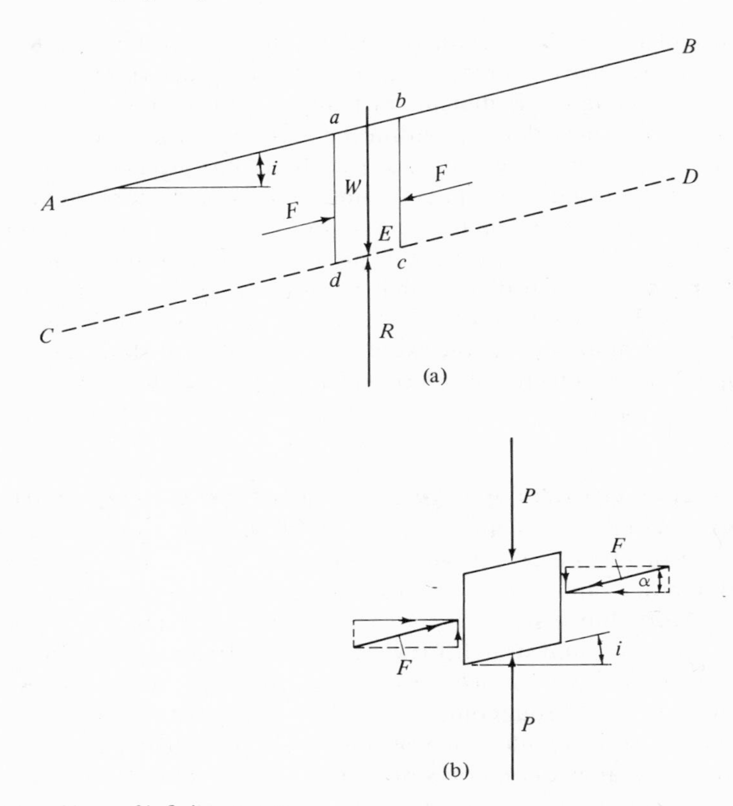

Fig. 20-1. Slope of infinite extent.

and lateral stresses F which are parallel to the top and bottom planes and act on the sides.

20.3. CONJUGATE RATIO. In the theory of elasticity it is known that, if the stress on a given plane at a given point is parallel to a second plane, the stress on the second plane at the same point must be parallel to the first plane. Such planes are said to be conjugate to each other, with respect to the stresses on them; and the stresses are called conjugate stresses. The stresses on the element under consideration in Fig. 20-1(b) comply with these conditions and are, therefore, conjugate stresses. Rankine has shown that conjugate stresses are related to each other by a ratio which is a function of the obliquity α of the lateral stresses and the maximum obliquity α_m. This ratio is called the conjugate ratio, or Rankine's lateral pressure ratio, and is expressed by the formula

$$\frac{F}{P} = K = \frac{\cos \alpha - \sqrt{\cos^2 \alpha - \cos^2 \alpha_m}}{\cos \alpha + \sqrt{\cos^2 \alpha - \cos^2 \alpha_m}} \qquad (20\text{-}1)$$

It is shown in Fig. 20-1(b) that each lateral stress can be resolved into a normal stress and a shearing stress, and that the obliquity of the resultant is equal to the slope angle i. Also, in a cohesionless soil the maximum obliquity is equal to the angle of friction ϕ. Therefore, Eq. (20-1) can be rewritten to apply to the infinite slope of cohesionless soil

$$\frac{F}{P} = K = \frac{\cos i - \sqrt{\cos^2 i - \cos^2 \phi}}{\cos i + \sqrt{\cos^2 i - \cos^2 \phi}} \qquad (20\text{-}2)$$

in which

F = lateral stress parallel to the slope;
P = vertical stress on a plane parallel to the slope;
i = slope angle;
ϕ = angle of friction of the soil; and
K = conjugate ratio.

It is obvious from Eq. (20-2) that

$$F = KP \qquad (20\text{-}3)$$

20.4. MAGNITUDES OF VERTICAL AND LATERAL STRESSES. As shown in Fig. 20-2, the vertical stress P is equal to the weight of a unit column of soil divided by the area of the column cut by a plane parallel to the

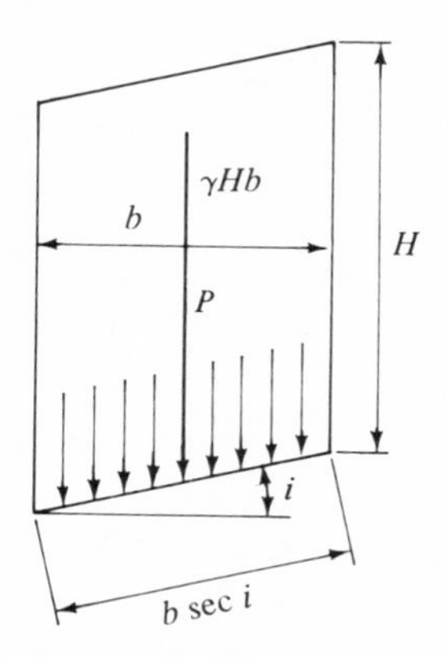

Fig. 20-2. Vertical stress on plane parallel to sloped surface.

slope. This stress can be determined by the formula

$$P = \gamma H \cos i \tag{20-4}$$

in which

γ = unit weight of soil;
H = depth of plane below surface; and
i = slope angle.

Substituting this value of P in Eq. (20-3), we obtain

$$F = \gamma H K \cos i \tag{20-5}$$

This is the expression for the lateral stress at any depth in a mass of cohesionless soil bounded by a sloping surface of infinite extent. The lateral stress in this case is oriented parallel to the sloping surface. It will be noted in Eq. (20-5) that F is a linear function of the depth H. Accordingly, the lateral stress increases at a uniform rate with the depth, as shown in Fig. (20-3). In this respect, the lateral pressure in cohesionless soil is analogous to hydrostatic pressure; and, for this reason, the conjugate ratio K is sometimes called the hydrostatic pressure ratio. Also, since the unit lateral pressure increases uniformly with depth below the surface, the pressure-distribution diagram is triangular in shape and the resultant pressure R acts at a distance $\frac{1}{3}H$ above the bottom of a vertical plane, as shown in Fig. 20-3.

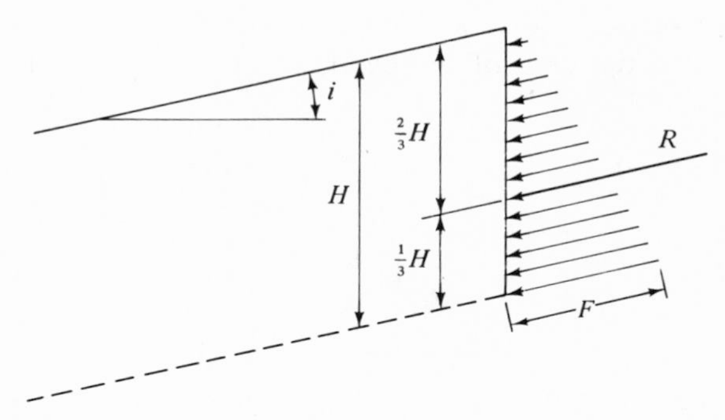

Fig. 20-3. Distribution of lateral pressure.

20.5. NORMAL AND SHEARING FORCES ON INCLINED SURFACE. Consider the forces acting on the base of any vertical slice, as in Fig. 20-4. The weight of the slice may be resolved into two components, which are, respectively, normal and parallel to the base. The parallel component is

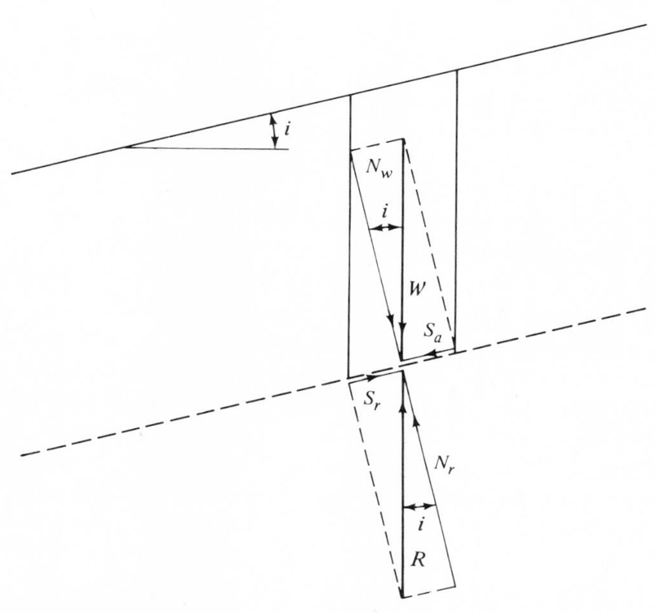

Fig. 20-4. Normal and shearing forces on an inclined surface.

a shearing force which tends to cause the slice to slide down hill. It is equal to the normal force multiplied by the tangent of the slope angle. Thus

$$N_w = W \cos i \tag{20-6}$$

$$S_a = N_w \tan i = W \sin i \tag{20-7}$$

in which

W = weight of slice;
N_w = normal forcc on base of slice;
S_a = shearing force on base of slice; and
i = slope angle.

The reaction to the weight is an upward force of equal magnitude. It may also be resolved into a normal component and a parallel component. The parallel component is the resisting force which prevents the slice from sliding down hill. Hence,

$$N_r = R \cos i \tag{20-8}$$

$$S_r = R \sin i = N_r \tan i \tag{20-9}$$

As long as S_r is equal to S_a, the slice will be in equilibrium and the slope will be stable. However, there is a limit to the resisting force which can be developed in a cohesionless soil, and this limit depends on the friction characteristics of the soil. Its value is

$$S_{r\,max} = N_r \tan \phi \qquad (20\text{-}10)$$

in which ϕ is the friction angle of the soil.

If the slope angle i is greater than the friction angle ϕ, then the shearing force S_a will be greater than the maximum shearing resistance $S_{r\,max}$ and failure will occur along the base of the slice. From this statement it may be concluded that the slope angle or angle of repose of a bank of cohesionless soil cannot exceed the internal friction angle. This conclusion is theoretically correct. However, in actual practice the angle of repose which can be relied upon is somewhat less than the friction angle of the soil. It should be noted that "angle of repose" has no meaning for a cohesive soil because cohesion also contributes to stability of a slope.

20.6. ANALYSIS OF MASS RESTING ON INCLINED LAYER OF IMPERMEABLE SOIL. One of the simplest of all practical problems in slope stability is the analysis of a sloping mass of soil of finite extent which rests upon an inclined layer of a clayey soil or rock that is less pervious than the soil above, as shown in Fig. 20-5. Water seeping down through the upper mass will tend to flow downward, saturating and weakening soil along the plane of contact. Under these circumstances

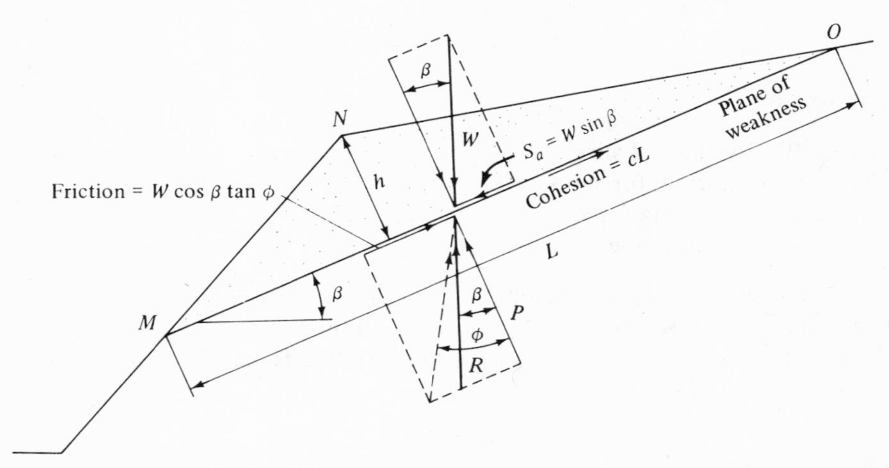

Fig. 20-5. Sloping mass on inclined layer of clayey soil.

the contact surface is a plane of weakness on which failure will occur if a slide develops. Therefore, an analysis of the shearing stresses on this plane and comparison with the shearing strength gives an indication of the stability of the slope.

The shearing stress on the plane of weakness MO is a variable quantity, its amount depending on the distribution of vertical pressure along the plane. However, no attempt will be made to determine the maximum shearing stress. Rather, it will be assumed that failure will occur when the total shearing force on the plane MO equals or exceeds the total shearing strength. In Fig. 20-5 the angle which the plane of weakness makes with the horizontal is designated as β. The total vertical pressure W on the plane is equal to the weight of the triangle of soil MNO and it acts through the centroid of the triangle. Thus,

$$W = \tfrac{1}{2}\gamma h L \tag{20-11}$$

The shearing stress is equal to the component of W which is parallel to the plane of weakness MO. Hence,

$$S_a = W \sin \beta \tag{20-12}$$

The total shearing strength $S_{r\,max}$ on MO is the sum of two components, namely, the cohesion C and the friction F. The cohesion component is equal to the unit cohesion c multiplied by the length L. The friction component is equal to the normal component of the reaction multiplied by the coefficient of friction of the soil, or $\tan \phi$. Thus,

$$C = cL \tag{20-13}$$

$$P = R \cos \beta = W \cos \beta \tag{20-14}$$

$$S_{r\,max} = cL + W \cos \beta \tan \phi \tag{20-15}$$

When $S_a < S_{r\,max}$, the slope is stable. When $S_a \geqq S_{r\,max}$, failure may be expected to develop along the plane of weakness. The factor of safety F_s of the slope is the ratio of the shearing strength to the shearing force, or

$$F_s = \frac{S_{r\,max}}{S_a} = \frac{cL + W \cos \beta \tan \phi}{W \sin \beta} \tag{20-16}$$

20.7. ANALYSIS OF HOMOGENEOUS SOIL. When a mass having a sloped boundary surface consists of soil which is reasonably homogeneous, the shearing-strength characteristics are the same on all interior surfaces within the mass. Failure, if it occurs, develops along a surface on which the shearing stress is greater than the shearing strength, and the most likely failure surface is the one on which the ratio of strength to stress is a minimum. This surface is called the maximum stressed surface, although the term is only relative and refers to the stress in relation to strength along the surface. If failure does not occur, the factor of safety of the slope is the minimum strength-stress ratio on the maximum stressed surface. Therefore, an analysis of a slope of this kind involves the determination of the location of the maximum stressed surface. A solution based upon the assumption that the maximum stressed surface is a plane passing through the toe of the slope is known as the Culmann method. Although there is considerable evidence that failure usually occurs on a curved surface instead of a plane, the Culmann method has certain advantages and may prove useful in some cases. The accuracy of this method is reasonably good for steep slopes or vertical banks, but the method underestimates F_s if applied to flatter slopes.

In Fig. 20-6, AB represents a slope of height H in homogeneous soil. Let θ be an angle of arbitrary value which defines a trial plane AC passing through the toe of the slope. The shearing force on this plane depends on

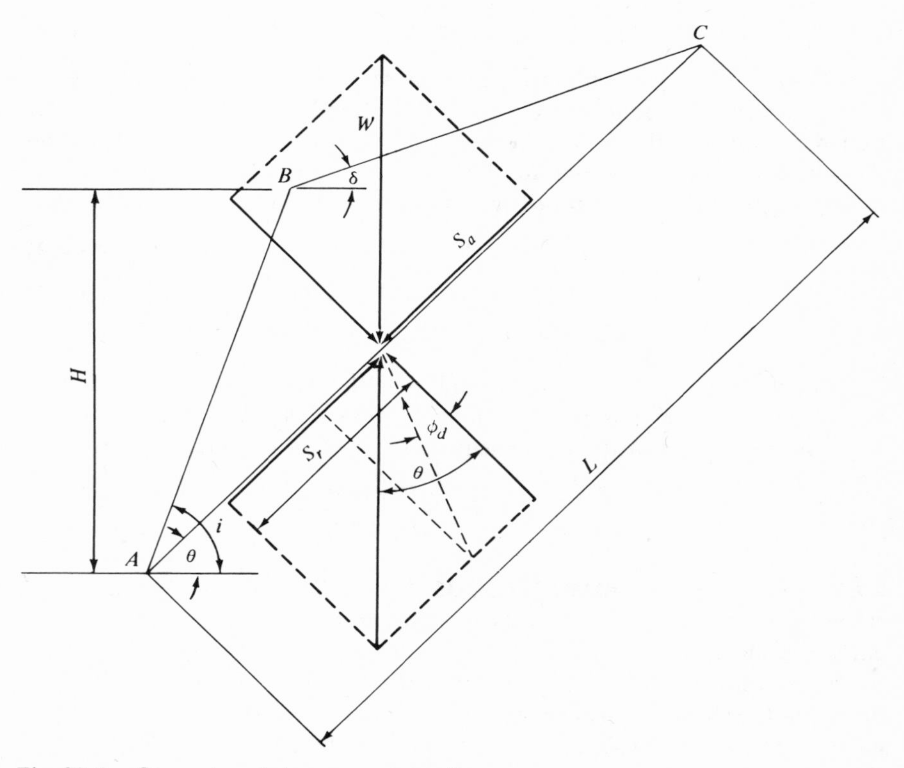

Fig. 20-6. Steep slope in homogeneous soil.

the weight of soil above and the angle θ. Thus,

$$S_a = W \sin\theta \tag{20-17}$$

The shearing resistance on the plane AC is equal to the shearing force at stresses less than failure; and it consists of some value of cohesion, which is less than the cohesion component of the strength of the soil, plus a friction force which is less than the friction component of soil strength. These actual values are referred to as developed cohesion and developed friction.

Let

c_d = the developed cohesion which is less than c; and
ϕ_d = the developed friction angle which is less than ϕ.

Then

$$W \sin\theta = c_d L + W \cos\theta \tan\phi_d \tag{20-18}$$

The weight of the soil in the triangle ABC is

$$W = \tfrac{1}{2}\gamma HL \csc i \sin(i - \theta) \tag{20-19}$$

Substituting this value of W in Eq. (20-18) and solving for c_d, we obtain

$$c_d = \tfrac{1}{2}\gamma H \csc i \frac{\sin(i - \theta)\sin(\theta - \phi_d)}{\cos\phi_d} \tag{20-20}$$

For any arbitrarily chosen value of ϕ_d that is less than ϕ, the maximum stressed plane will be the one defined by a critical angle θ_c which gives the maximum value of c_d. The critical angle θ_c may be obtained by trial by substituting various values θ in Eq. (20-20); or by applying the principle of maxima and minima, that is, setting the first derivative of c_d with respect to θ equal to zero and solving for θ. This operation gives

$$\theta_c = \tfrac{1}{2}(i + \phi_d) \tag{20-21}$$

Substituting in Eq. (20-20), we obtain

$$c_d = \frac{\gamma H[1 - \cos(i - \phi_d)]}{4 \sin i \cos\phi_d} \tag{20-22}$$

Also†

$$H = \frac{4\, c_d \sin i \cos\phi_d}{\gamma[1 - \cos(i - \phi_d)]} \tag{20-23}$$

†This equation is usually attributed to Culmann, who published it in "Die Graphische Statik" in 1866. However, Golder points out that it was first developed by Francaise in 1820. See *Geotechnique* **1** (No. 1), 66 (June 1948).

For a given soil material for which the unit weight γ, the cohesion c, and the angle of friction ϕ are known, and for any chosen values of factors of safety, it is evident from Eq. (20-23) that the greater the slope angle i, the smaller the height of slope H must be. It is pointed out that factors of safety must be applied separately to the friction component and the cohesion component.

Let

F_c = factor of safety with respect to cohesion; and

F_ϕ = factor of safety with respect to friction.

Then, since

$$F_c = \frac{c}{c_d} \qquad \text{and} \qquad F_\phi = \frac{\tan\phi}{\tan\phi_d}$$

$$c_d = \frac{c}{F_c} \tag{20-24}$$

$$\tan\phi_d = \frac{\tan\phi}{F_\phi} \tag{20-25}$$

EXAMPLE 20-1. Assume that a vertical cut ($i = 90°$) is to be made through a soil whose properties are as follows: $\gamma = 110$ pcf; $c = 600$ psf; $\phi = 24°$. Determine the safe depth of cut in this soil with a factor of safety of 2.

SOLUTION. Here,

$$c_d = \frac{600}{2} = 300 \text{ psf}$$

$$\tan 24° = 0.445$$

$$\tan\phi_d = \frac{0.445}{2} = 0.222$$

$$\phi_d = 12°\,30'; \cos\phi_d = 0.976; \cos(i - \phi_d) = 0.216$$

Substituting in Eq. (20-23), we obtain

$$H = \frac{4 \times 300 \times 1 \times 0.976}{110(1 - 0.216)} = 13.6 \text{ ft}$$

20.8. GRAPHIC SOLUTION OF EQ. (20-23). Equation (20-23) may be solved graphically by the construction procedure shown in Fig. 20-7. Beginning at point A, representing the toe of the slope, draw line AC making the angle ϕ_d with the horizontal. Lay off AB and BC each equal

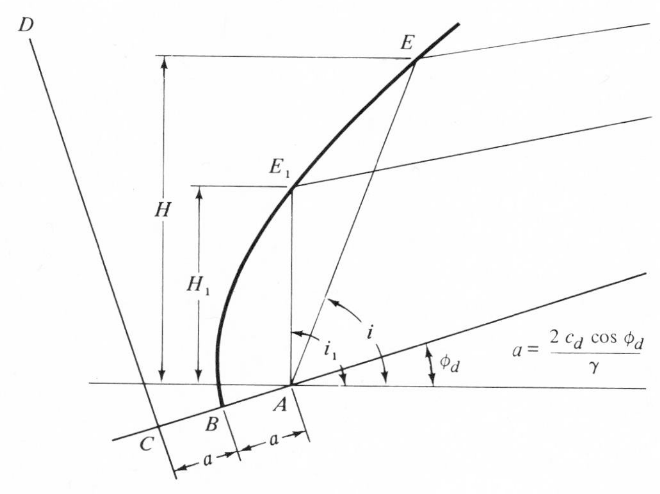

Fig. 20-7. Graphic solution of Eq. (20-23).

to

$$a = \frac{2c_d \cos \phi_d}{\gamma} \tag{20-26}$$

This quantity has the dimension of distance. Therefore, AB and BC may be laid off on any convenient scale, which establishes the scale of the diagram. Next, draw CD perpendicular to AC, and construct a parabola with point A as the focus and line CD as the directrix. Finally, draw a line beginning at point A and making the slope angle i with the horizontal. The intersection E of this line with the parabola marks the maximum height at which this slope will be stable, in accordance with the chosen factors of safety. It can be proved by trigonometry that the vertical distance from E to the horizontal line through A is equal to H, the height of slope as defined by Eq. (20-23).

20.9. OUTLINE OF GRAPHIC ANALYSIS OF STABILITY OF SLOPE. A graphical procedure for analyzing the stability of a slope, which is more accurate than the Culmann method, is the method of slices. Subsequent to 1920 a Swedish Geotechnical Commission made an exhaustive study of numerous slope failures on railroad cuts and fills in Sweden. Extensive borings into the soil involved in the slides indicated, in general, that failure usually occurs along a curved surface within the soil, rather than along a plane surface as assumed in the Culmann method. Also, it was found that analysis by the method of slices gave results which were in reasonably good agreement with the facts obtained in the field investigations. Because of the extensive use which this Commission made of the slices method, it is sometimes referred to as the Swedish method.

In applying the method of slices it is first necessary to choose arbitrarily a trial curved surface on which the total shearing force and the total shearing strength are determined for comparison, to establish the factor of safety for that particular surface. Then another surface is selected and analyzed, and so on, until the designer is satisfied that he has investigated the weakest surface and has thus determined the degree of stability of the slope. The shape of the curved failure surface probably approaches that of a circle in homogeneous soil, but may deviate from a circular arc in stratified or otherwise nonhomogeneous soil. However, the method is applicable to any shape of curved surface.

20.10. DETAILS OF METHOD OF SLICES. A cross section of the slope to be analyzed is drawn to scale, and the trial curved surface is drawn as shown in Fig. 20-8(a). The soil above the curve is then divided into vertical slices of any convenient equal width. The weight of each slice can be computed by multiplying the volume of the slice by the unit weight of the soil. Usually the thickness of the cross section is taken as unity in the direction normal to the page; and the weight of a slice is then equal to the area times the unit weight of the soil. The weight of a slice is a vertical downward force which is assumed to intersect the base of the slice at its

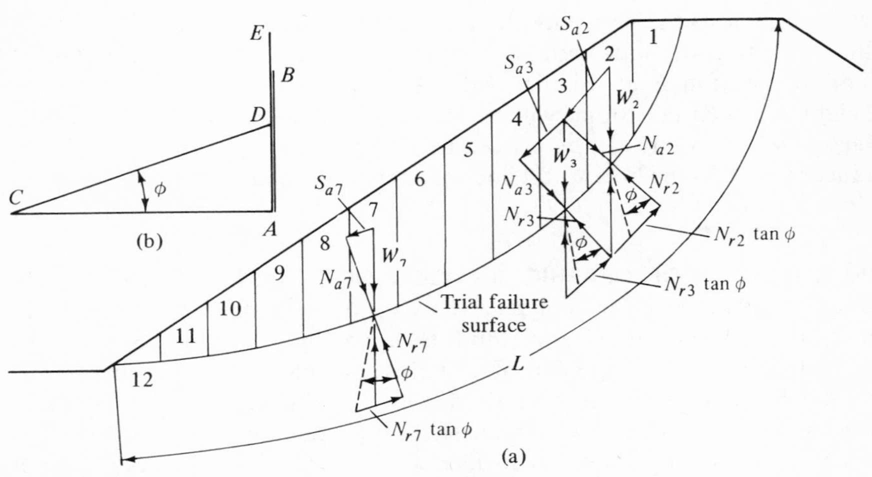

Fig. 20-8. Method of slices.

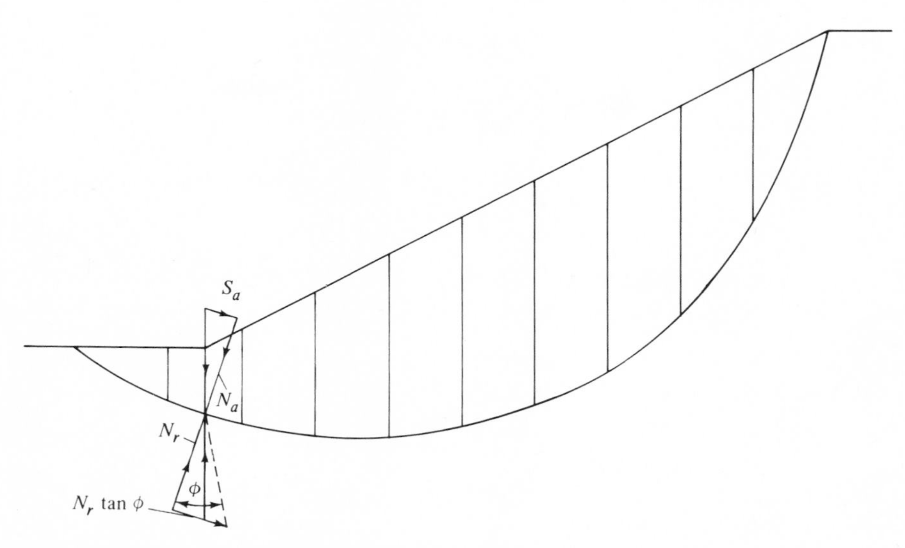

Fig. 20-9. Upward slope at lower end of surface.

center. Next, resolve the weight into two components, S_a and N_a, which are, respectively, parallel and normal to the base of the slice. The parallel component is the applied shearing force which tends to cause the slice to slide down hill. Repeat this process for each of the slices above the trial curved surface. The summation of the shearing forces on the bases of all the slices, or $S_{a1} + S_{a2} + S_{a3} + \cdots$, represents the total applied shearing force which acts on the curved surface and tends to cause the soil above that surface to slide downward. Attention is directed to the fact that the summation of shearing forces should be made algebraically. If the trial surface curves upward near its lower end, as illustrated in Fig. 20-9, the shearing component of the weight of the slices in this region will act in the opposite direction along the curve; and this fact needs to be taken into account in the determination of the total shearing stress.

The resistance to sliding along the curved surface is furnished by the cohesion and the friction of the soil. The total cohesion which the soil is capable of developing is equal to the unit cohesion multiplied by the total length of the curved surface, or cL. The friction component of the shearing strength at the base of each slice is equal to the normal component of the reaction multiplied by the coefficient of friction of the soil, or $N_r \tan \phi$. Then the total of all friction components along the curved surface is $(N_{r_1} + N_{r_2} + N_{r_3} + \cdots) \tan \phi$. The relationships here noted may be expressed in formula form as follows:

$$S_a = \sum_1^n S_{an} \qquad (20\text{-}27)$$

$$S_{r\,max} = cL + \sum_1^n N_{rn} \tan\phi \qquad (20\text{-}28)$$

20.11. FACTOR OF SAFETY. The factor of safety on the trial surface is equal to the total shearing strength divided by the total shearing force. Therefore,

$$F_s = \frac{S_{r\,max}}{S_a} = \frac{cL + \sum_1^n N_{rn} \tan\phi}{\sum_1^n S_{an}} \qquad (20\text{-}29)$$

The facts shown in eq. (20-29) can easily be represented graphically, as shown in Fig. 20-8(b). After the vectors representing the weights of the slices have been drawn to a suitable scale, draw the parallel and normal components of each weight vector. Then, with a pair of dividers, lay off the parallel components S_a on a vertical line from point A to B. The length of this line represents the total shearing force $\Sigma_1^n S_{an}$ on the curved surface. Next, with a pair of dividers, lay off the normal components N_r on a horizontal line from point A to C. This line represents the summation of the normal components of the vertical reactions. Lay off the angle ϕ at C and draw line CD. Then AD is equal to the summation of normal components multiplied by $\tan\phi$ and is, therefore, the friction component of the shearing strength of the curved surface. From point D, scale the total cohesion component of strength, or cL, upward to point E. This makes the line AE a vector representation of the total shearing strength. If AE is greater than AB, the slope is safe with respect to the trial surface; and vice versa. Also,

$$F_s = \frac{AE}{AB} \qquad (20\text{-}30)$$

20.12. LIMITATIONS OF THE ORDINARY METHOD OF SLICES. Slope stability problems are statically indeterminant—that is, there are more unknown quantities than there are equations of equilibrium. Any solution therefore involves simplifying assumptions: In the Culmann analysis the failure surface was assumed to be planar; whereas in the ordinary method of slices, shearing stresses on the sides of each slice are omitted from the analysis. Since downslope movement of the slices in Fig. 20-8 necessarily involves rotation of each slice, shearing stresses will occur on the sides. Since such stresses help to resist sliding, this method overestimates the factor of safety. The amount of error is small if the failure is at shallow depth. The ordinary method of slices is the most commonly used method for analysis of slope stability because it is simple, adaptable to any shape failure surface involving any number of materials, and the answers are on the safe side. Ordinarily three trial surfaces will establish F_s within 0.1.

A modification of the method of slices by Bishop (1) gives many of the same advantages as the ordinary method, plus better accuracy. Computations are more complex and involve a rapid trial-and-error procedure for each trial surface. The method is beyond the scope of the present text. Its use is explained fully in Terzaghi and Peck (*10*) and Lambe and Whitman (*5*).

20.13. PORE WATER PRESSURE AND SEEPAGE FORCES. Slope failures almost invariably involve partial saturation of the sliding mass of soil by ground water. This has two effects: a tendency to buoy up the soil and reduce the values of W and hence N and resisting force S_r, and introduction of a seepage force which adds to actuating force S_a. Both effects tend to reduce stability of a slope. The essential concepts of effective stress and seepage force were introduced in Sections 12.20 and 12.21.

The effect of submergence on an infinite slope can be seen by referring to Fig. 20-4. Replacing W by the buoyant weight W' reduces both the actuating force S_a and normal force N_w proportionately, and also reduces the resisting forces R, S_r, and N_r by the same proportion. Thus at equilibrium Eq. (20-7) still equals (20-10), and the submerged angle of repose remains equal to ϕ.

Next let us consider the effect of seepage parallel to the slope. Not only are buoyant unit weights involved, there is an elevation head loss, h, between points A and B shown in Fig. 20-10. The hydraulic gradient i_h, or head loss per unit length along the travel path, is

$$i_h = \frac{h}{l} = \sin i$$

From Eq. (12-13) the seepage force S per unit volume of soil becomes

$$S = i_h \gamma_w = \gamma_w \sin i \tag{20-31}$$

where i is the slope angle and γ_w is the unit weight of water.

In order to solve for resisting force one must use the buoyant unit weight in Eq. (20-6), giving

$$N'_w = W' \cos i$$

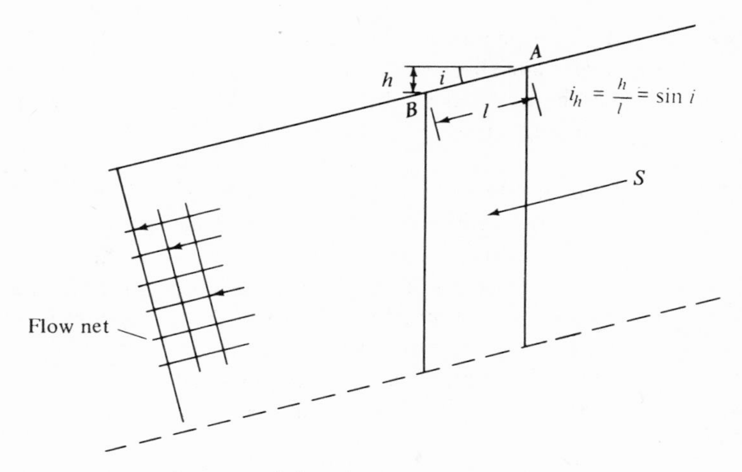

Fig. 20-10. Seepage force parallel to slope.

The resisting forces are then

$$N'_r = W' \cos i \tag{20-32}$$

and

$$S_{r\max} = W' \cos i \tan \phi$$

Dividing by volume of soil involved gives

$$\frac{S_{r\max}}{V} = \frac{W'}{V} \cos i \tan \phi$$

$$\frac{S_{r\max}}{V} = \gamma' \cos i \tan \phi \tag{20-33}$$

where γ' is the submerged unit weight of the soil. Since the seepage force S is assumed to act parallel to the failure surface it has no effect on resisting N_r or $S_{r\max}$. However, it does add to the actuating shearing force S_a, and Eq. (20-7) becomes

$$S_a = W' \sin i + VS$$

where V is the soil volume and S the seepage force per unit volume. Dividing by V and substituting for S

$$\frac{S_a}{V} = \gamma' \sin i + \gamma_w \sin i$$

$$= \sin i (\gamma' + \gamma_w)$$

$$= \gamma_t \sin i \tag{20-34}$$

where γ_t is the saturated unit weight.

At failure, S_a equals $S_{r\,max}$; therefore,

$$\gamma_t \sin i = \gamma' \cos i \tan \phi$$

$$\tan i = \frac{\gamma'}{\gamma_t} \tan \phi \tag{20-35}$$

Thus the angle of repose is reduced by seepage parallel to the slope.

EXAMPLE 20-2. A fine sand with a saturated unit weight of 130 pcf and $\phi = 30°$ stands on a 20° slope. The slope is submerged and then suddenly drained. Is the slope stable?

SOLUTION. Before and during submergence, $i = \phi$, and the slope is stable. After sudden drainage, from Eq. (20-35)

$$\tan i = \frac{130 - 62.4}{130} \tan 30° = 0.30$$

$$i = 16.8°$$

Since the angle of repose is less than the slope angle, the slope will fail.

The development of Eqs. (20-33) and (20-34) illustrates an important point: That seepage forces acting downslope parallel to a surface of sliding can be automatically taken into account by using the *saturated* unit weight to calculate actuating force, while using *submerged* unit weight to calculate resisting force. This is useful in slope stability calculations where seepage runs downhill roughly parallel to the failure surface. For resisting force it may be more convenient to calculate effective normal stress from total stress and pore water pressure rather than submerged unit weight. The equivalence of these two methods is discussed in Section 12.21.

EXAMPLE 20-3. Calculate the factor of safety for the trial surface in Fig. 20-11 by the ordinary method of slices.

SOLUTION. The unit weight of the alluvial clay is given. From a consideration of the relationships in Chapter 8, γ for the glacial till above the water table is 117 pcf, $e = 0.62$, $\gamma' = 63.5$ pcf, and $\gamma_t = 126$ pcf. Calculations for a section 1 ft thick are shown in Table 20-1. Note that the house weight per foot is added to slice 8. For slices 6, 7, and 8 cohesion is assumed to be the same as for the glacial till, but $\phi = 16°$.

EXAMPLE 20-4. Recalculate the F_s assuming drains are used to lower the water table to DD'.

TABLE 20-1. Calculations for Example 20-3

Slice	Average H (ft)	γ (pcf)	W (Kip) ($H\gamma X$)	i (deg)	S_a (Kip) (W sin i)	N (Kip) (W cos i)	l (ft)	u (Ksf) ($H_w \gamma_w$)	U (Kip) ($l \times u$)	N′ (Kip) (N − U)	(N − U) tan ϕ (Kip)
1 a	16	110	63	−40	−41	49	45	1.00	45	4	2
2 a	24	110	169								
b	16	126	129								
Total	—	—	298	−21	−107	278	70	2.50	175	103	48
3 a	12	110	66								
b	50	126	315								
Total	—	—	381	−13	−86	371	55	3.87	213	158	74
4	80	126	504	−3	−26	503	50	5.00	250	253	118
5 a	20	117	234								
b	86	126	1084								
Total	—	—	1318	9	206	1301	100	5.37	537	764	356
6 a	10	117	117								
b	78	126	983								
Total	—	—	1100	28	516	971	118	4.87	575	396	114
7 a	14	117	164								
b	60	126	756								
Total	—	—	920	33	501	771	118	3.74	441	330	95
8 h	—	—	10								
a	28	117	328								
b	40	126	504								
Total	—	—	842	40	541	645	130	2.50	325	320	92
					1504		686				899

$$F_s = \frac{S_\gamma}{S_a} = \frac{899 + 0.95(686)}{1504} = 1.03$$

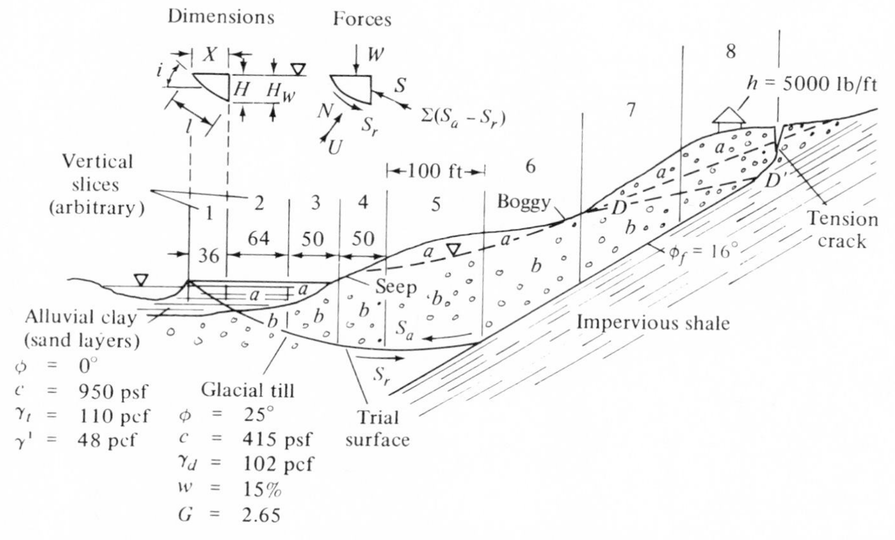

Fig. 20-11. Geologic section and trial surface for Example 20-3.

SOLUTION. Drain tile DD' will reduce the values of W in slices 7 and 8, reducing S_a and N. However, u and U also are reduced, such that N' and $N' \tan \phi$ are substantially increased. Cohesion is assumed to remain the same. The answer is 1.1.

In these examples the actuating forces S_a in slices 1 through 4 are negative, since the weight component parallel to the shear surface is acting to restrain the slide. This subtraction will result automatically if one assumes uphill slopes to be negative, as shown in Table 20-1. Since saturated unit weight is used to calculate S_a in these slices, seepage also is assumed to aid in restraining the slide, i.e., is assumed to be downward parallel to the shearing surface. This of course is impossible since seepage would converge from both directions at the bases of slices 4 and 5. A better procedure would be to draw a flow net or use piezometers and evaluate N_r' and S_a from true seepage directions.

The more general case with seepage force not necessarily parallel to a failure surface is shown in Fig. 20-12. Here the head loss h divided by the seepage distance l_s gives $i_h = \sin s$, where s is the seepage angle measured from horizontal. The seepage force per unit volume of soil is $S = \gamma_w \sin s$. Since S is a frictional loss, it cannot be negative. As shown in Fig. 20-11, it

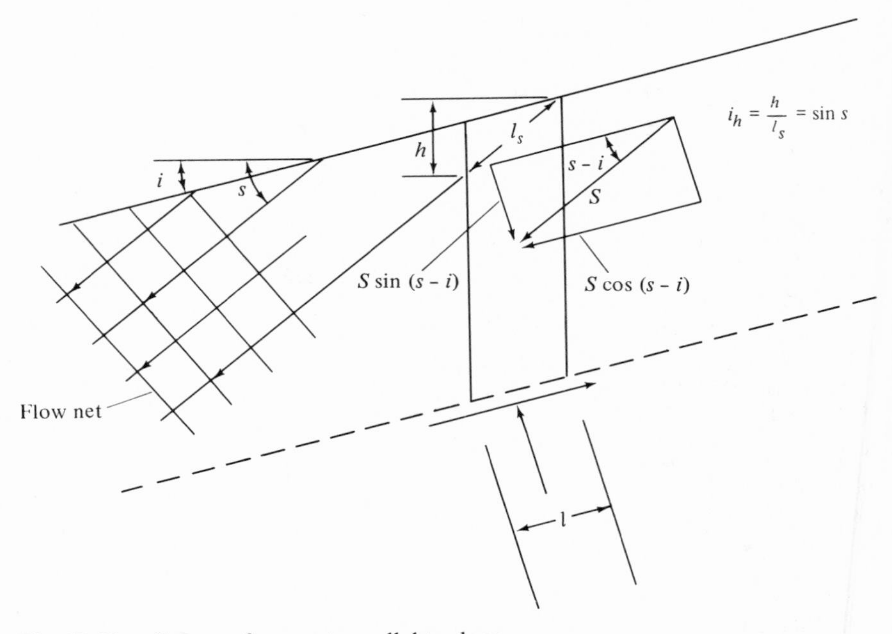

Fig. 20-12. Seepage force not parallel to slope.

adds components to both the normal and shearing forces. The normal force becomes

$$N'_r = W' \cos i + VS \sin(s - i)$$

where V is the volume of soil involved with seepage. Therefore for calculation of the resisting force, it can be shown that

$$N'_r = V[\gamma' \cos i + \gamma_w \sin s \sin(s - i)] \qquad (20\text{-}36)$$

Similarly, the total acting force including seepage is

$$S_a = V[\gamma' \sin i + \gamma_w \sin s \cos(s - i)] \qquad (20\text{-}37)$$

EXAMPLE 20-5. Calculate S_a and N' for slice 1 in Fig. 20-11, using the flow net in Fig. 20-13.

SOLUTION. From Fig. 20-13 the flow line nearly parallels the failure surface, both being uphill in slices 1 through 4. Therefore in Eqs. (20-36) and (20-37), i is negative and $s - i = 0°$. Since $\sin 0° = 0$ and cosines of angles between 0 and $-90°$ are positive, Eq. (20-36) reduces to (20-32),

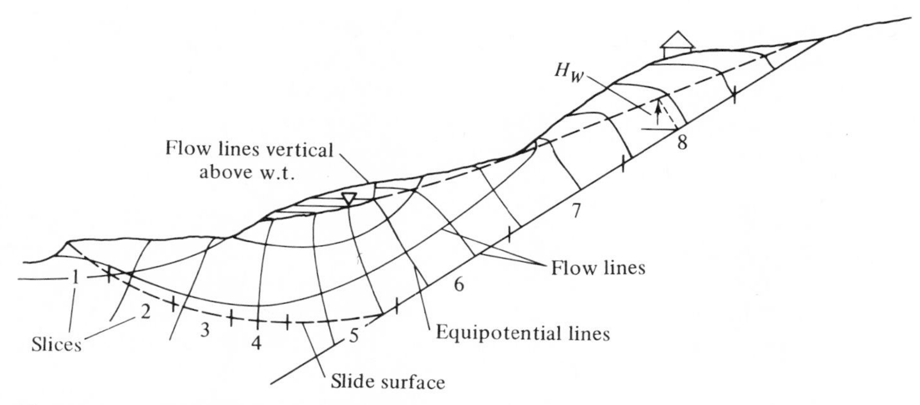

Fig. 20-13. Flow net for Example 20-5.

N_r' and $S_{r\,max}$ are unchanged. However Eq. (20-37) gives

$$S_a = (16)(36)(110 - 62.4)\sin(-40) + 62.4\sin(40)\cos(0)$$
$$= 575 - 30.7 + 40.1 = +5.4 \text{ Kip}$$

Therefore because of the seepage force the net S_a in slice 1 is not resisting but rather is contributing to sliding.

The correction illustrated in Example 20-5 ordinarily is not made since angle i is usually very low at the toe of a slope, giving low values for S_a. However it may be important for analysis of changes which will occur if the seepage direction is changed. As an extreme example, if seepage could be made normal rather than parallel to the shearing surface, $s - i = 90°$, and Eqs. (20-36) and (20-37) reduce to the use of submerged unit weight to calculate S_a and saturated unit weight to calculate N_r'. This is just the reverse of when seepage parallels the slope.

The calculation of effective normal stresses independent of seepage forces also may be improved through use of a flow net. Effective stresses for all slices should be determined as shown for slice 8 in Fig. 20-12: A line (dotted) is drawn parallel to the equipotential line to intercept the phreatic surface, and H_w is measured vertically from this intercept.

20.14. STABILIZATION OF LANDSLIDES. The three classical methods for stopping landslides are: unload the top, load up the toe, or drain. The effects of loading are easily seen in analysis by the method of slices. Drainage usually involves diversion of surface water and use of drain tile, but may involve the use of wellpoints or electroosmosis. An advantage of the latter two methods is that they may be used more effectively to reduce S_a and increase N_r, by changing the direction of seepage. A disadvantage is that if stabilization depends on altering seepage direction, sliding may resume if pumping is stopped. Tile drains are difficult to install in an active landslide because the trenches require shoring, and the tile lines often malfunction due to faulting.

Chemical stabilization constitutes a fourth method of treatment. This was successfully tried by Handy and Williams (*3*), lime being introduced in borings spaced at 5-ft intervals throughout the slide area. Although sliding stopped immediately because the drying action of the quicklime increased apparent cohesion, the main benefit was derived from a slow cementation reaction between lime, clay minerals, and water. Such reactions, called *pozzolanic reactions,* are most effective if the soil contains montmorillonite, although kaolinite also is reactive. Limited data indicate that other clay minerals or allophane are nonreactive, and the method should not be used with such soils. The cementitious product of pozzolanic reaction is a hydrated calcium silicate called *tobermorite,* which also is the main cementitious compound in portland cement concrete.

20.15. PROGRESSIVE FAILURES AND CREEP. Any failure surface usually involves a sufficient variety of soils or stress conditions that the peak strength is seldom reached simultaneously in soils all along the surface. Therefore one zone within the slide fails first, causing strength in this zone to decline to the residual strength, throwing additional load on other zones. The failure zones therefore gradually enlarge until the entire slide activates, explaining why landslides often start with slow creeping movements which gradually speed up to a sudden catastrophe. No quantitative methods are yet available to analyze this behavior except to apply a generous factor of safety, particularly with respect to cohesion. As previously discussed, both the residual strength and, in the case of creep, the yield strength cohesion are nearly zero, whereas ϕ often remains essentially unchanged.

For the same reason, i.e., peak strength exceeding residual strength, landslides sometimes occur in a number of successive slips, starting at the toe and working progressively headward until the slope is a series of stairsteps, each tilted backward and tending to pond water [Fig. 20-14(d)]. The first slide is an important warning, and a slope analysis should be immediately performed using the residual strength for soil in the active slide.

20.16. ANALYSIS OF SPECIAL CASES. Vertical cuts such as shown in Fig. 20-14(a) are common in trenching operations and in nature because of gullying or lateral erosion of river banks. The maximum stable height

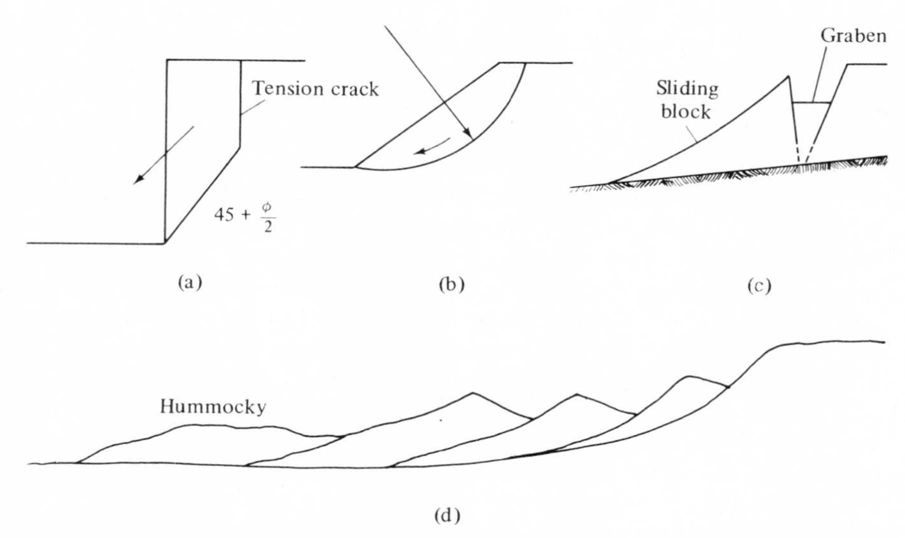

Fig. 20-14. Some types of landslides. (a) Fall; (b) rotational; (c) compound; and (d) successive.

of such a cut depends in part on the depth of tensile cracks behind the face, the worst condition being analogous to an unconfined compression (in plane strain) on soil at the bottom of the bank. Therefore the height H of such a cut should be such that the load $H\gamma$ does not exceed the unconfined compressive strength q_u of soil at the bottom, that is,

$$H_c\gamma = q_u \tag{20-38a}$$

From Eq. (19-22a), $q_u = 2c \cot(45 - \phi/2) = 2c \tan(45 + \phi/2)$, and

$$H_c = 2\frac{c}{\gamma} \tan\left(45 + \frac{\phi}{2}\right) \tag{20-38b}$$

where H_c is the critical height, c the soil cohesion, γ its unit weight, and ϕ is the soil angle of internal friction. Other expressions for critical height assume varying geometries of failure. For example, Eq. (20-23) for a vertical slope and planar failure surface becomes

$$H_c = 4\frac{c}{\gamma}\frac{\cos\phi}{1 - \sin\phi} \tag{20-39a}$$

or

$$H_c = 4\frac{c}{\gamma} \tan\left(45 + \frac{\phi}{2}\right) \tag{20-39b}$$

Because of the frequent observation of vertical tensile cracks running parallel to an open trench, Terzaghi (*9*, p. 154) derived a similar expression which takes into account tensile cracking whereby the multiplying constant becomes 2.67 instead of 4. However, field observations also show that the surface of sliding is distinctly curved rather than straight, as assumed in the above analyses. A recent study by Ellis (*2*) utilizes a curved cycloidal failure surface which more closely approaches field observations. Ellis found that failure would be incipient along the most critical cycloidal surface when the condition of equation (20-38b) is satisfied, that is when the multiplier is 2.0. Since an initiating failure will involve remolding and a loss of strength, a multiplier of 2.0 probably should not be exceeded for design of open trenches, particularly where life or property are involved. Furthermore, this assumes no seepage forces and no surcharge loading from excavated soil adjacent to the trench. The depth of a surcharge loading should be subtracted from the critical height unless the soil is piled outside of the potential failure, i.e., back from the excavation a distance greater than one-half the depth of the cut. Since the stability of open sewer trenches depends mainly on soil cohesion which is time dependent, pipe installation and backfilling must proceed without delay.

EXAMPLE 20-6. A sewer pipe is to be laid at depths of up to 12 feet in soil with $\gamma = 130$ pcf and $q_u = 10$ psi. What is the maximum depth of trench that will stand with vertical walls left unsupported?

SOLUTION. From Eq. (20-38a),

$$H_c = \frac{q_u}{\gamma} = \frac{10(144)}{130}$$

$$= 11 \text{ ft}$$

The use of shoring is indicated, or a flatter angle of cut could be used. A procedure for selecting the angle is given in (*2*).

PROBLEMS

20.1. Explain why a mass of soil bounded by a sloping surface is subjected to shearing stresses on interior surfaces. On what factors does the magnitude of these shearing stresses depend?

20.2. Define conjugate stresses and the conjugate ratio.

20.3. The slope of a mass of soil makes an angle of 18° with the horizontal and the angle ϕ of the soil is 25°. Compute the value of the conjugate ratio.

20.4. Determine the conjugate ratio for the soil of Problem 20.3 if the mass is bounded by a horizontal surface.

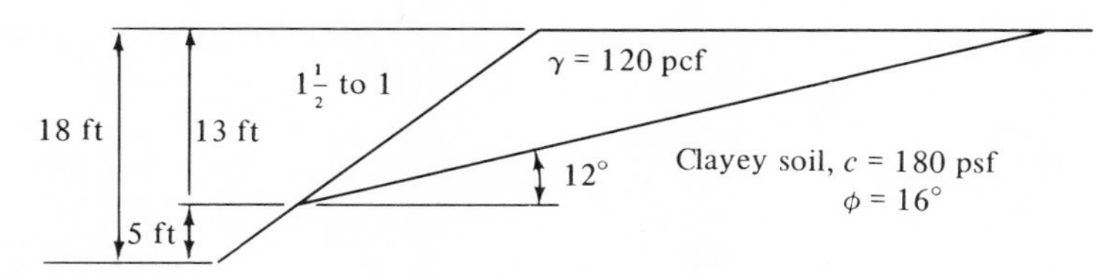

Fig. 20-15. Conditions for Problem 20.5.

20.5. A deep railroad cut intersects the plane of contact of two soil strata, the lower one of which is a highly impervious and cohesive clay. The plane of contact makes an angle of 12° with the horizontal. Other details of the situation are shown in Fig. 20-15. Assume that the shearing strength between the two strata is c = 180 psf and ϕ = 16°. Determine whether a slide will be likely to occur along this plane. If a slide is not likely, what is the factor of safety against sliding?

20.6. A soil has the following properties: γ = 120 pcf; c = 400 psf; ϕ = 25°. A trench with vertical sides is to be dug 10 ft deep in this soil. Using the Culmann method, determine whether or not the sides of this trench are safe against caving. If they are safe, and ϕ_d is assumed to be 15°, what is the factor of safety with respect to cohesion?

20.7. The stability of an embankment slope is being analyzed by the method of slices. On a particular curved surface through the soil mass, the shearing component of the weight of each slice and the normal component of the reaction at the base of each slice are as tabulated below. The length of the curved surface is 98 ft. If the ϕ angle of the soil is 12° and the cohesion is 50 psf, what is the factor of safety of the slope along this particular curved surface?

Slice No.	*Shear Component (lb)*	*Normal Component (lb)*
1	980	560
2	1020	1090
3	860	2040
4	530	3120
5	410	3360
6	320	3250
7	270	2740
8	200	1600
9	150	1260
10	40	830

20.8. An earth embankment is 70 ft high, has a top width of 20 ft, has side slopes of 2 to 1, and rests on a foundation which is very strong compared with the material in the embankment itself. The soil has the following properties: γ = 125 pcf; c = 560 psf; ϕ = 20°. Use the method of slices to determine the minimum factor of safety of the side slopes of this embankment.

20.9. Assume the embankment in Problem 20.8 is constructed on foundation soil which has the following properties: γ = 100 pcf; c = 60 psf; ϕ = 10°. Are the 2 to 1 slopes of this embankment sufficiently flat to provide a factor of safety of 1.5? What is the actual factor of safety?

20.10. Solve Eqs. (20-36) and (20-37) for N_r' and S_a if seepage is induced upslope parallel to the slope by use of electricity or vacuum.

REFERENCES

1. Bishop, A. W. "The Use of the Slip Circle in Stability Analysis of Earth Slopes." *Geotechnique* **5**, 7–17 (1955).
2. Ellis, H. B. "Use of Cycloidal Arcs for Estimating Ditch Safety." *ASCE J. Soil Mechanics and Foundation Eng.* **99** (SM2) (in press).
3. Handy, R. L., and W. W. Williams. "Chemical Stabilization of an Active Landslide." *ASCE* **37** (No. 8), (August 1967).
4. Highway Research Board, National Research Council. *Landslides and Engineering Practice.* Special Report No. 29, 1958.
5. Lambe, T. William, and Robert V. Whitman. *Soil Mechanics.* John Wiley, New York, 1969.
6. Lohnes, R. A., A. Millan, and R. L. Handy. "In-Situ Measurement of Soil Creep." *ASCE J. Soil Mechanics and Foundation Eng.* **98** (SM1), 143–147 (1972).
7. Skempton, A. W. "Long Term Stability of Clay Slopes." *Geotechnique* **14** 75–102 (1964).
8. Taylor, Donald W. *Fundamentals of Soil Mechanics.* John Wiley, New York, 1948.
9. Terzaghi, Karl. *Theoretical Soil Mechanics.* John Wiley, New York, 1943.
10. Terzaghi, Karl, and Ralph B. Peck. *Soil Mechanics in Engineering Practice,* 2nd ed. John Wiley, New York, 1967.

21

Embankments, Levees, Earth Dams

I. CONSTRUCTION

21.1. NEED FOR EARTH STRUCTURES. The necessities and desires of man have prompted him throughout the ages to engage in earth-moving operations in connection with the construction of engineering facilities. The function of roads, railroads, airports, and canals as arteries of transportation is greatly enhanced by cutting down the hills and filling the valleys in order to reduce the rise and fall of grade lines. The devastation of agricultural lands and urban areas by flood waters may be reduced by the construction of levees near river channels; and earth dams are widely employed to impound excess run-off water for irrigation, flood control, power development, navigation, water supply, and recreational purposes.

21.2. EARTH-MOVING EQUIPMENT. In ancient times the movement of the soil from cut to fill areas was largely accomplished by means of baskets carried on the backs of men; and, in some regions of the world where human labor is plentiful and cheap, this primitive method is still employed. Later, though as recently as the eighteenth century and the early part of the nineteenth century, wheelbarrows were widely used in earth-moving operations. Many of the canals and some of the early railroads in the United States were constructed in this manner. Next, horses and mules were hitched to slip scrapers, wheel scrapers, and fresno scrapers; and both the amount of human toil and the cost of moving a cubic yard of earth were greatly reduced. At the same time the practical length of haul between cut and fill areas was increased. As a result there was a great expansion in the potential benefit to be derived from engineering projects which required the movement of larger quantities of earth. Still later, the invention of the elevating grader and bottom dump wagons, both powered by horses and mules, made possible further benefits of such projects.

An outstanding technical development of the second and third quarters of the twentieth century has been the application of internal-combustion engine power to earth-moving operations. Equipment manufacturers have developed power shovels, drag lines, dump wagons and dump trucks, bulldozers, blade graders, carry-all scrapers, sheep's-foot rollers, pneumatic-tired rollers, and other items of specialized equipment, all of which contribute to the ease and efficiency of moving soil from one place to another. Much of this equipment rolls on pneumatic-tired wheels, which greatly increase the speed and the economical length of haul from cut to fill. As a result the real unit cost of moving earth has decreased very greatly, and engineers can now design and construct earth embankments and earth dams which are wider and higher and contain more cubic yards of material than was thought possible 25 to 50 years ago. These larger earth structures have brought new problems into focus. At the same time the versatility of modern earth-moving equipment makes possible a greater degree of control and closer adherence to design details and construction specifications.

21.3. SELECTION OF MATERIAL. The design of a soil structure should be preceded by a soil survey of both the cut areas and the fill areas, as outlined in Chapter 5. The results of this survey and of laboratory tests of the soil materials should be incorporated in a soil profile, which is made a part of the construction plans. The location and extent of each body of soil that will be encountered on the project should be clearly shown; and its color and other physical characteristics should be stated so that the earth-moving contractor can readily identify the various soils as they are revealed in the cut section or borrow area. Also the disposition of each soil in the embankment should be clearly shown. In planning the project, the available soils should be carefully selected and placed in the embankment in accordance with their various characteristics rather than solely on the basis of their proximity to the fill or the order in which they are encountered in excavation.

As a general rule, the poorest soils, that is, those which are fine-grained and have high shrinkage and swell characteristics, should be placed at the bottom and toward the center of an embankment. In this location the fluctuations in moisture content will be a minimum and the weight of overlying material will tend to inhibit the high volume changes which these poorer soils ordinarily undergo. The better soils, which are those containing higher percentages of sand and gravel, should be used in the outer portions and at the top of an embankment. These soils will be more stable against climatic variations and are usually stronger and more resistant to the internal stresses within the embankment. Finally, the topsoil from the cut sections, which contains higher percentages of organic matter and will more readily support vegetation, may be used to top out the side slopes and shoulder areas of an embankment. This soil will facilitate the growth of erosion-resistant grasses, which should be planted after completion of the fill. In this process of soil selection it may be necessary to reject completely certain of the very poorest soils, such as peats and mucks. If so, these facts should be clearly indicated on the plans, and the areas where they are to be wasted should be shown.

21.4. NECESSITY FOR ARTIFICIAL COMPACTION OF SOIL. Usually when soil is excavated, and then transported and dumped in a fill without being compacted artificially, it is loosened up to such an extent that the void ratio of the fill material is relatively high and the density of the fill is correspondingly low. With the passage of time, the filled soil gradually settles and compacts to a higher density. This residual compaction and accompanying settlement may be sufficient to lower the grade of the top of the embankment and cause a pavement or other structure thereon to settle and warp unevenly; as a result the pavement may crack extensively and the riding surface may deteriorate. Furthermore, the shearing strength of uncompacted soil is relatively low, and its permeability is relatively high. For these reasons, the practice of artificially compacting embankments is widespread; and, in the case of earth dams, controlled compaction of the soil during construction is essential.

21.5 METHODS OF OBTAINING COMPACTION. For reasons discussed in Section 19.38, compaction of an embankment during construction involves spreading the soil in thin layers, by means of the hauling equipment supplemented by blade graders and bulldozers. After each layer has been rolled and compacted to the required density, another layer is placed; and the operations are repeated until the embankment is completed. The compacted thickness of a layer is usually in the neighborhood of 6 in. although it may be as much as 9 to 12 in. or possibly more in situations where especially heavy equipment is used. On the other hand, in the case of a rather wet, rebounding type of soil, the thickness may be reduced to about

3 in. The main objective in this regard is to obtain the desired density of the soil throughout the full depth of the compacted layer.

The layers should be deposited approximately horizontal, but there should always be sufficient cross slope, as 1 ft in 20 to 50 ft, to permit the surface to shed water readily in case of rain. If surface drainage is not provided for during construction, water may pond in some areas of the fill and so saturate the soil that compaction becomes impossible. Under no circumstances should an embankment be constructed by end-dumping, that is, by beginning at one side of a valley and building the fill outward by dumping soil over the end of the new fill. This practice results in very poor and uneven compaction of the soil.

21.6. COMPACTION EQUIPMENT. The equipment for compaction of soil layers consists of rollers and tampers. Tamping equipment is used in areas, such as those around structures, which are too small for, or inaccessible to, rolling equipment. There are hand tampers, pneumatic tampers, and vibrator tampers. Rollers are of four general types: smooth rollers, pneumatic-tired rollers, tamping or sheep's-foot rollers, and vibratory rollers.

21.6.1. Smooth Steel Rollers. Generally, smooth rollers are of the three-wheel type, and the weight ranges from 5 to 12 tons. Pressures exerted range from 100 to 200 lb/in. of width of tread for the guide roller and 150 to 400 lb/in. for the rear or drive rollers. The pressure and weight classes of three-wheel rollers suited for compacting different soils are shown in Table 21-1.

TABLE 21-1. Pressure and Weight Classes of Three-Wheel Rollers Suited for Compacting Different Soils

Type of Soil	*Weight Group (tons)*	*Pressure (lb/in. of Width of Rear Rolls)*
Clean, well-graded sands, uniformly graded sands, and some gravelly sands having little or no silt or clay	Cannot be compacted satisfactorily with 3-wheel rollers	
Friable silt-sand and clay-sand soils which depend largely on their frictional qualities for developing bearing capacity	5 to 6	150 to 225
Intermediate group of clayey silts and lean clayey soils of low plasticity (minus No. 10 sieve)	7 to 9	225 to 300
Well-graded sand-gravels containing sufficient fines to act as filler and binder	10 to 12	300 to 400
Medium to heavy clayey soils	10 to 12	300 to 400

21.6.2. Pneumatic-Tired Rollers. Pneumatic-tired rollers are usually towed vehicles mounted on two axles with six tires per axle. Frequently the wheels are mounted in such a way that the tires wobble laterally as the roller is towed forward. This wobbling imparts to the soil a kneading action which facilitates compaction. Desirable inflation pressures for pneumatic-tired rollers operating on several types of soil are given in Table 21-2.

TABLE 21-2. Desirable Inflation Pressures of Pneumatic-Tired Rollers

Type of Soil	*Inflation Pressure (psi)*
Clean sands and some gravelly sands	20 to 40 (the greater pressure with the larger tires)
Friable silty and clayey sands which depend largely on their frictional qualities for developing bearing capacity	40 to 65
Clayey soils and very gravelly soils	65 and up

21.6.3. Sheep's-Foot Roller. The sheep's-foot, or tamping, roller is the most prevalent type used for compacting embankments. It consists of a steel cylindrical drum with steel projections or "feet" extending in a radial direction outward from the cylindrical surface. The tamper feet are 6 to 8 in. long and are so spaced in staggered rows as to obtain a contact of about two tamping feet per square foot of tamped area. The roller is drawn by a tractor back and forth over the layer to be compacted. On the first pass of the roller, the feet project into the loose soil and compact it near the bottom of the layer. On subsequent passes, the soil is compacted from the bottom upward until finally the feet project into the soil only an inch or two and the roller is said to have "walked out" of the soil. Additional passes of the roller beyond this stage will not compact the soil any appreciable amount.

The sheep's-foot roller is so designed that its weight may be varied by filling the cylindrical drum with different amounts of water or sand. The

TABLE 21-3. Desirable Characteristics of Sheep's-Foot Rollers

Type of Soil	*Area of Tamping Feet (sq. in.)*	*Contact Pressure (psi)*
Friable silty sands and clayey sands which depend largely on their frictional qualities for developing bearing capacity	7 to 12	75 to 125
Intermediate group of clayey silts, clayey sands, and lean clay soils with low plasticity	6 to 10	100 to 200
Medium to heavy clays	5 to 8	150 to 300

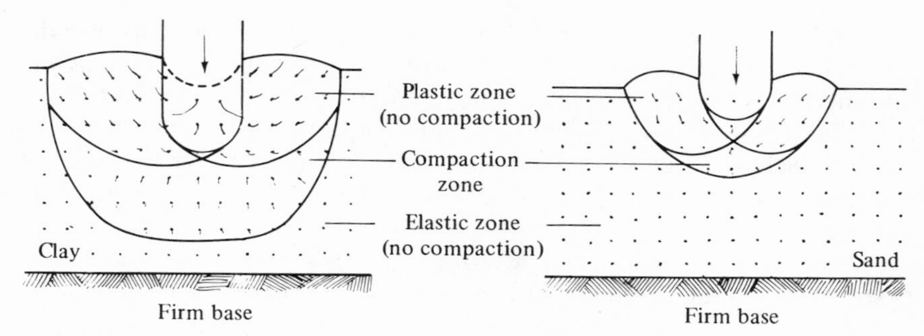

Fig. 21-1. Model studies simulating action of a sheep's-foot roller. After Butt *et al.* (*1*).

tamping feet are removable and may be obtained in several sizes, ranging from 5 to 12 sq. in. of foot area. Table 21-3 gives approximate areas of tamping feet and ground pressures which are used for three broad types of soils. It is necessary to consider various combinations of these features in order to secure efficient performance of tamping rollers. A roller that is most suitable for the compaction of clays will not walk out of a sandy soil; and one that is most suitable for the compaction of sandy soil will not produce the required density in clay soil. It is sometimes necessary to experiment with several different combinations of roller weight and foot size in order to obtain the best results with a particular soil.

Figure 21-1 illustrates the action of one foot of a sheep's-foot roller in clay and in sand. These graphs were obtained by photographing the soil movements in a glass-fronted box and indicate a much larger zone of compaction in clay than in sand. The greater effectiveness in clayey soils is also shown by data in Table 21-4, which gives the number of passes to achieve compaction of several soils in a field experiment conducted in England.

TABLE 21-4. Test Data: Number of Sheep's-Foot Roller Passes Required to Achieve 95 or 100% Standard Density[a]

Soil Type	*95%*	*100%*
Heavy clay	6	10
Silty clay	7	13
Sandy clay	14	38
Gravel-sand-clay	32	64

[a] Compacted in 9 in. lifts; contact pressure 115 psi. From Williams and MacLean, quoted by Johnson and Sallberg (*8*).

In this connection it may be mentioned that very sandy soils which have practically no cohesion cannot be compacted satisfactorily with a static roller. One method for compacting soil of this kind is to wet it thoroughly with large quantities of water and to operate a track-type tractor back and forth over the layer. The vibration of the tractor, coupled with a high degree of moisture content, will usually produce the desired compaction. A more effective method of vibration is to use a vibratory roller designed for this purpose.

21.6.4. Super-Compactors. Recent years have witnessed the development of exceptionally heavy pneumatic-tired rollers weighing up to 100 tons. One type, which is being used rather extensively on the construction of large earth dams, consists of four parallel articulate cells, each mounted on a single 21 by 24 in. tire. The cells can be loaded with wet or dry sand to provide a desired weight. Articulation permits the tires to pass over low or high spots in the soil surface without the load being shifted from one tire to another. Layers of soil up to 24 in. in thickness can be compacted satisfactorily by this equipment with many fewer passes than with a sheep's-foot roller.

Other developments include self-propelled sheep's-foot rollers which are capable of higher speeds than are tractor-drawn rollers. Heavier sheep's-foot rollers require larger contact areas. A similar action is obtained by self-propelled segmented-wheel rollers such as shown in Fig. 21-2.

Fig. 21-2. Self-propelled 25-ton segmented-wheel compactor.

21.6.5. Vibratory Compactors. Compaction to depths of up to 24 in. may also be achieved by lighter-weight compactors if they are vibrated, because of the additional force from the vibration. Vibration is particularly effective for densifying granular soils, but heavy vibratory compactors have been found to perform efficiently with cohesive soils as well. Vibratory compactors are of two general types: plates and rollers. Plate vibrators are especially adapted to small areas, rollers to quantity production, although there is considerable overlap in use of the two types.

Vibrations are commonly produced by rotating an off-center weight, producing a radial oscillation. A vertical vibration may be obtained by use of two counterrotating weights as shown in Fig. 21-3, such that horizontal forces at any instant oppose and cancel out. In either case the

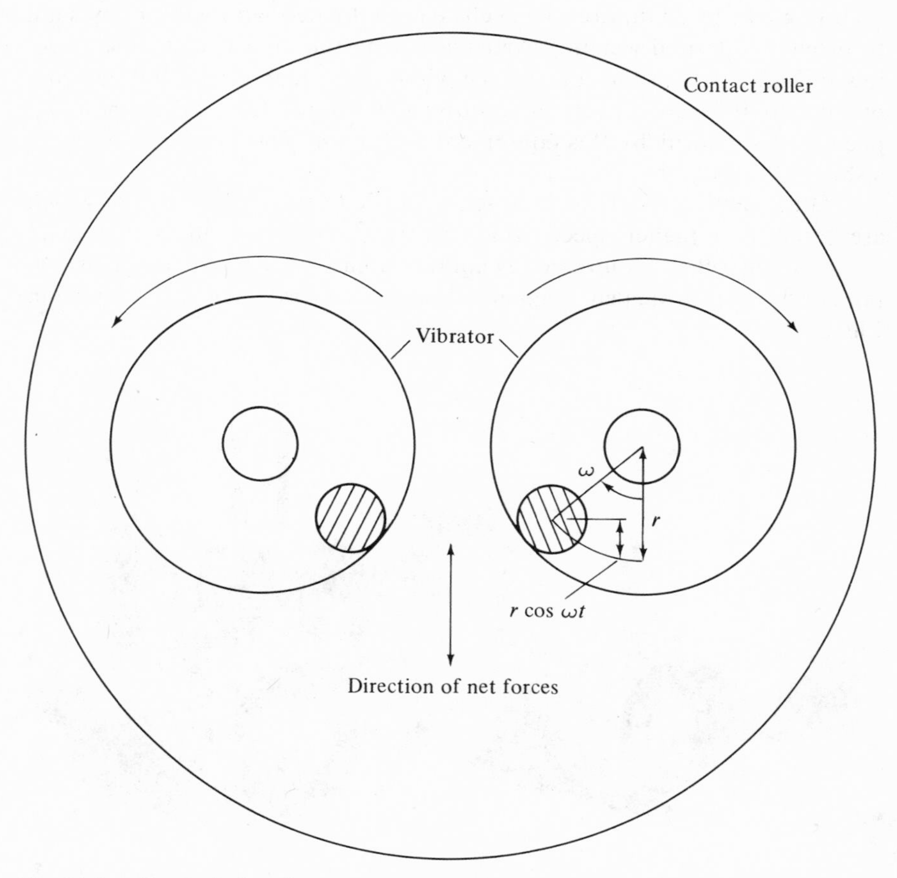

Fig. 21-3. Vertical vibrations produced by synchronized counterrotating cams.

vibrator thus is alternately light and heavy against the soil. In plate vibrators two counterrotating shafts are slightly out of phase so an oscillating horizontal component is developed which coincides with the light-heavy cycles to drive the machine forward.

A number of studies have shown an advantage in operating vibratory compactors at or near a resonant frequency for the system, particularly for granular soils. This is reasonable since effectiveness depends in part on how well successive vibrations reinforce one another and attenuate or carry into the soil. A vibrator on soil can be considered equivalent to a weight on a spring: If vertical motion is imparted to the weight, it will bounce up and down on the spacing at the natural frequency of the system. Operating a vibrator at such a frequency will cause overlapping or reinforcement of successive ups-and-downs, increasing amplitude of the vibrations.

By analogy with a weight on a spring, the natural frequency of a soil-vibrator system is

$$f_n = \frac{1}{2\pi} \sqrt{\frac{kg}{W + W'}} \tag{21-1}$$

where f_n is the natural frequency, k the soil "spring constant" or modulus of subgrade reaction which depends in part on size of the loaded area, g the acceleration of gravity, W the vibrating weight of the compactor, and W' the weight of the vibrating soil. Since both k and W' change as a soil compacts and these changes have not been evaluated, Eq. (21-1) is useful only as a guide, indicating that the lighter the vibrator, the faster should be the vibrations.

If one assumes that the vibrating soil weight W' is proportional to compactor vibrating weight W, Eq. (21-1) may be rewritten

$$f_n = \frac{c}{\sqrt{W(1 + b)}} \tag{21-2}$$

where b and c are factors. A plot of f_n vs. W is shown in Fig. 21-4 with experimental cost data obtained by Lewis (*10*) for a number of vibratory compactors operating at optimum conditions. In most cases the closer the points to the lines, the lower the cost for compaction. Plate vibrators probably have a lower resonant frequency for a given weight because of changes in factors b and c in Eq. (21-2).

Another important factor affecting vibratory compaction is the force of the vibration. This may be calculated from Newton's second law,

$$F = ma = W \frac{a}{g} \tag{21-3}$$

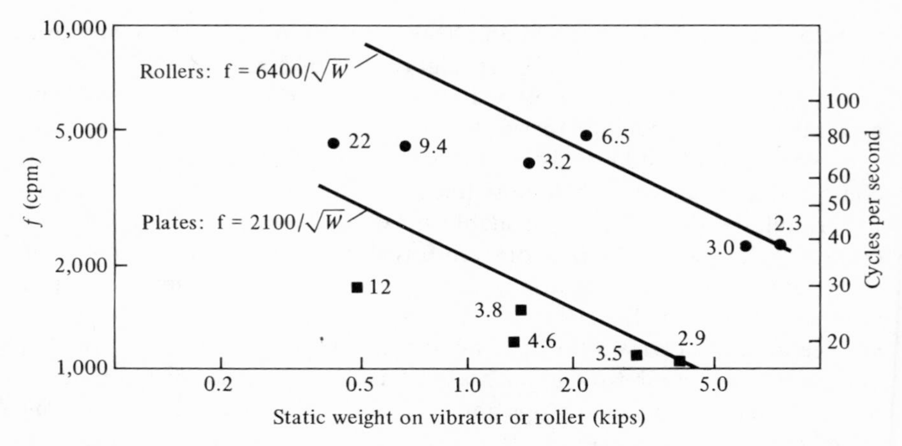

Fig. 21-4. Frequency, gross weight, and cost data for various vibratory compactors. Cost in cents per cubic yard under optimum operating conditions. Key: ●, vibrating rollers; ■, vibrating plate compactors. Data from Lewis (*10*).

where F is force, m is mass, a is acceleration, W is weight, and g is the acceleration of gravity. The quantity a/g is called the acceleration ratio, expressed as g's; force F therefore is proportional to weight of the vibrating mass and to the acceleration ratio.

Since F alternately acts upward and downward, the maximum and minimum total forces exerted by the vibrator are

$$W \pm F = W\left(1 \pm \frac{a}{g}\right) \tag{21-4}$$

When a/g exceeds 1.0, the vibratory force is sufficient to lift the vibrator off the ground during each cycle. This appears to be a minimum requirement for effective densification, and a/g commonly exceeds 1.5 to 2 in commercial vibrating compactors. The maximum acceleration a_m from harmonic motion may be calculated from the equation

$$a_m = (2\pi f)^2 \gamma \tag{21-5}$$

where f is the frequency and γ the amplitude, or maximum vertical movement measured either up or down from equilibrium. The amplitude γ therefore is one-half the total up-and-down movement.

EXAMPLE 21-1. Is the maximum acceleration ratio of the relative density test, ASTM D 2049-69 (Table 8-2), sufficient to lift the surcharge weight?

SOLUTION.

$$a_m = \left(2\pi \times 60 \frac{1}{\text{sec}}\right)^2 (0.025 \text{ in.})$$

$$= 3553 \text{ in/sec}^2 = 296 \text{ ft/sec}^2$$

$$\frac{a_m}{g} = \frac{296}{32.2} = 9.20$$

Since $a_m/g \gg 1$, yes, despite energy losses through the sample.

When a_m/g greatly exceeds 1.0, the energy which momentarily lifts the compactor off the ground is partly stored, since the compactor gains velocity for greater impact from the subsequent free fall. This is referred to as tampering action, and is especially effective for compacting clay where force is more important than resonance.

Hand or foot tamping is the oldest means for soil compaction, and as previously indicated a mechanized version is used for standard laboratory moisture-density tests. Mechanical drop-weight tampers or air-driven rammers also are used in the field where close quarters prevent the use of larger equipment, as in trench backfills, etc. Since these do not simulate harmonic motion, Eq. (21-5) does not apply, and the acceleration to be used in Eq. (21-3) represents deceleration after impact, which depends on hardness of the soil—as the soil hardens, the deceleration rate and force gradually increase. The compactive effort of tampers commonly is expressed in foot-pounds per cubic foot of soil.

21.7. ADJUSTING MOISTURE CONTENT. When the soil taken from a cut or borrow pit contains less moisture than the optimum for that soil, compaction is greatly facilitated by adding moisture, either by sprinkling the layers on the fill area or by wetting the cut. In some regions and at certain times of the year the soil may contain considerably more moisture than the optimum as it comes from the borrow area. This situation presents a very difficult problem from the standpoint of compaction control, as it is impossible to compact a soil that is too wet, no matter how many passes of a roller are made. The only practical alternative is to permit a wet soil layer to dry out on the fill before an attempt is made to compact it. Drying may be hastened by disking and scarifying the layer in order to gain maximum exposure to the sun and wind, but at best it is a slow process. Some soils dry much more rapidly than others. When a soil that dries very slowly is encountered, it may be advantageous to reduce the layer thickness to 3 or 4 in. in order to speed up the drying process.

Artificial drying of the soil is ordinarily far too expensive to be practicable. However, in the case of one major construction project, namely, the Mud Mountain flood-control dam in the State of Washington, the soil placed in the core of the dam was dried in rotary kilns similar to those used for burning clinker in the manufacture of portland cement. When dried to optimum moisture content, the soil was placed in the fill area under a huge tent to protect it from rainfall during compaction. The combination of high soil moisture in the borrow pit and continuously wet weather during the construction season made this procedure necessary on this particular project.

21.8. PRECAUTIONS AGAINST OVERCOMPACTION. As previously discussed, excessive foot pressure or excessive passes of a sheep's-foot roller, coupled with a moisture content on the high side of optimum, may cause shear-failure surfaces to develop adjacent to the contacts between the roller feet and the soil. These failure surfaces, referred to as "slickensides," constitute a definite weakness in the soil mass. In order to avoid overcompaction of cohesive soil, it is well to roll the material on the dry side of the optimum moisture content and to stop rolling the material as soon as the desired density is attained. Also, it may be necessary to reduce the foot pressure of the roller. Heavy hauling equipment should not be permitted to travel in the path of previous vehicles, but should be required to follow diverse haul-routes over the dump area. An embankment inspector should watch the action of the soil while it is being rolled. Excessive weaving under the roller or other equipment is generally indicative of excess moisture and possible development of slickensides.

21.9. SHRINKAGE. As a rule, the volume of soil which must be excavated from a cut section or borrow pit is greater than the volume of the embankment which can be constructed from the soil. This excess of yardage in cut over yardage in fill is called shrinkage and is usually expressed as a percentage of the fill quantity. Thus, if a fill of 100 cu yd is to be constructed and the shrinkage is 10%, 110 cu yd of excavation will be required. It is important, in planning an earth-moving project, to estimate the probable shrinkage which will occur as accurately as possible in order to balance the cut quantities and the fill quantities and to estimate the amount of right-of-way which will be required for borrow pits. When payments to a contractor are made on a unit-price basis, such as so much per cubic yard, the volume of soil excavated is most frequently used in computing the total payment and the volume of soil in the finished embankment is disregarded.

21.10. CAUSES OF SHRINKAGE. Shrinkage may be due to several causes. Probably the most influential factor is that density of the soil in a compacted fill is usually greater than the natural or cut density before the soil is excavated. Since soil consists of solid material plus void space, a greater volume of a lighter weight soil must be excavated in order to produce sufficient solids to make a given volume of a heavier soil. Another reason for shrinkage is the fact that the natural ground surface usually settles under the weight of an embankment and additional yardage is required to make up for this subsidence. Still another reason is the actual loss of soil which falls off the hauling equipment along haul roads which are not over the fill area. The amount of shrinkage is usually about 10 to 15% of the fill quantity, but may be much higher when the foundation of the embankment is highly compressible or when the void ratio of the natural soil in cut areas is unusually high.

An estimate of the amount of shrinkage to allow for a particular embankment can be made by comparing the cut density of the borrow material with the specified density of the soil in the finished embankment, and by estimating the amount of subsidence of the natural ground surface under the embankment.

21.11. DENSITY SPECIFICATIONS. Soil compaction is essentially a manufacturing process and requires specifications and controls to obtain a uniform product. The degree of compaction is defined as the ratio, in percent, of the embankment density to a specified standard density of the soil. Thus,

$$\text{Degree of compaction} = \frac{\text{Embankment density}}{\text{Specified standard density}} \times 100 \qquad (21\text{-}6)$$

The specified standard density may be the standard A.A.S.H.O. density, the modified A.A.S.H.O. density, or some other definite density. In highway work the degree of compaction of embankments usually varies from 90 to 100% of the standard A.A.S.H.O. density. In earth-dam construction the degree of compaction runs from 85 to 100% of the modified A.A.S.H.O. density.

To avoid overcompaction an upper limit may be placed on moisture content, usually 1 or 2% above optimum since extra water must be added to compensate for drying before compaction.

For the same reason 100% of standard or modified density is infrequently specified, the most common criteria being a 90 or 95% minimum requirement.

A recent trend which applies only to cohesionless materials is for the specification to be based on relative density, the advantage being to remove an erroneous implication of high density. For example, the center scale in Fig. 21-5 indicates that this sand may be readily compacted to 95% A.A.S.H.O. density with practically no control on moisture content,

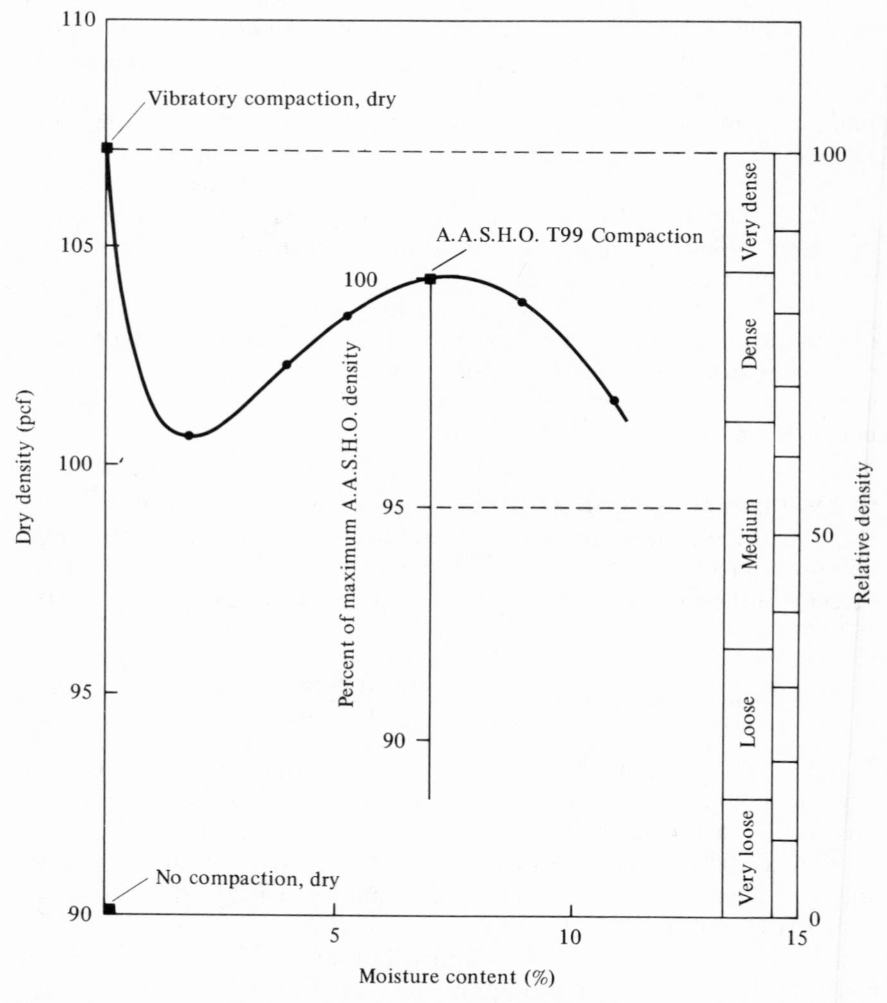

Fig. 21-5. A.A.S.H.O. T99 and relative density data for a sand.

but this density would coincide with a relative density of only 54%, which is a medium sand. A relative density of 80%, corresponding to a dense sand, would be a better specification which would equal 99% Standard A.A.S.H.O. (T99) density.

Relative density may be calculated from void ratios by Eq. (8-29), or more conveniently from densities by the equivalent equation:

$$D_r = \frac{\gamma_{max}(\gamma - \gamma_{min})}{\gamma(\gamma_{max} - \gamma_{min})} \times 100 \qquad (21\text{-}7)$$

where γ is the measured dry density and γ_{max} and γ_{min} are the maximum and minimum dry densities obtained from the laboratory relative density test, described in Section 8.16.

Control of the degree of compaction during construction may be obtained by actually measuring the density of the soil in each layer after rolling, to see if it complies with the specification. An alternate procedure is to determine by trial the number of passes of the roller which are required to densify a layer of soil. Then subsequent layers are given this same number of passes. This procedure minimizes the number of density determinations which must be made and is satisfactory when both the soil and its moisture content are reasonably uniform throughout the embankment.

21.11.1. Statistical Density Control. Elementary statistical theory is an aid to understanding of density test data as well as helping to arrange and interpret an effective testing program. For example, statistical theory explains why a completed compacted section should not be rejected on the basis of a single low test, since variations in density are entirely legitimate because of all the variables that affect the test. In practice a low test is usually repeated and the compacted density often becomes satisfactory on second evaluation; this also is explainable statistically, and neither test is "right" or "wrong," since they both represent field conditions. Where extensive density testing has been conducted on previously accepted and satisfactorily performing jobs, anywhere from 10 to 25% of the tests fall below the stated minimum acceptance requirement. In the A.A.S.H.O. Test Road (*5*), which was carefully controlled, 8.8% of the tests fell below the stated acceptance level of 95% compaction, although the average density was well above this.

Figure 21-6 shows the frequency distribution of 200 density tests on a compacted embankment in California. It can be seen that the data fit fairly closely to a theoretical normal distribution curve discussed in Chapter 6 and shown in the figure. Two parameters describe the distribution:

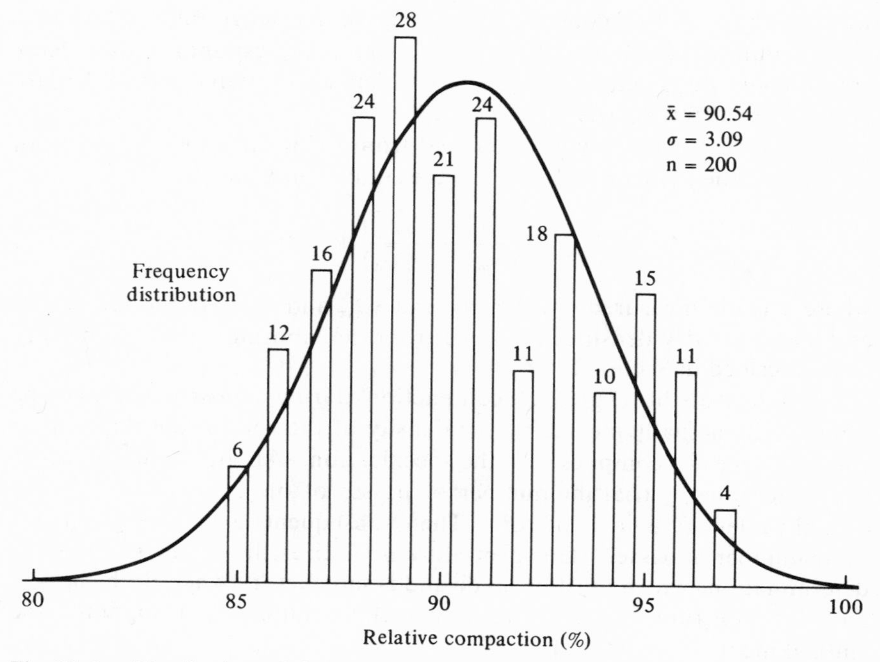

Fig. 21-6. Distribution of data from 200 density tests compared to a statistical normal distribution. From Sherman *et al.* (*14*).

the average or arithmetic mean, designated $\bar{x}$, and the standard deviation, designated σ. Since the $\bar{x}$ and σ given in the figure are only estimates of the true values because of the limited number of tests, $\bar{x}$ is called the sample mean, and σ the sample standard deviation. The standard deviation takes into account all random variations in compaction, material, and test procedure. The larger the random variations, the wider the bell-shaped normal distribution curve, and the larger is σ. Estimates for σ may be calculated by use of Eq. (6-8).

An easy way to test for a normal distribution is to convert the data to percent of the tests which meet each given percent compaction, and plot on probability paper to see if the result approximates a straight line. Probability paper was invented by an engineer, Hazen, and is similar to a cumulative curve such as used to show particle size distributions, Fig. 7-1, with the scales adjusted so that a normal distribution becomes a straight line. As in Fig. 7-1, the arithmetic mean or average may be read from the junction of the line with the 50% cumulative relative percent. In addition an estimate of the standard deviation, σ, may be read from the difference between the 50% and 16% cumulative relative frequencies.

The data of Fig. 21-6 are replotted on probability paper in Fig. 21-7 and indicate a fair compliance to a straight line, although there is a suggestion of two overlapping normal distributions (dotted lines). From this

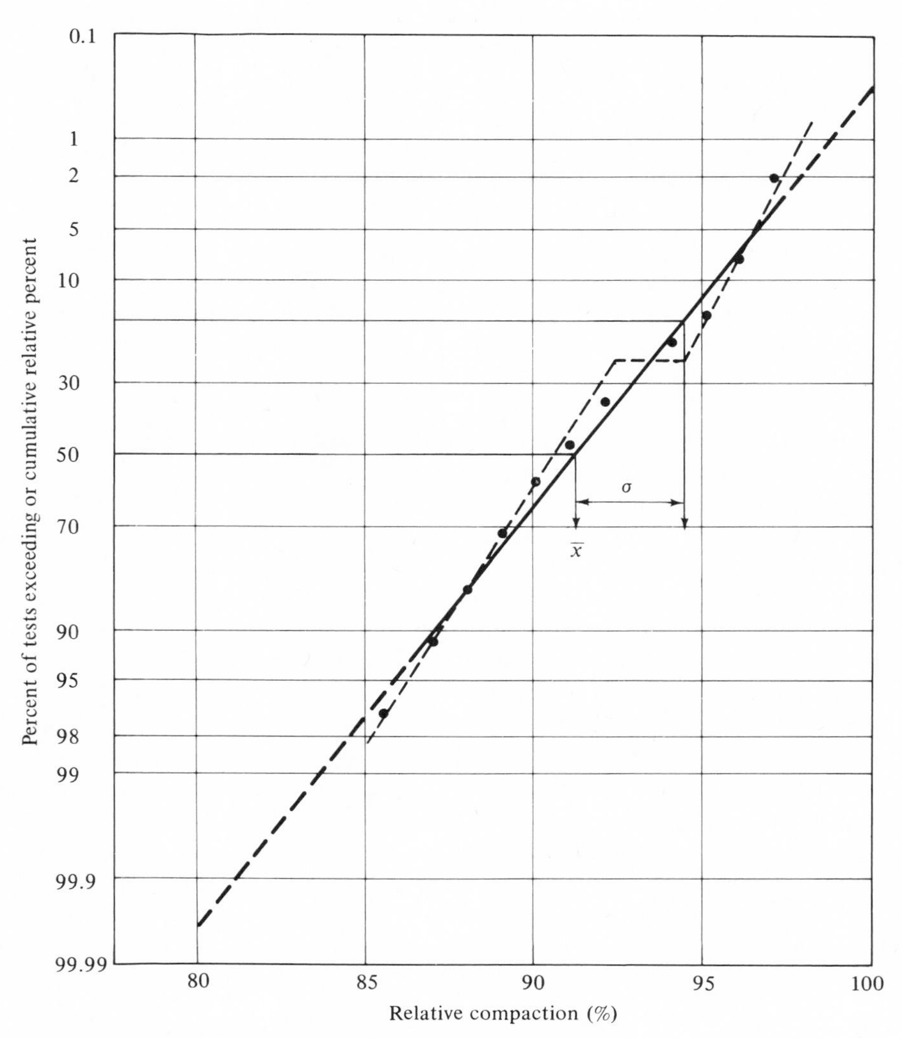

Fig. 21-7. Data of Fig. 21-6 plotted on probability paper.

graph and the assumption of a single normal distribution, about 64% of the embankment soil meets the specified minimum of 90% Proctor density, 12% meets 95% Proctor density, and less than 0.5% will be at 100% Proctor density. The σ of 3.1 is about average, the usual range being about 2 to 5.

The U.S. Bureau of Public Roads and the California Division of Highways have suggested that a realistic density specification should include, among other things: a statement of a desired mean which may be adjusted if the material varies; a numerical lower limit for percent compaction; the size area which would be reworked if a test does not pass; and the number of tests to be conducted before a decision is made. Sherman *et al.* (*14*) show that the widespread practice of rerunning low tests increases the probability of accepting unsatisfactory work.

21.11.2. Number and Spacing of Tests. Improperly arranged testing means improper data which may be worse than worthless since erroneous conclusions are worse than none at all. Sampling therefore is vital and should be directed and performed by trained, competent personnel.

Statistical methods and practical engineering concepts present two conflicting philosophies of sampling. Statistical theory assumes random sampling, meaning that test location should be determined only by chance, so as to minimize the possibility of systematic error. For example, if all density tests are conducted at 50 ft intervals down the centerline of a highway embankment we may introduce a systematic error or bias in favor of higher densities, since soil on the centerline is better confined than at the edges and also may receive more passes of the compactor. The statistician therefore will lay out his sampling sites on coordinates chosen from a table of random numbers, or he may utilize a deck of cards, flip a coin, or devise some other scheme to remove personal bias from the sampling program. The engineer, on the other hand, is concerned that all soil will be represented and prefers to lay out his tests at prescribed intervals so as not to miss any important areas. In practice the two principles are often combined: For example, tests may be at 50 ft intervals with lateral positioning determined by chance, with the result that neither school is so deeply offended.

Statistical theory provides methods for estimating the number of tests required for a 95% certainty on $\bar{x}$. According to A.S.T.M. Designation E 122-58,

$$N = \left(\frac{2\sigma}{E}\right)^2 \tag{21-8}$$

where N is the number of tests, σ is the estimated standard deviation, and E is the maximum allowable percent error in estimating the mean.

EXAMPLE 21-2. In carefully controlled compaction σ is about 2.0; if the average density is desired within $\pm 1\%$, one will need a minimum of

$$N = \left(\frac{2 \times 2}{1}\right)^2 = 16 \text{ tests.}$$

II. DESIGN

21.12. SIDE-SLOPE RATIO. An important feature of the design of earth embankments is the side-slope angle or ratio. This angle depends on the height of the embankment, on the shearing resistance of the soil, and on whether or not the embankment will be inundated by impounded water, as in the case of earth dams. Embankments and earth dams of considerable height, say in excess of about 50 ft, should be individually analyzed to determine the slope stability and the factor of safety against shear slides by the methods presented in Chapter 20. The side-slope ratios given in Table 21-5 may be used as a guide for the design of embankments less than 50 ft high.

The side-slope ratio may also depend on the shearing strength of the soil foundation. If the foundation is weak, the side slopes should be made flatter in order to spread the weight of the embankment over a wider area and thus reduce the concentration of shearing stresses in the foundation material. In order to investigate the stability of the foundation soil beneath an embankment, the results of some analyses of stresses and de-

TABLE 21-5. Minimum Side-Slope Ratios for Earth Embankments on Adequate Foundations

Side-Slope Ratio Horizontal: Vertical	*Kind of Soil and Other Conditions*
$\frac{1}{4}$:1	Unsaturated loess and permeable silt with some cohesion and no water table. Maximum safe height, 20 ft. Used to prevent erosion.
$1\frac{1}{2}$:1	All sand fills whether inundated or not; fills of cohesive soils less than 5 ft high and not subject to inundation.
2:1	Fills of cohesive soils more than 5 ft but less than 50 ft high and not subject to inundation.
3:1	All fills of cohesive soils not exceeding 50 ft in height and subject to total or partial inundation.

flections in a semi-infinite elastic medium with a trapezoidal load acting at the surface may be employed as a guide. Analyses of this kind give only approximate results, since the soil is not an elastic, homogeneous, isotropic material, as was discussed in Chapter 17. However, the procedure has value as a general guide, particularly when the foundation material is reasonably homogeneous. More complicated problems can be investigated by a technique known as finite-element analysis. This method consists of dividing the cross section of a structure into triangular elements and writing equations of equilibrium for each intersection or node. A computer is used to solve the equations. This is an advanced method and is beyond the scope of this text [see Clough and Woodward (*2*)].

21.13. DISTRIBUTION OF SHEARING STRESSES UNDER EMBANKMENT. Studies by Holl (*7*) show that the shearing stress on a horizontal plane in the foundation under a symmetrical trapezoidal load is zero along the vertical central axis through the load. As the distance away from the cen-

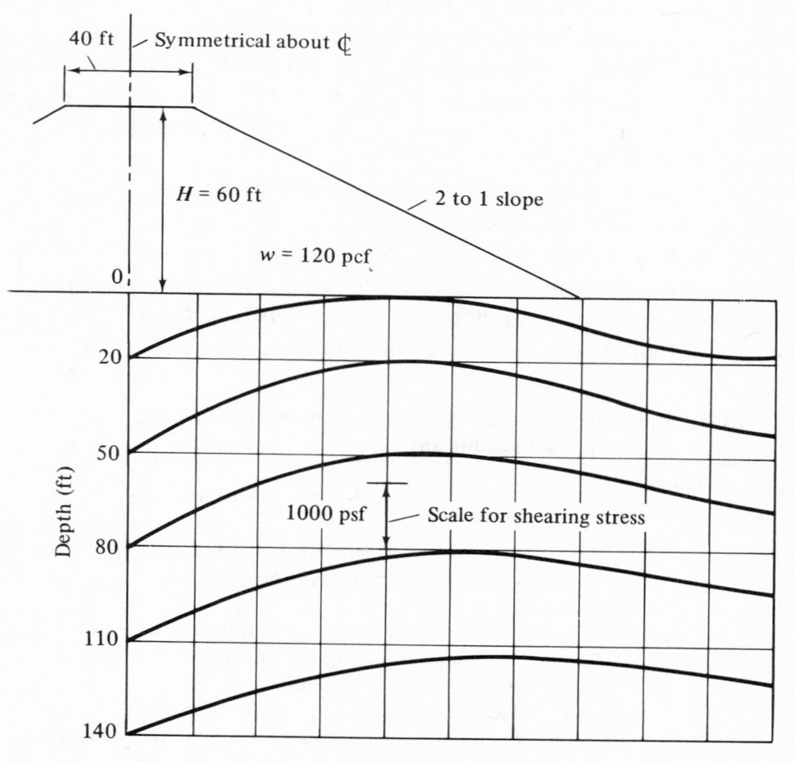

Fig. 21-8. Shearing stresses on various horizonal planes in the foundation material under an embankment.

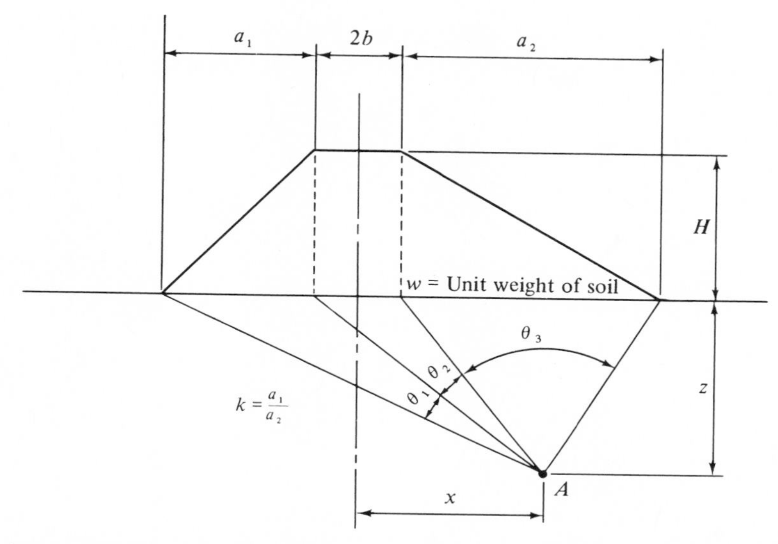

Fig. 21-9. Nomenclature for computing shearing stresses in the foundation under an embankment.

tral axis becomes greater, the shearing stress increases to a maximum value and then decreases again, approaching zero at some distance beyond the toe of slope of an embankment which produces the trapezoidal load. The locus of points of maximum shearing stress on the various horizontal planes lies nearly on a vertical line extending downward from the center of the side slope of the embankment. Also, the maximum shearing stress increases with depth below the foundation surface to a maximum value and then decreases. These facts are shown qualitatively in Fig. 21-8. The factor of safety of the foundation under an embankment can be investigated by calculating the shearing stresses at several points in the vicinity of the vertical line through the center of the side slope and comparing the stresses with the shearing strength of the soil.

The horizontal shearing stress at any point in the under soil in Fig. 21-9 may be computed by the formula

$$S_a = -0.318\gamma H \frac{z}{a_1} (\theta_1 - k\theta_3) \qquad (21\text{-}9)$$

in which S_a is the unit horizontal shearing stress at point A. Fig. 21-9; and the meanings of the other symbols are as shown in that diagram.

21.14. SHEARING STRENGTH OF FOUNDATION SOIL. The shearing strength at point A, Fig. 21-9, is equal to the cohesion of the soil plus the vertical intergranular pressure at the point times the tangent of the angle of friction. The vertical pressure is equal to the weight of a unit column of soil extending from point A to the foundation surface plus the vertical pressure due to the embankment load. This latter component may be computed by

$$p_2 = 0.318\gamma H \left[(\theta_1 + \theta_2 + \theta_3) + \frac{b}{a_1}(\theta_1 + k\theta_3) + \frac{x}{a_1}(\theta_1 - k\theta_3)\right] \quad (21\text{-}10)$$

in which p_2 is the vertical stress at point A due to the embankment load and the meanings of the other symbols are as shown in Fig. 21-9.

EXAMPLE 21-3. In an embankment like that illustrated in Fig. 21-10,

$$H = 60 \text{ ft} \qquad 2b = 40 \text{ ft}$$
$$\gamma = 125 \text{ pcf} \qquad a_1 = a_2 = 120 \text{ ft}$$

Side slopes: 2:1

The foundation has the following characteristics:

$$\gamma' = 65 \text{ pcf (submerged unit weight)} \qquad \phi = 20^\circ$$
$$c = 400 \text{ psf} \qquad \tan\phi = 0.364$$

Determine the factor of safety of the foundation after complete consolidation.

SOLUTION. First, investigate the factor of safety at point A which is selected at a depth $(a/2) + b = 80$ ft below the surface and beneath the center of one side slope. Then $\theta_1 = 18°40' = 0.33$ rad, $\theta_2 = 14°30' = 0.25$

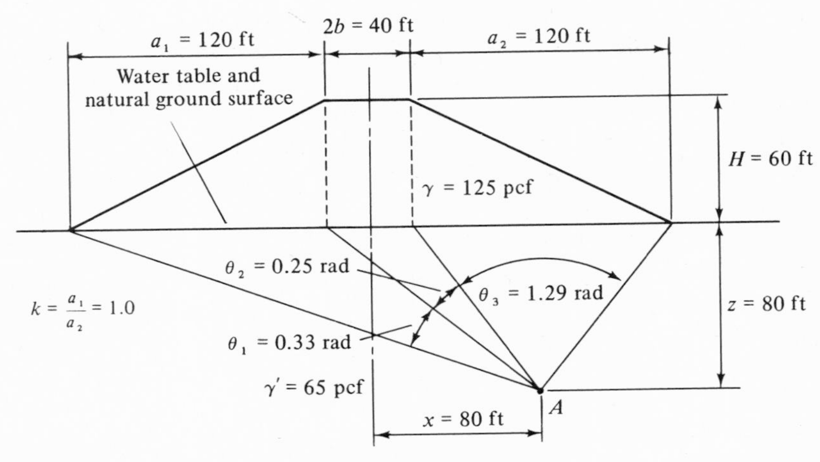

Fig. 21-10. Conditions for Example 21-3.

rad, and $\theta_3 = 73°40' = 1.29$ rad. According to Eq. (21-9), the horizontal shearing stress at point *A* is

$$S_a = -0.318 \times 125 \times 60 \times \frac{80}{120} \times [0.33 - (1 \times 1.29)] = 1530 \text{ psf}$$

The shearing strength at this point may be computed as follows: The pressure p_1 due to the submerged weight of the foundation material above point *A* is

$$80 \times 65 = 5200 \text{ psf}$$

Also, the pressure p_2 due to the weight of the embankment is, by Eq. (21-10),

$$p_2 = 0.318 \times 125 \times 60 \left\{ [0.33 + 0.25 + 1.29] + \frac{20}{120} [0.33 + (1 \times 1.29)] + \frac{80}{120} [0.33 - (1 \times 1.29)] \right\} = 3580 \text{ psf}$$

Then

$$p_1 + p_2 = 5200 + 3580 = 8780 \text{ psf}$$

and

$$S_r = 400 + (8780 \times 0.364) = 3600 \text{ psf}$$

The factor of safety is

$$\frac{S_r}{S_a} = \frac{3600}{1530} = 2.35$$

In order to get a more complete idea of the factor of safety of the foundation, it is recommended that similar computations be made for points beneath the center of the side slope and at depths of $(a/4) + b = 50$ ft and $(3a/4) + b = 110$ ft. If the factor of safety is not so large as desired, the side slopes should be made flatter and the computations should be repeated.

21.15. SLIDING BLOCK METHOD. When the foundation under an embankment or an earth dam is stratified or contains horizontal planes of cleavage which may also be planes of weakness, the shearing stresses may be studied by employing an approximate analysis which is sometimes referred to as the sliding block method. In this method the active horizontal pressure acting upon the vertical face of the downstream triangle of soil lying above the plane of cleavage is considered to be resisted by the shear-

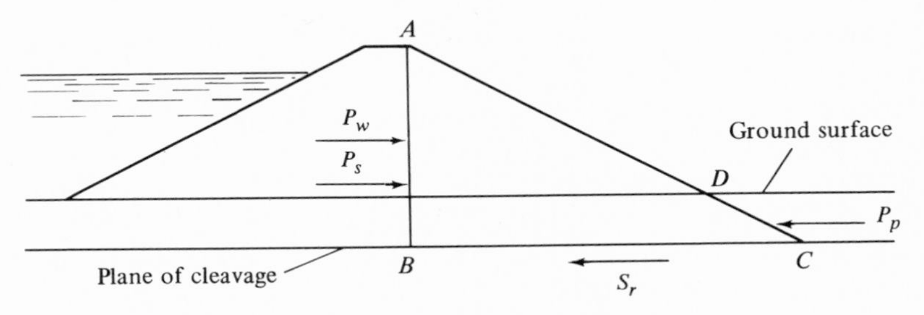

Fig. 21-11. Sliding block method of foundation analysis.

ing resistance developed along the base of the triangle plus the passive resistance which may develop at the downstream toe of the slope, as illustrated in Fig. 21-11. Here the sliding block is the triangle ABC which is pushed to the right by the water pressure P_w and the active soil pressure P_s acting on the vertical plane AB. Sliding is resisted by the total shearing resistance S_r along the base BC of the triangle plus the passive resistance pressure P_p acting on the sloping surface CD. In many cases the designer may not feel that the passive resistance pressure is sufficiently reliable to be depended upon, and it may be ignored. When this is true, the relationship between horizontal forces is

$$P_w + P_s = S_r \tag{21-11}$$

When S_r is equal to the total shearing strength along the base BC, the factor ofsafety F_s is

$$F_s = \frac{S_r}{P_w + P_s} \tag{21-12}$$

21.16. SUBSIDENCE. Even though the foundation soil is perfectly safe against shear failure, it will compress under the weight of an embankment. This causes subsidence of the natural ground surface on which the embankment is built. The subsidence is greatest under the central portion of the embankment, and decreases to approximately zero at the toes of the side slopes. It is approximately proportional to the height of fill at all points along a transverse cross section. The amount of the subsidence per foot of height of fill varies widely, depending upon the compressibility of the foundation soil. Obviously, if the foundation were solid rock, the subsidence would be essentially zero. It may range upward from this lower limit to as much as 1 in. or more per foot of height of fill.

Subsidence is an important consideration in connection with embankment construction because it adds to the amount of soil required to build the fill up to specified grade and constitutes a source of shrinkage or loss of the soil taken from the borrow pit. Also, subsidence must be taken into account in the design of a culvert or other conduit which is supported on the natural ground beneath an embankment. For example, if a culvert is built on a straight grade line from inlet to outlet and the natural ground subsides, the culvert conforms to the new contour and the distance from inlet to outlet will be lengthened. The combination of this distortion and the horizontal shearing strains in the foundation soil produced by the shearing stresses discussed in Sec. 21-13 causes the culvert to pull apart longitudinally at construction joints in order to conform to the new contour. The probability of this occurrence needs to be taken into account in the design of the culvert. In Fig. 21-12 is illustrated the settlement of a monolithic arch culvert in western Iowa due to subsidence of the natural ground under a 57-ft fill. The construction joints in this culvert opened up appreciably, as much as $5\frac{1}{2}$ in. in the two joints under the highest portion of the fill, but faulting and misalignment at the joints were prevented by heavy bells which were constructed at the upstream end of each independent section.

Structures of this kind, which are constructed monolithically from end to end without construction joints, are frequently pulled apart at several points, even though they are heavily reinforced in the longitudinal direction.

If the foundation soil is coarse-grained and has a relatively high co-

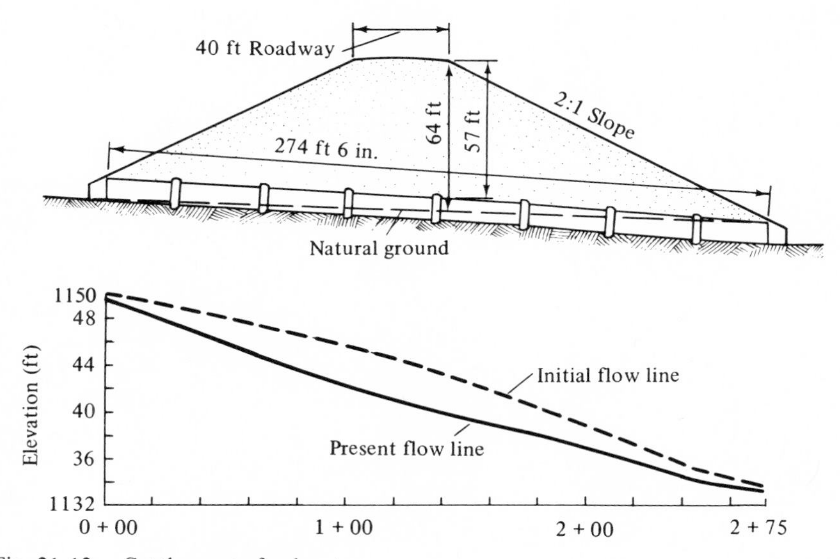

Fig. 21-12. Settlement of culvert.

efficient of permeability, or if the water table is at a very low elevation, most of the subsidence will occur almost immediately while the embankment is being constructed. On the other hand, if the soil is fine-grained and the water table is near the surface, the subsidence will occur very slowly as the soil consolidates and the water is squeezed out of the soil pores. In case residual subsidence due to consolidation is anticipated, it is desirable, if materials are at hand, to place a layer of 2 to 4 ft of sand or gravel at the base of the embankment. This layer will facilitate the removal of the water which squeezes out of the foundation soil and will prevent its being absorbed by the embankment.

21.17. USE OF SAND DRAINS. Vertical sand drains, or sand piles, are sometimes used to hasten the consolidation and settlement of highly compressible wet soils under embankments. They are constructed by boring vertical holes about 2 to $2\frac{1}{2}$ ft in diameter down through the compressible soil layer and filling the holes with clean coarse sand before the embankment is constructed. The piles are spaced from 5 to 20 ft apart in both the transverse and longitudinal directions, as illustrated in Fig. 21-13. They speed up the consolidation and stabilization of the compressible undersoil by greatly reducing the distance through which the pore water has to flow in order to escape. When the water reaches a sand pile, it can rise in the sand and flow away rapidly because of the high permeability of the sand. Vertical drains of this kind have been known to increase the rate of consolidation to about 20 to 25 times the rate for untreated soil.

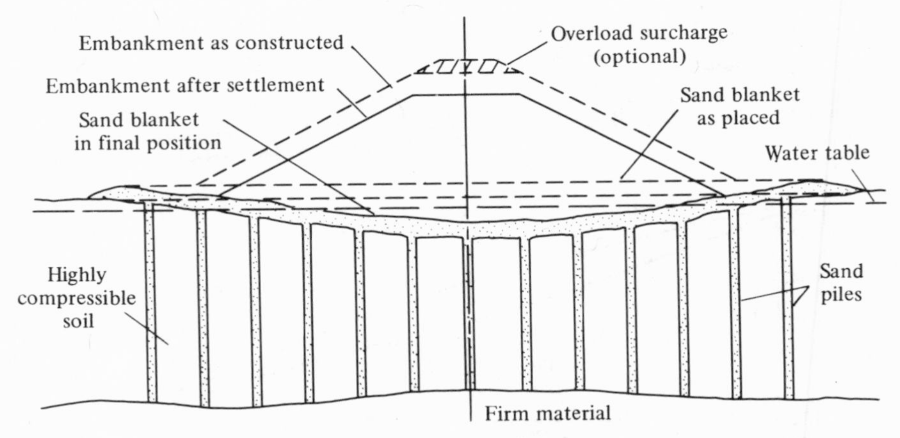

Fig. 21-13. Foundation stabilization by means of vertical sand drains (sand piles).

21.18. EFFECT OF HYDROSTATIC EXCESS PRESSURE IN PORE WATER. When an embankment is constructed on a highly compressible soil foundation having a low coefficient of permeability, the shearing resistance of the soil may be drastically reduced by the development of hydrostatic excess pressure in the pore water during construction. As consolidation proceeds with the passage of time, the shearing strength is gradually recovered; and, when consolidation is nearly complete, the shearing strength will probably exceed the original strength. The critical period for shear failure of the foundation is, therefore, during construction and shortly after completion of the embankment.

The magnitude of hydrostatic excess pressure in the soil water can be measured during construction by installing a vertical piezometer or standpipe and observing the elevation to which water rises in the pipe, as illustrated in Fig. 21-14. The standpipe should be perforated for several feet near the bottom end, and should be driven to the elevation at which it is desired to observe the hydrostatic excess pressure. The increase in elevation of water in the pipe, compared to the normal elevation of the water table in the soil, is an indication of the hydrostatic excess pressure in the pore water and evidence of a reduction in shear strength of the foundation. The hydrostatic excess pressure represents the portion of the total pressure due to the embankment load which is carried by the pore water. Since the coefficient of friction of water is zero, this portion of the pressure

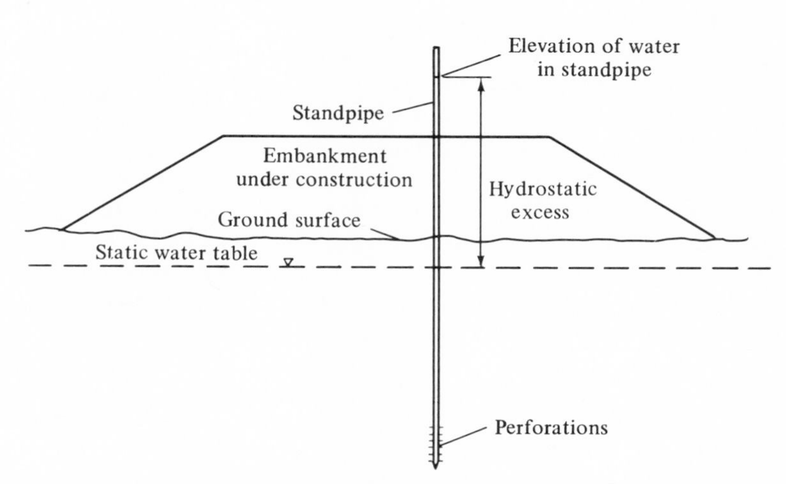

Fig. 21-14. Standpipe for observation of hydrostatic excess pressure in foundation during embankment construction.

is ineffective as a source of shearing strength of the soil. If, during construction, the hydrostatic excess pressure rises to a dangerous amount, it may be advisable to stop construction of the embankment for a period of time during which consolidation may proceed. When the excess pressure has been reduced to a safe value, construction may be resumed. This kind of control of the rate of construction may avoid a shear failure of the foundation during the most critical period in the life of the embankment and may insure its stability after completion.

21.19. REMOVAL OF PEAT OR MUCK FROM BENEATH EMBANKMENT. When an embankment is to be built over very poor soil, such as peat or muck, special procedures must be employed to insure stability of the earth structure and prevent uneven and long-continued settlement of the grade line. All such procedures are directed toward removal of the peat from beneath the embankment down to more stable soil below. If the peat bed is relatively shallow, say up to 4 or 5 ft deep, it may simply be excavated and wasted before the embankment of mineral soil is constructed. Where the bed is too deep to be excavated prior to fill construction, the peat may be displaced from beneath the fill by blasting. First the bottom 6 to 10 ft of the embankment is built over the surface of the peat. Then holes are drilled through this layer down into the peat, and dynamite charges are exploded in the bottoms of the holes. Because of the weight of the mineral soil overburden, the peat is displaced laterally and the soil layer is allowed to sink down to a more stable stratum. Then the embankment is completed in the normal manner.

Another method which has gained considerable favor is to displace the peat by temporarily overloading the embankment. In this method the fill is carried up to an elevation substantially above the height of the final grade line. The weight of the embankment plus the weight of the overload layer causes the peat to fail in shear and flow outward; and mounds or ridges are raised alongside the embankment which are sometimes higher than the embankment itself. After the peat flow has stopped, as evidenced by no further settlement of the embankment grade line, the overload may be removed by cutting the grade down to its final elevation. When this method is used, a year or more may be required for the foundation to become stabilized. If the embankment is for a highway which is to be paved, it is advisable to provide only a temporary surface of a flexible type during the period when the overload is in place.

21.20. DESIGN OF EARTH DAMS. Earth dams may be successfully constructed of a wide variety of earthen materials, and several different kinds of material may be combined in a number of different ways in any individual dam in order to accomplish the purpose of the structure in the most economical manner. In planning an earth dam it is necessary to make an extensive study of all the available materials, their physical properties, their proximity and ease of transportation to the site, and all other pertinent factors. Then these materials should be combined in such a way that the dam will be strong enough to withstand all the forces to which it will be subjected and will also be sufficiently water-tight to function satisfactorily. No earth dam can be made completely water-tight, but the amount of seepage water can be controlled by proper design and it can be controlled in such a way that the water will cause no harm to the structure. In all probability it will be desirable to make several designs, using various combinations and arrangements of the available soils, in order to determine the best and most economical combination.

Justin, *et al.* (*9*) have set forth a number of criteria which must be met if an earth dam is to function satisfactorily. These criteria provide a good check list by which the quality and performance of a dam can be judged. An earth dam should be so designed that:

1. There is no danger of overtopping (that is, there is sufficient spillway capacity and sufficient freeboard).
2. The seepage line is well within the downstream face.
3. The upstream-face slope is safe against sudden draw-down of the reservoir.
4. The upstream and downstream slopes are flat enough, with the materials utilized in the embankment, to be stable and to have a satisfactory factor of safety.
5. The upstream and downstream slopes are flat enough to make the shearing stresses induced in the foundation less than the shearing strength of the foundation soil and to provide a satisfactory factor of safety.
6. There is no opportunity for the free passage of water from the upstream face to the downstream face.
7. When water which passes through and under the dam reaches the discharge surface, its pressure and velocity are so small that it is incapable of eroding the material of which the dam or its foundation is composed.
8. The upstream face is properly protected against wave action, and the downstream face is protected against the erosive action of rain.

21.20.1. Spillway Capacity. The determination of spillway capacity which will be adequate to insure that an earth dam will not be overtopped during maximum flood flow is a problem in hydrology rather than soil engineering. Nevertheless, it is mentioned here because of its extreme importance in connection with earth dams. More failures of structures of this kind have occurred because of overtopping than from all other causes combined. When a dam is overtopped, the overflow water rushing over the downstream slope quickly erodes the soil structure; and failure is practically inevitable. In selecting the proper spillway capacity it is not enough to determine the greatest flood flow in recorded history and to apply an arbitrary factor of safety to this flow. It is better practice to study the maximum storms which have occurred within the climatic region in which the drainage area of the stream lies. Then, by transposing these storms to the drainage area under study, an idea of the maximum possible flood flow can be obtained. It is advisable to provide spillway capacity sufficiently large to handle the maximum possible flood flow thus determined with a generous margin of safety.

21.20.2. Freeboard. Closely associated with the problem of safe spillway capacity is that of determining adequate freeboard. Gross freeboard is defined as the vertical distance from the top of the spillway to the top of the dam. Net freeboard is the vertical distance from the impounded water surface to the top of the dam at the time the spillway is discharging the greatest flood for which it was designed. Freeboard is provided in order to prevent overtopping of the dam by tides, seiches, and waves or rollers. The freeboard should equal the sum of the estimated heights of these phenomena plus an arbitrary margin of safety based upon judgment.

21.20.3. Precautions against Seepage. The amount of seepage water passing through an earth dam and its foundation can be controlled to a considerable extent by the use of cut-off walls, impervious soil cores, upstream blankets of impervious soil, and combinations of these elements. The amount of seepage can be estimated for any particular arrangement by sketching a flow net in accordance with the principles set forth in Chapters 11 and 12. A cut-off wall is a relatively impervious element which extends downward through the foundation soil from the base of a dam. Its function is to reduce the amount of seepage water flowing through the foundation. If feasible, a cut-off wall should be carried down to solid rock or other very impervious material. If it extends only part way, its effectiveness is greatly reduced, as seepage water will flow downward and beneath the cut-off. Cut-off walls may be constructed of steel sheet piles; of plain or reinforced concrete; of dense, well-graded soil; or of fine-grained, impervious soil. Most of these require excavation in permeable soil or a cut-off would not be necessary, so dewatering is a difficult problem.

The "slurry trench" method recently has been devised for construction of concrete cut-off walls in soils below the water table. In this method the excavation is kept filled with a bentonite slurry, much as is used for drilling mud in weak strata. The slurry exerts pressure and helps stabilize the walls of the excavation, and allows clamshells to operate and remove soil from the bottom. When the excavation is completed, reinforcing steel may be lowered if desired, and concrete is introduced at the bottom through a canvas tube or "tremie." Soil acts as the form, and as the excavation fills with concrete the tremie is raised, and the clay slurry, being lighter, is displaced and may be saved for excavation of another wall section. The wall sections may be keyed together by use of an end-form (usually a vertical pipe) which is pulled prior to the next excavation.

A core wall is an impervious element which extends upward from the base of the dam to the top. It is practically always constructed of impervious soil and may simply be an extension of the cut-off wall upward to the top of the dam. If a cut-off wall of concrete or sheet piling is used, it should extend well up into the core wall to insure adequate bond and water-tightness at the junction of the cut-off wall and the core wall. The core wall is often located on the longitudinal center-line of the dam, but it may be located anywhere on the upstream side of the center.

An upstream blanket is a layer of impervious soil, about 5 or more ft thick, which is placed on the bottom of the reservoir area upstream from the dam. It is usually connected with the base of the core wall. It serves to reduce the seepage through the foundation soil and may be used instead of a cut-off wall. It reduces the seepage by increasing the distance through which water must flow in order to escape under the dam through the foundation material. Some typical arrangements of the various elements of earth dams are shown in Figs. 21-15 to 21-20, inclusive.†

21.20.4. Line of Saturation. The upper surface of the zone of saturation caused by seepage water is called the line of saturation, or the phreatic surface. It is the surface at which the pressure in the seepage water is atmospheric, and it is analogous to a ground water table. As pointed out in Chapter 12, the line of saturation is the upper boundary flow line of a flow net. It is important to keep this line of saturation well below the downstream slope surface of the dam. If the line of saturation is too high, it will intersect the downstream slope; and the seepage water will break out on the slope. This position produces a wet marshy condition in the vicinity of the downstream toe of the dam. If the velocity of the water below the break-out point is great enough, erosion and slough-

† Reproduced by permission from *Engineering for Dams*, Vol. 3, by J. D. Justin, J. Hinds, and W. P. Creager, published by John Wiley & Sons, Inc., 1945.

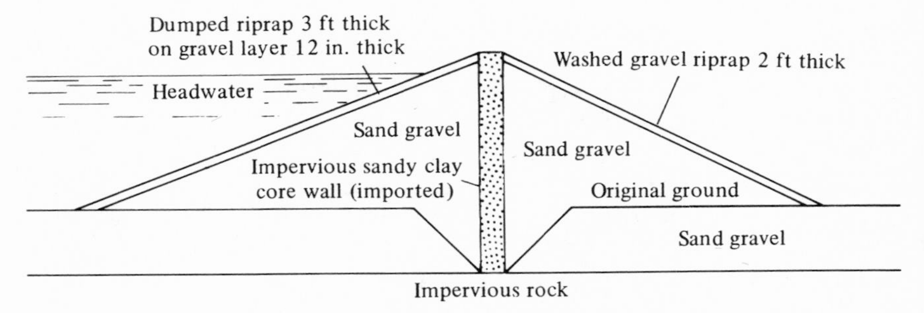

Fig. 21-15. Typical cross section of dam at a site where sand and gravel are readily available and impervious core material must be hauled a long distance [Justin *et al.* (*9*)].

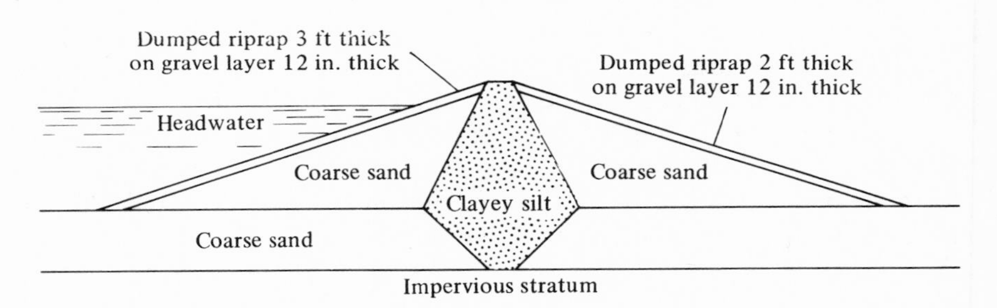

Fig. 21-16. Typical cross section of dam at a site where both coarse sand and impervious core material are readily available [Justin *et al.* (*9*)].

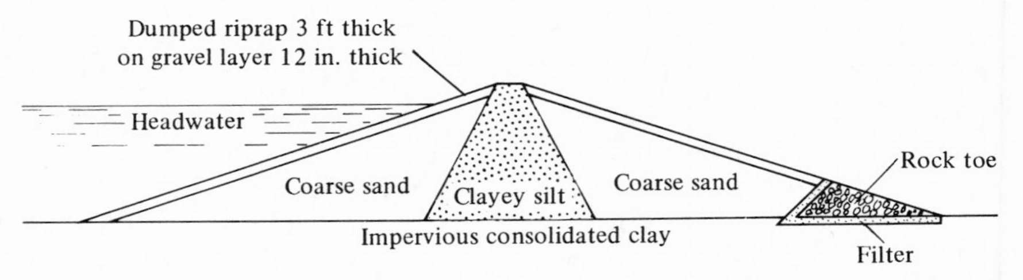

Fig. 21-17. Typical cross section of dam at a site where coarse sand and clayey silt are available and foundation material is impervious [Justin *et al.* (*9*)].

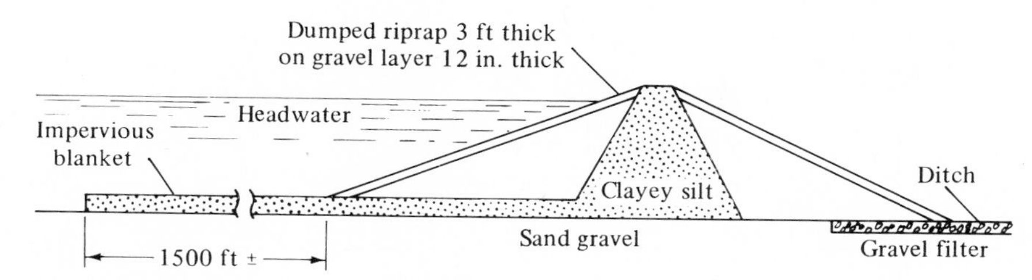

Fig. 21-18. Typical cross section of dam at a site where coarse sand and clayey silt are available and foundation material is pervious to a great depth [Justin *et al.* (*9*)].

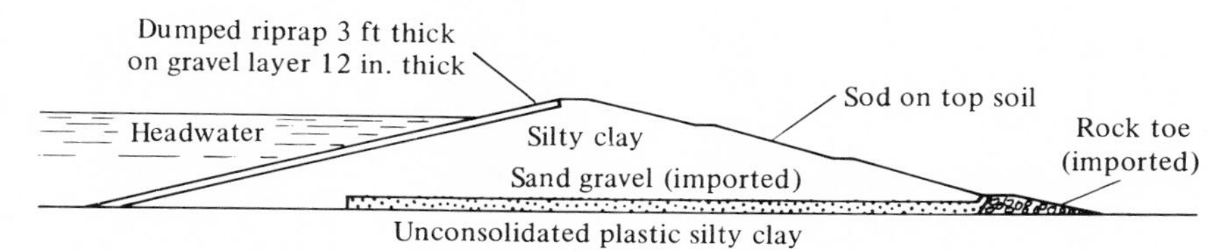

Fig. 21-19. Typical cross section of dam at a site where only available material is silty clay and foundation consists of plastic silty clay; slopes are flattened out to reduce shearing stresses in foundation soil [Justin *et al.* (*9*)].

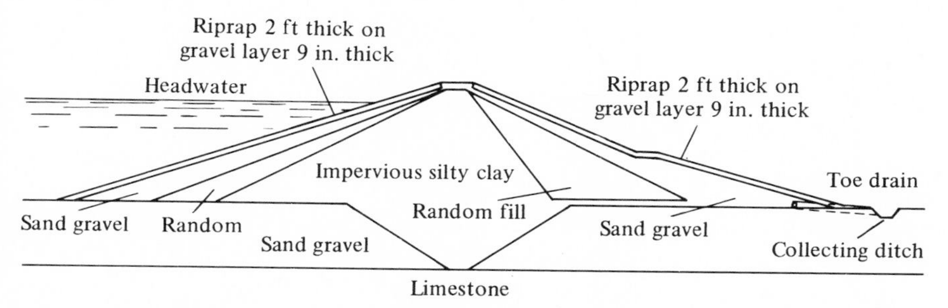

Fig. 21-20. Typical cross section of Corralville Dam, Iowa City, Iowa (Rock Island District, Corps of Engineers, U.S. Army).

ing of the soil may develop and the dam may eventually fail. But, even if erosive velocities do not develop, such a wet condition at the downstream toe is highly undesirable and should be avoided.

The position of the line of saturation can be controlled by providing artificial drainage for seepage water in the downstream section of the dam. Many arrangements of highly pervious elements constructed of rock or coarse gravel or sand in combination with tile drains are possible, the best choice depending on the local situation and the kinds of material most readily available. The purpose of the drainage structure is to provide a point of release well inside the downstream face toward which the seepage water will flow. It also serves as a collection gallery for the seepage water and provides an outlet for the water to flow harmlessly away from the dam.

21.20.5. Stability of Slopes. The slopes of both the upstream face and the downstream face of a dam should be flat enough to keep the shearing stresses, in the dam itself and in the foundation, below the shearing strengths of the soils which compose these elements of the structure; and the factor of safety should be ample. The factor of safety of an earth dam should, in general, be greater than that which is required for highway and railway embankments, because of the greater damage to life and property which may result from the release of large quantities of water in case of a dam failure. The principles of slope stability presented in Chapter 20 may be employed for analyzing the stresses in the dam. The discussion of stresses in the foundation under an embankment given earlier in this chapter is also appropriate to earth dams.

If, for any reason, the water level of the reservoir behind an earth dam is lowered very rapidly, a severe stress situation is created in the upstream slope of the dam, and possibly in the natural slopes bounding the reservoir, particularly if such slopes are composed of fine-grained soil which drains slowly. High stresses are induced in the soil by reason of the removal of hydrostatic pressure from the outside of the slope after a sudden draw-down, while there remains on the inside a hydrostatic pressure due to the saturated condition of the soil. Furthermore, the soil begins to drain by gravity as soon as the reservoir level is lowered and the forces due to seepage velocity augment the tendency of the slope to slide downward. Although sudden draw-down causes a severe strain on the stability of the upstream slope, it is a condition which rarely occurs except in the event of failure and washout of some other portion of a dam. If it is considered possible that a sudden draw-down condition may develop, the best defense is to use coarse, free-draining soil in the upstream shell of the dam. If the soil can drain as rapidly as the reservoir level drops, no undue stresses in the slope will be created.

Damage caused by sudden draw-down may also occur in river banks as the stage recedes subsequent to a flood, or in steep banks near the sea following a seiche or tidal wave.

21.20.6. Channels through Dam. It is very important in earth-dam construction to make sure that no planes or paths are created along which water may find a more or less direct passage through the dam. For example, it is essential that a good bond be secured between the soils in the dam and the foundation. This is usually accomplished by removing the top foot or so of the foundation surface to get rid of soil containing roots and other vegetable material. Then the surface should be scarified before the first layer of soil for the dam is laid and compacted. Another important precaution is to build baffle walls or diaphragms at comparatively short intervals along the length of any conduit which extends through the dam. These should completely encircle the conduit and should project outward about 4 to 6 ft from the barrel of the structure. After a dam is put in service, measures should be taken during maintenance to prevent burrowing animals from creating a passageway for water.

If seepage water emerges from the dam under sufficient pressure and with sufficient velocity to carry away particles of soil, a condition known as piping develops. This is a sort of internal erosion by which small channels grow in length from the downstream end; and, if unchecked, they may extend completely through the dam and cause it to fail. In a somewhat similar manner seepage water from the foundation may cause boils to develop at points downstream from the dam. The possibility of serious piping may be prevented by making the paths of percolation sufficiently long in relation to the head, thus reducing the hydraulic gradient, and by providing properly designed and constructed filters and drains so that a dangerous escape gradient will be avoided. For safety against piping in dams of fine-grained material the ratio of length l of path to head h should be not less than 5. In the case of a highly pervious foundation without a cut-off wall, the ratio l/h should be not less than 8 or 10. If downstream boils tend to develop after a dam is completed, the situation can usually be cured by placing a layer of coarse sand and gravel over the affected area and then adding layers of coarser stone until the weight of this overburden overcomes the tendency toward boiling, even at maximum reservoir head. Relief wells may also be installed to combat boiling.

21.20.7. Protection against Erosion. The upstream slope of a dam needs to be protected against erosion by wave action. Stone riprap is frequently used for this purpose. It may be dumped in a layer about 3 ft thick or hand-placed in a layer about 1 ft 6 in. in thickness. Regardless of the manner of placing the riprap, a 9 to 18 in. layer of either bank-run gravel or crushed stone having a maximum size of $1\frac{1}{2}$ in. should be placed beneath the riprap in order to prevent washing away of the soil as water rushes through the interstices between the riprap stones during wave action. If the water level in the reservoir is to be held fairly constant, the riprap may be terminated about 10 ft below the lowest water elevation. However, if the reservoir level is expected to fluctuate considerably, as in the case of a flood-control dam, it may be necessary to carry the riprap throughout the full height of the upstream face. The downstream face of the dam only needs to be protected against erosion by run-off water falling on the dam during rainstorms. Usually a good growth of sod will be sufficient for the purpose.

Soil-cement, or soil which has been mixed with portland cement and compacted under controlled conditions (Section 16.25), is becoming extensively used for riprap in areas where suitable rock is scarce. The soil-cement is usually mixed, spread and compacted in a series of overlapping horizontal strips running along the upstream face of the dam, giving it a

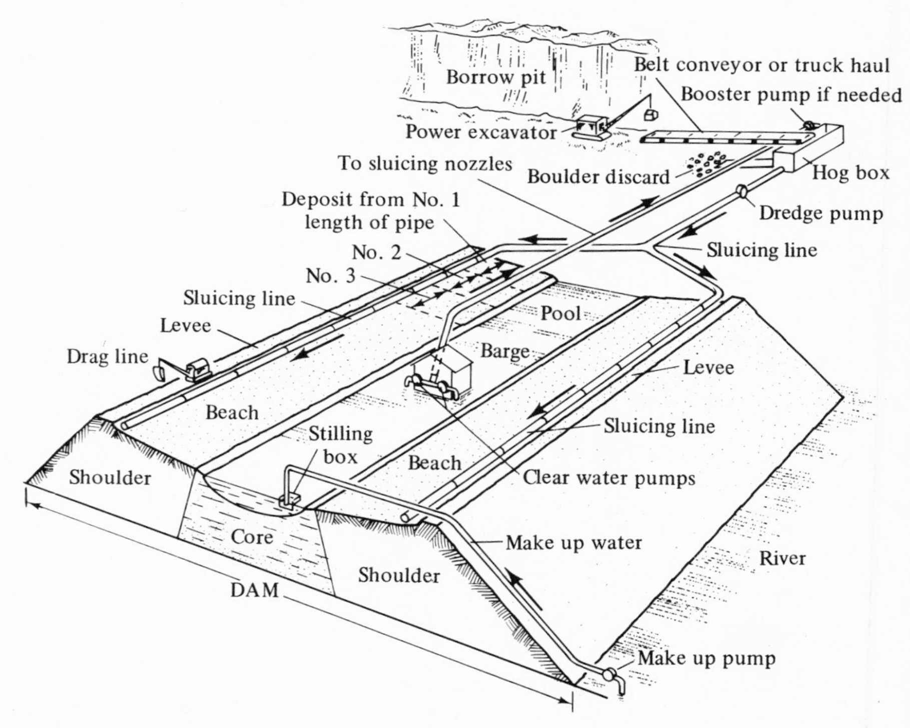

Fig. 21-21. Hydraulic-fill construction.

stair-step appearance. Bituminous concrete may be similarly used, but tends to be more costly.

21.20.8. Hydraulic-Fill Construction. There are two principal methods of constructing earth dams: the rolled-fill method and the hydraulic-fill method. The rolled-fill method was discussed earlier in this chapter in connection with suggestions for embankment construction. In hydraulic-fill construction, the soil is mixed with water and pumped through sluice pipes located along the outer edges of the fill. The soil-water mixture flows out of the sluice pipes and inward toward the longitudinal center of the dam. As it flows, the coarse particles of soil are deposited first, and then succeedingly smaller particles are released as the water flows over the beaches thus created. The fine particles of silt and clay are carried in suspension to a core pool at the center. Since the velocity of water in the core pool is practically zero, the silt and clay are permitted to settle out. Thus, the structure is built up with the coarse material in the outer shells and with fine-grained material making up the impervious core wall at the center. The diagram in Fig. 21-21 illustrates the hydraulic-fill method of earth-dam construction. Hydraulic fill construction becomes economical only for large fill operations with a source of cheap power, and therefore is rarely used.

PROBLEMS

21.1. State some general principles of soil selection in the design and construction of embankments.

21.2. Describe a sheep's-foot roller and explain how it functions in soil-compaction operations.

21.3. What is the best method for compacting cohesionless sand?

21.4. What is meant by overcompaction? How may it be avoided?

21.5. Would you expect the resonant frequency of a vibratory compactor to increase or decrease as weight is added to the compactor?

21.6. Should resonant frequency remain constant, increase, or decrease as the soil compacts? Why?

21.7. A loose sand is found to be at 90% maximum A.A.S.H.O. T99 density without any compaction. Explain.

21.8. The following percents of compaction were obtained from 5 density tests: 91, 97, 94, 90, 93. (a) Calculate the mean and standard deviation. (b) From data in Table 6-1, what are the 95% confidence limits on $\bar{x}$? (c) Based on the estimate of σ, predict how many additional tests will be needed to establish $\bar{x}$ within $\pm 2\%$.

21.9. From the data in Fig. 21-6, what are the 95% confidence limits on $\bar{x}$?

21.10. What is "shrinkage" in connection with embankment construction? What are the causes of shrinkage?

21.11. An earth dam 40 ft high is to be constructed of cohesive soil. The foundation material has a reasonably high shearing strength. What side slopes will be appropriate?

21.12. A highway embankment has a top width of 40 ft and side slopes of 2:1, and its height is 70 ft above the horizontal natural-ground surface. Assume that the unit weight of the embankment soil is 125 pcf and the weight of the foundation soil is 110 pcf. Determine the horizontal shearing stress at the following points in the undersoil (see Fig. 21-10): (a) $x = 100$ ft and $z = 50$ ft; (b) $x = 50$ ft and $z = 100$ ft; (c) $x = 100$ ft and $z = 100$ ft.

21.13. Calculate the vertical pressure at points (a), (b) and (c) in Problem 21.12 due to the weight of the overlying embankment. Add to these values the pressure due to the foundation soil.

21.14. Under what conditions would it be advisable to place a blanket layer of coarse-grained material between an embankment and its foundation?

21.15. What are vertical sand drains, and what is their function? Describe how they work.

21.16. Describe three methods of constructing an embankment over a peat bed.

21.17. Give the function of each of the following elements in an earth dam: a core wall; a cut-off wall; an upstream soil blanket. What kinds of materials are suitable for these elements?

21.18. Plans for an earth dam call for a 3 ft layer of dumped stone riprap on an 18-in. layer of crushed stone on the upstream face. What is the function of the stone riprap and of the crushed-stone layer?

21.19. Describe the two principal methods of earth-dam construction.

REFERENCES

1. Butt, G. S., T. Demirel, and R. L. Handy. "Soil Bearing Tests Using a Spherical Penetration Device." *Highway Research Rec.* **243,** 62–74 (1968).
2. Clough, R. W., and R. J. Woodward, III. "Analysis of Embankment Stresses and Deformations." *ASCE J. Soil Mechanics and Foundation Eng.* **93** (SM4), 529–549 (1967).
3. D'Appolonia, D. J., R. V. Whitman, and E. D'Appolonia. "Sand Compaction with Vibratory Rollers." *ASCE J. Soil Mechanics and Foundation Eng.* **95** (SM1), 263–284 (1969).
4. Highway Research Board. "Compaction of Embankments, Subgrades, and Bases." Highway Research Board Bulletin, 58 (84), 1952.
5. Highway Research Board. "The A.A.S.H.O. Road Test. Report 2, Materials and Construction." Highway Research Board Special Report, 61B, 1962.
6. Highway Research Board. "Vertical Sand Drains for Stabilization of Embankments." Highway Research Board Bulletin 115, 1955.
7. Holl, D. L. "Plane-Strain Distribution of Stress in Elastic Media." Bulletin 148, Iowa Engineering Experiment Station, Ames, Iowa, 1941.
8. Johnson, A. W. and J. R. Sallberg. "Factors Influencing Compaction Test Results." Highway Research Board Bulletin 319, 1960.

9. Justin, Joel D., Julian Hinds, and William P. Creager. *Engineering for Dams*, Vol. III. John Wiley. New York, 1945.
10. Lewis, W. A. "Recent Research into the Compaction of Soil by Vibratory Equipment," *Proc. 5th Intern. Conf. on Soil Mechanics and Foundation Eng.*, Vol. II, 261–268, 1961.
11. Lohnes, R. A., and Handy, R. L. "Slope Angles in Friable Loess." *J. Geol.* **76** (3), 247–258 (1968).
12. Meehan, R. L. "The Uselessness of Elephants in Compacting Fill." *Can. Geotech. J.* **4** (3), 358–360 (1967). (Repeated passes by elephants ineffective for compaction because they step in their old tracks.)
13. Plummer, Fred L., and Stanley M. Dore. *Soil Mechanics and Foundations*. Pitman, New York, 1940.
14. Sherman, G. B., R. O. Watkins, and R. H. Prysock. "A Statistical Analysis of Embankment Compaction." *Highway Research Board Rec.* **177,** 157–185 (1967).
15. Spangler, M. G. "Influence of Compression and Shearing Strains in Soil Foundations on Structures Under Earth Embankments." Highway Research Board Bulletin 125, 170 (1956).
16. U.S. Bureau of Reclamation. *Design of Small Dams*. U.S. Government Printing Office, Washington, D.C., 1960. 611 pp.

22

Pressure on Retaining Walls

22.1. KINDS OF LATERAL SOIL PRESSURE. Soil is neither a solid nor a liquid, but it has some of the characteristics of both of these states of matter. One of its characteristics which is similar to that of a liquid is the tendency to exert a lateral pressure against an object with which it comes in contact. This property of soil is highly important in engineering practice, since it influences the design of retaining walls, revetments, bulkheads, abutments, underground conduits, and many other structures of a similar nature.

There are two distinct kinds of lateral soil pressure, and a clear understanding of the nature of each is essential. First, consider a retaining wall which holds back a mass of clean, dry, cohesionless sand. The sand exerts a push against the wall by virtue of its tendency to slip laterally and seek its natural slope or angle of repose. This kind of pressure is known as the *active lateral pressure* of the soil. In this case the soil is the actuating element; and, if stability is maintained, the structure must be able to withstand the pressure exerted by the soil. Next, imagine that in some manner the retaining wall is caused to move toward the soil. When this situation develops, the retaining wall or other type of structure is the actuating element, and the soil provides the resistance for maintaining stability. The pressure which the soil develops in response to movement toward it is called the *passive resistance pressure*. It may be very much greater than the active pressure.

22.2. LIMITING MAGNITUDES OF ACTIVE AND PASSIVE RESISTANCE PRESSURES. The limiting values of both the active pressure and the passive resistance pressure for any given soil depend on the amount of movement of the structure against which the soil pressure impinges. In the case of active pressure, the movement of the structure is directed away from the soil and permits the development of strains within the soil mass. These strains mobilize shearing stresses which are directed in such a way that they help to support the mass and thus reduce the amount of pressure exerted by the soil against the structure. In the case of passive resistance pressure, internal shearing stresses also develop; but they act in the opposite direction from those in the active case and must be overcome by the movement of the structure. This difference in the direction of internal shearing stresses accounts for the difference in magnitude between active pressure and passive resistance pressure. The direction of movement and the direction of internal shearing stresses in the active and passive cases are shown in Figs. 22-1 and 22-2, respectively.

22.3. EARTH PRESSURE AT REST. Active pressures are accompanied by movements directed away from the soil; and passive resistance pressures, which are much larger, are accompanied by movements toward the soil. There must therefore be an intermediate pressure situation when the retaining structure is perfectly stationary and does not move or strain in either direction. The pressure which develops at zero movement is called *earth pressure at rest.* Its value is somewhat larger than the limiting value of active pressure, but it is considerably less than the maximum passive resistance pressure. The graph in Fig. 22-3 illustrates these relationships between earth pressures and movements of a retaining structure.

22.4. RANKINE THEORY OF LATERAL EARTH PRESSURE. The problem of determining the lateral pressure against retaining walls is one of the oldest in modern engineering literature. As early as 1687 a French military engineer, Marshal Vauban, set forth certain rules for the design of revetments to withstand the lateral pressure of soil. Since then many theories of earth pressure have been proposed by various investigators, and numerous experiments have been conducted in this field. Of the many theories which have been suggested, those by Coulomb and Rankine have stood the test of time and are often called the classical earth pressure theories. They have served the engineering profession well and are reliable, although it is recognized that they do not exactly fit all conditions which are encountered in practice. Nevertheless, they are basic to the problem and should be understood by all engineers who deal with earth pressures. As presented in this text, both theories are developed to apply only to cohesion-

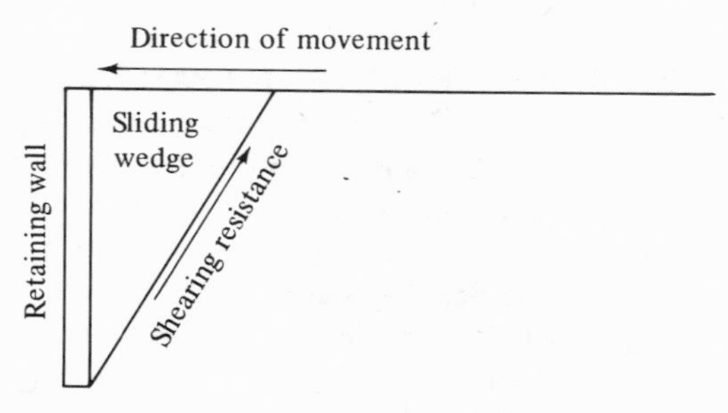

Fig. 22-1. Direction of movement and shearing resistance; active pressure case.
sure case.

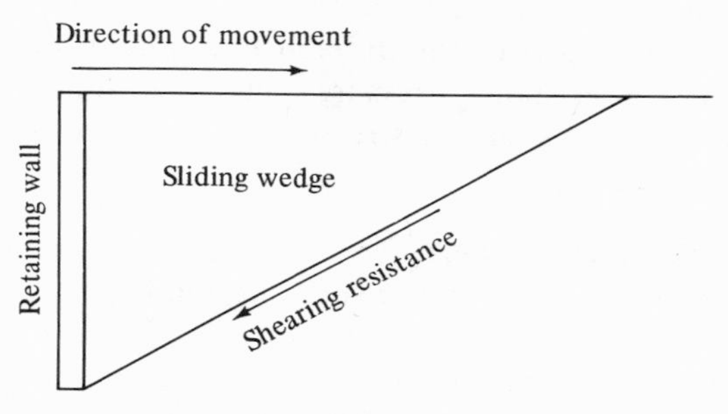

Fig. 22-2. Direction of movement and shearing resistance; passive resistance pressure case.

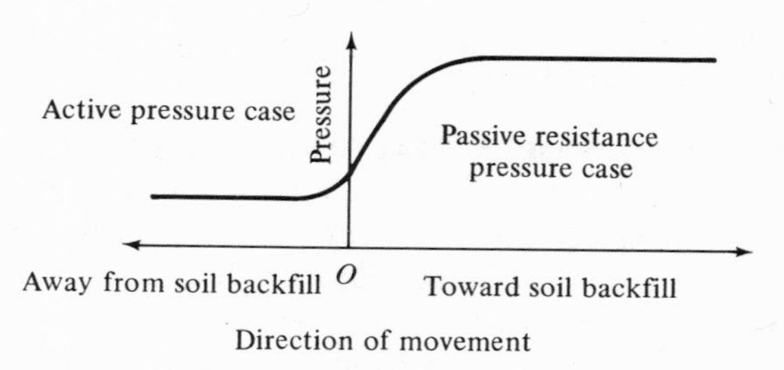

Fig. 22-3. Relation between soil pressure and direction of movement.

less soil backfill, since this is the situation most frequently assumed in retaining-wall design in practice.

Although the Rankine theory of lateral earth pressure was proposed in 1860, nearly a century later than the Coulomb theory, the principles ad-

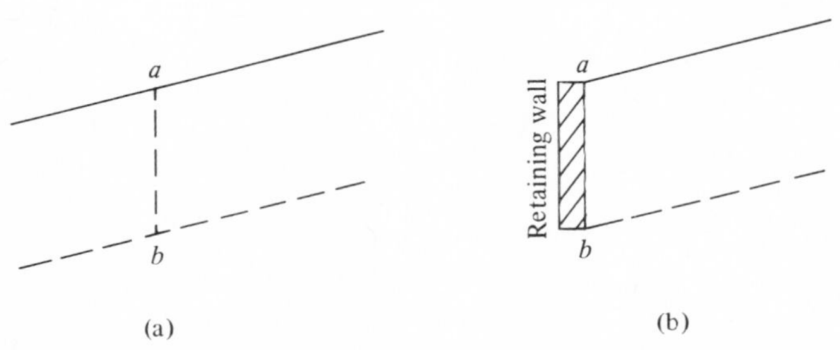

Fig. 22-4. Basic concept of Rankine theory of lateral earth pressure. (a) Infinite slope; (b) soil to the left of plane *a-b* replaced by retaining wall.

vocated by Rankine will be presented first. The theory is based upon the concept that a conjugate relationship exists between the vertical pressure and the lateral pressure at a point in a soil mass, as was discussed in connection with a slope of infinite extent in Chapter 20. As a matter of fact, Rankine's idea can be visualized rather easily by imagining first a slope of infinite extent, as shown in Fig. 22-4(a). Now assume that the soil to the left of the vertical plane *ab* is removed and a retaining wall is substituted for it, as indicated in Fig. 22-4(b). Then the pressure on the retaining wall is considered to be the same as that which existed on the plane *ab* before the wall replaced the soil. In accordance with Eq. (20-5), the unit pressure on the wall at any depth is

$$p = \gamma h K \cos i \tag{22-1}$$

in which

p = unit lateral pressure at any depth h;
γ = unit weight of soil;
h = depth below the top of the wall;
K = conjugate ratio; and
i = angle between the backfill surface and the horizontal.

The unit pressure p has its maximum value at the bottom of the wall where $h = H$.

22.5. HYDROSTATIC PRESSURE RATIO. If we continue with the analysis of a slope of infinite extent in accordance with the Rankine theory, we fine that the unit pressure on a wall increases linearly with depth and that the pressure-distribution diagram is a triangle with the apex at the top of the wall. This means that the resultant of pressure acts at the height $\frac{1}{3} H$ above the base of the wall. On the basis of triangular distribution, the total resultant pressure on a retaining wall is

$$P_R = \tfrac{1}{2}\gamma H^2 K \cos i \qquad (22\text{-}2)$$

in which

P_R = resultant pressure on the wall;
γ = unit weight of the soil;
H = height of the wall;
K = the conjugate ratio; and
i = angle of slope of the backfill.

For the special case of a horizontal backfill, which is probably the case most usually encountered in practice, the angle $i = 0$. Then $\cos i = 1$ and the conjugate ratio becomes

$$K = \frac{1 - \sin\phi}{1 + \sin\phi} = \tan^2\left(45^\circ - \frac{\phi}{2}\right) = \frac{\sqrt{\mu^2 + 1} - \mu}{\sqrt{\mu^2 + 1} + \mu} \qquad (22\text{-}3)$$

in which ϕ is the angle of internal friction of soil; and $\mu = \tan\phi$ is the coefficient of friction for soil.

In this form, K is often called Rankine's ratio of lateral pressure to vertical pressure. Also, it is sometimes called the hydrostatic pressure ratio. Equation (22-3) is derived from a Mohr circle later in this chapter.

22.6. VERTICAL COMPONENT OF EARTH PRESSURE. According to the Rankine theory, the line of action of the resultant pressure on a retaining wall is parallel to the slope of the backfill surface. This means that the resultant is horizontal when the backfill is level; but it has a vertical component, and is therefore inclined with the horizontal, when the backfill is on a slope. Experimental evidence confirms the existence of a vertical component of the resultant, but its magnitude appears to depend on the coefficient of friction between the soil and the wall, rather than on the slope of the backfill. The magnitude of the vertical component is

$$V = P \tan\phi_1 \qquad (22\text{-}4)$$

in which

V = vertical component of resultant pressure;
P = component of resultant pressure normal to back face of wall; and
ϕ_1 = angle of friction between soil and retaining wall.

The angle ϕ_1 can never be greater than the angle ϕ of internal friction of the soil. In practice ϕ_1 is frequently assumed to be about two-thirds of ϕ.

It is to be noted that the vertical component of the resultant pressure usually decreases the calculated overturning moment on a retaining wall,

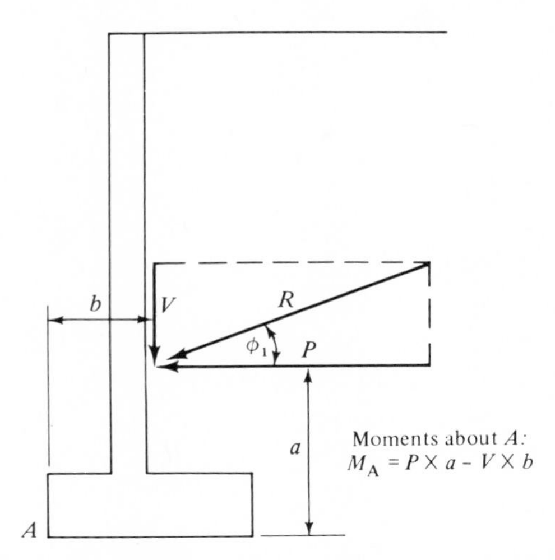

Fig. 22-5. Overturning moment with respect to front-face toe of a retaining wall.

as illustrated in Fig. 22-5. For this reason it is frequently neglected in retaining-wall design, as an added factor of safety.

22.7. PLANE OF FAILURE ACCORDING TO COULOMB THEORY. The Coulomb theory, which was proposed in 1773, is based upon the concept of a plane of failure extending diagonally upward and backward through the earth backfill, as shown in Fig. 22-6. The triangular mass of soil between this plane of failure and the back face of the retaining wall is sometimes referred to as the sliding wedge. It is reasoned that, if the retaining wall were suddenly removed, the soil within the sliding wedge would slump downward. Therefore, an analysis of the forces acting on the sliding wedge at incipient failure will reveal the horizontal thrust which it is necessary for the wall to withstand in order to hold the soil mass in place.

If a plane of failure is arbitrarily located so as to make an angle θ with the horizontal, such as plane BC in Fig. 22-6(a), the forces acting on the sliding wedge are as shown in Fig. 22-6(b). They consist of the weight W of the soil within the wedge, which acts through the centroid of the triangle ABC; a thrust N normal to the plane of failure, which is exerted by the soil to the right of plane BC; and a shearing force S, which acts upward along BC and which, at the limit of equilibrium, is equal to $N \tan \phi$. These forces must be balanced by the thrust P, which is assumed to act horizontally and to be concurrent with W, N, and S. The equal and opposite reaction to P is the lateral force which the wall must be designed to

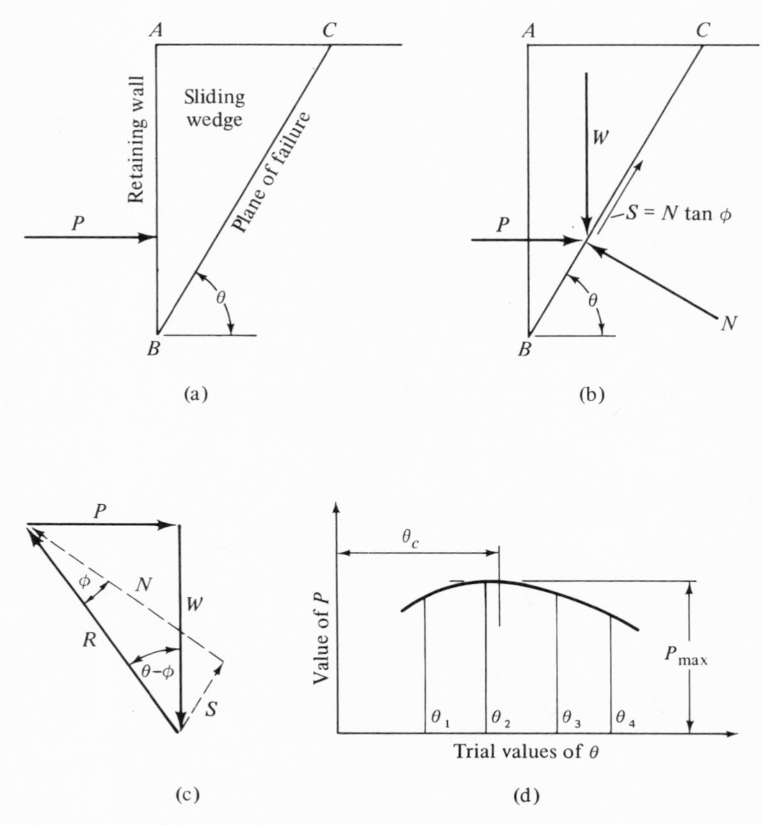

Fig. 22-6. Coulomb theory of lateral earth pressure. (a) If retaining wall were suddenly removed, force P would be required to hold sliding wedge in equilibrium; (b) forces acting on sliding wedge; (c) polygon of forces; (d) determination of maximum value of P.

withstand. If the unit weight of the soil and the friction angle are known, the force W acting through the centroid of the sliding wedge can be calculated. The magnitudes of N and S are not known, but the line of action of their resultant R makes the angle ϕ with a normal to the plane of sliding. It is therefore possible to construct the polygon of forces in Fig. 22-6(c). Then, by choosing a series of failure planes having different values of θ and obtaining the thrust on the retaining wall corresponding to each value, the maximum thrust can be determined. This is easily done graphically by plotting values of P against θ and measuring the maximum ordinate to the curve thus obtained, as shown in Fig. 22-6(d).

22.8. CALCULATION OF THRUST. The value of P may be derived analytically as follows. The forces acting on the sliding wedge are W, N, S, and P, as stated in Section 22.7 and as shown in Fig. 22-6(b). The forces N and S may be replaced by their resultant R, which acts along a line making the angle ϕ with the normal to the failure plane. It also makes the angle $(\theta - \phi)$ with the vertical. Since W, P, and R are three concurrent forces which are in equilibrium when failure is impending along the plane BC, they may be represented by the triangle of forces in Fig. 22-6(c). In this triangle,

$$P = W \tan(\theta - \phi) \tag{22-5}$$

But

$$W = \tfrac{1}{2}\gamma H^2 \cot \theta \tag{22-6}$$

Hence,

$$P = \tfrac{1}{2}\gamma H^2 \cot \theta \tan(\theta - \phi) \tag{22-7}$$

Since γ, H, and ϕ are constant for any particular wall and soil backfill, this equation shows that P varies with the angle θ, as was indicated in the graphical procedure of Section 22.7. To obtain the critical value of θ which yields the maximum value of P, equate $dP/d\theta$ to zero and solve for θ. This procedure gives

$$\theta_c = 45° + \frac{\phi}{2} \tag{22-8}$$

Substituting this value of θ_c in Eq. (22-7), we get

$$P = \tfrac{1}{2}\gamma H^2 K \tag{22-9}$$

in which

$$K = \frac{1 - \sin \phi}{1 + \sin \phi} = \tan^2\left(45° - \frac{\phi}{2}\right) \tag{22-10}$$

It will be noted that Eq. (22-9) is identical with the Rankine formula for a level backfill, or Eq. (22-2). Equation (22-9) gives the resultant horizontal pressure on a retaining wall with a vertical back face due to a cohesionless backfill bounded by a horizontal surface. A more general solution involving a sloping backfill, a battered wall, and a resultant thrust which makes an angle with the normal to the wall is given by

$$P_R = \frac{\gamma H^2}{2} \times \left[\frac{\sin^2(\beta - \phi)}{\sin^2 \beta \sin(\beta + \phi_1) \left(1 + \sqrt{\frac{\sin(\phi + \phi_1) \sin(\phi - i)}{\sin(\beta - i) \sin(\beta + \phi_1)}}\right)^2} \right] \tag{22-11}$$

The derivation of Eq. (22-11) is beyond the scope of this text. The nomenclature for this equation is shown in Fig. 22-7.

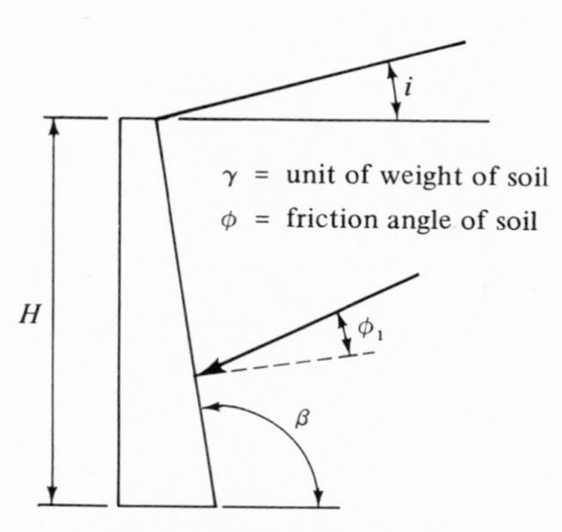

Fig. 22-7. Nomenclature for general Coulomb equation (22-11).

22.9. FAILURE SURFACE. It is assumed in the Coulomb theory that the plane of failure in the soil behind a retaining wall intersects the backfill surface at a distance back of the wall equal to H cot[45° + $\phi/2$]. But actual observations of retaining-wall failures indicate that the soil usually ruptures along a surface which passes through the heel of the wall and curves upward until it is practically vertical at the intersection with the backfill surface, as illustrated in Fig. 22-8. Also, these observations in-

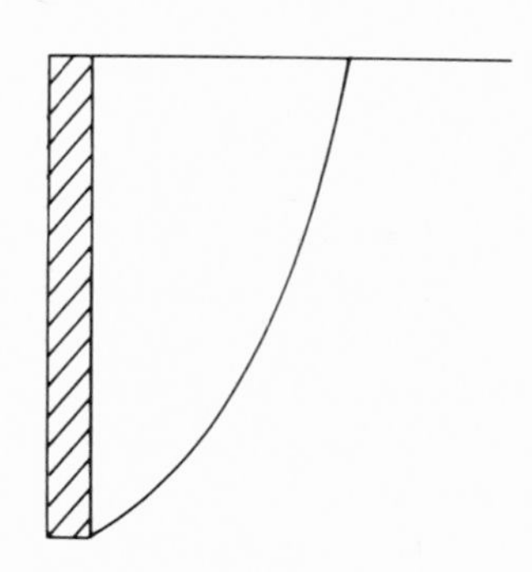

Fig. 22-8. Curved failure surface.

dicate that the distance from the retaining wall to the failure surface is rather consistently in the neighborhood of $0.4H$ to $0.5H$, which is much less than that assumed in the Coulomb theory. An analysis of the forces acting on a sliding wedge bounded by a curved surface, which more nearly coincides with actuality, can be made easily if it is assumed that the failure surface has the shape of a logarithmic spiral.

22.10. DETERMINATION OF THRUST ON WALL. The logarithmic, or equiangular, spiral is a curve expressed in polar coordinates by the equation

$$\rho = \rho_0 e^{m\alpha} \tag{22-12}$$

in which

ρ = a radius vector;
ρ_0 = any other radius vector;
e = base of natural logarithms;
m = a constant; and
α = angle between ρ and ρ_0, in radians.

This spiral has the interesting characteristic that the radius vectors of the curve make a constant angle with normals to the tangents at every point on the curve. Because of this property, the spiral is particularly useful in determining the earth pressure on a retaining wall.

To determine the horizontal pressure, first draw the retaining wall and backfill to any convenient scale, as in Fig. 22-9. Also, on a sheet of transparent paper, construct a logarithmic spiral whose equation is

$$\rho = \rho_0 e^{\alpha \tan \phi} \tag{22-13}$$

in which ϕ is the friction angle of soil.

Next, lay the transparent paper over the drawing and rotate the spiral until it passes through the heel of the wall and is approximately vertical at the point where it intersects the surface of the backfill. Transfer the curve within the limits of the backfill to the drawing and locate the position of the origin of the spiral on the drawing. On each increment of length of the curve which represents the failure surface within the soil mass, there is a normal force N; and at the limit of equilibrium the shearing force is $S = N \tan \phi$. The resultant of these two forces N and S makes the angle ϕ with the normal to the curve at the center of the increment. Therefore, in accordance with Eq. (22-13), the resultant force on each increment of length is collinear with its radius vector; and the lines of action of all resultants on all increments pass through the origin of the spiral. If moments are taken with respect to this origin, it is seen that the moment of all the resultants is zero and the moment of the weight of the failure wedge is equal to the moment of the thrust on the wall. From this

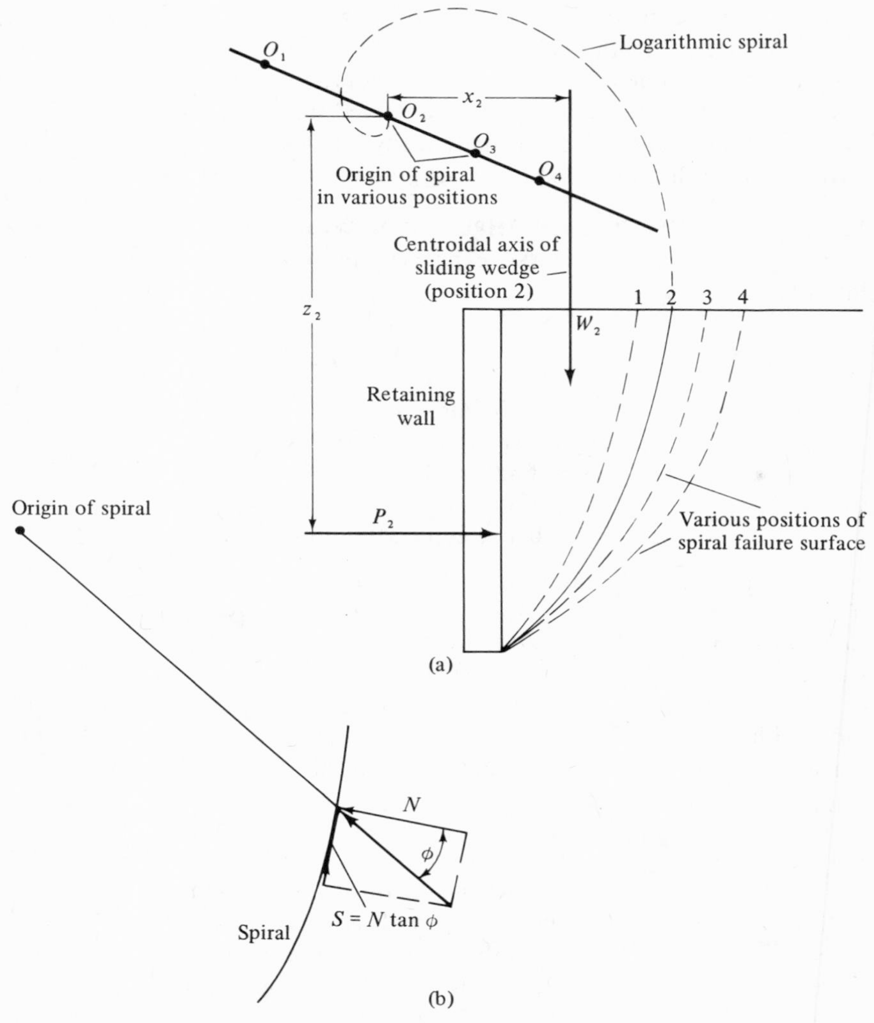

Fig. 22-9. Logarithmic spiral method of determining resultant pressure on a retaining wall.

fact, the thrust can be determined by the equation

$$P = \frac{Wx}{z} \tag{22-14}$$

in which

P = resultant horizontal pressure;
W = weight of sliding wedge;

x = horizontal distance from origin of spiral to centroid of sliding wedge; and

z = vertical distance from origin of spiral to resultant horizontal pressure.

By making several thrust calculations with the spiral-shaped failure surface in different locations, the maximum thrust on the wall can be determined in a manner similar to that outlined in Section 22.7.

Since the curved failure surface usually intersects a level backfill surface at a distance from the back of a retaining wall that is between $0.4H$ and $0.5H$, where H is the height of the wall, the thrust determined with the spiral in this location will be very close to the maximum value. Comparison of the results obtained by the Coulomb theory and the logarithmic-spiral method reveals no significant difference between the two. Therefore, it appears that the assumption of a plane surface of failure is justified, even though the curved shape more nearly coincides with observed facts.

22.11. PASSIVE RESISTANCE PRESSURE. When passive-resistance pressure conditions prevail, that is, when a structure is caused to move toward the soil backfill, the forces acting on the sliding wedge may be analyzed in substantially the same manner as in the active-pressure case. The essential differences are: The shearing force S is directed downward toward the heel of the wall and the plane of failure is more nearly horizontal, as shown in Fig. 22-10(a). After the normal force N and the shearing force S are replaced by their resultant R, the triangle of forces shown in Fig. 22-10(b) may be constructed. From this triangle,

$$P_p = W \tan(\theta + \phi) \tag{22-15}$$

or

$$P_p = \tfrac{1}{2} \gamma H^2 \cot \theta \tan(\theta + \phi) \tag{22-16}$$

Here again, as in the active-pressure case, the critical value of θ can be determined by the principle of maxima and minima. It is found to be

$$\theta_c = 45° - \frac{\phi}{2} \tag{22-17}$$

and the maximum passive resistance pressure is

$$P_p = \tfrac{1}{2}\gamma H^2 K_p \tag{22-18}$$

in which

$$K_p = \frac{1}{K} = \frac{1 + \sin \phi}{1 - \sin \phi} \tag{22-19}$$

Thus, it is seen that the limiting value of passive resistance pressure is much larger than the active pressure. As an example, suppose that the

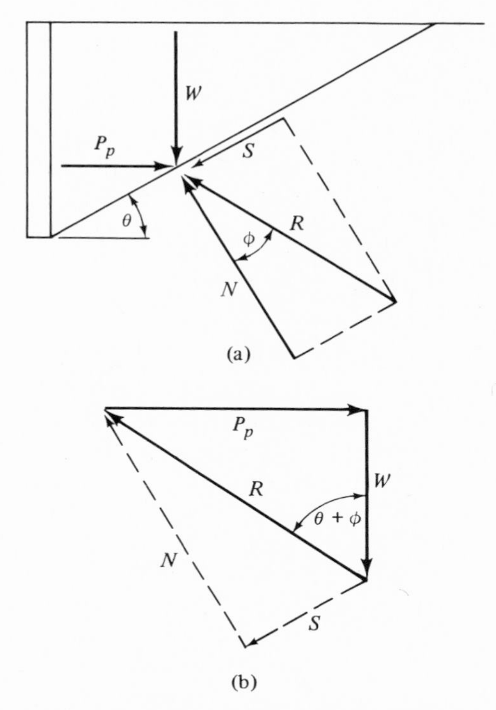

Fig. 22-10. Forces acting on a sliding wedge; passive resistance pressure case.

angle of internal friction of a soil is $\phi = 30°$. Then,

$$K = \frac{1 - \sin 30°}{1 + \sin 30°} = 0.33$$

and

$$K_p = \frac{1}{0.33} = 3.0$$

Accordingly, the limit of passive resistance pressure, that is, the capacity for resisting movement against this particular soil is nine times the active pressure. The ratio of passive resistance pressure to active pressure increases as ϕ increases. An example of passive pressure is soil at the toe of a retaining wall, as shown in Fig. 22-13.

22.12. EFFECT OF COHESION COMPONENT OF SOIL SHEARING STRENGTH. It is obvious from the role played by internal shear-forces in the backfill in both the Rankine analysis and the Coulomb analysis that a cohesion component of soil shearing strength tends to decrease the active pressure and to increase the passive resistance pressure on a wall, as compared with pressures computed for a soil having the same angle ϕ

and zero cohesion. This conclusion is confirmed by the commonly observed fact that a vertical bank of cohesive soil will in many instances stand unsupported for a considerable height and for a considerable length of time, a condition which would be impossible if the soil were cohesionless.

From the standpoint of economy in design it would appear to be logical to take cohesion into account in the determination of the overturning and translating forces which a retaining wall must resist. However, most designers are reluctant to do so because of the difficulty and uncertainty of determining the cohesion of the disturbed and manipulated soil which usually constitutes the backfill behind a wall. Also, the cohesion property of a disturbed soil is somewhat tenuous and may not be dependable under all climatic conditions and over a long period of time. It is rather widespread practice to assign a value of ϕ to each of a number of general soil types and calculate pressure on a wall by means of the Coulomb or Rankine theory, without attempting to determine a specific value of cohesion. One tabulation of such values of ϕ is shown in Table 22-1. They are recommended for design purposes, except in the case of very high walls for which extensive study of backfill soil characteristics may be justified.

TABLE 22-1. Usual Values of Unit Weight and Friction Angle for Soil Backfill

Backfill Material	*Unit Weight* γ *(pcf)*	*Internal Friction Angle* ϕ *(deg)*
Soft plastic clay	105–120	0–10
Wet, fine silty sand	110–120	15–30
Dry sand	90–110	25–40
Gravel	120–135	30–40
Loose loam	75– 90	30–45
Compact loam	90–100	30–45
Compact clay	90–110	25–45
Cinders	40	25–45
Compact sand-clay	115–125	40–50
Water	62.4	0

In the case of excavations made in cohesive clay soil, the cohesion should be taken into account, since in this case the soil behind the restraining sheeting is not disturbed and the cohesion is not destroyed. No satisfactory theory has been developed for calculating earth pressures against the bracing of an open cut in undisturbed clay for the many conditions which may be encountered. Terzaghi and Peck (7) have offered an empirical formula of the form

$$K_c = 1 - m\frac{4c}{\gamma H} \qquad (22\text{-}20)$$

in which

K_c = hydrostatic pressure ratio for use in Eqs. (22-2) and (22-9) for the case of excavations in cohesive soil;
m = a constant;
c = shear strength (cohesion) of soil;
γ = unit weight of soil; and
H = depth of cut.

For a number of measurements of strut loads in excavations in Japan, England, and Chicago the values of K_c fell within a range of plus or minus 30% of the value with $m = 1.0$. However in six observations made in Oslo the value of m was in the neighborhood of 0.4, and K_c was substantially greater.

It is emphasized that Eq. (22-20) is expected to give a pressure envelope for strut loads, rather than a statement of the actual earth pressure to be supported by the sheeting, for reasons discussed in the next section.

22.13. EFFECT OF MOVEMENT OF WALL. The discussion relative to the influence of the amount of movement of a retaining wall upon lateral pressure given in Section 22.2 and the facts illustrated in Fig. 22-3 indicate that a certain amount of yield in the direction away from the backfill soil is necessary for the pressure to be reduced to the value of active lateral pressure. This is true because a certain amount of strain must develop within the soil mass in order that the shearing forces which help to support the soil may be fully mobilized. The amount of movement required for active pressure to develop is quite small, on the order of 0.001 to $0.002H$ where H is the height of the wall. In special circumstances where movement is restricted, as in bridge abutments, pressures have been found to exceed the active case. The amount of movement toward the soil which is required to fully mobilize the passive resistance pressure of soil is, in general, much greater than the movement required in the case of active pressure. However, knowledge of this subject is so meager at the present time that no quantitative statement of the relationship between movement and pressure can be made. Further discussion of this relationship will be given in Chapter 26 in connection with the design of flexible pipe culverts.

The character of the movement of a retaining wall may influence the distribution of the active lateral pressure. If the wall yields by simple outward rotation or tipping about its base, the triangular distribution indicated by the classical theories will prevail and the line of action of the resultant strikes the wall at one-third of its height. If the wall yields

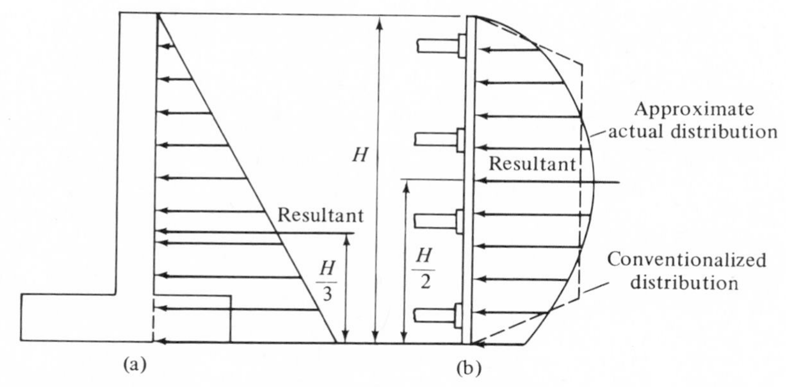

Fig. 22-11. Position of resultant pressure. (a) Pressure distribution on a retaining wall; (b) pressure distribution on sheeting of open cut.

by horizontal translation, that is, by sliding along the contact between its base and the foundation soil, a vertical arching action may develop in the backfill. In this case, the sliding wedge may be partially supported by arch action which brings about a redistribution of the lateral pressure on the wall and causes a reduction in the unit pressure near the base and an increase in pressure near the mid-height. This redistribution causes the position of the resultant to move upward and it may approach the mid-height. Such a condition seldom develops where a retaining wall is constructed independently and the backfill is then deposited against the wall; but it frequently occurs in the case of timbering or sheeting used to support the face of an open cut. Construction men have often observed that the pressure near the mid-height of such sheeting is considerably greater than that at the base of the cut [see Fig. 22-11(b)]. The total pressure on the restraining structure is about the same, regardless of the character of the yield, provided that it is directed away from the soil.

22.14. PRACTICAL SUGGESTIONS. The foregoing discussion of the theoretical aspects of pressures on retaining walls can be summarized by the following practical suggestions for earth-pressure computations.† The horizontal pressure on a retaining wall with a vertical back face and level backfill may be computed by the Rankine and Coulomb formulas for active pressure, or Eqs. (22-2) and (22-9). These formulas are valid when conditions are such that the top of the wall can move outward, either by rotation about the toe of the footing or by deflection of the wall, a distance equal to or more than about 0.1% of the wall height. This is a very small movement, being of the order of $\frac{1}{4}$ in. at the top of a 20-ft wall, and

† Most of these are adapted from Feld (*1*).

TABLE 22-2. Values of Rankine's Lateral Pressure Ratio

Internal Friction Angle ϕ *(deg)*	*tan* ϕ	*Lateral Pressure Ratio*		
		Active K	*Passive* $K_p = 1/K$	K_p/K
0	0	1.00	1.00	1.00
10	0.176	0.70	1.42	2.03
20	0.364	0.49	2.04	4.17
25	0.466	0.41	2.47	6.02
30	0.577	0.33	3.00	9.00
35	0.700	0.27	3.69	13.66
40	0.839	0.22	4.40	20.00
45	1.000	0.17	5.83	34.30
50	1.192	0.13	7.55	58.07
60	1.732	0.07	13.90	198.57

it can practically always be expected to develop. Values of the coefficient of friction and Rankine lateral pressure ratio for various angles of friction for both the active and passive cases are given in Table 22-2.

The general Coulomb formula, or Eq. (22-11), can be used to compute the lateral pressure on a retaining wall having a battered back face and sloping backfill, and when it is desired to take into account the friction between the soil and the wall. However, comparisons with experimental evidence indicate that the results are somewhat too small for negative fill slopes and somewhat too large for positive fill slopes, although the differences are no greater than 10%. It is sufficiently accurate for work of this kind (in which the properties of the soil are somewhat variable and it is difficult to determine them accurately) to use a table of ratios by which the pressure due to a level backfill on a vertical wall may be modified to fit the actual situation. Suggested ratios are given in Tables 22-3 and 22-4. In computing the ratios in Tables 22-3 and 22-4 the angle of frction ϕ_1 be-

TABLE 22-3. Modification Ratios for Battered Walls and Level Backfill

Wall Batter[a]	*Ratio*
+1:6	1.07
+1:12	1.03
Vertical	1.00
−1:6	0.95
−1:12	0.90

[a] Positive batter: toward backfill. Negative batter: away from backfill.

TABLE 22-4. Modification Ratios for Vertical Walls and Sloped Backfill

Slope Angle i (deg)	*Ratio*[a]		
	ϕ = 20°	ϕ = 30°	ϕ = 40°
+10	1.09	1.10	1.23
0	1.00	1.00	1.00
−10	0.88	0.83	0.74

[a]ϕ = Friction angle of soil.

tween the soil and the wall was assumed to be 30°, or equal to the angle ϕ of internal friction when ϕ is less than 30°.

22.15. LATERAL PRESSURE DUE TO SUBMERGED BACKFILL. The lateral pressure produced by a saturated submerged soil is equal to the full hydrostatic pressure of the water in the soil pores plus the pressure exerted by the soil solids. The most convenient way to determine the lateral pressure due to a submerged backfill is to compute these two components separately and to add the results together. The hydrostatic pressure can be computed by applying Eq. (22-9) and taking γ as 62.4 pcf and K as 1.0 (ϕ = 0°). The pressure due to the soil solids also can be computed by applying Eq. (22-9) and using the buoyant unit weight of the solid particles and a value of K corresponding to the friction angle ϕ of the soil in the saturated state. Note that this is not the same as using the saturated unit weight of the soil and applying the same K to the soil and water collectively, which would be incorrect.

22.16. POSITION OF LINE OF ACTION OF RESULTANT PRESSURE. The position of the line of action of the resultant pressure on a retaining wall has an important influence on the overturning moment exerted by the backfill; and, for this reason, careful consideration should be given to this detail of the problem. Both theoretical and experimental evidence indicate that the resultant on a wall of height H acts at a distance $\frac{1}{3}H$ above the base, as shown in Fig. 22-11(a), when the backfill is level or is a liquid material.

TABLE 22-5. Position of Line of Action of Resultant Pressure on Retaining Wall

Slope Angle i (deg)	*Position of Resultant Above Base*
All negative slopes	0.33H
0 (Level backfill)	0.33H
10	0.40H
20	0.42H
30	0.44H

When the backfill is on a slope, there is some conflict between theory and experiments. The relationships between slope angle and position of the resultant given in Table 22-5 are based largely on experimental evidence and are recommended for use in retaining-wall design. When the design of struts for supporting an open-cut excavation is being considered, the position of the resultant should be assumed to be at the mid-height, as shown in Fig. 22-11(b). For further discussion of this subject, see Ref. (*7*).

22.17. DRAINAGE OF FILL. Any retaining wall which is not designed to withstand the full hydrostatic pressure of submerged soil should be provided with adequate drainage facilities to insure that the backfill soil will not become saturated and submerged. Weep holes 4 to 6 in. in diameter should be placed in the wall at intervals along its length and at the lowest elevation at which free outlet drainage can be maintained. If a vertical layer of crushed rock or coarse gravel about 9 to 12 in. thick is placed directly behind the wall, the effectiveness of the weep holes will be greatly enhanced. In case the backfill soil is fine-grained material, a graded filter should be placed between the soil and the rock in order to prevent clogging of the drainage system by migration of the soil fines into the coarse layer. Where it is not practicable to install weep holes through a wall or if there is a question whether they will be adequately maintained throughout the life of the structure, a longitudinal drain may be placed behind the wall at an elevation a little above that of the footing. Water collected by this drain may be carried to outlets at the ends of the wall; or, in the case of a very long wall, it may be discharged through headers extending through the wall at intervals of several hundred feet. Seepage stresses act downward toward the drains and are usually small compared to precision of the analytical procedure.

22.18. EXPANSIVE CLAY. Often the fill behind retaining walls is not selected, and therefore may sometimes involve expansive clay soils. If an expansive soil is placed behind the wall dry and later becomes wet and expands, the wall usually moves, sometimes sufficiently to cause structural damage to the wall. Since the soil shear strength is no longer acting to hold the soil back from the wall, but instead is pushing against the wall, the passive pressure condition applies. However, designing for the passive condition is not sufficient because in the expansion process the soil structure is reconstituted, increasing cohesion and decreasing the friction angle. It is usually impractical to design a wall to restrain an expanding clay. If expansive clay must be used, a better procedure may be to insure that it is fully expanded when it is being backfilled, by designing for saturated conditions and wetting during backfilling. Loose backfilling of dry clay will not solve the problem, since the entire soil skeleton made up of individual clay clods will expand and exert pressure on the wall (see also Section 8.21).

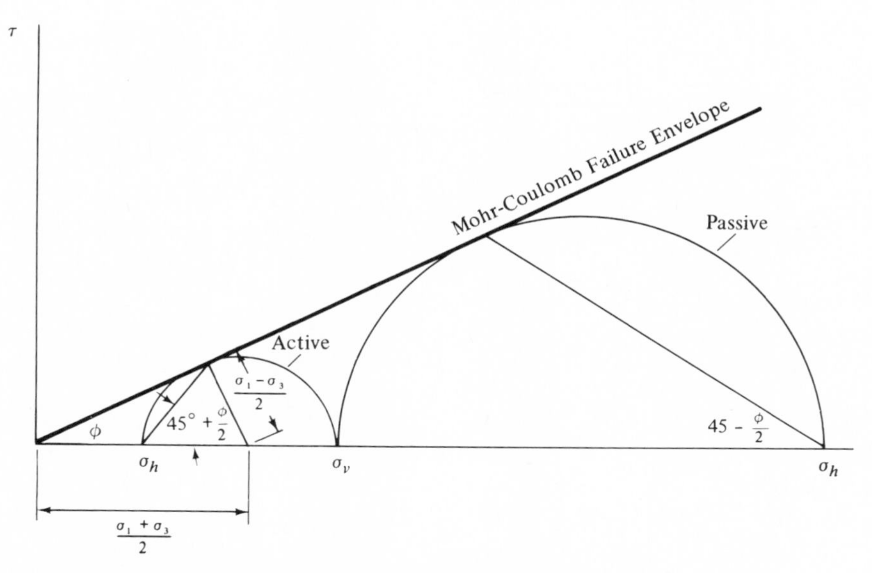

Fig. 22-12. Mohr circle represention of active and passive states.

22.19. MOHR CIRCLE REPRESENTATION OF ACTIVE AND PASSIVE PRESSURES. The Rankine coefficient K in Eq. (22-21) is easily derived with the aid of a Mohr circle, shown to the left in Fig. 22-12. Since the distance from the origin to the center of the circle is $(\sigma_1 + \sigma_3)/2$ and the radius of the circle is $(\sigma_1 - \sigma_3)/2$, if $c = 0$,

$$\sin\phi = \frac{\sigma_1 - \sigma_3}{\sigma_1 + \sigma_3}$$

$$\sigma_1 \sin\phi + \sigma_3 \sin\phi = \sigma_1 - \sigma_3$$

$$\sigma_1 - \sigma_1 \sin\phi = \sigma_3 + \sigma_3 \sin\phi$$

$$\sigma_1(1 - \sin\phi) = \sigma_3(1 - \sin\phi)$$

$$\frac{\sigma_3}{\sigma_1} = \frac{1 - \sin\phi}{1 + \sin\phi} = K \tag{22-21}$$

The contrast between active and passive states is shown by two circles tangent to the failure envelope in Fig. 22-12. In the active case σ_h is less than σ_v and therefore represents the minor principal stress. If the wall is pushed against the soil, σ_h increases until it is sufficiently large to cause passive failure, shown by the second circle in which σ_h is the major principal stress. In this case

$$K_p = \frac{\sigma_h}{\sigma_v} = \frac{\sigma_1}{\sigma_3}$$

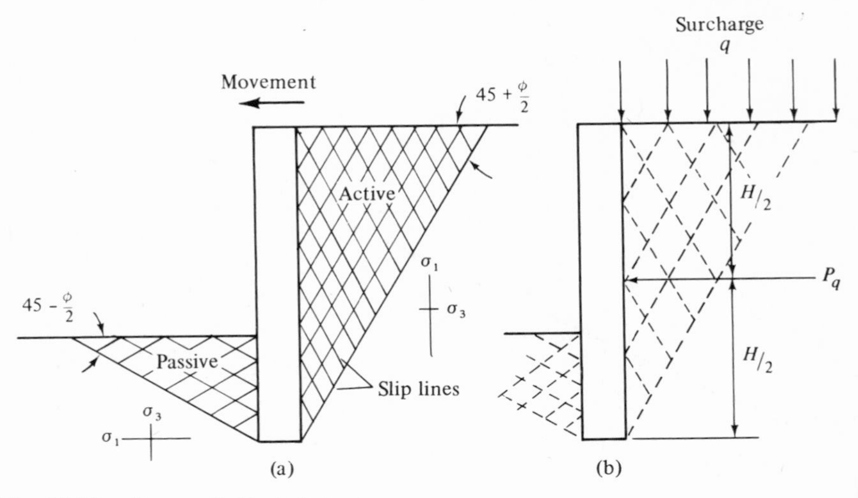

Fig. 22-13. Zones of plastic behavior.

which, comparing to Eq. (22-21), is the reciprocal of K.

If the wall is vertical, the respective σ_h points in Fig. 22-12 are origins of planes. Thus in the active case the slip planes are $45° + (\phi/2)$ with horizontal, which is identical to the angle obtained from the wedge analysis, Eq. (22-8). In the passive case the angle is $45° - (\phi/2)$, identical to that obtained in Eq. (22-17).

The Mohr circle representation is important because it shows that soil in either the active or passive states is on the verge of failure throughout the sliding wedges, as shown in Fig. 22-13(a). The wedges have a theoretically infinite number of parallel slip planes. If there is no volume change such a condition is referred to as an *ideal plastic state*. In contrast to an ideal elastic state, in a plastic state stress is not proportional to strain, but is such as to give a constant stress ratio equal to K. The stress ratio cannot exceed K because the soil would fail first; whereas if the stress ratio is less than K, the soil is no longer at failure, i.e., is no longer plastic.

22.20. EFFECT OF SURCHARGE LOADS. Retaining walls often are called upon to support surface loads on soil behind the wall in addition to forces induced by the weight of the soil itself. Experimental data indicate that pressures from soil resting against retaining walls are in close accord with those predicted from plastic theory. This is not true, however, for surcharge loads placed on top of the soil behind the wall. One reason may be that such loads are usually applied later, after the soil has had the opportunity to regain cohesion and is no longer in a state of incipient shear.

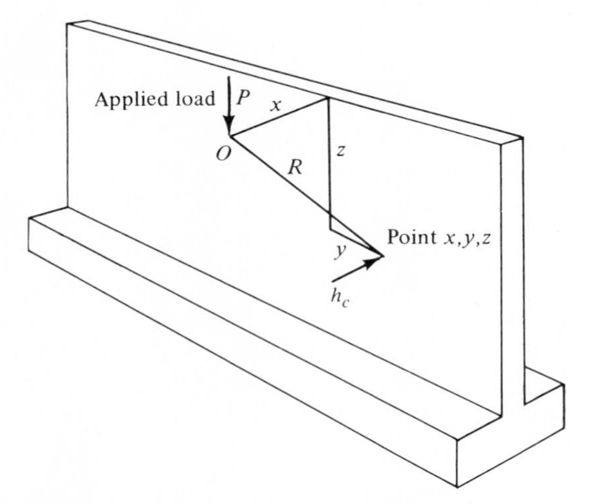

Fig. 22-14. Concentrated load on a surface of backfill at distance x from back of retaining wall.

Nevertheless plastic analysis is quite simple for surcharge loads extending from the face of the wall beyond the limit of the sliding wedge, as shown in Fig. 22-13(b). In this case if the soil behaves plastically the horizontal pressure is qK, where q is the vertical surcharge pressure. Since pressure qK is assumed to act upon the full height of the wall, the resultant is

$$P_q = qHK \tag{22-22}$$

acting at the mid-height of the wall. This is added to P or P_p from the weight of the soil, Eq. (22-9) or (22-18), whichever applies. As previously discussed, P or P_p act at the one-third height of the wall.

A much closer approximation to the actual stresses is obtained by assuming the soil behaves elastically—that is, stress is proportional to strain and the soil is no longer in plastic failure. Then the magnitude and distribution of lateral pressures against a retaining wall caused by surcharge loads applied at or near the surface of the backfill may be estimated by means of formulas which are modifications of the Boussinesq solution for distribution of stresses in an elastic medium due to a concentrated surface load. Experimental evidence indicates that the unit pressure on a relatively rigid concrete retaining wall due to a truck-wheel load applied at the surface of the backfill, as indicated in Fig. 22-14, may be computed by the formula

$$h_c = P\frac{x^2 z}{R^5} \tag{22-23}$$

in which

h_c = horizontal unit pressure at any point on the wall, in pounds per square foot;
P = applied truck-wheel load, in pounds;
x = horizontal distance from wall to load, in feet;
y = lateral distance from point to load, in feet;
z = vertical distance from point to load, in feet; and
$R = \sqrt{x^2 + y^2 + z^2}$.

Equation (22-23) gives a value of h_c which is approximately twice the pressure indicated by the classical Boussinesq formula, or Eq. (17-4). This increase in pressure is caused by the fact that a concrete retaining wall is

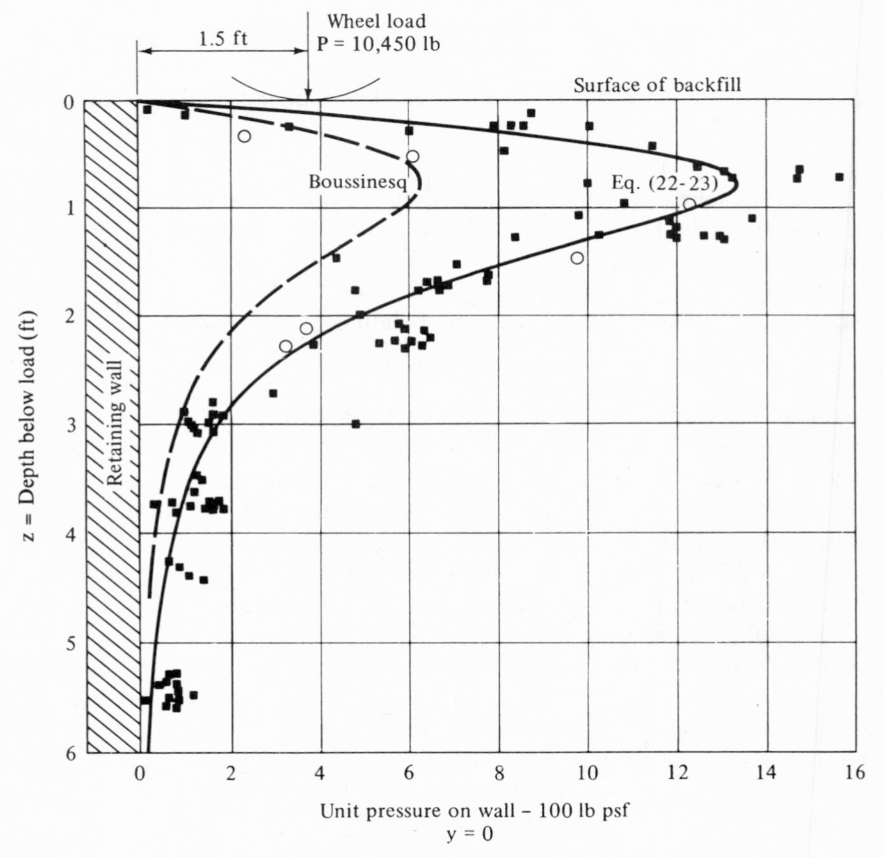

Fig. 22-15. Measured pressures on retaining wall caused by wheel load on backfill surface. Key: ■, Spangler (*4*); ○, Gerber (*2*).

very rigid in comparison with a vertical plane through the soil that would be considered if the backfill were continuous and not interrupted by a rigid wall. If the retaining structure is quite yielding (as yielding as the soil itself would be if a wall were not present), the pressure on the wall would be equal to the Boussinesq value or approximately one-half that indicated by Eq. (22-23). If the wall is of intermediate stiffness, then the pressure will be somewhere between these two limits. A comparison of experimental data with these two relationships is shown in Fig. 22-15, which supports use of the upper limit.

Experimental evidence also indicates that it is feasible to integrate Eq. (22-23) in the x direction and y direction between appropriate limits to obtain the lateral pressure at a point on a wall caused by a surcharge load which is uniformly distributed on a narrow strip or over an area on the backfill surface.

22.21. PARALLEL LINE OR NARROW STRIP LOAD OF FINITE LENGTH. If the surcharge load is applied uniformly along a narrow strip or line which is parallel to the retaining wall and of finite length, as indicated in Fig. 22-16, the expression in Eq. (22-23) may be integrated in the y direction. The horizontal unit pressure at a point at one end of the load, as in Fig. 22-16(a), may be found as follows:

$$h_s = p_s \int_0^{y_1} \frac{x^2 z}{R^5}\, dy \tag{22-24}$$

$$h_s = p_s \frac{x^2 z}{3R_1^4}\left[\frac{R_1^2 y_1}{(R_1^2 + y_1^2)^{1.5}} + \frac{2y_1}{(R_1^2 + y_1^2)^{0.5}}\right] \tag{22-25}$$

in which

h_s = horizontal unit pressure on the wall at any depth z opposite one end of a parallel strip load on the backfill surface;
p_s = load per unit length of strip;
x = distance from wall to strip load;
y_1 = length of strip load;
R_1 = $\sqrt{x^2 + z^2}$; and
3 = approximation for π.

Maximum pressures on the wall due to a strip load of finite length will occur at points directly opposite the center of the length. In Fig. 22-16(b), $y_1 = y_1'$. To calculate these maximum pressures, let y_1 in Eq. (22-25) equal one-half the length of the strip load, and multiply the result

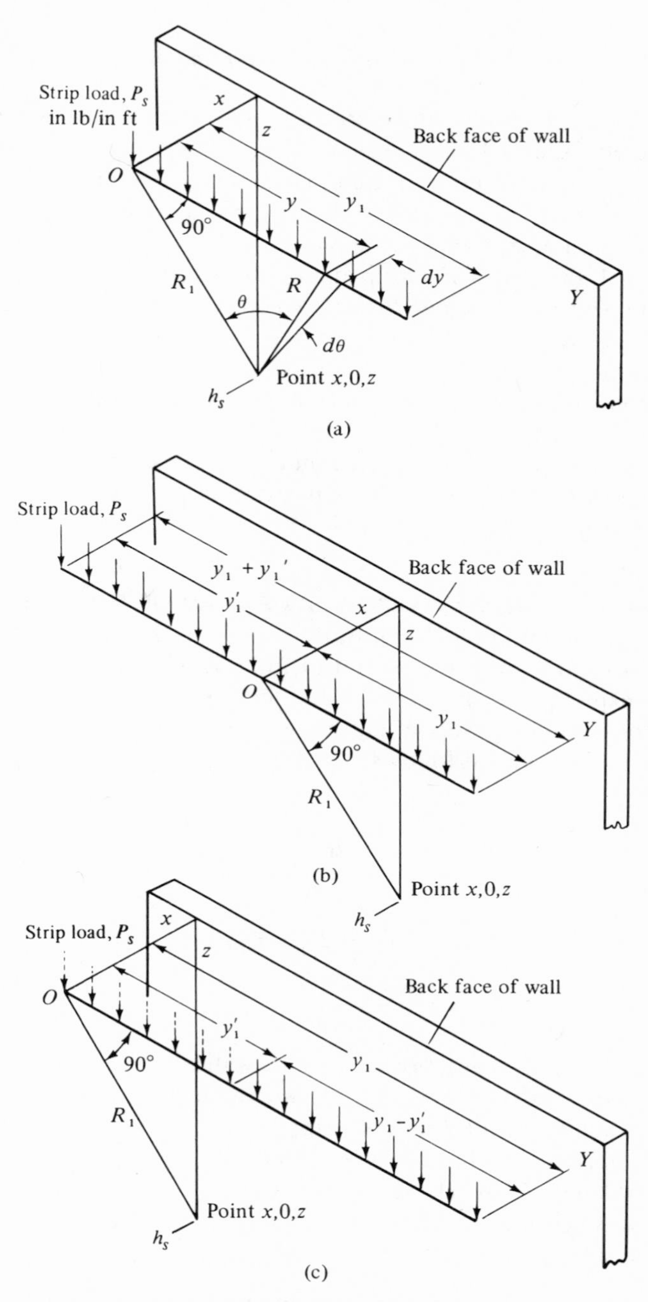

Fig. 22-16. Strip load of finite length parallel to wall. (a) Point at end of load; (b) intermediate point; (c) point beyond end of load.

by 2. Pressure at a point which is not opposite the center of the length may be calculated in two steps. First, let y_1 equal the distance from the point to one end of the load; then let y_1' equal the distance to the other end. The results of these calculations should be added together to obtain the pressure due to the complete strip load.

Similarly, for a point on the wall which lies beyond the end of the strip load, as in Fig. 22-16(c), imagine that the load is extended to a point opposite the point in question. Calculate the pressure due to the total strip load, including the imaginary extension. Next, calculate the pressure due to that extension alone, and subtract this value from the total-load pressure.

EXAMPLE 22-1. It is desired to estimate the horizontal pressure on a retaining wall at a depth of 3 ft, caused by a strip load of 2000 lb per lin/ft which is 8 ft long and which rests on the backfill along a line 2 ft back of the wall. The pressure is desired (a) opposite the center of the load; (b) opposite a point within the load and 2 ft from one end; (c) opposite an end of the load; and (d) opposite a point 2 ft beyond one end of the load.

SOLUTION. (a) By Eq. (22-25) in which $p_s = 2000$, $x = 2$, $z = 3$, $R_1^2 = 13$, $R_1^4 = 169$, and $y_1 = 4$, we obtain

$$h_s = 2000 \times \frac{4 \times 3}{3 \times 169}\left[\frac{13 \times 4}{(13 + 16)^{1.5}} + \frac{2 \times 4}{(13 + 16)^{0.5}}\right] = 86.2 \text{ psf}$$

The required pressure is $2 \times 86.2 = 172.4$ psf

(b) In this case, $y_1 = 6$ and $y_1' = 2$. The partial pressures are as follows:

$$\begin{aligned} &\text{for} \quad y_1 = 6, \qquad h_s = 92.0 \text{ psf} \\ &\text{for} \quad y_1 = 2, \qquad h_s = 63.6 \text{ psf} \end{aligned}$$

The total pressure is $92.0 + 63.6 = 155.6$ psf

(c) When $y_1 = 8$, $h_s = 93.6$ psf.

(d) Here, let $y_1 = 10$ and $y_1' = 2$. The corresponding pressures are

$$\begin{aligned} &\text{for} \quad y_1 = 10, \qquad h_s = 94.4 \text{ psf} \\ &\text{for} \quad y_1 = 2, \qquad h_s = 63.6 \text{ psf} \end{aligned}$$

The desired pressure is

$$94.4 - 63.6 = 30.8 \text{ psf}$$

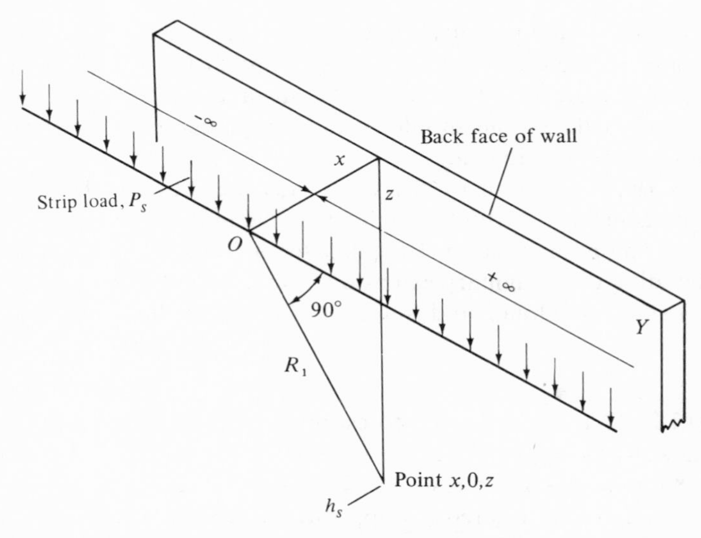

Fig. 22-17. Strip load of infinite length parallel to wall.

22.22. PARALLEL STRIP LOAD OF INFINITE LENGTH. If a parallel strip load is very long, it may be considered to extend from any point on the wall to $+\infty$ and $-\infty$, as illustrated in Fig. 22-17. For this case, the horizontal pressure at the point may be found by the relationship

$$h_s = p_s \int_{-\infty}^{+\infty} \frac{x^2 z}{R^5}\, dy \tag{22-26}$$

which yields

$$h_s = \frac{4p_s}{3}\frac{x^2 z}{R_1^4} \tag{22-27}$$

The letters have the same meanings as in Eq. (22-25).

22.23. AREA LOAD OF FINITE LENGTH. If a surcharge load is applied to a backfill uniformly over a rectangular area whose sides are parallel and perpendicular to the back face of the retaining wall, as shown in Fig. 22-18, the horizontal unit pressure at a point may be found by integrating the expression in Eq. (22-23) opposite an end of the loaded area, as in Fig. 22-18(a), in both the x direction and the y direction. For a point

$$h_a = p_a \int_{x_1}^{x_2} \int_0^{y_1} \frac{x^2 z}{R^5}\, dy\, dx \tag{22-28}$$

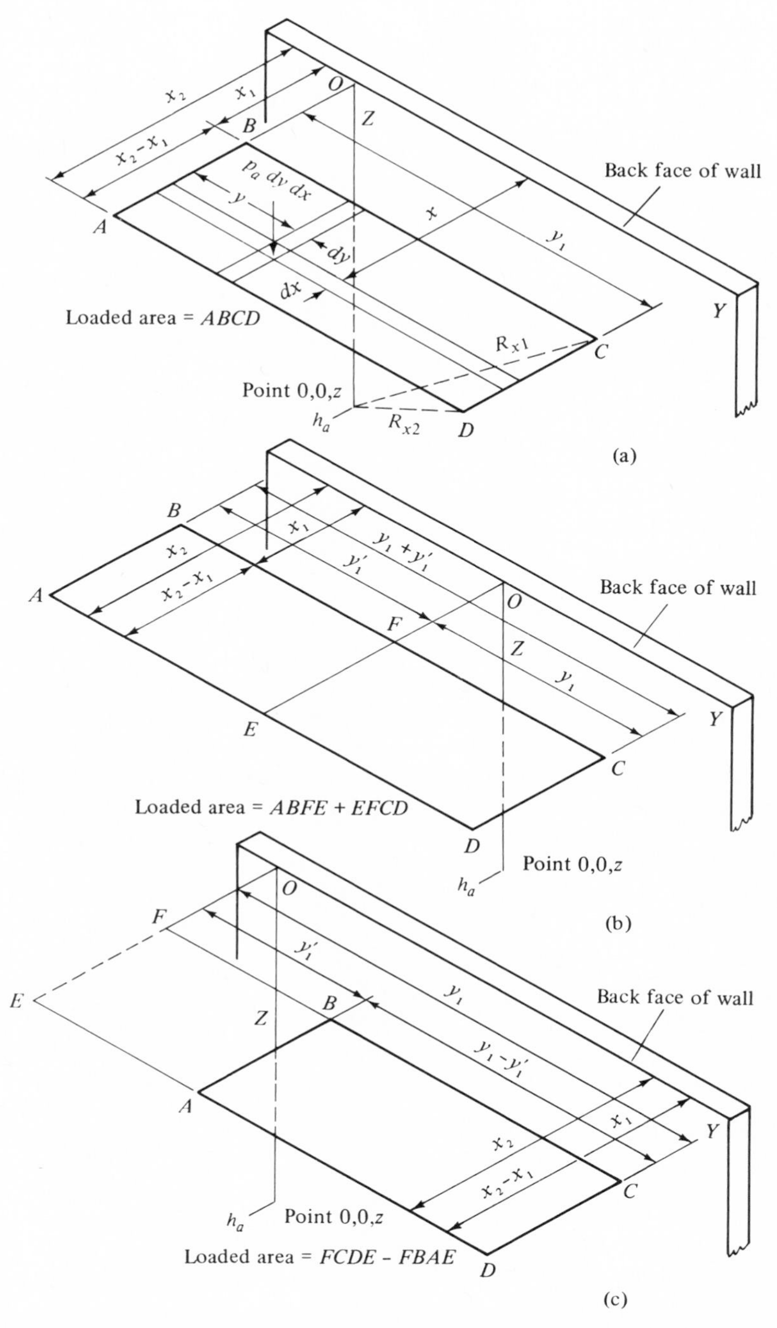

Fig. 22-18. Area load of finite length. (a) Point at end of load; (b) intermediate point; (c) point beyond end of load.

or

$$h_a = \frac{p_a}{3}\left[\arctan\frac{x_2 y_1}{z R_{x2}} - \frac{x_2 y_1 z}{(x_2^2 + z^2) R_{x2}} - \arctan\frac{x_1 y_1}{z R_{x1}} + \frac{x_1 y_1 z}{(x_1^2 + z^2) R_{x1}}\right] \tag{22-29}$$

in which

h_a = horizontal unit pressure on a retaining wall at any depth z opposite one end of a uniformly distributed area load on the backfill surface;
p_a = surcharge load per unit area;
x_1 = distance from wall to near side of load;
x_2 = distance from wall to far side of load;
$x_2 - x_1$ = width of load in direction normal to wall;
y_1 = length of loaded area;
R_{x1} = $\sqrt{(x_1^2 + y_1^2 + z^2)}$;
R_{x2} = $\sqrt{(x_2^2 + y_1^2 + z^2)}$; and
3 = approximation for π.

Again, as in the case of a parallel strip load, maximum pressures on the wall will occur at points opposite the center of an area load of finite length. In Fig. 22-18(b), $y_1 = y_1'$. To calculate these maximum pressures, let y_1 in Eq. (22-29) equal one-half the length of the area load, and multiply the result by 2. Pressure at a point which is not opposite the center of the load may be calculated in two steps. First let y_1 equal the distance from the point to one end of the load; then let y_1' equal the distance to the other end. These results should be added together to obtain the pressure due to the whole area load.

Similarly, for a point on the wall which lies beyond the end of the area load, as in Fig. 22-18(c), imagine that the load is extended to a point opposite the point in question. Calculate the pressure due to the total area load, including the imaginary extension. Next, calculate the pressure due to the extension alone, and subtract this value from the total-load pressure.

EXAMPLE 22-2. Assume that a railroad track rests on the backfill behing a retaining wall 20 ft high and the distance from the wall to the center of track is 8 ft. Other data are

γ = unit weight of soil = 120 pcf;
ϕ = friction angle of soil = 30°;
length of railroad ties = 8 ft;
loads from Cooper's E-72 locomotive: four driver axles at 5 ft centers, each weighing 72,000 lb;

impact = 75%; and
weight of track structure = 200 plf.

It is required to estimate the total lateral pressure on the wall due to both the soil backfill and the railroad traffic and to determine the position of the line of action of the resultant pressure.

SOLUTION. The earth pressure at the base of the wall is found by Eq. (22-1). From Table 22-2, $K = 0.33$. Hence,

$$p = 120 \times 20 \times 0.33 = 800 \text{ psf}$$

The distribution of the earth pressure is represented in Fig. 22-19 by line *A*.

To find the track load, assume that the rails and the ties distribute the weight of a locomotive in such a manner that it may be considered as a uniform load over an area equal to the length of ties multiplied by the length occupied by the driver axles. This area is 8 × 20 = 160 sq ft. The total load on it is as follows:

Drivers, 4 × 72,000	= 288,000 lb
Impact, 0.75 × 288,000	= 216,000 lb
Track, 20 × 200	= 4,000 lb
Total	= 508,000 lb

The load per unit area is 508,000 ÷ 160 = 3180 psf.

In Eq. (22-29), $p_a = 3180$, $x_1 = 4$, $x_2 = 12$, and $y_1 = 10$. The values of the horizontal pressures at various depths due to the track and traffic loads are listed in Table 22-6. Thus, the pressure at a depth of 4 ft is found as follows:

$$z = 4, \qquad R_{x1} = 11.48, \qquad \text{and} \qquad R_{x2} = 16.12$$

$$h_a = \frac{3180}{3}\left[\arctan\frac{12 \times 10}{4 \times 16.2} - \frac{12 \times 10 \times 4}{(144 + 16)16.12} - \arctan\frac{4 \times 10}{4 \times 11.48} + \frac{4 \times 10 \times 4}{(16 + 16)11.48}\right]$$

$$= 1060[1.077 - 0.186 - 0.716 + 0.436] = 647.7 \text{ psf}$$

$$2 \times 647.7 = 1295 \text{ psf}$$

The distribution of the pressure due to the track and traffic loads is represented in Fig. 22-19 by curve *B*.

To obtain the total pressure, add the earth pressure to the track-load pressure. The results are shown in Table 22-6 and are plotted in Fig. 22-19 as curve *C*. The area enclosed by curve *C* is the total pressure on the wall

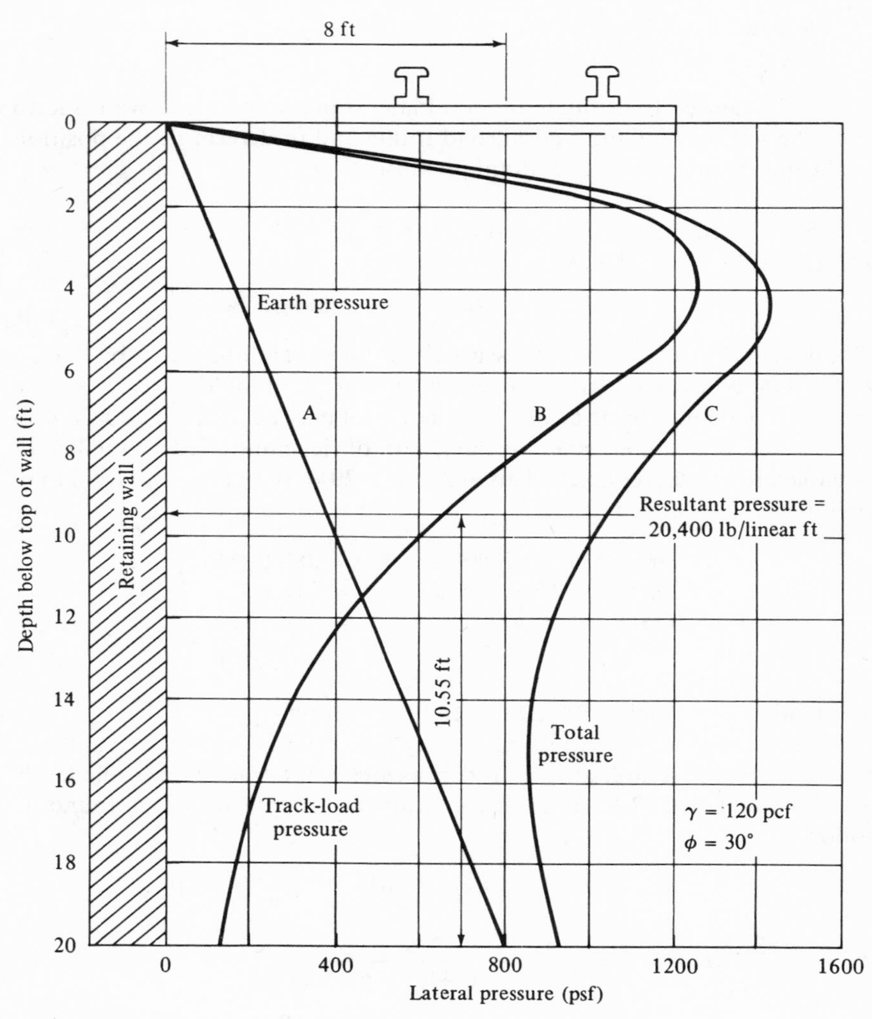

Fig. 22.19. Pressures in Example 22-2.

per foot of length. The resultant pressure on the wall acts through the centroid of this area.

22.24. AREA LOAD OF INFINITE LENGTH. If the area load is very long in the direction parallel to the wall, it may be considered to extend from any point on the wall to $+\infty$ and $-\infty$, as illustrated in Fig. 22-20. The horizontal pressure at a point due to this type of load may be calculated by integrating the expression in Eq. (22-27) in the x direction between the limits

TABLE 22-6. Pressures on Retaining Wall in Example 22-2

Depth Below Top of Wall (ft)	*Lateral Pressure (psf)*		
	Track Load	*Earth Load*	*Combined Load*
0	0	0	0
1	636	40	676
2	1089	80	1169
3	1288	120	1408
4	1295	160	1455
5	1216	200	1416
6	1085	240	1325
7	947	280	1227
8	814	320	1134
9	700	360	1060
10	588	400	988
11	496	440	936
12	420	480	900
13	358	520	878
14	299	560	859
15	250	600	850
16	214	640	854
17	180	680	860
18	155	720	875
19	130	760	890
20	117	800	917

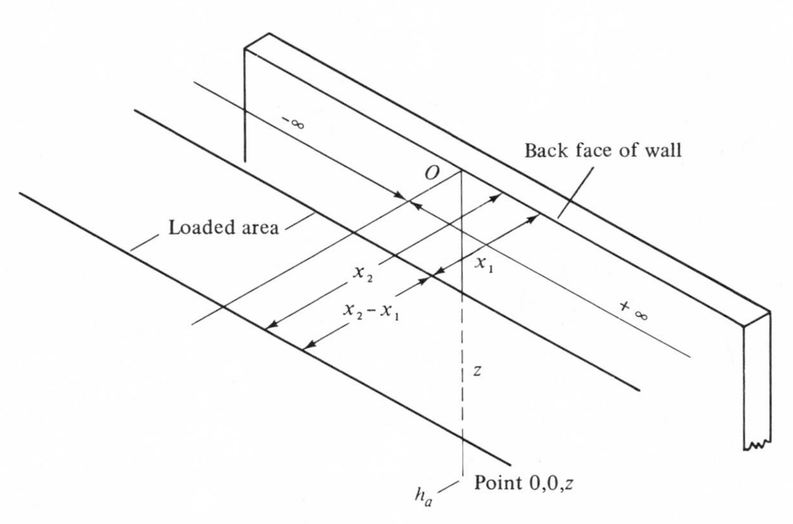

Fig. 22-20. Area load of infinite length.

x_1 and x_2. Thus,

$$h_a = \frac{4p_a}{3} \int_{x_1}^{x_2} \frac{x^2 z}{R_1^4}\, dx \tag{22-30}$$

which yields

$$h_a = \frac{2p_a}{3} \left[\arctan \frac{x}{z} - \frac{xz}{(x^2 + z^2)} \right]_{x_1}^{x_2} \tag{22-31}$$

22.25. STRIP LOAD OF FINITE LENGTH PERPENDICULAR TO WALL. The horizontal pressure on a wall due to a narrow strip load oriented perpendicular to the wall, as illustrated in Fig. 22-21, can be investigated by

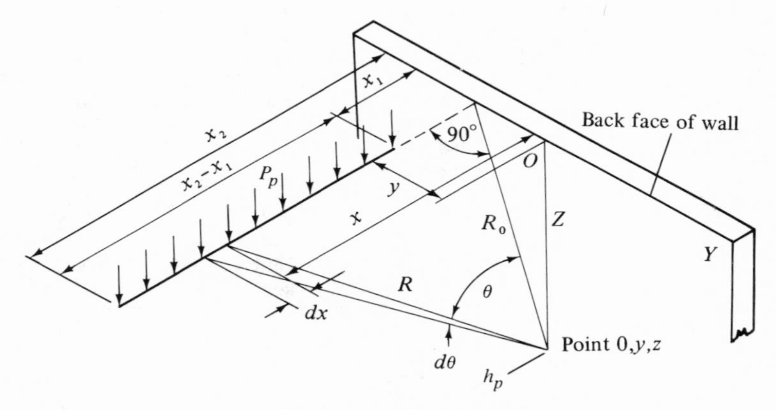

Fig. 22-21. Strip load of finite length perpendicular to wall.

integrating the expression in Eq. (22-23) in the x direction. Thus,

$$h_p = p_p z \int_{x_1}^{x_2} \frac{x^2}{R^5}\, dx \tag{22-32}$$

or

$$h_p = \frac{p_p z}{R_0^2} \left[\frac{\sin^3 \theta}{3} \right]_{\arctan(x_1/R_0)}^{\arctan(x_2/R_0)} \tag{22-33}$$

in which

h_p = horizontal pressure at any depth z on a wall due to a perpendicular strip load;

p_p = strip load per unit of length; and

x_1 = distance from wall to near end of load;

x_2 = distance from wall to far end of load;
$R_0 = \sqrt{y^2 + z^2}$;
$\theta = \arctan \frac{x}{R_0}$; and
3 = approximation for π.

22.26. PERPENDICULAR STRIP LOAD OF INFINITE LENGTH. When a strip load extends from the wall to a great distance, as shown in Fig. 22-22, x_1 = 0 and $x_2 = \infty$, and Eq. (22-33) becomes

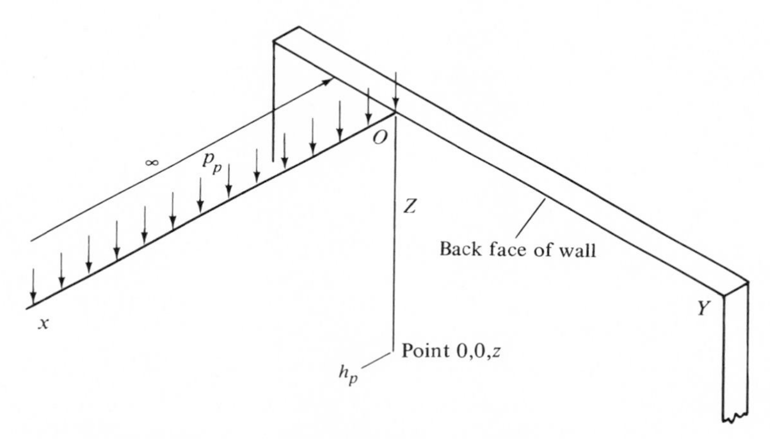

Fig. 22-22. Strip load of infinite length perpendicular to wall.

$$h_p = p_p \frac{z}{3R_0^2} \tag{22-34}$$

The maximum pressures on the wall will occur at various values of z in the vertical element opposite the strip load, that is, when $y = 0$. For this case, $R_0^2 = z^2$ and Eq. (22-34) reduces to

$$h_p = p_p \frac{1}{3z} \tag{22-35}$$

22.27. ANCHORED BULKHEAD. A type of restraining structure which is widely used, particularly in connection with water-front construction, is the sheet pile anchored bulkhead, represented in Fig. 22-23. It consists of interlocking steel sheet piling driven into the soil and anchored near the top by steel tie-rods or cables carried back into the soil and made fast to anchor piles or deadmen placed in firm soil. The fill behind the bulkhead tends to push it forward, and it is restrained by the anchors near the top

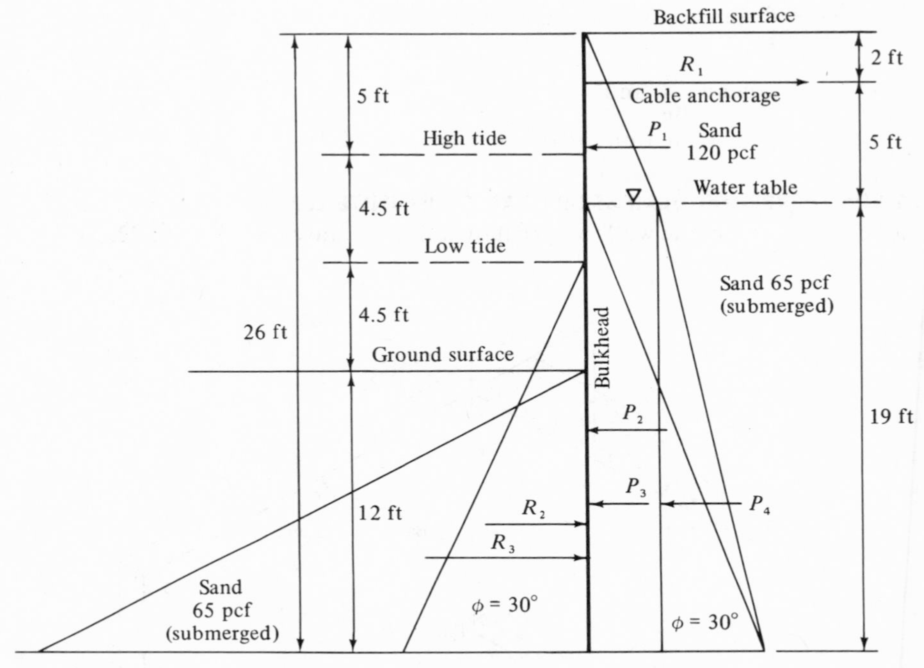

Fig. 22-23. Forces acting on an anchored bulkhead. *Actuating forces:* P_1 = active pressure of soil above water table; P_2 and P_4 = active pressure of soil below water table; P_3 = pressure of water below water table. *Resisting forces:* R_1 = tension in cable anchorage; R_2 = pressure of water below low tide; R_3 = passive resistance pressure of soil in front of bulkhead.

and by the mass of soil in front of the sheet piling below the ground surface. There is a tendency for the bottom of a retaining wall of this type to push forward and displace a wedge-shaped mass of soil immediately in front of it. The soil resists this tendency for displacement by virtue of its ability to exert passive resistance pressure.

In Fig. 22-23 are shown the actuating and resisting forces on a typical anchored bulkhead at a site adjacent to tidewater. The actuating forces consist of the active pressures of the soil and water back of the wall. The resisting forces consist of the tension in the rod or cable anchorage, the water pressure on the front side, and the passive resistance pressure of the soil within the penetration depth of the sheet piling. The stability factor of safety is the ratio of the total resisting force to the total actuating force. Obviously, the factor of safety of a bulkhead can be increased by increasing the depth of penetration of the sheet piling. The student is cautioned to remember that the resisting force R_3 is a passive force and only develops to the extent to which it is necessary to resist the active pressures which tend to push the bulkhead toward the soil.

EXAMPLE 22-3. Design the sheet pile bulkhead shown in Fig. 22-23, having a stability factor of safety of 1.25.

SOLUTION.

Actuating Forces × Factor of Safety
(All forces in lb/per lin ft of bulkhead)

$$P_1 = \tfrac{1}{2} \times 120 \times 7^2 \times 0.33 \times 1.25 = 1{,}225 \text{ lb}$$
$$P_2 = 19 \times 120 \times 7 \times 0.33 \times 1.25 = 6{,}650$$
$$P_3 = \tfrac{1}{2} \times 62.4 \times 19^2 \times 1 \times 1.25 = 14{,}160$$
$$P_4 = \tfrac{1}{2} \times 65 \times 19^2 \times 0.33 \times 1.25 = 4{,}850$$

Total actuating force 26,885 lb

Resisting Forces

$$R_2 = \tfrac{1}{2} \times 62.4 \times 16.5^2 \times 1.0 = 8{,}500 \text{ lb}$$

To determine R_3, write moments about R_1:

$$P_1 = 2.67 \times 1225 = +\ 3{,}270 \text{ ft lb}$$
$$P_2 = 14.5 \times 6650 = +\ 96{,}400$$
$$P_3 = 17.67 \times 14{,}160 = +251{,}000$$
$$P_4 = 17.67 \times 4850 = +\ 85{,}700$$

+436,370 ft lb

$$R_2 = 18.5 \times 8500 = -157{,}300$$

net moment +279,070 ft lb

$$R_3 = \frac{279{,}070}{20} = 13{,}950 \text{ lb}$$

Passive resistance capability of soil penetrated by sheet piling:

$$R_3 = \tfrac{1}{2} \times 65 \times 12^2 \times 3 = 14{,}050 \text{ lb}$$

14,050 is greater than 13,950; therefore penetration is satisfactory.
To determine R_1, $\Sigma H = 0$

$$R_1 = 1225 + 6650 + 14{,}160 + 4850 - 8500 - 13{,}950 = 4435 \text{ lb.}$$

22.28. REINFORCED EARTH. A novel method for horizontally reinforcing granular soil by means of corrugated steel or aluminum friction strips was developed in France by H. Vidal (*9*) and has been used for a number of high and difficult retaining walls. Since the usual soil loads act to give

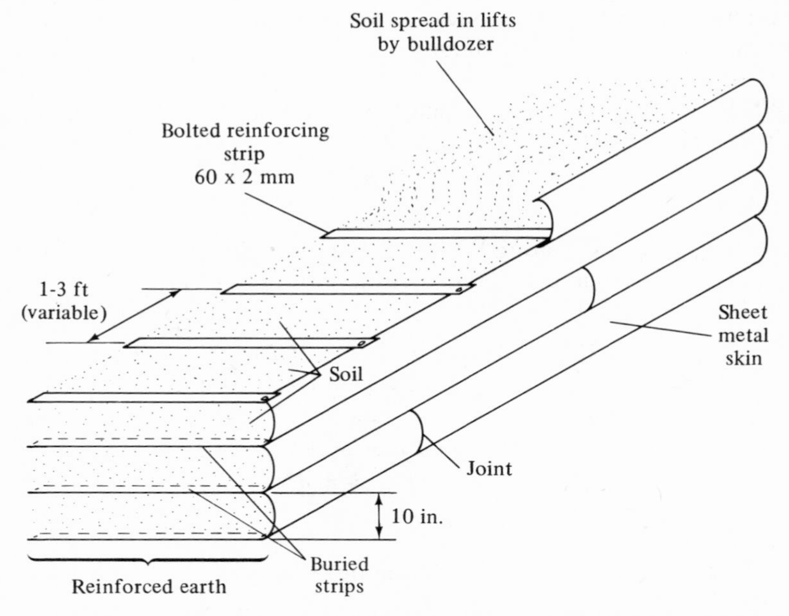

Fig. 22-24. Retaining wall by the reinforced earth method. Heights of 60 ft have been attained.

vertical compression, metal strips strung horizontally at intervals in the soil are brought into tension and resist horizontal expansion. For retaining walls, ends of the strips are bolted to horizontal sheet metal wall elements during assembly and filling (Fig. 22-24). The method differs from anchored bulkheads in that anchoring is by friction of backfill soil on the metal strips.

Model tests indicate that the length of the metal strips in reinforced earth should be at least 80% of the wall height to attain sufficient friction on the strips, and the method does not appear to be easily applicable to cohesive soils with low internal friction. Minimum tensile strength of the strips is calculated from the active coefficient of earth pressure, K. A simple 1:10 scale model to illustrate the principles can be constructed with paper, tape, and sand in the geometry shown in Fig. 22-24.

PROBLEMS

22.1. Define active lateral pressure and passive resistance pressure. Explain in detail the differences between them.

22.2. What is meant by earth pressure at rest?

22.3. State the basic philosophy underlying the Rankine and the Coulomb theories of lateral earth pressure.

22.4. A concrete retaining wall 24 ft high holds in place a backfill of gravel. The surface of the backfill is level with the top of the wall. Determine the unit pressure at the base of the wall and at its mid-height. What is the total resultant pressure on a section of the wall 1 ft long? Calculate the overturning moment about the base of the wall. Assume γ = 130 pcf, ϕ = 35°.

22.5. The space behind a concrete retaining wall 16 ft high is backfilled with compact loam soil. The surface of the backfill is level with the top of the wall. Determine the resultant pressure on a 1 ft length of the wall by Coulomb's graphical method.

22.6. Determine the resultant pressure on the retaining wall of Problem 22.5 by means of the logarithmic-spiral graphical method.

22.7. The space behind a concrete retaining wall 14 ft high is backfilled with a soil which weighs 120 pcf and has an angle of friction ϕ equal to 25°. The back face of the wall is battered 1 in. in 12 in., and the backfill slopes upward from the top of the wall on a 6:1 slope. Assume that the angle of friction between the soil and the wall is ϕ' = 20°. Calculate the resultant pressure on a 1 ft length of the wall by means of Eq. (22-11), and compare the result with that obtained by means of Tables 22-3 and 22-4.

22.8. A temporary cut is to be made through a cohesive soil which weighs 100 pcf and has an unconfined compressive strength of 325 psf. If the cut is 20 ft deep and the sides are vertical, estimate the total pressure, in pounds per linear foot, on timber sheeting used to restrain the cut faces. What will be the approximate location of the resultant pressure?

22.9. The space behind a retaining wall 10 ft high is backfilled with cohesionless sand, which will be saturated for the lower 6 ft of the wall. Assume that the angle ϕ for the sand is 22°, the unit weight above the line of saturation is 125 pcf, and the buoyant unit weight is 70 pcf. Determine the total lateral pressure on the wall per foot of length, and draw a diagram representing the distribution of pressure from top to bottom of the wall.

22.10. A retaining wall 26 ft high holds in place a level backfill of soil which weighs 115 pcf and has an angle ϕ equal to 20°. A continuous building footing 6 ft wide is to be constructed on the surface of the backfill at a clear distance of 10 ft from the back of the wall. The load on this footing will be 2 tons/sq ft. Compute the total pressure on a 1-ft section of the wall due to both the backfill and the footing; and determine the position of the resultant pressure.

22.11. A concentrated load of 20,000 lb is applied to the backfill surface at a distance of 6 ft back from a 10 ft high retaining wall. Calculate the magnitude and distribution of lateral pressure caused by the superimposed load on a vertical element of the wall which is 1 ft long and directly opposite the load.

22.12. A sheet-pile bulkhead is to be constructed as shown in Fig. 22-25. Estimate the factor of safety of the structure and the stress in the anchorage per lineal foot of bulkhead. (Note: Consider water on both sides of bulkhead balanced without reference to factor of safety.)

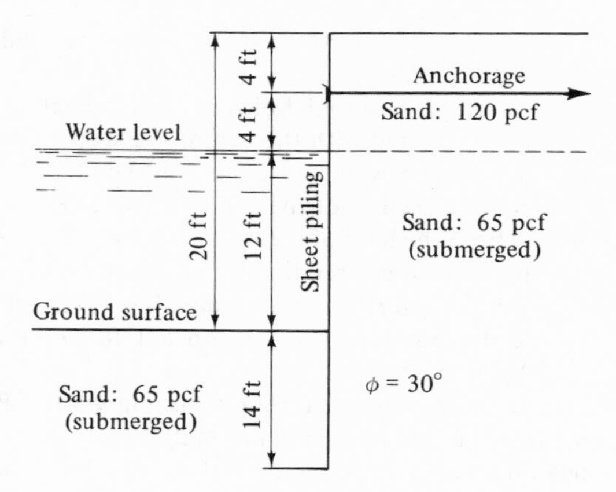

Fig. 22-25. Conditions for problem 22.12.

REFERENCES

1. Feld, Jacob. "Abutments for Small Highway Bridges," *Proc. Highway Research Board* **23** (1943).
2. Gerber, Emil. "Untersuchungen über die Druckverteilung im örtlich belasteten sand." Diss. A.-G. Gebr. Leemann. Zurich. 1929.
3. Mindlin, Raymond D. Discussion: "Pressure Distribution on Retaining Walls." *Proc. Intern. Conf. Soil Mechanics and Foundation Eng.* **3,** 155 (1936).
4. Spangler, M. G. "Horizontal Pressures on Retaining Walls Due to Concentrated Surface Loads." Bul. 140, Iowa Engineering Experiment Station, Ames, Iowa. 1938.
5. Spangler, M. G., and Jack L. Mickle. "Lateral Pressures on Retaining Walls due to Backfill Surface Loads." *Highway Research Board* **141** (1956).
6. Terzaghi, Karl. "General Wedge Theory of Earth Pressures." *Trans. Am. Soc. Civil Eng.* **106** (1941).
7. Terzaghi, Karl, and Ralph B. Peck. *Soil Mechanics in Engineering Practice,* 2nd ed. John Wiley, New York, 1967.
8. Tschebotarioff, Gregory P. *Soil Mechanics, Foundations and Earth Structures.* McGraw-Hill, New York, 1952.
9. Vidal, H. "La Terre Armée." *Ann. Institut Technique du Bâtiment et des Travaux Publics, Ser.: Materials* **38,** No. 259–260 (July/August 1969).

23

Bearing Capacity

23.1. FOUNDATION BED. The term *foundation* will here be used to refer to the lowest part of a structure. The function of a foundation is to transmit the weight of a structure, together with the effects of live loads and wind loads, to the material on which the structure rests in such a manner that the underlying material is not stressed beyond its safe bearing capacity. If the underlying material is solid rock, the design of a foundation is greatly simplified, since the bearing capacity of rock in natural beds is relatively very great and the area of contact between the foundation structure and the rock can be held to a minimum. Unfortunately, however, only a small percentage of structures can be founded on solid rock because of the great depth to ledge at most sites, and a foundation usually must be designed to rest on the soil deposit between the structure and ledge rock. The decision whether to found a structure on rock or to design the foundation to rest on the soil is largely dependent on the relative cost of the two alternatives, although a foundation on rock is worth considerably more and should be chosen if the increase in cost is not excessive. The material on which a foundation rests will here be called the *foundation bed.*

23.2. BEARING CAPACITY OF SOIL. Ultimate bearing capacity of a foundation bed is synonymous with the terms ultimate bearing value and ultimate soil pressure. It may be defined as the maximum unit pressure to which a foundation bed may be subjected without permitting a structure to settle excessively or in a manner which is detrimental to its function or structural integrity. The safe bearing capacity, or allowable bearing value, of a foundation bed is this maximum unit pressure divided by an appropriate factor of safety. Factors of safety in foundation design usually range from 3 to 5 or more. The value in any case depends on the importance of the structure and its character with reference to its ability to resist differential settlement, that is, the ability of the structure to withstand settlement of one part with respect to another part. The choice of an appropriate factor of safety may also depend on the extent of the designer's knowledge of the soil profile and of the behavior under load of the soil strata beneath the foundation.

23.3. GROSS AND NET BEARING CAPACITY. Bearing capacity is often thought of in terms of a gross value and a net value. Gross bearing capacity is the total unit pressure which can be supported at the base of the foundation. Net bearing capacity is the gross bearing capacity minus the vertical pressure that is produced on a horizontal plane level with the base by an adjacent surcharge. In other words, net bearing capacity is the total unit pressure minus the weight of the soil between the elevation of the bottom of a footing and the finished grade alongside. If a foundation rests directly on the surface of the ground, the gross bearing capacity and the net bearing capacity are identical.

When the soil surcharge above the bottom of a footing is at different elevations on the two sides, as indicated in Fig. 23-1, the weight of the shallower surcharge is used to determine the net bearing capacity. When a concrete floor or pavement rests on a soil surcharge, it is appropriate to compute an equivalent surcharge which is the height of a column of soil

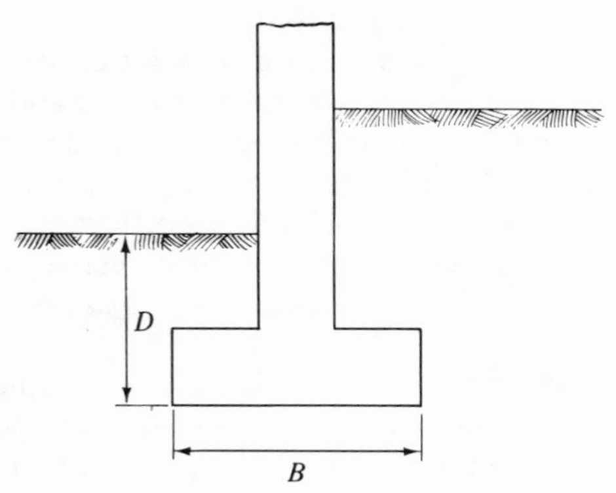

Fig. 23-1. Illustration of soil surcharge.

equal in weight to that of the actual soil and pavement above the bottom of the footing.

23.4. BEARING-CAPACITY REQUIREMENTS. In establishing the safe bearing capacity of a soil on which a foundation rests, it is necessary to consider two separate and distinct types of action by the soil when subjected to load and the influence of these actions on the supported structure. First, the bearing pressure must be low enough to insure that the compression strain in the soil will not be sufficient to cause detrimental settlement of the structure. The amount of settlement which can be tolerated varies widely. It depends on the type of structure, its ridigity or flexibility, its function, and its elevation with reference to adjacent buildings and street grades. A structure can usually tolerate a fair amount of settlement if the settlement is uniform throughout the structure, but differential settlement may cause substantial damage and should be held to a minimum.

A guide to the amount of allowable settlement is shown in Table 23-1. In general the limitations on total settlement derive from the maximum allowable differential settlement, which often is as much as 0.5 to 1.0 times the total settlement [Bjerrum, quoted in Lambe and Whitman (*6*)]. For example, Table 23-1 limits differential settlement to 0.002 L for several typical structures, where L is the column spacing. If L = 20 ft = 240 in., the maximum allowable differential settlement becomes 0.5 in. If we assume that this is one-half the maximum total settlement, the latter becomes

TABLE 23-1. Typical Limiting Settlements (Compiled from U.S.S.R. Building Code and Other Sources)

Limiting Factor or Type of Structure	*Maximum Allowable Settlement*	
	Differential[a]	*Total*
Drainage of floors	0.01–0.02 L	6–12 in.
Stacking, warehouse lift trucks	0.01 L	6 in.
Tilting of smokestacks, silos	0.004 B	3–12 in.
Framed structure, simple	0.005 L	2–4 in.
Framed structure, continuous	0.002 L	1–2 in.
Framed structure with diagonals	0.0015 L	1–2 in.
Reinforced concrete structure	0.002–0.004 L	1–3 in.
Brick walls, one-story	0.001–0.002 L	1–2 in.
Brick walls, high	0.0005–0.001 L	1 in.
Cracking of panel walls	0.003 L	1–2 in.
Cracking of plaster	0.001 L	1 in.
Machine operation, noncritical	0.003 L	1–2 in.
Crane rails	0.003 L	
Machines, critical	0.0002 L	

[a] L is the distance between adjacent columns, B is the width of base.

1.0 in., a figure frequently adopted as a maximum for major buildings. Soil engineers are sometimes asked to design a foundation for zero settlement; this is of course impossible unless the net bearing capacity is taken as zero, i.e., the weight of the structure just equals the weight of soil removed from the excavation, referred to as a floating foundation. Where settlement is critical it also can be minimized to a small fraction of an inch by founding on rock or by use of piles, discussed in the next chapter. The extra cost of such measures must of course be measured in terms of the advantages.

The second criterion for bearing capacity is that the soil will not exhibit shear failure as shown in Fig. 23-2(b) and (c). Such excessive settlement or tipping of a structure caused by shear failure of the supporting soil is often referred to as "breaking into the ground." The failure may occur under an individual footing or may involve an entire building resting on a mat foundation. If the failure occurs under a footing which is prevented from tilting, the failure may be symmetrical, pushing soil to both sides, as in Fig. 23-2(b). If there is no lateral restraint on tilting, the failure is usually one-sided, causing the structure to tilt, as in Fig. 23-2(c). So long as the footing is symmetrically loaded, the analysis of both cases is the same, since soil on the left side in Fig. 23-2(c) also was on the verge of failure when the right side failed, and thus both sides participate in support of the structure up to the instant of failure. The situation is analogous to a plank horizontally suspended by a rope at each end; both ropes support the plank until the load on the plank becomes too much, when ordinarily only one rope breaks. This analogy also serves to illustrate that if the load is not symetrically applied, the side receiving the larger portion of the load will fail, such that the total load at failure will be reduced by as much as one-half compared to the load at failure when loading is at the center. It therefore is essential that a factor of safety of at least 2 be applied for bearing capacity analysis of permanent structures, a value of 3 or more being normally used.

The two types of action just described, i.e., compression strain and shear failure, are nearly independent of each other and should be investigated separately. Obviously the safe bearing capacity finally adopted will be the least of the two results obtained from these studies.

23.5. SETTLEMENT INVESTIGATION. The general principles of foundation settlement due to compression strain or consolidation of the supporting soil were outlined and discussed in Chapter 18. In the case of a costly structure resting on fine-grained soil or beneath which a clay stratum may exist at some depth, a complete study of the consolidation characteristics of the underlying material may be justified. Such a study will serve as a valuable guide to the foundation designer. The first question which must

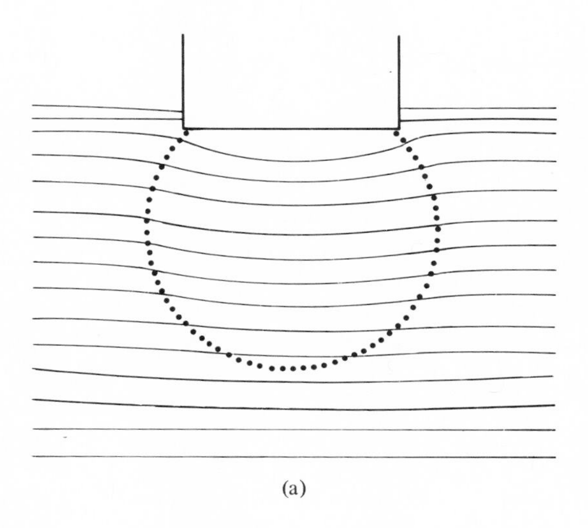

(a)

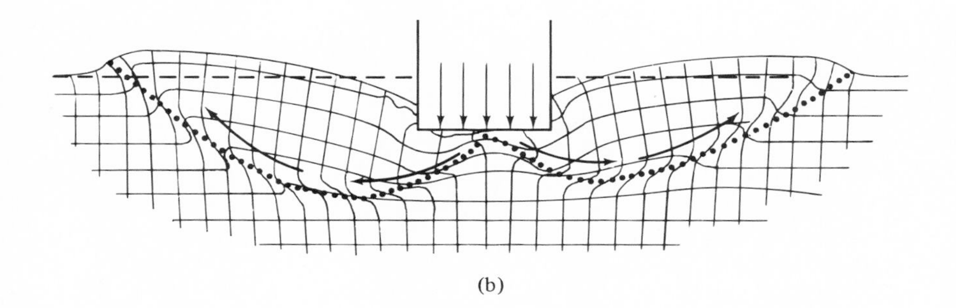

(b)

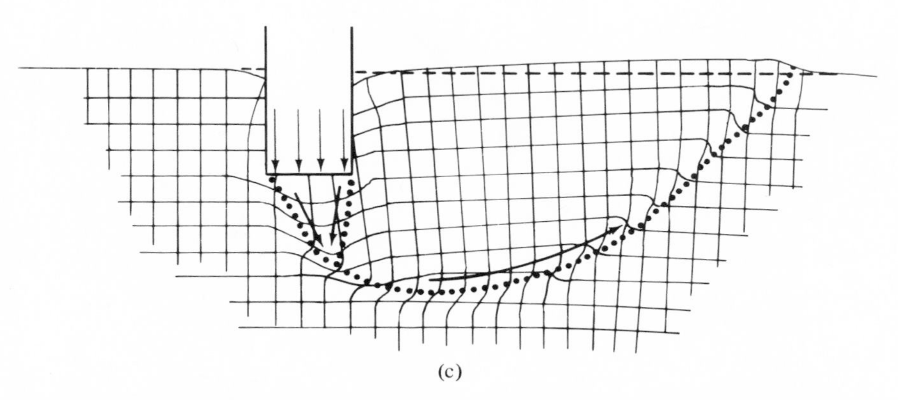

(c)

Fig. 23-2. Foundation settlement. From model studies by Jumikis (*5*) and by Selig (*9*). (a) Settlement due to compression; (b) settlement due to symmetrical shear failure; (c) settlement due to unsymmetrical shear failure.

be answered in such a case is how much settlement can be tolerated by the structure throughout its functional life. Then, with consolidation characteristics of the soil at hand, a safe bearing capacity which will limit the settlement to this tolerance can be established.

23.5.1. Settlement of Footings on Sand. Because of the difficulty in obtaining undisturbed samples of cohesionless sand, settlement predictions are usually made on the basis of tests performed in the field, most com-

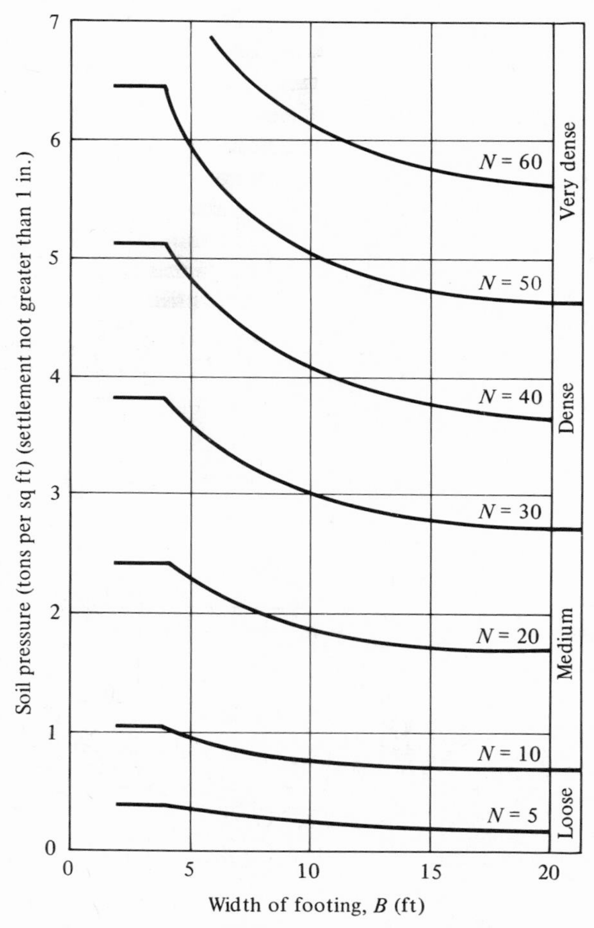

Fig. 23-3. Soil pressure corresponding to 1-in. settlement of footings on sand. The chart is based on a water table not closer than B below base of footing. (Reproduced by permission from *Foundation Engineering* by Peck, Hanson, and Thornburn, published by John Wiley and Sons, Inc., 1953.)

monly standard penetration tests. The relationship between the standard penetration test blow count N and relative density of sands was discussed in Chapter 6.

The relationship of settlement to relative density of sand can be seen by reference to the stress–strain curves in Fig. 19-6, in particular to the slopes of the initial portions of the curves, since the soil presumably will not be loaded to failure. Generally the looser the sand the flatter the stress–strain curve, i.e., the greater the strain or displacement for a given stress level. Tests conducted on the same sand with different levels of normal stress show another important relationship—the early part of the stress–strain curve becomes steeper with an increase in normal stress or confining pressure. Potential settlement therefore depends on both relative density and lateral confining pressure, and empirical correlations are possible between the blow count and anticipated settlement under various conditions of loading.

The most commonly used relationship is one by Terzaghi and Peck (*11*) shown in the form of the chart reproduced in Fig. 23-3, in which the bearing capacity which will permit 1 in. of settlement is shown in terms of width of footing and standard penetration values for the case where the water table is at an elevation B or more below the footing, B being the width of the footing. If the water table rises to near or above the footing, the bearing values given in the chart should be divided by 2 to compensate for the decrease in lateral confining pressure due to the decrease in vertical pressure caused by buoyancy.

Values of settlement larger or smaller than 1 in. can be deduced from Fig. 23-3 by assuming that settlement is directly proportional to load pressure, based on linearity of the stress–strain curve in the low stress region. Tests have shown that Fig. 23-3 tends to overestimate settlement by a factor of 2 or more, and that results at shallow depths are more reliable if the N values are corrected to 40 psi overburden pressure as discussed in Section 6.6. A chart for making this correction is shown in Fig. 23-4, which is a plot of Eq. (6-5). A rather high unit weight was used in drawing the chart to avoid overcorrection.

23.5.2. Settlement of Footings on Clay. As in the case of footings on sand, it is necessary to investigate the probable settlement characteristics of footings resting on clay in accordance with the discussion in Section 23.5 and the principles given in Chapter 18. In this connection, attention is directed to the usefulness of the liquid limit of a normally consolidated soil as an approximate indicator of its compression-strain characteristics, in accordance with Eq. (18-7).

Preconsolidation due to removal of overburden by geological erosion

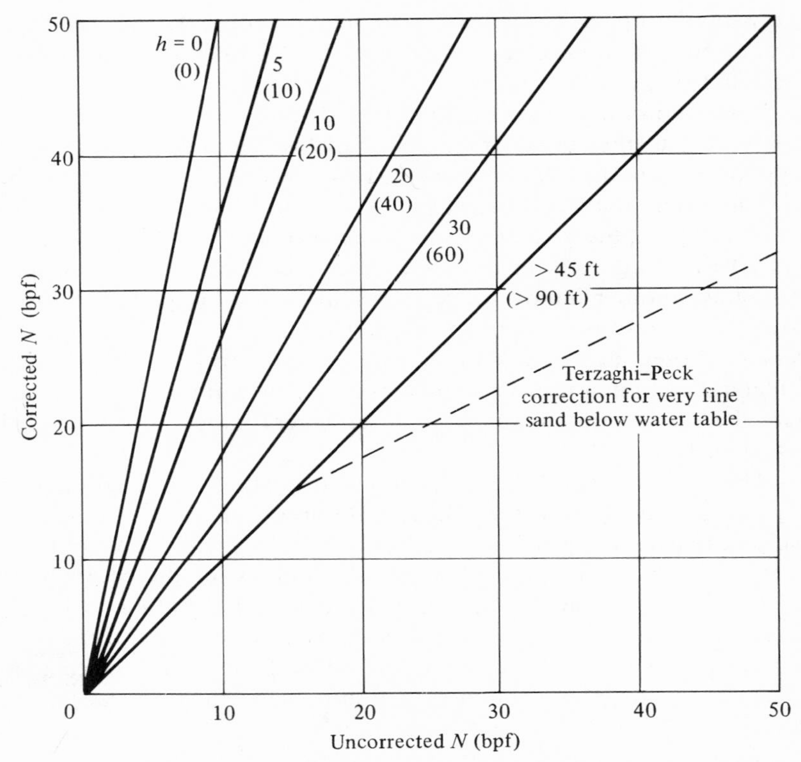

Fig. 23-4. Gibbs-Holtz surchange correction to standard penetration test blow count in sand. γ = 130 pcf; *h* is depth in feet. If below water table, use numbers in parentheses.

or by man, consolidation under the weight of glacial ice, etc., reduces the amount of potential settlement such that one relatively simple design practice is to assume that the net bearing capacity equals the preconsolidation pressure, which may be determined in accordance with the procedure outlined in Section 18.13.

The most common method for prediction of settlement on clays and soils with a clay binder combines one-dimensional consolidation test results with stress levels predicted from elastic theory, as discussed in detail in Chapter 18. More recent trends have included the use of pressuremeters to measure an effective modulus of elasticity (Section 6.5), and laboratory three-dimensional consolidation tests. The latter are performed in a triaxial chamber, and allow a measurement of deformations under all simulated past and anticipated vertical and horizontal stresses as predicted from elastic theory. This is referred to as the stress path method (Section 18.30).

Settlement studies should be made on individual footings and on the whole group of footings on which a structure rests, particularly if the footings are close together and if the compressible stratum primarily responsible for settlement extends to a considerable depth. It is easily possible for the settlement of any one footing to be influenced by the compression strains caused by loads on other adjacent footings.

23.6. PROPORTIONING FOOTINGS FOR EQUAL SETTLEMENT. When a structure is founded on separate column footings or pads, which are frequently referred to as spread footings, it is important that the settlement of the footings be as nearly uniform as possible. This objective is accomplished by proportioning the bearing areas of the footings according to the sums of the total dead load and some fraction of the live and wind loads. The fraction selected depends on the function of the structure and an estimate of the percentage of the live load which will be effective over substantial periods of time. Also each footing must be sufficiently large to insure that the safe bearing capacity of the soil is not exceeded when the total live and wind loads are effective.

For simplicity the settlement of individual footings is often assumed to be proportional to applied pressure regardless of the size of the footing. This is not strictly true, since as the size increases the pressure bulb extends deeper and involves more soil (See Fig. 23-10). If the footings are all nearly the same size, the error in this assumption is not appreciable.

Peck *et al.* (*8*) have presented rules of procedure for proportioning spread footings. They are as follows.

1. Determine the dead load for each footing, including the estimated weight of the footing.
2. Determine the maximum live load, including wind and earthquake load, that may act on the footing. This value is usually established by the building code.
3. Determine the ratio of the maximum live load to the dead load for each footing.
4. Select the footing for which this ratio is the largest, and determine the required area of this footing by dividing the sum of the dead load and maximum live load by the allowable soil pressure. This footing may be called the index footing.
5. To the dead load of the index footing add the live load that will actually be present often enough or long enough to govern the settlement. This live load is termed the reduced live load.

6. Divide the sum of the dead load and the reduced live load on the index footing by the area of the footing to obtain the reduced allowable soil pressure.
7. Use the reduced allowable soil pressure for determining the area of each other footing, considering the dead load and reduced live load for the footings.

If the weight of a footing determined from its actual required dimensions is appreciably different from the weight assumed in step 1, it may be necessary to redesign the footing. The more nearly correct dimensions should then be used for estimating the dead load.

Step 5 requires the exercise of judgment to a considerable extent. For example, the reduced live load for a warehouse or storage building will probably be a much greater percentage of the maximum live load than will the reduced live load for a school building where maximum live loads occur less frequently and are effective only for relatively short periods of time.

An alternate procedure, which is sometimes used, is to permit the use of a bearing capacity that is greater than the safe bearing capacity by some arbitrarily estimated percentage when dead load plus full live and wind loads are considered.

23.7. DETERMINATION OF SAFE BEARING CAPACITY. The evaluation of a material for a foundation bed from the standpoint of shear failure and the assignment of a safe bearing capacity for use in the design of a foundation are much more matters of good judgment based upon experience and logic than matters of technology. Although understanding of the engineering behavior of soil materials and scientific methods of analysis have advanced the art of foundation design immeasurably since about 1925, judgment still remains the principal ingredient in the process of selecting a safe bearing capacity for a particular case.

A number of different procedures for estimating safe bearing capacity are in rather widespread use. Several of these procedures will be outlined in the next few sections to give the student a basis for development of judgment in this field.

23.7.1. Soil Explorations. In the case of a small structure of relatively light weight, the safe bearing capacity of the soil on which the foundation is to rest is usually determined by visual inspection and judgment after the excavation for the footings has been completed. In the case of a larger and more important structure, the character of the foundation bed should be thoroughly explored prior to the preparation of the final design by means of borings, standard penetration tests, test pits, soil identification and strength tests, study of the performance of other structures resting on similar materials, and any other suitable methods available to the engineer. On the basis of such information, tentative decisions are made relative to the type of foundation, depth of footings, safe bearing capacity, and size of footings, and it is decided whether or not to use piling beneath the footings. These tentative decisions should be checked after excavation is completed and the results of the preliminary soil exploration have been verified.

It is never acceptable practice to proceed with the design or construction of even a moderately heavy structure without first making adequate borings which furnish specific knowledge in regard to the character of the underlying soil within the zone of influence of stresses produced by the foundation. No matter how familiar an engineer may be with subsurface conditions in a general area, the chances of encountering unusual and unexpected conditions at a specific site are so great that borings should always be made. Numerous examples could be cited in which an engineer believed his knowledge of conditions was sufficient to preclude the necessity of making expensive borings, but in which subsequent performance of the structure required costly remedial measures to be taken. Work and expense could have been avoided if specific knowledge had been obtained before-hand. It has been said facetiously that the foundation soil is always investigated; if not before construction, then certainly after a structure gets into trouble. Although this observation is not literally true, there is sufficient truth in it to justify the habitual practice of making soil borings at a proposed site.

23.7.2. *Specified Values of Safe Bearing Capacity.* The building departments of many of the larger cities and other organizations, such as state highway departments, have established tables of safe bearing capacities for various broad classes of foundation soils. These tables have been incorporated in building codes and standard design specifications. They are based upon the experience of the municipalities and highway organizations in their respective geographical areas, and in that sense they are extremely valuable as a guide in choosing a design value of the bearing capacity for a particular foundation bed. As a rule, however, they are greatly oversimplified in that they specify one value or a narrow range of values of safe bearing capacity for each class of soil, thus giving the impression that soil type is the only factor which influences bearing capacity. An intelligent decision in regard to safe bearing capacity requires the consideration of a number of details of the foundation structure itself, such as size and depth, proximity to other structures, and elevation of water table, in addition to soil type.

The Boston Building Code furnishes a good example of a table of safe bearing capacities. Sections 725 and 726 of the 1970 edition of this Code are as follows:

725.0 BEARING PRESSURES OF SOILS AND ROCKS. All applications for permits for the construction of new buildings or structures, and for the alteration of permanent structures which require changes that may affect their foundation, shall be accompanied by a statement describing the soils in the bearing strata, including sufficient records and data to establish their character and load-bearing capacity. Such records shall be certified by a licensed professional engineer.

725.1 SATISFACTORY FOUNDATION MATERIALS. The foundations of every permanent structure shall be supported by satisfactory bearing strata which shall mean:

a. Natural strata of rock, gravel, sand, inorganic silt, inorganic clay, or any combination of these materials with the limitations stated in section 725.2.3.
b. Compacted fills which satisfy the provisions of section 725.2.1.d.
c. Natural strata or artificial fills which can be changed into satisfactory bearing materials by pre-consolidation with a temporary surcharge in accordance with the provisions of section 725.2.1.e.

725.1.1. Where footings are supported at different levels, or at different levels from footings of adjacent structures, foundation plans shall include vertical sections showing to true scale all such variations in grade. The effect of such differences in footing levels on the bearing materials shall be considered in the design.

725.1.2. Foundations shall be constructed so that freezing temperatures will not penetrate into underlying soils that contain more than five (5) percent (by weight), passing a No . 200 mesh sieve. The foundations and grade beams of permanent structures, except when founded on sound rock, and except as otherwise provided in section 725.1.3. shall be carried down at least four (4) feet below an adjoining surface exposed to natural freezing. No foundation shall be placed on frozen soil. Foundations shall not be placed in freezing weather unless adequately protected.

725.1.3. Foundations of detached garages or similar accessory structures not exceeding eight hundred (800) square feet in area and not over one (1) story high, and grade beams of all structures, need not be carried more than one (1) foot below an adjoining surface exposed to natural freezing if the underlying soil to a depth of at least four (4) feet beneath the surface, and extending at least four (4) feet outside the building, is sand, gravel, cinders, or other granular materials containing not more than five (5) per cent (by weight) passing a No. 200 mesh sieve.

725.1.4. Foundations subject to hydrostatic uplift shall have adequate provisions to prevent heaving.

725.1.5. Basements and cellars shall be waterproofed in a manner consistent with their proposed use up to the maximum probable ground-water level. Under boilers, furnaces, and other heat-producing apparatus, suitable insulation shall be installed to protect the waterproofing against damage from heat as specified in articles 10 and 11. Foundations under heat-producing units shall be so insulated as to prevent evaporation of moisture from any underlying soil that is subject to shrinkage, and to protect the heads of wood piles against damage from heat.

725.2. CLASSIFICATION OF BEARING MATERIALS AND ALLOWABLE BEARING PRESSURES. ***725.2.1. Classification of Bearing Materials.*** The terms used in this section shall be interpreted in accordance with generally accepted engineering nomenclature. In addition, the following more specific definitions are used for bearing materials in the Greater Boston area:

a. Rocks
Shale: A soft, fine-grained sedimentary rock.
Slate: A hard, fine-grained metamorphic rock of sedimentary origin.
Conglomerate: A hard, well-cemented metamorphic rock consisting of fragments ranging from sand to gravel and cobbles set in a fine-grained matrix (locally known as Roxbury Puddingstone).

b. Granular Materials
Gravel: A mixture of mineral grains at least seventy (70) per cent (by weight) of which is retained on a No. 4 mesh sieve and possessing no dry strength.
Sand: A mixture of mineral grains at least seventy (70) percent (by weight) of which passes a No. 4 mesh sieve and which contains not more than fifteen (15) per cent (by weight) passing a No. 200 mesh sieve.
Coarse Sand: A sand at least fifty (50) percent (by weight) of which is retained on a No. 20 mesh sieve.
Medium Sand: A sand at least fifty (50) per cent (by weight) of which passes a No. 20 mesh sieve and at least fifty (50) percent (by weight) is retained on a No . 60 mesh sieve.
Fine Sand: A sand at least fifty (50) per cent (by weight) of which passes a No. 60 mesh sieve.
Well-Graded Sand and Gravel: A mixture of mineral grains which contains between twenty-five (25) per cent and seventy (70) per cent (by weight) passing a No. 4 mesh sieve, between ten (10) and forty (40) per cent (by weight) passing a No. 20 mesh sieve, and containing not more than eight (8) per cent (by weight) passing a No. 200 mesh sieve.

c. Cohesive Materials
Glacial Till: A very dense, heterogeneous mixture ranging from very fine material to coarse gravel and boulders and generally lying

over bedrock. It can be identified from geological evidence and from the very high penetration resistance encountered in earth boring and sampling operations.

Clay: A fine-grained, inorganic soil possessing sufficient dry strength to form hard lumps which cannot readily be pulverized by the fingers.

Hard Clay: An inorganic clay requiring picking for removal, a fresh sample of which cannot be molded by pressure of the fingers.

Medium Clay: An inorganic clay which can be removed by spading, a fresh sample of which can be molded by a substantial pressure of the fingers.

Soft Clay: An inorganic clay, a fresh sample of which can be molded with slight pressure of the fingers.

Inorganic Silt: A fine-grained, inorganic soil consisting chiefly of grains which will pass a No. 200 mesh sieve, and possessing sufficient dry strength to form lumps which can easily be pulverized with the fingers.

NOTE: Dry strength is determined by drying a wet pat of soil and breaking it with the fingers.

d. Compacted Granular Fill

A fill consisting of gravel, sand-gravel mixtures, coarse or medium sand, crushed stone, or slag, containing not more than eight (8) per cent (by weight) passing a No. 200 mesh sieve and having no plasticity, shall be considered satisfactory bearing material when compacted in nine (9) inch thick layers, measured before compaction, with adjustment of water content as necessary to achieve required compaction by applying to each layer a minimum of four (4) coverages of one of the following:

1. A vibratory roller with a steel drum with minimum weight of two (2) tons with a speed not exceeding one and one-half (1-½) miles per hour;
2. A rubber-tired roller having four (4) wheels abreast and weighted to a total load of not less than thirty-five (35) tons;
3. With the treads of a crawler type tractor with total load of not less than thirty-five (35) tons;
4. Other types of materials, compaction equipment, and procedures as may be approved by the building official on the basis of sufficient evidence that they will achieve compacted fllls having satisfactory properties.

The building official will require a competent inspector, qualified by experience and training and satisfactory to him, to be on the project at all times while fill is being placed and compacted. The inspector shall make an accurate record of the type of material used, including grain-size curves, thickness of lifts, type of compaction equipment and number of coverages, the use of water and other pertinent data. Whenever the building official or the inspector questions the suitability of a material, or the degree of compaction achieved, bearing tests shall be performed on the compacted material in accordance with the requirements of section 727.0. A copy of all these records and test data shall be filed with the building official.

e. Preloaded Materials

1. The building official may allow the use of certain otherwise unsatisfactory natural soils and uncompacted fills for the support of one (1) story structures, after these materials have been preloaded to effective stresses not less than one hundred and fifty (150) per cent of the effective stresses which will be induced by the structure.

725.2.3. TABLE 7-1. Allowable Bearing Pressures of Foundation Materials

Class of Material	*Allowable Bearing Pressure in Tons per Square Foot**
1. Massive igneous rocks and conglomerate, all in sound condition (sound condition allows minor cracks)	100
2. Slate in sound condition (minor cracks allowed)	50
3. Shale in sound condition (minor cracks allowed)	10 ‡
4. Residual deposits of shattered or broken bedrock of any kind except shale	10
5. Glacial Till	10
6. Gravel, well-graded sand and gravel	5
7. Coarse sand	3
8. Medium sand	2
9. Fine sand	1 to 2 ‡
10. Hard clay	5
11. Medium clay	2 †
12. Soft clay	1 †
13. Inorganic silt, shattered shale, or any natural deposit of unusual character not provided for herein	‡
14. Compacted granular fill	2 to 5 ‡
15. Preloaded materials	‡

*The allowable bearing pressure given in this section, or when determined in accordance with the provisions of section 727 will assure that the soils will be stressed within limits that lie safely below their strength. However, such allowable bearing pressure for Classes 9 to 12, inclusive, do not assure that the settlements will be within the tolerable limits for a given structure.

†Alternatively, the allowable bearing pressure shall be computed from the unconfined compressive strength of undisturbed samples, and shall be taken as 1.50 times that strength for round and square footings, and 1.25 times that strength for footings with length–width ratios of greater than four (4); for intermediate ratios interpolation may be used.

‡Value to be fixed by the building official in accordance with sections 726.0. and 727.0.

2. The building official may require the loading and unloading of a sufficiently large area, conducted under the direction of a competent engineer, approved by the building official, who shall submit a report containing a program which will allow sufficient time for adequate consolidation of the material, and an analysis of the preloaded material and of the probable settlements of the structure.

725.2.2. Bearing Values. The maximum pressure on soils under foundations shall not exceed values specified in section 725.2.3, table 7-1, except when determined in accordance with provisions of section 727.0 and in any case subject to the modifications of subsequent sections of this article.

726.0. SUBSURFACE EXPLORATIONS. ***726.1. Where Required.*** Where borings or tests are required, they shall be made at a sufficient number of locations and to such depths, and they shall be supplemented by such field or laboratory tests and engineering analyses, as are necessary in the opinion of the building official. When it is proposed to support the structure directly on bedrock, the building official may require drill holes or core borings to be made into the rock to a sufficient depth to prove that bedrock has been reached.

726.2. Soil Samples and Borings Reports. Samples of the strata penetrated in test borings or test pits, representing the natural disposition and conditions at the site, shall be available for examination of the building official. Wash or bucket samples shall not be accepted. Duplicate copies of the results obtained from all completed and uncompleted borings, plotted to a true relative elevation and to scale and of all test results or other pertinent soil data shall be filed with the building official.

23.8. THE KREY METHOD. A graphical procedure, known as the Krey method or the ϕ-circle method, may be employed to analyze the forces acting on the soil beneath a foundation and to determine the factor of safety against shear failure or breaking into the ground. This method is not widely used in practice, but has value in assisting the student to visualize the force relationships and the manner in which the shearing strength of the soil is mobilized to support a foundation load.

Consider a mass of cohesionless soil on which a foundation rests, as in Fig. 23-5. If a shear failure develops, it will occur along a surface having the approximate shape BCD. A trial surface of failure is established by arbitrarily choosing a center A of a circular arc BC. The radius r of this arc is the distance from A to the far edge B of the base of the footing. Complete the surface of failure by drawing the line CD that is tangent to the circle and makes an angle $[45° - (\phi/2)]$ with the surface of the ground. It is considered that the foundation load plus the weight of the soil in the

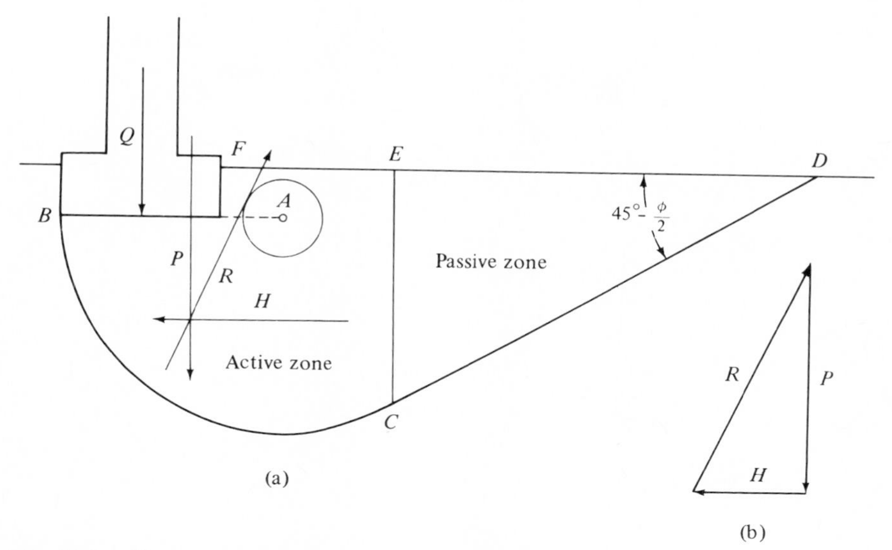

Fig. 23-5. Krey method; forces acting on active zone, excluding cohesion.

prism *BCEF* tends to cause rotation along the arc *BC*. This tendency produces an active horizontal thrust at the plane *CE*. This thrust is resisted by the passive resistance pressure of the triangle *CDE;* and the ratio of the ultimate passive resistance at the limit of equilibrium to the active thrust is the factor of safety against failure of the foundation soil.

23.8.1. Determination of Active Horizontal Thrust. The active horizontal thrust at the plane *CE* in Fig. 23-5(a) may be determined as follows: Find the magnitude and the line of action of the resultant vertical load *P*, which is the resultant of the foundation load and the weight of the active prism of soil *BCEF*. Draw the line of action of the active horizontal thrust *H*, which passes through a point one-third of the distance *CE* above point *C*. Then on each increment of length of the arc *BC* there is a normal force and a shearing resistance, as shown in Fig. 23-6. The resultant of each pair of these forces makes the angle ϕ with the radius of the circular arc *BC*. Then the distance from the center *A* of the arc to the line of action of this incremental resultant is $r \sin \phi$. Since this relationship is true for each incremental resultant, it is also true, for all practical purposes, of the resultant of all normal and shearing forces on the arc *BC*. It is concluded, therefore, that the line of action in Fig. 23-5 of the resultant force *R* on arc *BC* must be tangent to a circle whose center is at *A* and whose radius is $r \sin \phi$. This circle is called the ϕ-circle or the friction circle.

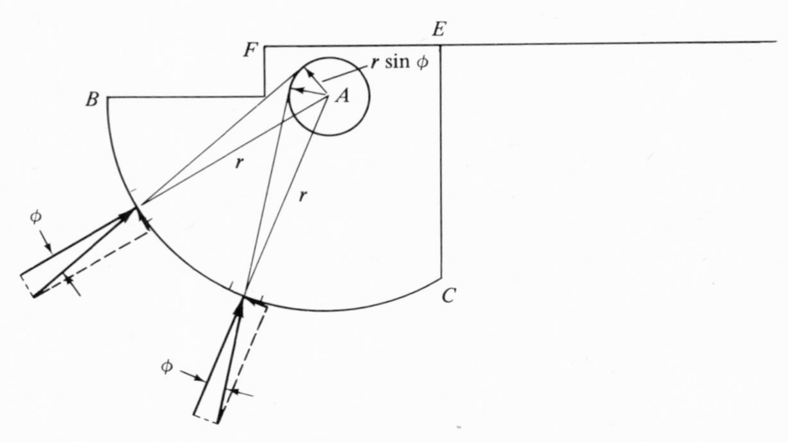

Fig. 23-6. Resultants of normal and shearing forces on arc *BC* are tangent to a circle whose radius is $r \sin \phi$.

The resultant R is concurrent with the vertical load P and the active horizontal thrust H. Therefore, the line of action of R is along a line which passes through the intersection of P and H and is tangent to the ϕ-circle. Since the magnitude of P is known, the magnitude of H can be determined by drawing the triangle of forces, as shown in Fig. 23-5(b).

23.8.2. Determination of Resultant Passive Resistance Pressure. The active horizontal thrust is resisted by the resultant passive resistance pressure supplied by the triangle of soil *CDE*, as shown in Fig. 23-7(a). The value of this resultant can be determined as follows. First compute the weight P_p of the soil in the triangle. This weight acts through the centroid of the triangle. The forces acting on the plane *CD* at the limit of equilibrium are a normal force and a shearing force which is directed toward *C*. The resultant R_p of these forces acts at the angle ϕ with the normal to *CD*, and it passes through the point of intersection of P_p and the line *CD*. The resultant passive resistance pressure H_p acts horizontally through this same point, and its magnitude can be determined by drawing a triangle of forces, as shown in Fig. 23-7(b).

23.8.3. Minimum Factor of Safety. The ratio of the resultant passive resistance pressure H_p to the active horizontal thrust H is the factor of safety against shear failure along the arbitrarily chosen trial surface *BCD* in Fig. 23-5. Next, select additional points, designated as A', A'', and so on, to represent centers of ϕ-circles, and repeat the foregoing procedure

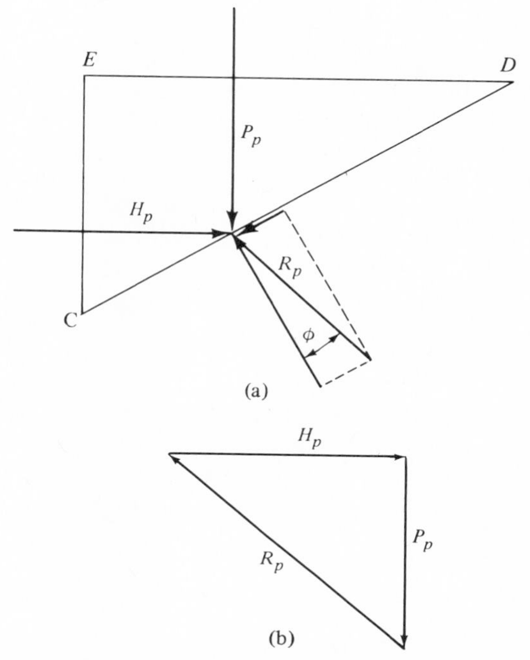

Fig. 23-7. Forces acting on passive zone, excluding cohesion.

for each failure surface thus established. From these various trials the minimum factor of safety for the foundation can be determined. If this value is less than desired, the foundation should be carried deeper or made wider, or modified in both ways, in order to increase the factor of safety.

23.9. INFLUENCE OF SURCHARGE, WIDTH OF FOOTING, AND POSITION OF WATER TABLE. A review of the Krey method of analysis indicates clearly the influence of a surcharge on the bearing capacity of the underlying soil. If the base of the footing were at a greater depth below the ground surface *FED* in Fig. 23-5, the weight P_p of the soil in the passive zone *CED* and the resultant passive pressure H_p would be increased. The resultant vertical pressure P would also be increased somewhat, but there would be a net gain in bearing capacity. This analysis illustrates the advantage of a deep foundation over a shallow one on the same type of soil.

Similarly, the width of footing has an influence on bearing capacity. If the footing shown in Fig. 23-5 were wider, the arc *BC* would penetrate deeper into the soil. The volume of soil in the passive zone *CED* and the length of the shear surface *BCD* would be increased, and there would be a net increase in the bearing capacity of the soil.

It is also clear that any condition which reduces the shearing resistance along the surface *BCD* reduces the bearing capacity of the soil. For example, if the water table rises to an elevation near the ground surface, the effective weight of the soil will be greatly reduced because of the buoyant effect of the ground water. This decrease in weight reduces the normal forces on the shear surface, and the bearing capacity of the soil will be reduced by about 50%. Foundations of structures in flooded areas often fail by breaking into the ground because of this reduction in unit weight and shear strength of the underlying soil during high water.

23.10. HOUGH'S PRESUMPTIVE BEARING CAPACITY. B. K. Hough (*2*) has published two diagrams giving the relationship between the safe bearing capacity of a number of generalized types of soil and the value of the standard penetration resistance *N* of the soil. These diagrams are based on

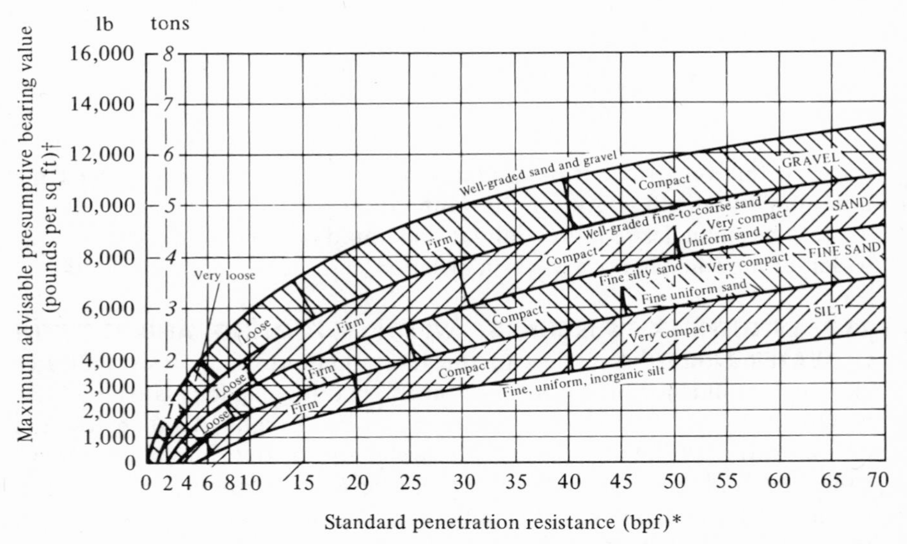

Fig. 23-8. Presumptive bearing values for granular soils. (Reproduced by permission from *Basic Soils Engineering*, 2nd ed., by B. K. Hough, published by the Ronald Press Co., © 1957, 1969.)
*Number of blows of 140-lb pin-guided drive weight falling 30 in. per blow required to drive a split-barrel spoon with a 2-in. outside diameter 12 in.
†Values must be corrected for effect of weak substrata, high ground water, and surcharge.

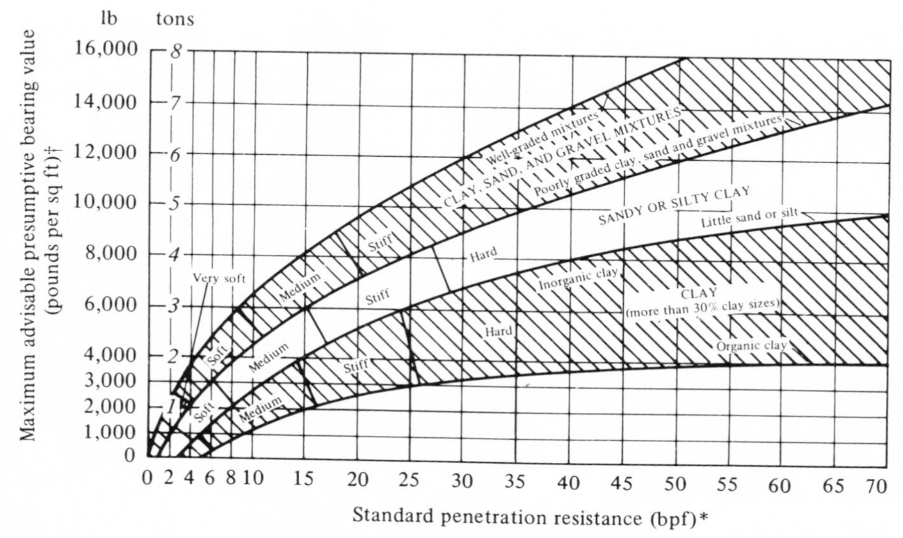

Fig. 23-9. Presumptive bearing values for clay and mixed soils. (Reproduced by permission from *Basic Soils Engineering*, 2nd ed., by B. K. Hough, published by the Ronald Press Co., © 1957, 1969.)
*See explanation in Fig. 23-8.
†Higher value may be used for precompressed (or compacted) clays of low sensitivity than for normally loaded or extrasensitive clays.

a study of a number of municipal building codes and of several hundred typical job records. They are reproduced in Figs. 23-8 and 23-9. The maximum advisable presumptive bearing values shown on these diagrams are for the design of conventional spread footings at ordinary depths. They should be adjusted for special conditions, such as the existence of relatively weak substrata, group action of footings, high water table, surcharge in connection with granular soils, and precompressed or preloaded clays of low sensitivity.

Since these graphs are based on experience, N values probably should *not* be corrected upward as shown in Fig. 23-4. The most common corrections are for high water table, where the bearing capacity should be reduced by a factor of 2, and surcharge pressure, which may be added to the bearing capacity. Preloaded clays are safe to the extent of their preloading. These charts are particularly useful for small structures in areas where there is no applicable building code.

23.11. LOAD TESTS. Making load tests has long appealed to both engineers and laymen as an ideal method for determining the safe bearing capacity of a foundation soil, and certainly there are situations for which this procedure is very valuable. On the other hand, there are some situations in which load tests may be not only misleading, but actually dangerous, because they may indicate a safe bearing capacity very much greater than the actual value.

Consider, for example, a case in which a relatively thin layer of hard clay lies above a thick stratum of soft clay, as illustrated in Fig. 23-10. If a load test is conducted by using a relatively small test plate, the significant bulb of pressure will lie wholly within the hard-clay stratum, and a high value of bearing capacity will be indicated by the test results. However, the actual bulb of pressure under a full-sized prototype footing may penetrate a considerable distance into the softer underlying layer, and the actual bearing capacity may well be much less than that indicated by the test procedure. Before an engineer recommends a program of load testing, he should be certain that conditions are such that the indicated results will be appropriate to the problem at hand.

Load tests are frequently used to check calculated settlement or bearing capacity on sand, or for evaluating bearing capacity of clays with fissures, slickensides, or other discontinuities that prevent an adequate evaluation from laboratory samples. Standard load tests are made with a steel circular plate 12 in. to 30 in. in diameter loaded by means of a 50 ton hydraulic jack acting upward against a dead load (ASTM Designation: D1194-57). The plate should be seated directly on the soil stratum to be tested, or a thin layer of sand may be used to insure a more uniform distribution of pressure. Where a pit is required to reach a particular depth, its diameter should exceed four times the diameter of the test plate.

Test loads are applied in increments, and the corresponding soil de-

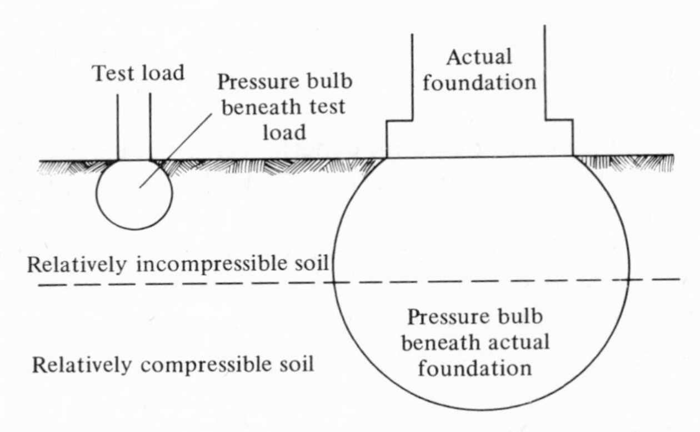

Fig. 23-10. Small test load area may not reveal influence of a buried compressible stratum on settlement of an actual structure.

formations are measured. Each increment should be left in place for sufficient time to permit the deformation to become practically constant before the next increment is applied. A load-deformation curve may be plotted, and the failure stress may be selected at the point at which the curve breaks more or less sharply. If the soil borings indicate a possibility that the material is weaker at greater depths, it is advisable to conduct additional tests at several elevations below the bottoms of the footings. The least bearing capacity revealed by these tests should be adopted for design purposes. It is not necessary to apply test loads to the soil at depths below the base of the footing greater than the width of the footing.

The reason for repeating plate tests on the same soil with two or more different sizes of plates is to ascertain the influence of size of the bearing area, since most footings are much larger than the largest plate which can be used. In sands the larger the plate the larger the allowable unit stress on the plate, because a larger plate improves confinement of the sand. For example, sand which will fail under the pressure of a man's heel may be adequate for support of a much larger unit stress over a larger area.

The failure stress from a plate bearing test is usually taken at a break

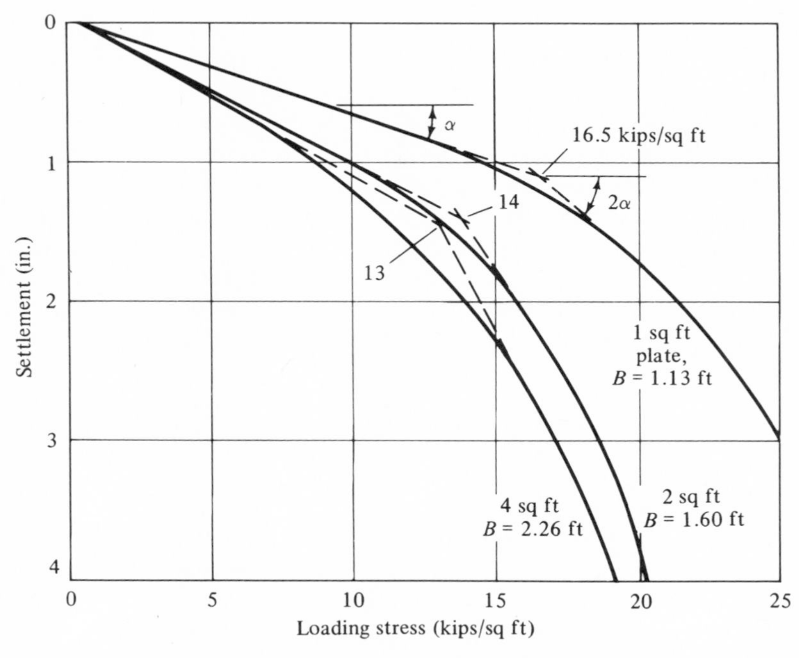

Fig. 23-11. Load-settlement curves for three different plates on a compressible clay soil. Data after Housel (*3*).

downward in the load-settlement curve. Since the break is seldom sharp, it may be more or less arbitrarily defined by intersecting tangents, one method being shown in Fig. 23-11. One procedure for analysis is to plot failure stress vs. plate width, extrapolate to the footing width, and apply the usual factor of safety of 3 or more to determine safe bearing capacity. This may be performed algebraically by a linear equation:

$$q_0 = aB + N_0 \tag{23-1}$$

in which

q_0 = bearing capacity in pounds per square foot;
B = bearing area width in feet;
N_0 = failure constant for a given soil; and
a = a constant.

This equation is a special case of the Terzaghi equation presented later in the chapter, the constant a varying from zero for an ideal clay to 1.0 for an ideal sand, and N_0 at failure representing roughly three times the unconfined compressive strength of the soil.

23.12. HOUSEL METHOD FOR CLAYS. In practice, plate bearing tests in clay soils sometimes indicate that instead of increasing or staying the same with increasing plate size, the failure stress decreases. This trend is shown by the data in Fig. 23-11. In this case the constant a in Eq. (23-1) is negative, and the equation is no longer rational. W. S. Housel (*3*) of the University of Michigan has suggested that this behavior relates to cohesive shearing resistance of soil immediately below the perimeter of the plate. Since the perimeter-to-area ratio is largest for smaller plates, smaller plates on such clays can support a larger loading stress at failure. The load is assumed to be carried by two components, shearing stress around the perimeter of the plate, and compressive stress in soil beneath the plate. Therefore

$$W = qA = Pm + An \tag{23-2}$$

in which

W = total load in pounds;
q = bearing stress in pounds per square foot;
A = plate or foundation area in square feet;
P = plate or foundation perimeter in feet;
m = perimeter shear in pounds per foot; and
n = compressive stress on soil column beneath the foundation in pounds per square foot.

Dividing by A,

$$q = \frac{P}{A} m + n \tag{23-3}$$

For a circular or square bearing area $P/A = 4/B$, and

$$q = \frac{4m}{B} + n \tag{23-4}$$

where B is the width. The latter equation closely resembles Eq. (23-1) except that q now is inversely instead of directly proportional to B, and the equation applies for any stress condition up to failure. An example of the use of the Housel method to predict stress at a certain settlement follows.

EXAMPLE 23-1. The data in Fig. 23-11 are to be used to design a 6-ft square footing. What total stress on the soil will cause an allowable settlement of the footing of 0.5 in.?

SOLUTION: Equation (23-4) is written for two different plates and solved simultaneously for m and n. From Fig. 23-11, at 0.5 in. settlement the stress on the smallest and largest plates is 7.5 and 4.5 kips/sq ft, respectively. Substituting,

$$7.5 = \frac{4m}{1.13} + n = 3.56\,m + n$$

$$4.5 = \frac{4m}{2.26} + n = 1.78\,m + n$$

$$3.0 = 1.78m$$

$$m = 1.69 \text{ kips/ft}$$

$$n = 1.50 \text{ kips/ft}^2$$

The equation now may be written for the full-size foundation 6 ft square:

$$q = \frac{4(1.69)}{6} + 1.50 = 1.12 + 1.50$$

$$= 2.6 \text{ kips/ft}^2$$

$$W = 2.6\,(6)\,(6) = 93.5 \text{ kips}$$

This q is considerably below the stress levels causing 0.5 in. settlement of the plates, which is reasonable because of the difference in pressure bulb.

The values of m and n change with different amounts of settlement, depending on the relative amount of stress carried by compression and by perimeter shear. Housel defines the maximum bearing capacity on the basis of m, n, and settlement—either settlement divided by n becoming a minimum, or $m \div n$ reaching a maximum. A factor of safety is then applied to obtain an allowable bearing capacity.

23.13. ANALYTICAL METHODS. In recent years there has been a trend away from empirical and semiempirical methods towards more sophisticated analyses utilizing principles of soil mechanics. The Krey or ϕ-circle method is such an analytical method, but applies only to cohesionless soils. An even simpler illustration still applying only to cohesionless soils draws an analogy between soil under a foundation and soil behind a retaining wall (Fig. 23-12). Line *EF* is analogous to the retaining wall with *W* the surcharge load, such that if we remove wedge II, wedge I fits the active case. However in order for wedge I to move downward, wedge II must be moved laterally, so wedge II fits the passive case. At failure the vertical and horizontal stresses are therefore related through the Rankine coefficients K and K_p. For the left-hand square in Fig. 23-12,

$$\sigma_h = \sigma_v K_p$$

and for the right-hand square

$$\sigma_h = qK = q/K_p$$

Equating and solving for q, the ultimate bearing capacity q_o in force per unit area is

$$q_o = \sigma_v K_p^2$$

Since the vertical pressure σ_v equals the depth D times the unit weight of the soil γ,

$$q_o = D\gamma K_p^2 = D\gamma \left(\frac{1 + \sin \phi}{1 - \sin \phi}\right)^2 \qquad (23\text{-}5)$$

Equation (23-5) is overly conservative because it neglects friction between the soil and foundation and between the two soil wedges. It nevertheless is useful as a quick check for the bearing capacity of foundations designed for granular soils on the basis of settlement. The ϕ may be evaluated by means of the corrected standard penetration test blow count, Section 6.6, and Fig. 23-17.

Terzaghi in 1925 suggested a somewhat similar approach for cohesive soils in which $\phi = 0$. In this case to determine the role of cohesion, the two stress squares in Fig. 23-12 may be visualized as two triaxial test specimens, number II being on its side. The vertical surcharge stress $D\gamma$ on this

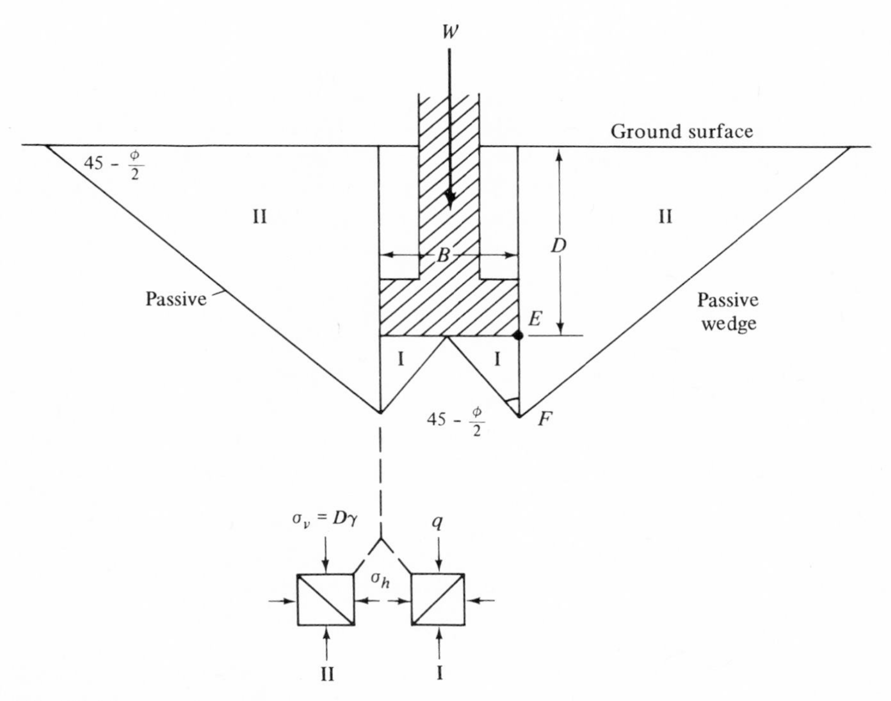

Fig. 23.12. Rankine solution for bearing capacity.

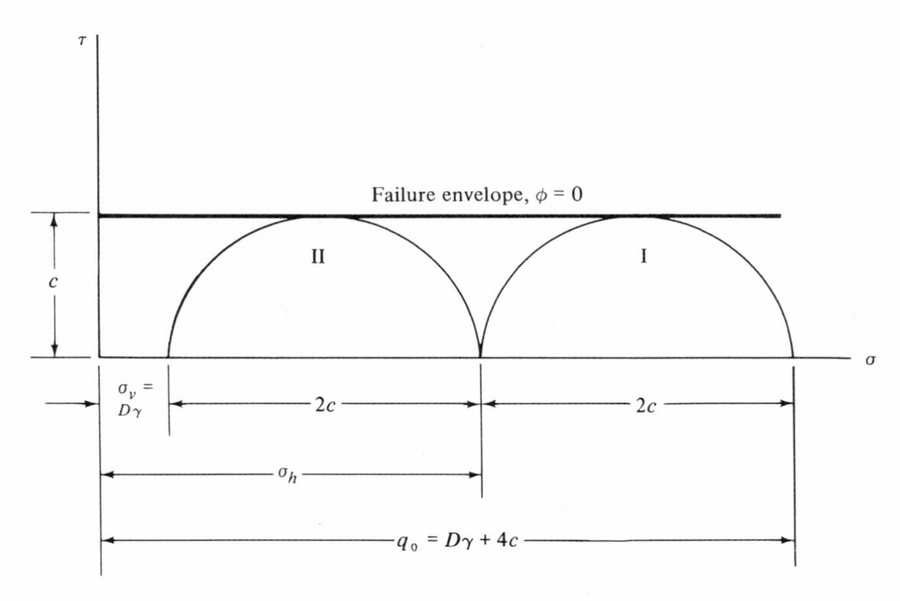

Fig. 23-13. Mohr circles showing stress conditions in a ϕ = o soil at failure—I: under a foundation; II: adjacent soil at the same level (see Fig. 23-12.).

square is the minor principal stress. The major principal stress may be found by referring to Mohr circle II in Fig. 23-13. It can be seen that if ϕ = 0, the major principal stress $\sigma_n = D\gamma + 2c$, where c is the soil cohesion. This becomes the minor principal stress for the square I in Fig. 23-12, and the major principal stress at failure becomes $q_o = D\gamma + 4c$. Subtracting surcharge, the net bearing capacity is therefore

$$q_o = 4c = 2q_u \tag{23-6}$$

where q_o is the bearing capacity and q_u is the unconfined compressive strength. This equation also is conservative for the same reasons as Eq. (23-5), but supports the common use of unconfined compressive strength for bearing capacity of small structures on clay, the nominal factor of safety being 2.

A similar approach, subject to the same criticisms, may be used for soils with both cohesion and internal friction. These and other analytic methods are skillfully reviewed in a historic context by Jumikis (*4*).

23.14. TWO-DIMENSIONAL ANALYTICAL METHODS BASED ON OBSERVED SHEAR PLANES. Although the assumed geometry of shear in Fig. 23-12 somewhat resembles the observed failure in Fig. 23-2(b), there are some major discrepancies, and the Rankine or principal stress solutions therefore would not be realistic even if boundary friction were included. Where a potential bearing capacity failure involves shearing, the problem may be analyzed by using trial surfaces and one of the methods suggested for analysis of slope stability, such as the method of slices. However, this is laborious and not precise because of assumptions made in that analysis, although the method of slices might be preferred over other methods for nonhomogeneous soil conditions.

One of the major contributions of Terzaghi was to adapt a theoretical solution by Prandtl to bearing capacity problems in soils. Prandtl was concerned with resistance of metal to a punch. He assumed zero friction between the metal and the punch, and that the metal is weightless. Terzaghi assumed essentially the same failure geometry, but added the effects of bearing friction and soil weight. The Terzaghi–Prandtl system for a long, continuous foundation is shown in Fig. 23-14(a). An essential element is a central wedge of soil under the foundation, which moves vertically as the foundation settles and thus pushes the adjacent soil aside. Such wedges have been observed in model studies (see Fig. 23-2c), and are known to form during driving of pile, explaining why flat-ended pile drive about as easily as cone-ended ones.

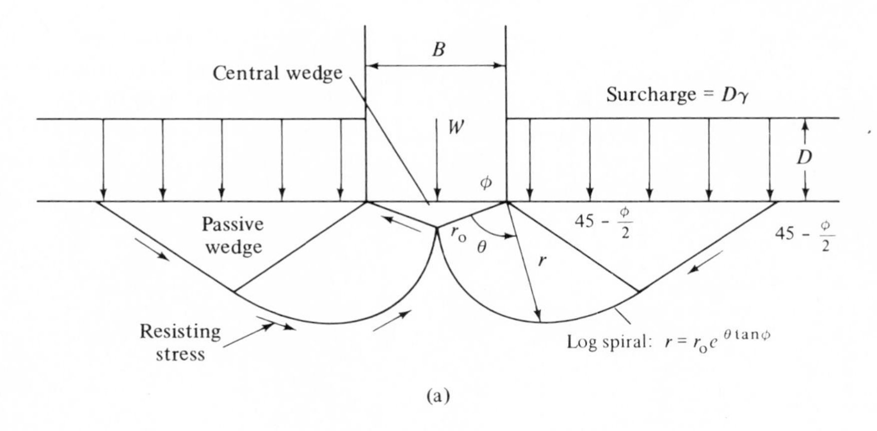

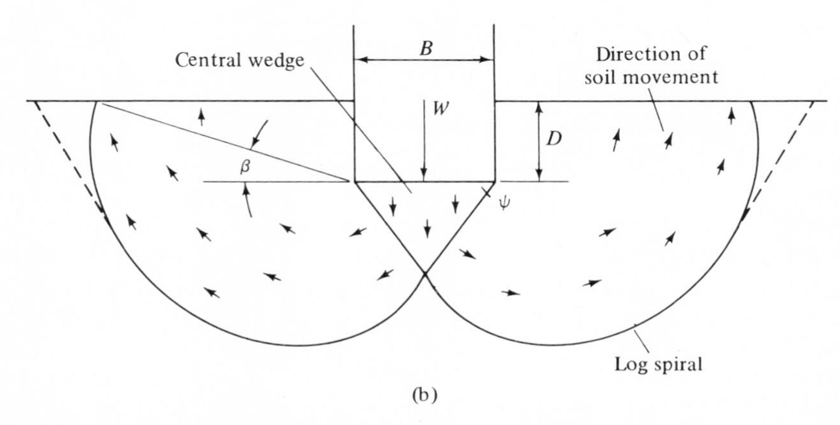

Fig. 23-14. Two idealized bearing capacity failure geometries. Both are drawn for $\phi = 10°$. (a) Prandtl–Terzaghi. (b) Meyerhof: $\phi < \psi < 45 + \phi/2$, ψ solved for minimum bearing capacity.

Terzaghi expressed his solution in terms of dimensionless bearing capacity factors designated N_γ, N_c, and N_q.

$$q_o = \underbrace{\frac{\gamma B}{2} N_\gamma}_{\text{width}} + \underbrace{cN_c}_{\text{cohesion}} + \underbrace{\gamma D N_q}_{\text{surcharge}} \tag{23-7}$$

where q_o is the bearing capacity in force per unit area, γ is the soil unit weight, B the width of the foundation, c the soil cohesion, and D the depth of the foundation. This is perhaps the most important equation in foundation engineering and served as a basis for Eq. (23-1) in the discussion of the plate bearing test. All three "N" bearing capacity factors are functions of the internal friction angle ϕ and the assumed shape of the failure zone. N values calculated by Terzaghi are shown in Fig. 23-15. Another useful advantage of this equation is that it allows one to see separately the effects of foundation width, soil cohesion, and surcharge.

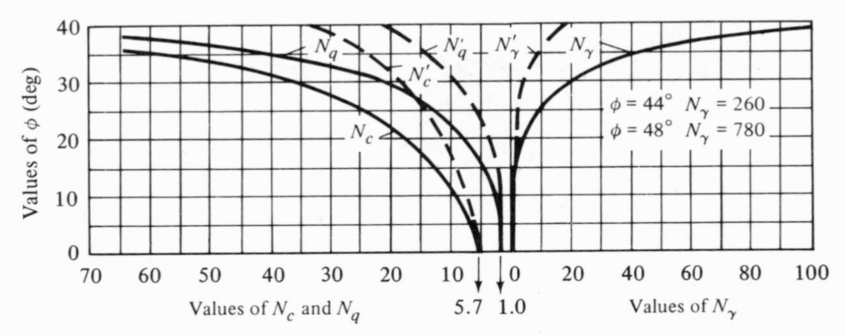

Fig. 23-15. Terzaghi bearing capacity factors for use in Eq. (23-7). Dashed lines are for local shear, discussed in Section 23.20. From Terzaghi, *Theoretical Soil Mechanics*, John Wiley and Sons, Inc., ©1943.

EXAMPLE 23-2. A clay soil has $\phi = 10°$, $c = 400$ psf, and $\gamma = 100$ pcf. Evaluate the relative influences of footing width and depth on total supporting capacity of a footing 4 ft wide and 5 ft deep.

SOLUTION. From Fig. 23-15, $N_\gamma = 1$, $N_c = 9$, and $N_q = 3$. Therefore according to Eq. (23-7),

$$q_o = \frac{100(4)}{2}(1) + 400(9) + 100(5)(3)$$

$$= 200 + 3600 + 1500 \text{ psf}$$

Multiplying by the width, 4 ft, gives supporting capacity Q in pounds per lineal foot:

$$Q = Bq_o = \underset{\text{width}}{800} + \underset{\text{cohesion}}{14{,}400} + \underset{\text{surcharge}}{6000} = 21{,}200 \text{ lb/ft}$$

23.15. EFFECT OF SUBMERGENCE. As discussed in previous chapters, the effect of submergence due to a high water table or flooding can be evaluated either by a consideration of pore pressure and effective stress, or by use of submerged unit weight. The latter is much more convenient in bearing capacity analysis because the soil unit weight appears in the first and third terms in Eq. (23-7). In cohesionless soils, where the second term becomes zero, submergence reduces the soil unit weight and therefore the bearing capacity by a factor of about one-half, which is the factor previously indicated in the discussion of presumptive bearing capacity of sands.

In cohesive soils if $\phi = 0$, then N_γ and N_q are low compared to N_c in Eq. (23-7). Since the soil unit weight does not appear in this term, theoretically the bearing capacity of $\phi = 0$ clays is not influenced by submergence. The bearing capacity of soils intermediate between sands and ideal clay will be affected by a factor between one-half and one.

EXAMPLE 23-3. Calculate the bearing capacity of a clayey sand, $\phi = 30°$, $c = 144$ psf, $\gamma = 120$ pcf, for a footing 4 ft wide by 5 ft deep, before and after submergence.

SOLUTION. From Fig. 23-15, $N_c = 36$, $N_q = 21$, and $N_\gamma = 20$. For the nonsubmerged case, Eq. (23-7) becomes

$$q_o = \frac{120\,(4)}{2}\,(20) + 144\,(36) + 120\,(5)\,(21)$$

$$= 4800 + 5184 + 12{,}600 = 22{,}584 \text{ psf}$$

After submergence, substituting (120 − 62.4) for the soil unit weight gives

$$q'_o = 2304 + 5184 + 6048 = 13{,}536 \text{ psf},$$

a reduction of 40%.

23.16. DEEP FOOTINGS. The surcharge term in Eq. (23-7) reflects only pressure and not strength of the adjacent soil which is lifted by a bearing capacity failure. Meyerhof found experimentally that when depth D exceeds width B, the bearing capacity factors increase rapidly with increasing depth because of shearing resistance of the surcharge. He also found that a more critical failure plane exists involving a deeper central wedge, as shown in Fig. 23-14(b). These two effects oppose and tend to balance out, the Meyerhof factors being somewhat lower than Terzaghi's for shallow foundations and higher for deep foundations. Meyerhof's factors vary depending on angle β in Fig. 23-14(b). For further details the reader is referred to Meyerhof (*7*). Use of the Terzaghi factors for deep foundations is on the safe side.

23.17. LOCAL SHEAR. The analysis of Section 23.14 assumes that soil along the entire shear plane reaches a limiting stress state at the same time, when actually shearing will begin first adjacent to the foundation and gradually extend downward. In loose sands and sensitive clays even the beginnings of a general shear failure could be disastrous, so in addition to the usual factor of safety of 3 or more, Terzaghi suggested that in these circumstances both ϕ and c also be reduced by one-third. This is taken into account by multiplying $c \times \frac{2}{3}$ and using lower bearing capacity factors shown by the dashed lines in Fig. 23-15. Analytical methods devised by Fröhlich also may be used, and these are extensively treated by Jumikis (*5*).

23.18. SAND: COMBINED GENERAL AND LOCAL SHEAR. Peck *et al.* (*8*) presented a compromise between the Terzaghi general shear and local shear bearing factors for noncohesive soils as shown in Fig. 23-16. These compromise factors emphasize local shear at low values of ϕ and general shear at high values, and are widely used in practice. Because of the arbitrary nature and close agreement between N_γ and N_q, an average may reasonably be used (Table 23-2). If $c = 0$, Eq. (23-7) becomes

$$q_o = N_a\left(\frac{\gamma_1 B}{2} + \gamma_2 D\right) \tag{23-8}$$

where

N_a = the average of N_γ and N_q;
γ_1 = unit weight of soil below bottom of footing;
γ_2 = unit weight of surcharge above bottom of footing;
B = width of footing;
D = least depth of surcharge, and
q_o = unit bearing capacity.

If the soil density above and below the footing is the same,

$$q_o = N_a\gamma\left(\frac{B}{2} + D\right) \tag{23-9}$$

Equation (23-8) may be quickly evaluated by the use of graphs shown in Figs. 23-17 and 23-18. In these graphs C is defined as the *surcharge ratio*, or ratio of surcharge depth to footing width:

$$C = \frac{D}{B} \tag{23-10}$$

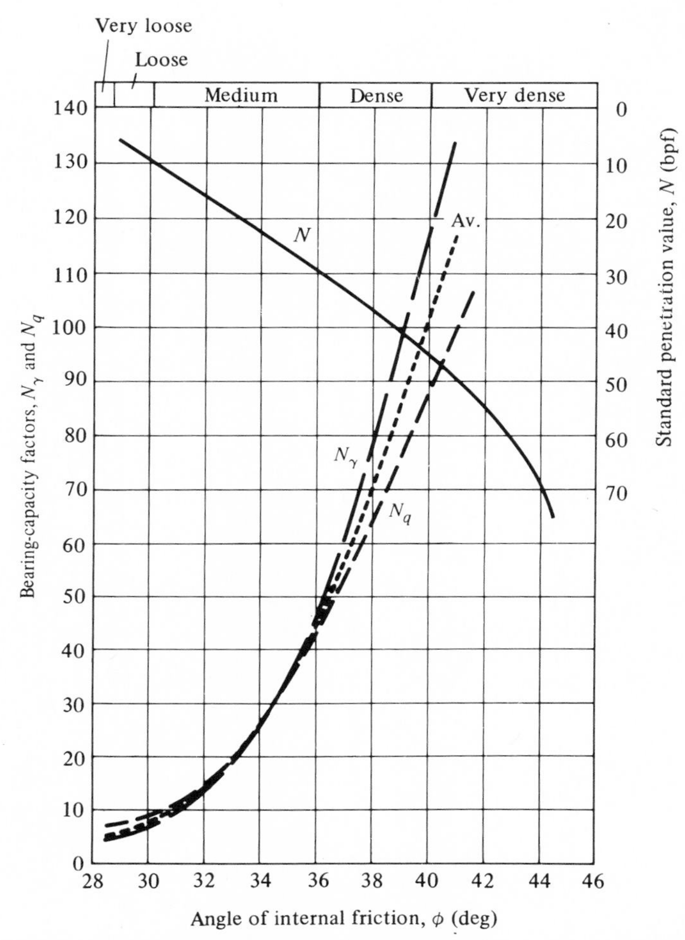

Fig. 23-16. Relationship between ϕ, bearing-capacity factors, and standard penetration values N of sand (Average line inserted by authors.) Reproduced by permission from *Foundation Engineering* by Peck, Hanson, and Thornburn, published by John Wiley & Sons, Inc., 1953.)

TABLE 23-2. Relative Density, ϕ, and Average Bearing Capacity Factors for Various Corrected Standard Penetration Values N' of Sand

N' *(bpf)*	*Relative Density (%)*	ϕ *(deg)*	*Failure Criterion*	$N_a = (N_\gamma + N_q)/2$
5	22	28.5	Local	6.9
10	37	30.0	Shear	7.8
15	46	31.5	.	13.7
20	53	33.0	.	20.8
25	59	34.5	.	30.7
30	64	36.0	.	45.2
35	69	37.5	.	63.7
40	73	38.0	.	84.5
45	77	40.0	General	105
50	80	41.0	Shear	122

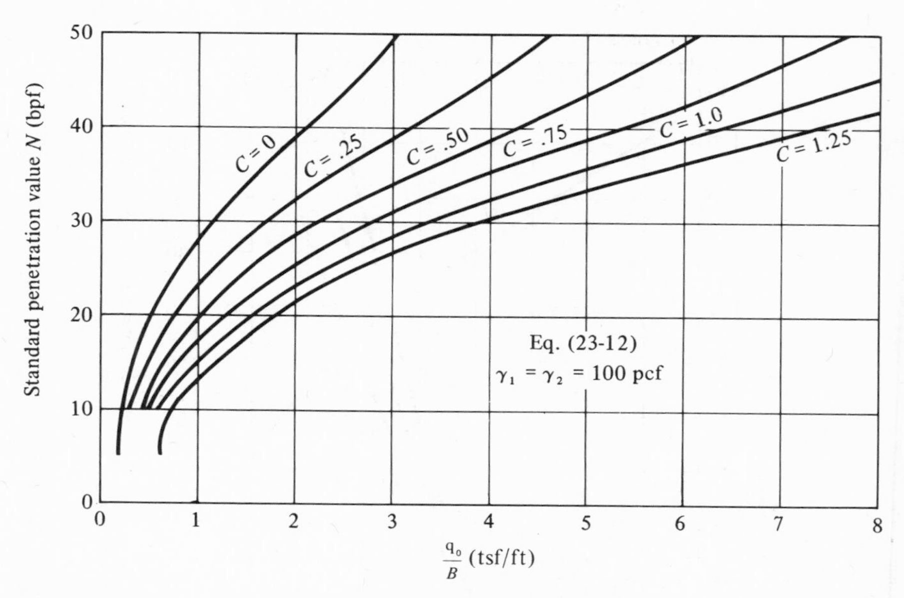

Fig. 23-17. Gross ultimate bearing capacity per foot width of footing. Note: Chart is for water table at distance B or more below base of footing. When water table is at top of surcharge, divide bearing capacity by 2.

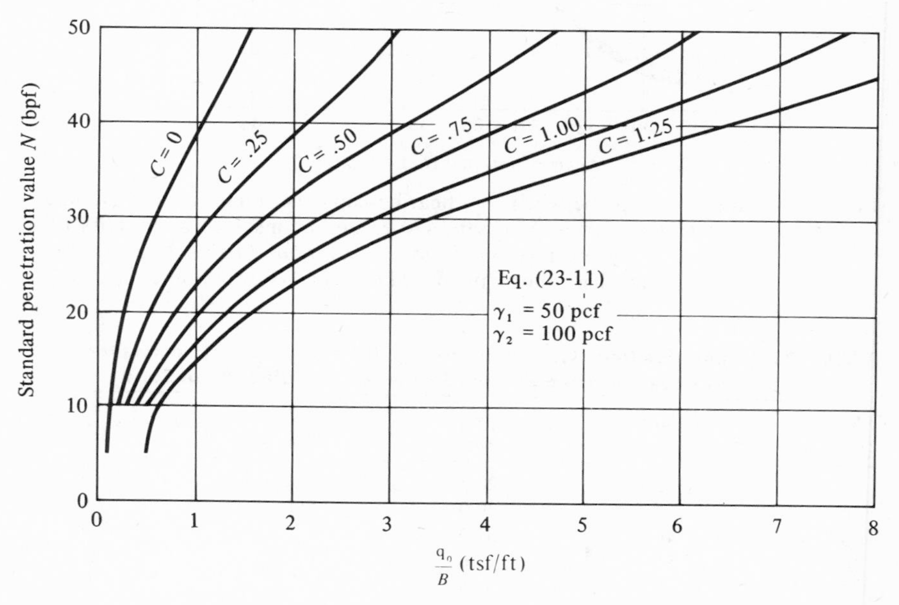

Fig. 23-18. Gross ultimate bearing capacity per foot width of footing. Note: Chart is for water table at base of footing.

Substituting for D, the sand bearing capacity equations become

$$q_o = N_a B \left(\frac{\gamma_1}{2} + C\gamma_2 \right) \tag{23-11}$$

or if $\gamma_1 = \gamma_2$,

$$q_o = N_a B \gamma (C + 0.5) \tag{23-12}$$

Figure 23-17 is drawn for Eq. (23-12) with $\gamma = 100$ pcf. If the soil and surcharge are subject to submergence, bearing capacity from this graph should be divided by 2. Figure 23-18 solves Eq. (23-11) for $\gamma_1 = 50$ pcf and $\gamma_2 = 100$ pcf, corresponding to a water table level at the base of the footing. The graphs also use the relation shown in Fig. 23-16 between friction angle ϕ and standard penetration test blow counts N, corrected according to Fig. 23-4. Since the graphs give ultimate bearing capacity, the result must be divided by a factor of safety to obtain an allowable bearing capacity.

To use the charts in Figs. 23-17 and 23-18, first assume a width of footing **B** and calculate the surcharge coefficient C. Then enter the chart at the standard penetration value N and read the corresponding value of q_o/B. Multiply this value by the width of footing to obtain the gross ultimate bearing capacity, and divide this result by a factor of safety to get the safe bearing capacity. If the assumed value of the footing width is larger or smaller than the width required to carry the footing load, select a different value of **B** and repeat the process until a combination of footing width and safe bearing capacity is found which will be both safe and economical.

EXAMPLE 23-4. Determine the size of footing resting on sand which will carry a column load of 145 tons. The corrected average standard penetration value of the sand below the footing is 25 blows/foot. The bottom of the footing will be 5 ft below the ground surface, and the water table is 10 ft below the footing. Use a factor of safety of 3.

SOLUTION. Assume a footing 5 ft square and 2 ft thick. The weight of the footing is $5 \times 5 \times 2 \times 150 = 7500$ lb, or 3.75 tons. Then the total footing load is 148.75 tons. The surcharge ratio is $C = 5/5 = 1.0$. For $N = 25$, it is found from the chart in Fig. 23-17 that $q_d/B = 2.3$; hence, $q_d' = 5 \times 2.3 = 11.5$ tons/sq ft. The safe trial bearing capacity is $11.5 \div 3 = 3.83$ tons/sq ft. So the required area of the footing is $148.75 \div 3.83 = 38.85$ sq ft. Since a footing 5 ft square provides only 25 sq ft, the first trial footing was too small.

Try a footing 6 ft square and 2.5 ft thick. Its weight is $6 \times 6 \times 2.5 \times 150 = 13{,}500$ lb, or 6.75 tons. Then $145 + 6.75 = 151.75$ tons and $C =$

$5/6 = 0.83$. From Fig. 23-17, $q_o/B = 2.05$. Hence, $q_o = 12.3$ tsf, $12.3 \div 3 = 4.1$ tsf, and $151.75 \div 4.1 = 37.0$ sq ft. This is slightly larger than the area of a footing 6 ft square, but is close. A designer would probably use a footing 6 ft 3 in. square without further effort toward refinement.

23.19. RECTANGULAR AND CIRCULAR FOOTINGS. The above two-dimensional solutions have been extended to three-dimensional problems by the use of models. Since a larger shear plane is involved, the cohesion factor N_c is increased; but since a short bearing area means less confinement of the soil, the width factor N_γ is decreased. The surcharge factor remains the same. For a square or circular footing Terzaghi (*10*) recommends multiplying $N_c \times 1.3$ and $N_\gamma \times 0.6$. However, based on more recent tests Leonards recommends multiplying $N_c \times 1.25$ and $N_\gamma \times 0.9$ for most soils. Thus little or no correction is needed for sands.

For clays ($\phi = 0$), Terzaghi and Peck recommend multiplying N_c by a factor $[1 + 0.3(B/L)]$ where B is the width of the footing and L is the length. For circular or rectangular footings the general bearing capacity Eq. (23-7) thus becomes

$$q_o = \frac{\gamma B}{2} N_\gamma + cN_c \left(1 + 0.3 \frac{B}{L}\right) + \gamma D N_q \tag{23-13}$$

For the special case of clay with $\phi = 0$, from Fig. 23-15, $N_\gamma = 0$, $N_c = 5.7$, and $N_q = 1.0$. Substituting,

$$q_o = 0 + 5.7c \left(1 + 0.3 \frac{B}{L}\right) + \gamma D$$

Since cohesion c equals one-half the unconfined compressive strength q_u,

$$q_o = 2.85\, q_u \left(1 + 0.3 \frac{B}{L}\right) + \gamma D \tag{23-14}$$

This equation is often used for design of footings on clay on the basis of unconfined compressive strength alone. As before, the gross ultimate bearing capacity should be divided by a factor of 3 or more to obtain a safe bearing capacity.

23.20. OTHER FAILURE THEORIES. Figure 23-19 shows several bearing capacity analyses solved for the special case of a cohesive ($\phi = 0$) soil and zero surcharge. Perhaps most pertinent is that despite various assumed geometries the results for either the general shear or local shear categories are quite close. Probably none of the assumed geometrics perfectly describes actual failure conditions, which are difficult to ascertain even from

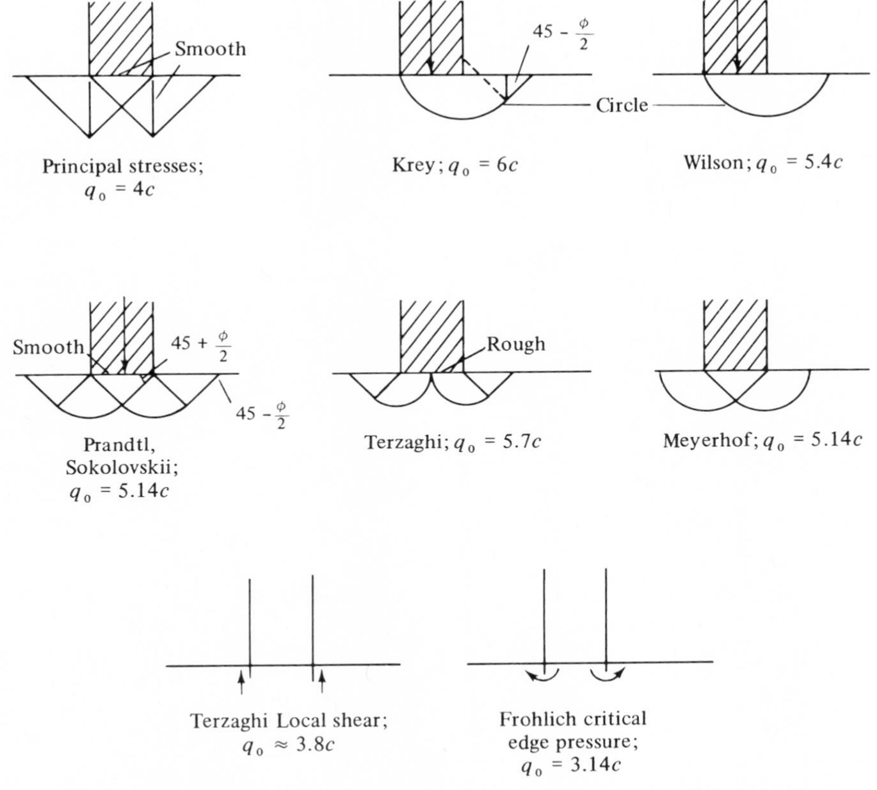

Fig. 23-19. Several bearing capacity theories solved for the special case of $\phi = 0$ and $D = 0$.

model tests because models involve inevitable distortions. Wilson solved for the most critical case of one-sided circular arc failure, the results being very close to the two-sided geometrics of Prandtl, Terzaghi, etc.; see Casagrande and Fadum (*1*).

23.21. SUMMARY: BEARING CAPACITY ANALYSIS. This chapter presents a number of commonly used alternatives for prediction of settlement and of ultimate bearing capacity of shallow foundations. In general they are arranged in order from the least to most complex and expensive. On the other hand the least complex methods of analysis are necessarily more conservative and therfore increase building costs, to cover possibilities not taken into account in the analysis. For example, bearing capacity on sands may be simply analyzed by assuming that cohesion is zero. This assump-

TABLE 23-3. Applicability of Various Methods to Evaluate Bearing Capacity, in Approximate Order of Increasing Investigative Detail

Criteria			
Settlement		*Ultimate Bearing Capacity*	
Sand	*Clayey Soils*	*Sand*	*Clayey Soils*
Soil identification from borings, referred to existing building codes			
SPT[a] (Fig. 23-3)	Preconsolidation stress, or compression index from liquid limit (Chapter 18)	SPT[a] and Fig. 23-8 or 23-9 (Presumptive bearing values)	
—	—	SPT[a] and Fig. 23-17 or 23-10 (c = 0 analysis)	Unconfined compressive strength and Eq. (23-14), or vane shear for soft clays. (ϕ = 0 analysis)
Plate load tests			
	Consolidation tests or stress path method (Chapter 18).	Triaxial, direct shear, bore-hole shear, or other tests to evaluate c and ϕ or c' and ϕ' for bearing capacity analyses	

[a] SPT = Standard penetration test or equivalent test (N-value).

tion is necessary if cohesion is not measured because in some sands cohesion *is* zero. Similarly, bearing capacity on clays is simply analyzed by assuming the friction angle is zero. These simpler methods of analysis will be sufficient and accurate for foundations on sand or on clay, but will be overly conservative for the vast range of intermediate soils which have both cohesion and friction. Another factor is cost of the structure—if the structure costs little, there is little to be saved by an extensive boring and testing program and analysis, whereas structures in the multi-million dollar category require the most extensive foundation soil investigation, including check solutions by several methods of analysis.

A third factor involved in the decision about the extent of investigation is whether or not unusual soil conditions are encountered in the preliminary borings which should be made for any structure. Conditions such as land fills, quick clays, quicksand, landslides, excavation adjacent to existing structures, etc., often warrant a detailed site study even for inexpensive structures because safety is involved.

These factors are summarized in Table 23-3; the final decision on methods used of course involves engineering judgment.

PROBLEMS

23.1. Define bearing capacity of a foundation soil. Define allowable bearing value.

23.2. Explain the difference between gross and net bearing capacity.

23.3. What two types of action of the soil comprising a foundation bed must be considered when the safe bearing capacity is being determined?

23.4. What is meant by the term differential settlement?

23.5. A square footing is to be founded on a stratum of medium clay. What is the allowable bearing capacity according to the Boston Building Code? If the unconfined compression strength of the clay is 1.5 tsf, what bearing capacity may be allowed?

23.6. If the footing in Problem 23.5 is under a long wall and has a length-width ratio of 8, what is the allowable bearing capacity?

23.7. A square column footing, founded on sand at a depth of 4 ft below ground level, will carry a load of 100 tons. The minimum average standard penetration value of the sand is 20 blows/foot. Determine the size of footing that will be required to be safe against a shear failure with a factor of safety of 3, (a) if the water table is 10 ft below ground level, (b) if the water table is 4 ft below ground level, and (c) if the water table is at ground level.

23.8. Will the footings of sizes determined in Problem 23.7 probably settle more or less than 1 in.?

23.9. The standard penetration of a stiff silty clay is 20. What is Hough's "presumptive bearing value" for this material?

23.10. Determine the perimeter–area ratio of each of the following footings: 28 ft square, 5 ft square, 9 ft in diameter (circular), 12 ft × 24 ft (rectangular), and octagon 15 ft between parallel sides.

23.11. A footing 12 ft square is to be constructed on a homogeneous soil. The maximum settlement of the footing is not to exceed $\frac{3}{4}$ in. Two test areas are loaded at the site, one being 3 ft by 3 ft and the other 5 ft by 5 ft. The load on the smaller area, which causes a settlement of $\frac{3}{4}$ in., is 34,000 lb, and that on the larger area is 66,500 lb. What is the bearing capacity of the soil, and what load will the 12-ft square footing carry?

23.12. The standard penetration test blow count on a cohesionless sand near the ground surface is 5. Assuming γ_{sub} = 50 pcf. (a) Estimate N'; (b) Estimate ϕ; (c) Calculate ultimate bearing capacity of a 4 ft wide footing in the ground surface by the following methods: Rankine, Terzaghi general shear, Terzaghi local shear, compromise general and local shear. Check the results with Fig. 23-17 or 18. Select which is most applicable, select a suitable factor of safety and define a safe bearing capacity. (d) How does the answer in (c) compare with Hough's presumptive bearing capacity obtained with N and with N'? Which N appears to be most valid for this use?

23.13. The unconfined compressive strength of a clay is 0.5 ton/sq ft. (a) Determine a safe column load on a footing 4 ft square located 10 ft below the ground surface. Assume γ = 100 pcf. (b) An excavation adjacent to the footing removes all but 1 ft of surcharge. What is the effect on the factor of safety?

23.14. Shear and density tests give the following data: γ = 120 pcf, c = 400 psf, ϕ = 10°. (a) Find the ultimate bearing capacity of a linear footing 6 ft wide at a depth of 22 ft below ground elevation, and 5 ft below the basement level, assuming a low water table. (b) What are the contributions of footing width, cohesion, and surcharge?

23.15. Calculate the minimum cohesion for a clay soil to support a D8 crawler tractor weighing 51,000 lb with a track width of 22 in. and a total contact area of 5445 sq in. Assume γ = 100 pcf, ϕ = 0, and zero depth of sinking.

23.16. (a) Will the tractor of Problem 23.15 overstress a loose, cohesionless sand, ϕ = 25°, γ = 100 pcf? If so, predict how deep the tractor will sink if there is no change in ϕ. (b) Repeat, assuming that the water table is at the ground surface. (c) Would (b) qualify as quicksand?

REFERENCES

1. Casagrande, A., and R. E. Fadum. "Application of Soil Mechanics in Designing Building Foundations." *ASCE Trans.* **109,** 383–416 (1944). Discussion by Terzaghi.
2. Hough, B. K. *Basic Soils Engineering*, 2d ed. Ronald Press, New York, 1969.
3. Housel, W. S. "A Generalized Theory of Soil Resistance." *ASTM Spec. Tech. Publ.* **206,** 13–29 (1956).
4. Jumikis, A. R. *Soil Mechanics.* Van Nostrand Co., Princeton, New Jersey, 1962.
5. Jumikis, A. R. *Theoretical Soil Mechanics.* Van Nostrand Reinhold, New York, 1969.
6. Lambe, T. W., and R. V. Whitman. *Soil Mechanics.* John Wiley, New York, 1969.

7. Meyerhof, G. G. "The Ultimate Bearing Capacity of Foundations." *Geotechnique* **2,** 301–332 (1951).
8. Peck, R. B., W. E. Hanson, and T. H. Thornburn. *Foundation Engineering.* John Wiley, New York, 1953.
9. Selig, E. T. "A Technique for Observing Structure-Soil Interaction." *ASTM Materials Res. and Standards* **1** (9), 717–719 (Sept. 1961).
10. Terzaghi, Karl. *Theoretical Soil Mechanics.* John Wiley, New York, 1943.
11. Terzaghi, K., and R. B. Peck. *Soil Mechanics in Engineering Practice,* 2nd. ed. John Wiley, New York, 1967.

24

Piles and Pile-Driving Formulas

24.1. TYPE OF PILES. Bearing piles are frequently used under foundations for the purpose of transmitting the foundation load to a deeper soil stratum having a higher bearing capacity than that immediately below, or to gain the increased bearing capacity which is associated with greater depth of load application. In the case of piers and abutments of bridges, an additional function of piles may be to prevent failure due to undercutting of the foundation by scour of the stream bed during flood flow. Also, piles may be driven into a stratum of loose sand for the purpose of compacting or densifying the material. Densification is accomplished by the vibration of driving and by decrease in void ratio to compensate for the volume of soil displaced by the driven piles. This operation increases the bearing capacity of the sand without reference to the load-carrying capacity of the piles.

Piles may be made of many different kinds of materials, and they vary widely in length and cross section. Some of the more common types in present-day use are shown in the classification diagram in Fig. 24-1. Bearing piles which transmit their load to the soil by friction between the pile surface and the soil are called friction piles. Those which transmit their load to a stratum with a high bearing capacity at the lower end are called point bearing piles. Actually, most piles transmit the load to the soil by a combination of these actions. Point bearing piles may be thought of as long slender columns. However, they are not free columns, since they de-

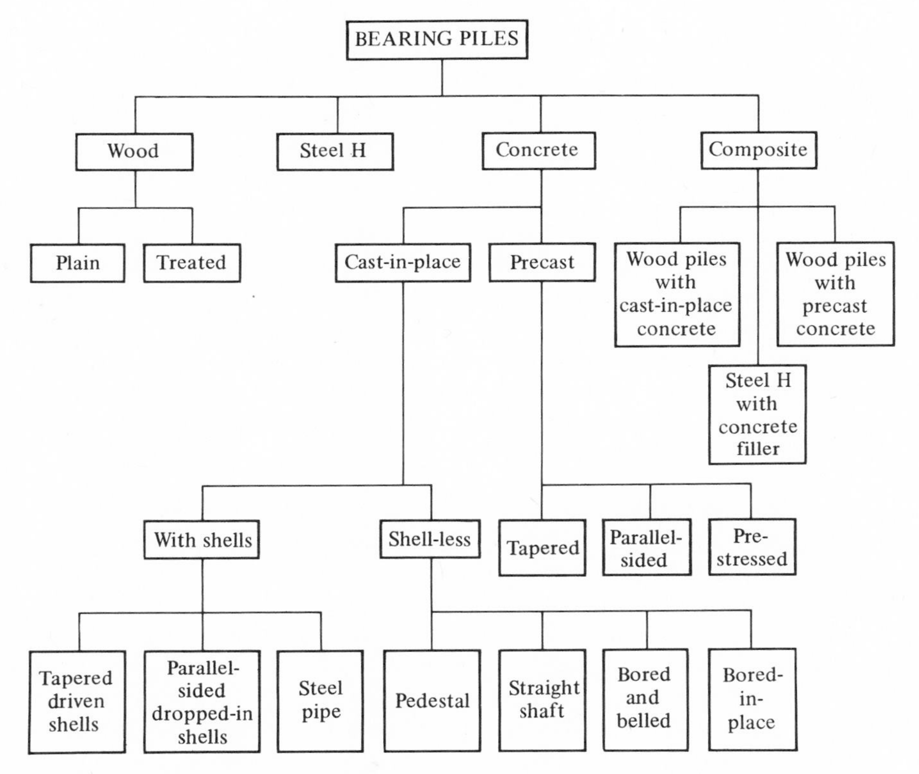

Fig. 24-1. Types of bearing piles (adapted from Cummings).

rive considerable lateral restraint against buckling from the relatively soft soil through which they are driven. Failure of a point bearing pile by column buckling is practically an unheard-of occurrence, unless the piles extend for long distances through air or water, which do not provide any lateral support against buckling. Piles driven into a soil stratum with a substantial length extending through an overlying newly placed fill may be subjected to *negative skin friction* or down-drag shear as the fill consolidates under its own weight. This friction or shear adds to the load which the pile will be required to carry.

Piles may be driven vertically or they may be driven on a batter to resist horizontal components of a foundation load. Batter piles are sometimes referred to as spur piles. The batter of a spur pile is expressed as a fraction in which the horizontal leg of the slope triangle is the numerator and the vertical leg is the denominator. Thus, a batter of $\frac{1}{3}$ means a slope at the rate of 1 ft horizontal to 3 ft vertical.

24.2. WOOD BEARING PILES. Plain wood piling is the oldest and perhaps the most commonly used type, because of its comparative cheapness and availability. It may not be the most economical in the long run, however, and careful study of the relative overall cost should be made before the pile type is finally selected. Plain or untreated wood is very susceptible to decay when subjected to alternate wetting and drying. Therefore, unless the pile is completely buried below the lowest elevation of the permanent ground water table, its life may be comparatively short. There are many instances on record in which untreated wood piles under buildings in urban areas have deteriorated prematurely because the ground water table has receded to an elevation below the tops of the piles. When such circumstances are even remotely possible, consideration should be given to alternate types of piling, such as creosoted wood, concrete, or composite piles of wood and concrete.

24.3. CONCRETE PILES. The use of concrete piles is increasing, largely because the diminishing supply of timber and high transportation costs have improved their economic position in comparison with that of wood piles. As indicated in Fig. 24-1, concrete piles are of two general classes, namely, cast-in-place and precast. Cast-in-place piles may or may not have a steel shell left in the ground into which the concrete is poured. The purpose of the shell which is left in the ground is to prevent mud and water from mixing with the fresh concrete and to provide a form for the pile. The shell-less piles may be used in drier, more stable locations where a hole in the soil will hold its shape and the fresh concrete may be poured directly in contact with the surrounding soil.

24.4. DRIVEN SHELLS. When steel shells are left in the ground to receive cast-in-place piles, they may be tapered driven shells, parallel-sided dropped-in shells, or steel pipes. One type of tapered driven-shell pile is made as follows: A thin corrugated steel shell, closed at the bottom with a steel boot, is placed on the outside of a steel mandrel or core; and the shell and the core are driven together into the ground, as indicated in Fig. 24-2(a). When sufficient driving resistance has been developed, driving is stopped and the mandrel is withdrawn from the shell. The steel-lined hole in the ground can be inspected with a lamp on a drop cord or a flashlight, as indicated in Fig. 24-2(b); and the shell is then filled with concrete, as shown in (c). A relatively thin shell is used because the heavy steel driving mandrel supports the shell against collapse during the driving operation. Shells of this kind may be tapered uniformly from the bottom boot to the top, or they may be step-tapered. The step-tapered shell consists of a series

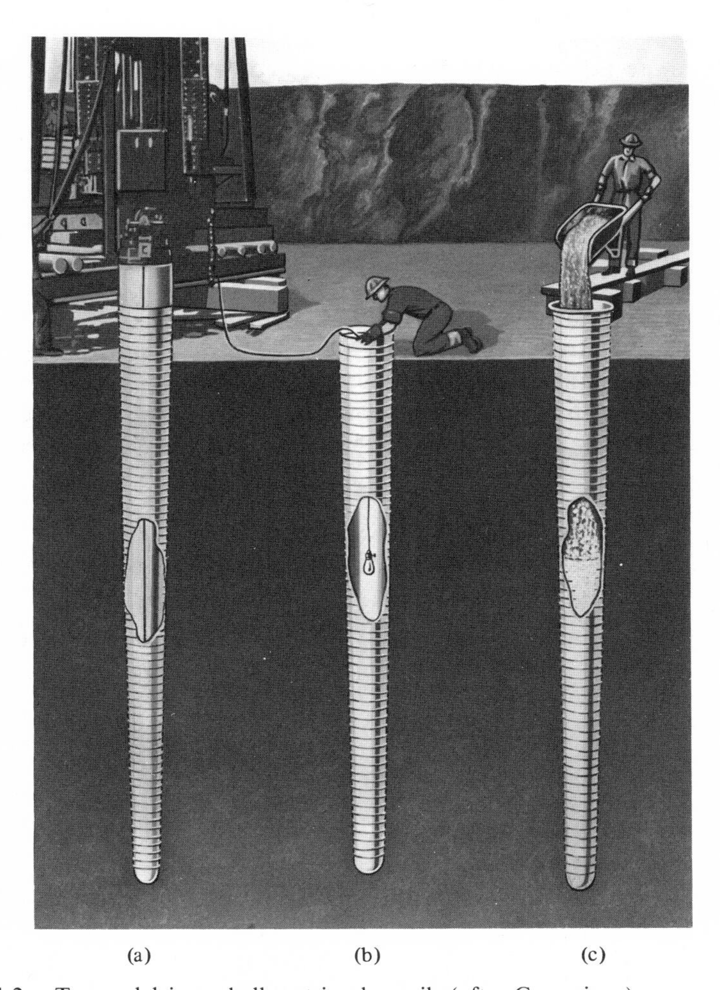

(a) (b) (c)

Fig. 24-2. Tapered driven-shell cast-in-place pile (after Cummings).

of parallel-sided tubes of different diameters, each about 8 ft long. The tube of smallest diameter, with the bottom-boot closure, is placed at the lower end of the shell. Then tubes of successively larger diameter are added to this first section until the desired length of shell is obtained. The driving mandrel is step-tapered in a similar manner, and the shoulder formed at each point of change of diameter bears directly on the top end of each shell tube. Thus, the driving force is well distributed throughout the length of the shell.

Another type of driven shell has a scalloped cross section which is formed by a series of vertical flutings running the full length of the shell. They are uniformly tapered and are closed at the bottom by a steel boot. The shells are driven into the ground and then filled with concrete to form cast-in-place piles. Shells of this kind are driven without the aid of a driving mandrel; therefore, they must be heavy enough to withstand the driving stresses. The thinnest shells which can be used for this type of pile are $\frac{1}{8}$ in. thick, but thicknesses of $\frac{3}{16}$ in. and $\frac{1}{4}$ in. are much more common, the selected thickness depending on the length and diameter of the pile.

The parallel-sided, dropped-in shell pile is constructed by driving a heavy steel pipe, usually about 16 in. in diameter and $\frac{1}{2}$ in. thick, with the aid of a steel mandrel, as indicated in Fig. 24-3(a). When suitable driving resistance is encountered, driving is stopped and the mandrel is removed from the pipe. A thin metal shell, such as a corrugated-metal culvert pipe, which is slightly smaller in diameter than the drive pipe is dropped down

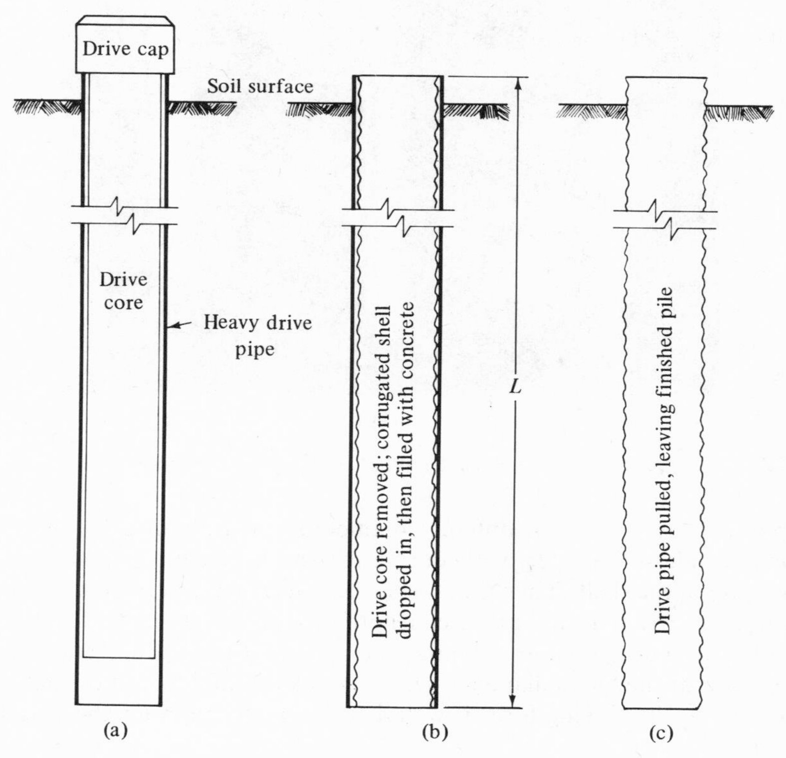

Fig. 24-3. Steps in construction of parallel-sided, dropped-in shell pipe.

inside, as represented in Fig. 24-3(b), and filled with concrete. Then the drive pipe is pulled out of the ground, and the finished pile is left, as shown in (c). The bearing capacity of this kind of pile depends almost wholly on point resistance, since the surrounding soil may not close in sufficiently to develop any appreciable friction along the sides of the pile.

24.5. STEEL-PIPE PILES. A third kind of cast-in-place pile with a steel shell left in the ground is the steel-pipe pile with either a closed end or an open end. The closed-end pipe pile is simply a steel pipe which is closed at the bottom end with a heavy boot, driven into the ground, and then filled with concrete. Sometimes a pipe is driven all in one piece or it may be made of several pieces of pipe either welded together or fitted together with special internal sleeves.

The open-end steel pipe is usually driven to point bearing on rock. While the pipe is being driven, it is open at the bottom and fills with soil as driving progresses. When driving is completed, this soil is removed from the pipe either with air and water jets or with a miniature orange-peel dredging bucket. After the soil is removed, the pipe is given a few blows with the pile-driving hammer to make sure that it is properly seated in the bedrock; any water remaining in the pipe is pumped out; and the pipe is filled with concrete. A pile installed in this manner is, in effect, a caisson of relatively small diameter and it has a high bearing capacity.

24.6. SHELL-LESS PILES. There are several kinds of shell-less concrete piles, such as pedestal piles, straight-shaft piles, bored and belled piles, and bored-in-place piles. The pedestal pile is formed by first driving a heavy steel pipe containing a mandrel or core into the ground until satisfactory resistance is encountered, as indicated in Fig. 24-4(a). Then the mandrel is removed and a small quantity of conrete is poured into the pipe, as indicated in (b). The mandrel is replaced in the pipe and lowered to the concrete, after which the pipe is pulled upward 2 or 3 ft, as indicated in (c). Then the hammer is used to drive the mandrel downward, and thus force the concrete out of the lower end of the pipe where it forms into a bulb-shaped mass, as shown in (d). Then the mandrel is removed once more and the pipe is filled with concrete up to the top. Finally the mandrel is placed on the concrete to hold it down while the steel drive pipe is pulled out of the ground, as represented in (e). The completed pile is shown in Fig. 24-4(f). The straight-shaft shell-less concrete pile is made with the same equipment and in essentially the same manner, but the pipe is completely filled with concrete after the mandrel is removed the first time.

The bored and belled pile is primarily suited to locations where the soil is relatively firm and a hole bored to receive the concrete will stand

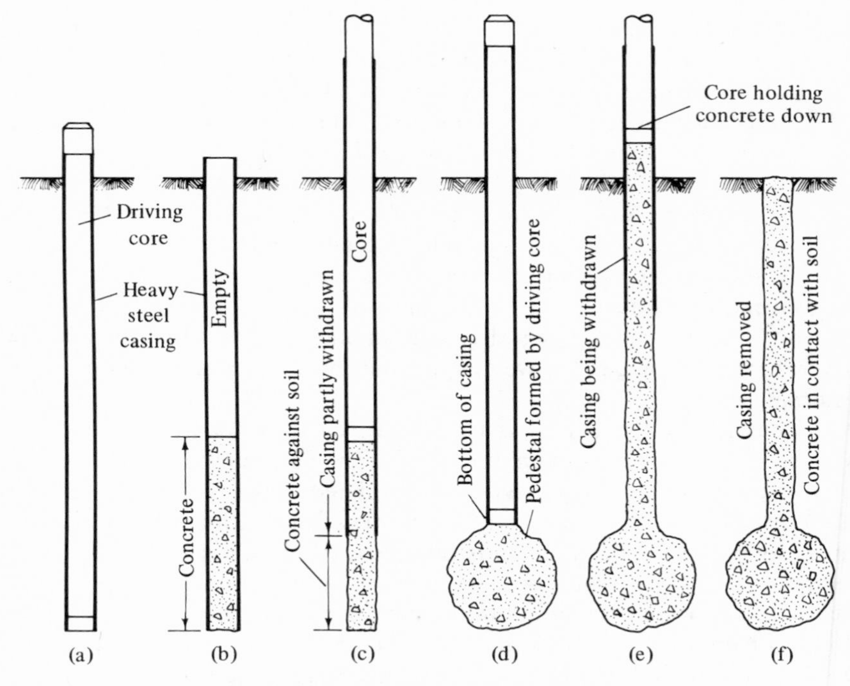

Fig. 24-4. Construction of concrete pedestal pile (after Cummings).

without caving or slumping. The procedure is to bore a hole to a predetermined depth. Then the auger is removed, an underreamer is inserted, and the bottom of the hole is reamed out to form a bell-shaped base as shown in Fig. 24-5(a). The underreamer consists of expandable blades which open out on an angle of about 30° with the vertical to form the belled base. When cutting is complete the underreamer is retracted and removed from the hole. Then the hole is cleaned out and filled with concrete.

The bored-in-place pile is more suitable for use in less stable soils. It is formed by boring a hole down and into a high-bearing stratum with a continuous flight auger having a hollow central shaft, as illustrated in Fig. 24-5(b). Then as the auger is slowly withdrawn, high-strength concrete mortar is pumped under pressure through the hollow shaft to form the pile. The pumping pressure establishes intimate contact between pile and soil, and a rough contact surface is created by filling small horizontal depressions. This roughness contributes to a reasonably high friction component of the total bearing capacity of the pile. The bored-in-place pile may be particularly adaptable in locations where vibrations of conventional pile driving operations might cause damage to existing structures.

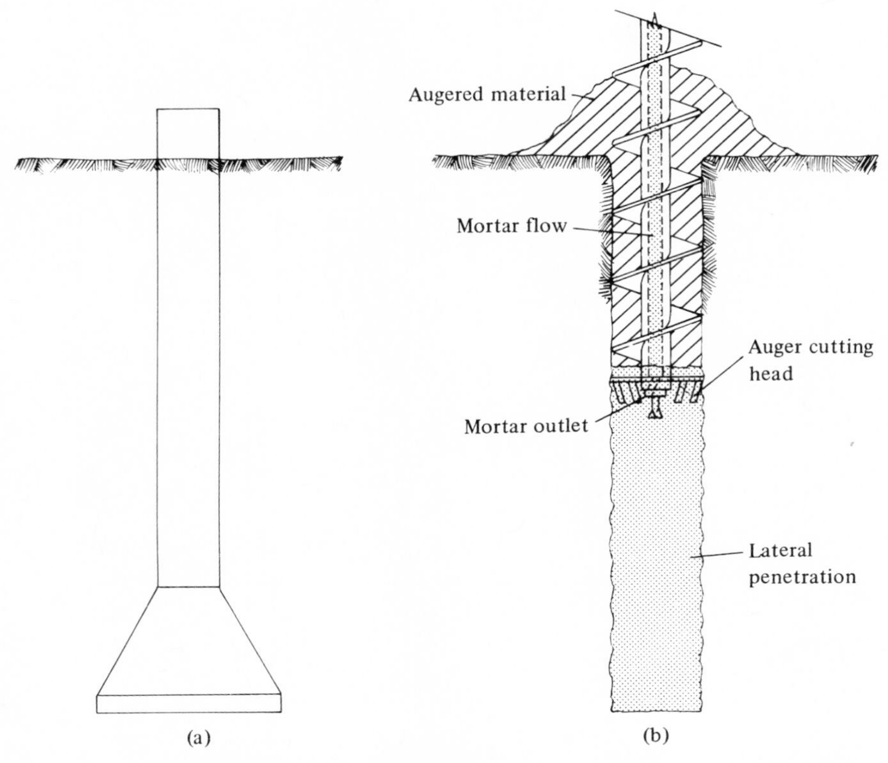

Fig. 24-5. (a) Bored and belled pile; (b) bored-in-place pile (courtesy Lee Turzillo Contracting Co.).

The shell-less types are the cheapest kinds of all concrete piles, since they are made of plain concrete only without steel either in the form of reinforcement or a left-in-place shell. They require greater care in construction, however, because driving operations of subsequent piles may distort and injure newly made piles before the concrete has an opportunity to set.

24.7. PRECAST CONCRETE PILES. Precast concrete piles are made either square or octagonal in cross section, and they may be tapered uniformly from tip to butt or tapered only for a length of 4 or 5 ft at the lower end and parallel-sided for the rest of their length. Three forms are shown in Fig. 24-6. The piles are usually cast in a horizontal position; and for this reason they are made square or octagonal instead of circular. Nonprestressed piles are reinforced with longitudinal bars and transverse steel, which may

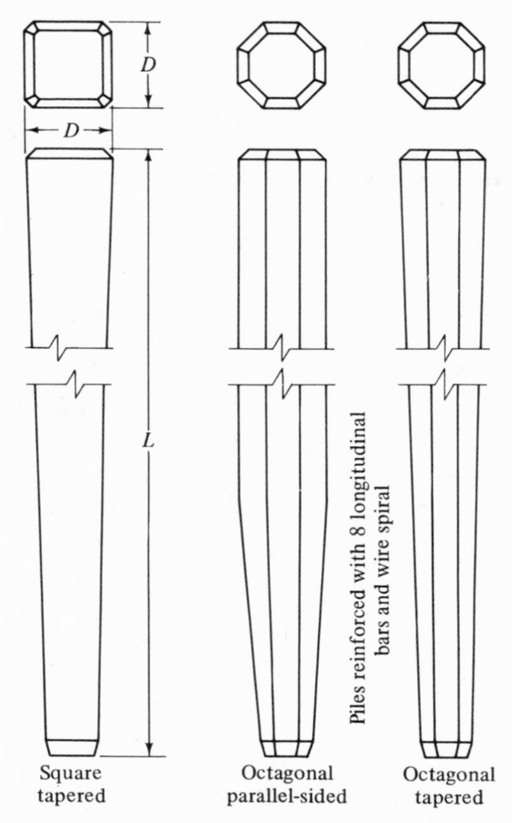

Fig. 24-6. Three types of precast concrete piles.

be in the form of separate loops or continuous spirals. The transverse steel is always spaced closer together for a length of 3 or 4 ft at each end of the pile than near the midsection, because the stresses in the pile during driving are greatest near the ends.

Handling and driving stresses in precast concrete piles are very severe and for this reason, prestressed concrete piles are coming into greater usage. They may be either pretensioned or posttensioned. The level of prestress is relatively low, about one-fourth to one-third the usual level for more orthodox structural members. This seems to provide sufficient strength to withstand the usual handling and driving stresses. It is doubtful if prestressing contributes to the strength or efficiency of the pile after it is in place.

Precast piles, whether prestressed or not, must be handled very carefully prior to driving, to prevent injury from longitudinal bending stresses due to their own weight. They should always be lifted by slings under the third-points to reduce bending stresses to a minimum. Uniformly tapered precast piles are limited in length to about 40 ft because of the small cross-sectional area of the lower end of such a pile. When greater lengths are needed, the parallel-sided pile with tapered end is more appropriate because of the greater cross-sectional area available to resist handling stresses. Piles of this design well over 100 ft in length have been successfully used. Octagonal precast piles up to 36 in. in diameter have been made, but their weight is excessive. This objection can be remedied somewhat by casting the pile with a 10-in. or 12-in. cylindrical hole centered on the longitudinal axis.

Precast piles are usually manufactured in a temporary casting yard near the site where they are to be used, because transportation and handling costs are high. When very hard driving conditions are anticipated, each pile may be equipped with a steel driving shoe. This is a pointed steel cap which fits over the tip of the pile. It is anchored into the concrete at the time of casting by steel straps.

24.8. COMPOSITE PILES. Composite piles of timber and concrete combine the low cost of wood piles with the advantages of concrete piles, which do not have to be buried below the ground water level. Where a composite pile is used, the wood section is driven down until its upper end is about at the ground surface. Then it is topped off with a concrete pile, which is used as a follower to drive the head of the wood pile down below the ground water level. When driving is completed, the concrete section extends from below the water level up to the cut-off grade, as shown in Fig. 24-7.

The most important element of the composite pile is the joint between the wood and concrete sections. It should be designed to provide good positive bearing between the concrete and wood, and should have some strength in both tension and bending. Also, it should be simple in design and easily constructed in the field. A satisfactory joint can be made, as indicated in Fig. 24-7, by cutting the upper end of the wood section to form a tenon about 9 in. in diameter and 18 in. long. The shoulders of the tenon should be firm and square. A few heavy spikes driven into the tenon will aid in bonding the concrete to the timber. Next, a length of corrugated-metal culvert pipe, whose inside diameter is the same as that of the butt of the wood pile, is placed over the tenon and filled with concrete. The concrete around the tenon will have a relatively thin section, and a few hoops of heavy steel wire in this region will provide insurance against damage when driving is resumed. The use of high-early-strength cement in the con-

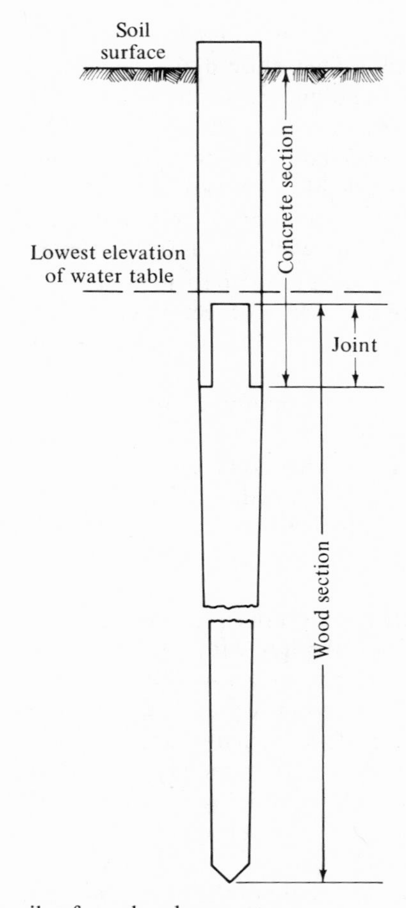

Fig. 24-7. Composite pile of wood and concrete.

crete section will reduce the time interval between pouring the concrete and finish driving of the composite pile.

The method of constructing a composite pile just outlined has the distinct advantage that the joint between the wood and the concrete is made above ground, where it can be observed and inspected. It has the disadvantage that considerable time must elapse between pouring the concrete section and final driving. Several other methods may be employed in forming a composite pile. One is first to drive a heavy steel casing and driving mandrel down below water level; and then to remove the mandrel, to insert in the casing a wood pile with a tenon cut at the head, and to drive it down below the water level. The mandrel in this case should be equipped with a drive cap, which fits over the tenon and bears solidly on both the shoulder and the end of the tenon. After the wood pile is driven to the proper elevation, a corrugated metal pipe is inserted in the casing down over the tenon, and the pipe is filled with concrete to complete the pile.

24.9. STEEL H-BEAMS. Steel H-beam sections are comparatively new in the field of bearing piles. They have proved to be especially useful for

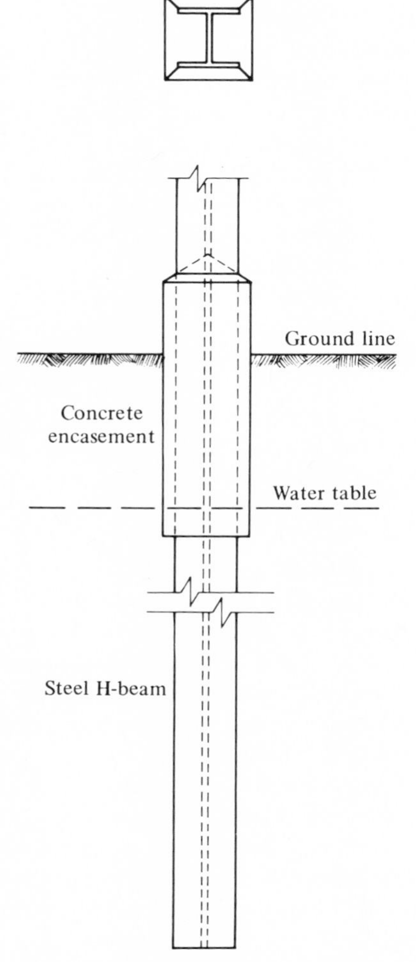

Fig. 24-8. Steel H-beam pile encased in concrete near ground line to protect the steel against corrosion.

trestle structures in which the piles extend above the ground line and serve both as bearing piles and as columns. Because of their small cross-sectional area, they can often be driven into a river bed of sand and gravel into which, even with the aid of jetting, it would be very difficult to drive a displacement pile such as one of timber or concrete. However, this ability of the H-beam to penetrate more easily into dense material often works to its disadvantage in softer material. Bearing piles frequently support their load mainly by friction along the sides of the pile, and this friction is increased by compaction of the soil as it is displaced laterally during pile driving. Since H-beams compact the soil very little, greater lengths of this type of pile are sometimes necessary to support the same load which can be carried by a shorter length of a displacement pile.

Steel H-beams which are intended to serve as point bearing piles are sometimes fitted with a steel base plate or shoe. This increases the bearing area between the pile and the relatively stiff stratum to which it is driven. The displaced cross-sectional area of softer soil through which such piles are driven is much greater than that of the pile itself, and friction between the soil and the pile is practically nil.

When a steel pile is exposed to the air or to alternate wetting and drying above a ground water table, it is subject to corrosion the same as any steel structure under similar conditions. In some installations, H-beam piles are protected from corrosion by frequent painting with an asphalt compound. Sometimes, when an H-beam is used both as a bearing pile and as a column of a bent, it is protected from corrosion by encasing it in concrete for several feet above and below the ground line, as shown in Fig. 24-8.

24.10. PILE-DRIVING HAMMERS. A pile is driven by means of a hammer which applies high-energy blows to the butt or head of the pile. There are four types of hammers in common use: the drop hammer, the single-acting steam hammer, the double-acting or differential-acting steam hammer and the diesel-powered hammer. The drop hammer is simply a heavy steel casting with grooves on two sides which fit over vertical guide rails in the pile-driver leads. It is lifted to the desired height by means of a rope and pulley and is allowed to drop onto the pile head. The hammer may be released by means of a trip hook which disconnects the rope from the hammer and allows it to fall freely; or it may be released by disengaging the clutch on the hoisting drum. in which case the hammer drags the hoisting line as it falls.

A steam pile-driving hammer is one that is automatically raised and dropped a comparatively short distance by the action of a steam cylinder and piston supported in a frame which rests upon and follows the pile as it is driven. In the single-acting steam hammer the striking part is raised by steam power to a certain height where the steam is automatically released and the hammer falls to the pile head by gravity. At this point the steam pressure is automatically turned on; and the cycle of lift and drop continues at a rapid rate. In the double-acting steam hammer the striking part is not only raised by steam, but its gravity fall is augmented by steam power on the downward stroke. Steam hammers are capable of applying from 50 to 110 blows/min. Hammers may also be operated with compressed air or diesel fuel; in the latter, each fall of the hammer compresses an air-fuel mixture, causing it to ignite and lift the hammer.

24.11. PROTECTION OF PILE HEAD. When a wood pile is being driven, the blows of the hammer tend to crush the wood fibers at the pile head and may cause the pile to split vertically for some distance. This crushing of the wood is referred to as brooming. It is necessary to cut the damaged portions of the piles off before the foundation is placed on the piles, and the piles should be of sufficient length to allow for this cut-off. In order to reduce the amount of brooming and splitting, a heavy steel ring is often placed over the head of the pile during driving. In the case of a precast concrete pile a metal drive cap is placed over the pile head with laminated layers of wood between the cap and the pile. Also, a hardwood block is placed above the cap to help cushion the blow of the hammer and protect the pile head.

24.12. BEARING CAPACITY OF PILES. The determination of the bearing capacity of a pile is a very difficult problem. The only way that it can be determined with certainty is to test-load a pile after it is driven; but this is an expensive and time-consuming operation. Nevertheless, test-loading of a number of representative piles is justified in the case of a large and expensive structure.

A preliminary estimate of the bearing capacity of an individual pile may be obtained as follows: The total supporting capacity is the sum of end-bearing plus side friction (Fig. 24-9):

$$Q = Q_p + Q_s \tag{24-1}$$

End-bearing Q_p may be estimated by use of a bearing capacity formula corrected for a rectangular or circular bearing area, Eq. (23-13), with Tezaghi factors (Fig. 23-15). The Meyerhof factors are larger and more accurate for deep foundations in dense soils, but the lower Terzaghi factors may be preferred since end-bearing and side frictional support will seldom peak out at the same amount of deflection of the pile. The Terzaghi factors also are close to those derived by Vesić (*9*) for a limited failure zone.

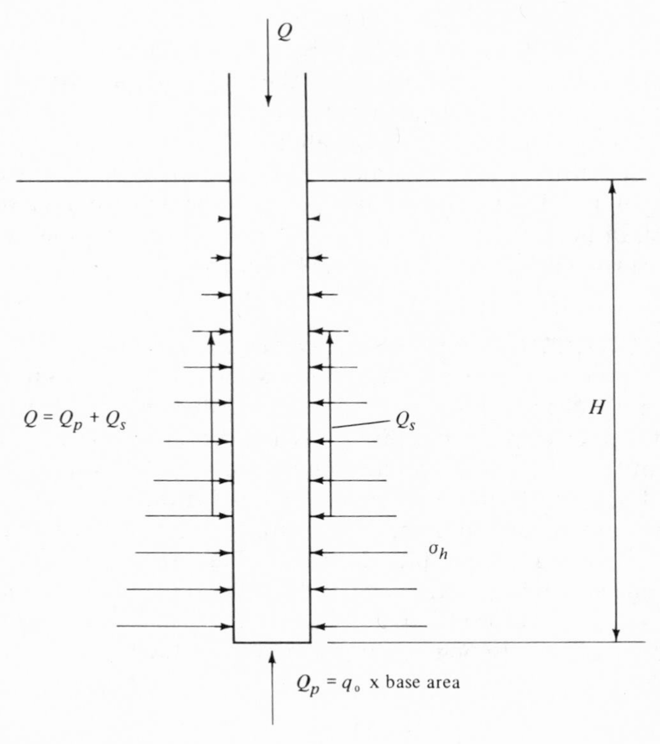

Fig. 24-9. Forces acting on a pile.

Side friction may be estimated by the formula

$$Q_s = CH(c_d + \bar{\sigma}_h \tan \phi_d)$$
$$Q_s = CH(c_d + \tfrac{1}{2} K_d H \gamma \tan \phi_d) \quad (24\text{-}2)$$

in which

Q_s = total support from side friction;
C = circumference of the pile;
H = depth of pile penetration;
c_d = developed cohesion between the soil and pile;
$\bar{\sigma}_h$ = average horizontal pressure for the length of the pile;
K_d = ratio of horizontal to vertical earth pressure;
γ = unit weight of the soil; and
ϕ_d = angle of friction developed between the soil and pile.

A rectangular circumference is assumed for a steel *H* pile since the spaces between the flanges fill with soil. The upper limits of c_d and ϕ_d are c and ϕ for the soil. One of the most important unknowns is the value of K_d. The amount of test data available to evaluate K_d is very limited, primarily because of lack of adequate data on the full depth of the soil involved. When a number of piles are driven close to one another, K_d should exceed 1.0 and may approach the coefficient of passive resistance, K_p [Eq. (22-19)]. For bored pile K_d may approximate K_o, the coefficient of earth pressure at rest. For bored and cast-in-place pile a reasonable assumption is to calculate $\bar{\sigma}_h$ from the unit weight of the fluid concrete, i.e., use $K_d = 1.0$ and $\gamma = 150$ pcf.

EXAMPLE 24-1. Estimate the bearing capacity of a 30-ft, 12 in. diameter cast-in-place pile in soil with $c = 1.4$ psi, $\phi = 15°$, and $\gamma = 110$ pcf, and a water table at the ground surface.

SOLUTION. $Q = Q_p + Q_s$
Solving first for Q_p, from Eq. (23-13) and Fig. 23-12,

$$q_o = \frac{\gamma B}{2} N_\gamma + cN_c \left(1 + 0.3 \frac{B}{L}\right) + \gamma D N_q$$

where B and L are width and length of the bearing area, both 1.0 ft. Using submerged unit weight of the soil, and the surcharge depth D as equal to the length of pile,

$$q_o = \frac{47.6\,(1)}{2}(2) + (1.4 \times 144)(13)(1.3) + 47.6\,(30)(5)$$

$$= \underset{\text{width}}{47.6} + \underset{\text{cohesion}}{3407} + \underset{\text{depth}}{7140} = 10{,}595 \text{ psf}$$

$$Q_p = q_o \pi (\tfrac{1}{2})^2 = 8321 \text{ lb}$$

Solving Eq. (24-2) for Q_s by using K_d and γ for fluid concrete,

$$Q_s = \pi(1)(30)(1.4 \times 144) + \tfrac{1}{2}(1)(30)(150) \tan 15°$$

$$= \underset{\text{cohesion}}{19{,}001} + \underset{\text{friction}}{56{,}853} = 75{,}834 \text{ lb}$$

$$Q = 8{,}321 + 75{,}834 = 84{,}155 \text{ lb} = 84 \text{ Kip}$$

A check should also be made that this does not exceed the unconfined compressive strength of the concrete in the pile.

An advantage of the above type of preliminary analysis is that it allows one to see the various contributing factors for both end-bearing and side friction. For example, the above example apparently is primarily a friction pile, and for added safety the end-bearing might be neglected, giving an ultimate capacity of 75 Kip. This could be used as a guide for a load test or divided by an appropriate factor of safety if used for design.

24.13. ACTION OF PILE GROUPS. Even if every pile were test-loaded, there would remain a question of the bearing capacity of all the piles in a group because of the overlap of the stressed zones of soil around adjacent individual piles. It is a widely recognized fact that the supporting power of a group of piles will be less than the sum of the bearing capacities of the individual piles, particularly in the case of friction piles. There is very little quantitative information available in regard to the relationship between group and individual-pile bearing capacity. One rule-of-thumb procedure, known as the Feld rule, is to reduce the bearing capacity of each pile by one-sixteenth on account of the effect of the nearest pile in each diagonal or straight row of which the pile is a unit. Thus for the 9-pile group in Fig. 24-10, the reduction will be $\frac{3}{16}$ for each corner pile; $\frac{5}{16}$ for the center pile on each side; and $\frac{8}{16}$ for the pile in the center of the group. This is a reduction of 28% below the gross bearing capacity obtained by multiplying the capacity of one pile by the total number of piles. The *efficiency* of the group is, therefore, 100 minus 28 or 72%.

Also, in the case of friction piles driven into a deep stratum of cohesive soil, it is necessary to investigate the bearing capacity and stability of the whole mass of soil which encompasses the pile group. This may be done by assuming that the supporting strength is equal to the shearing strength around the periphery of the mass plus the bearing capacity at the

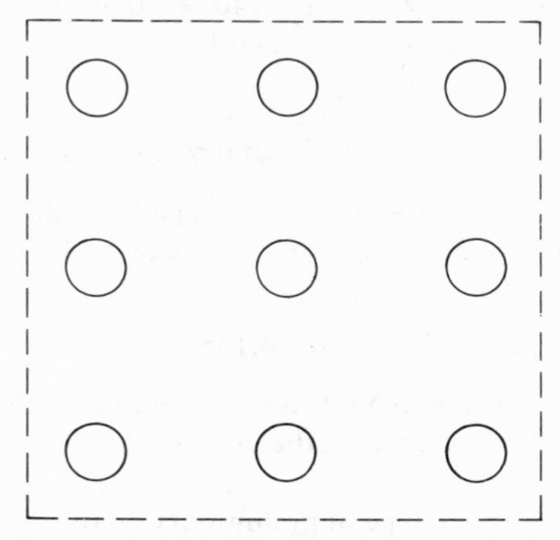

Fig. 24-10. Efficiency of group = 72 percent.

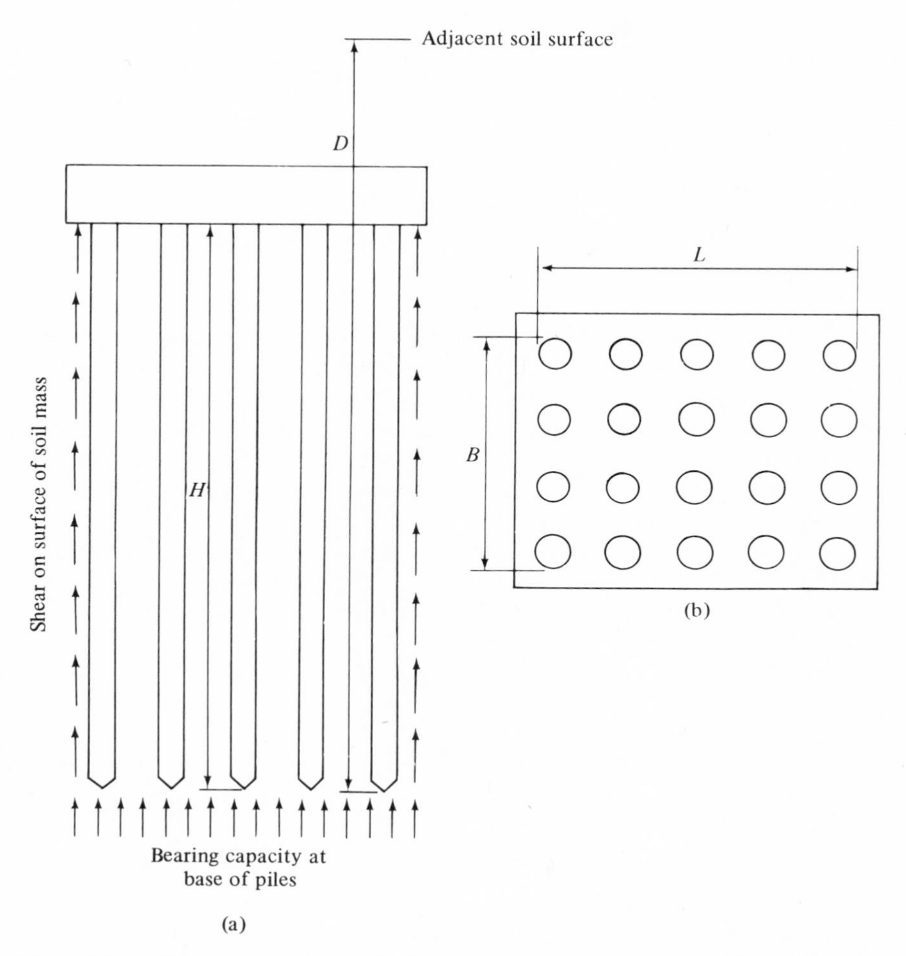

Fig. 24-11. Bearing capacity = shear on sides of soil mass + bearing value at base of enveloping soil.

base of the mass, acting as a pier. Thus, referring to Fig. 24-11 and Eq. (23-14), the gross ultimate bearing capacity of the pile group is [see Refs. (*5*) and (*8*)]

$$Q_g = c\left[(2B + 2L)H + 5.7BL\left(1 + 0.3\,\frac{B}{L}\right)\right] + BL\gamma D \qquad \text{(24-3)}$$

in which

Q_g = gross ultimate bearing capacity of pile group;
c = cohesion of soil;

= one-half unconfined compression strength $= q_u/2$;
B = breadth of pile group;
L = length of pile group;
H = depth of pile penetration;
γ = unit weight of soil; and
D = depth of surcharge.

The result of Eq. (24-3) should be divided by an appropriate factor of safety, say 2 or 3.

24.14. BASIC ENERGY RELATIONSHIP. For many decades, engineers have attempted to devise formulas for determining the static bearing capacity of piles from facts observed during the process of driving. This procedure involves expressing dynamic forces and effects in terms of static forces; and it is a very difficult thing to do because of energy losses during the driving process and the resiliency of the pile and the soil into which it is being driven. There are probably several dozen different formulas of this kind in the engineering literature, but none of them is satisfactory in all respects. The basis for practically all these formulas is the simple energy relationship which may be stated by either of the following equations:

$$WH = RS \tag{24-4}$$

or

$$R = \frac{WH}{S} \tag{24-5}$$

in which

W = weight of hammer;
H = height of fall;
WH = energy of hammer blow;
R = resistance to penetration;
S = pile penetration under one hammer blow; and
RS = resisting energy of pile.

24.15. ENGINEERING NEWS FORMULA. Practically all pile-driving formulas are modifications of the basic energy relationship. These modifications are the results of efforts by various investigators to take account of the many and varied conditions which actually exist in a pile-driving operation and which cause the actual energy relationships to deviate from those given in Eqs. (24-4) and (24-5). The simplest dynamic formula of this kind, and the one which is probably used most widely in the United States, is known as the Engineering-News formula. This formula has the following forms:

$$R = \frac{2WH}{S + 1} \qquad \text{(for drop hammers)} \qquad (24\text{-}6)$$

$$R = \frac{2WH}{S + 0.1} \qquad \text{(for single-acting steam hammers)} \qquad (24\text{-}7)$$

in which

R = bearing capacity of the pile, in tons;
W = weight of hammer, in tons;
H = hammer drop, in feet; and
S = penetration of pile, in inches.

The Engineering-News formula is almost wholly empirical in character. It was developed many years ago when practically all piles were of wood and were driven by drop hammers of relatively light weight. It ignores many factors which influence bearing capacity, such as pile length, weight, cross section, taper, kind of material, and character of soil. Nevertheless, it has a good background of experience and, *on the average*, gives reasonably good results.

The theoretical factor of safety of the Engineering-News formula is supposed to be 6. However, comparisons of calculated bearing capacities and the results of load tests indicate that, on the average, the actual factor of safety is much less than the theoretical value. Also, such comparisons reveal a wide spread between minimum and maximum actual factors of safety. For example, in one series of tests, Mumma (*6*) determined the actual factor of safety, that is, the quotient obtained by dividing the ultimate test load by the calculated working load of 58 individual piles driven and tested by the U.S. Army Corps of Engineers. The test piles represented a wide variety of types, lengths, sizes, soil conditions, and geographical locations.

Another series of tests conducted by the Michigan State Highway Commission (*4*) indicated the factor of safety of a similar number of piles. These tests were conducted at three sites selected for their distinctive soil types. At one site the soil was stiff and cohesive and required hard driving to overcome high resistance to pile penetration. At a second site, deep soft cohesive soil offered weak resistance. At the third site the soil consisted of deep granular deposits with intrabedded organic materials. Four types of piles were used: H-section, pipe, flute-tapered monotube, and step-tapered shell.

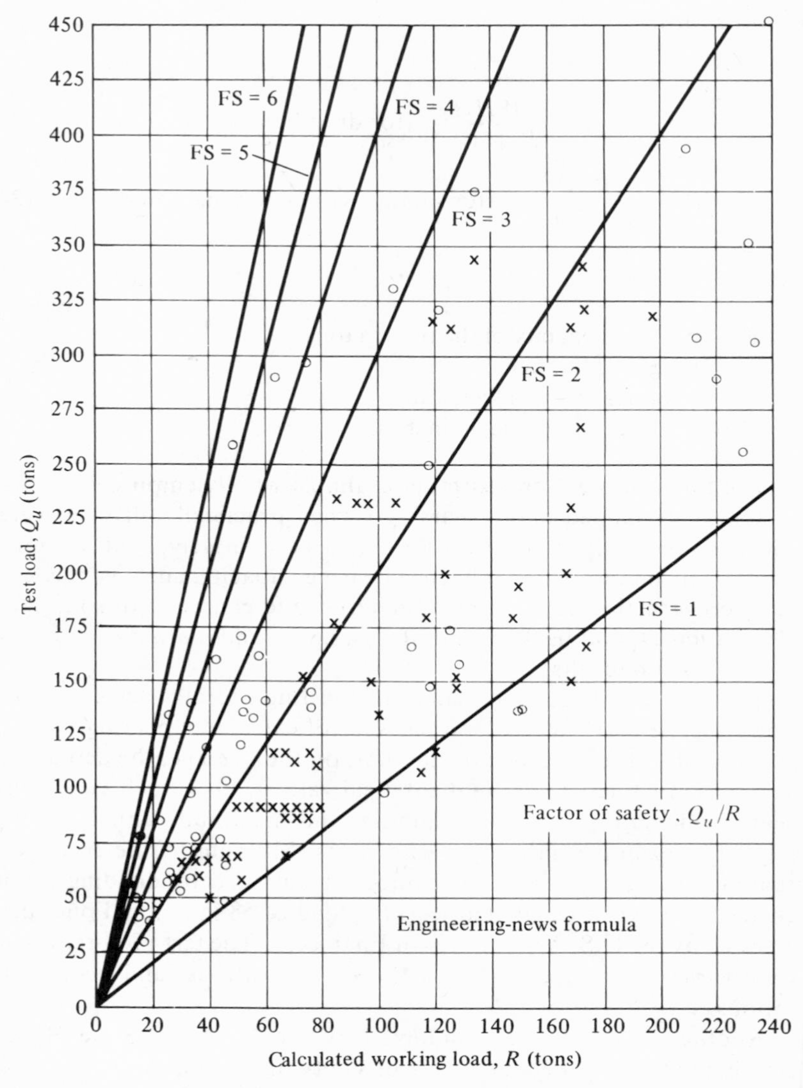

Fig. 24-12. Comparison of test loads and calculated loads on piles (Engineering News Formula). Key: o—U.S. Army Corps of Engineers Tests (Mumma); x—Michigan State Highway Commission Tests.

The relationships between ultimate test loads from both the Corps of Engineers and the Michigan tests and working loads calculated by the Engineering-News formula are shown in Fig. 24-12. It will be noted that there is a wide spread in actual factors of safety, and that eight of the test piles indicated values less than 1.0.

24.16. RABE FORMULA. An empirical formula which takes into account practically all the factors that influence pile bearing capacity has been developed by W. H. Rabe (*6*) of the Ohio Department of Highways. It is based upon facts observed in regard to pile type, length, and cross sec-

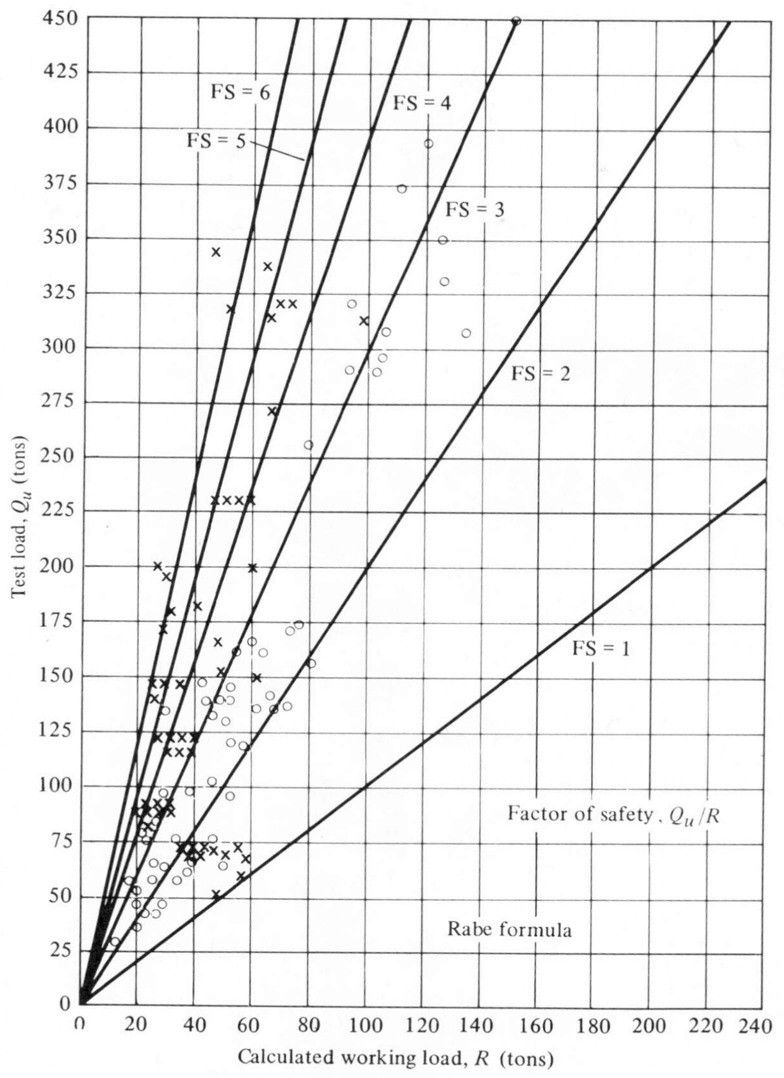

Fig. 24-13. Comparison of test loads and calculated loads on piles (Rabe formula). Key: ○—U.S. Army Corps of Engineers Tests (Mumma); x—Michigan State Highway Commission Tests.

tion, soil conditions, hammer characteristics, and test-load results on 101 pile-driving jobs. It represents the opposite extreme from the standpoint of simplicity as compared with the ENR formula. The results of the Rabe formula agree with test loads more closely, as indicated in the diagram of Fig. 24-13. Factor of safety plots for a number of other pile driving formulas are available (*4, 7*).

The Rabe formula is

$$R = \frac{MFD}{S + C} \times \frac{W}{W + \frac{P}{2}} \times B \tag{24-8}$$

in which

R = bearing capacity of pile, in pounds;
M = hammer efficiency factor;
F = WH for a drop hammer or a single-acting steam hammer, in foot-pounds, or the manufacturer's rated energy for a differential-acting or double-acting steam hammer, in foot-pounds;
W = weight of striking parts of hammer, in pounds;
H = height of fall of striking parts, in feet;
D = correction factor for battered piles;
S = average penetration of pile during last few blows, in inches per blow;
$C = C_1 + C_2 + C_3$ = temporary compression loss, in inches;

TABLE 24-1. Values[a] of the Soil Factor B_s

Kind of Soil Penetrated	B_s
Very wet plastic clay or silt; muck and loam; very poor soil	0.25
Soft clay or silt; rather poor soil	0.50
Medium clay or silt; average quality soil	0.70
Hard clay or silt; loose sand or gravel; good soil	0.85
Dense sandy silt; moderately compact sand or gravel; very good soil	1.00
Very compact sand or gravel; shale or hardpan; excellent foundation material	1.25

[a]1. Interpolate values of B_s between those given if necessary.

2. B_s depends only on the kind of soil penetrated. Do not increase B_s if the point comes in contact with rock.

3. If wet clay or silt has good adhesive quality, B_s may be increased; but the increase must not exceed 25%.

4. If there are boulders in the soil, reduce B_s by an amount which depends on the possibility that some piles will be stopped short by contact with boulders; but the reduction should not exceed 25%.

5. If several kinds of soil are penetrated (as is usual), estimate a weighted average overall value of B_s for the entire depth of penetration, giving considerably more weight to the soil near the point than to that above.

P = weight of pile or pile apparatus, including driving cap, in pounds (as it is constituted when S is measured); and

B = $B_s \times B_l \times B_c$ = correction factor for type of soil and pile length and cross section.

In order to apply the Rabe formula, it is necessary to determine several factors. Three of these factors are the following: B_s, which is called the soil factor and is obtained from Table 24-1; B_l, which is called the pile length factor and is found from Fig. 24-14; and B_c, which is called the pile cross-section factor and is found from Fig. 24-15.

Another factor is the temporary compression loss, which is the sum of the following three losses: One part of the compression loss is that in the cap and cushion and is found by the formula

$$C_1 = \frac{R \times \dfrac{W + \dfrac{P}{2}}{W}}{6{,}000{,}000 \times B} \times \frac{H}{3} \qquad (24\text{-}9)$$

in which C_1 is the compression loss in the cap and cushion, in inches.

The value of C_1 determined by Eq. (24-9) appears to be in foot-pounds. However, this equation is entirely empirical and the result is to be considered a distance in inches. The term $H/3$ needs to be evaluated only for a drop hammer. It may be taken equal to 1 for any steam hammer.

The second part of the compression loss is that in the pile and is found by the formula

$$C_2 = \frac{VRL}{AEB} \qquad (24\text{-}10)$$

in which

C_2 = compression loss in the pile, in inches;

V = a factor depending on the pile taper and on the vertical arrangement of the soil with reference to the center of resistance to driving;

L = length of the pile, in inches;

A = average equivalent cross-section area of pile, in square inches; and

E = modulus of elasticity of the pile material, in pounds per square inch.

The value of V is given in Table 24-2. For a precast concrete pile, A is taken as the area of the concrete plus 10 times the area of the longitudinal steel reinforcement; and for a steel H-beam with concrete filler, A is taken as the area of the steel plus 0.1 times the product of the area of the concrete and the fractional length of the filler. The value of E is 30,000,000 psi

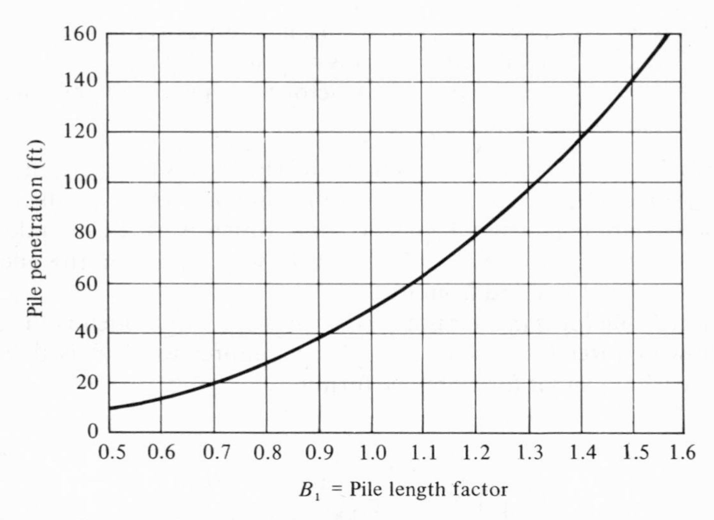

Fig. 24-14. Pile length factor in Rabe formula [after Rabe (*6*)].

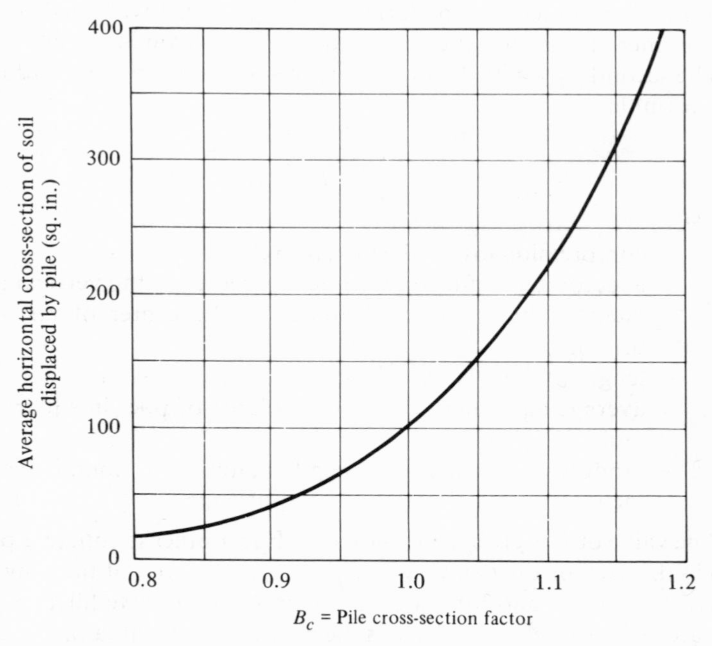

Fig. 24-15. Pile cross-section factor in Rabe formula [after Rabe (*6*)].

TABLE 24-2. Values of Factor V

	Pile Characteristics							
			Length Corresponding to Taper of 1 in					
Vertical Arrangement of Soil with Reference to Center of Resistance to Driving	*Steel H with Filler Near End*	*Non-tapered*	*Over 20 ft*	*20 ft*	*16 ft*	*12 ft*	*8 ft*	*4 ft*
1. Point bearing; rock or other hard material at point; poor soil above	1.00	1.00	1.00	1.00	1.00	1.00	1.00	1.00
2. Point bearing; rock or other hard material at point; fairly good soil above	0.95	0.95	0.95	0.93	0.92	0.91	0.89	0.87
3. Point bearing; rock or other hard material at point; very good soil above	0.90	0.88	0.85	0.82	0.79	0.76	0.72	0.68
4. Abrupt increase in firmness of soil near point, but not reaching rock or other hard material	0.88	0.80	0.75	0.70	0.66	0.62	0.55	0.49
5. Uniform firmness; full penetration	0.85	0.75	0.70	0.63	0.57	0.52	0.44	0.36

for steel, 3,000,000 psi for concrete, and 1,500,000 psi for timber. To evaluate C_1 or C_2, it is necessary to estimate R in advance. If the estimated value of R differs from the value computed by the Rabe formula by more than 5%, determine the values of C_1 and C_2 which correspond to that computed value and repeat the calculation for R.

The third part of the compression loss is that in the soil. This part of the loss is designated as C_3 and may be taken equal to 0.04 in. for all conditions.

Where sufficient data are not available for computing C_1 and C_2, no serious error will be introduced if the sum of C_1, C_2, and C_3 is assumed to be 0.15 in. for a single-acting or double-acting steam hammer or 0.25 in. for a drop hammer.

Another factor introduced in the Rabe formula is a correction factor for battered piles. This factor is designated as D and is found by the formula

$$D = \frac{1 - UG}{\sqrt{1 + G^2}} \tag{24-11}$$

In this equation, U is a coefficient of friction, which is obtained from

TABLE 24-3. Values of Various Factors in Rabe's Formula Which Depend on Hammer Type

Type of Hammer	*For Use in Eq. (24-8)*	*For Use in Eq. (24-11)*	*For Use in Eq. (24-12)*			
			Minimum F = WH		*Minimum F = WH*	
	M	*U*	*N*	*J*	*N*	*J*
Drop hammer which drags hoisting line	4.00	0.20	0.30	1.4	0.80	2.2
Drop hammer released by trip	4.75	0.20	0.30	1.4	0.80	2.2
Single-acting steam (Vulcan type)	5.00	0.10	0.18	1.2	0.45	1.8
Differential-acting steam (Vulcan type)	5.25	0.05	0.16	1.2	0.40	1.8
Double-acting steam (McKiernan–Terry type)	6.00	0.05	0.16	1.2	0.40	1.8

Table 24-3; and G is the rate of batter of the pile, as $\frac{1}{3}$ or $\frac{1}{4}$. For a vertical pile, $G = 0$, and $D = 1$.

Still another factor is the hammer efficiency factor, which is designated as M and is obtained from Table 24-3.

24.17. MOST EFFECTIVE SIZE OF HAMMER. The size of hammer and the energy of the blow delivered to the pile head are important factors to be considered in connection with any pile-driving job. If the hammer and the energy of the blow are too small, the pile will penetrate very slowly and tediously, and there will be excessive strain on the driving equipment. If the hammer weight and energy per blow are too large, the pile may be driven more rapidly than the soil can be displaced and readjusted around the pile. There will then be extraordinary resistance to penetration and the pile may be damaged. Damage from this cause may take the form of shattering of the head of a precast concrete pile or excessive brooming and splitting of the head of a wood pile. A more insidious type of damage may occur near the lower end of the pile, especially of a wood pile where it cannot be inspected. Also, failure of the pile by shear or bending and buckling may develop below the ground as a result of over-driving. A pile should be watched carefully as it is being driven. If the rate of penetration suddenly increases toward the end of driving, there is real danger of over-driving and the weight and height of fall of the hammer should be checked.

Rabe has also devised an empirical formula for size of hammer, which is helpful in choosing the correct driving equipment for a particular pile-driving job. This formula is

$$F = WH = \frac{NCR}{B\left[1 - J\dfrac{\frac{P}{2}}{W + \frac{P}{2}}\right]} \tag{24-12}$$

Values of N and J are given in Table 24-3. All other nomenclature is the same as for Eq. (24-9). Rabe recommends that the height of fall H for a drop hammer should not be more than 10 ft and preferably not more than 7 ft. Any combination of hammer size and height of fall which gives a value of F between the maximum and minimum values obtained by Eq. (24-12) and Table 24-3 should prove satisfactory on a pile-driving job.

24.18. LOAD TESTS. Load tests are widely accepted as the most reliable method of determining the bearing capacity of a pile. A common method

Fig. 24-16. Pile load test (courtesy of Rock Island District, Corps of Engineers, U.S. Army).

of conducting a load test is to drive anchor piles adjacent to the test pile, support a beam on these piles, and then jack against the beam, as illustrated in Fig. 24-16. As test loads are applied, pile settlements are measured, usually by means of a surveyor's level and a target rod.

Test loads may be applied continuously or in cycles. In the continuous method the test load is applied in increments until failure occurs or until sufficient load has been applied to satisfy the building-code or specification requirement. Each increment of load should be held constant until the rate of settlement becomes essentially zero. The load-settlement curve thus obtained is known as a total-settlement curve or gross-settlement curve.

In the cyclic method, a load increment is applied, and the settlement is observed. Then the load is released, and the pile is permitted to rebound. Repeating this process for several increments will permit data to be obtained for plotting a net-settlement curve or a plastic-settlement curve. Examples of total-settlement and net-settlement curves are shown in Figs. 24-17 and 24-18.

The determination of the failure load from a settlement curve requires the exercise of a considerable amount of judgment. Various organizations have developed empirical methods of determining the failure load. One typical method is as follows: Observe the load at which the total settlement, including elastic deformation of the pile, is not over 0.01 in./ton of test load. Divide this load by 2 to obtain the working load. It is assumed that the total settlement is obtained when no further settlement has occurred in 24 hr.

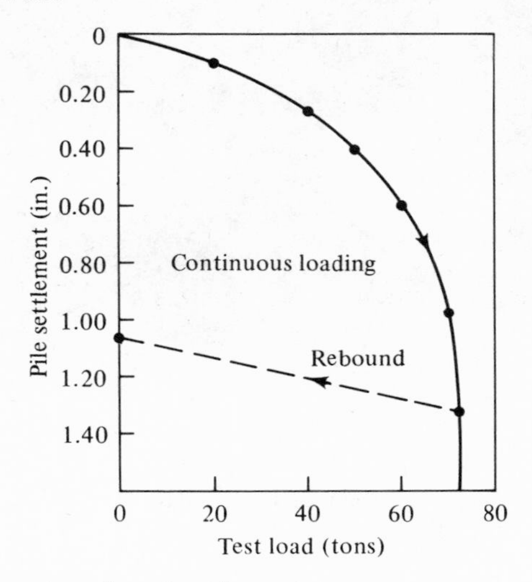

Fig. 24-17. Construction of total-settlement curve.

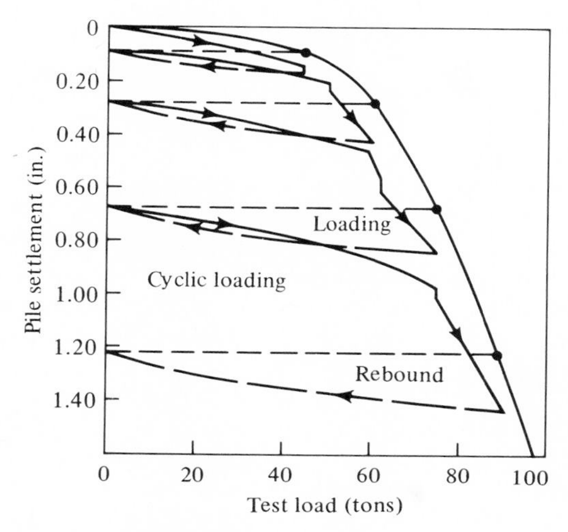

Fig. 24-18. Construction of net-settlement curve.

The accuracy of load testing can be seriously affected by the procedure used in performing the test. In some soils, such as loose sand, the driving of anchor piles or supporting piles nearby may densify the sand to such an extent that the bearing capacity of the test pile will be increased. In other soils, such as plastic clay, the driving of anchor piles may lower the bearing capacity of the test pile by further disturbing the structure of the clay and thereby lowering its shear strength. It is usually specified that the distance from the test pile to an anchor pile shall be not less than 5 pile diameters.

PROBLEMS

24.1. What are friction piles? Point bearing piles? Describe the distinguishing characteristics of each. What is negative skin friction?

24.2. What are the principal materials of which foundation piles are made?

24.3. What are batter or spur piles and why are they used?

24.4. A bridge pier in the center of a stream may rest on bearing piles. What function do the piles perform in addition to increasing the load carrying capacity of the pier?

24.5. If there is danger that timber piles beneath a footing may be subjected to wetting and drying during their functional life, what precautions should be taken to insure their permanence?

24.6. What is a bored-and-belled pile and how is it installed?

24.7. A group of 24 wood piles 1 ft in diameter are driven in four rows of six piles each. They are spaced 3 ft 6 in. center to center. If each individual pile has a bearing capacity of 24 tons, what is the bearing capacity of the group?

24.8. If the piles in Problem 24.7 are driven into cohesive soil having an unconfined compression strength of 400 psf, check to determine the stability of the whole group.

24.9. A pile is driven by a drop hammer weighing 4000 lb. The height of fall is 10 ft and the average penetration of the last blows is 0.05 in. What is the bearing capacity of the piles according to the Engineering-News formula?

24.10. A pile is driven by a single-acting steam hammer which weighs 2000 lb. The height of fall is 4 ft and the average penetration under the last few blows is 0.33 in. What is the bearing capacity of the pile by the Engineering-News formula?

24.11. A wood pile 7 in. in diameter at the point and 12 in. in diameter at the butt is driven 26 ft into medium to hard clay. The weight of the pile and driving cap is 1500 lb; and the weight of the drop hammer, which is released by a trip, is 2000 lb. The height of fall of the hammer is 5 ft. The average penetration of the pile under the last few blows of the hammer is 0.33 in. Determine the bearing capacity of the pile by both the Engineering-News formula and the Rabe formula.

24.12. Check the suitability of the hammer size and height of fall in Problem 24.11.

24.13. Estimate the minimum length of 12-in. diameter friction pile to support 20 tons in soil with $c = 3$ psi and $\phi = 0$. Disregard end bearing. What is the maximum compressive stress in the pile?

24.14. Caissons 36 in. in diameter are to bear on a sand stratum which lies underneath 80 ft of soft clay. If the standard penetration blow count in the sand is 24 bpf, estimate end-bearing for each caisson. The saturated unit weight of the clay is 124 pcf and the water table is at the ground surface.

24.15. What would be the effect of lowering the water table in Problem 24.14 after the caissons are in?

24.16. Standard penetration test blow counts in a sand average 23 bpf after correction to 40 psi overburden pressure. Relative density tests give maximum and minimum dry densities of 120 pcf and 90 pcf, respectively. What spacing of 12-in. diameter driven pile would be required to increase ϕ by 5°?

24.17. Evaluate the effect of the pile in Problem 24.15 on skin friction if the piles are 15 ft long, assuming $K = 2.0$ and $\phi_d = \phi$ for the sand.

24.18. Evaluate the effect of the pile in Problem 24.15 on allowable soil pressure under a warehouse floor, the maximum allowable settlement being 1 in.

24.19. Calculate the maximum and net uplift forces on a 12-in. diameter pile extending through 20 ft of expansive clay, 7 ft of which are below the permanent water table. Assume $c_d = 2$ psi, $\phi_d = 0$.

REFERENCES

1. Chellis, Robert D. In *Foundation Engineering,* Leonards, G. A., Ed. Chapter 7. McGraw-Hill, New York, 1962.

2. Cummings, A. E. "Dynamic Pile Driving Formulas." *J. Boston Soc. C.E.*, (January 1940).
3. Cummings, A. E. "Lectures on Foundation Engineering." *Circular Series No. 60,* Engineering Experiment Station, University of Illinois, 1949.
4. Michigan State Highway Commission. "A Performance Investigation of Pile Driving Hammers and Piles." Lansing, Michigan, 1965.
5. Peck, Ralph B., Walter E. Hanson, and Thomas H. Thornburn. *Foundation Engineering.* John Wiley, New York, 1953.
6. Rabe, W. H. "Dynamic Pile-Bearing Formula with Static Supplement." *Engineering News-Record* (December 26, 1946).
7. Spangler, M. G., and H. F. Mumma. "Pile Test Loads Compared with Bearing Capacity Calculated by Formulas." *Proc. Highway Research Board* **37** (1958).
8. Terzaghi, Karl, and Ralph B. Peck. *Soil Mechanics in Engineering Practice,* 2nd ed. John Wiley, New York, 1967.
9. Vesić, A. B. "Bearing Capacity of Deep Foundations in Sand." *Highway Research Rec.* **39,** 112–153 (1963).

25

Loads on Underground Conduits

25.1. THEORY OF LOADS ON UNDERGROUND CONDUITS. Underground conduits of the types used for sewers, drains, culverts, water mains, gas lines, and the like have served to improve the standard of living of mankind since the dawn of civilization. Remnants of structures of this kind have been found among the earliest examples of the practice of the engineering arts, but only within the past several decades has it been possible to design conduits on a rational basis with a degree of precision comparable with that obtained in the design of bridges and buildings. The loads to which buried conduits are subjected in service and their supporting strength under various installation conditions may be determined by means of the Marston Theory of Loads on Underground Conduits.

The Marston theory is named for Anson Marston who developed it in the early years of the twentieth century. Marston was the first Dean of Engineering at Iowa State University and Director of the Engineering Research Institute, serving in those capacities from 1904 to 1932. In the early 1900s a great deal of land drainage was being done in northern Iowa and engineers began using larger diameter pipelines and deeper cuts with the result that structural failures became widespread. They turned to Marston for help and the development of the theory began. Since then numerous investigations of failures of sewers, culverts, and other structures of this type have verified the general correctness of the Marston approach.

The basic concept of the theory is that the load due to the weight of the soil column above a buried pipe is modified by arch action in which a part of its weight is transferred to the adjacent side prisms, with the result that in some cases the load on the pipe may be less than the weight of the overlying column of soil. Or, in other cases, the load on the pipe may be increased by an inverted arch action in which load from the side prisms is transferred to the soil over the pipe. The key to the direction of load transfer by arch action lies in the direction of relative movement or tendency for movement, between the overlying prisms of soil and the adjacent side prisms, as illustrated in Figs. 25-1.

The transfer force associated with arch action at the plane of relative movement is the resultant of vertical and horizontal components of force. The Marston theory of loads deals with these components rather than the resultant itself.

25.2. CLASSES OF UNDERGROUND CONDUITS. For purposes of load computation, underground conduits are divided into two major classes, known as *ditch conduits* and *projecting conduits*, the classification being based on the construction or environmental conditions which influence

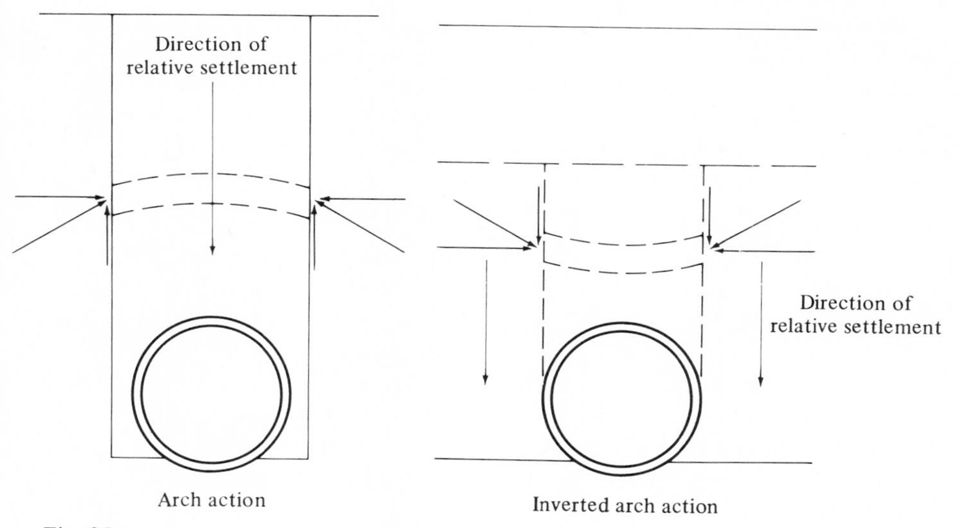

Fig. 25-1.

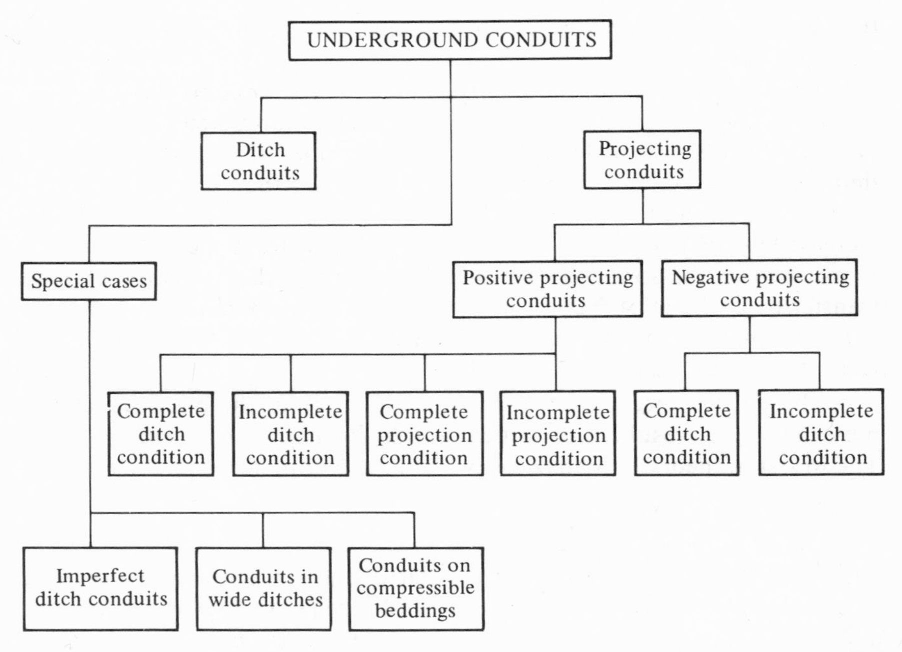

Fig. 25-2. Construction conditions which influence loads on underground conduits.

the load. Projecting conduits are further subdivided into *positive projecting conduits* and *negative projecting conduits*. Also there are several special cases having characteristics which are similar to those of both of the major classes. The accompanying chart shows the various construction conditions which influence loads on underground conduits.

25.3. DEFINITIONS. A *ditch conduit* is defined as one which is installed in a relatively narrow ditch dug in passive or undisturbed soil and which is then covered with earth backfill. Examples of this class of conduits are sewers, drains, water mains, and gas mains. A *positive projecting conduit* is one which is installed in shallow bedding with its top projecting above the surface of the natural ground and which is then covered with an embankment. Railway and highway culverts are frequently installed in this manner. A *negative projecting conduit* is one which is installed in a relatively narrow and shallow ditch with its top at an elevation below the natural ground surface and which is then covered with an embankment. This is a very favorable method of installing a railway or highway culvert, since the load produced by a given height of fill is generally less than it would be in the case of a positive projecting conduit. This method of construction is most effective in minimizing the load if the ditch between the top of the conduit and the natural ground surface is backfilled with loose uncompacted soil. The *imperfect ditch conduit*, sometimes called the *induced trench conduit*, is an important special case which is somewhat similar to the negative projecting conduit, but is even more favorable from the standpoint of reduction of load on the structure. Obviously, this method of construction should not be employed in embankments which serve as a water barrier, since the loosely placed backfill may en-

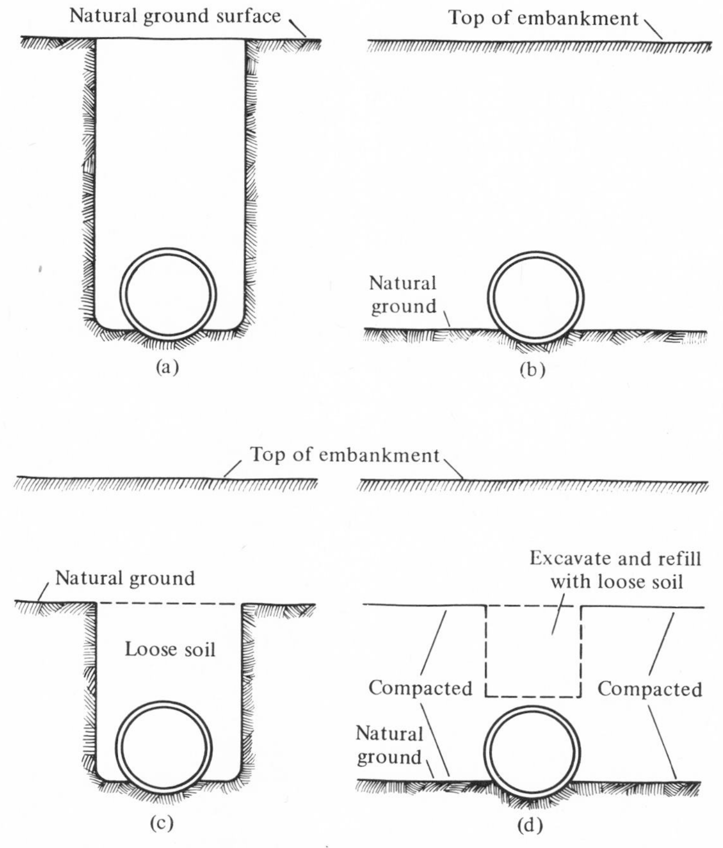

Fig. 25-3. Various classes of conduit installations. (a) Ditch conduit; (b) positive projecting conduit; (c) negative projecting conduit; and (d) imperfect ditch conduit.

courage channeling of seepage water through the embankment. The essential elements of these four main classes of conduits are shown in Fig. 25-3.

25.4. CHARACTERISTICS OF LOAD ON DITCH CONDUIT. In the case of a ditch conduit, the backfilling material has a tendency to consolidate and settle downward. This action plus the settlement of the conduit into its soil foundation, causes the prism of soil within the ditch and above the pipe to move downward relative to the undisturbed soil at the sides. This relative movement mobilizes along the sides of the ditch certain shearing stresses or friction forces which act upward in direction and which, in association with horizontal forces, create an arch action that partially supports the soil backfill. The difference between the weight of the backfill and these upward shearing stresses is the load which must be supported by the conduit at the bottom of the ditch. If it is assumed that cohesion between the backfill material and the sides of the ditch is neglıgible, then the magnitude of the vertical shearing stresses is equal to the active lateral pressure exerted by the soil backfill against the sides of the ditch multiplied by the tangent of the angle of friction between the two materials.

This assumption of negligible cohesion is justified on two accounts. Even when a ditch is backfilled with cohesive soil, considerable time must elapse before effective cohesion between the backfill and the sides of the ditch can develop after backfilling. Also, the assumption of no cohesion yields the maximum probable load on the conduit. This maximum load may develop at any time during the life of the conduit as a result of heavy rainfall or some other action which may eliminate or greatly reduce cohesion between the backfill and the sides of the ditch.

25.5. VERTICAL PRESSURE ON TOP OF CONDUIT. In the mathematical derivation of the formula for loads on ditch conduits, the following notation will be employed:

W_c = load on conduit, in pounds per linear foot;
γ = unit weight (wet density) of filling material, in pounds per cubic foot;
V = vertical pressure on any horizontal plane in backfill, in pounds per linear foot of ditch;
B_c = horizontal breadth (outside) of conduit, in feet;
B_d = horizontal width of ditch at top of conduit, in feet;
H = height of fill above top of conduit, in feet;

h = distance from ground surface down to any horizontal plane in backfill, in feet;
C_d = load coefficient for ditch conduits;
μ = $\tan \phi$ = coefficient of internal friction of fill material;
μ' = $\tan \phi'$ = coefficient of friction between fill material and sides of ditch;
K = ratio of active lateral unit pressure to vertical unit pressure; and
e = base of natural logarithms.

The coefficient μ' may be equal to or less than μ, but cannot be greater than μ. The value of K will be taken as Rankine's ratio and may be found by Eq. (22-3), which is

$$K = \frac{\sqrt{\mu^2 + 1} - \mu}{\sqrt{\mu^2 + 1} + \mu} = \frac{1 - \sin \phi}{1 + \sin \phi} = \tan^2\left(45^\circ - \frac{\phi}{2}\right) = \frac{1}{\tan^2\left(45^\circ + \frac{\phi}{2}\right)}$$

Let Fig. 25-4 represent a section of a ditch and ditch conduit 1 unit in length; and consider a thin horizontal element of the fill material of height dh located at any depth h below the ground surface. The forces acting on

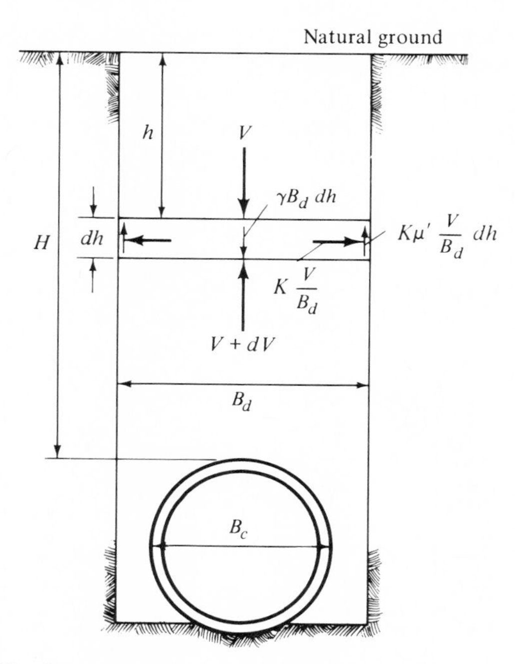

Fig. 25-4. Free-body diagram for ditch conduit.

this element at equilibrium are: V, the vertical force on the top of the element; $V + dV$, the vertical force on the bottom of the element; $\gamma B_d dh$, the weight of the element; and $K(V/B_d)dh$, the lateral pressure on each side of the element, it being assumed that the vertical pressure on the element is uniformly distributed over the width B_d. Since the element has a tendency to move downward in relation to the sides of the ditch, these lateral pressures induce upward shearing forces equal to $K\mu'(V/B_d)dh$. Equating the upward and downward vertical forces on the element, we obtain

$$V + dV + 2K\mu'\frac{V}{B_d}\,dh = V + \gamma B_d dh \tag{25-1}$$

This is a linear differential equation, the solution for which is

$$V = \gamma B_d^2 \frac{1 - e^{-2K\mu'(h/B_d)}}{2K\mu'} \tag{25-2}$$

At the elevation of the top of the conduit, $h = H$; and, by substituting this value in Eq. (25-2), we obtain an expression for the total vertical pressure at the horizontal plane through the top of the conduit. The portion of this total pressure which is carried by the conduit depends on the rigidity of the conduit in comparison with that of the fill material between the sides of the conduit and the sides of the ditch. In the case of a very rigid pipe, such as a burned-clay, concrete, or heavy-walled cast-iron pipe, the side fills may be relatively compressible and the pipe itself will carry practically all of the load V. On the other hand, if the pipe is a relatively flexible, thin-walled pipe, and the soil is thoroughly tamped in at the sides of the pipe, the stiffness of the side fills may approach that of the conduit and the load on the structure will be reduced by the amount of load which the side fills are capable of carrying.

25.6. MAXIMUM LOADS ON DITCH CONDUITS. For the case of a rigid ditch conduit with relatively compressible side fills, the load on the conduit will be

$$W_c = C_d \gamma B_d^2 \tag{25-3}$$

in which

$$C_d = \frac{1 - e^{-2K\mu'(H/B_d)}}{2K\mu'} \tag{25-4}$$

Evaluation of W_c is made easy by the use of the computation diagram in Fig. 25-7, in which values of C_d for various values of H/B_d are plotted for several kinds of filling materials having different coefficients of internal friction.

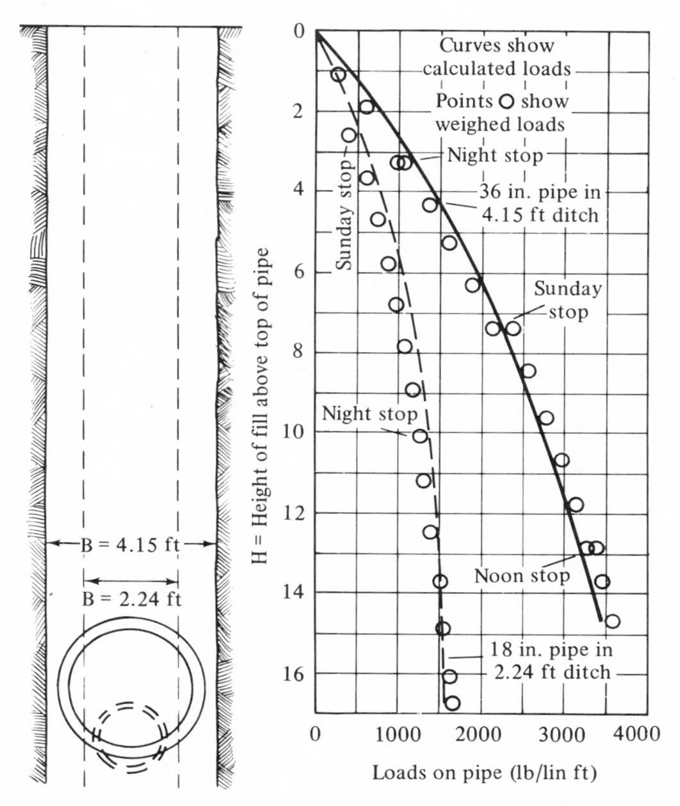

Fig. 25-5. Weighed loads vs. calculated loads by Eq. (25-3) on 18 in. and 36 in. rigid pipes in ditches (*1*).

Comparisons between actual weighed loads on rigid pipes in ditches and calculated loads in two of Marston's early experiments (*4*) are shown in Fig. 25-5.

For the case of a flexible pipe conduit and thoroughly tamped side fills having essentially the same degree of stiffness as the pipe itself, the value of W_c given by Eq. (25-3) might be multiplied by the ratio B_c/B_d. The load on the flexible pipe would then be

$$W_c = C_d \gamma B_c B_d \qquad (25\text{-}5)$$

It is emphasized that for Eq. (25-5) to be applicable, the side fills must be compacted a sufficient amount to have the same resistance to deformation under vertical load as the pipe itself. This equation should not be used merely because the pipe is a flexible type.

In actual practice, it is probable that the load on a pipe lies somewhere

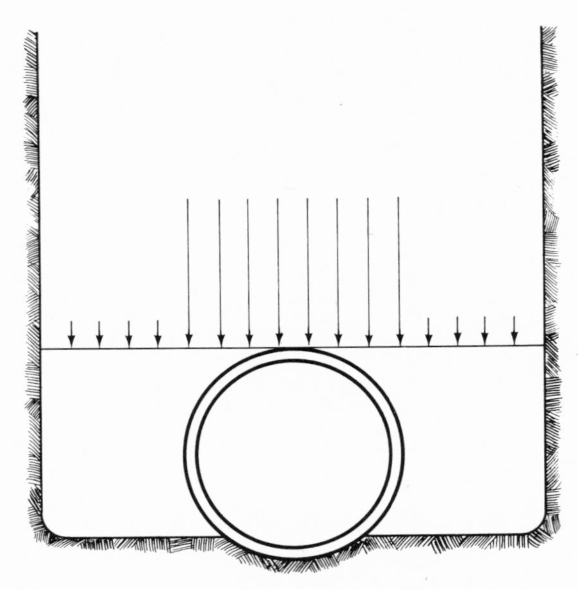

Fig. 25-6. Load distribution at level of top of pipe when sidefill soil columns are substantially less strain resistant than the rigid pipe.

between that indicated by Eq. (25-3) and Eq. (25-5), depending upon the relative rigidity of the pipe and the sidefill columns of soil, as illustrated in Fig. 25-6. Research is needed to establish the relationship between load on a pipe and the total backfill load under various conditions of relative rigidity. Some recent laboratory tests by a private research agency indicated that the load imposed on a 10-in. flexible sewer pipe under a dropped-in sand backfill, with no compaction of the sidefills, was about 70 to 90% of the backfill load on a similarly installed rigid pipe.

EXAMPLE 25-1. Assume that a rigid sewer pipe with an outside diameter of 24 in. is to be laid in a ditch which is 3.5 ft wide at the top of the pipe and is to be covered with 28 ft of clayey soil backfill which weighs 120 pcf. Determine the load on the sewer.

SOLUTION. In this case,

$$H = 28\ \text{ft} \qquad \gamma = 120\ \text{pcf}$$

$$B_d = 3.5\ \text{ft} \qquad B_d^2 = 12.25$$

$$H/B_d = 8.0$$

From the curve in Fig. 25-7 marked "ordinary maximum for clay," C_d

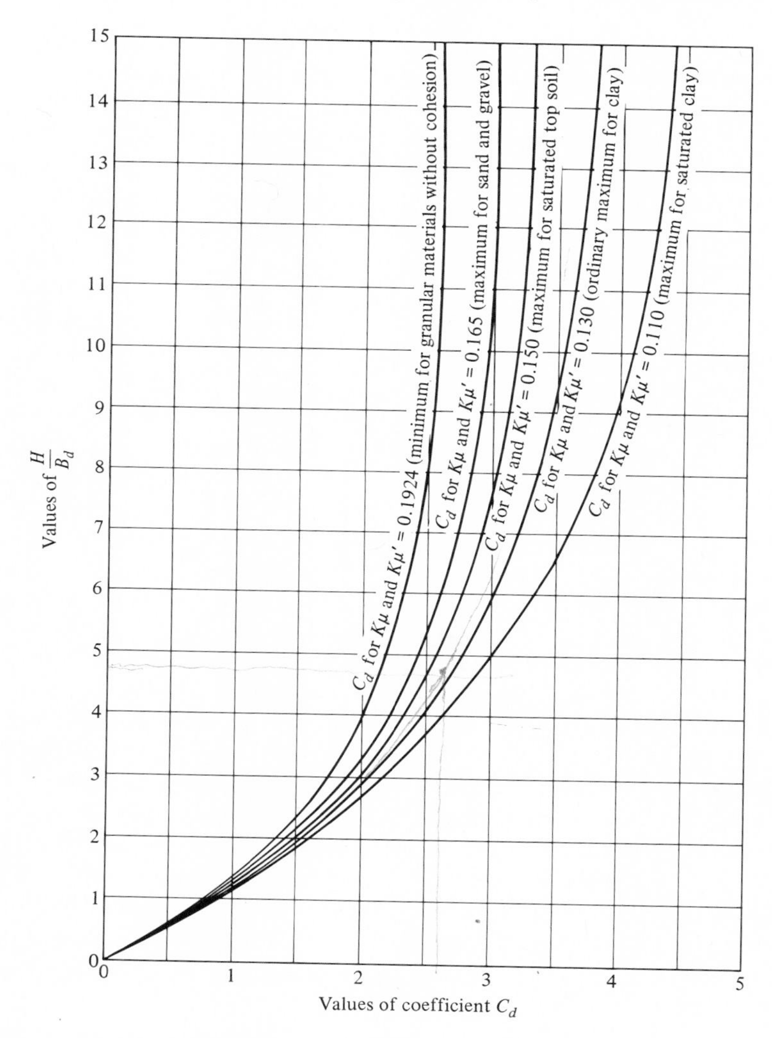

Fig. 25-7. Diagram for coefficient C_d for ditch conduits.

is 3.35. Substituting in Eq. (25-3), we get

$$W_c = 3.35 \times 120 \times 12.25 = 4930 \text{ lb/linear ft of pipe}$$

25.7. DEVELOPMENT OF MAXIMUM LOAD. The preceding ditch-conduit formulas, with proper selection of the physical factors involved, give the maximum loads to which any particular conduit may be subjected in service. However, because of the development of cohesion, any particular conduit may escape the maximum load for a long time, sometimes until it is removed for some reason other than load failure. Experiments and field observations show that the load on a conduit at the time the fill is completed may be less than it will be at some later time. That is, the load keeps building up for a period of time after the maximum height of fill is reached. This lag characteristic has been observed in extreme cases to amount to as much as 20 to 25% of the total load; and its development may require several years. It accounts for the fact that sewers and other conduits which have been observed to be structurally sound immediately upon completion are sometimes found to be cracked several months or several years later.

The theoretical loads found by Eqs. (25-3) and (25-5) are working values which should be used in the design of sewers, drains, and other ditch conduits to prevent cracking of the pipe. This should be the goal of the sewer designer; for, although it is true that a cracked sewer may often continue to function as a conduit for an indefinitely long time without collapsing (as a result of the development of passive resistance pressures against the sides of the pipe), sewers and drains should not be designed to allow such cracks to develop, since there is no assurance that a cracked pipe will always stand up. Long continued, unusually wet periods may so weaken the soil that it will no longer support the sides of the pipe, even though it may have done so for a long time previously. Also, the soil at the sides of the pipe may be dangerously softened, or even washed away, by water forced through the cracks of the pipe if, because of extraordinary circumstances, the pipe is forced to operate under head. Furthermore, a cracked sewer pipe will act as a drain, and an excessive amount of ground water will be allowed to infiltrate into the sewer, overtaxing its capacity and greatly increasing the cost of sewage treatment. Engineers have long recognized the necessity of tightly sealing the joints of sanitary sewer pipes to prevent ground water infiltration, but such efforts are of no avail if the pipes crack. It is obvious that pressure pipes, such as water and gas mains, cannot function if they are cracked.

25.8. LOAD ON CONDUIT IN DITCH WITH SLOPING SIDES. The width of ditch B_d in the load formulas previously derived is the width of a normal ditch with vertical sides. In case the ditch is constructed with sloping sides, as indicated in Fig. 25-8, experiments have shown that the width of the ditch at or slightly below the top of the conduit is the proper width to use for B_d in calculating the load. If it is desired to dig either a ditch with sloping sides or one which is very wide in comparison with the size of the conduit, it is good practice to lay the conduit in a relatively narrow sub-ditch at the bottom of the wider ditch. Then, in accordance with the principle just

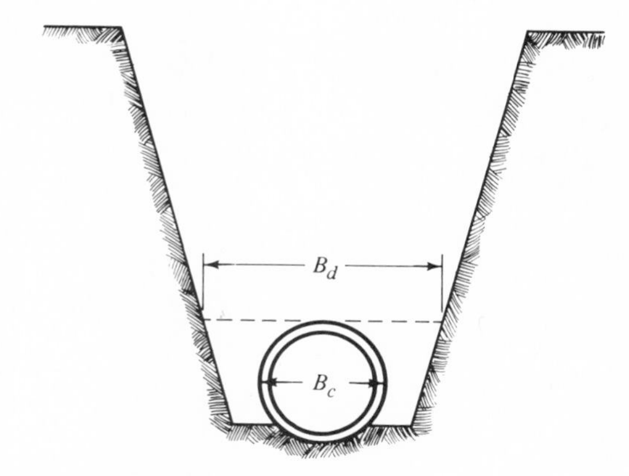

Fig. 25-8. Effective width of ditch with sloping sides.

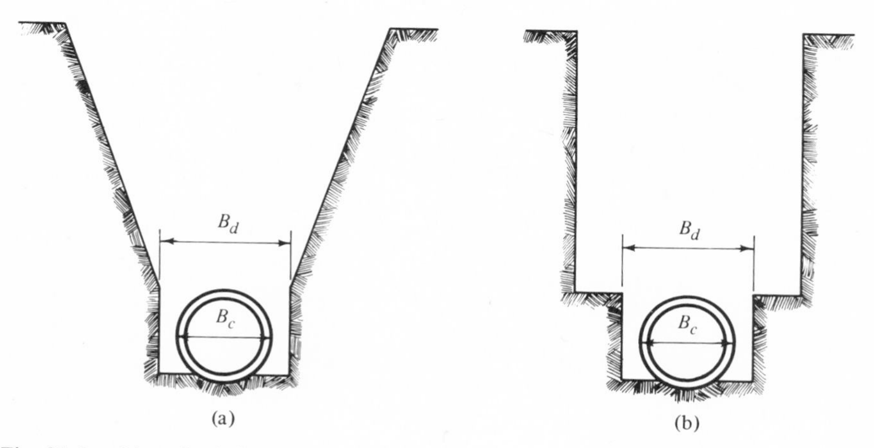

Fig. 25-9. Use of subditch.

stated, the load on the conduit will be held to a reasonable minimum value. The use of a sub-ditch for this purpose is illustrated in Fig. 25-9.

25.9. SHEARING STRESSES IN EMBANKMENT OVER POSITIVE PROJECTING CONDUIT. A positive projecting conduit, as defined and also as the name implies, is installed with its top projecting some distance above the natural ground surface. It may be of any shape, such as circular, rectangular, or elliptical; it may be made of any material, such as concrete, burned clay, cast iron, corrugated metal, wood, plastic, etc.; and it may possess any degree of rigidity, from the very rigid concrete pipes and

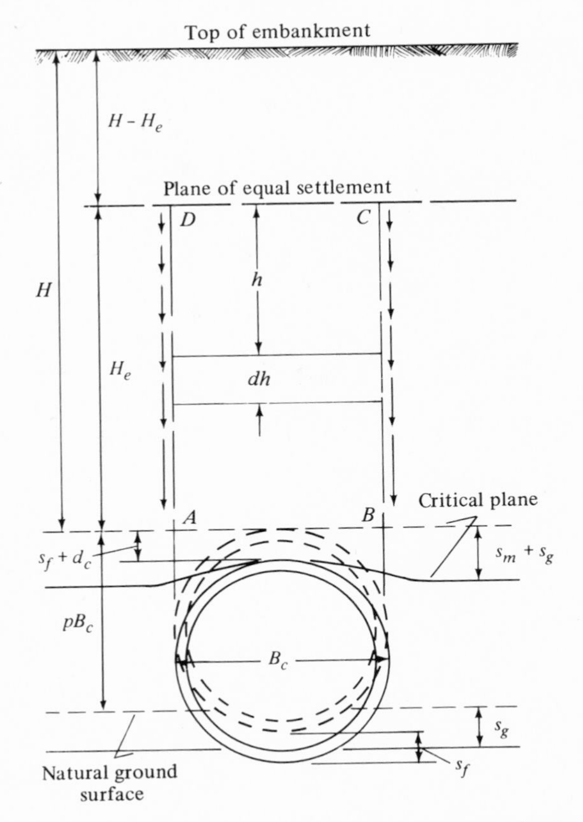

Fig. 25-10. Settlements which influence loads on positive projecting conduits (incomplete projection condition). Key: dashed line, initial elevation ($H = 0$); solid line, final elevation.

monolithic box culverts to the relatively flexible, light-weight, corrugated-metal pipes.

When a conduit is installed as a positive projecting conduit, shearing forces again play an important role in the production of the resultant load on the structure. In this case the planes along which relative movements are assumed to occur and on which shearing forces are generated are the imaginary vertical planes extending upward from the sides of the conduit, as indicated in Figs. 25-10 and 25-11. The width factor in the development of an expression for load is the outside breadth of the conduit, designated as B_c. The vertical distance from the natural ground surface to the top of the structure is expressed as pB_c, in which p is the *projection ratio*.

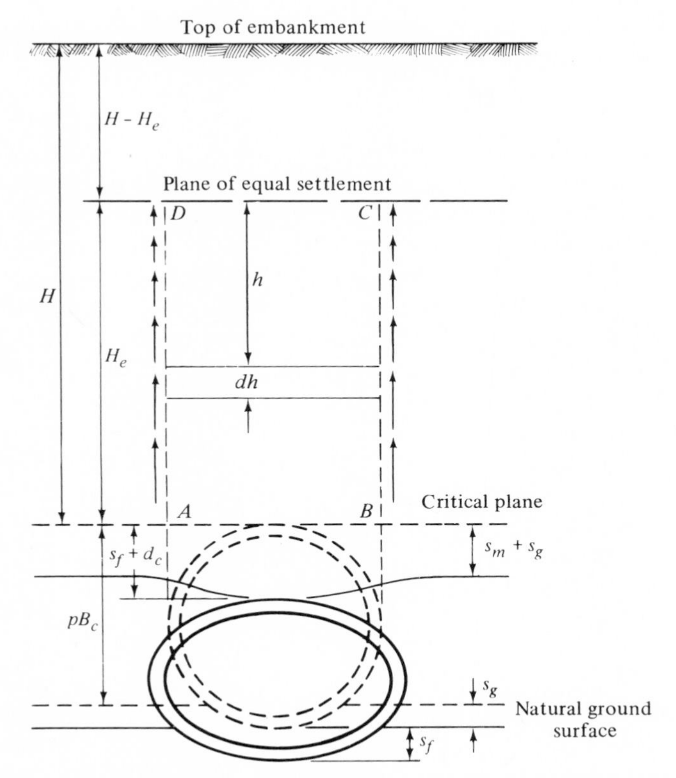

Fig. 25-11. Settlements which influence loads on positive projecting conduits (incomplete ditch condition). Key same as in Fig. 25-10.

25.10. SETTLEMENT RATIO. The magnitudes and directions of relative movements between the interior prism *ABCD*, Fig. 25-10 or Fig. 25-11, and the adjacent exterior prisms are influenced by the settlement of certain elements of the conduit and the adjacent soil. These settlements are combined into an abstract ratio, called the *settlement ratio*, according to the formula

$$r_{sd} = \frac{(s_m + s_g) - (s_f + d_c)}{s_m} \tag{25-6}$$

in which

r_{sd} = settlement ratio;
s_m = compression strain of the side columns of soil of height pB_c;

s_g = settlement of the natural ground surface adjacent to the conduit;
s_f = settlement of the conduit into its foundation; and
d_c = shortening of the vertical height of the conduit.

25.11. CRITICAL PLANE. In connection with settlement of a conduit, it is convenient to define a critical plane, which is the horizontal plane through the top of the conduit when the fill is level with its top, that is, when $H = 0$. During and after construction of the embankment, this plane settles downward. If it settles more than the top of the pipe, as illustrated in Fig. 25-10, the settlement ratio is positive; the exterior prisms move downward with respect to the interior prism; the shearing forces on the interior prism are directed downward, and the resultant load on the structure is greater than the weight of the prism of soil directly above it. This is known as the projection condition.

If the critical plane settles less than the top of the conduit, as in Fig. 25-11, the settlement ratio is negative; the interior prism moves downward with respect to the exterior prisms; and the shearing forces on the interior prism are directed upward, and the resultant load is less than the weight of the soil above the structure. This is called the ditch condition, because the shearing forces act upward, or in the same direction as in the case of a ditch conduit.

25.12. PLANE OF EQUAL SETTLEMENT. In the case of a ditch conduit the shearing forces extend all the way from the top of the pipe to the ground surface. In a projecting-conduit installation, however, if the embankment is sufficiently high, the shearing forces may terminate at some horizontal plane in the embankment which is called the *plane of equal settlement*. A plane of equal settlement develops because a part of the vertical pressure in the exterior prisms is transferred by shear to the interior prism, or vice versa. This transfer of pressure causes different unit strains in the interior and exterior prisms, and at some height above the conduit the accumulated strain in the exterior prism plus the settlement of the critical plane will just equal the accumulated strain in the interior prism plus the settlement of the top of the structure. Above the plane of equal settlement the interior and exterior prisms settle equally. Since there are no relative movements between the adjacent prisms, no shearing forces are generated in the zone above this plane.

When the height of the plane of equal settlement above the top of the conduit, which height is designated as H_e, is less than the height H of the embankment, the plane of equal settlement is real. This is called either the incomplete ditch condition or the incomplete projection condition, because the shearing forces do not extend completely throughout the total height of embankment. If H_e is greater than H, the plane of equal settlement is

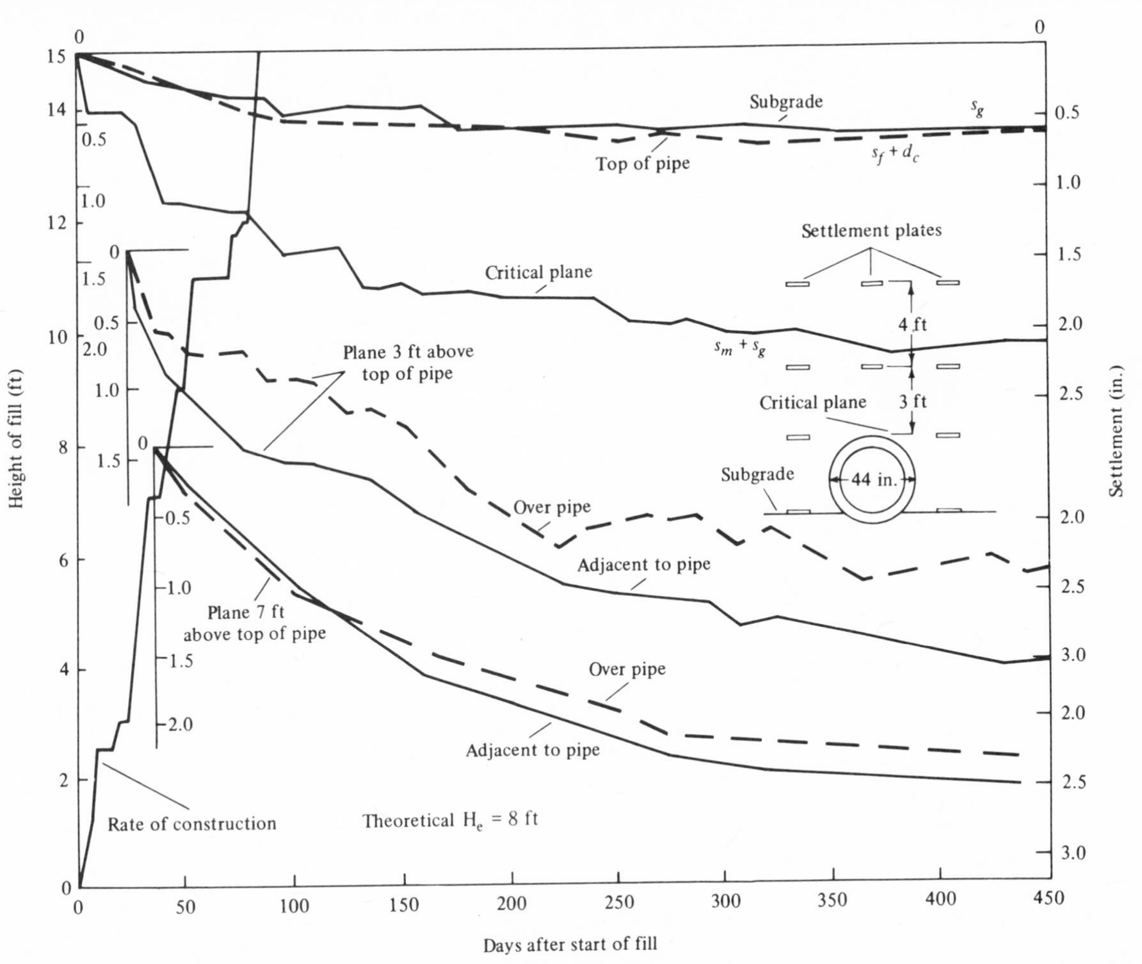

Fig. 25-12. Settlement measurements at planes within a 15-ft high embankment over a 44-in.-concrete pipe culvert. These measurements demonstrate the existence of a plane of equal settlement (*10*).

imaginary. This is referred to as either the complete ditch condition or the complete projection condition, because the shearing forces do extend completely to the top of the embankment.

The actual existence of a plane of equal settlement in a 15 ft high embankment over a 44-in. diameter concrete pipe culvert was demonstrated by measurements of settlements as shown in Fig. 25-12. In this experiment settlements were measured at the subgrade level, in the critical plane at the top of the pipe, and in the planes 3 ft and 7 ft above the top, both within the prism of soil directly over the structure and at points laterally removed from this prism. As shown in the figure, the points adjacent to the pipe settled a greater amount in the critical plane and at an elevation 3 ft above the pipe, but at the 7-ft level, the settlements at points over the pipe and adjacent thereto were nearly the same. The theoretical height of the plane of equal settlement, calculated from measured values of the settlement ratio, was approximately 8 ft.

25.13. MARSTON'S FORMULA FOR LOAD ON POSITIVE PROJECTING CONDUIT. By a process similar to that employed in the case of ditch conduits which is described in Section 25.5, Marston derived a formula for the vertical load on a positive projecting conduit. For the complete ditch or projection condition, the formula is

$$W_c = C_c \gamma B_c^2 \tag{25-7}$$

in which

$$C_c = \frac{e^{\pm 2K\mu(H/B_c)} - 1}{\pm 2K\mu} \tag{25-8}$$

The plus signs are used for the complete projection condition, and minus signs are used for the complete ditch condition.

Also, for the incomplete ditch or projection condition,

$$C_c = \frac{e^{\pm 2K\mu(H_e/B_c)} - 1}{\pm 2K\mu} + \left(\frac{H}{B_c} - \frac{H_e}{B_c}\right) e^{\pm 2K\mu(H_e/B_c)} \tag{25-9}$$

The plus signs are used for the incomplete projection condition, and the minus signs are used for the incomplete ditch condition.

In Eqs. (25-7) to (25-9),

W_c = load on conduit, in pounds per linear foot;
γ = unit weight of embankment soil, in pounds per cubic foot;
B_c = outside width of conduit, in feet;
H = height of fill above conduit, in feet;
H_e = height of plane of equal settlement, in feet;
K = Rankine's lateral pressure ratio;

μ = $\tan\phi$ = coefficient of friction of fill material; and
e = base of natural logarithms.

A formula for evaluating H_e is derived by equating an expression for the sum of the total strain in the interior prism plus the settlement of the top of the conduit to a similar expression for the sum of the total strain in an exterior prism plus the settlement of the critical plane. This formula is

$$\left[\frac{1}{2K\mu} \pm \left(\frac{H}{B_c} - \frac{H_e}{B_c}\right) \pm \frac{r_{sd}p}{3}\right] \frac{e^{\pm 2k\mu(H_e/B_c)} - 1}{\pm 2K\mu} \pm \frac{1}{2}\left(\frac{H_e}{B_c}\right)^2$$

$$\pm \frac{r_{sd}p}{3}\left(\frac{H}{B_c} - \frac{H_e}{B_c}\right)e^{\pm 2K\mu(H_e/B_c)} - \frac{1}{2K\mu} \cdot \frac{H_e}{B_c} \mp \frac{H}{B_c} \cdot \frac{H_e}{B_c} = \pm r_{sd}p\frac{H}{B_c} \qquad (25\text{-}10)$$

Use the upper signs for the incomplete projection condition, for which the settlement ratio is positive, and use the lower signs for the incomplete ditch condition, for which the settlement ratio is negative.

25.14. LOAD-COMPUTATION DIAGRAM FOR POSITIVE PROJECTING CONDUITS. It is both difficult and time-consuming to solve Eqs. (25-9) and (25-10). Fortunately the results can be given in a relatively simple diagram from which values of the load coefficient C_c can be obtained for substitution in Eq. (25-7). Such a diagram is shown in Fig. 25-13. It will be noted that C_c is a function of the ratio of the height of fill to the width of the conduit, or H/B_c, and of the product of the settlement ratio and the projection ratio, or $r_{sd}p$, as well as of the friction characteristics of the soil. However, Marston pointed out that the influence of the coefficient of internal friction μ is relatively minor in this case, and it is not considered necessary to differentiate between various soils as for ditch conduits. Therefore, in constructing Fig. 25-13, it was assumed that $K\mu = 0.19$ for the projection condition, in which the shearing forces are directed downward, and that $K\mu = 0.13$ for the ditch condition, in which the shearing forces are directed upward. This diagram gives reasonable maximum loads the accuracy of which is within the degree of precision of the assumptions upon which the analysis is based.

The ray lines in Fig. 25-13 represent values of C_c versus H/B_c according to Eq. (25-9), whereas the envelope curves correspond to Eq. (25-8). The ray lines intersect the envelope curves at points where $H_e = H$. Therefore, this diagram can be used to estimate the height of the plane of equal settlement in a particular case, as well as to estimate the load on the conduit. Also, the rays are straight lines which can be represented by equations for use when the value of H/B_c exceeds the limit of the diagram. These equations are given in Table 25-1.

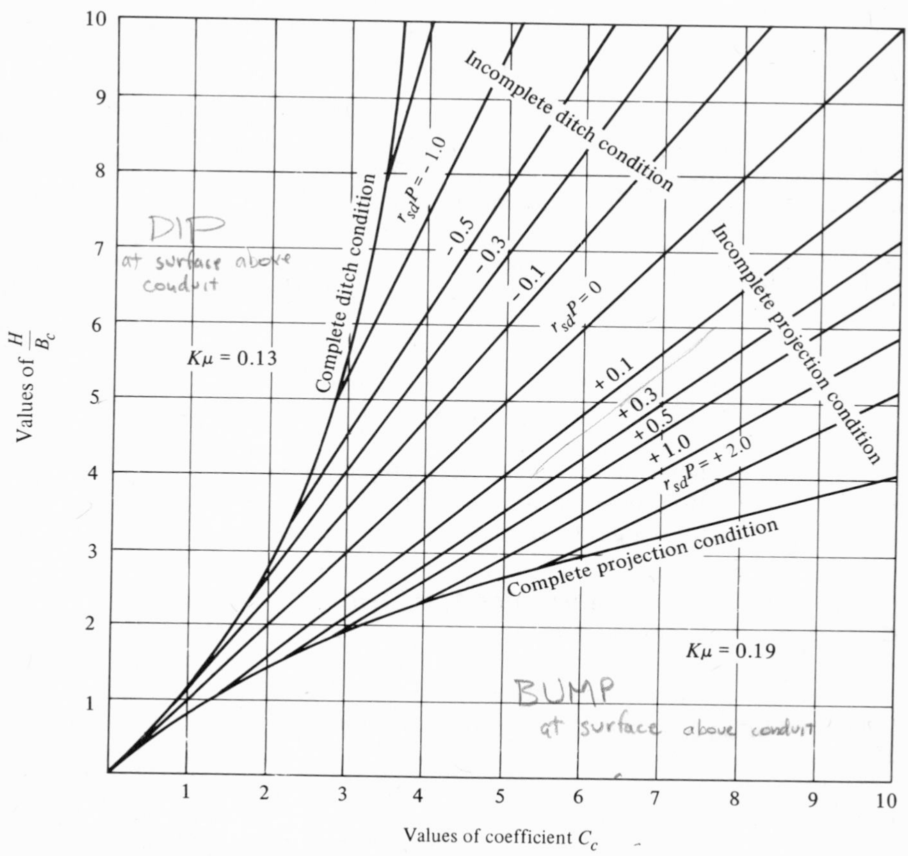

Fig. 25-13. Diagram for coefficient C_c for positive projecting conduits.

TABLE 25-1. Values[a] of C_c in Terms of H/B_c

Incomplete Projection Condition $K\mu = 0.19$		Incomplete Ditch Condition $K\mu = 0.13$	
$r_{sd}p$	Equation	$r_{sd}p$	Equation
+0.1	$C_c = 1.23H/B_c - 0.02$	−0.1	$C_c = 0.82H/B_c + 0.05$
+0.3	$C_c = 1.39H/B_c - 0.05$	−0.3	$C_c = 0.69H/B_c + 0.11$
+0.5	$C_c = 1.50H/B_c - 0.07$	−0.5	$C_c = 0.61H/B_c + 0.20$
+0.7	$C_c = 1.59H/B_c - 0.09$	−0.7	$C_c = 0.55H/B_c + 0.25$
+1.0	$C_c = 1.69H/B_c - 0.12$	−1.0	$C_c = 0.47H/B_c + 0.40$
+2.0	$C_c = 1.93H/B_c - 0.17$		

[a] From Ref. (*1*).

25.15. NATURE OF THE SETTLEMENT RATIO. Although the settlement ratio r_{sd} is a rational quantity in the development of the load formula, it is difficult, if not impossible, to predetermine the actual value which will be developed in a specific case. Therefore, it is more practicable to consider this ratio as an empirical quantity and to determine working values for design purposes from observations of the performance of actual culverts under embankments. Such observations have been made, and the values recommended in Table 25-2 are based on them.

TABLE 25-2. Design Values of Settlement Ratio

Conditions	*Settlement Ratio*
Rigid culvert on foundation of rock or unyielding soil	+1.0
Rigid culvert on foundation of ordinary soil	+0.5 to +0.8
Rigid culvert on foundation of material that yields with respect to adjacent natural ground	0 to +0.5
Flexible culvert with poorly compacted side fills	−0.4 to 0
Flexible culvert with well-compacted side fills[a]	−0.2 to +0.8

[a]Not well established.

25.16. THE CASE FOR WHICH $r_{sd}p = 0$. Examination of the load-computation diagram in Fig. 25-13 indicates that when the product of the settlement ratio r_{sd} and the projection ratio p equals zero, then $C_c = H/B_c$. When this value of C_c is substituted in Eq. (25-7), the load formula reduces to $W_c = H\gamma B_c$; that is to say, the load is equal to the weight of the prism of soil directly above the conduit. The settlement ratio is equal to zero when the critical plane settles the same amount as the top of the conduit, that is, when $s_m + s_g = s_f + d_c$. The projection ratio is equal to zero when the structure is installed in a narrow and shallow trench so that its top is approximately level with the adjacent natural ground. This is a transition case between positive and negative conduits and is sometimes referred to as a zero projecting conduit.

25.17. MEASURED VERSUS THEORETICAL LOADS. Correlation between actual measured loads on a 44-in. diameter concrete pipe under a 15-ft embankment and the loads calculated by the Marston theory shows very close agreement as indicated in Fig. 25-14. This load-measuring experiment was continued for a period of 21 yr, with the results shown in Fig. 25-15. The average load on the 16-ft long culverts increased rapidly during the first 6 months after completion of the embankment, then fluctuated up and down within a range of about 10% of the maximum load during the balance of the 21-yr period. Also it is indicated that the

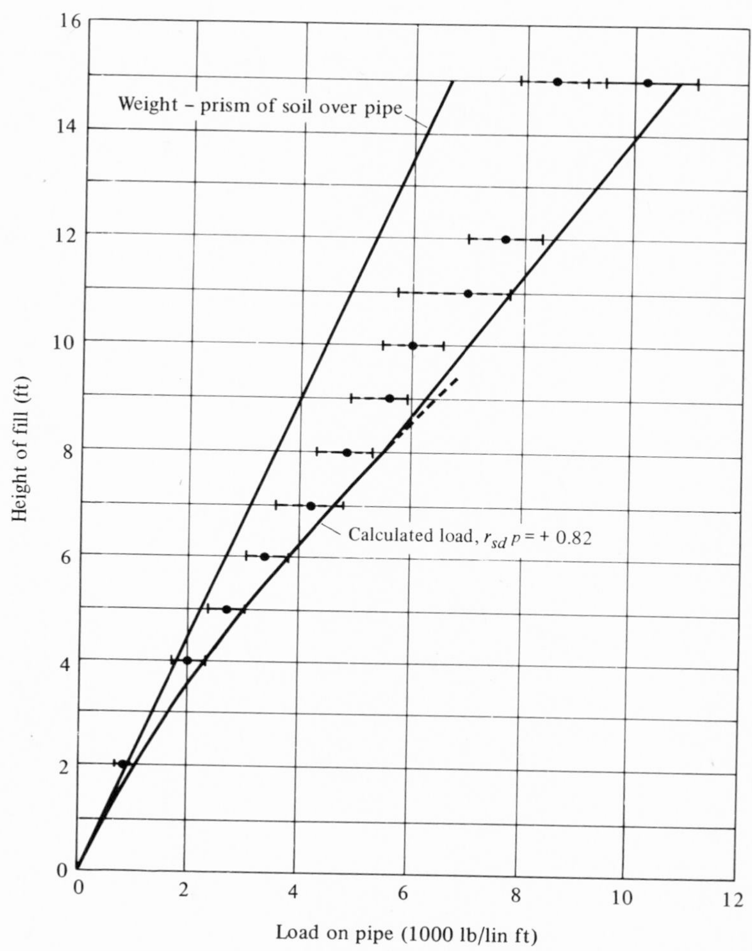

Fig. 25-14. Comparison between measured and theoretical loads on a 44-in. concrete pipe culvert under a 15-ft embankment (*10*). Key: dashed line, during construction of fill; solid line, maximum load after completion of fill (maximum, minimum, and average load on four 4-ft long sections.

load on the concrete pipe was consistently greater by about 50% than that on a parallel corrugated steel pipe of approximately the same diameter. This difference in load reflected the influence of the difference in vertical deflection of the two kinds of pipe, which in turn influenced the settlement ratios of the installations and the magnitude of the shear stress components of the resultant loads on the culverts.

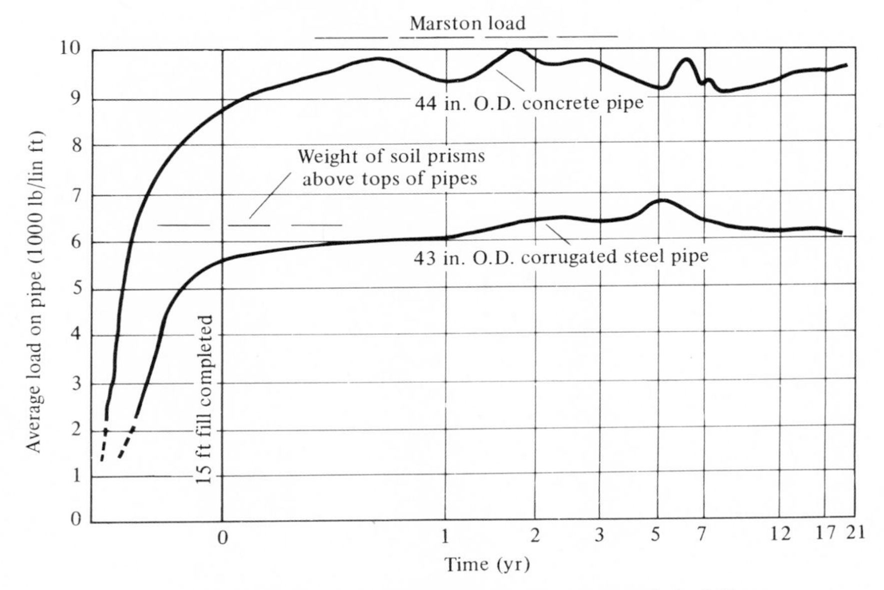

Fig. 25-15. Measured loads on rigid and flexible pipe over period of 21 yr.

EXAMPLE 25-2. Suppose that it is desired to determine the load on a 6 × 6 ft box culvert under a 50-ft fill which has a unit weight of 120 pcf. Assume that the outside width of the barrel is 7.67 ft, the projection ratio is 0.5, and the settlement ratio is +0.6.

SOLUTION. In this case,

$$H = 50 \text{ ft} \qquad p = 0.5$$
$$B_c = 7.67 \text{ ft} \qquad r_{sd} = +0.6$$
$$H/B_c = 6.52 \qquad r_{sd}p = +0.3$$
$$\gamma = 120 \text{ pcf}$$

From Fig. 25-13 the value of C_c is 9.0. Substituting in Eq. (25-7), we obtain

$$W_c = 9 \times 120 \times 7.67^2 = 63{,}500 \text{ lb/linear ft of barrel}$$

25.18. LOAD ON IMPERFECT DITCH CONDUIT DUE TO EARTH FILL. In the imperfect ditch conduit construction procedure, illustrated in Fig. 25-3(d), the pipe is first installed as a positive projecting conduit. Then the soil backfill at the sides and over the pipe is compacted up to some speci-

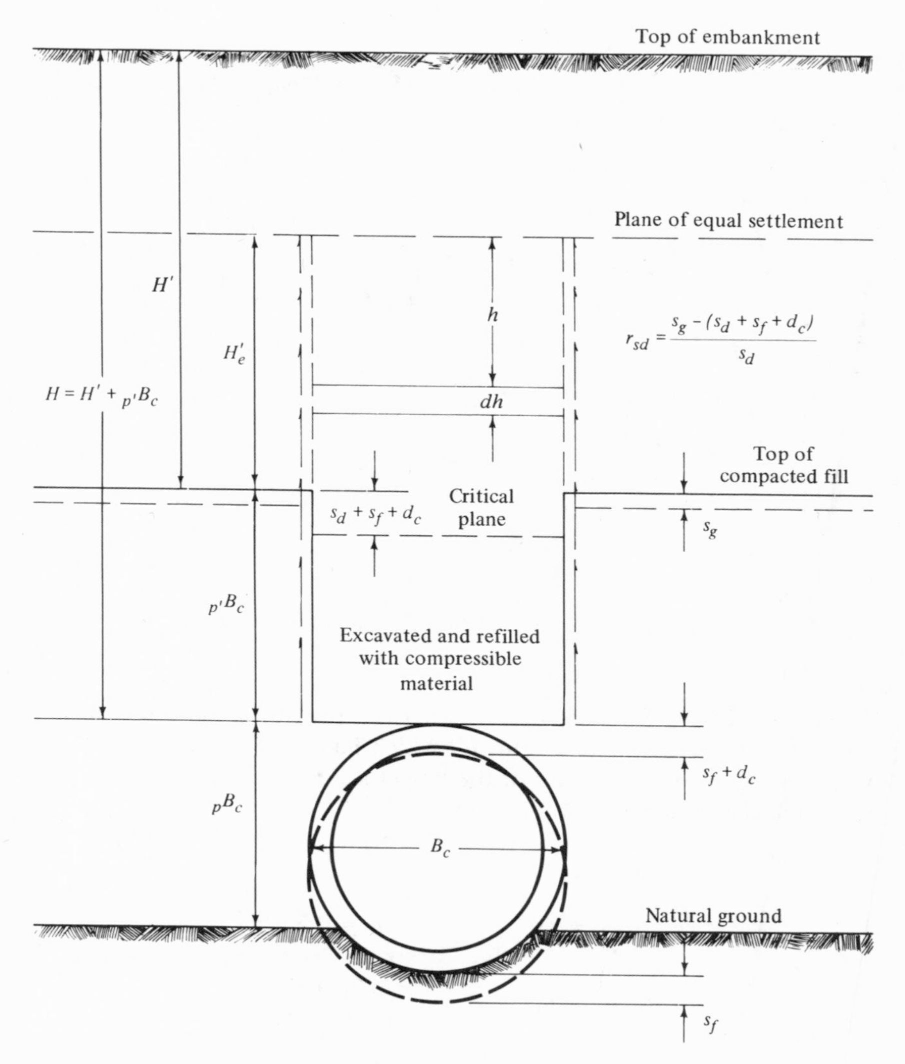

Fig. 25-16. Settlements which influence loads on imperfect ditch conduits. Key: solid line, initial elevation ($H = 0$); dashed line, final elevation.

fied elevation above its top. Next, a trench of the same width as the outside horizontal dimension of the pipe is excavated down to the structure and refilled with very loose, compressible material. This may be simply the excavated and loosened soil, or it may be material whose compressibility is augmented by the addition of straw or hay or other bulky material. The objective of this method of construction is to insure that the interior prism of soil will settle more than the exterior prisms, thereby generating friction forces which are directed upward on the sides of the interior prisms. The resultant load on the conduit is thereby reduced.

A formula for the load on an imperfect ditch conduit may be derived from the settlements and geometrical considerations illustrated in Fig. 25-16. The load formula is

$$W_c = C_n \gamma B_c^2 \tag{25-11}$$

in which C_n is a load coefficient which is a function of the ratio of the height of fill to the width of ditch, H/B_c; of the projection ratio p'; and of the settlement ratio r_{sd}.

The projection ratio p' in an imperfect ditch installation is equal to the depth of the ditch divided by its width and is considered to be a positive quantity. The critical plane is defined as the horizontal plane in the ditch filling material at the level of the surface of the compacted backfill before settlement occurs. The settlement ratio r_{sd} is defined as the result obtained by dividing the difference between the settlement of the compacted fill surface and the settlement of the critical plane by the compression of the column of soil of depth $p'B_c$. Thus

$$r_{sd} = \frac{s_g - (s_d + s_f + d_c)}{s_d} \tag{25-12}$$

in which

r_{sd} = settlement ratio for imperfect ditch conduits;
s_g = settlement of surface of compacted soil, in feet;
s_d = compression of fill in ditch within height $p'B_c$, in feet;
s_f = settlement of flow line of conduit, in feet;
d_c = deflection of conduit, i.e., shortening of its vertical dimension, in feet; and
$(s_d + s_f + d_c)$ = settlement of critical plane, in feet.

The settlement ratio is always a negative quantity in this case. The value of the coefficient C_n may be obtained from one of the diagrams in Figs. 25-17 to 25-20. As was mentioned in the case of the projecting-conduit diagram of Fig. 25-13, the ray lines in these diagrams intersect the envelope curves at points where $H_e = H$. Therefore, the height of the plane of equal settlement can be estimated by multiplying the value of H/B_c at this intersection by B_c to obtain H_e.

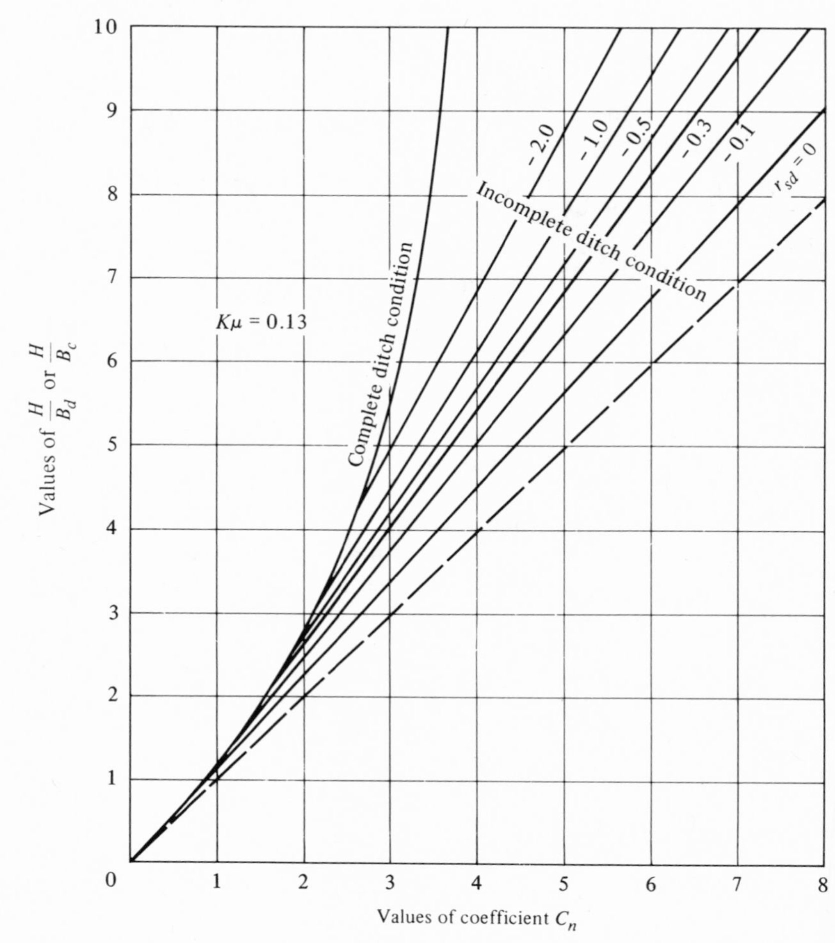

Fig. 25-17. Diagram for coefficient C_n for negative projecting conduits and imperfect ditch conditions when $p' = 0.5$.

TABLE 25-3. Values[a] of C_n in Terms of H/B_d or H/B_c

r_{sd}	*Equation*
0	$C_n = 0.88H/B + 0.03$
−0.1	$C_n = 0.77H/B + 0.09$
−0.3	$C_n = 0.71H/B + 0.14$
−0.5	$C_n = 0.67H/B + 0.17$
−1.0	$C_n = 0.61H/B + 0.23$
−2.0	$C_n = 0.53H/B + 0.33$

[a]Negative projecting conduits and imperfect ditch conduits (*1*), $p' = 0.5$.

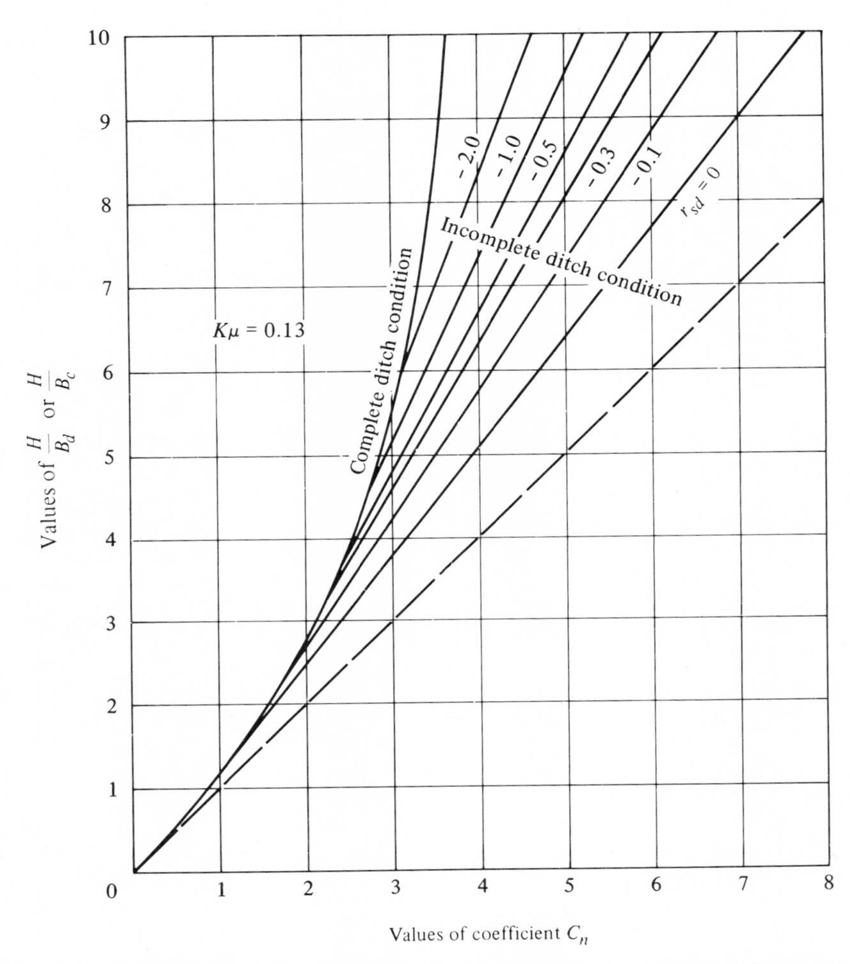

Fig. 25-18. Diagram of coefficient C_n for negative projecting conduits and imperfect ditch conduits when $p' = 1.0$.

TABLE 25-4. Values[a] of C_n in Terms of H/B_d or H/B_c

r_{sd}	*Equation*
0	$C_n = 0.77H/B + 0.11$
−0.1	$C_n = 0.65H/B + 0.25$
−0.3	$C_n = 0.58H/B + 0.34$
−0.5	$C_n = 0.53H/B + 0.41$
−1.0	$C_n = 0.47H/B + 0.52$
−2.0	$C_n = 0.40H/B + 0.69$

[a] Negative projecting conduits and imperfect ditch conduits (*1*). $p' = 1.0$.

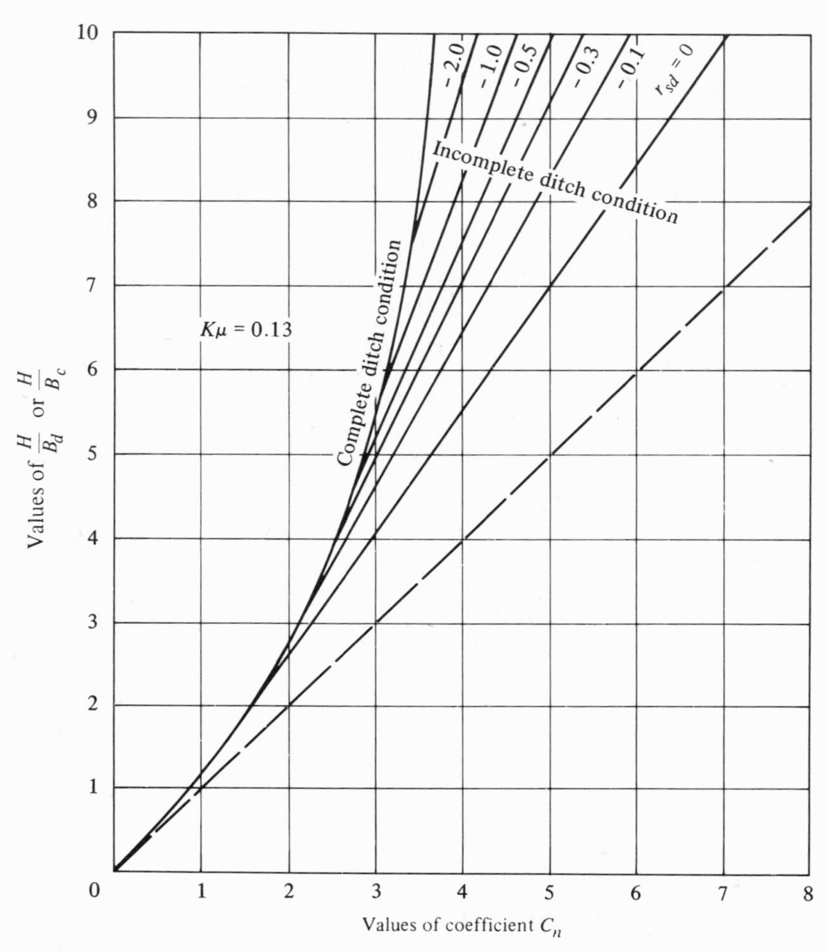

Fig. 25-19. Diagram of coefficient C_n for negative projecting conduits and imperfect ditch conduits when $p' = 1.5$.

TABLE 25-5. Values[a] of C_n in Terms of H/B_d or H/B_c

r_{sd}	*Equation*
0	$C_n = 0.68H/B + 0.23$
−0.1	$C_n = 0.55H/B + 0.44$
−0.3	$C_n = 0.48H/B + 0.58$
−0.5	$C_n = 0.44H/B + 0.66$
−1.0	$C_n = 0.38H/B + 0.81$
−2.0	$C_n = 0.31H/B + 1.15$

[a] Negative projecting conduits and imperfect ditch conduits (*1*), $p' = 1.5$.

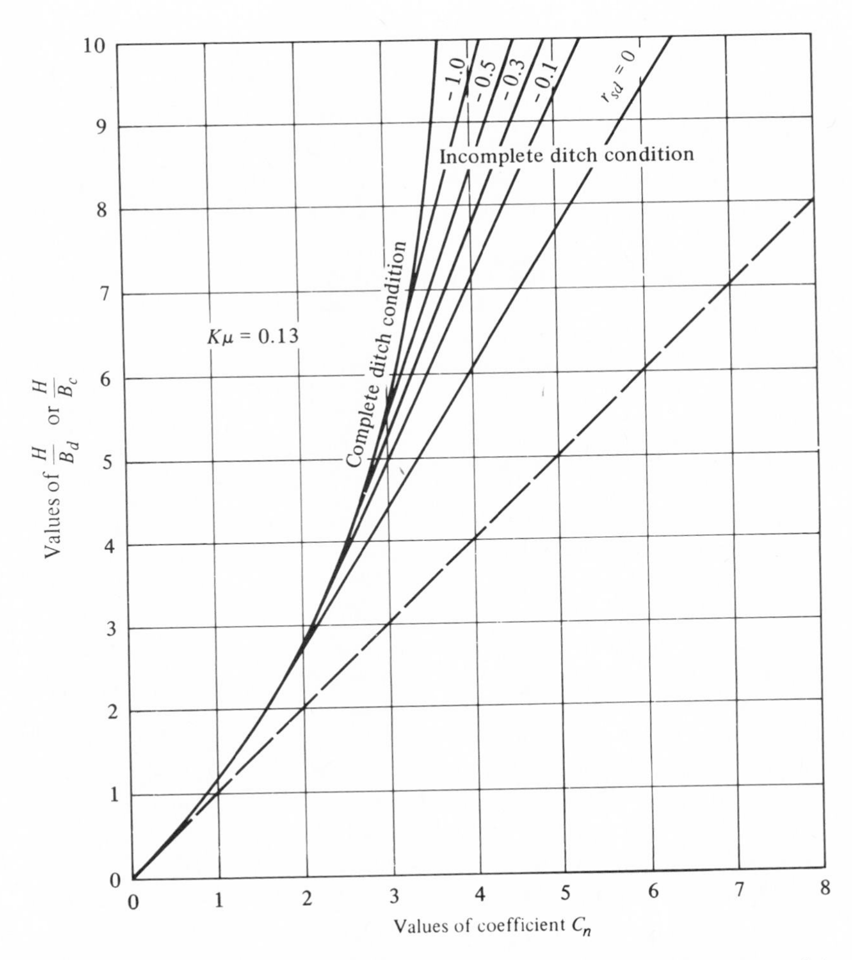

Fig. 25-20. Diagram of coefficient C_n for negative projecting conduits and imperfect ditch conduits when $p' = 2.0$.

TABLE 25-6. Values[a] of C_n in Terms of H/B_d or H/B_c

r_{sd}	*Equation*
0	$C_n = 0.59H/B + 0.37$
−0.1	$C_n = 0.47H/B + 0.65$
−0.3	$C_n = 0.40H/B + 0.82$
−0.5	$C_n = 0.36H/B + 0.92$
−1.0	$C_n = 0.31H/B + 1.11$
−2.0	$C_n = 0.24H/B + 1.52$

[a]Negative projecting conduits and imperfect ditch conduits (*1*). $p' = 2.0$.

Research directed toward the determination of loads on imperfect ditch conduits has not progressed so far as it has in connection with the other classes of conduits. In the absence of extensive factual data relative to probable values of the settlement ratio for conduits of this class, it is tentatively recommended that this ratio be assumed to lie between -0.3 and -0.5 for the purpose of estimating loads. Recent research reported by Taylor (*14*) of the Illinois Division of Highways indicated that the measured settlement ratio of a 48 in. reinforced concrete pipe culvert, installed as an imperfect ditch conduit under 30 ft of fill varied from -0.25 to -0.45.

25.19. NEGATIVE PROJECTING CONDUIT. An analysis of loads on negative projecting conduits, as illustrated in Fig. 25-3(c), follows the same procedure as that for imperfect ditch conduits, except that width factor is B_d, the width of the shallow ditch in which the pipe is installed, instead of the width of the imperfect ditch, B_c. The same load coefficient diagrams are applicable to both case.

25.20. MODIFIED IMPERFECT DITCH PROCEDURE. A modification of the original imperfect ditch construction procedure, which has been developed by practicing engineers, is to install a conduit and place the sidefills up to a foot or so above the pipe (see Fig. 25-21). Then baled straw is placed directly over the structure and the embankment soil is compacted for a substantial width, say 2 or 3 pipe diameters, along the sides of the straw bales. When the fill has reached the top of the bales, the wires are cut and a second layer of bales installed. Again the fill is placed and compacted up to the top. This procedure can be repeated until the desired negative projection ratio, p', is accomplished, after which construction of the embankment is continued in the normal manner.

Several reinforced concrete pipe culverts installed in the above manner in Humboldt County, California, (*3*) performed very favorably in comparison with culverts in the same area which were installed as ordinary projecting conduits, as indicated in Fig. 25-23. The engineer in charge of these installations also reported that the baled straw method simplified construction procedures. A photograph of the operation is shown in Fig. 25-22.

Another project involving an 18-ft diameter corrugated steel pipe culvert under 83 ft of fill was recently constructed near Wolf Creek, Montana (*6*), using this baled straw procedure. Strain gauges placed at the spring line of the pipe indicated that the vertical load on the structure was equal to about one-half the dead weight of the overlying prism of embankment material. Also it is of interest to note that the original 36-in. depth of baled straw was compressed to 11 in. under the pressure of

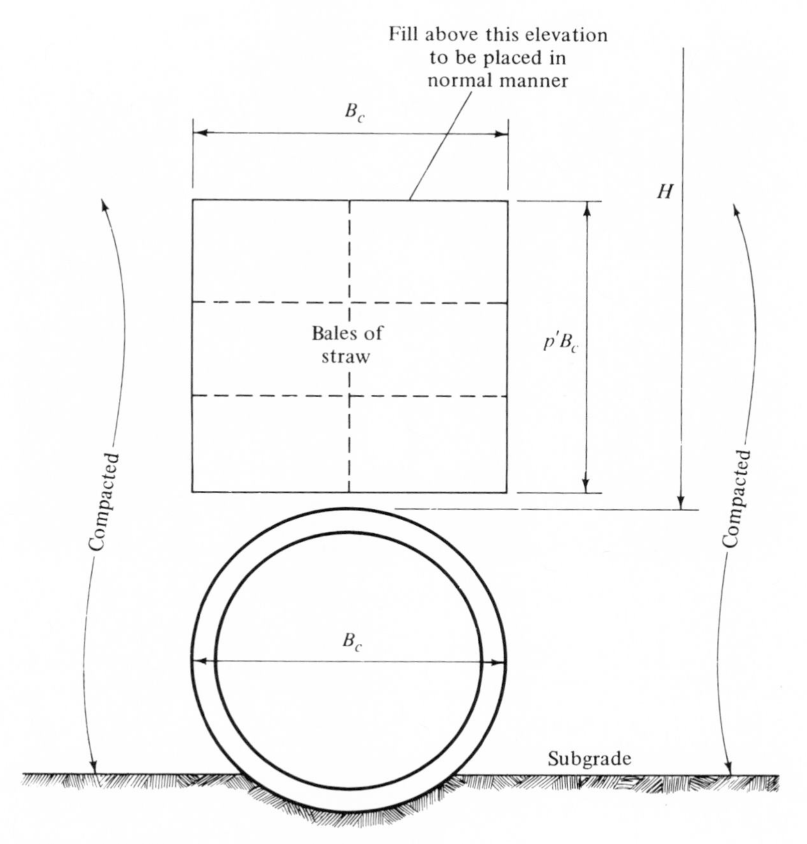

Fig. 25-21. Modified imperfect ditch construction.

the prism of soil above. In spite of this substantial amount of compression, there was no visible sag in the grade of the embankment, indicating that a plane of equal settlement had developed at some elevation below grade.

EXAMPLE 25-3. Suppose that a 60-in. reinforced concrete pipe, 6 ft in outside diameter, is installed as an Imperfect Ditch Conduit with $p' = 0.5$. The height of fill above the top of the pipe will be 50 ft. Determine the load on the structure and the height of the plane of equal settlement.

SOLUTION. In this case,

$$H = 50 \text{ ft} \qquad p' = 0.5$$
$$B_c = 6 \text{ ft} \qquad r_{sd} = -0.3 \text{ (assumed)}$$
$$H/B_c = 8.33 \qquad \gamma = 120 \text{ pcf}$$

Fig. 25-22. Modified imperfect ditch construction using baled straw (*3*).

From Fig. 25-17, $C_n = 5.8$. Substituting in Eq. (25-11), we obtain

$$W_c = 5.8 \times 120 \times 6^2 = 25{,}000 \text{ lb/linear ft}$$

The ray line for $r_{sd} = -0.5$ in the diagram intersects the envelope curve at $H/B_c = 2.8$. Therefore the height of the plane of equal settlement H_e is $6 \times 2.8 = 16.8$ ft. This is well below the top of the embankment, indicating that there is no danger of sag in the top of the embankment.

25.21. CONDUITS PLACED IN WIDE DITCHES. Equation (25-3) for determining the load on a ditch conduit indicates that this load is a function of the width of the ditch in which the conduit is laid; that is, the wider the ditch, the greater is the load on a conduit laid in it. Obviously, there is a limiting width beyond which this principle does not apply; since, in a ditch which is very wide relative to the conduit, the sides of the ditch will be so far away from the conduit that they cannot possibly affect the load on it.

Studies by Schlick (*7*) of the effect of the width of ditch on the load transmitted to a rigid conduit indicate that it is safe to calculate the load by means of the ditch-conduit formula for all widths of ditch below that which gives a load equal to the load indicated by Eq. (25-7) for a positive projecting conduit. In other words, as the width of the ditch increases,

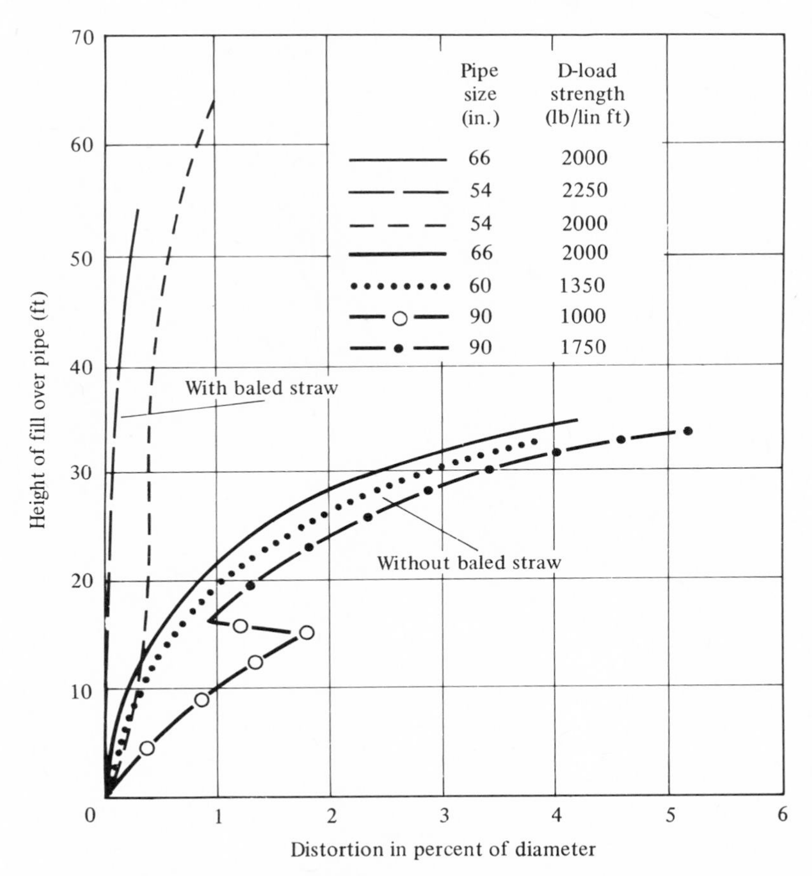

Fig. 25-23. A comparison of the performance of culverts placed with baled straw and without straw. Distortion is horizontal diameter minus vertical diameter.

other factors remaining constant, the load on a rigid conduit increases in accordance with the theory for a ditch conduit until it equals the load determined by the theory for a projecting conduit. The width at which this load equality develops is called the *transition width*. For greater widths, the load remains constant regardless of the width of the ditch.

The diagram in Fig. 25-24 shows values of the ratio of width of ditch to width of conduit, or the ratio B_d/B_c, at which the loads on a rigid conduit are equal by both the ditch conduit theory and the projecting conduit theory. For values of B_d/B_c less than those given in the diagram, the load on a rigid conduit may be determined by the ditch conduit theory. For greater values of this ratio, use the projecting conduit theory.

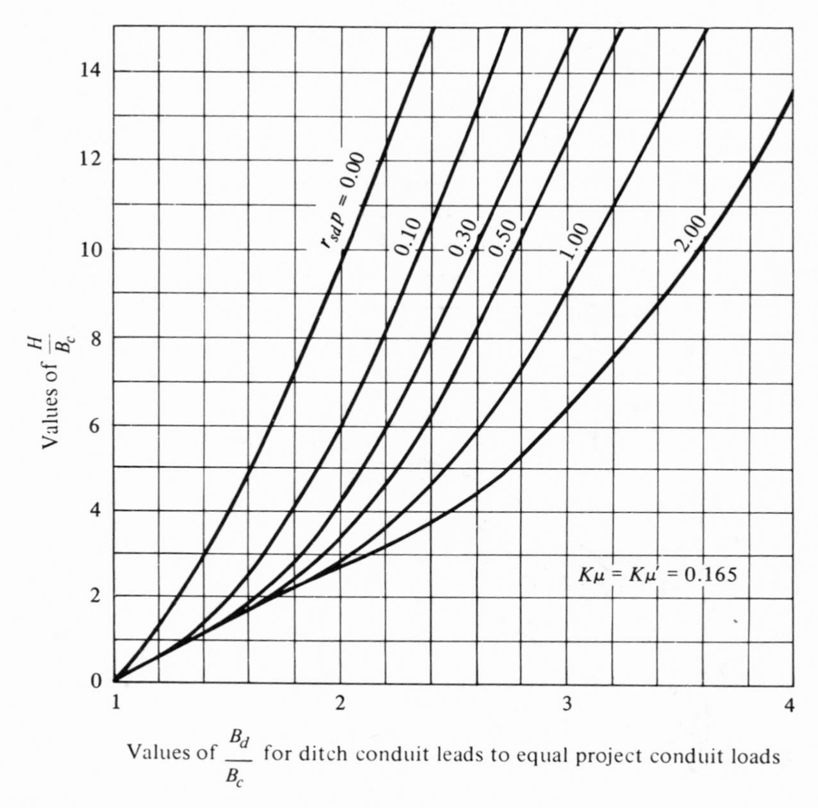

Fig. 25-24. Curves for transition-width ratio.

It is difficult to determine an appropriate value of $r_{sd}p$, the product of the settlement ratio and the projection ratio, in the application of the transition width concept. In the absence of specific information relative to this physical factor, $r_{sd}p = 0.5$ is suggested as a reasonably good working value.

25.22. REDUCTION OF LOAD BY USE OF YIELDING FOUNDATION. In certain situations involving rock foundations and rock fills over the conduits, it may not be practicable to use the imperfect ditch method of construction. Under such conditions, it may be advantageous to apply another special method of constructing projecting conduits so as to make certain that the vertical load will be less than that which would be normally developed. The basic feature of this method is to place the conduit on a very yielding foundation by excavating in the rock foundation a trench having a width somewhat greater than the outside width of the conduit and refilling this trench with loose, highly compressible soil on which the conduit is constructed.

This method produces a less severe loading condition by insuring an abnormally high settlement of the top of the conduit in relation to the settlement of the critical plane in the embankment; and thereby reducing the value of the settlement ratio and, consequently, the load on the structure. Another favorable feature of this method of construction, especially in the case of a circular pipe conduit, is the opportunity afforded to obtain a wider distribution of the reaction between the pipe and its bedding; and thus to increase its load-carrying capacity, as will be discussed in the next chapter. A disadvantage of this type of construction is that the flow line of the structure is permitted to settle more than it would ordinarily. This disadvantage can be neutralized to an appreciable extent, however, by constructing the conduit on a camber.

25.23. LOADS ON CONDUITS DUE TO LOADS APPLIED TO SURFACE OF FILL. In addition to being subjected to external loads because of the filling material, underground conduits are also subject to loads due to highway, railway, or airplane traffic or to other types of loads applied at the surface of the fill and transmitted through the soil to the underground structure. Such loads are of major importance when a conduit is placed under a traffic way with a relatively shallow covering of earth.

Extensive experiments on both ditch and projecting conduits have indicated that a static concentrated surface load, such as a truck wheel, is transmitted through the soil covering to the underground structure substantially in accordance with the Boussinesq solution (see Chapter 17) for stress distribution in a semi-infinite elastic solid, as indicated in Fig. 25-25. These experiments also indicated the magnitude of impact loads produced by moving wheel loads. From the facts revealed by these tests, Marston (5) proposed the following formula for live loads on underground conduits:

$$W_t = \frac{1}{A} I_c C_t P \tag{25-13}$$

in which

W_t = average load per unit length of conduit, due to wheel load, in pounds per foot;
A = length of conduit section on which load is computed, in feet;
I_c = impact factor;
C_t = load coefficient; and
P = concentrated wheel load on surface of fill, in pounds.

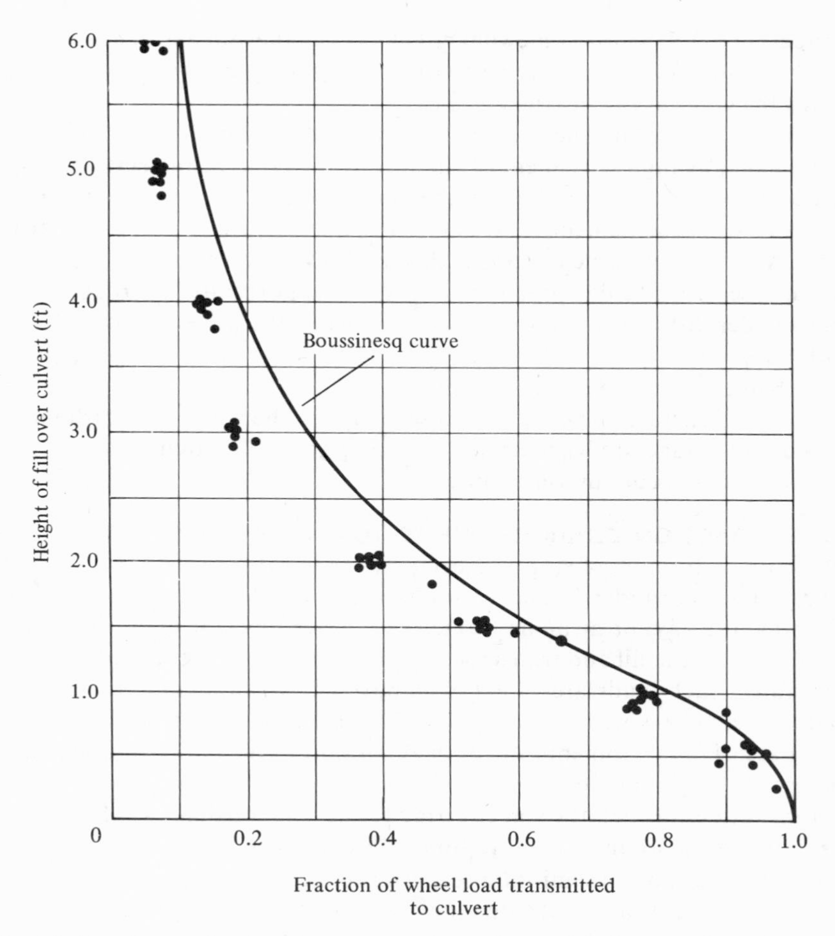

Fig. 25-25. Static wheel loads transmitted to a 2 ft × 3 ft 6 in. section of culvert (*13*).

The impact factor I_c is equal to unity when the surface load is static. When it is moving, as in the case of a truck or airplane wheel, the value of I_c depends on the speed of the vehicle, its vibratory action, the effect of wing uplift, and, most importantly, the roughness characteristcs of the roadway surface. The experiments referred to showed that the value of the impact factor is independent of the depth of cover over a culvert, and they indicated that design values of I_c should range from 1.5 to 2.0 for trucks operating on an unpaved roadway. No quantitative experimental evidence is available for paved roadway conditions or for airplane traffic.

The value of the coefficient C_t is dependent on the length and width of the conduit section and on the depth of cover over the conduit. It is based upon the Boussinesq law of stress distribution and may be evaluated in accordance with the principles discussed in Chapter 17. When estimating the live load on a circular or arch-shaped conduit, experimental evidence indicates that it is valid to calculate the load on a rectangular area which is the vertical projection of the conduit section on a horizontal plane through the top of the structure.

The length of conduit A is the actual length of an individual precast section of sewer or culvert pipe which is 3 ft or less in length. For continuous structures of monolithic concrete, corrugated metal, or steel, or for long sections of cast-iron pipe, A is the *effective length*. Effective length is defined as the length of pipe over which the average live load produces the same effect on stress or deflection as does the actual load, which is of varying intensity along the pipe. No research information is available at this time concerning the effective length of continuous pipe lines of various materials and diameters. Until factual information on this subject becomes available, an effective length of 3 ft is suggested for use in design of pipes greater than 3 ft in length.

PROBLEMS

25.1. Define and describe Ditch Conduits; Positive Projecting Conduits; Imperfect Ditch Conduits.

25.2. Compute the load on a clay sewer pipe, having an inside diameter of 21 in. and wall thickness of 2 in., which is installed in a ditch whose width is 3 ft 9 in. at the elevation of the top of the pipe. The height of backfill over the top of the pipe is 18 ft. Assume the soil is a clayey material having a coefficient of friction of 0.2 and unit weight of 120 pcf.

25.3. Suppose the pipe in Problem 25.2 is installed in a ditch which is 4 ft 9 in. wide. What will be the percentage increase in load?

25.4. A 5 × 5 reinforced concrete box culvert with 10 in. sidewalls is installed with its top 3 ft above the adjacent natural ground level. What is the projection ratio for this installation?

25.5. A circular pipe culvert is installed as a positive projecting conduit. As the fill over it is constructed, the tabulated settlements develop. Calculate the settlement ratio for this installation.

Element	*Settlement (ft)*
Critical plane	0.54
Top of pipe	0.18
Natural ground at sides of pipe	0.09

25.6. An elliptical concrete pipe has outside dimensions of 7.0 ft on the major axis and 4.1 ft on the minor axis. When installed in a culvert with the major axis vertical (VE), the top of the pipe projects 5 ft above the adjacent natural ground surface. What is the projection ratio for this installation?

25.7. Suppose the pipe in the preceding problem is installed with the major axis horizontal (HE) and the top of the pipe is 3 ft above the adjacent ground. What is the projection ratio?

25.8. An 84 in. diameter flexible pipe is installed with its top 5 ft above the adjacent ground level. It is covered with an embankment 18 ft above its top. The soil weighs 120 pcf. Assume a settlement ratio of −0.2. Estimate the load on the pipe.

25.9. A reinforced concrete pipe culvert having an outside diameter of 5.83 ft is installed in such a manner that it projects 4.6 ft above the ground level. The foundation soil is described as very unyielding in character. If this culvert is covered with a fill 24 ft high which has a unit weight of 120 pcf, what is the load on the structure?

25.10. Estimate the height of the plane of equal settlement in Problem 25.9.

25.11. A rigid pipe with an outside diameter of 5.83 ft is installed as a positive conduit. Soil is thoroughly compacted to a distance of 12 ft on each side and to an elevation 7 ft above the top of the pipe. A trench 5.83 ft wide is excavated directly over the pipe to a depth of 6 ft and refilled with loose compressible soil. Then an embankment is constructed to a height of 56 ft above the top of the pipe. Assume the unit weight of the soil is 125 pcf and the settlement ratio is −0.35. Estimate the load on the pipe and the height of the plane of equal settlement.

25.12. An 18-in. o.d. pipe is installed in a ditch which is 60 in. wide. Assume $r_{sd}p = 0.3$ and the height of fill is to be 20 ft. What is the transition width? Should the load on this pipe be calculated by the ditch conduit theory or the projecting conduit theory?

25.13. A concrete box culvert 4 ft wide (outside) is covered with 2 ft of fill which weighs 125 pcf. A truck wheel weighing 10,000 lb stops directly over the center of the culvert. Determine the total dead and live load on a 3-ft length of the culvert which is centered directly beneath the wheel.

REFERENCES

1. Clarke, N. W. B. *Buried Pipelines*. Maclaren and Sons, London, 1968.
2. Deen, Robert C. "Performance of a Reinforced Concrete Pipe Culvert under Rock Embankment." *Highway Research Board Rec.* **262** (1969).
3. Larson, Norman G. "A Practical Method for Constructing Rigid Conduits under High Fills." *Proc. Highway Research Board,* **41,** 273 (1962).
4. Marston, Anson, and A. O. Anderson, "The Theory of Loads on Pipes in Ditches and Tests of Cement and Clay Drain Tile and Sewer Pipe." Bul. 31, Iowa Engineering Experiment Station, Ames, Iowa, 1913.
5. Marston, Anson. "The Theory of External Loads on Closed Conduits in the Light of the Latest Experiments." Bul. 96, Iowa Engineering Experiment Station, Ames, Iowa, 1930.

6. Scheer, Alfred C., and Gerald A. Willett, Jr. "Rebuilt Wolf Creek Culvert Behavior." *Highway Research Board Rec.* **262** (1969).
7. Schlick, W. J. "Loads on Pipe in Wide Ditches." Bul. 108, Iowa Engineering Experiment Station, Ames, Iowa, 1932.
8. Schlick, W. J. "Loads on Negative Projecting Conduits." *Proc. Highway Research Board*, **31,** 308 (1952).
9. Spangler, M. G. "Long Time Measurement of Loads on Three Pipe Culverts." 52nd Annual Meeting, Highway Research Board, Washington, D.C., Jan. 1973. *Highway Research Board Rec.* **443** (1973).
10. Spangler, M. G. "Field Measurements of the Settlement Ratios of Various Highway Culverts." Bul. 170, Iowa Engineering Experiment Station, Ames, Iowa, 1950.
11. Spangler, M. G. "A Theory of Loads on Negative Projecting Conduits." *Proc. Highway Research Board,* **29,** 153 (1950).
12. Spangler, M. G. "A Practical Application of the Imperfect Ditch Method of Construction," *Proc. Highway Research Board,* **37,** 271 (1958).
13. Spangler, M. G., Robley Winfrey, and Clyde Mason. "Experimental Determination of Static and Impact Loads Transmitted to Culverts." Bul. 79, Iowa Engineering Experiment Station, Ames, Iowa, 1926.
14. Taylor, R. K., "Induced Trench Method of Culvert Installation." 52nd Annual Meeting, Highway Research Board, Washington, D.C. Jan. 1973. Discussion by Spangler. *Highway Research Board Rec.* **443** (1973).

26

Supporting Strength of Underground Conduits

26.1. FACTORS AFFECTING SUPPORTING STRENGTH OF CONDUIT. Underground conduits are constructed in a wide variety of shapes and of many different structural materials. In general, the load on a conduit is independent of its shape and the material of which it is made, except for the effect these properties may have on the settlement of the top of the conduit. On the other hand, the supporting strength or load-carrying capacity of a conduit is intimately dependent on its shape and the kind and quality of material of which it is made.

Monolithic types of reinforced-concrete structures, such as arch and box culverts, may be satisfactorily designed by any of the current procedures for analysis of rigid-frame structures. Numerous measurements of the distribution of the vertical load on both rectangular and circular culverts due to the earth overburden have indicated that, essentially, it is uniformly distributed over the width of the conduit and may be so considered for design purposes. Such structures may or may not be subjected to active lateral pressure on all or a portion of their side-wall areas, the loading depending on the local situation.

It is impossible to overemphasize the importance of good quality bedding and backfilling practices in connection with the installation of underground conduits of circular or elliptical shape, in order to enable such pipes to develop their maximum potential of load-carrying capacity.

The shape of the bedding influences the distribution of the vertical reaction on the bottom of the pipe. In contrast to the vertical earth load on top, the lateral distribution of the bottom reaction depends primarily upon the character of the bedding on which the pipe is installed, and bending moment in the pipe wall is greatly influenced by this distribution. It is important to note that good lateral distribution of the upward re-

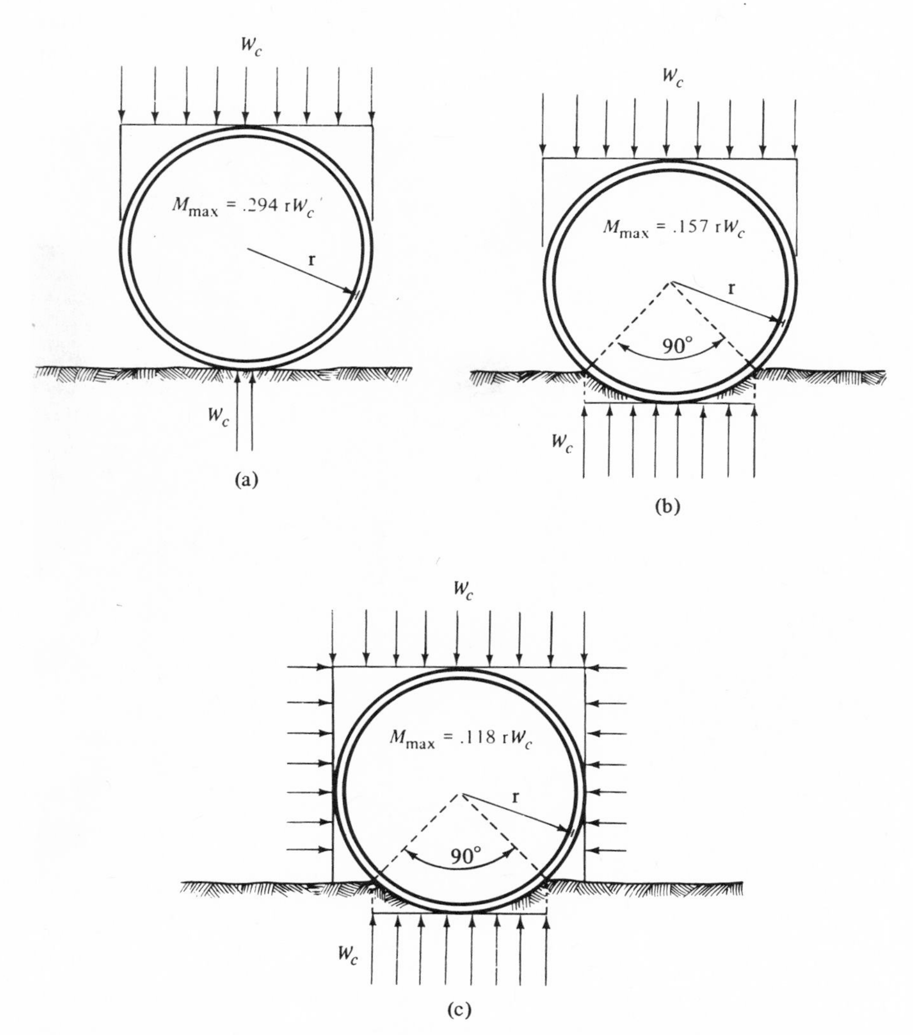

Fig. 26-1. Influence of good bedding and backfilling practice on bending moment in pipe wall.

action is most readily achieved by shaping the bedding material by means of a template cut to fit the contour of the pipe for the width specified. The practice of placing a pipe on a flat bed and then tamping soil material within the triangular-shaped spaces beneath the haunches of the pipe does not yield the desired result in terms of efficient distribution of the bottom reaction.

Backfilling practices influence the development of active lateral pressures on the sides of the pipe, which, when they can be relied upon, cause bending moments in the opposite direction from those produced by the

Fig. 26-2. Failure in the invert of a large diameter reinforced concrete pipe, caused by poor quality bedding and unfavorable conditions for development of lateral pressures against the sides of the pipe.
Note: This pipeline was repaired by drilling holes through the walls at the lower quarter points and injecting pressure grout to improve the bedding. Then the shattered concrete was removed and the steel protected by gunite. The pipe has served satisfactorily for more than 20 yr and gives promise of a long life.

vertical load and reaction, and therefore decrease the bending moment stress in the pipe wall and increase its supporting strength.

The importance of these principles is illustrated numerically in Fig. 26-1. In (a) is shown a circular pipe placed on a flat surface without any attempt to shape the bedding to fit the contour of the pipe, and without effective lateral pressure. This load condition causes the bottom reaction to be highly concentrated and the maximum bending moment is 0.294 rW_c. In (b) the bedding is preshaped to fit the pipe over a width of 90° which distributes the reaction approximately uniformly over this width, and the maximum bending moment is 0.157 rW_c. Thus the pipe in (b) theoretically can support 1.87 times, or nearly double the load on pipe (a) without an increase in wall stress.

The illustration in (c) shows the influence of active lateral pressure on stress in the pipe wall. If it is assumed that the pipe bedding is the same as in (b) and that lateral pressure of an intensity equal to about one-third the vertical pressure of the overlying soil acts against the portion of the pipe which projects above the subgrade, the bending moment is still further reduced to 0.118 rW_c. The supporting strength is increased to 2.49 times that of pipe (a) and 1.33 times that of pipe (b). These illustrations indicate the very great influence of bedding and backfilling practices on the structural performance of a pipe in the ground. The quality of these procedures and the care taken in the installation of a pipe may very well spell the difference between success and structural failure of a buried conduit. An illustration of the type of damage which can result from poor bedding and backfilling practices is shown in Fig. 26-2.

From the above considerations it is indicated that the field supporting strength of this type of structure is dependent upon three factors: (1) the inherent strength of the pipe; (2) the quality of the bedding as it affects the distribution of the bottom reaction; and (3) the magnitude and distribution of active lateral pressures which may act on the sides of the pipe.

26.2. TESTS FOR INHERENT STRENGTH OF PIPES. Rigid circular or elliptical pipes, usually precast of such materials as plain or reinforced concrete, burned clay, asbestos cement, and cast iron, are not readily analyzed by principles of mechanics; and, since they are usually relatively small structures, their inherent supporting strength can be most easily determined by testing a representative group of specimens in the laboratory. Several methods of testing sections of pipes have been devised. Four methods of supporting and loading pipes are: the two-edge bearing, the three-edge bearing, the sand bearing, and the Minnesota bearing. The details of these methods are shown in Fig. 26-3. Of these tests, that with

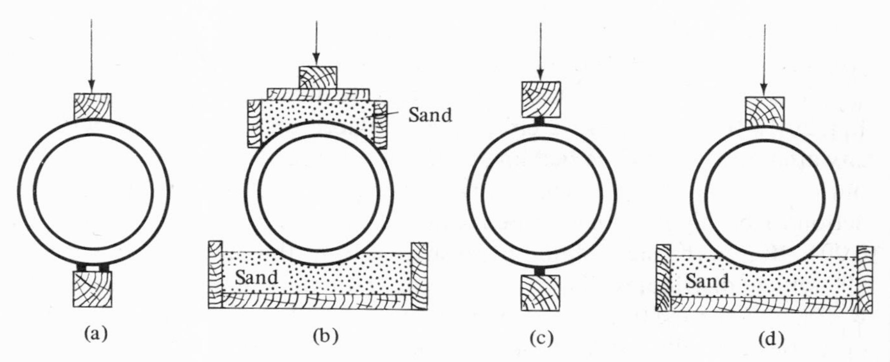

Fig. 26-3. Four types of bearing for laboratory tests of pipe. (a) Three-edge bearing; (b) sand bearing; (c) two-edge bearing; and (d) Minnesota bearing.

the three-edge bearing is the simplest and most easily performed; and it also gives accurate and uniform results. For these reasons it is widely employed in pipe-strength determinations, although some engineers prefer the sand bearing test because of the wider distribution of both the applied load and the reaction.

As will be noted in Fig. 26-3 the test load and the reaction on the pipe are distributed differently in each of the types of tests, and the breaking load or supporting strength will likewise be different. It is convenient to express the supporting strength of a pipe in terms of its strength when tested by the three-edge bearing method. The ratio of the strength of a pipe under any stated condition of loading, whether in the field or in the laboratory, to its strength by the three-edge bearing test is called the *load factor* or *strength ratio* for the stated condition and is designated as L_f. Numerous tests of rigid pipes have indicated the following values of load factor for the other three test methods shown in Fig. 26-3.

Sand bearing test	1.5
Two-edge bearing test	1.0
Minnesota bearing test	1.1

Furthermore, when a pipe is placed and loaded in a field installation, the distribution of the load and reaction will be much different from that in the test loading; and in certain cases lateral pressures will be exerted against the sides of the pipe, and its ability to support vertical loads will thus be increased. Load factors for various bedding conditions and lateral pressure situations are discussed below.

26.3. BEDDING CONDITIONS FOR DITCH CONDUITS. A wide variety of bedding conditions affecting the load and reaction distribution and the lateral-pressure situation may be encountered in conduit-construction practice, and a wide range of supporting strengths of a given conduit may be obtained simply by varying the installation conditions. It is feasible to establish and define classifications of bedding conditions covering a range of practical attainments and to determine a load factor for each classification which, when multiplied by the three-edge bearing laboratory

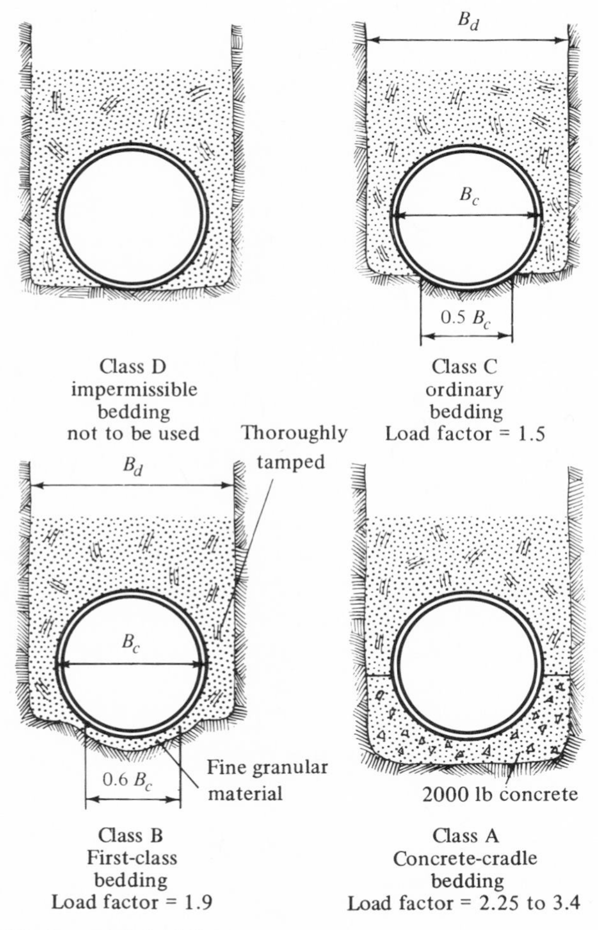

Fig. 26-4. Ditch conduit beddings.

strength of the pipe, will give the supporting strength for pipes installed in accordance with the definition of that classification.

For ditch conduits, the following bedding classifications have been defined and are illustrated in Fig. 26-4.

Impermissible bedding (Class D) is that method of bedding a ditch conduit in which little or no care is exercised to shape the foundation to fit the lower part of the conduit exterior or to refill all spaces under and around the conduit with granular materials at least partially compacted.

Ordinary bedding (Class C) is that method of bedding a ditch conduit in which the conduit is bedded with "ordinary" care in an earth foundation preshaped to fit the lower part of the conduit exterior with reasonable closeness for a width of at least 50% of the conduit breadth; and in which the remainder of the conduit is surrounded to a height of at least 0.5 ft above its top by granular materials that are shovel-placed and shovel-tamped to completely fill all spaces under and adjacent to the conduit. All this work must be done under the general direction of a competent engineer.

First-class bedding (Class B) is that method of bedding a ditch conduit in which the conduit is carefully bedded on fine granular materials in an earth foundation that is carefully preshaped to fit the lower part of the conduit exterior for a width of at least 60% of the conduit breadth; and in which the remainder of the conduit is entirely surrounded to a height of at least 1.0 ft above its top by granular materials that are carefully placed to completely fill all spaces under and adjacent to the conduit and that are thoroughly tamped on each side and under the conduit as far as practicable in layers not exceeding 0.5 ft in thickness. All work must be done under the direction of a competent engineer represented by a competent inspector who is constantly present during the operation.

Concrete-cradle bedding (Class A) is that method of bedding a ditch conduit in which the lower part of the conduit exterior is bedded in plain or reinforced concrete of suitable thickness under the lowest part of the conduit and extending upward on each side of the conduit for a greater or less proportion of its height. (See Section 26.6.)

The load factors for these bedding classes have been determined experimentally to be as follows:

Impermissible bedding (Class D)	1.1
Ordinary bedding (Class C)	1.5
First-class bedding (Class B)	1.9
Concrete-cradle bedding (Class A)	2.2–3.4

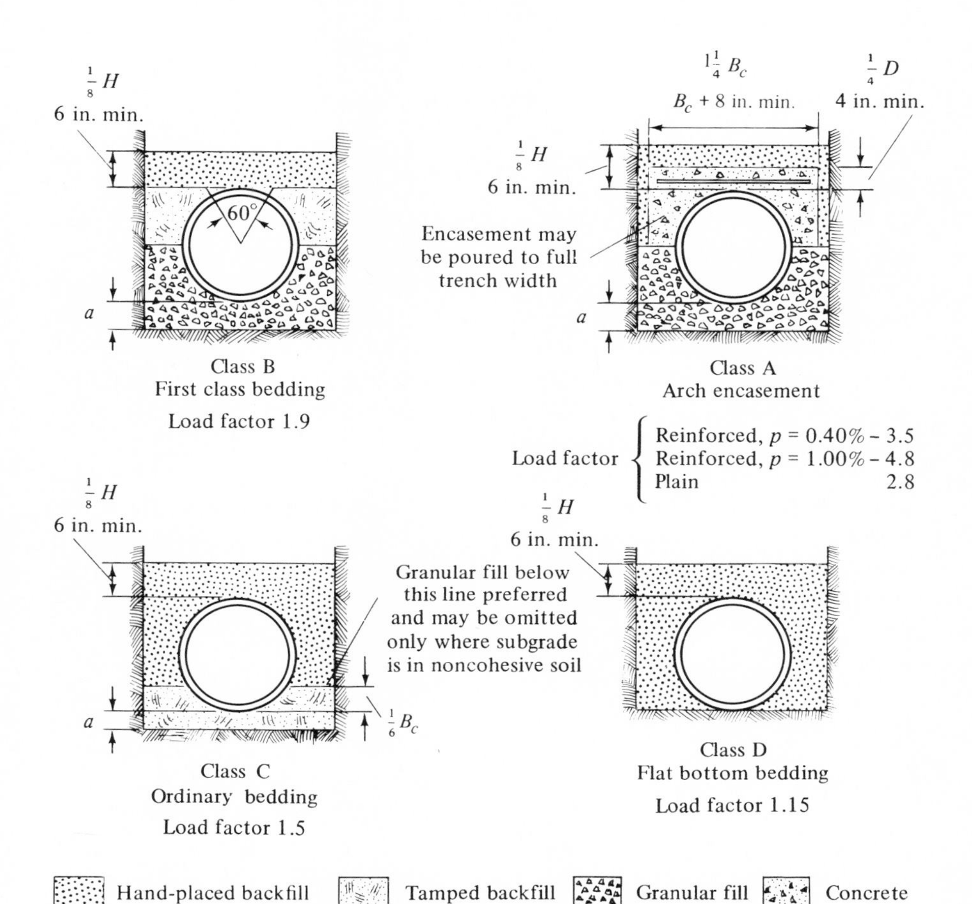

Fig. 26-5. Beddings and load factors for rigid conduits in trenches. Notes: *Granular fill* is to be crushed stone or pea gravel with not less than 95% passing $\frac{1}{2}$ in. and not less than 95% to be retained on a No. 4 sieve; to be placed in not more than 6 in. layers and compacted by slicing with a shovel. *Tamped backfill* shall be finely divided, job excavated material free from debris, organic material, and stones, compacted to 95% maximum density as determined by A.A.S.H.O. T-99. Granular fill may be substituted for all or part of tamped backfill except that, where used below pipe, it shall be carried to not less than $\frac{1}{6}$ B_c above pipe bottom. *Hand placed backfill* shall be finely divided material free from debris and stones. *Legend:* B_c—outside diameter of pipe, H—backfill cover above top of pipe, D—nominal pipe size, a—fill below pipe (see table), and p—area of transverse steel expressed as a percentage of area of concrete at crown.

Table of Fill Depths below Pipe

D (in.)	*a (in.) Min. Soil*	*a (in.) Min. Rock*
27 & smaller	3	6
30–60	4	9
66–larger	6	12

The factor for concrete-cradle bedding depends on the design of the cradle and the quality of the concrete.

The bedding classes just described were developed a number of years ago, when hand labor was extensively employed in the installation of conduits in trenches. More recently, practicing engineers have developed methods of bedding sewer pipes which require the use of selected granular materials, but much less hand labor. Typical of such more modern beddings are those shown in Fig. 26-5. This information was supplied through the courtesy of Mr. Henry Benjes of Black and Veatch, Consulting Engineers of Kansas City, Missouri (*1*).

26.4. FACTOR OF SAFETY. As is true of all engineering structures, the structural design of underground conduits requires the application of a reasonable factor of safety. For rigid pipes the following relationship is appropriate:

$$W_c = \frac{S_{eb} \times L_f}{F_s} \tag{26-1}$$

or

$$S_{eb} = \frac{W_c \times F_s}{L_f} \tag{26-2}$$

in which

W_c = design supporting strength = load on conduit;
S_{eb} = three-edge bearing strength; and
F_s = factor of safety.

Factors of safety used in practice vary widely. For pipe without steel reinforcement, suggested values based on minimum test strength of a representative number of test specimens are from 1.3 to 1.5. For reinforced concrete pipe, the suggested factor of safety is 1.0, based on the minimum 0.01-in. crack test strength. In this case the residual strength of the pipe between the 0.01 in. crack and the ultimate strength, plus the passive resistance pressure which develops against the sides as the pipe deflects after cracks form, provide a generous factor of safety against collapse. The 0.01 in. crack test strength is defined as the maximum load applied to the pipe at the time of the development of a crack having a width of 0.01 in. measured at close intervals throughout a length of 1 ft or more. It is usually expressed as a D-Load, that is, the test strength per linear foot of pipe divided by the internal diameter in feet.

The 0.01 in. crack width represents a test criterion only and is not to be regarded as a failure situation. All reinforced concrete structures are expected to crack as the tensile stress is transferred to the reinforcement, because the modulus of elasticity of steel is much greater, and the steel will stretch more than the concrete at working stresses. Unless cracks open up a sufficient amount to permit or promote oxidation and corrosion of the steel, reinforced concrete pipes will continue to perform their load-carrying function indefinitely, even though fine cracks have developed.

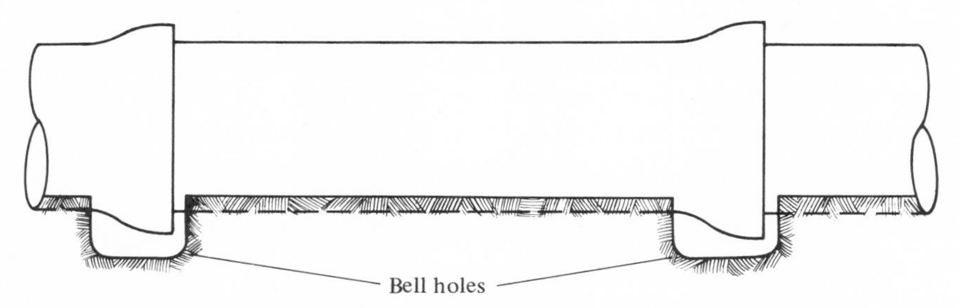

Fig. 26-6. Bell holes must be deep enough and wide enough to insure that all the bottom reaction acts on the barrel of the pipe.

Fig. 26-7. Clay pipe for which adequate bell holes were not provided.

26.5. BELL HOLES FOR BELL AND SPIGOT PIPE. The seat of strength of a bell and spigot pipe lies in the barrel, and care must be exercised to be sure that all of the bottom reaction is carried by the barrel. This is accomplished by excavating bell holes in the pipe bedding which are deep enough and wide enough to insure that all of the reaction is confined to the barrel. Some laboratory research by the National Clay Pipe Institute has indicated that the load factor for pipes which bear heavily on the bells is in the range from 0.5 to 0.75. A recent (1970) investigation of a 30-in. bell and spigot pipe sewer indicated that, because of inadequate bell holes, the bells of some pipe sections had settled to near contact with bedrock. This caused a high concentration of reaction on the bottom of the bells and these sections cracked top and bottom. The estimated load factor in this situation was 0.8. In other words, the pipe failed under an earth load which was less than the three-edge bearing strength of the pipe. These experiences emphasize the need for providing adequate bell holes, as illustrated in Fig. 26-6. A photograph of an 18-in. clay pipe for which adequate bell holes were not provided and which settled a sufficient amount to develop a high concentration of reaction on the bell is shown in Fig. 26-7.

26.6. BEDDING CONDITIONS FOR PROJECTING CONDUITS. As in the case of ditch conduits, it is convenient to name and define several classes of bedding conditions for projecting conduits and to determine a value of N, a bedding factor, for each class. These classes of bedding are illustrated in Fig. 26-8 and are defined as follows:

Impermissible projection bedding is that method of bedding a positive projecting conduit in which little or no care is exercised to preshape the foundation surface to fit the lower part of the conduit exterior or to fill all spaces under and around the conduit with granular materials. This type of bedding also includes the case of a conduit on a rock foundation in which an earth cushion is provided under the conduit but is so shallow that the conduit, as it settles under the influence of vertical load, approaches contact with the rock.

Ordinary projection bedding is that method of bedding a positive projecting conduit under an embankment in which the conduit is bedded with "ordinary" care in an earth foundation preshaped to fit the lower part of the conduit exterior with reasonable closeness for at least 10% of its overall height; and in which the remainder of the conduit is surrounded by granular materials that are shovel-placed to completely fill all spaces under and adjacent to the conduit. All this work must be done under the general direction of a competent engineer. In the case of a rock founda-

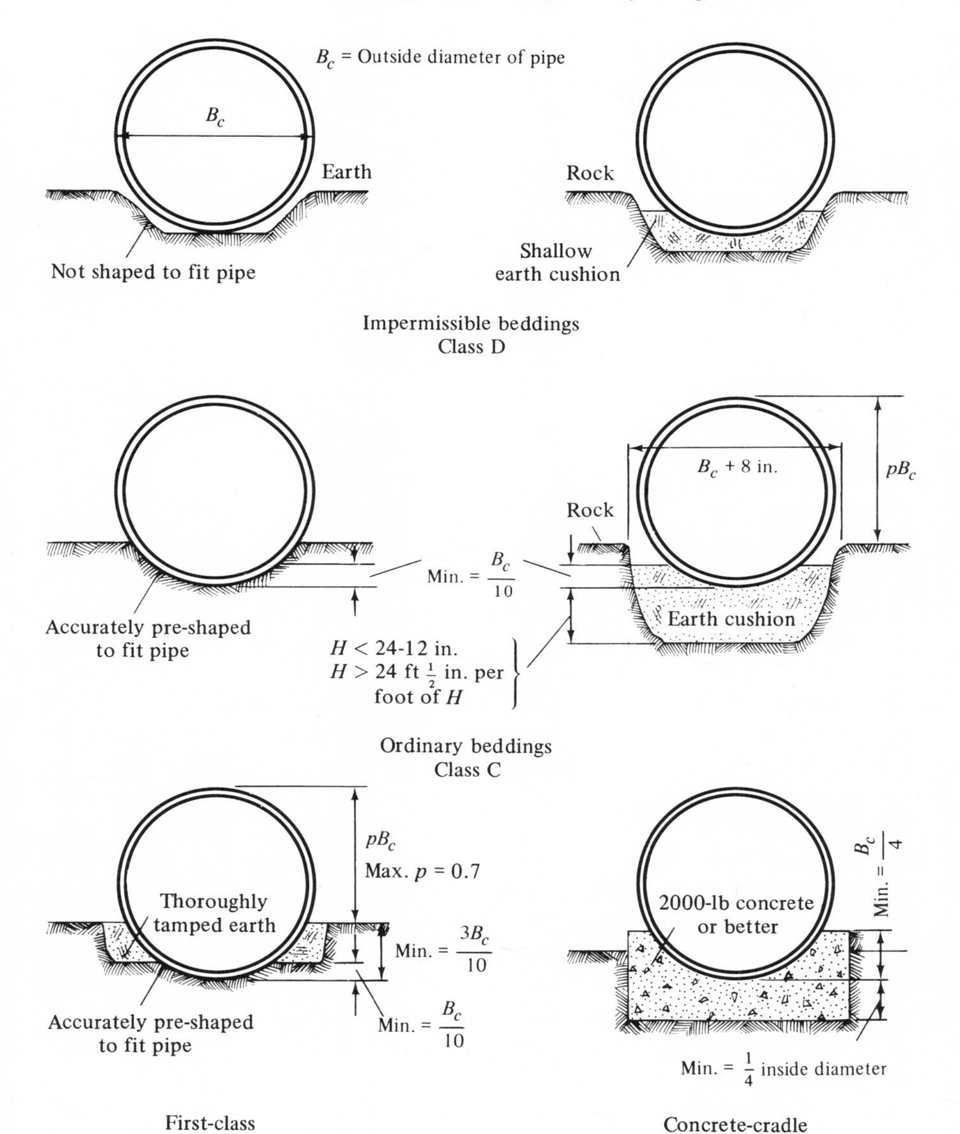

Fig. 26-8. Bedding for positive projecting conduits.

tion, the pipe is bedded on an earth cushion, which is preshaped in accordance with the foregoing specifications for an earth foundation and has a thickness under the pipe of not less than $\frac{1}{2}$ in./ft of height of fill and not less than 12 in.

First-class projection bedding is that method of bedding a positive projecting conduit, in which the projection ratio is not greater than 0.70; in which the conduit is carefully bedded on fine granular materials in an earth foundation that is carefully preshaped to fit the lower part of the conduit exterior for at least 10% of its overall height; and in which earth filling material is thoroughly rammed and tamped, in layers not exceeding 0.5 ft in depth, around the conduit for the remainder of the lower 30% of its height. All work must be done under the direction of a competent engineer, represented by a competent inspector who is constantly present during the operation.

Concrete-cradle projection bedding is that method of bedding a positive projecting conduit, in which the lower part of the conduit exterior is bedded in a cradle that is constructed of 2000-lb concrete or better, has a minimum thickness under the pipe of one-fourth its nominal internal diameter, and extends up the sides of the pipe for a height equal to one-fourth its outside diameter.

26.7. LOAD FACTOR FOR POSITIVE PROJECTING CONDUITS. In culvert construction practice, a rigid pipe is very often installed as a positive projecting conduit; and the fill material may exert an active lateral pressure against those portions of the sides of the pipe which project above the natural ground surface. These lateral pressures contribute significantly to the supporting strength of the structure as illustrated in Fig. 26-1. The supporting strength of a positive projecting conduit is, therefore, a function not only of the distribution of the vertical load and the vertical reaction on the pipe but also of the magnitude and distribution of the active lateral pressure on those portions of its sides which are exposed to the embankment filling material. A formula for calculating the load factor for a positive projecting conduit (*8*), which takes into account these side pressures, is

$$L_f = \frac{A}{N - xq} \tag{26-3}$$

in which

L_f = load factor;
A = a parameter which is a function of the shape of the conduit (circular conduit, A = 1.43; horizontal elliptical (HE), A = 1.34; and vertical elliptical (VE), A = 1.02);

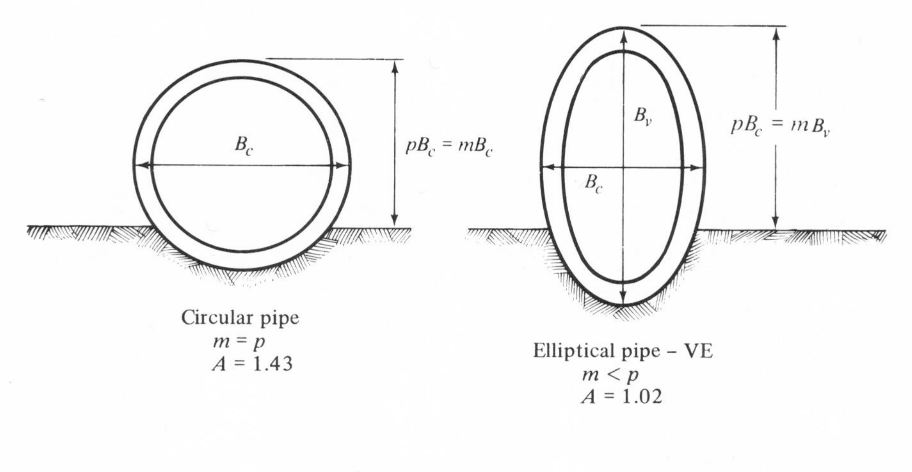

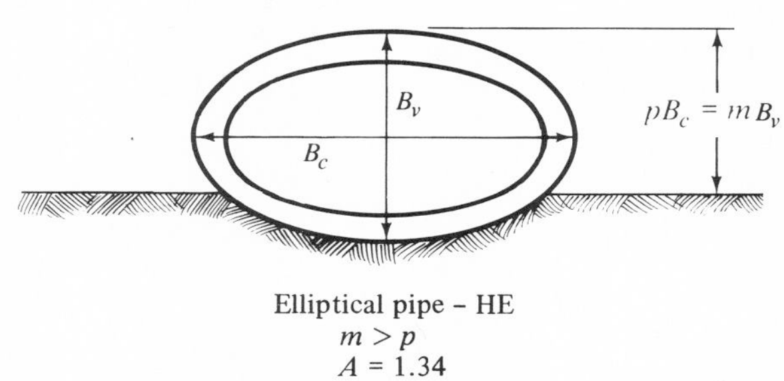

Fig. 26-9. Illustration of ratio m.

N = bedding factor = a parameter which is a function of the distribution of the vertical reaction;

q = ratio of the *total* lateral pressure to the *total* vertical load (For those situations in which lateral pressure cannot be relied upon to be effective, $q = 0$.);

x = a parameter which is a function of the area of the vertical projection of the pipe on which active lateral pressure of the fill material acts. It is expressed as a function of m; where

m = the ratio of the vertical projection above the subgrade to the vertical dimension of the conduit (see Fig. 26-9). Note: For circular conduits, $m = p$, the projection ratio.

The value of q may be found by the formula

$$q = \frac{pK}{C_c}\left(\frac{H}{B_c} + \frac{p}{2}\right) \tag{26-4}$$

in which

p = the projection ratio;
K = ratio of active lateral unit pressure to vertical unit pressure;
C_c = load coefficient for projecting conduits;
H = height of fill above top of conduit, in feet; and
B_c = horizontal breadth (outside) of conduit, in feet.

When the load and reaction situation causes a circular pipe to crack first at the top, which is usually the case when pipes are bedded in a concrete cradle, N' and x' should be substituted for N and x in Eq. (26-3). Values of N, N', x, and x' for circular pipe are given in Table 26-1. Values of N and x for elliptical pipe are given in Table 26-2.

TABLE 26-1. Values of x (or x′) and N (or N′) for Circular Pipe

Value of $p = m$	*Value of* x	*Value of* x'	*Type of Bedding*	*Value of* N	*Value of* N'
0.3	0.217	0.743	Impermissible (Class D)	1.310	—
0.5	0.423	0.856	Ordinary (Class C)	0.840	—
0.7	0.594	0.811	First class (Class B)	0.707	—
0.9	0.655	0.678	Concrete cradle (Class A)	—	0.505
1.0	0.638	0.638			

TABLE 26-2. Values of N and x for Elliptical Pipe (9)

Type of Pipe	*Value of* m	*Value of* x	*Type of Bedding*	*Value of* N
Horizontal	0.3	0.146	Ordinary (Class C)	0.763
elliptical (HE)	0.5	0.268		
	0.7	0.369	First class (Class B)	0.630
	0.9	0.421		
Vertical	0.3	0.238	Ordinary (Class C)	0.615
elliptical (VE)	0.5	0.457		
	0.7	0.639	First class (Class B)	0.516
	0.9	0.718		

EXAMPLE 26-1. Assume that a 60-in. circular reinforced concrete pipe is to be installed by the Imperfect Ditch method of construction. The outside diameter of the pipe is 6.0 ft and when installed in a Class C bedding will project 4.2 ft above the adjacent subgrade. The imperfect ditch will be excavated to a depth of 6.0 ft and refilled with very loose soil. Assume a settlement ratio of −0.3. The height of fill will be 50 ft and the soil weighs 120 pcf.

Calculate the load on the pipe and the required three-edge bearing

strength with a factor of safety of 1.0 based upon the 0.01 in. crack strength.
Load calculation data:

$$H = 50 \text{ ft} \qquad p' = 6/6 = 1.0$$
$$B_c = 6.0 \text{ ft} \qquad r_{sd} = -0.3$$
$$H/B_c = 8.33 \qquad \gamma = 120 \text{ pcf}$$
$$C_n = 5.2 \text{ (Fig. 25-18)}$$
$$W_c = 5.2 \times 120 \times 6^2 = 22{,}500 \text{ pcf} \quad (4500\text{D})$$

Supporting strength calculation data:

$$K = 0.33 \qquad A = 1.43$$
$$p = 4.2/6.0 = 0.7 \qquad N = 0.840 \text{ (Table 26-1)}$$
$$x = 0.594 \text{ (Table 26-1)}$$

$$q = \frac{0.7 \times 0.33}{5.2}\left(8.33 + \frac{0.7}{2}\right) = 0.385 \qquad \text{[Eq. (26-4)]}$$

$$L_f = \frac{1.43}{0.840 - (0.594 \times 0.386)} = 2.34 \qquad \text{[Eq. (26-3)]}$$

$$\text{required three-edge strength} = \frac{22{,}500 \times 1}{2.34} = 9600 \text{ plf}$$

$$\text{D-load strength} = \frac{9600}{5} = 1920\text{D}$$

ASTM Specification C76 Class IV pipe required:

$$\text{Class IV pipe} = 2000\text{D at 0.01 in. crack}$$

$$\text{actual factor of safety} = \frac{2000\text{D} \times 2.34}{4500\text{D}} = 1.04$$

EXAMPLE 26-2. Assume that an elliptical reinforced concrete pipe is to be installed as a positive projecting conduit with the major axis horizontal (HE). The horizontal outside diameter of the pipe is 7.42 ft and the vertical diameter is 5.08 ft. The pipe will be laid in a Class C bedding. The top of the pipe projects 3.7 ft above the adjacent subgrade and the settlement ratio is assumed to be +0.4. The height of the fill over the pipe will be 18 ft and the soil weights 120 pcf.

Calculate the load on the pipe and the required three-edge bearing strength, using a factor of safety of 1.0 based upon the 0.01 in. crack strength.

Load calculation data:

$H = 18$ ft
$B_c = 7.42$ (inside span $= 6.33$ ft)
$H/B_c = 2.43$
$C_c = 3.3$ (Fig. 25-13)

$p = 3.7/7.42 = 0.5$
$r_{sd} = +0.4$
$r_{sd}p = +0.2$
$\gamma = 120$ pcf

$$W_c = 3.3 \times 120 \times 7.42^2 = 21{,}800 \text{ plf} \quad (3444\text{D})$$

Supporting strength calculation data:

$K = 0.33$
$m = 3.7/5.08 = 0.7$
$x = 0.369$ (Table 26-2)

$A = 1.34$
$N = 0.763$ (Table 26-2)

$$q = \frac{0.5 \times 0.33}{3.3}\left(2.43 + \frac{0.5}{2}\right) = 0.134$$

$$L_f = \frac{1.34}{0.763 - (0.369 \times 0.134)} = 1.88$$

$$\text{required three-edge strength} = \frac{21{,}800 \times 1}{1.88} = 11{,}600 \text{ plf}$$

$$\text{D-load strength} = \frac{11{,}600}{6.33} = 1833\text{D}$$

ASTM Specification C507 Class HE-IV pipe required:

$$\text{Class HE-IV pipe} = 2000\text{D at 0.01 in. crack}$$

$$\text{actual factor of safety} = \frac{2000\text{D} \times 1.88}{3444\text{D}} = 1.09$$

EXAMPLE 26-3. Assume that an elliptical reinforced concrete pipe is to be installed as a positive projecting conduit with the major axis vertical (VE). The horizontal outside diameter is 5.08 ft and the vertical diameter is 7.42 ft. The pipe will be laid in a Class B bedding. The top of the pipe projects 5.2 ft above the adjacent subgrade and the settlement ratio is assumed to be +0.6. The height of fill over the top of the pipe is 28 ft and the soil weighs 120 pcf.

Caclulate the load on the pipe and the required minimum three-edge bearing strength, using a factor of safety of 1.2 based upon the 0.01 in. crack strength.

Load calculation data:

$H = 28$ ft
$B_c = 5.08$ ft (inside span $= 4.0$ ft)

$p = 5.2/5.08 = 1.0$
$r_{sd} = +0.6$

$H/B_c = 5.5$ $\qquad r_{sd}p = +0.6$

$C_c = 8.5$ (Fig. 25-13) $\qquad \gamma = 120$ pcf

$$W_c = 8.5 \times 120 \times 5.08^2 = 26{,}400 \text{ plf} \quad (6600\text{D})$$

Supporting strength calculation data:

$K = 0.33$ $\qquad A = 1.02$

$m = 5.2/7.42 = 0.7$ $\qquad N = 0.516$ (Table 26-2)

$x = 0.639$ (Table 26-2)

$$q = \frac{1.0 \times 0.33}{8.5}\left(5.5 + \frac{1.0}{2}\right) = 0.235$$

$$L_f = \frac{1.02}{0.516 - (0.639 \times 0.235)} = 2.79$$

$$\text{required three-edge strength} = \frac{26400 \times 1.2}{2.79} = 11{,}350 \text{ plf}$$

$$\text{D-load strength} = \frac{11{,}350}{4.0} = 2840\text{D}$$

ASTM Specification C507 Class VE-V pipe required:

$$\text{Class VE-V pipe} = 3000\text{D at } 0.01 \text{ in. crack}$$

$$\text{Actual factor of safety} = \frac{3000\text{D} \times 2.79}{6600\text{D}} = 1.27$$

26.8. LOAD FACTORS FOR SPECIAL CONDITIONS. The proper load factor to use in determining the field supporting strength of a conduit in a particular type of installation needs to be given careful study in special cases of loading. For example, analytical considerations indicate that the load factors for live loads produced by wheel loads applied at the roadway surface and for all classes of bedding are nearly the same, varying from about 1.5 to 1.7. In the case of a negative projecting conduit or a zero projecting conduit, the structure is installed in a ditch and the situation is similar to that typical of a ditch conduit, except that working conditions in the trench may be drier and more favorable. If such favorable conditions are anticipated and the soil at the sides can be well compacted, some lateral pressure against the sides of the pipe may be assumed to be effective in determining a load factor. When a conduit is installed and loaded by the imperfect-ditch method of construction, the structure will be subject to active lateral pressures on its sides, much the same as a positive projecting conduit, and this fact needs to be taken into account in determining a proper load factor.

26.9. EFFECT OF INTERNAL PRESSURES. Buried conduits such as water and gas mains, and natural gas and petroleum products transmission pipelines, may be subjected to both external earth and traffic loads and to internal fluid pressure. When a circular conduit is loaded externally, the pipe wall is subjected to bending-moment stresses which tend to overstress the pipe at the top, at the bottom, and at the two sides. When the conduit is subjected to internal fluid pressure, the pipe wall is stressed in tension. When it is subjected to both external loads and internal pressure, these stresses combine in the pipe wall, with the result that a conduit will not carry so much external load when subjected to internal pressure as otherwise. The converse is also true; that is, a pipe will not withstand so much internal pressure when subjected to external load as when no external load is applied.

Two general classes of pipe material are considered in connection with this problem: cast iron pipe, which is a relatively brittle, nonelastic type; and steel or ductile iron pipe which are elastic in character. The supporting strength of cast iron pipe under external load and internal pressure was investigated in depth by Schlick (*2*).

26.10. CAST IRON PIPE UNDER COMBINED LOADING. If the bursting strength and the three-edge bearing strength of a pipe are known, the relationship between the internal pressures and external loads which will cause failure may be computed by means of the formula

$$s = S\sqrt{\frac{T - t}{T}} \qquad (26\text{-}5)$$

in which

s = three-edge bearing load at failure under combined internal and external loading, in pounds per linear foot;
S = three-edge bearing strength with no internal pressure, in pounds per linear foot;
T = bursting strength of pipe with no external load, in pounds per square inch; and
t = internal pressure at failure under combined internal and external loading, in pounds per square inch.

As an example, suppose that a 24-in. cast-iron water pipe has a three-edge bearing strength of 3500 lb/linear ft and a bursting strength of 540 psi. The graph of Eq. (26-5) for this pipe is shown in Fig. 26-10. If the pipe were subjected in service to an internal pressure of 180 psi (including an allowance for water hammer), this graph shows that it would have a three-edge strength of 2850 lb/linear ft; for an internal pressure of 300 psi, the

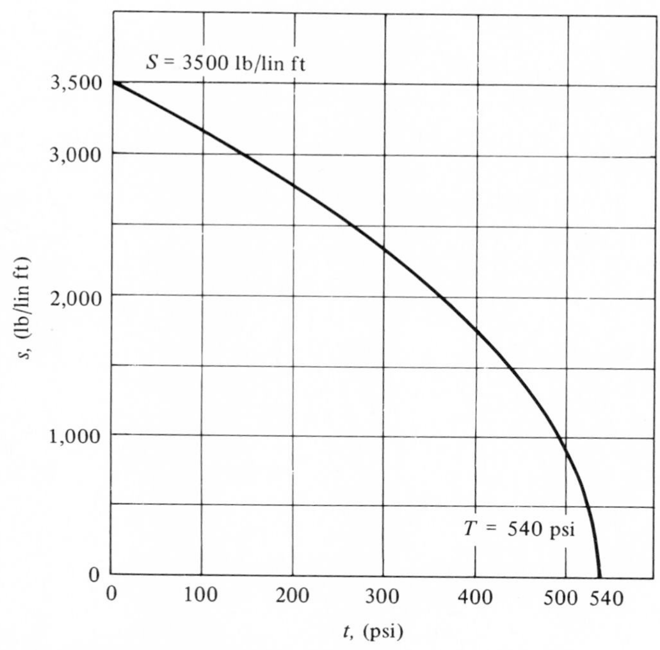

Fig. 26-10. Graph of Eq. (26-5) for $S = 3500$ and $T = 540$.

three-edge strength would be 2325 lb/linear ft; and so on. The three-edge bearing strength indicated by the graph must then be multiplied by an appropriate load factor to obtain the supporting strength of the pipe as it is actually installed.

26.11. BEDDING CONDITIONS AND LOAD FACTORS FOR WATER AND GAS PIPES. The methods of installing cast-iron pipes for water and gas service frequently produce bedding conditions which are considerably different from those commonly encountered in the case of sewers and culverts. Six field conditions which influence the bedding condition, and therefore the field supporting strength, of cast-iron pressure mains are described in Table 26-3. The load factors or strength ratios for pipes of different diameters installed under the various field conditions are shown in Fig. 26-11. Thus, if the 24-in. pipe of Section 26-10 is subjected to an internal pressure of 300 psi and is installed in accordance with condition E, its field supporting strength is $1.52 \times 2325 = 3540$ lb/lin ft. If it were installed in accordance with condition C, the field supporting strength would be $0.85 \times 2325 = 1980$ lb/linear ft.

TABLE 26-3. Bedding Conditions for Pressure Mains

Field Condition	*Description*
A	Flat-bottomed trench; backfill untamped
B	Flat-bottomed trench; backfill firmly tamped under and around pipe
C[a]	Pipe laid on blocks; backfill untamped
D[a]	Pipe laid on blocks; backfill firmly tamped under and around pipe
E	Shaped-bottom trench; backfill untamped
F	Shaped-bottom trench; backfill firmly tamped under and around pipe

[a] Field conditions C and D are obsolete.

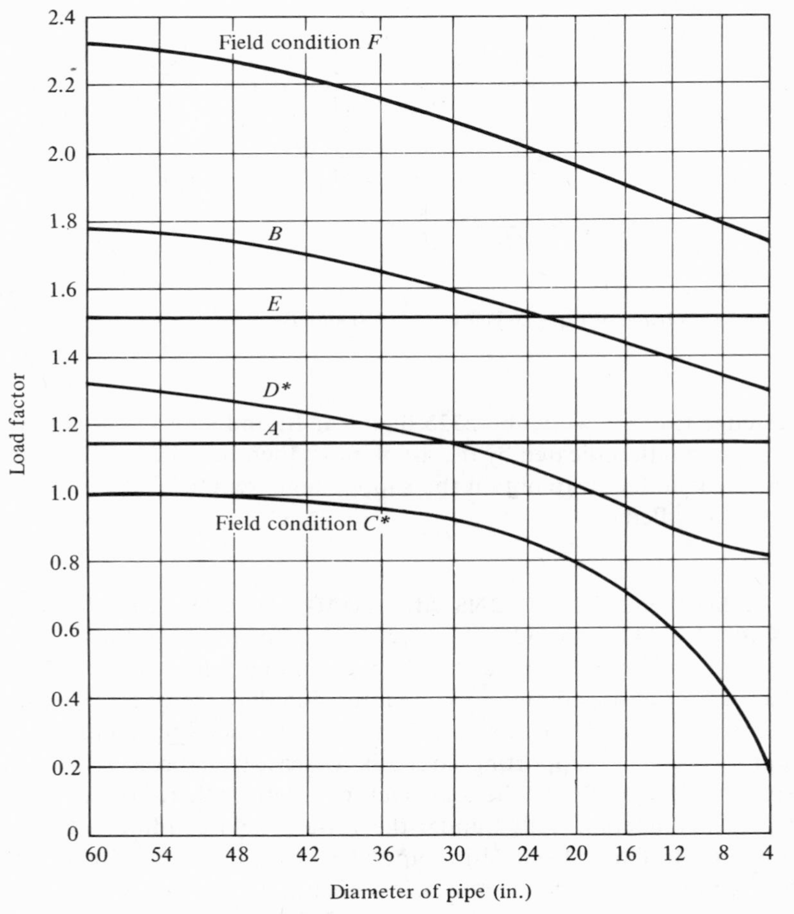

Fig. 26-11. Working values of load factor for cast-iron pipe in six field conditions. *Field conditions *C* and *D* are obsolete.

26.12. TOTAL STRESS IN PRESSURED PIPELINES OF STEEL OR DUCTILE IRON. Pipelines of these kinds are usually laid in trenches and then backfilled before they are pressurized. The pipe deflects an amount Δx under the influence of the earth load and assumes an elliptical shape with the major axis horizontal and the minor axis vertical (see Fig. 26-12).

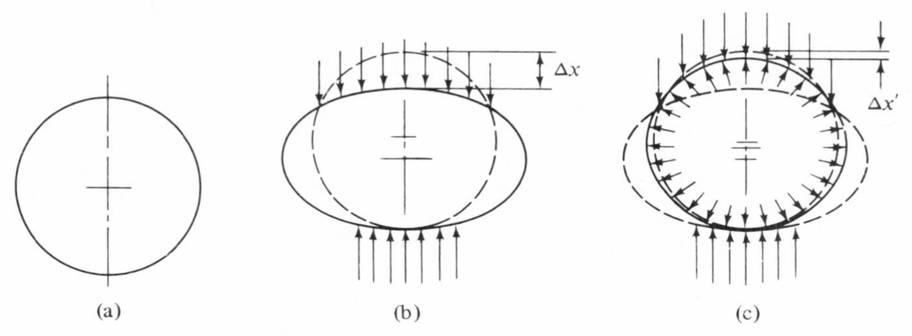

Fig. 26-12. Deflections of steel or ductile iron pipe under various conditions of external load and internal pressures. (a) No external load, no internal pressure; (b) external load only, no internal pressure; (c) external load plus internal pressure.

When internal pressure is introduced into the pipe, the resultant of the vertical components of this pressure will be greater than the resultant of the horizontal components because of the elliptical shape. This excess vertical internal pressure against the upper half of the pipe acts in opposition to the vertical load and combines with the resilience or stored energy in the deflected pipe to resist the external load.

Under these conditions, equilibrium of forces on the pipe will prevail when the sum of the vertical excess pressure and the resilience in the deflected pipe is equal to the external load. Therefore, when internal pressure is introduced into a circular pipe which has been deflected by external load, the deflection will decrease to some equilibrium value, $\Delta x'$, and the shape of the pipe will be stabilized as an ellipse intermediate between a circle and the deflected shape under external load alone. The bending stresses in the pipe wall corresponding to this equilibrium deflection are less than the stresses due to external load alone, and these bending stresses are assumed to be algebraically additive to the tensile hoop stress due to internal pressure.

On the basis of the foregoing hypothesis, the maximum combined stress in a pipe due to earth load and internal pressure may be expressed by the formula

$$S = S_1 + S_2 = \frac{p(D - 2t)}{2t} + 0.117 \times \frac{C_d \gamma B_d^2 Etr}{Et^3 + 2.592pr^3} \qquad (26\text{-}6)$$

in which

S = maximum combined stress, in pounds per square inch;
S_1 = hoop stress due to internal pressure, in pounds per square inch;
S_2 = bending stress due to fill load, in pounds per square inch;
C_d = calculation coefficient for fill load on ditch conduits (see Fig. 25-7);
γ = unit weight of soil backfill, in pounds per cubic foot;
B_d = width of ditch at level of top of pipe, in feet;
E = modulus of elasticity of pipe metal, in pounds per square inch;
D = outside diameter of pipe, in inches;
t = thickness of pipe wall, in inches;
r = mean radius of pipe = $(D - t)/2$, in inches; and
p = internal pressure, in pounds per square inch.

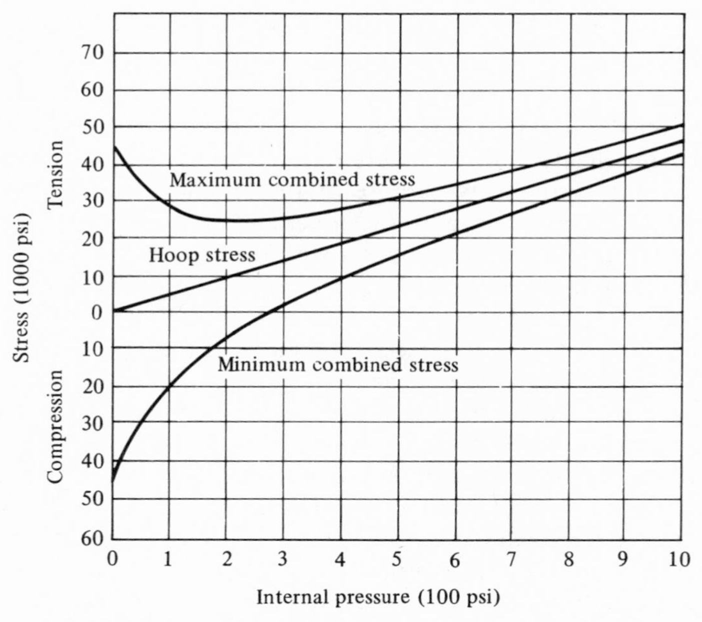

Fig. 26-13. Combined stress external load plus internal pressure. Values are for 36 in. × $\frac{3}{8}$ in. steel pipe; H = 6 ft; B_d = 5 ft. and γ = 120 pcf. The bedding contact angle is 30°.

In Eq. (26-6) all factors are in inch units except γ and B_d, which are in feet. However, the coefficient of the term for S_2 has been adjusted so that the result S is in pounds per square inch.

If the pipeline is subjected to vertical loads due to surface traffic vehicles, the stress from this source should be added to that indicated by Eq. (26-6).

A typical diagram of combined stress in a 36 in. by $\frac{3}{8}$ in. steel pipeline subjected to earth load and to various internal pressures is shown in Fig. 26-13. It will be noted that the combined tensile stress decreases to some minimum value as internal pressure is increased, and then increases with further increase in pressure. The combined stress curve becomes asymptotic to the internal pressure curve at infinity.

26.13. SOURCES OF SUPPORTING STRENGTH OF UNDERGROUND CONDUITS. In general it may be said that all of the underground conduits derive their ability to support the earth above them from two sources: first, the inherent strength of the pipe to resist external pressures; second, the lateral pressure of the soil at the sides of the pipe, which produces stresses in the pipe ring in opposite directions to those produced by the vertical loads and thereby assists the pipe in supporting the vertical loads. In rigid pipes, such as those made of concrete, cast iron, and burned clay, the inherent strength of the pipe is the predominant source of supporting strength. The only lateral pressure that can be safely depended upon to augment the load-carrying capacity of the pipes is the active lateral pressure of the soil, since the rigid pipes deform very little under the vertical load and, consequently, the sides do not move outward enough to develop any appreciable passive resistance pressure in the enveloping soil, at least until after the pipe is cracked or broken.

In flexible pipes, such as corrugated-metal culverts and thin-walled steel or plastic conduits, the situation is reversed. The pipe itself has relatively little inherent strength, and a large part of its ability to support vertical load must be derived from the passive pressures induced as the sides move outward against the soil. The ability of a flexible pipe to deform readily and thus utilize the passive soil pressure on the sides of the pipe is its principal distinguishing structural characteristic and accounts for the fact that these relatively light-weight, low-strength pipes can support earth fills of considerable height without showing evidence of structural distress. It is apparent from these considerations that any attempt to analyze the structural behavior of the flexible conduits must take into account the soil at the sides as an integral part of the structure, since such a large proportion of the total supporting strength is attributable to the side material.

26.14. FAILURE OF FLEXIBLE CONDUITS. Another major difference between the rigid types of conduits and the flexible types is that the latter usually fail by deflection rather than by rupture of the pipe walls, as do the former. A flexible pipe, installed in the ordinary manner without vertical struts or other prestressing devices, will deflect under the vertical earth load, the vertical diameter becoming less and the horizontal diameter becoming greater by appreciable amounts. The outward movement of the sides of the pipe against the enveloping fill material brings into play the passive resistance of the soil, which acts horizontally against the pipe and keeps the actual deflection of the pipe considerably below the amount the pipe would deflect if acted upon by the vertical earth loads alone.

This action continues, as the embankment is built higher, until the top of the pipe becomes approximately flat. Additional load may then cause the curvature of the top portion of the pipe to reverse direction, and the top may become concave upward. When this occurs, the sides of the pipe will pull inward; and the side supports of the pipe will be eliminated, since they are passive forces that cannot follow the inward movement. The deflection of the pipe will therefore proceed as rapidly as the earth above can follow the downward movement of the top of the pipe and exert pressure on the structure. Finally, complete collapse and failure may result. The whole action is one of large deflection change, accompanied by high bending moment in the pipe wall.

A hypothetical sequence of the development of pipe deflection is

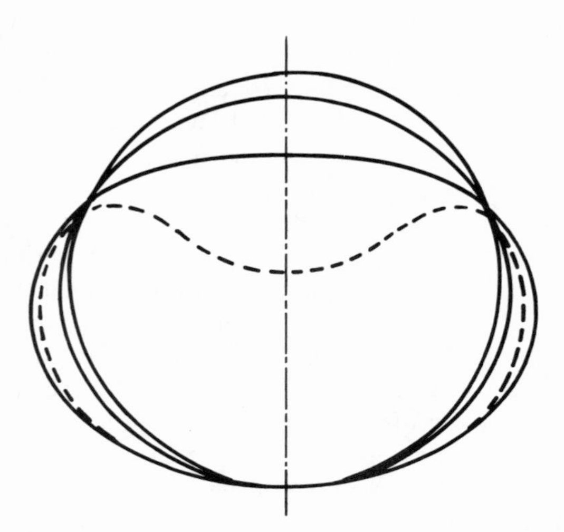

Fig. 26-14. Stages of deflection of a flexible-pipe culvert.

Fig. 26-15. Failure by excessive deflection of an 84-in. corrugated steel pipe.

shown in Fig. 26-14. A complete deflection failure of an 84-in. corrugated steel pipe is illustrated in Fig. 26-15.

26.15. STRUCTURAL CHARACTERISTICS OF FLEXIBLE CONDUITS. A number of field loading experiments on corrugated-metal pipe culverts, in which the vertical and lateral pressures on the pipes and the deflections of the pipes were measured, have led to the following conclusions regarding structural characteristics of flexible conduits:

1. The vertical load may be determined by Marston's theory of loads on conduits and is distributed approximately uniformly over the breadth of the pipe.
2. The vertical reaction is equal to the vertical load and is distributed approximately uniformly over the width of bedding of the pipe.
3. The horizontal pressure on each side of the pipe is distributed para-

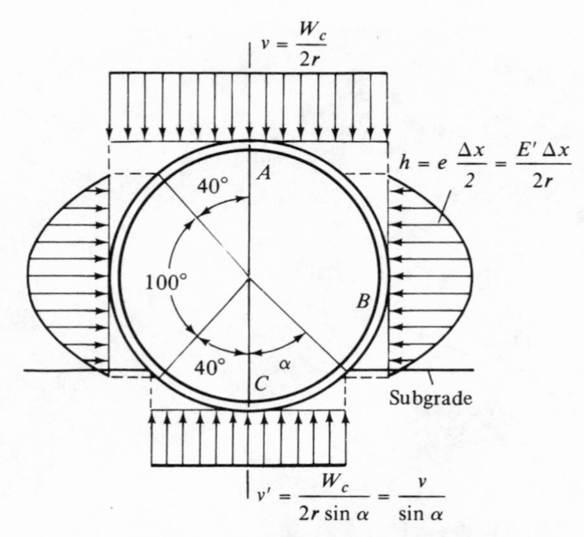

Fig. 26-16. Assumed distribution of pressure on flexible culvert pipe.

bolically over the middle 100° of the pipe; and the maximum unit pressure, which occurs at the ends of the horizontal diameter of the pipe, is equal to the modulus of passive resistance of the fill material multiplied by one-half the horizontal deflection of the pipe, or since $er = E'$, the modulus of soil reaction divided by the radius and multiplied by one-half the horizontal deflection.

This assumed loading is shown graphically in Fig. 26-16. The deflection of a flexible culvert pipe resulting from the load system just described is very often augmented by the continued yielding of the soil at the sides of the pipe in response to the horizontal pressures over a considerable period of time after the maximum vertical load has developed. This yielding results in a continuation of the pipe deformation to a value beyond that which is primarily attributable to the vertical load. Therefore, when it is desired to estimate the maximum ultimate deflection of a flexible pipe culvert, it may be necessary to introduce a quantity which has been called the deflection lag factor. The deflection lag factor cannot be less than unity and has been observed to range upward toward a value of 2.0. It appears to depend upon the quality of the soil at the sides of the pipe. A well-graded dense soil will permit very little, if any, residual deflection, and the lag factor can safely be ignored; while a loosely placed soil may induce a relatively large deflection lag. Except in the case of very high quality, well-compacted backfill soil, a deflection lag factor of about 1.25 is recommended for design purposes. For best results, the backfill soil should be compacted for a width of one or two pipe diameters on each side of the pipe.

26.16. CALCULATION OF DEFLECTION OF FLEXIBLE CULVERT. A formula for computing the deflection of a flexible pipe culvert is

$$\Delta x = D_l \frac{K W_c r^3}{EI + 0.061 E' r^3} \tag{26-7}$$

in which

Δx = horizontal deflection of the pipe, in inches (it may be considered the same as the vertical deflection);
D_l = deflection lag factor;
K = a bedding constant, its value depending on the bedding angle; α, in Fig. 26-16;
W_c = vertical load per unit length of the pipe, in pounds per linear inch;
r = mean radius of the pipe, in inches;
E = modulus of elasticity of the pipe material, in pounds per square inch;
I = moment of inertia per unit length of cross section of the pipe wall, in inches^4 per inch;
E' = er = modulus of soil reaction, in pounds per square inch; and
e = modulus of passive resistance of the enveloping soil, in pounds per square inch per inch.

It is recommended that the deflection of a corrugated-metal pipe culvert should not exceed about 5% of the nominal pipe diameter.

Values of the bedding constant K for various values of the bedding angle are shown in Table 26-4. The bedding angle α is defined as one-half the angle subtended by the arc of the pipe ring which is in contact with the pipe bedding, and over which the bottom reaction is distributed, as shown in Fig. 26-16.

TABLE 26-4. Values of Bedding Constant

Bedding Angle, α (deg)	*Bedding Constant, K*
0	0.110
15	0.108
$22\frac{1}{2}$	0.105
30	0.102
45	0.096
60	0.090
90	0.083

26.17. STIFFNESS FACTOR. The stiffness factor EI in Eq. (26-7) may be evaluated by testing the pipe metal to determine its modulus of elasticity and by calculating the moment of inertia of the shape of the cross section of the pipe wall. In many cases, however, it will be easier to subject a representative section of the pipe to a laboratory three-edge bearing test to determine the relationship between the load and the change in diameter of the pipe. The effective values of the product EI may then be obtained by substituting the measured loads and deflections in the following formulas:

$$EI = 0.149 \frac{Wr^3}{\Delta y} \qquad (26\text{-}8)$$

$$EI = 0.136 \frac{Wr^3}{\Delta x} \qquad (26\text{-}9)$$

in which Δy and Δx are the vertical and horizontal deflections of the pipe ring, respectively; W is the three-edge bearing test load, in pounds per linear inch; and the meanings of r, E, and I are as given previously.

TABLE 26-5. Moments of Inertia and Values of *EI* per Inch Length of Steel Pipe with Standard Corrugations ($2\frac{2}{3}$ in. Pitch by $\frac{1}{2}$ in. Depth)

U.S. Gauge No.	*Moment of Inertia* *(in.^{4}/in.)*	*Stiffness Factor* EI *(lb/in.,* E = *29,000,000 psi)*
4	0.008275	239,975
6	0.006744	195,576
8	0.005512	159,848
10	0.004373	126,817
12	0.003317	96,193
14	0.002326	67,454
16	0.001848	53,592
20	0.001104	32,016
24	0.000733	21,257
30	0.000366	10,614

TABLE 26-6. Moments of Inertia and Values of *EI* per Inch Length of Steel Pipe with Structural Plate Corrugations (6 in. Pitch by 2 in. Depth)[a]

U.S. Gauge No.	*Moment of Inertia* *(in.^{4}/in.)*	*Stiffness Factor* EI *(lb/in.,* E = *29,000,000 psi)*
1	0.1659	4,811,100
3	0.1463	4,242,700
5	0.1270	3,683,000
7	0.1080	3,132,000
8	0.0961	2,786,900
10	0.0781	2,264,900
12	0.0604	1,751,600

[a] From *Armco Handbook of Drainage and Construction Products.*

Computed values of I, which is the moment of inertia of a cross section of the pipe wall per linear inch of pipe, and values of the stiffness factor EI for metal having a modulus of elasticity of 29,000,000 psi are given in Table 26-5 for pipes having standard corrugations and in Table 26-6 for pipes having structural plate corrugations. For smooth steel pipe, $I = t^3/12$, where t is the thickness of the pipe wall.

26.18. MODULUS OF PASSIVE RESISTANCE. The modulus of passive resistance of the side filling material is defined as the unit pressure developed as the side of a pipe moves outward a unit distance against the side fill. Little is known about the exact nature of this modulus. In the Rankine theory of lateral soil pressures, the limiting value of the ratio of passive horizontal pressure to vertical pressure which a granular soil without cohesion can develop is shown to be the reciprocal of the active pressure ratio. However, this theory does not give a clue in regard to the amount of movement required to develop the limiting value of passive pressure; and it would seem that the actual passive pressure may be any value less

TABLE 26-7. Values of E' for 18 Flexible Pipe Culverts

Item	*Location*	*Pipe Diam. (in.)*	*Soil Type*[a]	*Fill Height (ft)*	*Mod. of Passive Resist., e (psi/in.)*	*Value of $E' = er$ (psi)*
1[b]	Ames, Iowa	42	Loam top soil (U)	15	14	294
2[b]	Ames, Iowa	42	Well-graded gravel (U)	16	32	672
3[b]	Ames, Iowa	36	Sandy clay loam (T)	15	28	502
4[b]	Ames, Iowa	36	Sandy clay loam (U)	15	13	234
5[b]	Ames, Iowa	42	Sandy clay loam (T)	15	25	525
6[b]	Ames, Iowa	42	Sandy clay loam (U)	15	15	315
7[b]	Ames, Iowa	48	Sandy clay loam (T)	15	29	696
8[b]	Ames, Iowa	48	Sandy clay loam (U)	15	14	336
9[b]	Ames, Iowa	60	Sandy clay loam (T)	15	26	780
10[b]	Ames, Iowa	60	Sandy clay loam (U)	15	12	360
11[c]	Chapel Hill, N. C.	30	Sand	12	25	375
12[c]	Chapel Hill, N. C.	31.5	Sand	12	56	882
13[c]	Chapel Hill, N. C.	30	Sand	12	80	1200
14[c]	Chapel Hill, N. C.	20	Sand	12	35	350
15[c]	Chapel Hill, N. C.	21	Sand	12	82	861
16[c]	Culman Co., Ala.	84	Crushed sandstone (C)	137	190	7980
17[c]	McDowell Co., N. C.	66	Clayey sandy silt (C)	170	40	1320
18[d]	Wolf Creek, Mont. (reconstructed)	216	Graded crushed gravel (C)	83	58	6300

[a] U—untamped; T—tamped, C—compacted.
[b] Side pressure and pipe deflections measured.
[c] Side pressures estimated, pipe deflections measured.
[d] Load and pipe deflections measured.

than the maximum, the ratio depending on the soil characteristics and the amount of movement of the sides of the pipe.

Some recent research (*11*) has indicated that this modulus is strongly influenced by the size of the pipe, and that for a given type of soil in a given state of compaction, the product er of the modulus and the radius of the pipe, designated as E', is reasonably constant. That is to say, for the same soil, the modulus is inversely proportional to the pipe radius. Also, observations on a limited number of pipes in service, where sufficient information is available to make an approximate estimate, indicate that the value of E' varies widely. The range was from a minimum of 234 psi, in the case of a shovel-placed uncompacted sandy clay loam, to a maximum of 7980 psi, for a crushed sandstone soil which was compacted to Proctor density. A tabulation of estimated values of E' for 18 actual flexible pipe culvert installations is given in Table 26-7. This table may be used as a guide in the selection of an appropriate value of E' for use in the design of flexible pipe conduits.

26.19. STRESSES IN FLEXIBLE PIPE WALL. Although pipe deflection is the principal criterion for design of a flexible pipe conduit, it may also be important to determine the bending moment and tangential thrust stresses around the periphery of the pipe wall. This may be of particular importance in the design of longitudinal bolted seams of field-assembled structures. The seams are subjected to a combination of tangential thrust and bending moment. The thrust subjects the bolts to single shear stress and the moment creates a prying action which subjects one row of bolts to direct tension. Therefore the stress on a bolt may be a composite of shear and tension.

Equations for the moment and thrust at the bottom of a circular pipe (point C in Fig. 26-16) are

$$M_c = A W_c r \qquad (26\text{-}10)$$

$$R_c = B W_c \qquad (26\text{-}11)$$

with values of A and B given in Table 26-8.

Equations for the moment and thrust at the bottom of the pipe due to horizontal loads (Fig. 26-16) are

$$M_c = -0.166\,hr^2 \qquad (26\text{-}12)$$

$$R_c = 0.511\,hr \qquad (26\text{-}13)$$

in which

$$h = \frac{E'\,\Delta x}{2r}$$

TABLE 26-8. Values of A and B in Eq. (26-10) and (26-11) for Various Values of the Bedding Angle α

α	A	B	$\sin \alpha$
0	0.294	0.053	0
15	0.234	0.050	0.259
30	0.189	0.040	0.500
45	0.157	0.026	0.707
60	0.138	0.014	0.866

To determine the total stress situation at any point D on the periphery at counterclockwise angle ϕ from the bottom point C, it is most convenient to write the moment and thrust equations for vertical load and lateral pressure separately. Then the resultant stress at a point is obtained by algebraic combination of these stresses. Expressions for evaluating moments and thrusts are given in Eqs. (26-14) to (26-31).

Moment, thrust, and shear resulting from vertical load: when ϕ lies between 0 and α

$$M_D = W_c r\left[A + B(1 - \cos\phi) - 0.250\frac{\sin^2\phi}{\sin\alpha}\right] \tag{26-14}$$

$$R_D = W_c\left(0.500\frac{\sin^2\phi}{\sin\alpha} - B\cos\phi\right) \tag{26-15}$$

$$S_D = W_c\left(0.500\frac{\sin\phi\cos\phi}{\sin\alpha} - B\sin\phi\right) \tag{26-16}$$

when ϕ lies between α and 90°

$$M_D = W_c r[A + B(1 - \cos\phi) - 0.50\sin\phi + 0.25\sin\alpha] \tag{26-17}$$

$$R_D = W_c(0.500\sin\phi + B\cos\phi) \tag{26-18}$$

$$S_D = W_c(0.500\cos\phi - B\sin\phi) \tag{26-19}$$

when ϕ lies between 90° and 180°

$$M_D = W_c r[A + B(1 - \cos\phi) - 0.25(1 + \sin^2\phi - \sin\alpha)] \tag{26-20}$$

$$R_D = W_c(0.500\sin^2\phi + B\cos\phi) \tag{26-21}$$

$$S_D = W_c(0.500\sin\phi\cos\phi - B\sin\phi) \tag{26-22}$$

Moment, thrust, and shear resulting from horizontal loads: when ϕ lies between 0° and 40°

$$M_D = hr^2(0.345 - 0.511 \cos \phi) \tag{26-23}$$

$$R_D = 0.511\, hr \cos \phi \tag{26-24}$$

$$S_D = 0.511\, hr \sin \phi \tag{26-25}$$

when ϕ lies between 40° and 140°

$$M_D = hr^2(0.199 - 0.500 \cos^2\phi + 0.143 \cos^4\phi) \tag{26-26}$$

$$R_D = hr(\cos^2\phi - 0.568 \cos^4\phi) \tag{26-27}$$

$$S_D = hr(\sin \phi \cos \phi - 0.568 \sin \phi \cos^3\phi) \tag{26-28}$$

when ϕ lies between 140° and 180°

$$M_D = hr^2(0.345 + 0.511 \cos \phi) \tag{26-29}$$

$$R_D = 0.511\, hr \cos \phi \tag{26-30}$$

$$S_D = 0.511\, hr \sin \phi \tag{26-31}$$

PROBLEMS

26.1. When five specimens of clay sewer pipe are tested in three-edge bearings, the minimum test strength is found to be 1820 lb/linear ft. Applying a factor of safety of 1.4, what will be the design load for these pipes when installed in a ditch with Ordinary (Class C) Bedding? When installed with First Class (Class B) Bedding?

26.2. What will be the required minimum three-edge bearing strength of the sewer pipe in Problem 25.2 if it is installed with Ordinary (Class C) Bedding and with a factor of safety of 1.3?

26.3. What will be the required minimum three-edge bearing strength of the culvert pipe in Problem 25.9 if it is installed with First Class (Class B) Bedding? Assume the angle ϕ of the fill material is 30° and use a factor of safety of 1.0 based upon the 0.01 in. crack strength.

26.4. A 60-in. equivalent diameter elliptical reinforced concrete pipe has an outside major diameter of 7.38 ft and the minor diameter is 5.10 ft. The pipe is installed as a positive projecting conduit with the major axis vertical (VE) in a Class C bedding. The top of the pipe projects 6.0 ft above the adjacent natural ground surface and the settlement ratio is assumed to be +0.6. The height of fill over the top of the pipe is 35 ft and the soil weighs 130 lb/cu ft.

Calculate the load on the pipe and the required minimum three-edge bearing strength using a factor of safety of 1.0 based upon the 0.01 in. crack strength.

26.5. An 18-in. nonreinforced concrete pipe is to be installed under 12 ft of cover in a ditch which is 36 in. wide at the elevation of the top of the pipe. Assume a clay soil weighing 120 pcf Class C bedding, and a factor of safety of 1.5. Determine the required minimum three-edge bearing strength of the pipe.

26.6. A 60-in. reinforced concrete culvert pipe having 6 in. sidewalls is to be installed as a positive projecting conduit in a Class C bedding and covered with an embankment 24 ft high. Assume the projection ratio is 0.5, the settlement ratio is +0.8, and the unit weight of the soil is 120 pcf. Also assume the lateral earth pressure ratio $K = 0.33$, and that the earth pressure is effective over the full projection of the pipe. Determine the required minimum three-edge bearing strength of pipe which will not develop a crack wider than 0.01 in. Factor of safety is 1.0.

26.7. The same size pipe as in Problem 26.6 is to be installed as an imperfect ditch conduit on a Class C bedding under 50 ft of fill. Assume $p' = 1.0$, $r_{sd} = 0.4$, $\gamma = 120$ pcf, $K = 0.25$, and the lateral pressure is effective over the top 0.9 of the pipe. Determine the required minimum 0.01 in. crack three-edge bearing strength of the pipe. Factor of safety is 1.0.

26.8. A 60-in. 12-gauge metal pipe having standard corrugations ($\frac{1}{2}$ in. deep at $2\frac{2}{3}$ in. centers) is to be installed as a projecting conduit in 60° bedding and covered with an embankment 30 ft high. Assume the projection ratio is 0.7, the settlement ratio is zero, the unit weight of soil is 120 pcf, and the modulus of soil reaction is 600 psi. Estimate the long time deflection of the pipe, assuming the deflection lag factor is 1.25.

26.9. A 30-in. cast-iron water main has a bursting strength of 560 psi and a three-edge bearing strength of 8400 lb/linear ft. If the pipe is installed in a flat-bottomed trench, the backfill is untamped, and the pipe operates at a pressure of 120 psi plus an allowance of 50% for water hammer, how much external load will the pipe support? If the pipe is installed on a shaped trench bottom and the backfill is untamped, how much external load will it support?

26.10. A 30-in. o.d. by $\frac{3}{8}$ in. steel line pipe carrying natural gas at a maximum operating pressure of 600 psi is installed in a 3.5 ft wide trench under 6 ft of cover. Assuming the soil weighs 120 pcf and $E = 30 \times 10^6$ psi, determine the maximum unit tensile stress in the pipe wall.

REFERENCES

1. Benjes, Henry J. "Recommended Practice for the Design and Installation of Conduits in Trenches." Washington University Conference on Conduit Strengths and Trenching Requirements, St. Louis, Mo., 1958. Also private communication, 1970.
2. Schlick, W. J. "Supporting Strengths of Cast-Iron Pipe for Water and Gas Service." Bul. 146, Iowa Engineering Experiment Station, Ames, Iowa, 1940.
3. Sears, E. C. "Engineering Properties and Design of Ductile-Iron Pipe in Underground Pressure Service." Paper 62-WA-353. American Society of Mechanical Engineers, 1962.
4. Sears, E. C. "Ductile-Iron Pipe Design." *J. Amer. Waterworks Assoc.* **56,** 4 (1964).
5. Spangler, M. G. "Protective Casings for Pipelines." *The Petroleum Engineer* (April 1952).

6. Spangler, M. G. "Secondary Stresses in Buried High Pressure Pipelines." *The Petroleum Engineer* (November 1954).
7. Spangler, M. G. "The Structural Design of Flexible Pipe Culverts." Bul. 153, Iowa Engineering Experiment Station, Ames, Iowa, 1941.
8. Spangler, M. G. "The Supporting Strength of Rigid Pipe Culverts." Bul. 112, Iowa Engineering Experiment Station, Ames, Iowa, 1933.
9. Swanson, Harold, and Mason D. Reed. "Structural Characteristics of Reinforced Concrete Elliptical Sewer and Culvert Pipe." *Highway Research Board* **56,** Rec. 35 (1964).
10. Watkins, R. K., and A. P. Moser. "Response of Corrugated Steel Pipe to External Loads." *Highway Research Board Rec.* 373, 86 (1971). Also discussion by M. G. Spangler.
11. Watkins, R. K., and M. G. Spangler. "Some Characteristics of the Modulus of Passive Resistance of Soil: A Study in Similitude." *Proc. Highway Res. Board* **37,** 576 (1958).

Index

Printer and Binder: Haddon Craftsmen
78 79 80 81 8 7 6 5 4 3